Bioanalysis

Advanced Materials, Methods, and Devices

Volume 11

Series Editor
Tuan Vo-Dinh, Fitzpatrick Institute for Photonics, Duke University, Durham, NC, USA

The book series on BIOANALYSIS: Advanced Materials, Methods, and Devices is intended to serve as an authoritative reference source for a broad, interdisciplinary audience involved in the research, teaching, learning, and practice of bioanalytical science and technology. Bioanalysis has experienced explosive growth due to the dramatic convergence of advanced technologies and molecular biology research, which has led to the development of entirely new ways to probe biomolecular and cellular processes as well as biological responses to implanted biomaterials and engineered tissues. Novel optical techniques using a wide variety of reporter gene assays, ion channel probes, and fluorescent probes have provided powerful bioanalytical tools for cell-based assays. Fluorescent reporters allow the development of live cell assays with the ability for in vivo sensing of individual biological responses across cell populations, tracking the transport of biological species within intracellular environments, and monitoring multiple responses from the same cell. Novel classes of labels using inorganic fluorophors based on quantum dots or surface-enhanced Raman scattering labels provide unique possibilities for multiplex bioanalyses.

Laser-based technologies are important in the development of ultrasensitive bioanalytical techniques. Lasers are now used as excitation light sources in a wide variety of molecular bioassays. Today, single-molecule detection techniques using laser excitation provide the ultimate tools to elucidate cellular processes. The possibility of fabricating nanoscale materials and components has recently led to the development of devices and techniques that can measure fundamental parameters at the molecular level. With "optical tweezer" techniques, for example, small particles may be trapped by radiation pressure in the focal volume of a high-intensity, focused laser beam. Ingenious optical trapping systems have also been used to measure the force exerted by individual motor proteins.

Whereas the laser has provided a new technology for excitation, the miniaturization and mass production of sensor devices and their associated electronic circuitry has radically transformed the ways detection and imaging of biological species can be performed in vivo and ex vivo. Sensor miniaturization has enabled significant advances in imaging technologies over the last decade in such areas as microarrays and biochips for bioanalysis of a wide variety of species. The miniaturization of high-density optical sensor arrays has also led to the development of advanced high-resolution imaging methods at the cellular or molecular scales. With powerful microscopic tools using near-field optics, scientists are now able to image the biochemical processes and sub-microscopic structures of living cells at unprecedented resolutions.

Recently, nanotechnology, which involves research on and development of materials and species at length scales between 1 to 100 nanometers, has been revolutionizing important areas in bioanalysis at the molecular and cellular level. The combination of molecular nanotechnology and various sensing modalities (optical, electrochemical, etc) opens the possibility of detecting and manipulating atoms and molecules using nano-devices, which have the potential for a wide variety of bioanalyses at the cellular level. These new bioanalytical tools are capable of probing the nanometer world and will make it possible to characterize the chemical and mechanical properties of biomolecules and cells, discover novel phenomena and processes, and provide science with a wide range of tools, materials, devices, and systems with unique characteristics.

This book series will present the most recent scientific and technological advances in materials, methods and instrumentation of interest to researchers, students, and manufacturers. The goal is to provide a comprehensive forum to integrate the contributions of biophysicists, biomedical engineers, materials scientists, chemists, chemical engineers, biologists, and others involved in the science and technology revolution reshaping molecular biology and biomedicine.

More information about this series at http://www.springer.com/series/8091

Leonid V. Tanin · Andrei L. Tanin

Biomedical and Resonance Optics

Theory and Practice

Springer

Leonid V. Tanin
CJSC Holography Industry
Minsk, Belarus

Andrei L. Tanin
Belarusian Medical Academy
of Postgraduate Education
Minsk, Belarus

ISSN 2364-1118 ISSN 2364-1126 (electronic)
Bioanalysis
ISBN 978-3-030-60772-2 ISBN 978-3-030-60773-9 (eBook)
https://doi.org/10.1007/978-3-030-60773-9

This Springer imprint is published by the registered company Springer Nature Switzerland AG
The registered company address is: Gewerbestrasse 11, 6330 Cham, Switzerland

Dedicated to the memory of the dearest Teacher,
Valentina Konstantinovna Tanina, who instilled sense of love and respect for medicine in us,
which defined the future of our professions and allowed seeing it from the perspective of a physicist and a medic.

Leonid V. Tanin and Andrei L. Tanin

Foreword

Modern medicine is constantly in need of creation and implementation of new contactless, noninvasive methods of human organism study. During the last decades, many methods based on the X-ray radiation, electrophysiological and magnetic properties of tissues became habitual in clinical practice. In addition to that methods of computerized, positron-emission and magnetic resonance tomography, Doppler ultrasonography and many other methods that facilitate medical processes are widely used and improved in medicine.

In 1975, one of the authors of the monograph Leonid V. Tanin, the graduate of Leningrad Optical School: Physical Department of Leningrad State University (Optics and Spectroscopy Department, 1971), postgraduate course of postgraduate course of Leningrad Physics and Technology Institute named after A. F. Ioffe of the Academy of Sciences of the USSR (Plasma optics sector, 1974) is coming back to Belarus.

Being employed by the Institute of Physics of the Academy of Sciences of the BSSR, young scientist, Candidate of physical and mathematical sciences Leonid V. Tanin began conducting active scientific studies in the spheres of laser physics, holography and medicine. In 1978, Leonid V. Tanin, being the employee of the Institute of Physics of the Academy of Sciences of the BSSR, initiated and created the group "Coherent-optical studies of medical–biological systems" in the Institute of Neurology, Neurosurgery and Physiotherapy of the Ministry of Health of the BSSR. Leonid V.Tanin was in charge of this group of scientists and highly skilled specialists more than 20 years. Leonid V. Tanin stood at the origins and was one of the first creators of a new scientific school subsequently named biomedical optics. The perspectivity of this school in science was confirmed by the fact that few publications during the period from 1978 to 1982 today turned into grandiose international scientific congresses and conferences.

The results of original fundamental and applied studies conducted in the co-authorship with scientists of different spheres: medics, biologists, physicists, biophysicists, engineers, specialists in the sphere of electronics, programmers that provided the implementation of brand new coherence-optical methods of the diagnosing, treatment and prevention of pathologic disorders of cardiovascular and neuromuscular systems and also other diseases connected with the disorders of metabolic processes, into the clinical practice, are presented in the given monograph.

One of the first goals of Leonid V. Tanin at that time was to develop physical principles of a new generation of precision medical–biological devices based on the holography and speckle-optics for the remote noninvasive diagnostics of diseases and highly precise preventive examinations with sensitivity and big spatial resolution that allowed to increasing the objectivity, information value and reliability of the derived results. On the basis of these developments, laser speckle-optical microhematomyograph (laser specklometer), holographic microscope, laser-holographic complex (holographic cardiograph) for contactless defining surface blood flow with marking out the components of this integration process, were created.

In the result of the studies conducted under the guidance and direct participation of Leonid V. Tanin during several years in the Research Institute of Neurology, Neurosurgery and Physiotherapy of the BSSR together with the Institute of Physics of the Academy of Sciences of the BSSR, it was first found out that the change of muscle tonus accompanied by the changes of oscillations of muscles fiber and adjacent tissues deforms spectrum of fluctuations of speckle-field. Measurements of spectrum characteristics of dynamic laser speckle-field, formed at the reflection of light from the tissue over the contraction muscle, allowed significantly supplementing the study of the contractile function of skeletal muscles. Similar method of defining contractile activity of muscles was named a laser myography by authors. This new noninvasive, quite sensitive method was made the basis of the development and implementation into the medical practice of a laser device (Laser specklometer) aimed for the study of biomechanical characteristics of skeletal muscles and microcirculation of skin. It is necessary to point out that the idea of creation of a new generation diagnostic device on the basis of the principles of speckle-optics, called laser specklometer, and its realization belongs to Leonid V. Tanin. Nowadays, in the whole world, there are no analogs of this device that allows studying noninvasively, remotely, with high sensitivity practically simultaneously the activity of muscles and microhemodynamics of skin in the process of curing the diseases of peripheral and central nervous system. Thanks to the speckle-optical diagnostics, it became possible to derive the data about functional condition of skeletal muscle, root of spinal marrow and single nerve trunks that is very important for defining the treatment tactics of this category of patients.

After graduating from Minsk State Medical Institute in 1993 and after clinical studies on the basis of the Institute of Neurology, Neurosurgery and Physiotherapy of the Ministry of Health of the Republic of Belarus, Andrei L. Tanin joins to the conduction of these studies. He conducts studies in the field of biomedical optics and defends a Candidate's dissertation on the topic "Restoration of the functions of peripheral nerves after neurography under the influence of laser hemotherapy." Andrei L. Tanin made a great contribution to the creation and development of the brand new for medicine speckle-optical diagnostics and laser hemotherapy in the treatment of diseases of peripheral and central nervous systems. Being a first-class neurosurgeon, Associated Professor, Candidate of Medical Sciences and Director of

"Republican Theoretical and Practical Center of Neurology and Neurosurgery" of the Ministry of Health of the Republic of Belarus (in 2010–2015), Andrei L. Tanin developed and implemented into medical practice clinical and diagnostic complex, which includes neurologic, electrophysiological and thermal imaging methods, laser speckle-optical methods of studies that made it possible to derive more complete and objective information about the character of damage and the flow of regenerative innervations process in surgically restored nerve trunks and tissues, which are innervated by them. Due to these studies, speckle-optical method became widely used on the early stage of diagnostics of such diseases of nervous system as compressive–ischemic and traumatic damages of peripheral nerves and also for the observation of the dynamics of the restoration of the functions lost as a result of damage of peripheral nervous system during the treatment.

Laser microhematomyograph (laser specklometer) has been effectively used in the Institute of Neurology, Neurosurgery and Physiotherapy of the Ministry of Health of the Republic of Belarus for the diagnostics of the diseases of central and peripheral nervous systems, but its possibilities still have not been fully discovered. It is necessary to highlight that the sphere of its application extends not only the measurements of skin blood flow and muscle tonic state but also on the studies of other organs and tissues.

In the course of the studies, in charge of which was Leonid V. Tanin and in which he actively participated, for the first time there was created series of contactless coherent-optical methods on the basis of optical holography, holographic and speckle-interferometry, holographic microscopy, speckle-photography, laser diffractometry with simultaneous electrophysiological control for the diagnostics and treatment of different neurologic diseases, control over the process of the development of which can be quantitatively estimated both on the cellular level and on the whole organism that allows studying the mechanisms of appearing and development of the diseases of this type was created. With the help of these methods and specially developed devices, pulsation activity of nerve cells was studied, water wave properties of nerve fibers were observed for the first time, the dependence of the speed of muscle contraction on the time, the field of deformations of epithelial tissue were defined.

Starting from 1978 with the participation of Leonid V. Tanin methods of laser puncture (LP), methods of combined usage of LP, hyperbaric oxygenation, LP and classical acupuncture and also intravenous laser blood irradiation (ILBI) for the treatment of diseases of peripheral nervous system (PNS) were developed and implemented in the Institute of Neurology, Neurosurgery and Physiotherapy of the BSSR. These studies allowed maintaining and organizing at qualitatively new level physical therapy, reasonably and actively attracting laser radiation of different intensity, length, spectral composition and other physical and pharmacological means. On the basis of the profound studies of the mechanisms of influence of laser radiation on the manifestation of atherosclerotic process depending on the degree of its manifestation certain techniques were developed that allow determining the regimes of radiation of whole blood and vessels walls: the intensity and duration of laser radiation, frequency and time of influence that provide the best treatment effect.

There were created highly sensitive methods of holographic interference microscopy and on their basis devices for the study of the changes of the form of red blood cells and lymphocytes, the areas of inner and outer monolayers of cell membrane, which arise during different hypoxic conditions accompanying the disorder of the processes of metabolism during atherosclerosis. For the first time, holographic studies of structural–functional characteristics of nerve, muscle fibers during the influence of laser radiation, magnetic field and hyperbary were conducted. It is necessary to especially highlight the results connected with the detection of water wave properties of nerve fibers that contain world priority and allow having a new look at the mechanisms of information transmission in human body with the participation of neuromuscular system.

The appearing of new contactless, noninvasive diagnostic methods and devices for determining blood characteristics (sizes, forms of erythrocytes, speed of surface blood flow) and adaptive-tonic function of peripheral nervous system will have a contribution to realization of the program of human health resumption. Already today its implementation into the healthcare practice particularly for the diagnostics of blood microcirculation and defining of muscle tonic state considerably increases the possibilities of diagnostic methods and contributes to the increase of the effectiveness of the treatment. Together with the usage of modern achievements in the sphere of laser technology and coherent-optics, non-traditional approach to the problem of the diagnostics and treatment of the diseases connected with the disorder of motion activity open enormous possibilities of prediction of appearance of complications and selection of a risk group among this pathology. Implementation into everyday practice of highly sensitive laser methods of express diagnostics of these diseases in considerable degree supplements the possibilities of already existing diagnostic methods and devices.

The application of highly sensitive methods of holographic interferometry for the studies and diagnostics of vertebral diseases of peripheral nervous system, in pathogenesis of which considerable role belongs to the vascular factor, is quite perspective. The detection of hemodynamic disorders and establishment of the correlation of changes of the system hemodynamics with the indicators of regional blood flow in patients with neurologic manifestation of osteochondrosis will allow detecting the pathogenesis of a disease in a specific person that will give the possibility to conduct differentially therapy during this pathology. Besides, with the help of holographic and moiré methods of contouring the data concerning the intensity of anatomic changes of spine that is important when dealing with the question about reasonability of the development of the indicators of kinesitherapy including the manual therapy can be derived.

In 1992–1995 on the initiative and with active participation of Leonid V. Tanin, the studies of the development and production of laser-holographic complex (laser cardiograph) for the defining of the state of human cardiovascular system began together with the employees of the Institute of Cardiology of the Ministry of Health of the Republic of Belarus, the Institute of Electronics and State Scientific Institution “B. I. Stepanov Institute of Physics of the National Academy of Sciences (NAS) of Belarus” (the Institute of Physics of the NAS of Belarus). The results of these studies showed that the usage of holographic methods in the area of cardiology opens enormous possibilities for the conducting of contactless, noninvasive diagnostics at the pathology of cardiovascular system.

The created laser-holographic complex is a new generation device, which will allow conducting studies over the operation of cardiovascular system, respiratory apparatus, muscular system, etc. The principle of operation of the complex is based on the defining and analysis of the field of the shift of the human chest surface points, which are caused by the activity of mechanical waves arising during the heart contraction. The formed illustrative picture of a shift field in different phases of a cardiac cycle will allow deriving the information about the functional state of valves and cardiac muscle and also about hemodynamic based on the chest surface. The laser-holographic complex will allow conducting the study of regulation mechanism, the integration of a nervous, cardiovascular, muscle and osteoarticular systems at a new level. Apart from deriving new knowledge in the sphere of scientific medicine, the complex can be widely applied in the clinical practice, notably: conducting timely treatment of mostly widespread and hardly detected at the early stages of diseases.

The usage of the given developments is planned in the sphere of neurology, neurosurgery, cardiosurgery in the clinical practice alongside with electrocardiography, angiocardiography and other methods.

Besides, Leonid V. Tanin presented the results of a number of priority studies conducted by him in the sphere of the development of the methods of resonance interferometry and fluorescence for the studies of plasma (notably in FT-1 tokamak device); dynamic resonance holography in the pairs of atomic sodium; developments of multiple long method of holographic contouring of the surface relief in statics and dynamics; study of the coherent properties of the laser sources with unstable mode structure; developments of the methods of holographic interference microscopy for the measurement of the deformations of microobjects: For the first time, the studies of the deformations of the resonator of the heterojunction laser mirrors with double interostructure in the system $GaAs$—$Al_xGa_{1-x}As$ were conducted. The perspectivity is shown of holographic microscopy application for the research of the parameters of semiconducting lasers in the wide range of currents, temperatures and mechanical loads.

The works of Leonid V. Tanin on the recording of dynamic holograms in a new class of detecting media—alkali metal vapors that for several years defined the analogous works of foreign authors are pioneering. He experimentally showed the perspectivity of laser application on the colorants with reconfigurable wavelength of radiation for the holographic interferometry (including resonance) and color holography that considerably widen the possibilities of given scientific directions. It is important to highlight that the results presented in Chap. 1 derived with the help of resonance methods with high sensitivity of measurements are presented in this monograph not accidentally. They give the possibility to develop on their basis new scientific approaches toward the study of spectral and functional conditions of medical and biological objects.

In conclusion, it is necessary to notice that the monography *Biomedical and Resonance Optics: Theory and Practice* prepared for the publishing by Leonid V. Tanin and Andrei L. Tanin is a fundamental scientific work at the meeting point of physics and medicine that opens perspectives for further studies in this sphere and undoubtedly causes great interest of scientists and specialists.

Minsk, Belarus A. N. Rubinov

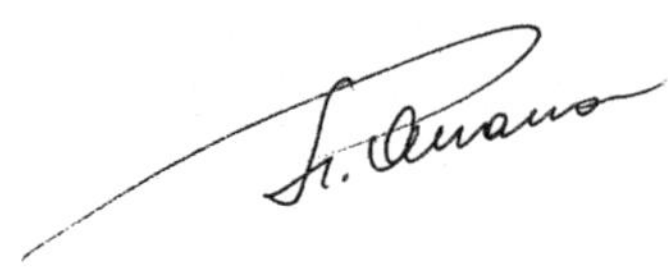

Minsk, Belarus P. A. Apanasevich

Minsk, Belarus I. P. Antonov

Minsk, Belarus G. I. Sidorenko

Preface

The creation of modern contactless, noninvasive methods for the human organism study and their introduction into medical practice with the help of a new generation of medical–biological devices based on the holography and speckle-optics is one of the actual objectives, and it became a reason for the creation of the given work.

The main idea of the studies of Leonid V. Tanin is the creation and development of a new direction in the medicine called biomedical optics. The author set himself a task to develop physical basis and principles of studies of medical–biological objects, starting from fundamental studies up to concrete practical applications. Enormous interest toward acquiring new knowledge in the sphere of biology and medicine constituted the search of modern scientific approaches and methods of studies with the usage of coherent and nonlinear optics, holography, resonant interferometry, speckle-optics, laser physics, etc., and their adaptation in the practical sphere.

In 1975, Leonid V. Tanin suggested beginning coherent-optical, holographic and speckle-optical studies of neuromuscular and cardiovascular systems that remain actually up to the present moment. A few years later in 1978 in the Institute of Neurology, Neurosurgery and Physiotherapy of the Ministry of Health of the BSSR, he created the scientific group "Coherent-optical studies of medical–biological objects," which was engaged into the development and usage of brand new contactless highly sensitive methods and devices and appliances for the diagnostics and treatment of diseases of peripheral and central nervous systems. In the same year, the author for the first time published scientific works in this area.

The book consists of five chapters. In Chap. 1, the results of the studies in the sphere of resonance optics including resonance dynamic holography, resonance interferometry and fluorescence are presented. Chapter 2 covers holographic interference microscopy for the study of such microobjects as nerve fiber, muscle fiber, lymphocytes, nerve, muscles in conditions of electrostimulation, influence of laser radiation, magnetic fields, hyperbaries with the usage of continuous and gated radiations of He–Ne laser, including the development and creation of an experimental sample of holographic interference microscope operating on the transmission and in the reflected light.

Holographic methods of interferometry for the study of changing in the course of time conditions of diffuse objects (in particular, functional state of the human blood circulation system) with the usage of the regimes: multiple long-wave reconfigurable interval between components of a spectrum laser radiation on the colorants and dual monopulses of laser radiation at the ruby with reconfigurable period of their succession, on the basis of these studies the development and creation of an experimental sample of laser-holographic complex (holographic cardiograph) at the dual monopulse of ruby laser are presented in Chap. 3. In Chap. 4, the attention is paid to the speckle-optical methods and devices for the study of longitudinal shift, oscillation, deformations, speed of the movement of the objects with complex diffusing surface (skin, muscle tissue) that became the basis for the creation of experienced sample of laser diagnostic device of a new generation—specklometer (microhematomyograph).

The principles of operation of laser specklometer as well as the development and usage of the methods of speckle-optical diagnostics and laser hemotherapy in the treatment of the diseases of human peripheral nervous system are presented in Chap. 5. Andrei L. Tanin made a considerable contribution to these studies.

An enormous contribution to writing the last four chapters of the monograph, to the comprehension, analysis and interpretation of the derived biomedical results as well as to its preparation for the publishing made the Candidate of Medical Sciences, neurosurgeon of the highest category Andrei L. Tanin. The main result in this monograph (in particular, studies conducted in the area of biomedical optics) is its practical importance in modern medicine. That is why he, neurosurgeon of the highest category with long-termed practical experience in the sphere of neurology and neurosurgery, in considerable degree independently conducted the studies on the subject "Restoration of the functions of peripheral nerves after neurography under the influence of laser hemotherapy." Together with other specialists: Doctor of medical sciences, Prof. G. K. Nedzved, Doctor of medical sciences, Prof. N. I. Nechipurenko, Candidate of medical sciences L. A. Vasilevskaya, Candidate of medical sciences S. E. Rovdo, etc., he created and developed brand new in modern medicine speckle-optical diagnostics and laser hemotherapy in the treatment of the diseases of peripheral and nervous systems. The monograph *Laser Hemotherapy in the Treatment of Diseases of Peripheral and Central Nervous Systems* was published by him with participation of the following authors Leonid V. Tanin, N. I. Nechipurenko, G. K. Nedzved and S. E. Rovdo (Minsk 2004). Today the studies in this sphere, which were begun 10 years ago, continue developing in the State Establishment "Republican Theoretical and Practical Center of Neurology and Neurosurgery" of the Ministry of Health of the Republic of Belarus.

Authors express their gratitude to all co-authors of these studies who helped to bring the idea of the development of the field of biomedical optics to life. It is pleasant to realize that the efforts spent on the development and creation of laser specklometer—a new generation unique device—are not wasted but bring enormous benefits to the medical science today. Ignaty Petrovich Antonov (Academician, Doctor of Medical Sciences, Professor, former Head of the Institute of Neurology, Neurosurgery and Physiotherapy of the Ministry of Health of the

Republic of Belarus) deserves special words of gratitude for his wise scientific foresight and provided to us support in a difficult time, unique intuition toward everything new in science, for the paternal care shown not only to us, the authors of this book Leonid V. Tanin and Andrei L. Tanin, but also to the creation together with us a brand new scientific school of speckle-optical diagnostics and laser hemotherapy in the treatment of diseases of peripheral nervous system. We have a deep feeling of gratitude and appreciation toward Boris Ivanovich Stepanov (Academician, Doctor of physical and mathematical sciences, Professor, former Director of the Institute of Physics of the Academy of Sciences of the BSSR) and to Alexander Sergeevich Rubanov (Academician, Doctor of physical and mathematical sciences, Professor, Head of the Laboratory of Optical Holography of this Institute) for presenting to us opportunity to realize ourselves in the creation and development of a brand new school in modern scientific and biomedical optics. Besides, we would like to express our gratitude to the Candidate of medical sciences Ludmila Alexandrovna Vasilevskaya who during many years effectively, with great passion and professionalism, uses laser specklometer for the diagnostics of the diseases of peripheral and central nervous systems.

We express our gratitude to the people with whom we were lucky to communicate, work, study and gain experience. I would like to touch every chapter and point out each person who supported us:

Chapter 1: S. E. Frish, Yu. I. Ostrovsky, G. V. Ostrovskaya, A. N. Zaidel, Yu. N. Denisyuk, V. G. Sidorovich, G. V. Dreiden, V. I. Gladushchak, E. Y. Shreider, E. N. Shedova, I. I. Komissarova, V. N. Philippov, N. A. Pobedonostseva, U. V. Kovalchyk, I. A. Merkulov, G. T. Razdobarin, V. V. Semenov—Russia; B. I. Stepanov, A. N. Rubinov, V. A. Mostovnikov, M. M. Loiko, V. S. Motkin, S. A. Batische, S. A. Anufrik, V. S. Burakov, P. A. Apanasevich, N. V. Tarasenko, A. S. Rubanov, E. V. Ivakin, B. A. Sotsky—Republic of Belarus; M. S. Soskin, S. G. Odulov, V. B. Taranenko—Ukraine; K. K. Schwarz, A. O. Ozols—Latvia;
Chapter 2: B. I. Stepanov, A. S. Rubanov, I. A. Bulygin, V. N. Kalunov, G. V. Abramchik, S. V. Konev, E. A. Chernitsky, V. M. Mazhul, I. V. Markhvida, A. V. Goroshkov, L. E. Alikevich (Batai), B. Y. Vilner, N. I. Luschitskaya, I. P. Antonov, G. K. Nedzved, V. V. Panteleev, N. V. Solovei, R. M. Tanina, V. A. Lapina, U. I. Musienko, A. R. Gavrilova, L. A. Vasilevskaya, S. E. Rovdo, V. V. Dubovik, G. A. Govor, L. I. Rachkovsky, V. B. Shalkevich, G. G. Petrovsky, G. I. Ryabtsev, A. N. Kuzmin, V. F. Voronov—Republic of Belarus; O. S. Sotnikov, N. G. Vlasov, A. V. Sokolov, E. N. Lekhtsier, A. N. Metelkin—Russia, A. S. Davydov—Ukraine, Hert von Bally—Germany;
Chapter 3: P. A. Apanasevich, A. S. Rubanov, I. V. Markhvida, I. L. Drobot, N. M. Spornik, I. P. Antonov, S. D. Bezzubik, S. A. Aleksandrov, L. I. Rachkovsky, M. M. Loiko, S. K. Dik, V. K. Zabarovsky, S. A. Naumovich, A. S. Artushkevich, A. I. Golovko, G. I. Sidorenko, A. V. Frolov, V. V. Mironchik, V. A. Pilipovich, A. A. Kovalev, B. N. Tushkevich, U. F. Morgun, A. V. Agashkov, F. G. Drik, V. V. Manikalo—Republic of Belarus; Yu. I. Ostrovsky, M. M. Butusov, Yu. N. Denisyuk, N. G. Vlasov, A. I. Larkin, A. P. Kapitsa, B. G. Turukhano, N.

Turukhano, V. G. Komar, O. B. Serov, V. A. Vanin, M. M. Ermolaev, M. K. Shevtsov—Russia; V. B. Markov—Ukraine; Sh. D. Kakichashvili—Georgia; Hert von Bally—Germany; John Colfield, Hans Bjelkhagen—USA, Michel Grosmann —France, P. Gregush—Hungary;
Chapter 4: I. V. Markhvida, A. S. Rubanov, N. A. Fomin, I. A. Utkin, S. K. Dik, I. L. Drobot—Republic of Belarus; I. S. Klimenko, N. G. Vlasov—Russia; Marti Lopes Luis—Cuba;
Chapter 5: I. P. Antonov, A. F. Smeyanovich, A. S. Rubanov, I. V. Markhvida, A. A. Kumeisha, S. A. Aleksandrov, M. M. Loiko, S. P. Andreev, A. S. Uzukbadzhakov, V. M. Aranchuk, S. K. Dik, M. M. Korol, P. N. Bagrov, V. A. Dmitriev, G. K. Nedzved, L. A. Vasilevskaya, S. E. Rovdo, R. M. Tanina, V. A. Lapina, A. V. Astapenko, P. A. Vlasuk, L. N. Anatskaya, A. R. Gavrilova, L. I. Matusevich, V. A. Misnikova, E. P. Titovets, N. I. Nechipurenko, G. G. Petrovsky, G. I. Ovsyankina, G. V. Zobnina, S. A. Likhachev, Y. G. Shanko, V. I. Hodylev—Republic of Belarus; G. B. Semenov, E. M. Dianov, V. A. Kozlov, O. A. Vlasenko —Russia; T. Asakura—Japan.

We would like to express our gratitude and respect to the reviewers of the first edition of our book *Coherent and Biomedical Optics: Theory and Practice*, 2007, to the Doctor of Technical Sciences, Prof. Nikolai Georgievich Vlasov, Doctor of technical sciences, Prof. Nikolai Maksimovich Spornik and Doctor of physical and mathematical sciences, Prof. Boris Nikolaevich Tushkevich, to the talented scientists—generators of brand new ideas in the sphere of coherent-optics and holography, published in numerous scientific articles, patents and authors' inventions both in our country and abroad, to those who passed away early but managed contributed greatly to modern science by creating, developing and realizing on practicing a number of scientific schools such as iridescent holography, holographic and speckle-interferometry, holographic interferometry with the use of dual monopulses and many others.

We are grateful to the reviewers of the second extended edition of this book titled *Biomedical and Resonance Optics: Theory and Practice*, to the great scientist in the sphere of holography, laser physics, coherent and nonlinear optics, Doctor of physical and mathematical sciences Boris Ganievich Turukhano, Head of the Laboratory of Holographic Informational and Measuring Systems of St. Petersburg Institute of Nuclear Physics named after B. P. Konstantinov of Russian Academy of Sciences, to Doctor of physical and mathematical sciences Vladimir Victorovich Kabanov, Head of the Laboratory of Optical Holography of the State Scientific Institution "B. I. Stepanov Institute of Physics of the National Academy of Sciences of Belarus," to the Doctor of physical and mathematical sciences Alexei Leonidovich Tolstik, to the Professor of the Department of Laser Physics and Spectroscopy of Belarusian State University and Doctor of Medical Sciences Arnold Fedorovich Smeyanovich, Head of the Department of Neurosurgery of State Establishment "Scientific and Practical Center of Neurology and Neurosurgery" of the Ministry of Health of the Republic of Belarus who took the responsibility to consider the results of our studies.

In conclusion, we would like to express our special appreciation and gratitude for the help in the preparation of the manuscript for the publishing by the employees of Magic of Light, Ltd., A. V. Madiarov, A. A. Korolenko, A. A. Kazak, S. P. Zabreiko, A. V. Krivko-Krasko, secretaries S. I. Korotkova, O. I. Kocherga, K. P. Ionayskaite, M. G. Potashova and V. A. Lakhanskaya.

Minsk, Belarus

Leonid V. Tanin
Andrei L. Tanin

Contents

About the Authors

Leonid V. Tanin Academician of the International Academy of Engineering, Doctor of physical and mathematical sciences, honored Inventor of the Republic of Belarus, Chairman of the Board of Directors of CJSC HOLOGRAPHY INDUSTRY.

In 1971, he graduated from Physical Department of State Leningrad University (Department of Optics and Spectroscopy), in 1974—postgraduate course of Leningrad Physical and Technical Institute of A. F. Ioffe of the Academy of Sciences of the USSR (Sector of Plasma Optics). In 1978, he created a scientific group "Coherent-optical studies of medical and biological systems" in the Research Institute of Neurology, Neurosurgery and Physiotherapy of the Ministry of Health of the BSSR, and he was in charge of it more than 20 years. He is one of the founders of a future-oriented field in science–biomedical optics. He developed the basis of speckle-optical diagnostics and laser hemotherapy in the treatment of diseases of peripheral and central nervous systems, he created the diagnostic device "Laser specklometer" that does not have analogs in the whole world, for the study of biomechanical activity of muscles and surface blood flow. In 2012 for the monograph Biomedical and Resonant Optics: Theory and Practice, Leonid V. Tanin and Andrei L. Tanin were awarded the First Prize named after Yu. I. Ostrovsky "For the best scientific work in the field of optical holography and interferometry" of the Russian Academy of Sciences. In 2014, He successfully defended his doctoral thesis on the topic "Resonant holographic and speckle-optical studies of phase, diffuse and mirror objects."

He is the author of 300 scientific publications, including four monographs, 78 patents and certificates of authorship and patents that are registered both in our country and abroad, co-author of two teaching textbooks.

He was a member of the International Program Committees in the field of biomedical optics (1990, 1992, Germany; 1994, Japan; 2002, Belarus). He presented lecture courses in this sphere in the leading universities and world scientific centers.

In 1994, he was Elected Chairman of the Belarusian Society of Biomedical Optics in the Belarusian branch of SPIE (USA).

Andrei L. Tanin Associate Professor, from 2010 to 2015, Director of State Establishment "Republican Scientific and Practical Center of Neurology and Neurosurgery" of the Ministry of Health of the Republic of Belarus, Chief Neurosurgeon of the Ministry of Health of the Republic of Belarus.

After graduating from Minsk State Medical Institute, Andrei L. Tanin began studying at clinical residency and afterward he entered postgraduate studies in the neurosurgical department of the Research Institution of Neurology, Neurosurgery and Physiotherapy of the Ministry of Health of the Republic of Belarus. In 2003, he successfully defended candidate dissertation "Restoration of the function of peripheral nerves after neurography under the influence of laser hemotherapy".

He is one of those who developed a completely new in modern medicine speckle-optical diagnostics and laser hemotherapy in treatment of peripheral and central nervous systems. With his participation, the monograph Laser Hemotherapy in the Treatment of Diseases of Peripheral Nervous System was published.

Since 2015, he has been the associate professor of the department of neurology and neurosurgery of State Educational Establishment "Belarusian Medical Academy of Postgraduate Education", a leading researcher at the State Establishment "Republican Scientific and Practical Center of Neurology and Neurosurgery." He was awarded the badge

"Excellence in Health Care of the Republic of Belarus", diplomas of the Ministry of Health of the Republic of Belarus and the National Academy of Sciences of Belarus.

His main research interests are related to the problems of surgical treatment of intracerebral hypertensive hemorrhages, supratentorial gliomas and damage to peripheral nerves.

He is the author of more than 115 scientific publications, including three monographs, four patents and seven teaching textbooks.

Chapter 1
Resonance Methods for Increasing Sensitivity of Interferometry, Fluorescence, Dynamic Holography

Coherent-optical—holographic [1–4], interference [5–12] and fluorescent [13–15] methods make different demands to the light source depending on the tasks: By considerable energy output and power of the radiance, it can be its high spatial coherence and monochromaticity; tunable wavelength in a broad spectral range, two and more generation lines of the radiance with the controlled spectral interval, different width of pulse generation of the radiance, and also generation of two or several pulses with the controlled time interval, the radiance with given and changeable direction of vector polarization and so on.

At the heart of the studies, which are presented in the first chapter, are the phenomena of anomalous dispersion. In the studies of these phenomena, D. S. Rozhdestvensky [12, 16–26] discovered a lot of interesting things, particularly, his studies became the continuation of his school.

The results of the studies on resonance optics, which include dynamic resonance holography, resonance interferometry and fluorescence are given in the present chapter. Resonance methods [27–29], which enable to regulate the medium sensitivity to the recording of dynamic resonance gain-phase (bulk and plane) holograms are also proposed and experimentally carried out in this chapter. Thereby these resonance methods reveal a new class of detecting media for dynamic holograms—gaseous medium [30–35], and also they allow raising enormously the sensitivity of interferometry and fluorescence for concentration detection of normal and excited atom in plasma [13, 27]. For the first time, the author reveals the availability of using pulsed wavelength-tuned dye laser radiation [12, 36] for these purposes. The presence of coherent light source with gradually tuned wavelength during the recording of resonance interferograms enables to approach to the absorption line of the components of heterogeneity arbitrary close that considerably increases the sensitivity of this method to the definite type of particles [27, 37, 38]. As well as great sensitivity increasing it also enables to study space distribution of the given component without distorted influence of the rest of the components, so it becomes possible to determine the contribution to the refraction of different components of the mixture. The method sensitivity is considered, and the boundaries are determined of its applicability for

L. Tanin et al., *Biomedical and Resonance Optics*, Bioanalysis 11,
https://doi.org/10.1007/978-3-030-60773-9_1

the case of absorption line, which has dispersive profile in the assumption of the final width of the probing radiation line. Results of the studies visually confirmed the advantage of using dye lasers in this method. So, for example, during the experiment on the detection of the sodium atoms concentration Rhoda mine 6G laser with wavelength adapted to the resonance transition NaI ($\lambda = 589.66$ nm) was used [39]. Further in the works [40, 41], detecting the concentration of excited hydrogen atoms in the laser-induced spark or studying the laser flare on the target containing lithium generating dye solution replacement was made that was provided by another resonance line wavelength: hydrogen H_α ($\lambda = 656.3$ nm) or Li ($\lambda = 670.7$ nm). These joint studies made it possible to extend the number of studied plasma components and also increase the sensitivity of resonance interferometry method [27].

As the objects for testing of resonance methods of increasing interferometry sensitivity, fluorescence and dynamic holography sodium plasma (the flame of alcoholic lamp and arc discharge), high-temperature hydrogen plasma in the facility of tokamak and also atomic sodium vapors have been chosen.

The results of the present work on the application of the dye laser for resonance interferometry, fluorescence and holography opened the prospects and were used during resonance-holographic, resonance-interference and resonance-fluorescent plasma studies in the Plasma Optics Department of Leningrad Physics and Technology Institute named after A. F. Ioffe of the Academy of Sciences of the USSR (study supervisors A. N. Zaidel, Yu. I. Ostrovsky, G. V. Ostrovskaia, and the author who was one of the employees of this department at that time). Resonance fluorescence experiments carried out by the employees of the Institute of Plasma Physics Laboratory G. T. Razdobarin, V. V. Semenov and others in association with the employees of the Laser Plasma Diagnostics Laboratory of the Institute of Physics of the Academy of Sciences of the BSSR (supervisor V. S. Burakov) along with the author enabled to determine neutral hydrogen atom concentration under plasma excitation by dye laser radiation with lamp pumping [13] with resonance line wavelength H_α.

For efficient realization of these coherent-optical methods, it was necessary to develop and manufacture pulsed wavelength-tuned dye laser radiation with laser and lamp pumping and to determine its optimal operation.

During these studies, there was a need to study space and temporal coherence of dye laser radiation, the mode structure of which changes from pulse to pulse. In connection with the absence of the methods of measurement of spatial coherence radiation of such light sources, holographic step and holographic integral methods were proposed and developed [42, 43]. Also for spatial coherence function (SCF) measurement, the device of radiation laser source was proposed, developed and made. It was named a coherometer [44].

The necessity of the step method development was caused by the fact that usual holographic method because of its nonlinearity of the record, inherent to actual photomaterials, does not make it possible to measure mutual-coherence γ of laser edge points having essentially different radiation brightness. In this case, the use of the set of holograms photographed with different exposure (step attenuation) enables to overcome this problem and to recover the full form of spatial coherence function (SCF) between any laser edge points.

Though holographic and step methods enable to recover the full form of SCF, their use is quite difficult and time-consuming experimental task. So, for example, the number of measurements n necessary for SCF determining for points of laser end is equal to N^2 (where $N \approx 100$). At the same time in some cases, when behavior details of $|\gamma|^2$ are insignificant, it is convenient to determine averaged values $|\gamma|^2$ (e.g., to approximate $|\gamma|^2$ by homogeneous function) for the space coherent characteristic. For the measuring laser edge-averaged SCF distribution, less time-consuming integral method is proposed.

The possibilities of different methods, which are used in the radiation spatial coherence investigations, have been studied. They are interference, holographic, holographic with microphotometry of intensity initial distribution, integral under the spatial coherence measurement of pulsed organic dye laser radiation with unstable mode structure, particularly, rhodamine 6G laser with the laser pumping.

1.1 Rhodamine 6G Laser with Laser Pumping for Holography, Resonance Interferometry and Fluorescence

For the first time, dye solution generation was found in the USSR [45, 46], in the USA and in FRG in 1966 [47, 48]. B. I. Stepanov, A. N. Rubinov, and V. A. Mostovnikov contributed to the creation and development of these lasers; therefore, they were honored with the USSR State Prize [49, 50].

Dye lasers are a special class of optical quantum generators [50–52]. Having broad amplification band, organic dyes enable to carry out graduated frequency tuning in the broad spectral region (up to 100 nm in one dye type). During the experimental studies on resonance interferometry and dynamic resonance holography, "Raduga-3M" dye laser-type [39, 53], which was developed in the Institute of Physics of the Academy of Sciences of the BSSR, was assumed as a basis.[1] Using "Raduga-3M," it was possible to get quite high-power laser radiation in the spectral region of 360–1200 nm. In the present work for radiation generation, the effective dye

[1]The dye laser "Raduga-3M" with laser pumping was for the first time mounted by the author from the separately developed drawings into the instrument version in the Institute of Physics of the Academy of Sciences of the BSSR and was made in the optical mechanics departments of the Leningrad Physics and Technology Institute named after A. E. Ioffe of the Academy of Sciences of the USSR (1971). During the process of producing, several changes were added to this laser device. In particular, the length of the resonator was increased, and the Fabry–Perot interferometer was introduced in it with the basis of 100 μm, what gave the possibility to narrow the width of generator spectrum up to 0.01–0.03 nm and carry out the regime of one-and four-frequency generation with the pulse energy of 10^{-3} J and power of about 0.1 mW. These changes made it appropriate for the usage in holography purposes and resonance interferometry.

Rhodamine 6G with the wavelength-tuned radiation of 570–620 nm (that coincides with resonance sodium of natrium NaI—589 nm and 589.6 nm) was chosen.

In Fig. 1.1, the optical scheme of the laser facility can be seen. The optical pumping of the cell with the Rhodamine 6G alcoholic solution was made by the second ruby laser harmonic, working in Q-switched mode. Ruby laser resonator is formed by 100% dielectric mirror (1) and by the stack made of two sheet glasses (4) and the dye lasers—by the dielectric mirrors with the reflection index of 100 and 60%, accordingly (10).

Thereafter, the first mirror was replaced by the reflecting grating (7), which can be seen on the scheme (Fig. 1.1). During the substitution of 100% dielectric mirror for the diffraction grating (7) (1200 lines/mm with the first reflection index of 70%), the narrowing of the generation spectrum up to 0.6–2.0 nm occurred. The graduated frequency tuning in the spectral region of 570–620 nm was ensured by the diffraction

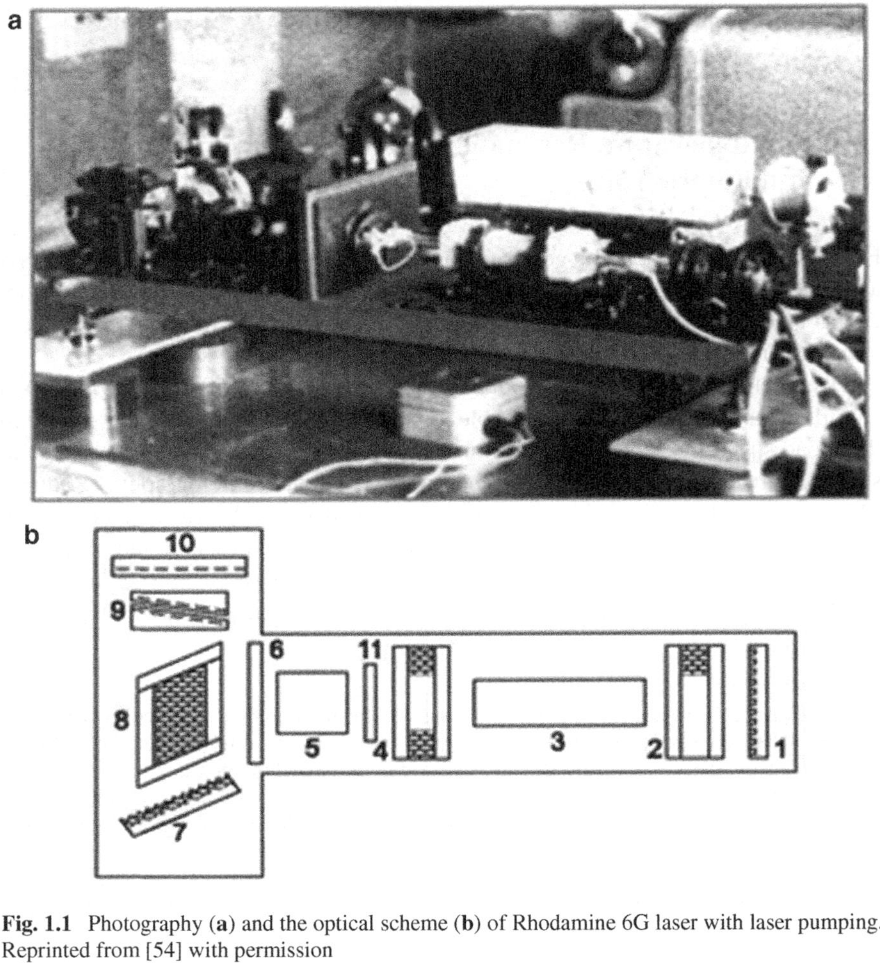

Fig. 1.1 Photography (**a**) and the optical scheme (**b**) of Rhodamine 6G laser with laser pumping. Reprinted from [54] with permission

grating rotation. Figure 1.2b shows the spectrogram where during its mounting some changes were made, in particular, the resonator length was increased and spatial coherence spatial coherence with the basis of 100 μm was included, and due to these changes it became possible to narrow the generation spectrum width of 0.1–0.3 J and to carry out single-frequency and four-frequency generation regime with the pulse energy of 10^{-3} J and with the pulse power of about 0.1 mW. These changes made it appropriate for the purposes of holography and resonance interferometry.

Spectogram where the spectra of two pulses were sequentially fixed during the generation wavelength tuning through grating rotation is shown in Fig. 1.2. In this case, maximal generation energy is 0.006 J.

For further generation spectrum narrowing into the Rhodamine 6G laser, the Fabry–Perot interferometer (9) with the basis of 100 μm (free spectral range is 1.8 nm) was included, the mirrors of which had dielectric covers with the reflection index of 70%. Great dispersion of the Fabry–Perot interferometer equal to 1.8 nm provides narrowing of generation lines up to 0.01–0.03 nm that corresponds to the

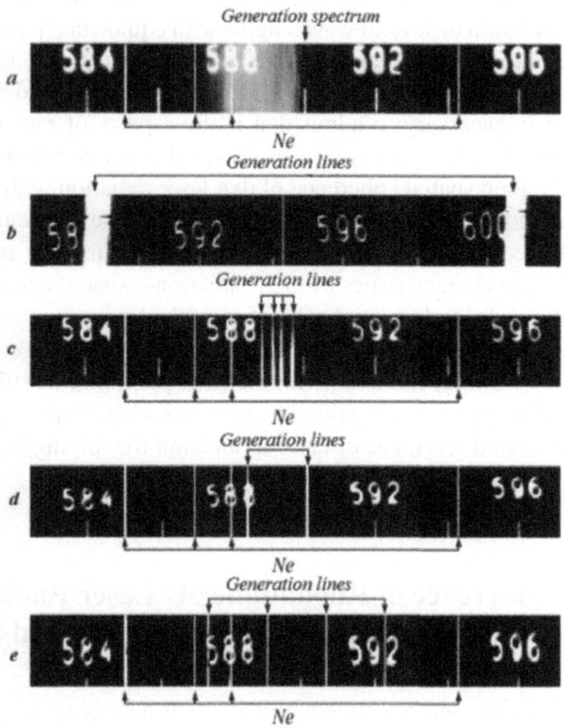

Fig. 1.2 Generation spectra of rhodamine 6G dye laser with laser pumping: **a** the non-selective resonator (the mirrors are $R = 100\%$, $R = 60\%$, the resonator length is 20 cm); **b** selective resonator (the diffraction grating, the exit mirror is $R = 60\%$); **c** selective resonator (the diffraction grating, the exit mirror is $R = 60\%$, Fabry–Perot interferometer, four impulses with Fabry–Perot interferometer sequential slope); **d** four-frequency generation mode. Reprinted from [54] with permission

radiation coherence length of 36 mm and allows using it for the holography and resonance interferometry purposes [41, 47]. Moreover, in this case, with the Rhodamine 6G concentration corresponding to the absorption constant of 12 cm^{-1} for $\lambda =$ 347.2 nm under the change of pumping energy from 0.03 to 0.05 J single-frequency or two-frequency generation mode was produced (Fig. 1.2c, d). The spectrogram (Fig. 1.2c) also shows radiation frequency tuning with the Fabry–Perot interferometer slope (four generation pulses with the interferometer sequential slope). Meanwhile, grating slope does not change. Wavelength-tuned interval in the case of single-frequency mode covered spectral region of 25 nm that was achieved by simultaneous slope of the grating and interferometer. In this case, the generation energy peak was 0.002 J. By sloping Fabry–Perot interferometer without changing the diffraction grating position, it was possible to tune laser frequency as well as to change the spectral intensity relations between the generation lines. The interferometer base increase resulted in the decrease of the interval between the generation lines.

By doubling the dye concentration in the solution and increasing pumping energy to 0.065 J, simultaneous generation mode of four wavelengths with 1.8 nm path was produced. Meanwhile, the width of each component was 0.03 nm (see Fig. 1.2e). Simultaneous generation of several wavelengths with equidistant path between them is of the utmost interest in the area of dye lasers practical use, in particular, in holographic interferometry to outline the relief of the time-dependent surfaces. At the same time, it is necessary to admit that if the aspects of radiation frequency selection (the control of time coherence of pulsed dye lasers radiation) have been studied quite well, then spatial coherence of dye laser radiation with laser pumping has not been studied. It is connected with the fact that the challenge of the laser source practical use is the absence of measuring and controlling methods of one of the most important characteristics of their radiation—spatial coherence. And the most difficult thing is to study spatial coherence of the laser radiation with unstable mode structure. The problem has appeared how to measure spatial coherence of the pulsed Rhodamine 6G laser radiation with the laser pumping. In 1974, L. V. Tanin published the studies on this laser radiation coherence characteristics using his own methods in order to find resources of laser application for solving holography tasks [39, 55]. This will be in detail discussed below.

1.2 Spatial Coherence of Rhodamine 6G Laser Radiation with Laser Pumping and Its Measuring Methods

In the holographic experimental practice, it is often necessary to know a quantitative variable, which describes interference pattern contrast produced during the use of radiation sources under study. As such, a radiation characteristic mutual-coherence function is taken. This function describes the correlation between radiation electrostatic field amplitude in arbitrary points of space r_1 and r_2 and time t and $t + \tau$ [56]:

$$\Gamma(r_1, r_2, \tau) = \langle \mathcal{E}(r_1, t + \tau)\mathcal{E}(r_2, t) \rangle \tag{1.1}$$

Averaging here is typical for such experiments, and it is equal to the time interval T (e.g., T is the exposure time) the quantity of which is supposed to be much larger than the period of electromagnetic oscillations. The quantity of the function Γ_{12} depends on the field correlation in the points r_1 and r_2 as well as on the quantity of these fields, i.e., on the light intensity. It is convenient to norm this function for the correlation studies and thereby to identify the degree of coherence as:

$$\gamma_{12}(r_1, r_2, \tau) = \frac{\Gamma_{12}(r_1, r_2, \tau)}{\sqrt{\Gamma_{11}(r_1, r_1, 0)\Gamma_{22}(r_2, r_2, 0)}} \tag{1.2}$$

The degree of coherence γ_{12} contains the information about field correlation in different space points r_1, r_2 and with different phase shift τ.

In most cases[2] [57], the degree of coherence can be expressed by two functions product, one of which is spatial coherence and another one is time coherence.

$$|\gamma_{12}(r_2, r_2, \tau)| = |\gamma_{12}(r_1, r_2, 0)||\gamma_{12}(r_1, r_1, \tau)| \tag{1.3}$$

The optical fields, for which spatial and time coherence effects can be separated from each other, have been named mutually spectroscopically pure [58]. In accordance with the above-mentioned spatial coherence, it identifies the correlation between the fields in different space points r_1 and r_2 with phase zero shift, and time coherence is provided by the field correlation in the present space point r_1 during different periods of time.

Under certain assumptions, spatial coherence is generally determined by the light spectral characteristics, and time coherence—by the geometrical adjectives of the source (by its extent and its angularity). Spatial and time coherence function module characterizes the interference pattern contrast, which appears under two light beams superposition and consequently can be detected from the interference and diffraction experiments.

For mutual-coherence function, E. Wolf [59] derived wave equation, which allows detecting a range of common characteristics of this function behavior. So, particularly on the basis of this equation, it is possible to determine the value of mutual-coherence function in the volume V, if the value Γ_{12} is set on the surface S, which limits this volume:

$$\Gamma(r_1, r_2, \nu) = \iint\limits_{SS'} \frac{\partial G}{\partial n}(r_1, S)\frac{\partial G^*}{\partial n'}(r_2, S')\Gamma\big(S, S', \nu\big)\mathrm{d}S\mathrm{d}S', \tag{1.4}$$

[2] It becomes possible when the path difference of the points r_1 and r_2 is considerably lower than the coherence length of the light emission $L = c\tau$, where c is the speed of light that can be easily carried out for the quasi-monochromatic light, the effective spectral bandwidth $\delta\nu$ of which is connected with the frequency ν by the relation: $\delta\nu/\nu \ll 1$.

here G is the Green's function of wave function for Γ; $\frac{\partial}{\partial n}$ is the normal derivative to the surface S.

Equation (1.4) is written for the private presentation of mutual-coherence function $\check{\Gamma}$:

$$\widetilde{\Gamma}(r_1, r_2, \tau) = \int_0^{\infty} \Gamma(r_1, r_2, \nu)\exp(-2\pi\nu\tau)\mathrm{d}\nu \tag{1.5}$$

Supposing that different points of the radiator surface give off uncorrelated light, i.e., on the surface:

$$\Gamma\left(S, S', \nu\right) = \delta_s\left(\vec{S} - \vec{S}'\right)\Gamma(S, \nu), \tag{1.6}$$

where δ_s is the delta function. And substituting (1.6) into (1.4), we find out that

$$\Gamma(r_1, r_2, \nu) = \int \frac{\partial G}{\partial n}(r_1, S)\frac{\partial G^*}{\partial n'}(r_2, S)\Gamma(S, \nu)\mathrm{d}S. \tag{1.7}$$

It follows that the field, statistically independent in different radiant points, during the propagation takes incomplete spatial coherence. Thus, high spatial coherence radiation can be produced from the usual thermal sources if they are situated at rather long distances. However, in this case, the light intensity $I \sim \frac{1}{r^2}$, where r is the radiator distance, for example, to the sun, considerably weakens.

Thermal source low spatial coherence is provided by the fact that the photon phases, which are emitted by different source atoms, are not correlated. Laser radiation spatial coherence is closely related to the number of transverse modes [59, 60]. Depending on the correlation degree of laser radiation in different modes, it can have high or low spatial coherence. If special synchronizing systems are not applied, then the laser radiation in different modes, as a rule, is not correlated. In this case for radiation spatial coherence improvement, transverse mode selection is usually carried out that can be achieved by including adjustable diaphragm resonator into [61].

As is known, in the case of thermal source radiation, low time coherence occurs. The fact is that the excited atom goes into the ground state during finite time $\tau_1 \sim 10^{-8}$ s. Wave packet emitted by elementary radiator is characterized by definite phase and amplitude, and, consequently, for the time interval $\tau < \tau_1$, thermal source radiation turns to be coherent. But the initial phases of different wave packets are arbitrary, that is why for $\tau > \tau_1$ radiation is practically incoherent. It is possible to raise radiation spatial coherence by including narrow-band filter. As is known, wave packet while passing through such a filter spreads in time $|\tau_1' > \tau_1|$ that leads to the spatial coherence increase. For more sophisticated treatment in this case, it is convenient to move to Fourier representation for time dependence of the electric field light wave.

Let us examine the radiation, which consists of the Gaussian-shape packages:

$$\mathcal{E} \sim \mathcal{E}_0 \exp\left(-\frac{|\vec{r} - \vec{c}t|^2}{ra}\right) \exp[i(k_0 r - \omega t)]. \tag{1.8}$$

From here, it is easily seen that coherence characteristic time of this radiation is $\tau = \frac{a}{c}$, where c is the speed of light, and the coherence length is $l = a$.

Fourier representation of such a wave package is given by:

$$\mathcal{E}(k) \sim \exp\left[\frac{(k - k_0)^2}{2\sigma}\right] \exp[i(kr - \omega t)], \tag{1.9}$$

where $\sigma = \frac{1}{a}$.

Thus, the package under consideration is characterized by wave vector spread $\Delta k \approx \sigma = \frac{1}{a}$, i.e. $\Delta kl \approx 1$; similarly for frequency spread we have $\Delta\omega\tau \approx 1$.

It can be easily seen from here that for the spatial coherence raise, it is necessary to reduce the half-width of the radiation line $\Delta\omega$. In this context, determination of the degree of temporal coherence is connected with radiation spectral width measuring. For the temporal coherence quantitative description, it is convenient to use the notion of radiation coherence length, which, as it is known [62], can be given by:

$$\ell = \frac{\lambda^2}{\Delta\lambda}, \tag{1.10}$$

where $\Delta\lambda$ is the radiation spectrum width.

For the spatial coherence degree measuring more often interference [63–66], diffraction [67, 68] and holographic [69–73] methods are used. For the wave fields characteristic when spatial coherence degree is revealed as spatially inhomogeneous function [72], it is necessary to know the full spatial coherence function, which includes mutual-coherence degree between all field points. This problem can be solved with the help of holographic methods [69, 73].

One of the most effective methods for the spatial coherence studies of dye lasers is the holographic method proposed by Yu. N. Denisyuk, D. I. Staselko and R. R. Gerke. The main idea of the holographic method [69] is as follows. A hologram is a photosensitive layer where the photographic record of the interference pattern is carried out and which appears under the superposition of reference and object waves. The contrast of this interference pattern is provided by the radiation spatial–temporal characteristics of the used source of light. In turn, the brightness of the image reconstructed by a hologram is in proportion to the interference structure contrast square. Thus, the brightness of the image reconstructed by a hologram contains radiation spatial–temporal coherence information. The scheme of hologram recording is diverse, and it depends on the temporal coherence studies [74] or on the spatial coherence studies of the radiation [42–44, 69].

For the spatial coherence studies, it was proposed [69] to record a hologram in such a manner that the laser edge is designed on a hologram (reference beam) and on the flat diffuse screen (object beam). If radiation contains several transverse modes and each of them has its own amplitude-phase distribution on the resonator edge and which differs from the distribution of the rest, then different modes will create its own independent interference patterns in the hologram plane. In other words, in those parts of the hologram where each mode interferes itself, there will be a contrast interference pattern, and in the region of overlap of different modes the interference pattern will be blurred. Meanwhile, the more modes there are in the radiation, the faster the contrast of the total interference pattern in the hologram plane is decreasing. A hologram will reconstruct the wave field only in the part where pattern contrast is distinct from zero, i.e., within the visibility of interference pattern. Thus, scanning the hologram by a narrow beam, for example, by the beam of He–Ne laser, in the reconstructed image of the diffuse screen only those parts of the laser edge will be seen, which during the hologram recording have been radiating this mode. Hereof, by reconstructing different hologram parts, it is possible to identify the number of modes generated by the laser. In this regard, it is important to note that for obtaining the image of the mode field full structure, the diffuse screen is chosen in such a way that the radiation from the whole points of the diffuse screen reaches each part of holograms. The variants of this method are presented below. They enable to determine quantitatively the radiation spatial coherence degree: just holographic and integral methods [42–44, 69]. Holographic procedure is convenient to be used as it enables to get a peculiar spatial coherence function image for one pulse that is very important during the studies of laser sources spatial coherence function, the radiation of which has unstable mode structure.

1.2.1 Holographic Step and Integral Methods of Radiation Spatial Coherence Measurement

In some cases, edge intensity distribution of laser sources, which operate in a single-mode regime as well as in a multimode regime, is quite heterogeneous. In this case, it is almost impossible to get the linear recording of the whole hologram surface. Therefore, the proposed method [73] for the quantitative measurement of spatial coherence full function cannot be used in this way. The only assumption in the proposed method [42], which can be named as a "step method," is quasi-linearity of the hologram recording.

It can be explained in the following way. Let the change of amplitude ratio of hologram transmission coefficient conditioned by the exposure interference part δH be linearly connected with δH:

$$T(H+\delta H) = T(H) + f(\delta H), \tag{1.11}$$

where $H = It$ is the average exposure; $\delta H = \delta It$ is the exposure version, which is provided by the intensity interference part δI, and $f = \frac{\mathrm{d}T}{\mathrm{d}H}$ is the linear amplitude transmission coefficient. The value of the ratio f depends on the reference beam intensity and depending on the value H can be considerably changed along the whole hologram surface. But in small hologram parts, it is assumed that f is a constant and does not depend on δH. Using the work studies [69] and (1.11), it is easy to get the following formula for the reconstructed image intensity:

$$I(R_i, r_j) = f^2(r_j) I_0(r_j) I_p(R_i) \left|\gamma(R_i, r_j, \tau_{ij})\right|^2 \tilde{I}(r_j), \quad (1.12)$$

where $I(R_i, r_j)$ is the image intensity in the point R_i during the reconstruction by hologram point with the coordinate r_j by the light intensity $\tilde{I}(r_j)$. $I_0(r_j)$ and $I_p(R_i)$ are the functions of r and R, respectively, and are determined by the field intensities in the hologram and object planes during the exposure.

After simple transformations, we finally get the following formula for spatial coherence function and for transmission linear coefficient:

$$\left|\gamma(r_i, r_j)\right|^2 = \left[\frac{I(r_i, r_j) I(r_j, r_i)}{I(r_i, r_i) I(r_j, r_j)}\right]^{1/2}, \quad (1.13)$$

$$\frac{f^2(r_i)}{f^2(r_j)} = \frac{I(r_i, r_j)}{I(r_j, r_i)}, \quad (1.14)$$

where $I(r_i, r_j)$ is the light intensity in the point r_i of the reconstructed image at illumination of the hologram point r_j.

From (1.14), it is seen that, generally speaking, $I(r_i, r_j) \neq I(r_j, r_i)$ as the transmission linear coefficient f can depend on the point position r on the hologram surface. The main idea of the step method consists in the following: The determination of the full form of spatial coherence function takes several steps with the help of the set of holograms recorded with different exposure. Knowing the distribution of relative values f on the hologram surface and the transmission T as the function of the coordinate r on each hologram, it is easy to single out linear recording area. This is the area where f does not depend on T, or the same is $I(r_i, r_j) = I(r_j, r_i)$. Notice that in the linear recording area, (1.13) can be rewritten in a simpler way:

$$\left|\gamma(r_i, r_j)\right|^2 = \frac{I(r_i, r_j)}{\sqrt{I(r_i, r_i) \times I(r_j, r_j)}}. \quad (1.15)$$

From the above, we can conclude that the step holographic method allows determining the full spatial coherence function in the case of sudden end edge laser variances.

But sometimes, it is convenient to use averaged values $|\tilde{\gamma}|^2$, for example, to approximate $|\gamma|^2$ by a homogeneous function. The homogeneous function depends only on the vector $\vec{r} = \vec{r}_1 - \vec{r}_2$, which characterizes relative position of these two points on the laser end. In this case for determining the averaged distribution $|\gamma|^2$ on the laser end, we propose to use a holographic method, which will be called an integral method.

The main idea of the integral method is that the recording of a hologram is carried out according to the scheme (Fig. 1.5) proposed in [69]. But unlike the methods developed in these works, we are going to measure not the differential (local) intensity of the reconstructed image $I(R, r)$ but the integral intensity of the whole image:

$$I_i(r) = \int\limits_S I(R, r) d^2 r, \tag{1.16}$$

what is achieved due to the registration of the whole image by the photocathode of the electron-multiplier phototube. Then according to (1.12), we have the integral relation, which connects the intensity of the reconstructed image $I_I(r)$ and squared absolute value of spatial coherence function $|\gamma(R, r)|^2$:

$$I_i(r) = \int I_0(R, r) d^2 r = f^2(r) I_0(r) I_b(r) \int I_p(R) |\gamma(R, r)|^2 d^2 R, \tag{1.17}$$

where $I_b(r)$ is the intensity of the reconstructing beam, $I_0(r)$ is the radiation intensity in the point r of the hologram plane, and $I_p(r)$ is the radiation intensity in the point R of the laser edge.

Consider that the recording conditions are linear. This assumption does not greatly limit the applicability of the method for the average value determination $|\gamma(R, r)|^2$.

Consider also that $|\gamma(r_1, r_2)| = |\gamma(r_1 - r_2)|$, i.e., we will approximate spatial coherence function by the homogeneous function (thus, our results will give precise description of $|\gamma|^2$ of the sources with homogeneous spatial coherence function and in the case of heterogeneous spatial coherence function our results will give for it the averaged approximating form). As the function depends only on $\vec{x} = \vec{r} - \vec{R}$, then passing in the right side of (1.17) from integration on d^2R to integration on d^2x, we get the following integral equation for $\gamma(x)$:

$$\frac{I_i(r)}{I_0(r)} = C \int I_p(\vec{r} - \vec{x}) |\gamma(x)|^2 d^2 x, \tag{1.18}$$

where $I_i(r)$ is the experimentally taken dependence of the integral on the position of the recovery point on a hologram, and $I_0(r)$ and $I_p(\vec{r} - \vec{x})$ are the light intensity in the points of a hologram and an image, which can be experimentally detected by the degree of the photographic-plate blackening.

The value of the constant C can be calculated, but as it can be easily seen, the normalizing condition $|\gamma(0)|^2 = 1$ makes the value of this constant negligible during the distribution construction γ. Equation (1.18) is the integral equation with the difference kernel, and it can be solved by decomposition in the Fourier integral. Then for the value $|\gamma(x)|^2$, we have

$$|\gamma(x)|^2 = C_2 \int d^2 p \exp(-ipx) \frac{\int \frac{I_i(r)}{I_0(r)} \exp(ipr) d^2 r}{\int I_p(r) \exp(ipr) d^2 r}, \tag{1.19}$$

where constant C_2 is directly detected from $|\gamma(0)|^2 = 1$.

Thus, for the calculation of the average distribution $|\tilde{y}(x)|^2$, it is necessary to determine experimentally the value of the integral intensity $I_i(r)$ and the intensity distribution on the laser edge $I_0(r)$. Further calculations can be made, for example, using computer or in the analytical form using function approximation $I_i(r)$ and $I_0(r)$.

1.2.2 Measuring Spatial Coherence of 6G Rhodamine Laser Pumping by Interference and Holographic Methods: Holographic, Holographic with Microphotometry of Initial Intensity Distribution and Integral

To measure dye laser spatial coherence function, the setup with the Mach–Zehnder interferometer was mounted. On the scheme below, we can see its components (Fig. 1.3).

For spatial coherence estimation of the dye laser, it is convenient to use the interference method, which was proposed by Yu. I. Ostrovsky [75]. The method is based on the depth measuring of localization area of interference fringes, which have been produced in the Mach–Zehnder double-beam interferometer. This method also enables to study radiation spatial coherence with low time coherence as it is possible to straighten optical path difference between interfering beams thoroughly.

As it is shown in [75], spatial coherence degree of heat homogeneous circular source is connected with the shift from the plane of the interference pattern localization in the following way:

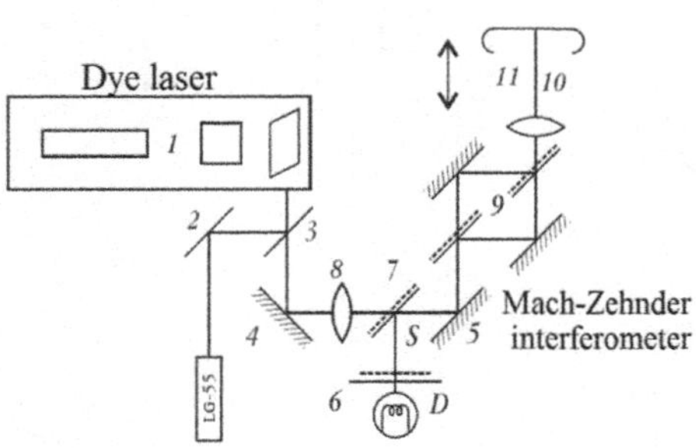

Fig. 1.3 Scheme of measurement of spatial coherence function by the interference method: 1—the laser; 2, 3, 4, 5, 7—the deflecting mirrors; 6—the heat source with diaphragm from 1 to 0.01 mm; 8—the lens for scroll display in front of input mirror of the Mach–Zehnder interferometer; 9—the Mach–Zehnder interferometer; 10—the lens for interference fringes localization; 11—the filmboard with the movement table. Reprinted from [54] with permission

$$|\gamma_{12}| = \frac{2J_1\left|\frac{\pi \nu_0 dpy}{f(p-y)}\right|}{\frac{\pi \nu_0 dpy}{f(p-y)}}, \tag{1.20}$$

where J_1 is the Bessel function of the first kind, d is the source diameter, f is the distance from the lens of interference pattern localization to its plane; p is the distance from the plane of the interference pattern localization to the both real source images formed by the source; y is the coordinate of the given point calculated from the plane of interference pattern localization (see Fig. 1.4a). From Fig. 1.4, it can be seen that when moving off the localization plane L the contrast of interference fringes will decrease, and it will decrease faster, the larger the length of the source is. Spatial coherence studies (Fig. 1.3) were carried out using two combinations of sort elements in the resonator of Rhodamine 6G laser with laser pumping: (1) diffraction grating (generation spectrum width is 1 nm), (the case of low time coherence); (2) diffraction grating, the Fabry–Perot interferometer (generation spectrum width is 0.03 nm), (the case of high time coherence).The Mach–Zehnder interferometer was tuned up in white light for horizontal fringes, spatial frequency of which is 30 cm^{-1}. For recording interferograms, the "Micrat 300" film was used. The recording was made during sequential shift of the filmboard along the optical axis.

When moving along the axis, it was necessary to determine the size of localization region of interference fringes, which were produced in the radiation of dye laser with

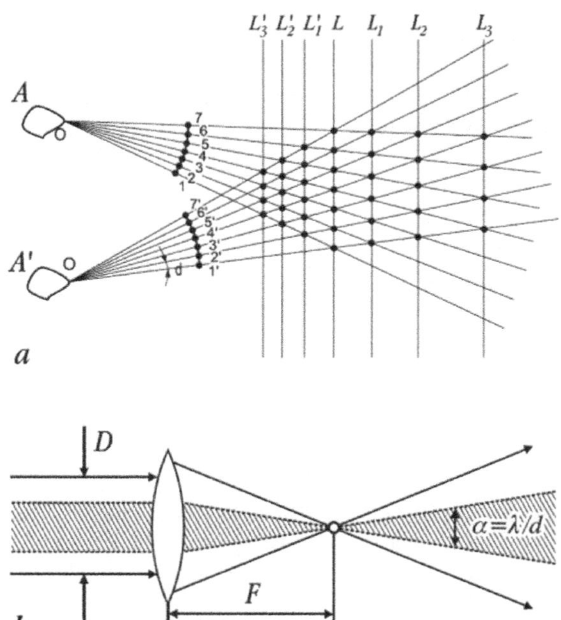

Fig. 1.4 To the question of establishment of interrelationship of radiation coherence degree between heat and laser sources. Reprinted from [54] with permission

laser pumping. Then, changing the diameter of the radiating surface of "reference" heat source using variable slit, the coincidence has been achieved of localization region sizes of interference patterns produced during radiation of "reference" heat source and the laser under consideration. In this case, spatial coherence of laser radiation corresponds to spatial coherence of radiation of heat homogeneous circular light source.

The presence of lens led to the increase of efficiency of angular sizes of laser source that is equivalent to spatial coherence degradation of light.

The results of the studies have shown that in both cases (in the first case, there was no Fabry–Perot interferometer in the resonator, and in the second case, the Fabry–Perot interferometer was included into the resonator) Spatial coherence of dye laser radiation focused by the lens (8) with the focal distance $F = 19$ cm in front of input mirror of the Mach–Zehnder interferometer ($v = 30\ \text{cm}^{-1}, f = 11.6\ \text{cm}, p = 3.6\ \text{cm}$) corresponds to spatial coherence of radiation of heat homogeneous arc source with the diameter d less than 0.1 mm. It can be easily shown that to the source of such a size coherence region of laser end corresponds, the diameter of which is (Fig. 1.4b):

$$D \approx F\frac{\lambda}{d}, \tag{1.21}$$

i.e., for our case $F = 19$ cm, $\lambda = 5.8 \times 10^{-5}$ cm, $d \leq 10^{-2}$ cm, $D \leq 1.2$ mm. It should also be noted that the present method can be used only for the study of light sources, which have stable mode structure, as spatial coherence function determination (as in any interference method) is carried out by the set of interferograms.

Comparison with heat source is reasonable when spatial coherence function under consideration (a priori) is homogeneous, in the opposite case averaged values $\tilde{y}$ will be received.

For measuring spatial coherence function of 6G laser radiation with the selective cavity (see Fig. 1.1) by a holographic method, the scheme similar to the mentioned one in the works [42–44, 69] was used. The installation diagram can be seen in Fig. 1.5.

Laser end with fivefold increasing was designed onto the hologram and diffusive screen. Near the laser end, the rectangular diaphragm (2) of $2 \times 3\ \text{mm}^2$ was put, which made it possible to link point coordinates of laser end on the hologram, diffusive screen and reconstructed image. The distance from the screen to hologram is 100 mm, and average angle between reference and object beams is about 12°. Hologram registration was made on the "Mikrat 900" film, and the hologram size is $10 \times 15\ \text{mm}^2$.

In Fig. 1.6a, the scheme of wave front reconstruction and intensity registration is shown. The hologram was exposed by narrow beam of He–Ne laser (with the diameter of 1 mm), which could move in two mutually perpendicular directions x and y. Photomultiplier tube, which was used for the measuring intensity along with the diaphragm (diameter 0.5 mm) put in front of it, could also move along the reconstructed image what made it possible to measure spatial coherence function of laser end points.

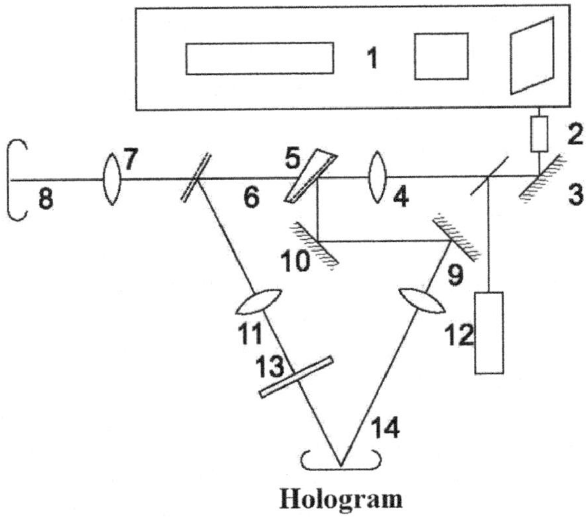

Fig. 1.5 Scheme of recording a hologram of diffusive screen for determining spatial coherence function: 1—the laser under investigation; 2—the rectangular diaphragm, 3, 5, 10—the deflecting mirrors; 4, 11, 12—the lenses, which design the laser end image and rectangular diaphragm onto the hologram and diffusive screen; 6, 9—the mobile mirrors system used for flattering of the path difference of object and reference beams; 7—the lens, which designs laser end intensity distribution onto the film (8); 13—the diffusive screen; 14—the hologram. Laser LG–55 is used for the scheme adjusting. Combination of dye laser beams and LG-55 has been achieved using the plates 15 and output mirror 10 (see Fig. 1.11). Reprinted from [54] with permission

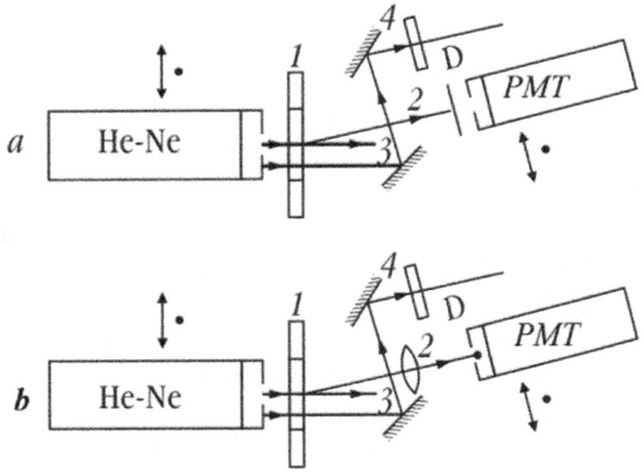

Fig. 1.6 Scheme of wave front reconstruction and intensity registration: **a** when determining SCF by a holographic method; **b** when determining SCF by an integral method. Reprinted from [54] with permission

To avoid the influence of instability of reconstructing source and photomultiplier tube, along with the measurement of point intensity of the reconstructed image (2) by the same photomultiplier tube, the light beam (3) incident upon a hologram which preliminary was weakened by the filter (4), was periodically under control.

From the measured intensity values of the reconstructed image, illumination constant signal, which was provided by hologram noise, was subtracted. Measuring systematic errors are substantially conditioned by a hologram and photomultiplier tube nonlinearity. In general, accidental errors are provided by inaccuracy of the points coordinates overlapping laser end in the hologram plane and in the plane of the reconstructed image. Scatter of results of our experiments was 10% at determination of $|\gamma|^2$ just by holographic method and 15% at microphotometry of the intensity distribution on laser end.

Figure 1.7 shows the results of calculations of normalized degree of laser end points, which lie on two mutually perpendicular lines l and h and which meet the laser end center (l is the line passing on the generating area of the end in crosswise direction, which coincides with the direction of pumping pulse propagation). Spatial coherence degree is given for the points concerning the laser end center. The measuring was made using holograms recorded by laser radiation under pumping energy $E = 0.04$ J and $E = 0.03$ J and dye concentration corresponding to absorption constant of 12 cm^{-1} for $\lambda = 347.2$ nm.

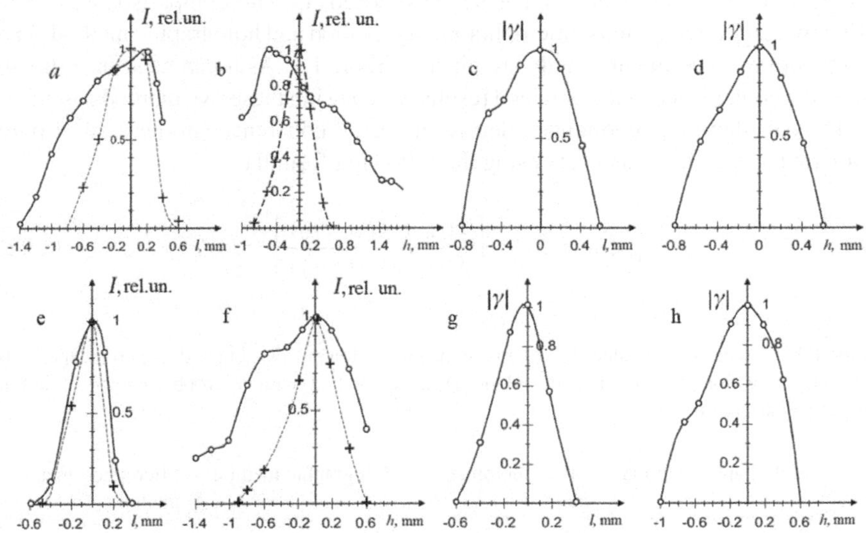

Fig. 1.7 Intensity distribution on the laser end for the points, which lie on two mutually perpendicular lines l and h and which meet the laser end center corresponding to the point of their crossing: dotted line—intensity $I_M(R)$ of laser end reconstructed image; solid line—initial intensity distribution on the laser end $I(R)$ received with the help of direct microphotometry; **a**, **b** corresponds to $E = 0.04$ J; **e**, **f** $E = 0.03$ J, where E—the pumping energy; **c**, **d**, **g**, **h** calculated using these intensities $I_M(R)$ and $I(R)$ normalized degree of spatial coherence of corresponding points of the laser end relative to its central point. Reprinted from [54] with permission

As it can be seen from Fig. 1.7 from the whole generating end area of $2 \times 3\ \text{mm}^2$ (pumping energy is 0.04 J), spatial coherence area (0.2 level) takes $1.3 \times 1.3\ \text{mm}^2$. If pumping energy is $E = 0.03$ J, then this area is narrowed to $0.8\ \text{mm}^2$ (the comparison is given for one-axis points, Fig. 1.7c, g). Spatial coherence function contour had one maximum. The coherence area sizes from pulse to pulse have not been changed (under pumping energy $E = 0.04$ J and dye concentration corresponding to absorption constant of $12\ \text{cm}^{-1}$ for $\lambda = 347.2$ nm). Dependence of this area on the concentration (in the studied concentrations interval of corresponding to absorption constants of $10–18\ \text{cm}^{-1}$ for $\lambda = 347.2$ nm) under constant level of pumping energy of 0.04 J was weakly evident. At the same time, we could observe strong dependence of coherence area sizes in the direction of pumping pulse propagation on pumping energy. When the pumping energy was $E = 0.04–0.05$ J, coherence area size was the largest and formed 1.3 mm, in the case when pumping energy was 0.01 J, coherence area size was decreasing to 0.2 mm. In Fig. 1.7, the dotted line shows intensity distribution in the reconstructed image of diffusive screen, which is also the characteristics of coherence, because, namely it characterizes the quantity of coherent light emitted by laser end points. It seems that asymmetry of the initial intensity distribution on the laser end (see Fig. 1.7a, b, e, f—solid line) is provided by unsteady optical inhomogeneity. It appears under transverse, concerning cavity axis, pumping pulse and also under the presence of the Fabry–Perot interferometer in the resonator. Along with the determination of spatial coherence function of dye laser, the comparison of values $|\gamma|$ for one of the generation pulses was carried out. This comparison was made using two different methods: microphotometry method and holographic method. The results of this experimental study are given in Table 1.1. As it can be seen from the table, the coincidence of the obtained results, when using these two methods, is good.

The calculation of normalized degree of spatial coherence (in the case of pure holographic method) was made using the following formula

$$|\gamma(r_0, r_j)|^2 = \left[\frac{I(r_0, r_j)I(r_j, r_0)}{I(r_0, r_0)I(r_j, r_j)}\right]^{1/2}, \tag{1.22}$$

Table 1.1 Comparison of values $|\gamma|$ for one of the generation pulses, obtained by two independent methods: pure holographic method and holographic method with microphotometry of initial intensity distribution

No.	*h* Holographic method	*h* Microphotometry method	*l* Holographic method	*l* Microphotometry method
1	0.53	0.47	0.51	0.52
2	0.82	0.81	0.93	0.93
3	1.0	1.0	1.0	1.0
4	0.81	0.81	0.91	0.92
5	0.63	0.60	0.70	0.69
6	0.45	0.44	0.58	0.58

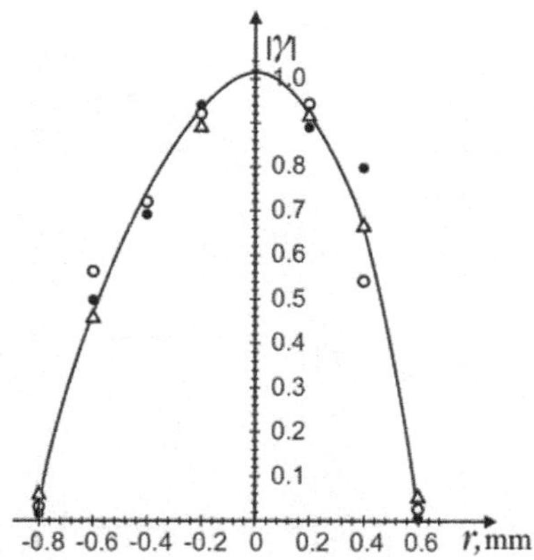

Fig. 1.8 Study of spatial coherence function homogeneity. Reprinted from [54] with permission

where $I(r_0, r_j)$ is the intensity of the image in the point r_0 during the reconstruction of a hologram along the point r_j; $I(r_j, r_0)$ is the intensity of the image in the point r_j during the reconstruction of a hologram along the point r_0; $I(r_j, r_j)$ is the intensity of the image in the point r_j during the reconstruction of a hologram along the point r_j; index 0 is the laser end point index, as relative to which spatial coherence function was determined, $j = 1, 2, 3$ is the running index of the laser end point.

As it can be seen from Fig. 1.8, laser spatial coherence function considerably depends on the distance between laser edge points and practically does not depend on the position of supporting point r_j.

Thus, spatial coherence function is almost homogeneous function. It proved the possibility to measure in this case SCF by interference method.

The measuring scheme of spatial coherence function of 6G Rhodamine laser by integral method differs from the previous scheme only on the reconstruction step.

Figure 1.6b shows the scheme of wave front reconstruction and intensities registration.

The image of the diffusive screen has been reconstructed by a small part of a hologram, the size of which was determined by the beam diameter of He–Ne laser (equal to 1 mm).

The whole reconstructed image was gathered by the lens photomultiplier tube PMT-12. By moving the beam of He–Ne laser along the hologram in two mutually perpendicular directions, integral intensity was measured of the whole reconstructed image of the diffusive screen. The measurements were carried out on the same holograms where γ was determined by pure holographic method and by holographic one with microphotometry of initial intensity distribution on the laser end (see Fig. 1.7). It has given the opportunity to predetermine about the correlation of the results, which were obtained using different methods.

In Fig. 1.9, line contours of equal intensity (integral along the whole image) are plotted for one of the generation pulses ($E = 0.03$ J). Initial intensity distribution was determined by photographic photometry method. To simplify the calculation,

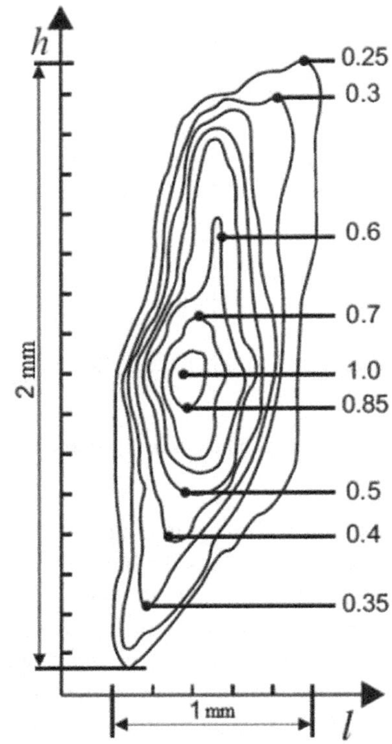

Fig. 1.9 Line contours of equal integral images intensity of the laser end reconstructed under the illumination of different holograms points. Reprinted from [54] with permission

distribution of $I_0(r)$ and $I_i(r)$ was approximated (see Figs. 1.7 and 1.10) by the Gaussian law:

$$I_i(r) = I_i(0)\exp\left[-\left(\frac{x^2}{x_i^2} + \frac{y^2}{y_i^2}\right)\right],$$

$$I_0(r) = I_p(r) = I_0\exp\left[-\left(\frac{x^2}{x_0^2} + \frac{y^2}{y_0^2}\right)\right], \tag{1.23}$$

where x_i, y_i, x_0, y_0 are the corresponding experimentally determined values. Then for $|\gamma(x)|^2$ we have

$$|\gamma(x)|^2 = \exp\left[-x^2\left(\frac{x_0^2 - x_i^2}{x_0^2(2x_i^2 - x_0^2)}\right) + y^2\left(\frac{y_0^2 - y_i^2}{y_0^2(2y_i^2 - y_0^2)}\right)\right]. \tag{1.24}$$

By substituting in (1.24) the values x_0 and x_i, found from the graphics, we get averaged value $|\gamma(x)|^2$ along the axis l.

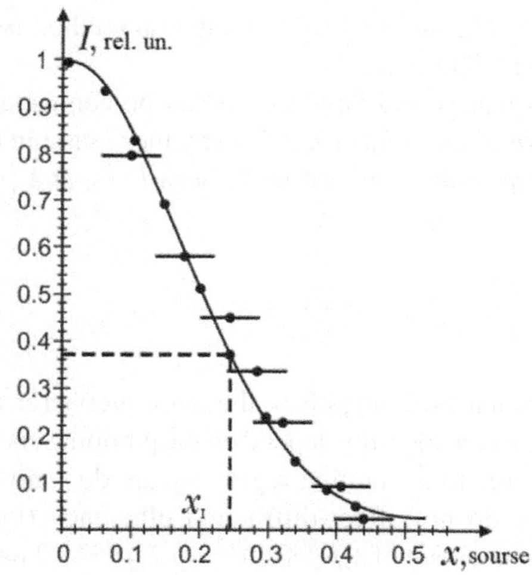

Fig. 1.10 Dependence of integral intensity on the diameter of contours equal to intensity along the axis *l*. Reprinted from [54] with permission

Figure 1.11 shows the averaged values $|\gamma(x)|^2$ determined by the integral method and also the values of spatial coherence function found with the help of pure holographic method (in both cases one hologram was processed, and spatial coherence function was determined relative to the same laser end points). It is seen that though some details in the behavior of $|\gamma|^2$ are not reflected in the integral curve, in general, both curves (a) and (b) in the Fig. 1.11 look similarly. So, the coherence area defined

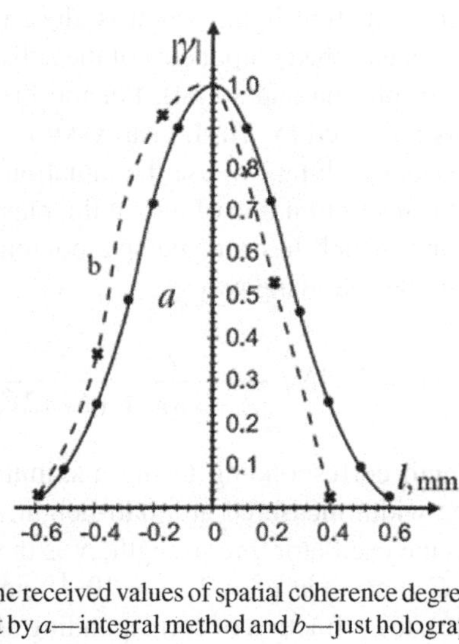

Fig. 1.11 Comparison of the received values of spatial coherence degree of the same laser end points concerning the central point by *a*—integral method and *b*—just holographic method. Reprinted from [54] with permission

from the condition $\gamma = 1/2$ and produced by integral method is 0.57 mm, and only by holographic method—0.60 mm.

The observed divergences to a large extent can be connected with the simplest approximation of distribution of $I_0(r)$ and $I_i(r)$ by the Gaussian law. In this simplest approximation, the connection between half-width I_i, I_0 and γ is specified by the relation:

$$x_j^2 = \frac{x_0^2(2x_i^2 - x_0^2)}{x_0^2 - x_i^2}. \tag{1.25}$$

At laser pumping by nanosecond pulses, the sizes of coherence areas in the direction of pulse distribution considerably depend on the pumping energy. But the value of relation of coherence area to generating region square does not practically change. Concentration changes do not affect diffraction efficiency (in the interval corresponding to absorption constant of 10–18 cm^{-1} for $\lambda = 347.2$ nm). During scanning hologram laser end by narrow beam of He–Ne laser, the form of the reconstructed image has not practically changed. That indicates the absence of the modes over TEM_{00}.

1.3 Resonance Method for Increasing Interferometry Sensitivity in the Studies of Low-Temperature Sodium Plasma

Resonance methods for increasing interferometry sensitivity are based on the interferograms recording in the radiation light, which is close in frequency connected with the absorption line of one of the components of the substance under study (e.g., sodium, lithium, potassium plasma and so on). For the first time, obtaining resonance interferograms was proposed by Yu. I. Ostrovsky in 1961 [37]. The idea of this method is that for plasma probing, we use the radiation with the wavelength is closely connected with the absorption line of one of the plasma components.

Near the absorption line, which has a dispersive contour [76], atoms and ions refraction are described in the following way:

$$n - 1 = C\lambda_0^3 N f \frac{\lambda - \lambda_0}{(\lambda - \lambda_0)^2 + (\Delta\lambda/2)^2}, \tag{1.26}$$

where λ_0 is the wavelength corresponding to the maximum absorption line, and $\Delta\lambda$ is the absorption line width measured at a half-height; λ is the wavelength of translucent radiation; f is the oscillator line strength; N is the atom concentration on the absorption level and $C = e^2/4\pi m_e c^2 = 2.24 \times 10^{-14}$ cm.

As it comes from (1.26), the refraction of corresponding atoms as it approaches to absorption lines increases sharply and can enormously exceed the refraction of the same atoms away from the absorption line. Thus, the usage of wavelength, which

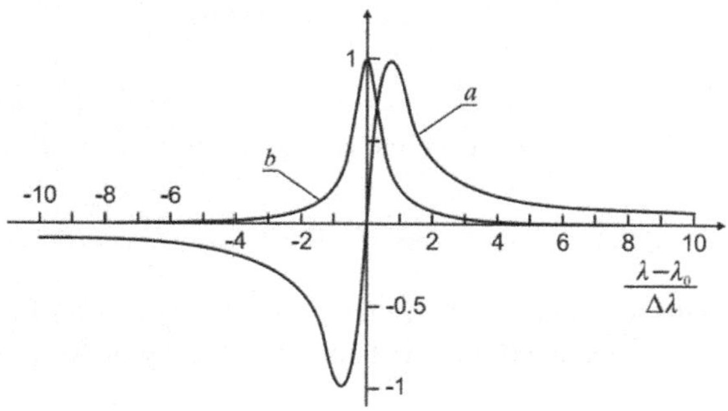

Fig. 1.12 Refraction path of $(n - 1)$ $(n - 1)_0$ (a) and absorption constant x/x_0 movement (b) in the coordinate system for the calculation of radiation near spectral line. Reprinted from [54] with permission

is closely connected with the absorption line, for producing interferograms, makes it possible to increase considerably the sensitivity of measuring of corresponding atoms concentration (Fig. 1.12).

The method of resonance interferometry is applicable for determining the concentration of atoms and ions, which have strong absorption lines. Indeed, absorption line contour, in the case of its dispersive form, is described in the following way

$$\varkappa = 2\pi c\lambda_0^2 fN\frac{\Delta\lambda}{(\lambda-\lambda_0)^2+\left(\frac{\Delta\lambda}{2}\right)^2}. \tag{1.27}$$

From the comparison of (1.26) and (1.27), it is possible to determine the link between the refraction maximum value $(n - 1)_0$, which can be achieved under $\lambda - \lambda_0 = \pm\,(\Delta\lambda/2)$, and absorption constant $\varkappa_0$ in the line center, i.e., under $\lambda = \lambda_0$

As it is shown in the study [27, 28], minimal atom concentration, which can be detected by the method of resonance interferometry, is determined in the following way:

$$N_{\min} = \frac{\Delta\lambda|n-1|_0}{C\lambda_0^3 f} \approx \frac{\Delta\lambda|k_{\min}|}{C\lambda_0^2 fl}, \tag{1.28}$$

where $k_{\min}$ is the minimal shift of interference fringe.

From (1.28), it is seen that atom limiting concentration is proportional to the absorption line width and depends on the measuring precision of interference fringes. Here it is assumed that the line of translucent radiation lies at a distance of half-width of absorption line having a dispersive contour.

Using radiation with arbitrary wavelength for producing interferograms, the calculation of concentration lower limit, which can be researched by the method of resonance interferometry ($K_{\min} = 0.1$), can be made using the following formula

$$N_{\min} = \frac{0.1}{C\lambda_0^2 fl} \frac{(\lambda - \lambda_0)^2 + (\Delta\lambda/2)^2}{|\lambda - \lambda_0|}. \tag{1.29}$$

If the distance from the absorption line is considerably higher than its half-width

$$|\lambda - \lambda_0| \gg \Delta\lambda/2,$$

then minimally detected concentration increases linearly as translucent line moves away from absorption line maximum and does not depend on its width

$$N_{\min} \approx \frac{0.1}{C\lambda_0^2 fl} |\lambda - \lambda_0|. \tag{1.30}$$

Maximum atom concentration, which can be studied by the method of resonance interferometry, is provided by absorption of probing radiation by plasma. It can be explained by the fact that during transmission of one of the interfering beams through the plasma, the relation of its intensities changes that leads to the contrast decay of interference pattern. Upper limit of measured concentrations can be detected in the following way:

$$N_{\max} \approx \frac{3}{\pi\lambda_0^2 fl} \frac{C(\lambda - \lambda_0) + (\Delta\lambda/2)^2}{\Delta\lambda}. \tag{1.31}$$

Here it is assumed that the minimal contrast of the interference pattern (enough for observing) is $\gamma = 0.1$; $l\varkappa \approx 6$, where $l\varkappa$ is the layer transmission density.

Figure 1.13a[3] shows limits of the measured values of atoms concentrations as a distance function of the probing line from the absorption line center (for $\lambda_0 = 5 \times 10^{-5}$ cm). The calculated curves are received under the assumption that the width of probing radiation line $\delta\lambda$ is much smaller than $\Delta\lambda$. From the figure, it is seen that the concentration interval measured by one interferogram expands as probing line moves away from absorption line. In other cases, as it is shown in Fig. 1.13b, when the line width of probing radiation is compared or bigger than absorption line, considerable decrease of the method and the narrowing of concentration interval take place.

In the present chapter, the experimental results of the possibility of sensitivity increase of the method of resonance interferometry using Rhodamine 6G laser with laser pumping are given [12, 36]. Sodium plasma was taken as the subject of inquiry. The simple model of such a plasma can be DC arc and alcohol lamp flame, which contain sodium vapors. The arc and flame plasma itself are studied enough. These objects have been chosen just for the demonstration of the possibility of the method of resonance interferometry. The purpose of this experimental study was to determine the concentration of normal sodium atoms in the alcohol lamp flame and in the DC arc.

[3]The idea of calculations of plots in Fig. 1.13a belongs to G. V. Ostrovskaya.

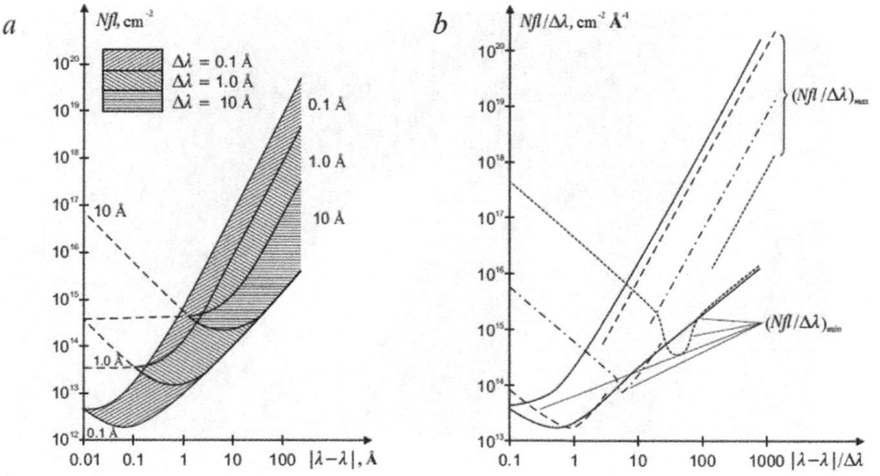

Fig. 1.13 Validity range of the method of resonance interferometry at monochromatic probe radiation and different widths $\Delta\lambda = 0.01$; 0.1; 0.1 nm of absorption line. On the *x*-axis, the distance between the absorption line and the line of probe radiation $|\lambda - \lambda_0|$ was put (**a**); sensitivity limits of the method as the function $|\lambda - \lambda_0|/\Delta\lambda$ at different relations of the widths of probe radiation $|\delta\lambda|$ and absorption line $|\Delta\lambda|$. Solid curve—$\delta\lambda/\Delta\lambda = 0$, dotted—$\delta\lambda/\Delta\lambda = 1$, dash-dotted—$\delta\lambda/\Delta\lambda = 10$, dots—$\delta\lambda/\Delta\lambda = 100$ (**b**). The idea of calculation of plots in **a, b** belongs to G. V. Ostrovskaya. Reprinted from [54] with permission

Experiments on resonance interferometry in the neighborhood of sodium D-doublet were carried out with the help of the studied wavelength-tuned radiation of Rhodamine 6G laser.

Photo (a) and the optical scheme (b) of the experimental device for obtaining resonance interferograms are presented in Fig. 1.14. Spectrum generation width is 0.03 nm. Figure 1.15 shows the generation frequency tuning near sodium resonance doublet, which was carried out by the slope of the Fabry–Perot interferometer located in the dye laser resonator. Spectrograph DFC-8 controlled the position of generation line.

For simultaneous interferogram, obtaining radiation with two wavelengths was used: $\lambda_1 = 589.7$ nm—dye laser wavelength distant from long-wave component of resonance doublet NaI, $\lambda = 589.593$ nm on 1 nm; $\lambda_2 = 694.3$ nm—generation wavelength of ruby laser, which carries out optical pumping by its second harmonic of the same dye laser.

Let us describe Fig. 1.14 with the help of the deflecting mirrors 11, 12, 14, dye laser radiation and the part of ruby laser radiation (reflected from the filter (6) FC-7) were directed to the Mach–Zehnder interferometer (17). Meanwhile, both probing beams coincided with the direction. The lens 13 was used to form probing beam of necessary divergence. The Mach–Zehnder interferometer was adjusted in white light on horizontal interference fringes, spatial frequency of which could be changed in wide ranges from 0 to 50 lines/mm. Interferograms registration was carried out

a

b

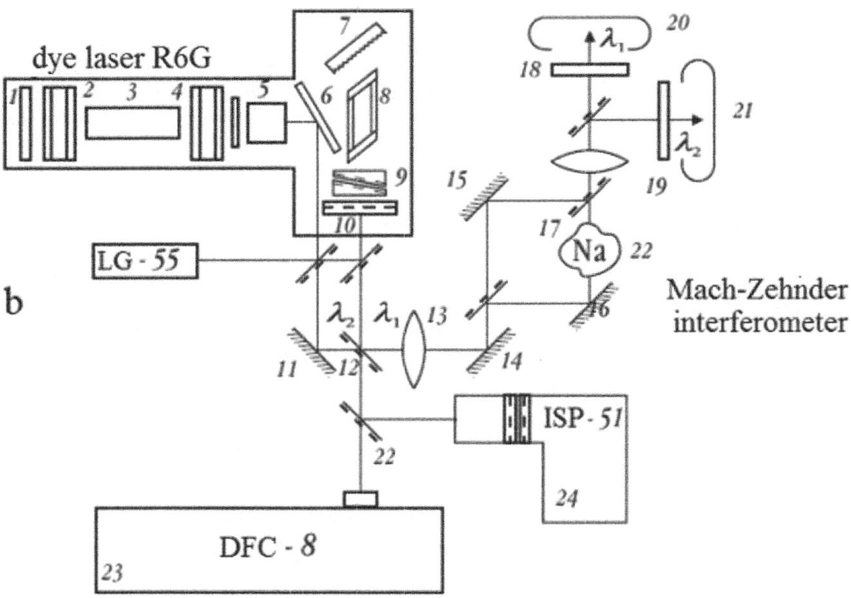

Fig. 1.14 Photo (**a**) and the optical scheme (**b**) of the device for obtaining resonance interferograms. Reprinted from [54] with permission

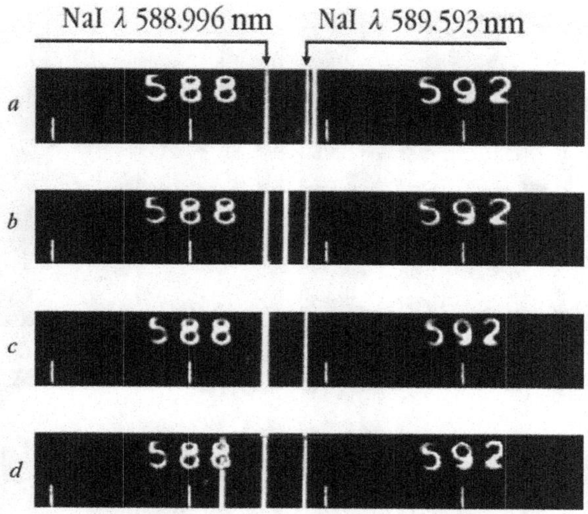

Fig. 1.15 Generation frequency tuning near sodium resonance doublet. Generation wavelength: **a** 589.7 nm, **b** 589.3 nm, **c** 590.0 nm, **d** 588.4 nm. Reprinted from [54] with permission

under three values of space frequency: 0 (infinite width line—Fig. 1.8), 5 lines/mm and 10 lines/mm. Interferograms were recorded simultaneously on two films 20, 21.

In front of one of them, the filter 3C-8 (18) was placed, which transmits only radiation with the wavelength of 589.7 nm, in front of the other—the filter KC-19 (19), which transmits only radiation with the wavelength of 694.3 nm.

DC arc or alcohol lamp flame 22 containing sodium vapors were included into the Mach–Zehnder interferometer arm 17.

Figures 1.16 and 1.17 show interferograms of alcohol lamp flame with sodium vapors corresponding to dye laser radiation ($\lambda_1 = 589.7$ nm and 589.0 nm) and to ruby laser radiation ($\lambda_2 = 694.3$ nm). Similar interferograms for DC arc with sodium packing are shown in Fig. 1.18. Spatial frequency of interference fringes for the presented interferograms is 5 lines/mm. Produced simultaneously in two wavelengths λ_1 and λ_2, interferograms allow estimating averaged on thickness of probing layer concentration of sodium atoms by fringe relative shift[4] (expressed in a number of fringes).

Formulas for interference fringes shift on interferograms produced in wavelengths λ_1 and λ_2 take the following form:

$$k_1 = \frac{\left[(n-1)_1 - (n_0 - 1)\right]l}{\lambda_1} \tag{1.32}$$

$$k_2 = \frac{\left[(n-1)_2 - (n_0 - 1)\right]l}{\lambda_2} \tag{1.33}$$

[4]Relative shift of interference fringes means the difference in shift of interference fringes on interferograms produced in different wavelength λ_1 and λ_2.

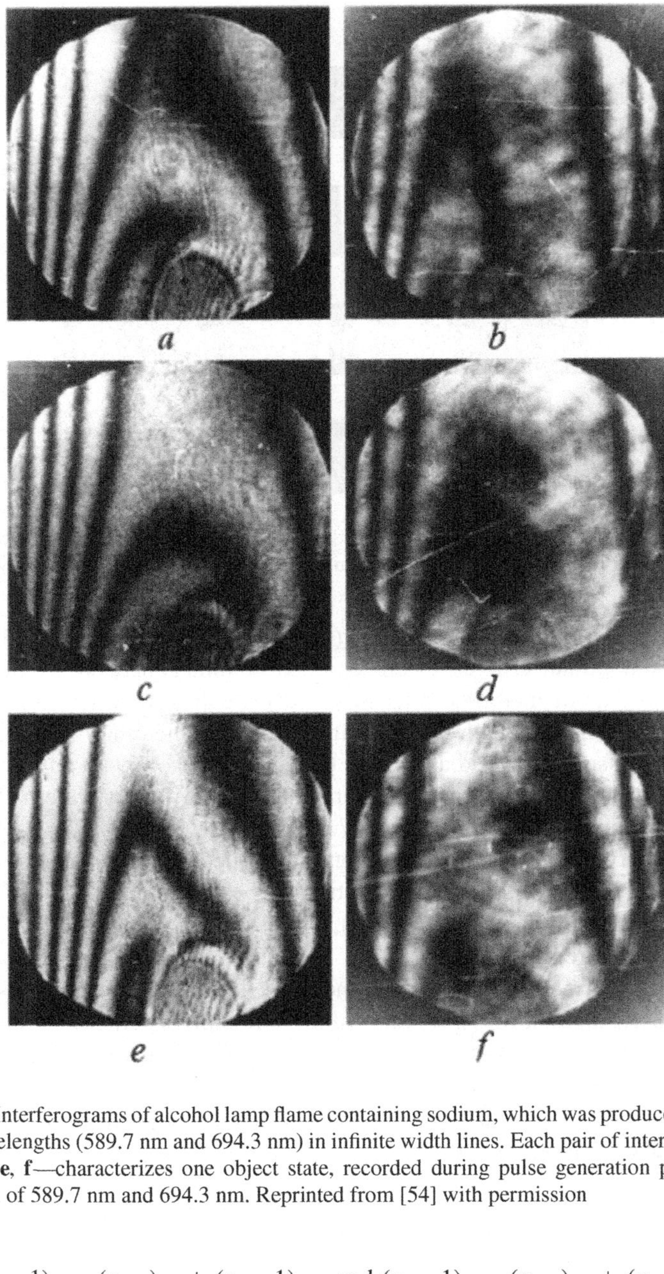

Fig. 1.16 Interferograms of alcohol lamp flame containing sodium, which was produced separately in two wavelengths (589.7 nm and 694.3 nm) in infinite width lines. Each pair of interferograms—**a**, **b**, **c**, **d**, **e**, **f**—characterizes one object state, recorded during pulse generation period in the wavelength of 589.7 nm and 694.3 nm. Reprinted from [54] with permission

where $(n-1)_1 = (n_{0\text{-}1})_{\text{air}} + (n-1)_{1Na}$ and $(n-1)_2 = (n_{0\text{-}1})_{\text{air}} + (n-1)_{2Na}$ are the refraction values provided by the change of density of air in perturbed area and by the content of sodium atoms for $\lambda_2 = 694.3$ nm and $\lambda_1 = 589.7$ nm; $n_{0\text{-}1}$ is the refraction of unperturbed air layer; l is the typical area thickness.

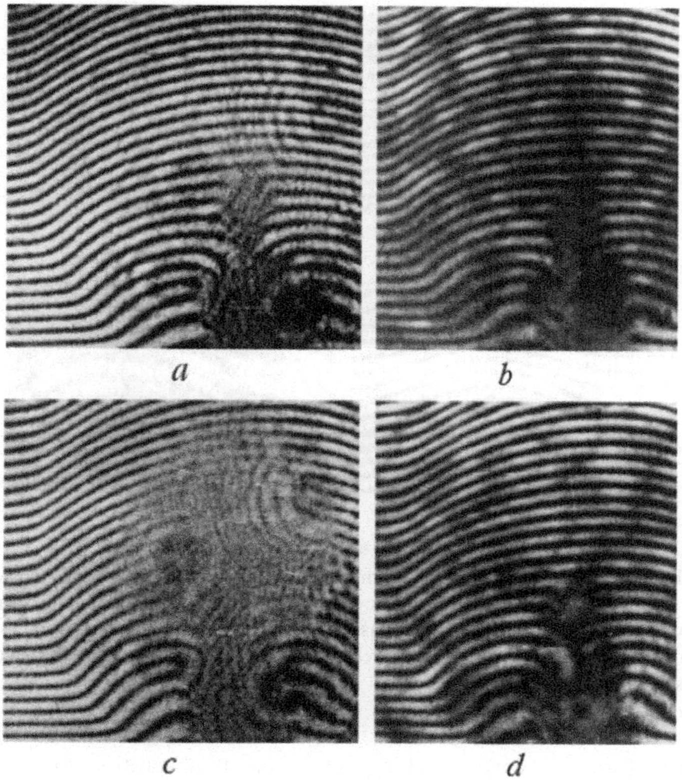

Fig. 1.17 Interferograms of alcohol lamp flame containing sodium and produced separately in two wavelengths (589.7 and 694.3 nm). Spatial band frequency is 5 lines/mm. **a**, **b** correspond to simultaneous recording of interferograms in the wavelength of 589.7 nm and 694.3 nm; **c**, **d** correspond to recording of interferograms in the wavelength of 589.0 nm and 694.3 nm. Reprinted from [54] with permission

As one of the radiation wavelengths (589.7 nm or 589.0 nm), which X-rays plasma, is closely connected with absorption sodium lines (589.996 nm or 589.996 nm), then measuring sensitivity relative to sodium atoms considerably increases and almost does not depend (accurate within the second component of the Cauchy formula) on measuring air refraction for both wavelengths of 589.7 nm and 694.3 nm, in radiation of which simultaneous interferograms recording takes place. On each interferograms, directly measured value reveals fringes shift from their unperturbed position. Equation (1.34) for relative shift of interference fringes comes from (1.32) and (1.33):

$$k_2\lambda_2 - k_1\lambda_1 = \left[(n-1)_{2Na} - (n-1)_{1Na}\right]l \tag{1.34}$$

The equation, which established the connection between relative shift of interference fringes and sodium atoms concentration, can be obtained using the Zelmeer

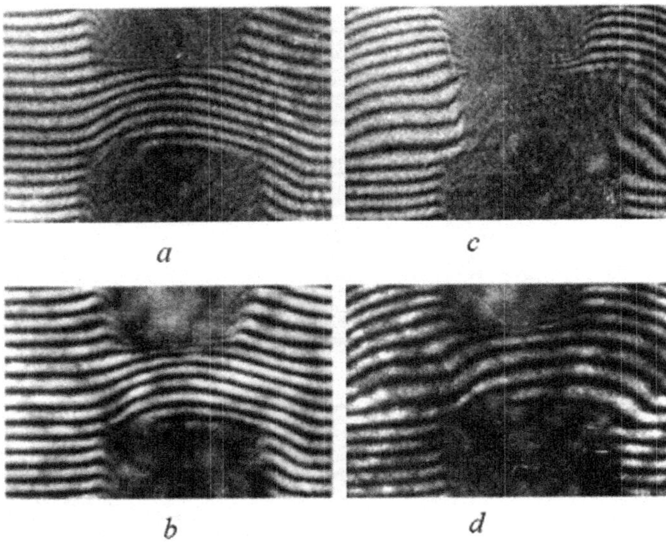

Fig. 1.18 DC arc interferograms with sodium packing produced separately in two wavelengths (589.7 and 694.3 nm). Spatial band frequency of 5 lines/mm. **a**, **b** Correspond to simultaneous recording of interferograms in the wavelengths of 589.7 and 694.3 nm; **c**, **d** correspond to simultaneous recording of interferograms in the wavelength of 589.7 and 694.3 nm. Reprinted from [54] with permission

approximate formula (without taking into account absorption line width) and neglecting sodium refraction by the wavelength of 694.3 nm:

$$N_{Na} = \frac{(k_2\lambda_2 - k_1\lambda_1)(\lambda_1 - \lambda_0)}{Cf_{02}\lambda_0^3 l}, \tag{1.35}$$

where f_{02} is the oscillator strength of absorption sodium fringe.

Determined averaged on layer thickness concentration of sodium atoms in DC arc was 7×10^{13} cm^{-3}, in the alcohol lamp flame is 6×10^{14} cm^{-3} that approximately corresponds to ultimate sensitivity of measurements for the distance of 0.1 nm between the wavelength of probing line and resonance sodium fringe.

As it is seen from Fig. 1.13a, the received concentration values of sodium atoms in arc plasm are also on the response limit of the method sensitivity that was observed experimentally. Under frequency detuning of probing radiation from the absorption line center on 0.5 nm, interference fringes shift is out of response limit of the methods.

In the case of mismatching of probing radiation from the absorption line center, less than 0.1 nm strong absorption of radiation by atomic sodium can be observed, like it is on interferograms (see Figs. 1.17c and 1.18b).

Thus, for sodium plasma under study, the optimal mismatch value between the wavelength of probing radiation and the wavelength of resonance sodium doublet

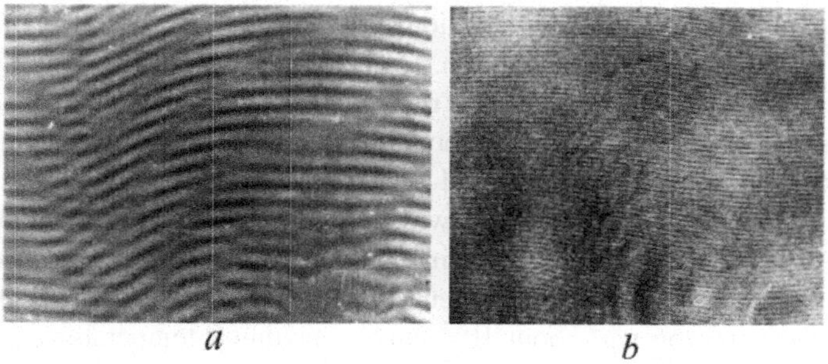

Fig. 1.19 Moire interferograms produced in the wavelengths of 589.7 and 694.3 nm; **a** 5 lines/mm; **b** 10 lines/mm. Reprinted from [54] with permission

component (589.695 nm), under which the method of resonance interferometry can function, is 0.1 nm under the values $\lambda = 0.01$ nm and $\lambda = 0.03$ nm.

In the alcohol lamp flame, differences in shift of interference fringes produced in λ_1 and λ_2 were almost absent under parameter mismatch $\xi = 0.9$ nm.

The type of the method of resonance interferometry is moiré comparison method of interferograms, one of which has been produced in radiation with the wavelength λ closely connected with absorption resonance line of homogeneity under study. And if one of the lines, for example λ, is closely connected with absorption line λ_0 of homogeneity under study and the second one λ_2 is placed at a large distance from it, then according to (1.26) measuring sensitivity is going to be changed depending on the value $\lambda_1 - \lambda_0$.

It is easy to see that the highest measuring sensitivity will be in the case if λ_1 and λ_2 are on the both sides from the absorption line at a distance $\Delta\lambda/2$.

The pattern, which was obtained in the result of overlapping of two interferograms (which is usually carried out by photographing of two interferograms on one photoplate), enables to determine plasma dispersion quantitatively.

As an example, Fig. 1.19 shows obtained interferograms of the alcohol lamp flame containing sodium and registered in radiation with the wavelengths of 589.7 and 694.3 nm (under space frequencies of interference fringes of 5 lines/mm, 10 lines/mm).

1.4 The Resonance Fluorescence Method for Hydrogen Plasma Diagnostics in the FT-1 Tokamak Device with the Use of Dye Lasers

For the diagnostics of high-temperature hydrogen plasma in FT-1 tokamak device for the first time, the method of resonance fluorescence [77] was applied with the use of dye laser with laser and lamp pumping [78, 79]. The idea of this method can be expressed in the following way. Laser radiation passing through plasma selectively excites atoms and ions, optical transition frequency of which coincides with the frequency of the present laser radiation. Intensity increase of atoms and ions glow (fluorescence signal) is registered by the receiving equipment. By fluorescence signal size, it is possible to determine the population of levels, which are in charge of the present optical transition and in the end—the concentration of the respective atoms and ions in plasma, and by Doppler width of fluorescence line it is possible to determine their temperature. The experiments took place in hydrogen plasma of the FT-1 tokamak device with hydrogen neutral atom concentration of 10^9–10^{10} cm^{-3} (electron concentration was about 10^{13} cm^{-3}, maximum temperature of electrons was 300 eV).

FT-1 tokamak represents toroidal magnetic trap with longitudinal current of 27 kA. Torus major diameter is 125 cm, minor diameter is 40 cm, and filament diameter is 30 cm. On this device, the studies on plasma heating under longitudinal current and microwave field [80] take place.

For exciting the signal of hydrogen neutral atoms, fluorescence in plasma dye laser with the generation in line region ($\lambda = 656.3$ nm) was used. As a dye laser, "Raduga-3M" device with laser pumping was used [52]. Generation line width was about 0.5 nm, generation pulse energy was 10^{-3} J, and its duration was 2×10^{-9} s.

During the studies, it was determined that the laser with lamp pumping should be considered more promising for the diagnostics of hydrogen plasma by the method of resonance fluorescence, as it has stable dye generation in the range of 440–700 nm. Generation line width in the range H_a reaches 0.8 nm while using the interferometer with the base of 10 microns as a selector. Generation pulse energy is 0.08 J with the duration of 2.5×10^{-6} s.

Figure 1.20 schematically shows the experimental setup. Its main components are the following: the laser with gradually tuned radiation frequency with the lamp pumping (1); the optical scheme of beam forming (2) including two lenses $F = 25$ cm, $F = 75$ cm and diaphragm of 0.2 cm; FT-1 tokamak equipped with the system of black diaphragm for input and output of laser radiation and by the light trap opposite

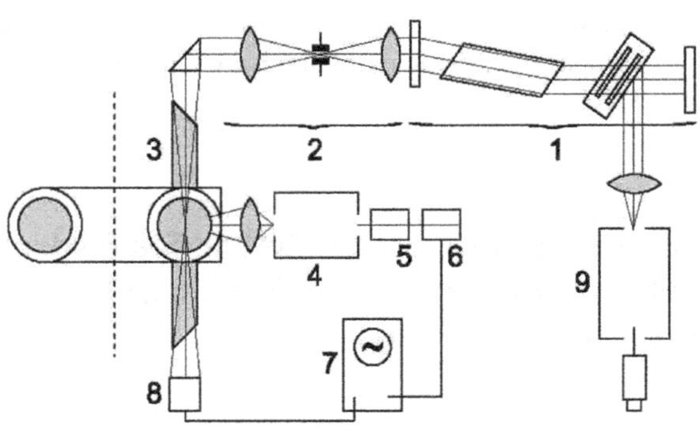

Fig. 1.20 Optical scheme of an experimental device for the diagnostics of hydrogen plasma by the method of resonance fluorescence. Reprinted from [54] with permission

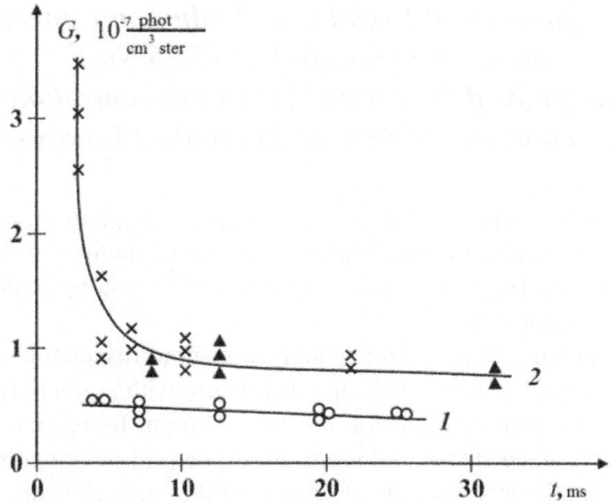

Fig. 1.21 Dependence of fluorescence intensity in time under the exciting of hydrogen plasma by dye laser radiation with lamp pumping. Reprinted from [54] with permission

the watch window (3); the monochromator MDR-2 (4); the radiation detector PMT-84 (5); the pulse amplifier (6); the oscilloscope C8-2 (7); the laser pulse energy meter (8); the monochromator DM-1 with the microscope for laser oscillation frequency control (9).

Beam cross section in the chamber was about 1 cm^2, solid angle of light acquisition was 5×10^{-3}. Fluorescence radiation was observed under the angle of 90° to the laser beam. Fluorescence volume was designed on the monochromator slot with the reduction of 1:5. For the detection of absolute value of hydrogen neutral atoms concentration, the experiments were carried out on Rayleigh scattering of gas laser radiation. Argon has been chosen as working gas. Gas pressure in the discharge chamber was about 1 atm. Figure 1.21 shows the results of measuring fluorescence intensity in time under plasma excitation by dye laser radiation with lamp pumping. The curve (1) is related to the discharge standard conditions (discharge current is 27 kA) after long training of the chamber by heating and by discharge. The curve (2) corresponds to the conditions with less thorough training of the chamber (discharge current is 17 kA). Hydrogen neutral atoms concentration near discharge maximum calculated by the amplitude of a signal is equal to 1×10^9 cm^{-3} and 2×10^9 cm^{-3}, respectively.

1.4.1 To the Question of Creation of Radiation Coherence Source with the Wavelength L_a *(121.6 Nm) for the Study of Hydrogen Atoms Concentration in Plasma by the Method of Resonance Fluorescence*

One of the possible methods of receiving coherence radiation in vacuum spectral region is the generation of the third harmonic in the medium with high third-order nonlinearity [81, 82]. High efficiency of transformation can be achieved under the condition of phase matching.

Near resonance lines, it is possible due to anomalous dispersion. Similar experiments were carried out in xenon [83] and its mixture with argon [84]. In our experiments as nonlinear medium, krypton has been chosen. Its resonance line is $\lambda =$ 123.6 nm, and it is at a distance of 2 nm from resonance line of hydrogen atom L_a.

During the experiments, the optical scheme of the setup illustrated in Fig. 1.23a was used. As an exciting source of the third harmonic generation in krypton optical quantum generator, "Raduga-6" on organic dye solution was used (Fig. 1.22). It enables to get high-power generation with gradually tuned spectrum in the range of 260–1200 nm [78]. OQG "Raduga-6" consists of the master laser on solutions (1), two amplifiers (2, 3) and the output crystal KDR (4), which enables to broaden spectral working range in ultraviolet region. Excitation of OQG and two dye amplifiers is carried out using the monopulse ruby laser (5) with three amplifiers (6, 7, 8). With the help of "Raduga-6", it becomes possible to get radiation on the wavelength of 369.6 nm with energy of 0.1 J and with the duration of $\approx 10^{-8}$ s. Radiation spectral width was $\approx 10^{-2}$ nm.

The cell (3) with the length of 11 cm was filled with spectroscopically pure krypton where impurity content was no more than 0.1%. Laser radiation in the cell was focused with the help of the lens (2) at a distance of 3 cm from the input window. The third harmonic isolation from the exciting radiation ($\lambda = 369.6$ nm) was carried

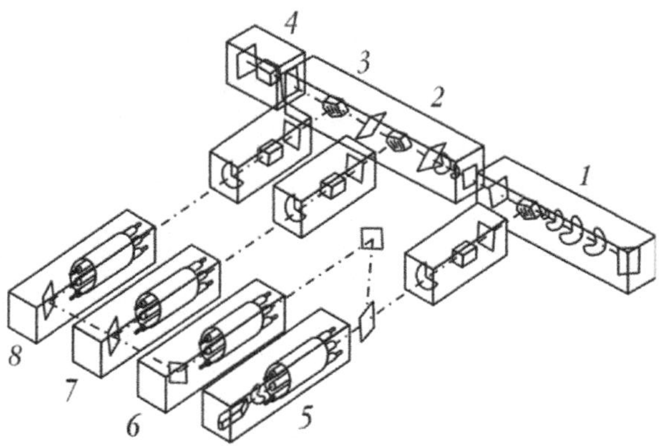

Fig. 1.22 Scheme of optical quantum generator "Raduga 6" on organic dye solution. Reprinted from [54] with permission

out with the help of the monochromator with the prism (4) made of MgF_2. Exciting light passed in the monochromator was absorbed by the filter (5). The third harmonic generation radiation was registered with the help of the electron multiplier IB (6), long-wave sensitivity limit of which is ≈200 nm.

Figure 1.23b shows the dependence of the third harmonic intensity on the pressure. Experimental curve (the third harmonic wavelength $\lambda = 123.2$ nm) has strong maximum under the pressure of 25–30 Torr (Fig. 1.23b). By preliminary estimation, generating energy on the wavelength of 123.2 nm is not less than 1 kJ. Near resonance line, $\lambda = 123.6$ nm krypton was the medium with anomalous dispersion. The maximum on the curve corresponds to the condition of phase matching of interacting waves. For generating power increasing, it is possible to make larger krypton pressure and to reconstruct phase matching condition by means of adding buffer gas with positive dispersion, for example, argon.

It is possible to assume that generating power increase due to adding of buffer gas and also the changes of some settings makes it possible to get generation on the wavelength of 121.6 nm.

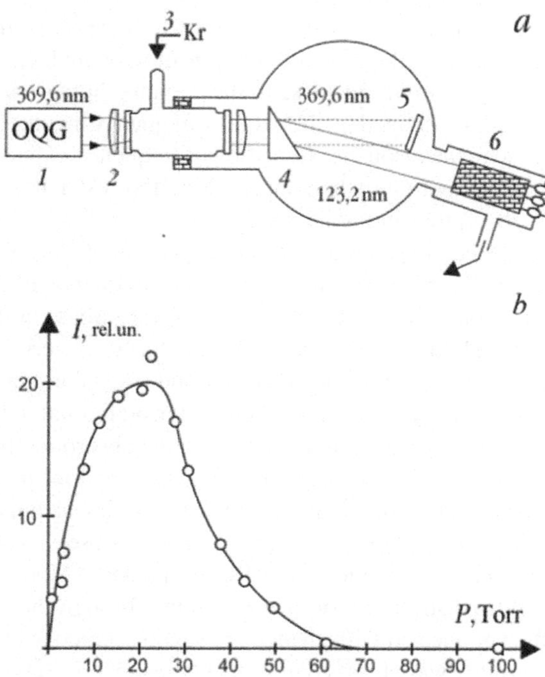

Fig. 1.23 **a** Scheme of the experimental setup for producing coherence radiation in vacuum spectral range (123.2 nm); **b** dependence of radiation third harmonic intensity generating near resonance krypton line on the pressure. Reprinted from [54] with permission

1.5 New Class of Detecting Media for Holography—Gaseous Media. The Study of the Conditions of Dynamic Gain-Phase Holograms Recording (Plane and Bulk) in Sodium Vapors

As it is known, medium is used as a recording material for the registering dynamic holograms. The optical properties of this medium can be changed quite fast and reversible under electromagnetic radiation. As a rule, liquids [85, 86], crystals [87–90] and semiconductors [91, 92] are used as such a medium. The idea to record a dynamic hologram in gaseous medium for the first time was proposed and experimentally carried out in the work [30]. The study of the possibilities of the dynamic hologram recording with low relaxation time in the gaseous resonance medium is of significant interest in terms of using a new sensitive medium for hologram recording. The studies using laser radiation-tuned frequency made it possible to determine the conditions of dynamic gain-phase hologram recording in a new class of detecting media—gaseous media (using sodium vapors). The results of our studies are foreground in the science of our and foreign countries [30–35].

It is clear that gain-phase relief, which is recorded in vapors in the case of laser radiation frequency tuning to the used atomic transition, is high enough that corresponds to the medium sensitivity increase to the dynamic hologram recording. The sensitivity of such media is defined by the value of frequency mismatch, which forms radiation hologram with the resonance transition of frequency of a gaseous medium. It is conditioned by the fact that when approaching the resonance, the changes in the refractive and absorption index increase enormously. Dye solution laser with laser pumping should be considered as a more convenient light source for holograms recording in such media. By varying oscillator frequency of such lasers, it is possible to tune in resonance with the used atomic transitions. In the works of A. M. Bonch-Bruevich with the employees [93, 94], induced scattering of light beams angularly propagated to each other in rubidium and potassium vapors was under study. When radiation density is 10^7–10^8 W/cm^2, then additional light beams can be seen in the spectrum short-wave region concerning the absorption line. Theoretical interpretation of the detected phenomena as the process of four-photon parametric scattering of monochromatic radiation, provided by nonlinear polarization on the same frequency, allows the authors [93, 94] to explain the observed effects including the absence of additional beams in long-wave spectrum parts. Moreover, in the work [93] change of angular distribution of beams passing through the vapors could be observed. This effect connected with changes of vapors refractive index under resonance photons absorption was observed in short-wave and long-wave spectrum parts and also between the doublet lines (except the center of gravity of doublet-point where vapors refractive index is 1).

In the described experiment [32], sodium vapors were used as detecting media for holograms recording. Sodium vapors were developed in the cell (4) (Fig. 1.24c, d) with optical transparent windows, from which beforehand with the help of a

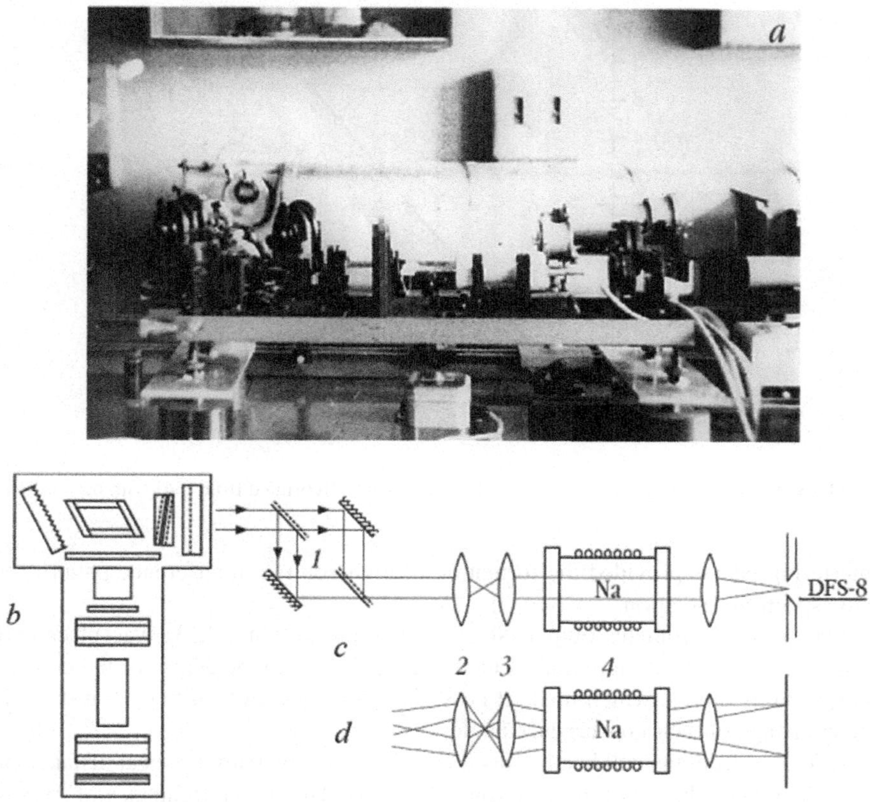

Fig. 1.24 Photo (**a**) and the basic diagram of the experimental setup for dynamic resonance hologram recording: **b** dye laser; **c** top view; **d** side view. Reprinted from [54] with permission

backing pump, the air was evacuated to the pressure 10 atm. The change of the optical density of sodium vapors in the cell was made with the help of a special furnace, the temperature of which was graduated depending on the current passing through it with the help of chromel–alumel thermocouple. The graphics of this dependence can be seen in Fig. 1.25. Knowing the tension of sodium vapors and the temperature, it was possible to determine the upper limit of sodium atoms concentration in the cell. In fact as sodium vapors were unsaturated, it was lower in several times.

The studied 6G Rhodamine laser with laser pumping was used as a light source. Radiation frequency of this laser was tuned for resonance sodium doublet (λ = 588.996 nm and 589.593 nm).

Half-width of laser generation fringe was about 1 nm (the resonator included a diffraction grating and an output mirror). Generation output energy was several milli joules at the pulse duration of about 15 ns on the half-power level. The scheme of holograms recording (see Fig. 1.24) included the Mach–Zehnder interferometer (1) tuned for horizontal fringes in white light. The absence of path difference between

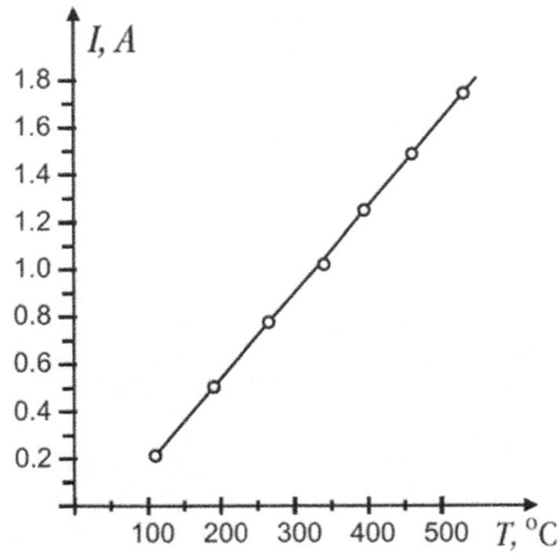

Fig. 1.25 Dependence of current value on the temperature. Reprinted from [54] with permission

interfering beams provided the formation of high contract interference pattern for wide spectrum radiation.

On the interferometer output, single telescopic system (2, 3) was placed. It designed the localization region of interference fringes to the center of the cell (4) with sodium vapors, which made it possible to save the shift of beams in the cell under change of their convergence angle.

Outgoing from the cell beams was focused by the lens with the focal distance of $F = 200$ mm on the slit of the spectrograph DFS-8. The spatial frequency of fringes changed from 3 to 20 lines/mm that corresponded to beam convergence angle of 2×10^{-3}–12×10^{-3} rad. Intensity relations of beams on the interferometer output vary from 1:1 to 1:30. Sodium atoms concentration in the cell changed from 10^{14} to 10^{16} cm^{-3}, and the layer thickness (along the beams motion) was about 2 cm. In the absence of sodium vapors in the cell in the focal plane of the spectrograph, two horizontal fringes can be observed. They correspond to the spectra of initial beams (see Fig. 1.26a). At space frequencies, which correspond to the recording of a thin hologram, at radiation density of 10^5 W/cm^{-2}, on spectrograms equidistant fringes could be observed. The position of these fringes corresponded to the expected diffraction grating pattern formed by initial beams. Figure 1.26b shows one of such spectrograms produced under the beams convergence angle of 4×10^{-3} rad. and vapor concentration of 10^{14} cm^{-3}.

Additional beams disappeared when additional path difference exceeding 1.5 mm was included into the interferometer. It corresponds to effective line width of 0.2–0.3 nm, radiation spectrum assisted in forming dynamic grating (see Fig. 1.26c). And when optical path difference increased from 0.1 to 1.5 mm, then additional beams disappeared at first from long-wave and then from short-wave parts of both spectrum components.

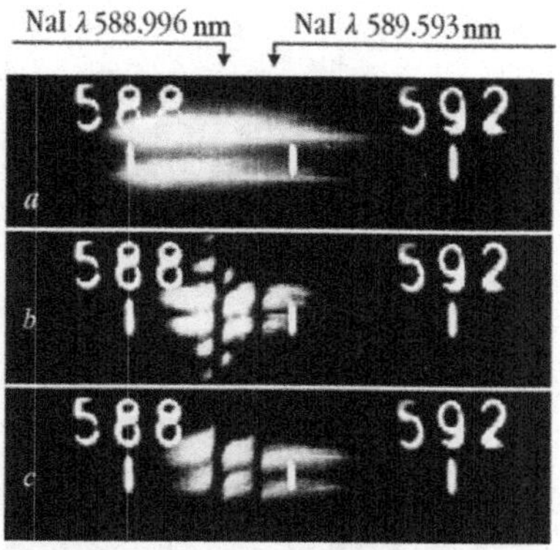

Fig. 1.26 **a** Beams intensity distribution under the absence of sodium vapors; **b** under sodium vapors concentration of 10^{14} cm^{-3}; **c** under the same sodium vapors concentration including additional path difference of 2 mm. Reprinted from [54] with permission

When changing beams convergence angle from 2×10^{-3} to 12×10^{-3} rad. on dynamic phase grating, the change of diffraction pattern (up to three procedures) took place (Fig. 1.27). And when the second cell with sodium vapors, optical density of which is closely connected with optical density in the first cell and what is equivalent to the exclusion of wavelength interval from the initial beams spectrum corresponding to sodium vapors absorption lines, is put before interferometer, then additional beams disappeared.

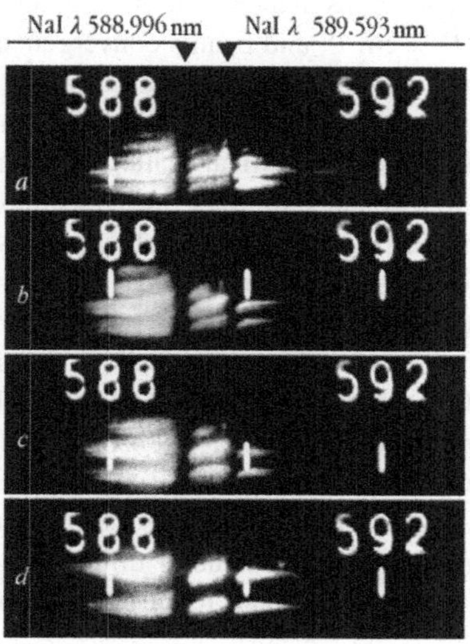

Fig. 1.27 Change of the diffraction pattern on dynamic phase grating at change of convergence angle between beams recording this grating. Reprinted from [54] with permission

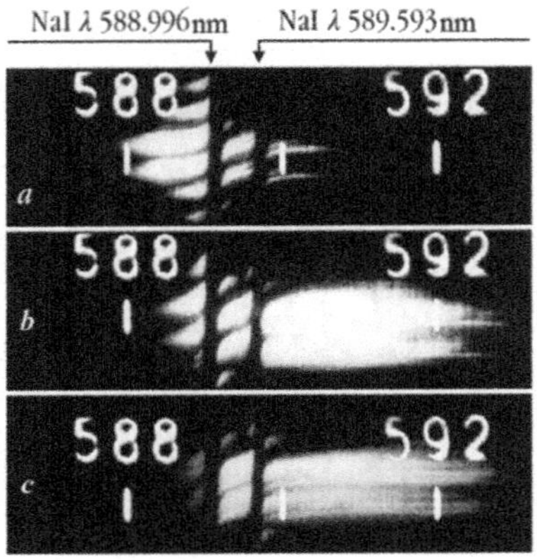

Fig. 1.28 Diffraction image on dynamic phase gratings under dye laser frequency tuning near sodium absorption lines (under intensity relations recording dynamic phase grating of 1:1). Reprinted from [54] with permission

It shows that in the present experiments, additional beams, which appeared in the result of diffraction of initial beams on the gain-phase grating, are provided by radiation absorption near contour centers of resonance sodium lines. Additional beams could be observed in a short-wave as well as in long-wave spectrum region concerning each doublet component (see Fig. 1.28).

The presence of additional beams from the long-wave side of resonance component of 589.593 nm is also confirmed by the results received under the shift of the radiation spectrum concerning resonance sodium doublet (see Fig. 1.28); sodium vapors concentration is 10^{14} cm^{-3}, and beams convergence angle is 4×10^{-3} rad. In the spectra of additional beams in the interval between the doublet line, characteristic dips could be observed. These dips were located in frequencies region, for which space modulation of medium optical features is minimal (see Fig. 1.28). In the parts corresponding to resonance doublet lines, clearly defined absorption lines could be observed. It indicates the absence of absorption saturation along the full region thickness occupied by sodium vapors. Near these lines, characteristic bends could be observed, which were provided by the presence of a lapse rate of sodium vapors in the cell. From the spectral intensity path in the first order of diffraction, shown in Fig. 1.29, it follows that the intensity of additional beams is maximum near the components of resonance sodium doublet.

Intensity distribution on the spectrum height for the wavelength $\lambda = 588.6$ nm is shown in Fig. 1.30. At inequality of intensities of interference beams, asymmetry is observed in the diffraction image. At large difference in intensities, only one diffraction beam arises, which corresponds to the diffraction of a strong beam on the dynamic grating.

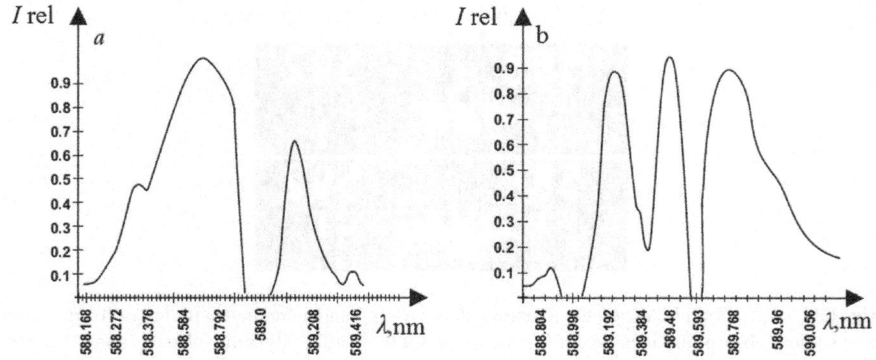

Fig. 1.29 Spectral intensity path in the first diffraction order: **a** when adjusting the laser radiation spectrum to the short-wave component of the resonant sodium doublet; **b** when adjusting the laser radiation spectrum to the long-wave component of the same doublet. Reprinted from [54] with permission

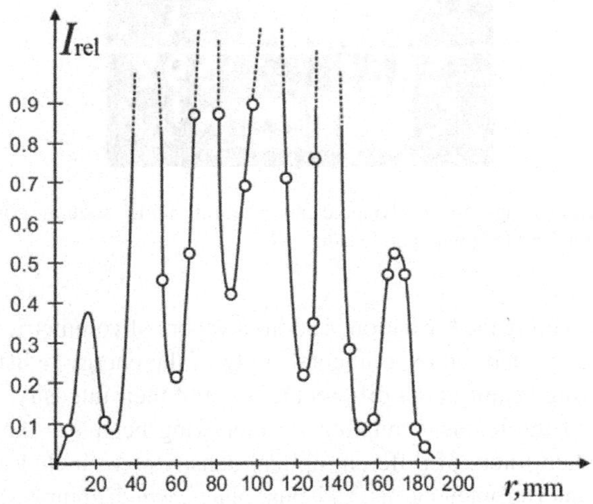

Fig. 1.30 Intensity distribution on the spectrum height of initial and additional beams for the wavelength of 588.6 nm. Reprinted from [54] with permission

Figure 1.31 shows a diffraction image, which corresponds to the intensity relation of 1:30. Under the given convergence angle of 4×10^{-3} rad. with the increase of the concentration from 10^{14} to 10^{16} cm^{-3}, the intensity of the additional beam could be observed (see Fig. 1.32).

In the studied interval of sodium vapors concentration, the additional beams intensity dramatically decreases under convergence angles more than 5×10^{-3} rad, which corresponds to the transition from plane to volume dynamic grating (see Fig. 1.27d).

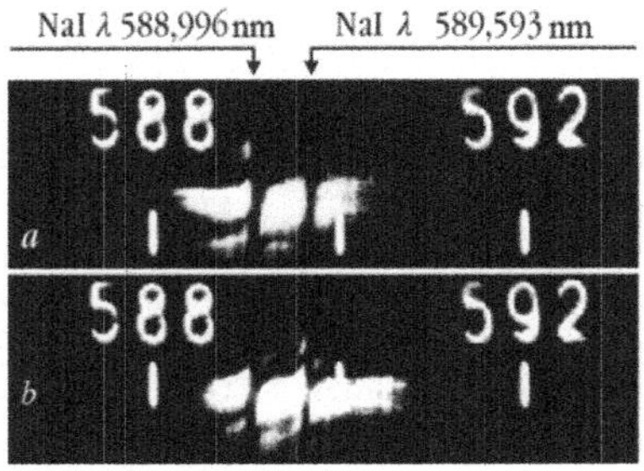

Fig. 1.31 Diffraction image on the dynamic phase grating under frequency tuning of the dye laser near sodium absorption lines (under intensities relation recording dynamic phase grating of 1:30). Reprinted from [54] with permission

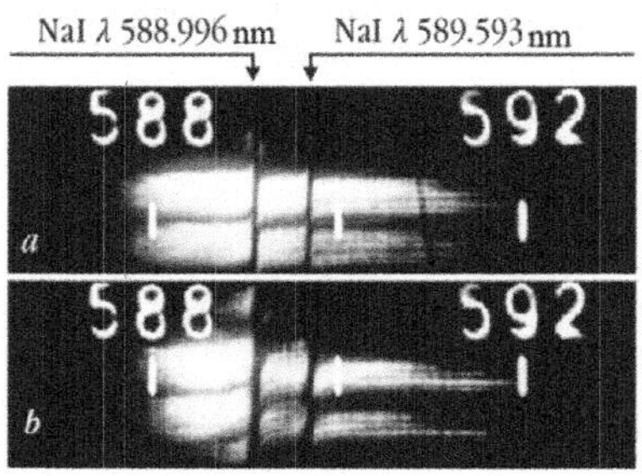

Fig. 1.32 Diffraction image on the dynamic phase grating under sodium atoms concentration changing. Reprinted from [54] with permission

In conditions providing the formation in sodium vapors of volumetric dynamic holograms, there was carried out experimental study of the energy redistribution effect between interfering beams under different relation of their intensity.

The change of the intensity relation of interfering beams to the value of 1:30 was achieved when filters of different optical density were included in one of the Mach–Zehnder interferometer arms, and glass plates, which compensate optical path difference—in the other one. The spectrograms obtained were processed by the method of photographic photometry.

Under the intensity values of initial beams, which considerably differ from each other, we detected essential (up to 4 times) increase of the relation of spectral intensities of weak and strong beams at the output of the cell with sodium vapors. Redistribution of energy between the beams was observed only for wavelengths lying in the range of contours of sodium vapors absorption line. A control measurement was carried out to estimate measurement errors, which connected with microphotometry spectrogram. These spectrograms were produced in similar conditions with homogeneous heat source. Figure 1.33 shows one of such spectrograms, which was photographed with the help of the spectrograph DFS-8, in front of the slit of which step attenuator was placed. The presence of step attenuator made it possible to

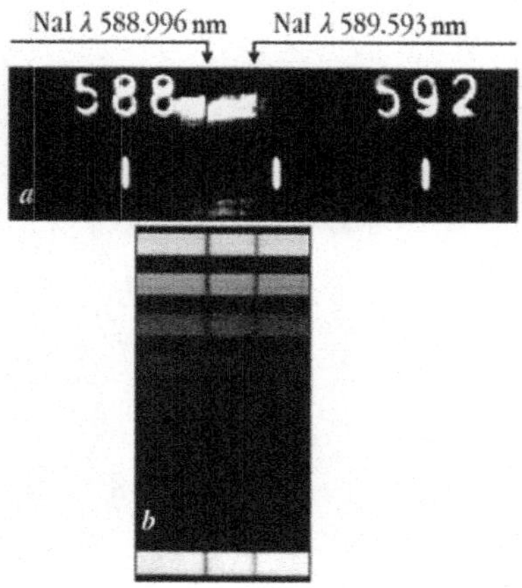

Fig. 1.33 Spectrograms produced in the result of passing through radiation sodium vapors: **a** dye laser; **b** homogeneous heat source. Reprinted from [54] with permission

compare the results of microphotometry with the same intensity relations, which were on spectrograms produced in dye laser radiation. On the control spectrograms, the effect did not appear, so it shows the absence of errors connected with microphotometric measurement.

The observed effect of energy redistribution from the strong wave to the weak one qualitatively matches to the summary of the theoretical treatment of 3D amplitude dynamic holograms. This treatment was held in the work [95]. The results of the diffraction on the volume amplitude dynamic grating are shown in Fig. 1.34. This figure shows the ratio of normalized intensities of a weak beam and strong one depending on the wavelength. The strongest effect could be observed under the sodium vapors concentration of 10^{16} cm^{-3}. Under the concentration of 10^{14} cm^{-3}, it almost disappeared.

1.6 Study of the Influence of Mismatch of the Polarization Planes of Beams Forming Bulk Diffraction Grating on the Self-diffraction Process

In the conditions, which provide the recording of 3D dynamic holograms in sodium vapors, several experiments with beams, polarization planes of which were turned relative to each other, were carried out. For this purpose into one of the Mach–Zehnder interferometer arms, the quartz plate was included. It turned the polarization plane of the initial beam at 120°. Into another interferometer arm glass plate and filter were included. It compensated the optical path difference and created intensity relation of interfering beams equal to 1:15. For matching the direction of polarization plane

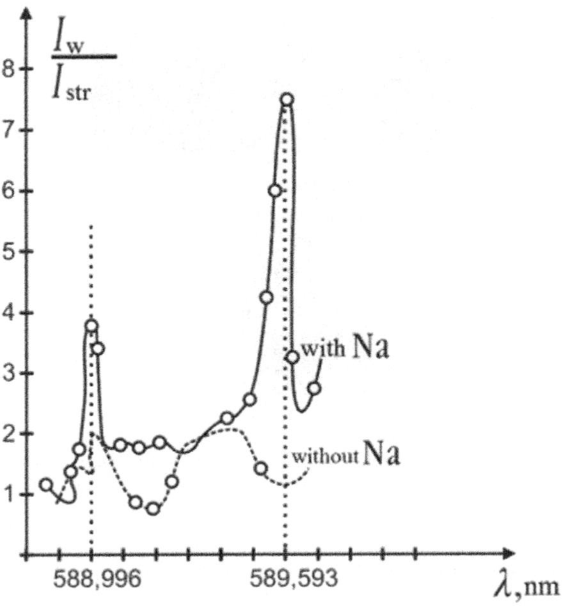

Fig. 1.34 Relation ratio of normalized intensities of weak beam I_w and strong one I_{str} depending on the wavelength. Reprinted from [54] with permission

of dye laser and He–Ne laser used for optical scheme adjustment, the Frank Ritter prism was placed at the output of both lasers.

At the cell output, the Frank Ritter analyzer prism was also placed, which in the absence of sodium vapors in the cell quenches plane-polarized radiation of a weak wave. Meanwhile, an orthogonally polarized component of a strong beam passed through it. It was natural to assume that parallel polarized (concerning initial polarization) component of a strong beam and linearly polarized radiation of a weak beam in sodium vapors will form volume dynamic grating (interference fringes frequency is $\approx$15 lines/mm). After this, at the cell output, the radiation of this polarization is damped by the analyzer. It will be possible to elicit a fact of recording volume dynamic grating in the case of getting diffraction radiation of the orthogonally polarized component of a strong beam in the direction of the weak one. Figure 1.35c shows the scheme of the polarization direction of beams interacting with analyzers.

The results of filling the cell with sodium vapors can be seen in Fig. 1.36. When only a weak beam passed through sodium vapors, its radiation was damped by the analyzer that could also be observed without sodium in the cell (see Fig. 1.36a). Figure 1.36b, c shows the spectrogram produced only in case if one strong beam falls on the cell.

In the case of simultaneous passing of weak and strong beams through sodium vapors, the analyzer transmitted a considerable part of radiation in the direction

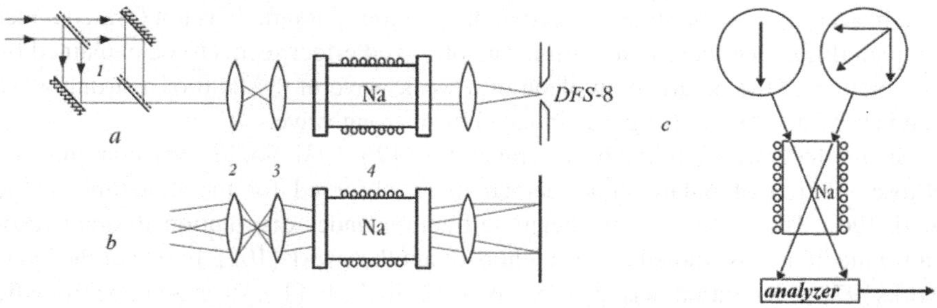

Fig. 1.35 Optical scheme of beams formation for recording dynamic diffraction gratings in the cell with atomic sodium vapors: **a** top view; **b** side view; **c** the diagram view of interacting beams, polarization planes of which are turned relative to each other. Reprinted from [54] with permission

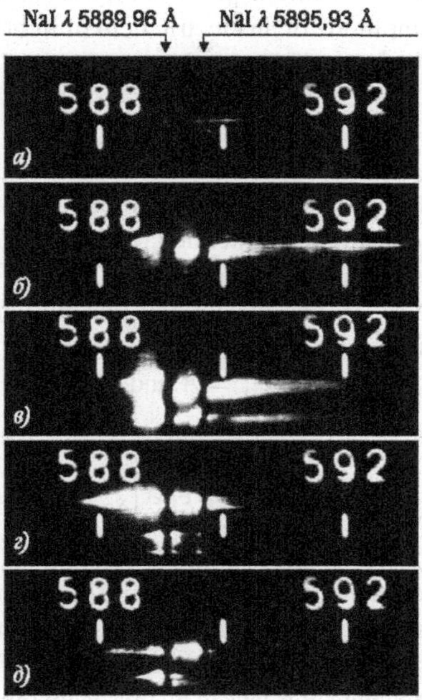

Fig. 1.36 Spectrograms produced when passing through sodium vapors of weak and powerful beams (intensity relation is 1:15), polarization planes of which are turned relative to each other (under vapor concentration of 10^{15} cm^{-3} and beam convergence angles of 9×10^{-3} rad). Reprinted from [54] with permission

of a weak beam. The observed effect could be observed especially strong near the resonator (see Fig. 1.36c).

Additional experiments (see Fig. 1.36d, e) have shown that the effect under review does not depend on the path difference (path difference is $\approx$5 mm) between interfering beams, so it is not conditioned by light diffraction on the volume dynamic grating, which, according to the studies stated in the previous paragraph, is not formed under the path difference more than 2 mm. The observed effect seems to be explained by the rotation of the polarization plane of a weak wave in the field of a strong wave that is provided by birefringence induced by a strong wave.

Such effects were studied by several authors [96–103]. So, in particular, induced ellipse rotation of polarization seemed to be observed for the first time in the work [96]. The effect of light beam self-action under distribution through resonance medium was studied by the authors of another work [102]. In one of the latest works [97], polarization twisting of weak waves in the field of strong monochromatic radiation was investigated.

In our case, unlike the work [97], polarization twisting of weak wave in the field of strong wave of the same frequency was observed. This effect is referred to nonlinear optics and can be described phenomenologically if we make an assumption about the dependence of dielectric medium permeability ε (in the region of resonance absorption line) on the amplitude of an electric field of light wave $\vec{\mathcal{E}}$. In quadratic on $\mathcal{E}$ approximation for tensor component, we have

$$\varepsilon_{ik} = \varepsilon_{ik}^{0} + \gamma_{iklm}\mathcal{E}_l^*\mathcal{E}_m, \tag{1.36}$$

where ε_{ik}^0 is the non-perturbed value of tensor components of dielectric constant (for gas $\varepsilon_{ik}^0 = \varepsilon^0\delta_{ik}$, where δ_{ik} is the Kronecker delta), and γ_{iklm} is the tensor components of fourth-rank tensor describing nonlinearity.

For reasons of symmetry, it is possible to show that this tensor has nonzero components of only three types γ_{xxxx}, γ_{xxyy}, γ_{xyxy}, and $\gamma_{xxyy} = \gamma_{xxxx} - \gamma_{xyxy}$. Then denoting as in the work [104] $\gamma_{xxyy} = \frac{1}{2}B$, $\gamma_{xyxy} = A$ and $\gamma_{xxxx} = A + \frac{1}{2}B$, we will rewrite (1.36) in the final form:

$$\varepsilon_{ik} = \left[\varepsilon^0 + \frac{1}{2}B\sum \mathcal{E}_n^*\mathcal{E}\right]\delta_{ik} + A\mathcal{E}_i^*\mathcal{E}_k \tag{1.37}$$

In the conducted experiment two beams, namely high-intensity and low-intensity beams are used. For simplicity, we are going to assume that displaying nonlinear medium properties is provided only by high-intensity beam. Also, we are going to examine the fact how the change of ε caused by this beam influences the conditions of passing low-intensity beam. When choosing the direction of propagation of powerful beam along z-axis, then

$$\vec{\mathcal{E}}^{(M)} = \left(\vec{\mathcal{E}}_x^{(M)}, \vec{\mathcal{E}}_y^{(M)}, 0\right).$$

As the angle between the directions of low-intensity and powerful-intensity beams is low ($\varphi \approx 10^{-3}$ rad.), it is possible to approximately assume that weak beam also propagates along z-axis:

$$\vec{\mathcal{E}}^{(C)} = \left(\vec{\mathcal{E}}_x^{(C)}, \vec{\mathcal{E}}_y^{(C)}, 0\right).$$

Medium anisotropy caused by powerful beam leads to the fact that along z-axis modes can propagate, which have the difference of wave vectors:

$$\Delta k = k_1 - k_2 = \frac{A\left|\vec{\mathcal{E}}^{(M)}\right|^2 \omega}{2\sqrt{\varepsilon} c}. \tag{1.38}$$

Components $\vec{\mathcal{E}}_x^{(M)}$, $\vec{\mathcal{E}}_y^{(M)}$ satisfy the equations:

$$\begin{aligned} &\mathcal{E}_x^{(M)} e_{2x} + \mathcal{E}_y^{(M)} e_{2y} = 0, \\ &-\mathcal{E}_y^{(M)} e_{1x} + \mathcal{E}_x^{(M)} e_{1y} = 0, \\ &\left(e_i e_j\right) = \delta_{ij}, \end{aligned} \tag{1.39}$$

where $\vec{e}_1$ and $\vec{e}_2$ are the unit polarization vectors.

By decomposing weak wave at the cell input ($z = 0$) on these modes for the dependence of total amplitude of the field in a weak beam, we get the following formula:

$$\vec{\mathcal{E}}(z) = \left(\vec{\mathcal{E}}^{(C)}\vec{e}_1\right)\vec{e}_1 \exp\left(\frac{i\Delta k}{2} z\right) + \left(\vec{\mathcal{E}}^{(C)}\vec{e}_2\right)\vec{e}_2 \exp\left(-\frac{i\Delta k}{2} z\right). \tag{1.40}$$

As it was mentioned above, a weak beam passes through the system of the first polarizer (Nicol prism)–second polarizer, which is crossed with the first one. Choosing plane polarization direction of a weak beam at the cell input behind x-axis for the component $\vec{\mathcal{E}}_y^{(C)}$ passing through the second Nicol, we get the following formula

$$\mathcal{E}_y^{(C)}(L) = \left(\mathcal{E}^{(C)} e_1\right) e_{1y} \exp\left(\frac{i\Delta k}{2} L\right) + \left(\mathcal{E}^{(C)} e_2\right) e_{2y} \exp\left(-\frac{i\Delta k}{2} L\right), \tag{1.41}$$

where L is the cell length.

As at the cell input $\mathcal{E}_y^{(C)}(z) = 0$, then (1.41) can be rewritten in the following way:

$$\mathcal{E}^{(C)}(L) = \left(\mathcal{E}^{(C)} e_1\right) e_{1y} 2i \sin\frac{\Delta k}{2} L. \tag{1.42}$$

Solving (1.39) and (1.42) and assuming that $\frac{\Delta k}{2} L \ll 1$, for light intensity passed through the second Nicol $I^{(C)}(L)$, we get the following formula:

$$I^{(C)}(L) \sim I^{(C)}(0)\left(I^{(M)}\right)^2 \sin^2\left(2\varphi\frac{A^2}{\varepsilon^0} L^2\right), \tag{1.43}$$

where $I^{(C)}(0)$ is the intensity of a weak beam at the output; $I^{(M)}$ is the intensity of a strong beam; $tg\varphi = \frac{\left|\mathcal{E}_y^{(M)}\right|}{\left|\mathcal{E}_x^{(M)}\right|}$; ε^0 is the dielectric medium permeability under the absence of powerful beam; A is the constant of nonlinearity; L is the cell length.[5]

As it is seen from (1.43), light intensity at the output is connected with the intensity of a weak beam by linear law, by the square of the intensity of powerful beam and square of cell length and strongly depends on powerful beam polarization. When deducing this formula, we neglected the effect of proper rotation, i.e., the change of polarization of powerful beam at the length of the cell.

But this effect does not make fundamental changes to the results but complicate calculations and the final answer. Moreover, it should be noted that polarization automatic changes do not appear in the case of only circular or plane polarization [97].

The conducted studies allow concluding that mismatching of polarization planes of initial interfering beams, which participate in the recording of volume dynamic gratings in the medium with resonance absorption leads to the effect of rotation of the plane of polarization of a weak wave in the field of a strong wave of the same frequency provided by birefringence, displaying of which does not allow detecting the diffraction contribution.

1.7 Results and Conclusions

The majority of the results of the studies presented in the first chapter were made by the author during his postgraduate study and during his work in Leningrad Physics and Technology Institute AS USSR named after A. F. Ioffe of the Academy of Sciences of the USSR (1971–1975).

Dye laser "Raduga-3M" with laser pumping for the first time was mounted by Tanin L. V. from the separately developed drawings in the Institute of Physics of the Academy of Sciences of the BSSR into the instrument version and was made in the optical mechanics departments of Leningrad Physics and Technology Institute named after A. E. Ioffe of the Academy of Sciences of the USSR (1971). During its mounting, some changes were made, in particular, the resonator length was increased and the Fabry–Perot interferometer with the basis of 100 μm was included. Due to this, it became possible to narrow the generation spectrum width to 0.01–0.03 nm with pulse energy of 10^{-3} J and power of about 0.1 mWt. These changes made it suitable for the purposes of holography and resonance interferometry [12, 36, 39].

[5]In the case of strong effects or very large length of a cell, the dependence is more complicated: $I^{(C)}L \sim \sin^2 A|E|^2 L/2\sqrt{\varepsilon}$.

Among the most considerable results, the following can be singled out:

1. During the studies of dynamic resonance holography, a new class of detecting media–gaseous media was detected. In particular, additional beams were produced. They appeared in the result of initial beams diffraction on the gain-phase grating formed in the result of radiation absorption near the contour centers of resonance sodium lines. Additional beams could be observed in short-wave as well as in long-wave spectrum regions. The observed effect of redistribution of energy from strong wave to weak wave in the result of scattering on the amplitude volume dynamic grating recorded in sodium vapors is of practical interest in the mechanism of the control of image contrast. It qualitatively corresponds to the conclusions of theoretical treatment proposed by V. G. Sidorovich.

The studies made it possible to determine the conditions of the recording of gain-phase holograms in sodium vapors: optimal values of sodium vapors concentration, convergence angles of the beams forming the grating, radiation density necessary for the hologram recording, and also to find out the influence of the change of the ratio of beam intensities, forming the grating, on a self-diffraction process.

Also, we studied the effect of rotation of the polarization plane of a weak wave in the field of a strong wave of the same frequency detected while investigating the influence of polarization planes mismatching of initial beams on the self-diffraction process. In sodium vapors, these beams form volume dynamic grating.

2. The experimental study on the possibility of increasing the sensitivity of resonance interferometry method using wavelength-tuned dye laser radiation was carried out.

Light sources with the fixed wavelength, used heretofore in the studies using the method of resonance interferometry, considerably constrained its sensitivity. Practically always, the necessity exists to select not the light source to the particular plasma absorption lines, but quite the contrary—the object of study—plasma to definite, fixed laser absorption lines. Because of this fact, it was impossible to use possibilities of increasing sensitivity of the method for the majority of the objects under consideration.

The basic requirements were determined to light source used in the studies by the method of resonance interferometry and the boundaries of this method applicability for the case of the absorption line with the dispersion contour under the assumption of finite width of the probe radiation line.

There was measured distribution of sodium atoms concentration in arc plasma and in the alcohol lamp flame containing sodium salt. It was established that measurement sensitivity of sodium atoms concentration considerably depends on the value of wavelength mismatching of probe radiation from the absorption line center of sodium resonance doublet and has its maximum when $\Delta\lambda \sim 0.1$ nm. In the case when mismatching of probe radiation from the absorption line center is less than 0.1 nm, strong radiation absorption by resonance line wings could be observed.

The present studies allowed increasing sensitivity of measurements by the method of resonance interferometry in comparison with the results of similar interference measurements made by other authors. As a probing source, they used stimulated Raman scattering of ruby laser radiation in nitrobenzene under the mismatch value of $\Delta\lambda \sim 0{,}7$ nm. Minimally detectable sodium atoms concentration in arc plasma was $\sim 10^{13}$ cm^{-3} with the layer thickness of 5 mm.

3. For the first time in the result of the studies on resonance fluorescence, the character was established of spatial and temporal hydrogen atoms distribution and metal impurity atoms in high-temperature plasma of tokamak. The main result of these experiments on tokamak FT-1 tokamak was practical mastering of the range of measuring concentrations of 10^9–10^8 hydrogen atoms by line H_a. To broaden spectral range of the fluorescence method for the first time coherent radiation on the hydrogen wavelength L_a (121.6 nm) was produced. Taking into consideration the influence of buffer gas, the optimal conditions of the generation process of the third harmonic in gaseous mixtures (the mixture of krypton and argon) were found. It made it possible to achieve conversion efficiency of 10^{-4} and to get coherent radiation in the range of vacuum ultraviolet radiation (121.6 nm) on the level of kilowatt power enough for resonance fluorescence excitation in the vacuum spectrum range.

The collaboration of several research teams assisted to the effectiveness of all this quite complicated experimental work. Among them are on the part of Russia—plasma optics departments (A. N. Zaidel, Yu. I. Ostrovsky, G. V. Ostrovskaya, V. I. Gladuschak, E. J. Shreider), plasma physics laboratories of the Physics and Technology Institute named after A. E. Ioffe of the Russian Academy of Sciences (G. T. Razdobarin, V. V. Semenov, L. V. Sokolova, I. P. Folomkin), holography laboratories (V. G. Sidorovich, D. I. Staselko) of the State Optical Institute named after S. I. Vavilov (Leningrad); on the part of Belarus—laser plasma diagnostics laboratories (V. S. Burakov, N. V. Tarasenko, P. A. Naumenkov, P. J. Misakov and others), generating organic compounds laboratories (A. N. Rubinov, V. S. Motkin, M. M. Lojko and others), laser systems and devices laboratories (V. A. Mostovnikov, S. A. Batische and others) and Optical Holography Laboratories (A. S. Rubanov) of the Institute of Physics of the Academy of Sciences of the BSSR (Minsk). A great role in the organization of the cooperation of these research teams from Russia and Belarus in order to carry out joint science studies was played by L. V. Tanin.

The studies on dynamic resonance holography [30–35], resonance interferometry [27–29], resonance fluorescence [13–15] using dye lasers [12, 36, 39, 78, 79] carried out for the first time with the active participation of the author of this book were successfully continued by his followers [105–112].

The results of these studies were used in the works of other authors and co-authors in the studies of potassium, cesium, lithium and hydrogen plasma. These results are also used in PIAS named after P. V. Lebedev RAS (Moscow), the Physics and Technology Institute named after A. E. Ioffe of Russian Academy of Sciences (St. Petersburg), the Institute of Physics of the NAS of Belarus and so on.

The studies on the use of sodium vapors as resonance detecting media (the first work was published by the author in 1974 [30]) are the priority in the USSR and in foreign countries (see A. Yariv, IEEE, J. Quantum Electr. QE-15,524,1978 [33]) and, as later studies of other authors showed (Appl. Phys. Lett. 32, 12, 813, 1978, JOSA. 68, N. 5, 685, 1978), these media with narrow resonance were the most effective under the wave surface processing and reversal of the wave fronts.

References

1. D.A. Gabor, Nature **6**, 777 (1948)
2. Y.N. Denisyuk, DAS USSR **144**(6), 1275 (1962). (in Russian)
3. Y.N. Denisyuk, Opt. Spectrogr. **15**(4), 522 (1963). (in Russian)
4. E. Leith, J. Upatnieks, JOSA **52**, 1123 (1962)
5. R. Brooks, L. Heflinger, R. Wuerker, Appl. Phys. Lett. **7**, 248 (1965)
6. J. Burch, Prod. Eng. **44**, 431 (1965)
7. M.H. Horman, Appl. Opt. **4**, 333 (1965)
8. R.I. Collier, E.T. Doherty, K.S. Pennington, Appl. Phys. Lett. **7**, 223 (1965)
9. R.L. Powell, K.A, Stetson, JOSA **55**, 1593 (1965)
10. K.A. Heines, B.P. Hildebrand, Phys. Lett. **19**, 10 (1965)
11. Yu.I. Ostrovsky, M.M. Butusov, G.V. Ostrovskaya, *Holographic Interferometry* (Nauka, Moscow, 1977), 339 p. (in Russian)
12. Yu.I. Ostrovsky, L.V. Tanin, Tech. Phys. J. **45**(8), 1756 (1975). (in Russian)
13. V.S. Burakov, P.Y. Misakov, P.A. Naumenkov, S.V. Nechaev, G.T. Razdobarin, V.V. Semenov, L.V. Sokolova, L.V. Tanin, et al., *Abstracts of All-Union Conference Lasers Based on Complex Organic Compounds and Their Usage*, Dushanbe, the TSSR, 27–30 Sept 1975, p. 317. (in Russian)
14. S.A. Batishche, V.S. Burakov, V.I. Gladuschak, V.A. Mostovnikov, P.A. Naumenkov, G.T. Razdobarin, A.S. Rubanov, A.N. Rubinov, V.V. Semenov, L.V. Tanin, Tech. Phys. Lett. **3**(14), 647 (1977). (in Russian)
15. S.A. Batishche, V.C. Burakov, V.I. Gladuschak, V.A. Mostovnikov, P.A. Naumenkov, G.T. Razdobarin, A.S. Rubanov, A.N. Rubinov, V.V. Semenov, L.V. Tanin, *Abstracts of All-Union Conference: Lasers Based on Complex Organic Compounds and Their Usage*, TSSR, Dushanbe, 27–30 Sept 1977, p. 142. (in Russian)
16. D.S. Rozhdestvensky, Phys. Chem. J. **42**, 87 (1910). (in Russian)
17. D.S. Rozhdestvensky, Zhurnal Russkogo fiziko-himicheskogo obschestva. Physics series **44**, 395 (1912). (in Russian)
18. D.S. Rozhdestvensky, *Simple Ratio in the Spectra of the Alkali Metals* (PROMEDIA, Leningrad, 1915), 111 p. (in Russian)
19. D.S. Rozhdestvensky, Proc. GOI **2**(13) (1921). (in Russian)
20. D.S. Rozhdestvensky, Proc. Acad. Sci. USSR **8**, 1119 (1934). (in Russian)
21. D.S. Rozhdestvensky, *Work on Anomalous Dispersion in the Vapors of Metals* (AS of the BSSR, Moscow, 1951), p. 406. (in Russian)
22. D.S. Rozhdestvensky, N.P. Penkin, J. Phys. USSR **5**, 319 (1941). (in Russian)
23. D.S. Rozhdestvensky, N.P. Penkin, Proc. Acad. Sci. USSR. Phys. Ser. **5**, 97 (1941). (in Russian)
24. Yu.I. Ostrovsky, N.P. Penkin, Opt. Spectrosc. **3**, 193 (1957). (in Russian)
25. Yu.I. Ostrovsky, N.P. Penkin, Opt. Spectrosc. **9**, 703 (1960). (in Russian)
26. Yu.I. Ostrovsky, L.V. Tanin, et al., *Abstracts of Optical Holography and Its Application*, RSFSR, Leningrad, 1974, p. 62. (in Russian)
27. G.V. Dreiden, A.N. Seidel, G.V. Ostrovskaya, Yu.I Ostrovsky, N.A. Pobedonostseva, L.V. Tanin, Plasma Phys. **1**(3), 462 (1975)

28. G.V. Dreiden, A.N. Zaidel, G.V. Ostrovskaya, Yu.I. Ostrovsky, N.A. Pobedonostseva, L.V. Tanin, V.N. Philippov, E.N. Shedova, *Abstracts of 4 All-Union Conference on Low Temperature Plasma Physics*, Kiev, the USSR, 1975, p. 134. (in Russian)
29. G.V. Ostrovskaya, Yu.I. Ostrovsky, L.V. Tanin, *Abstracts of the International Symposium on Applied Holography "Interkamera"*, Prague, Czechoslovakia, 1978, p. 229
30. Yu.I. Ostrovsky, V.G. Sidorovich, D.I. Staselko, L.V. Tanin, *Abstracts of Optical Holography and Its Application*, RSFSR, Leningrad, 1974, p. 62. (in Russian)
31. Yu.I. Ostrovsky, V.G. Sidorovich, D.I. Staselko, L.V Tanin, *Abstracts of I All-Union Conference "Lasers Based on Complex Organic Compounds"*. Minsk, the BSSR, 1975, p. 213. (in Russian)
32. Yu.I. Ostrovsky, V.G. Sidorovich, D.I. Staselko, L.V. Tanin, Lett. Tech. Phys. J. **1**(22), 1030 (1975). (in Russian)
33. A.I. Yariev, IEEE J. Quantum Electron. QE **15**(6), 524 (1979)
34. Yu.I. Ostrovsky, V.G. Sidorovich, D.I. Staselko, L.V. Tanin, *Abstracts of II All-Union Conference on Holography*, Kiev, the USSR, 1975, p. 22. (in Russian)
35. Yu.I. Ostrovsky, V.G. Sidorovich, D.I. Staselko, L.V. Tanin, *Abstracts of VIII All-Union Conference of Coherent and Nonlinear Optics*, Tbilisi, the GSSR, 1976, p. 310. (in Russian)
36. L.V. Tanin, *Proceedings of the Conference "Holographic methods of study,"* 1978, p. 194. (in Russian)
37. Yu.I. Ostrovsky, Copyright certificate No. 268732, 1961 (Bulletin of inventions 14 (1970)). (in Russian)
38. D.A. Lenard, J.C. Keck, ARS J. **32**, 142 (1962)
39. Yu.I. Ostrovsky, L.V. Tanin, *Abstracts of Optical holography and Its Application*, RSFSR, Leningrad, 1974, p. 34. (in Russian)
40. G.V. Ostrovskaya, N.A. Pobedonostseva, ZhTF **45**, 1462 (1975). (in Russian)
41. G.V. Dreiden, G.V. Ostrovskaya, N.A. Pobedonostseva, F.N. Philippov, Tech. Phys. Lett. **1**, 106 (1975). (in Russian)
42. L.V. Tanin, *Abstracts of IV Republican Conference of Young Scientists in Physics*, Minsk, Belarus 1977, p. 16. (in Russian)
43. I.A. Merkulov, L.V. Tanin, *in Abstracts of I All-Union Conference "Lasers Based on Complex Organic Compounds"*. Minsk, the BSSR, 1975, p. 65. (in Russian)
44. L.V. Tanin, Copyright Certificate No. 736 236, 29 Jan 1980. (in Russian)
45. B.I. Stepanov, A.N. Rubinov, V.A. Mostovnikov, ZhETF Lett. **5**, 144 (1967). (in Russian)
46. B.I. Stepanov, A.N. Rubinov, ZhPS **7**(4), 507 (1967). (in Russian)
47. P.P. Sorokin, J.R. Lankard, IBMJ. Res. Dev. **10**(2), 162 (1966)
48. F.P. Schafer, W. Schmidt, J. Volse, Appl. Phys. Lett. **9**, 306 (1966)
49. A.N. Rubinov, Optical lasers based on solutions of complex organic compounds, Abstract of the dissertation, the Institute of Physics of the AS of the BSSR, 1972. (in Russian)
50. B.I. Stepanov, A.N. Rubinov, UFN **95**(1), 45 (1968). (in Russian)
51. V.A. Mostovnikov, A.N. Rubinov, M.M. Loiko, ZhPS **9**(5), 776 (1968). (in Russian)
52. B.V. Bokut, N.S. Kazak, V.A. Mostovnikov, A.N. Rubinov, ZhETF Lett. **15**(1), 26 (1972). (in Russian)
53. M.M. Loiko, V.A. Mostovnikov, V.S. Motkin N.S. Kazak, A.N. Rubinov, B.V. Bokut, Preprint of IP of AS of the BSSR, ATO3692 (1971). (in Russian)
54. L.V. Tanin, A.L. Tanin, *Biomedical and Resonance Optics: Theory and Practice* (Belaruskaya navuka, Minsk, 2011), 687 p. (in Russian)
55. L.V. Tanin, Study of the coherence properties of pulsed dye laser and its use in holography and holographic interferometry, Abstract of the Dissertation, Minsk, 1976. (in Russian)
56. E. Wolf, L. Mandel, UFN **88**(4), 613 (1966). (in Russian)
57. E. Wolf, L. Mandel, UFN **88**(2), 347 (1966). (in Russian)
58. J. Perine (ed.), *Coherence of Light* (Mir, Moscow, 1974). (in Russian)
59. J. Klauder, E. Sudarshan, in *Principles of Quantum Optics*, ed. by S.A. Akhmanova (Mir. Moscow, 1970). (in Russian)
60. B.A. Sotsky, A.M. Goncharenko, Opt. Spectrosc. **19**, 731 (1965). (in Russian)

61. A.A. Mikaelian, M.A. Ter-Mikaelian, U.G. Turkov, *Optical Generators on the Solid Object* (Sovetskoye Radio, Moscow, 1967), 384 p. (in Russian)
62. R. Collier, K. Burkhart, L. Lin, in *Optical Holography*, ed. by Yu.I. Ostrovsky (Mir, Moscow, 1973), 686 p. (in Russian)
63. D.F. Nelson, R.J. Collins, J. Appl. Phys. **32**(4), 739 (1961)
64. M. Janossy, L. Csillag, K. Kantor, Phys. Lett. **18**(2), 124 (1965)
65. A. Javan, W. Benett, D.R. Herriott, Phys. Rev. Lett. **6**(3), 106 (1961)
66. L.G. Nazarova, Opt. Spectrosc. **23**(4), 757 (1970). (in Russian)
67. B.A. Sotsky, Opt. Spectrosc. **19**(1), 138 (1965). (in Russian)
68. E.S. Voropai, A.M. Sarzhevsky, B.A. Sotsky, ZhPS **7**(6), 834 (1967). (in Russian)
69. D.I. Staselko, Y.N. Denisyuk, Opt. Spectrosc. **28**(2), 323 (1970). (in Russian)
70. M. Lurie, J. Opt. Soc. Am. **58**(5), 614 (1968)
71. K. Murata, T. Asaka, H. Fujiwara, Opt. Acta **17**(1), 5 (1970)
72. J. Weingartner, JOSA **60**(4), 572 (1970)
73. D.I. Staselko, V.B. Voronin, A.G. Smirnov, Opt. Spectrosc. **34**(3), 561 (1973). (in Russian)
74. D.I. Staselko, Y.N. Denisyuk, A.G. Smirnov, Opt. Spectrosc. **26**(3), 413 (1969). (in Russian)
75. G.V. Dreiden, Y.I. Ostrovsky, E.N. Shedova, Opt. Spectrosc. **32**(2), 367 (1972). (in Russian)
76. A. Unzold, in *Physics of stellar atmospheres*, ed. by E.R. Mustel (Izdatelstvo inostrannoyi literatury, Moscow, 1949), 630 p. (in Russian)
77. D.E. Evans, Phys. Ion. Cases **641** (1976)
78. S.P. Andreev, S.A. Batishche, S.A. Mezen, V.S. Mostovnikov, *Abstracts of I All-Russian Conference "Laser Optics"*, Leningrad, RSFSR, 4–8 Jan 1977 p. 358. (in Russian)
79. L.V. Alkevich, A.F. Bohonov, V.S. Burakov, ZhPS **35**(4), 606 (1981). (in Russian)
80. I.P. Gladkovsky, V.E. Golant, Dyachenko, et al., ZhTF **43**, 1932 (1973). (in Russian)
81. G.C. Bjorklund, IEEE. J. Quant. Elect. QE **11**, 287 (1975)
82. S.C. Wallace, G. Zdasiuk, Appl. Phys. Lett. **28**, 449 (1976)
83. A.H. Kung, Appl. Phys. Lett. **25**, 653 (1974)
84. A.H. Kung, J.F. Yong, S.E. Harris, Appl. Phys. Lett. **22**, 301 (1973)
85. E.V. Ivakin, A.S. Rubanov, ZhPS **15**(3), 543 (1971). (in Russian)
86. E.V. Ivakin, I.P. Petrovich, A.S. Rubanov, Quantum Electron. **1**(13), 96 (1973). (in Russian)
87. D.L. Steabler, W. Phillips, Appl. Opt. **13**(4), 788 (1970)
88. D.L. Steabler, W. Phillips, Appl. Phys. Lett. **24**(6), 268 (1974)
89. K.K. Schwarz, Y.J. Kristapson, Registering media for holography: collected papers, 137 (1975). (in Russian)
90. K.O. Hill, Appl. Opt. **10**(7), 1695 (1971)
91. Y. Vaitkus, Y. Visschakas, K. Yaroshyunas, Quantum Electron. **2**(9), 2068 (1975). (in Russian)
92. YuYu. Vaitkus, K.Y. Yurashyunas, Lithuanian Phys. Proc. **2**, 355 (1974). (in Russian)
93. A.M. Bonch-Bruevich, V.A. Hodovoy, V.V. Khromov, JETP Lett. **11**, 431 (1970). (in Russian)
94. A.M. Bonch-Bruevich, S.G. Przhibelsky, V.A. Hodovoy, V.V. Khromov, ZhETF **65**(1), 61 (1973). (in Russian)
95. V.G. Sidorovich, D.I. Staselko, Opt. Spectrosc. **38**(6), 1202 (1975). (in Russian)
96. A.M. Bonch-Bruevich, V.A. Hodovoi, V.V. Khromov, *Abstracts of VI All-Union Conference on Non-linear Optics*, Minsk, the BSSR, 1972, p. 87. (in Russian)
97. V.M. Aratyunyan, T.A. Panazyan, G.G. Adonts, ZhETF **68**(1), 44 (1975). (in Russian)
98. A.I. Alekseev, V.I. Galitsky, ZhETF **57**(3), 1002 (1969). (in Russian)
99. A.A. Dienes, Phys. Rev. **174**(2), 400 (1968)
100. A. Dienes, IEEE. J. QE **5**(3), 162 (1969)
101. P.A. Apanasevich, V.G. Dubovets, ZhPS **17**(5), 796 (1972). (in Russian)
102. A.J. Mouradian, G.G. Adonts, V.G. Kolomiets, Proc. Acad. Sci. Armen. SSR. Phys. **8**, 331 (1973). (in Russian)
103. G.G. Adonts, Proc. Acad. Sci. Armen. SSR. Phys. **8**(4), 241 (1973). (in Russian)
104. N. Bloembergen, in *Nonlinear Optics*, ed. by S.A. Akhmanov, R.V. Khokhlov (Mir, Moscow, 1966), 424 p. (in Russian)

105. N.A. Pobedonostseva, Determination of the concentration of atoms in dense plasmas, Abstract of the Dissertation, Physical and Technical Institute named after A.F. Ioffe, Academy of Sciences of the USSR, 1975. (in Russian)
106. Y.V. Kovalchuk, Resonant interaction of laser radiation on the dye with the plasma of a laser spark, Abstract of the Dissertation, Physical and Technical Institute named after A.F. Ioffe, Academy of Sciences of the USSR, 1979. (in Russian)
107. A.E. Korolev, D.I. Staselko, Opt. Spectrogr. **57**(2), 299 (1984). (in Russian)
108. A.E. Korolev, D.I. Staselko, Opt. Spectrogr. **58**(1), 147 (1985). (in Russian)
109. A.E. Korolev, The study records the resonant dynamic holograms in the metal vapor with the saturation of homogeneously broadened absorption contours, Abstract of the Dissertation, GOI named after Vavilov, 1987. (in Russian)
110. V.S. Burakov, G.T. Razdobarin, N.V. Tarasenko, et al., ZhPS **53**(4), 547 (1990). (in Russian)
111. V.S. Burakov, S.N. Raikov, N.V. Tarasenko, ZhPS **64**(3), 281 (1997). (in Russian)
112. N.V. Tarasenko, Laser-fluorescence spectroscopy of atomic and molecular processes in a non-equilibrium plasma media, Abstract of the Dissertation, IMAF, 1999. (in Russian)

Further Reading

113. A.N. Rubinov, T. Sh, Efendiev. Quantum Electron. **3**(15), 129 (1973). (in Russian)
114. B.A. Bushuk, A.I. Rubinov, T.I. Smolskaya, ZhPS **17**(6), 1112 (1972). (in Russian)
115. N.A. Borisevich, V.V. Gruzinsky, N.M. Paltarak, et al., ZhPS **14**(1), 41 (1971). (in Russian)
116. V.I. Kravchenko, M.S. Soskin, A.A. Smirnov, ZhPS **11**(5), 796 (1969). (in Russian)
117. G.G. Dyadyusha, E.A. Tikhonov, M.T. Shpak, Ukr. Phys. J. **13**(9), 1529 (1968). (in Russian)
118. N.A. Naboykin, A.A. Ogurtsova, A.E. Podgorny, F.S. Pokrovskaya, Quantum Electron. **4**, 229 (1969). (in Russian)
119. L.M. Bonch-Bruevich, T.K. Razumova, I.O. Starobogatov, Optico-mech. prom-st. **9**, 23 (1975). (in Russian)
120. S.S. Anufrik, S.P. Zabirko, V.S. Motkin, A.M. Rusetsky, *Abstracts of I All-Union Conference "Lasers based on complex organic compounds"*, Minsk, the BSSR, 1975, p. 217. (in Russian)
121. D.J. Bradley, A.J.F. Durrant, C.M. Gale, et al., IEEE J. Quant. Electron. QE-4 11, 707 (1968)
122. V.P. Goncharov, L.Y. Minko, S.A. Mikhnov, V.S. Strizhnev, Sq. Electron. **4**, 112 (1971). (in Russian)
123. T.W. Hansch, J.S. Shahin, A.L. Schawlow, Phys. Rev. Lett. **27**(11), 707 (1971)
124. M.C. Peterson, M.J. Kurylo, W. Braun et al., J. Opt. Sac. Am. **61**(6), 746 (1971)
125. R.J. Thrash, H. von Weyssenhoff, J.S. Shirk, J. Chem. Phys. **55**(3), 4653 (1971)
126. V.C. Burakov, P.A. Naumenkov, G.A. Kolosovsky, ZhPS **16**(1), 54 (1972). (in Russian)
127. R.A. Keller, E.F. Zalewski, N.C. Peterson, J. Opt. Soc. Am. **62**(3), 319 (1972)
128. M.B. Denton, H.V. Malmstadt, Appl. Phys. Lett. **18**(11), 485 (1971)
129. J. Kuhl, G. Marowsky, Opt. Commun. **4**(2), 125 (1971)
130. P.B. Sackett, J.T. Yardley, Chem. Phys. Lett. **9**(6), 612 (1971)
131. G. Megie, O. de Witte, Rev. Phys. Appl. **6**(3), 341 (1971)
132. M.L. Bolshov, I.D. Guzeev, A.V. Zybin, et al., ZhPS **19**(5), 821 (1973). (in Russian)
133. P.P, Sorokin, J.R. Lankard, J. Chem. Phys. **51**(7), 2929 (1969)
134. D. Dimock, E. Hinnov, L.C. Johnson, Phys. Fluids **12**(8), 1730 (1969)
135. G.K. Tumakaev, V.R. Lazovskaya, ZhTF **43**(9), 2008 (1973). (in Russian)
136. A.A. Afanasiev, V.S. Burakov, V.V. Zheludok, S.V. Nechaev, Dokl. **15**(10), 889 (1971). (in Russian)
137. B. Stansfield, R. Nordwell, J. Meyer, Phys. Rev. Lett. **26**, 1219 (1971)
138. C.F. Burrell, H.J. Kunze, Phys. Rev. Lett. **28**(1), 1 (1972)
139. V.G. Sidorovich, *Abstracts of All-Union Conference "Optical Holography and Its Application"*, Leningrad, RSFSR, 1977, p. 4. (in Russian)

140. G.V. Dreiden, A.I. Seidel, Y.I. Ostrovsky, E.N. Shedova, ZhFT **43**(7), 1537 (1973). (in Russian)
141. G.V. Dreiden, YuI Ostrovsky, E.N. Shedova, A.N. Zaidel, Opt. Commun. **4**(3), 209 (1971)
142. G.V. Dreiden, G.V. Ostrovskaya, N.A. Pobedonostseva, V.N. Philippov, Tech. Phys. Lett. **1**(2), 106 (1975). (in Russian)
143. D.I. Staselko, V.L. Strigun, *Abstracts of All-Union Conference "Optical Holography and Its Application"*, Leningrad, RSFSR, 1974, p. 31. (in Russian)
144. B.H. Soffer, B.B. Mc Farland, Appl. Phys. Lett. **10**(10), 266 (1967)
145. A.M. Bonch-Bruevich N.N. Kostin, V.A. Hodovoi, Opt. Spectrogr. **24**(6), 1014 (1968). (in Russian)
146. B. Soep et al., Opt. Commun. **1**(9), 433 (1970)
147. I.G. Naumenko, A.M. Korobkov, M.I. Dzyubenko, Opt. Spectrogr. **34**(6), 1175 (1973). (in Russian)
148. I.L. Garlsten, T.J. Mellrath, Opt. Commun. **8**(1), 52 (1973)
149. D. Kato, T. Sato, Opt. Commun. **5**(2), 134 (1972)
150. H. Walther, J.L. Hall, Appl. Phys. Lett. **17**(6), 239 (1970)
151. D.J. Bradley, W.G.I. Caughey, J.I. Vukusic, Opt. Commun. **4**(2), 150 (1971)
152. F.C. Strome, J.P. Webb, Appl. Opt. **10**(6), 1348 (1971)
153. S.A. Myers et al., Opt. Commun. **4**(2), 187 (1971)
154. V.I. Revenko, V.B. Timofeev, PTE **6**, 168 (1972). (in Russian)
155. S.A. Batishche, V.S. Motkin, P.I. Myshalov, *Abstracts of I All-Union Conference "Lasers based on complex organic compounds"*, Minsk, the BSSR, 1975, p. 225. (in Russian)
156. C.V. Shank, I.E. Bjorkholm, K.H. Kogelnik, Appl. Phys. Lett. **18**(9), 395 (1971)
157. M.V. Belokon, A.N. Rubinov, Quantum Electron. **1**(7), 1651 (1974). (in Russian)
158. L.P. Kaminov, H.P. Weber, E.A. Chandross, Appl. Phys. Lett. **18**(11), 497 (1971)
159. V.I. Tomin, B.A. Bushuk, ZhPS **17**(2), 218 (1972). (in Russian)
160. V.N. Lukyanov, L.T. Semenov, N.V. Shelkov, S.D. Yakubovich, Quantum Electron. **2**(11), 2373 (1975). (in Russian)
161. L.E. Erickson, A. Szabo, Appl. Rhys. Lett. **18**(10), 433 (1971)
162. E.F. Zalewski, R.A. Keller, Arr1. Opt. **10**, 12 (1971)
163. C.-Y. Wu, J.R. Lombardi, Opt. Commun. **7**(3), 233 (1973)
164. A.A. Friesem, U. Ganiel, G. Neuman, Appl. Phys. Lett. **23**(5), 249 (1973)
165. F.V. Karpushko, A.S. Rubanov, DAN BSSR **16**(7), 600 (1972). (in Russian)
166. V.V. Kabanov, A.S. Rubanov, A.L. Tolstik, A.V. Chaley, Opt. Commun. **71**(3, 4), 219 (1989)
167. V.V. Kabanov, A.S. Rubanov, A.L. Tolstik, Modern Opt. Laser Phys. **313** (1993). (in Russian)
168. S.M. Karpuk, A.S. Rubanov, A.L. Tolstik, Quantum Electron. **24**(1), 52 (1997). (in Russian)
169. S.M. Karpuk, A.S. Rubanov, A.L. Tolstik, A.V. Chaley, Tech. Phys. Lett. **20**(12), 4 (1994). (in Russian)
170. A.S. Rubanov, A.L. Tolstik, S.M. Karpuk, O. Ormachea, Opt. Commun. **181**(1–3), 183 (2000)
171. O.G. Romanov, A.L. Tolstik, Opt. Spectrosc. **105**(5), 812 (2008). (in Russian)
172. O.G, Romanov, A.L. Tolstik, ZhPS **75**(4), 509 (2008). (in Russian)
173. V.V. Kabanov, A.S. Rubanov, A.L. Tolstik, A.V. Chaley, Opt. Quantum Electron. **19**(3), 351 (1987)
174. V.V. Kabanov, A.S. Rubanov, IEEE J. Quantum Electron. V. QE **26**(11), 1990 (1990)
175. V.V. Kabanov, Laser Appl. Opt. Metrol. **95** (2004)
176. V.V. Grigoriev, V.V. Kabanov, ZhPS **75**(2), 187 (2008). (in Russian)
177. V.V, Kabanov, A.S. Rubanov, A.L, Tolstik, A.V. Chaley, Opt. Quantum Electron. 19(3), 351 (1987)
178. E.V. Ivakin, S.M. Karpuk, A.S. Rubanov, A.E. Tolstik, A.V. Chaley, Proc. Acad. Sci. Phys. Ser. **56**(8), 41 (1992). (in Russian)
179. A.L. Tolstik, BSU **159** (2002). (in Russian)

180. S.M. Shandarov, N.I. Burimov, J.N. Kulchin, R.V. Romashko, A.L. Tolstik, V.V. Shepelevitch, Quantum Electron. **38**(11), 1059 (2008). (in Russian)
181. I.N. Agishev, A.L. Tolstik, Proc. Russ. Acad. Sci. Phys. Ser. **72**(12), 1735 (2008)
182. O.G. Romanov, A.L. Tolstik, ZhPS **76**(3), 395 (2009). (in Russian)
183. A.S. Bondarchuk, E.A. Melnikova, Proc. Russ. Acad. Sci. Phys. Ser. **70**(9), 1289 (2006). (in Russian)

Chapter 2
Holographic Microscopy of Phase and Diffuse Objects Under the Influence of Laser Radiation, Magnetic Fields, Hyperbary

The first works on the development of the directions of biomedical optics in Belarus were studied by L. V. Tanin in 1975. Soon the urgency of this direction became clear, and due to this in 1978, he formed the scientific team "The coherent-optical studies of medico-biological systems" in the Institute of Neurology, Neurosurgery and Physiotherapy of the Ministry of Health of the BSSR where these scientific studies are still carried out. Coherent-optical methods of peripheral neuromuscular and cardiovascular systems treatments and diagnostics were developed [1–77].

In the present chapter, there are proposed and studied crucially new coherent-optical methods. Namely, the direction of holographic interference microscopy was created and developed. In particular, the methods and devices, which allow producing 3D images and studying changes of their state, were developed [1–3, 5, 7–10, 13, 19, 54]. It can be also noticed that among the treatment methods the bases of such directions as laser acupuncture [4, 6, 11, 12] and intravascular blood irradiation [16, 18, 21, 26–28, 30–53] were taken into consideration. During the experiment and in the hospital, medical and physics specialists have developed new treatment methods. It was made as a result of studying the influence of electrophysiological factors, medical products, strong pulsed magnetic fields [14], hyperbary [15, 17] on the separate nerve and muscle fibers, nerves, muscles in its vital state and, finally, on neuromuscular and cardiovascular systems. Particularly, some laser radiation features (monochromaticity, power density, tuned wavelength, high spatial coherence, polarization) made it possible to use them in modern neurology. The stimulatory effect is known of low-temperature laser radiation (632.8 nm) on the activity of enzyme system of the nerve cell, on the increase of maximum pulse frequency passed by the nerve without transformation during irradiation. It is detected that laser radiation exercises a salutary influence over the regeneration process of the nerve fiber. All this arose the question of the laser radiation practical application in the hospital during the treatment of peripheral nervous system diseases [29, 34, 35, 37, 39–41, 43, 45, 48, 53, 75, 77].

L. Tanin et al., *Biomedical and Resonance Optics*, Bioanalysis 11,
https://doi.org/10.1007/978-3-030-60773-9_2

Since 1978 in the Institute of Neurology, Neurosurgery and Physiotherapy of the Ministry of Health of the BSSR, the employees of the team "Coherent-optical studies of medico-biological systems" under supervision and on the initiative of the author L. V. Tanin along with the employees of the first neurologic and physiotherapy departments under the supervision of Professor I. P. Antonov and Professor G. K. Nedzved carried out laser therapy by low-intensity laser radiation during the treatment of patients with peripheral nervous system diseases including intravascular laser blood irradiation (ILBI) [4, 6, 11, 12, 16, 18, 21, 26–28, 30– 40, 75, 77].

At the same time, the necessity appeared of more detailed study of structural and functional characteristics of neuromuscular tissue under the influence of laser radiation, magnetic field, increased pressure in connection with its medicinal use as the mean providing the formation of stable adaptive process at different organization levels [15, 17, 77].

Thus, the development and integration of the method of holographic interference microscopy with electrophysiological control is of utmost interest. It is used for the studies of the structure and function changes in isolated excited specimen of nerve and muscular tissue being in the conditions of increased pressure. When solving this problem, we based on the following conditions: noninvasivity, high measurement sensitivity, dynamic studies and the possibility of simultaneous electrophysiological control. There are some peculiarities when using coherent-optical methods in the studies of nerve and muscular tissues. These peculiarities are connected with the multiple light scattering in the object, its microstructure changes, the use of increasing optics under the considerable thickness of optical windows of hyperbaric cells, the deformation of cells useful capacity under pressure application. It generates the need for the developing additional experimental processes that allow using holographic methods in the present conditions [15, 17, 70, 71, 72]. These studies are covered in detail in Sect. 2.3.

Primarily, as a technical object, a semiconductor crystal of laser diode was chosen. It was chosen for the purpose of developing the method of holographic microscopy with regard to the studies of diffusely reflective medico-biological microobjects. This object was used to train the design features of the holographic interference microscope in reflected light, and the holographic method was adapted to the interference studies of diffusive objects. As a result, the integrated study of thermal characteristics of semiconductor crystal faces of laser diode was carried out. The studies of the deformation of laser diodes, which worked in continuous and pulsed regimes and which were carried out along with the employees of the Laboratory of Semiconductor Optics of the Institute of Physics of the NAS of Belarus using the method of holographic interference microscopy, are also of independent scientific and practical interest [55–69].

2.1 Holographic Microscopy for the Study of Phase, Diffusive and Mirror Microobjects

For the first time, the use of a holographic recording of wave front with its consequent reconstruction as a lensless method of microobjects studies was proposed by D. Gabor [78–82]. Though his idea of images enlargement due to the difference of wavelengths at the level of recording and during the reconstruction has not found wide application, the holographic principle itself considerably broadened and added the possibilities of classical microscopy approaches [83, 84]. Microscopy, using such possibilities of holography and holographic interferometry as microobjects 3D image recovery with its posterior processing, interferometric comparison of time-spaced microobject states of any irregular shape [85, 86], has a number of advantages over the classical interference microscopy.

As it is known, in the studies of medical and biological objects, in vivo microscopy plays a great role. But, along with noninvasiveness, high sensitivity, precision, resolution, possibility to observe the microobjects under study with rather high magnification and so on what differs in vivo microscopy advantageously, this method like any other method has some disadvantages. So, in optical microscopes a large increase inevitably leads to a decrease in sharpness. It complicates the study and photorecording of short-lived bulk structure, the sizes of which are beyond the limits of depth of microscope objective sharpness. The possibilities of classical in vivo microscopy are considerably limited by the object mobility. The objects with higher mobility, such as muscle tissue and so on, are not suitable for the microscopical study, because at insignificant contractions, the structures under consideration get outside the field of view, especially when working with high magnification. The same difficulty can be observed in cinemicrography. Moreover, in classical microscopy, the difference in depth planes of the object under study (it can be a cell or cell population) is recorded, for example, photographically with the objective overfocusing during each exposure. It is obvious that time consumed on overfocusing, microscope adjustment and single-frame photography of microobject section limits and sometimes excludes the possibility of research operation. Firstly, it concerns the study of fast processes or short-lived microobjects where there is a necessity to set volume correlation between structural formations in the present moment (in the studies of the process of cell or cell ensemble division or during microcirculation and so on).

Moreover, living organisms are used to be developing continuously. A morphologist, a physiologist or a cytochemist looking through the microscope in some cases cannot be sure that focusing on the new preparation part they see it in the same functional mode. Developing processes in some cases are so transient that they can escape the researcher's view because of the inertness of human visual analyzer. In this connection, the number of objects suitable for the studies by the method of vital microscopy is limited. The limitation of lifetime of the majority of microobjects leads to the fact that in the classical microscopy considerable part of the studies is connected with the estimation of microobjects quantity and geometrics in sections, smears and so on [87]. At the same time when changing a living microobject on a histologic specimen, considerable part of the information on the object of investigation is distorted and lost. Such a situation leads, for example, to attempts of using electronics and computer technologies for the synthesis of 3D images by the averaging of multiple photographs of sections of several homogeneous objects focused on different levels of morphogenesis. Such techniques are too bulky, complicate the work of the experimenter and exclude considerable part of structural changes.

In due course, the necessity of developing fixation and visualization histologic methods was detected by the absence of the possibility to process and study living specimens immediately and also by the necessity of their dead storage (up to several years). On the other hand, to date photographic recording of sections and smears is almost the only way to create disease classifiers and atlases of microobjects.

In its turn, the holographic method of recording and time reconstruction of full object optical information enables to solve the problems concerning microobjects certification and also the studies on their recognition in a new way. Living microobjects holograms producing can help when studying rare and typical diseases and allow classifying these unique pathologies. Taking into consideration the dynamics of the aspect of the disease and the possibilities of transformation of shape and sizes of micro objects (red cells, lymphocytes, nerve and muscle fibers and so on), holographic recording can assist the medical diagnostics. The use of holographic recording method in microscopy enables to reveal weak movement region in the microobject quite easily (for example, cytoplasmic flow in the cell [88] or the process of plasmolysis of the onion epidermis cells). If the motion is not an oscillating periodic process than on the level of the hologram recording, the moving part in the object does not practically influence its formation and during the reconstruction has smaller contrast than its fixed parts. Though in this case the quantitative determination of speed of movement is difficult, velocity profile can be easily determined.

The first works in this area appeared in 1966 when R. F. Wanlighten and Kh. Osterberg informed about their holographic microscope [89]. During the depth of field tuning up to 40 μm, the microobjects with the size up to 2 μm could be seen on the reconstructed image. A hologram preserved the information about 3D object.

G. Ellis [90] went further. He used prior processing of the reconstructed image, and due to this, he could show phase-contrast, dark field and interference effects. Diatoms with the size about 100 μm have been chosen as objects of study.

G. Knoh has not only demonstrated the possibilities of holography, but also studied moving plankton (size is about 1 mm) in movement using for the "fixation" pulsed ruby laser operated in a free-running lasing mode [91].

K. Snow and R. Wandervaken described holographic microscopes operated in reflected and transmitted light. Phase as well as diffusive specimens could be chosen as objects of study. And these microscopes were interference and operated in real time [92].

The first domestic work in this area appeared in 1972 [93] and was dedicated to the study of onion epidermis cells. At that time not only qualitative, but also quantitative data were produced, which were recorded from interferograms. Cell topogram was produced, and two-dimensional distribution of optical thickness was built. The path between two neighboring interference fringes was 4 μm. Maximum thickness of the cell was 24–28 μm. The recording of changes, which were made in the onion steam, enabled to determine, for example, such characteristics as growth rate—1.5 $\times 10^{-5}$ m/s [94]. And the method of holographic interferometry was the easiest and the simplest. In the next few years for the tasks of its type, holographic interference microscope was created [95].

At the same time, the distribution of holographic methods in the area of object dynamics studying can be clearly seen. Old methods are developing, and new methods appear. For example, Heflindger, G. L. Stuart and others [96] used the camera for observing the moving plankton. With the help of this camera, they recorded holograms. Then, reconstructed images with the positive resolution of microorganisms were studied under a microscope. In the other work, they showed that using the same equipment, the resolution up to 1 μm can be got [97].

In the works [98, 99], holographic interferometry and microscopy were used for recording fast periodic processes in microbiological objects.

In the works [1–3, 5, 7–10, 13, 15], we can find the results of preliminary studies of myelinated nerve fibers of frog sciatic nerve, cavy spleen lymphocytes and frog red blood cells.

Important feature of the holographic method is the possibility to study diffusely scattered objects, to carry out a quantitative comparison of time-spaced condition of living structures (differential interferometry) [100] and to get relief contour of static and time-dependent object surfaces using long-wave tuned lasing regime [70, 72, 101, 102].

The analysis of the literature showed that in the works, which are dedicated to this theme, mainly microobjects with the size from 1 μm up to 1 mm are used that shows the opportunity of effective application and development of the methods of holographic interference microscopy in biology and medicine.

Wide possibilities of the methods of holographic interference microscopy were also showed in the works on the research of dynamic distribution of the liquid in the narrow channel (width of 180 μm, depths of 90 μm) [103], on the study of refractive index distribution in a transparent medium near the wire with the diameter of 20 μm heated by electrical pulse [104], on the study of diaphragm deformation of 2 mm under small pressure disturbance [105], on the producing of the interferograms of onion epidermis cells [106], urchin ovules [107]. At the present time, we have gathered a considerable experience of using holographic methods for practical purposes. In the work [108], for example, using holographic interference microscope with the coherent noise averaging, plasmodium migration in the mold cells is under consideration. It is shown that the change of the protoplasm flow directivity and the change of the cytoplasmic channel thickness take place simultaneously. The authors of the work [109] on the holographic interferograms calculated the refractive index distribution in the section of light guide glass fiber was calculated. The methods of holographic interference microscopy were used to study the conditions of cells crystallization under low temperature with the purpose of determining optimal regime of their freezing for long-term preservation [110]. In particular, they helped to see the processes of crystallization and mechanical failure of red cells and to determine the optimal freezing rate [111]. In the works [112, 113], they detected the application for the growth condition monitoring and crystals homogeneity. In [114, 115], specially developed method of holographic interferometry of anisotropic media for liquid crystal studies was used.

It is known that hemoglobin oxygenation and deoxygenation rate in the red cell depends on the morphological (cell surface area) and physiological (for example, pH and inner concentration 2.3—diphosphoglycerate) factors [116]. So, red cells shape and surface area studies depending on their functional state are of utmost interest [117]. A number of works is known, which successfully used methods of holographic interferometry [118, 119] to study morphological features of erythrocytes. Interferometry, including holographic one, is one of the main methods for detecting hemoglobin content in separate red cells [120].

The information received using these methods enables to determine the relation between the areas of inner and outer monolayer of cell membrane and, consequently, can be helpful for the studies of mechanism of action of different medical products [121].

The monograph [122] shows the perspective application of holographic methods for the studies of microcirculation, red cells shape changing when moving along the bloodstream. In the work [123] using the method of holographic interferometry, the analysis of red cells shapes and sizes during hereditary spherocytosis was performed. In the work [124], it is noticed that morphological characteristics such as volume, diameter, deformability, aggregation degree and others are extremely sensitive to the external influence and physiological organism state.

Some new approaches in holographic interferometry can also add the methods, which have already been tested in microscopy. One of them is holographic heterodyne interferometry. Its main idea is the creation of small shift between interfering wave frequencies. In this case in any point of interference pattern, the intensity is changing sinusoidally that allows using radio aids measuring phase delay more precise than by traditional fringes deviation. For example, in the work [125] where the bases of holographic heterodyne interferometry are shown in detail, the accuracy is increased 450 times.

Among other developments in coherent optics used in microscopy, the microscope analyzing coherent antistokes radiation of Raman scattering [126] should be mentioned. It enables to study microobjects with spatially inhomogeneous chemical composition and has 10 μm resolution.

Definite achievements of the Doppler anemometry are realized in the microscope described in [127, 128]. It is used for the study of the Brownian movement of the particles and flows in the plants, for the research of blood cells movement in a human organism. The speed changing in such a system is determined by a half-width of autocorrelational function of the intensity of scattered light radiation.

A lot of attention is paid to the development of scanning microscopes where the object is scanned by focused laser beam, and the image is rendered on the TV screen. Such microscopes have stronger light-gathering power and often better resolution, less aberrations that allows processing images better [129]. New works have appeared, which inform about the possibility of phase reconstruction in such images, and creation of interference confocal microscope [130], operating almost in the same way as a dotty-diffractive interference microscope, which uses interference of directly transmitted light together with diffracted one [131]. The authors [132, 133] created the laser projection microscope with image brightness gain in copper vapors, which operates at the wavelength of $\lambda = 514$ nm. The methods of detecting microobjects characteristics and their ensembles directly by diffraction pattern and without microobjects increase are being developed. In biology, such methods are used during the studies of muscle specimen and during red cells studies.

Development and use of the methods of holographic interference microscopy for the studies of structure-functional changes of peripheral nerves and innervated by them tissues are the most important processes for the modern neurology. The study of the influence of some pharmacological and physical factors (laser, magnetic fields, hyperbary and so on) on the peripheral nervous system can become the basis for the development of new methods of diagnostics and treatment of its pathology. The studies of peripheral nerve fibers and cells in their lifetime play a special role during the investigation of the mechanism of stimulating action of low-energy laser radiation (used, for example, during the treatment of lumbar osteochondrosis, radiculitis, trigeminal nerves by laser acupuncture method [4, 6, 11, 12], during the study of transfer processes and information processing in a living organism). Information content and reliability of these studies can be improved by the use of independent electrophysiological and holographic methods [3, 15].

The use of common light microscopy and its methods such as polarized, phase-contrast, dark field, luminescent, the methods of electron microscopy and X-ray beams diffraction enabled to get a list of new information about the structure and ultrastructure of nerve fibers. There is a necessity to specify the mechanism of trophic function realization of the nervous system in order to retrace the process of transfer from hyperactivity to neuron damage, to study the dynamic of reconstructive regenerative process and many others. Thus, the study of the structure and function of living nerve cell, nerve fiber is still of certain importance.

According to the above-said, the development of coherent-optical, highly sensitive, contactless methods of a lifetime microobjects study, particularly, holographic microscopy, is the actual problem.

2.1.1 Holographic Interference Microscopes Operating on Transmission and in Reflected Light

The main types of holographic microscopes

A holographic method was proposed by Denis Gabor at the end of the 40-ies when he was trying to improve the electron microscope [78–82]. In the similar "lensless" microscope, the magnification can be received without optics during the recording and reconstruction [134, 135]. One of the possible schemes can be seen in Fig. 2.1.

The magnification in such a scheme can be defined by the expression:

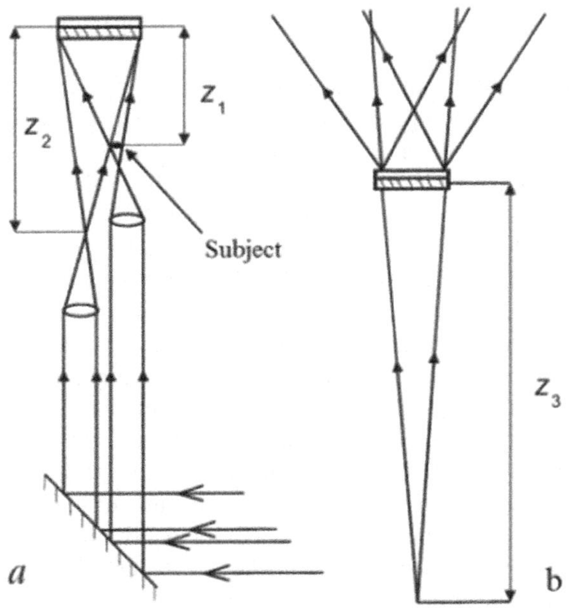

Fig. 2.1 Scheme of the recording (**a**) and reconstruction (**b**) of holograms "in a lensless" holographic microscope. Reprinted from [136] with permission

$$M = \left[1 \pm \frac{z_1 \lambda_1}{z_3 \lambda_2} - \frac{z_1}{z_2}\right]^{-1},$$

where z_1 is the distance from an object to a hologram, z_2 and z_3 are the distances from pinholes during the recording and reconstruction, respectively.

Minus refers to the real image and plus—to the virtual image. The expression shows that the magnification depends on the wavelength relation used during the recording of λ_1 and reconstruction of λ_2, wave front curvature and can be easily regulated.

But, "lensless" microscopes are not widely used because of the strong aberrated distortions provided by different longitudinal and lateral magnification. All aberrations are removed only in the case when $M = 1$.

Absolutely new what is brought by holography and holographic interferometry into microscopy is the possibility to study diffusely reflective microobjects with the accuracy of a light wave fractions.

Holographic recording, which is based on the two-stage recording process (conservation) and time reconstruction of full optical information about the object, enables to solve in a new way questions on study of microobjects. By recording the single hologram of an object, it is possible to study it many times using different visualization methods: interference, phase-contrast, light and dark fields [90].

For receiving enlarged image of a microobject in holographic microscopy, it is possible to single out two approaches [137]. In the first case, a hologram of a microobject is recorded, and 3D image reconstructed by it (an optical copy of a microobject) is studied with the help of a microscope with a resolution of 1 μm and by a large field of view, by sequential overfocusing of the objective. Sequence of holograms recorded during the certain period of time allows receiving spatial as well as time information about the object and thereby characterizing its structure-functional state. As a rule, during the "lensless" recording of a microobject at the level of its volumetric image reconstruction, the wave front [91, 138] is used, which is conjugated with a reference wave front. Then, in the space in front of a hologram from the side of an observer, the real image is formed, which can be fully studied on the whole depth of an object at consecutive overfocusing of a microobjective in spite of its quite short focal distance. Such a method can be used for the studies of thick layers and separate microobjects which are located in this layer at a great depth. Virtual images of microobjects reconstructed by a hologram are rarely used. It is due to the absence of long-focal strong microobjectives and thin substrates, which could be coated by holographic recording media. Thus, the deeper the object is, the more difficult it is to study its virtual images.

Another approach of magnification in holographic microscopy is the recording on a hologram of a microobject an image preliminary enlarged by the adjusted optical system and located between an object and a hologram [89]. Thus, in this case, holographing is made not of the objects itself, but of its image. The use of microobjectives as project optics, which convert the cone of scattered by the object rays with high numerical aperture into a cone of incident on a hologram rays with small numerical aperture, allows reducing in comparison with the first approach the requirements to the photolayer resolution, its density and its homogeneity, but it can be reached by considerable narrowing of the field of view.

One of the first holographic microscopes is described in the work [89] (Fig. 2.2a). During the reconstruction of a hologram by a beam identical to the reference one and observing virtual images, the authors show the possibility of producing the resolution close to the limit of ~1 μm with the depth tuning of ~40 μm. Some expansion of the field of view and the depth of the reconstructed scene can be produced by putting a hologram between an objective and an eyepiece [139] (Fig. 2.2b). The reconstructed

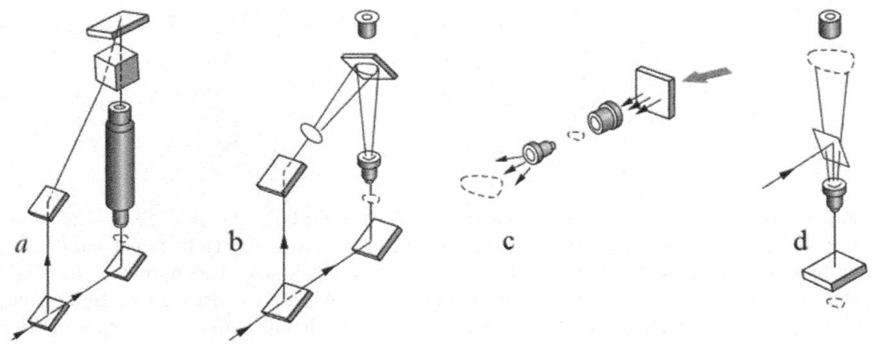

Fig. 2.2 Optical schemes of holographic microscopes. Reprinted from [136] with permission

image is observed through the eyepiece, which can be moved by the field of the recorded image. In order to produce a deep holographic image of high fidelity, it is necessary to expose a hologram by a beam conjugated with the reference beam (Fig. 2.2c, d). Then in the place where the object is at recording, its real image without aberrated distortions is formed. It was formed due to its compensation during wave distribution through the optical system in opposite direction. But this method of aberration compensation is quite sensitive to such factors as accuracy of setting a hologram and optical system, which makes an enlarged image of a microobject, reference and reconstructed beams mutual inclination, wavelength changing [140]. If necessary requirements are fulfilled, then this microscope scheme provides certain advantages. For example, as it is shown in [140], the realization of this scheme makes it possible to study three-dimensional scene with the depth of ~10 cm and the resolution of 2–3 μm (numerical aperture of microscope objective is 0.24).

The use of the methods of holographic interferometry in microscopy gives more interesting possibilities during the study of phase microscopic objectives. To detect the wavelength phase modulation passed through the transparent object, it should be converted into amplitude modulation [141]. Usually, it is fulfilled with the help of interference microscope (Fig. 2.3a).

In such a scheme, the optical paths of the reference beam and the beam of comparison should be carefully equalized, i.e., refractive index and glass element thickness should be equal. The magnification should be also the same.

At the same time, any above-described holographic microscope can be transformed into an interference one. There is no need to select doubled optical details in the holographic interference microscope precisely. Instead of it, the waves passed the same way during different periods of time are used (Fig. 2.3b).

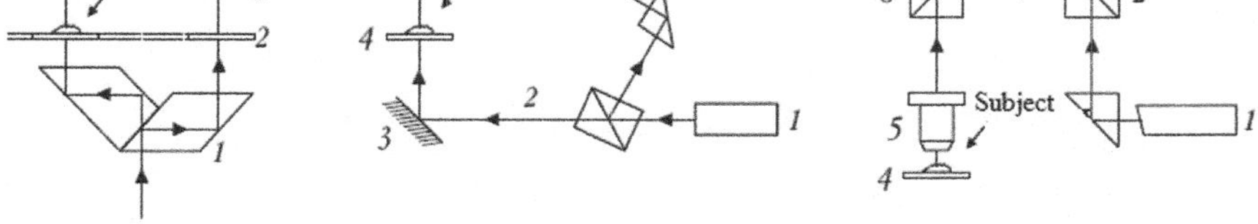

Fig. 2.3 Classical interference microscope (**a**): 1—the beam splitter, 2—the slide, 3—the microscope objectives, 4—the phase plates, 5—the light prism, 6—the ocular. Holographic interference transmission microscope (**b**): 1—the laser, 2—the dividing cube, 3—the rotary mirror, 4—the object under study, 5—the microobjective, 6—the hologram, 7—the ocular, 8—the camera; Holographic interference reflection microscope (**c**): 1—the laser, 2,8—the dividing cube, 3—the rotary mirror, 4—the object under study, 5—the microobjective, 9—the rotary prism, 6—the hologram, 7—the ocular. Reprinted from [136] with permission

It follows that holographic microscopes, based on different methods of hologram recording, can simultaneously operate as holographic interferometers using the method of two-base exposure or real time, multiple longwave or immersion [142]. And these holographic microinterferometers differ from the traditional ones by new principal possibilities such as the comparison of wave fronts scattered by an object during different periods of time (for example, microobject initial state and its strained state in the result of mechanical, chemical and electromagnetic effects), and they also enable to carry out the comparison of wave fronts of any complicated form (including diffusely scattered objects). Moreover, such microinterferometers enable to carry out interference comparison of waves of different frequencies that makes it possible to detect microobjects dispersion characteristics or to study wave interference scattered by microobjects in the range of hologram aperture. It should be noted that precise overlapping of the compared microobjectives state using holographic interferometry is doing automatically, while their identification in traditional interferometry is carried out with lower accuracy only during complicated comparison, for example, when photographing two different states of a microobject. There is also a set of other practical advantages. So, for the visualization of phase object relief, for the detection of deviation sign of the interference fringe and for the quantitative processing of microobjects interferograms making small phase incursions, the interferograms in fringes of final width appear to be convenient. For producing such an interferogram, the fringe period should be lower than a microobject sizes. Holographic interferometers without any difficulties enable to produce reference fringes with the frequency higher than the frequency of traditional industrial interference microscopes, and thus, using them makes it more convenient to get quantitative information about phase microobjects.

Figure 2.4a shows the interferogram with the limiting for traditional interference microscope band frequency, and Fig. 2.4b shows the interferogram of the same object produced in holographic microscope. These interferograms were recorded during the

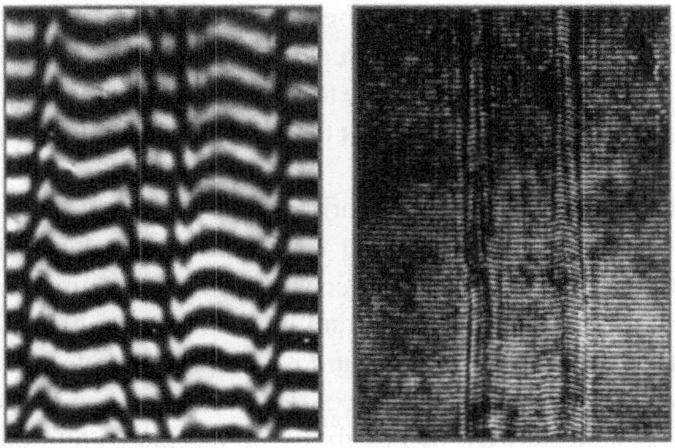

Fig. 2.4 Nerve fiber interferograms under traditional (**a**) and holographic (**b**) interference microscopes. Reprinted from [136] with permission

structure and optical characteristics of nerve fibers study experiments [1–3, 5, 8, 9, 15]. These studies were carried out with the use of the specially developed and mounted holographic microscope, the optical scheme of which is shown in Fig. 2.3b. Such schemes provide in holographic microinterferometry more fringes per unit object length than the schemes with lensless object holographing. Really, if two final width fringes are produced by inclining a hologram under an angle α, then in the scheme with preliminary increase within the reconstructed image area, period of reference fringes T is determined by formula $T = \frac{\lambda}{2}/\sin\frac{\alpha}{2}$, and in the scheme of "lensless" registration with further increase $T = \frac{\lambda}{2N}/\sin\frac{\alpha}{2}$, where N is the increase of a microobjective.

2.1.2 *Shortly About Coherent Noises in the Images of Diffusive Microobjects—Speckles and the Ways of Their Elimination*

Small space resolution (10 μm), reached in [143, 144] during the microobject holographing, is connected with the fact that high space frequencies in its spectrum and corresponding to small parts demand high efficiency of numerical aperture of the recording system and its high resolving power. High coherent radiation scattered on heterogeneous glasses, diaphragms edges, specks of dust and on the object leads to the formation of speckles chaotic interference structure which deteriorates image quality [145, 146]. To a large degree, the losses depend on the light scattering of coherent radiation by a microobject. First of all, it covers diffusely scattered microobjects where we can observe radiation dispersion from their surfaces as well as by the depth. Moreover, it is the result of the nonlinearity of the record and its own noises of the recording media. And yet, in spite of the difficulties, emerged during the study of microobjects, space resolution can be increased. For example, in [147], it is realized using the scheme in colliding beams. Just one this methodological approach decreases image speckle-pattern due to the reconstruction by white incoherent light source, on the one hand, and on the other hand—due to the microobject recording in immediate proximity from the photographic medium that corresponds to the hologram aperture magnification.

Partial speckles suppression can be carried out by multiple exposure on the one photoplate of the same object image, but with different speckle-pattern [148, 149] at including into the light beam, which reconstructs a hologram, randomly moving phase heterogeneity or during the hologram recording in the light of several wavelengths [150, 151]. The last one is of utmost interest, because at the same time, it is possible to decide an issue about color rendering in the microobject image. The authors [152] also showed the possibility of speckle-pattern decreasing in the microobject reconstructed image by the way of hologram recording of the focused image because such holograms make it possible to reconstruct their sources with low spatial coherence. In the works [153, 154], the scheme of holographic microscope for phase object studies is proposed. In this microscope, the coherent noises suppression is achieved with the help of including in light beam of revolving glass parallelepiped. During the studies of phase microobjects, which can change polarization state of scattered on them radiation, for coherent background elimination, the method of polarizing filtration can be quite effective [155]. It is possible that overcoherence of laser radiation also influences the quality of the microobject holographic image. Taking into consideration low holographing microobject volume (depth is 20 μm), it is reasonable to record reflecting holograms of phase and diffusely scattered microobjects in incoherent light source radiation (for example, lamps similar to DAC-50 with glowing body of 1 mm can be used as such sources). Attempts of holographing in incoherent light were carried out in the works [156, 157]. It also should be noticed that the latest developed recording media have resolving power up to several thousand lines/mm, which is enough for the recording of high space frequencies produced during microobject structures scattering. But, light sensitivity of such media is quite low. That is why it is very important, in particular during holographic study of biological microobjects, to reach compromise between maximum allowed microobject raying energy, light sensitivity of recording holographing medium and its resolving power.

2.1.3 Peculiarities of Microobject Holographic Interferograms Formation Connected with the Dependence of Interference Pattern Contrast on Defocusing

In holographic interferometry, which allows "remembering" and comparing wave fronts in different periods of time and of complicated forms up to random fields, there are a lot of possibilities. Using holographic interferometry in solving biomedical tasks, it is necessary to take into consideration object mobility and the presence of magnification optics. Though, at the moment, it can be said that principal problems of holography and holographic interferometry are solved. There are only actual problems such as the search of the best technical decision and the schemes of holographic devices, more effective media and radiation sources, the decrease of the laboriousness and the cost of holographic recording, the development of algorithms, which lower the laboriousness and improve the precision of holographic interferograms processing and searching new and effective fields of their application.

In different schemes of holographic and speckle-interferometry, there is an object longitudinal shift. Defocusing produced by axial shift leads to a considerable transformation of the wave spread through the optical scheme and influences the interference pattern characteristics, results and measurement range. Holographic interference microscopy with its possibility of interfering comparison of mirror-reflecting as well as diffusely scattering surfaces concerning deformations analysis was developed in the works [56, 57, 65, 158, 159]. Unlike other schemes of holographic interferometry, the presence of the microobjective is necessary here. The influence of the objective on interference pattern forming was studied by a number of authors [160, 161]. So, it becomes interesting to study the influence of defocusing on field amplitude correlations on which the form of holographic interferograms depends.

The objective can differently influence interference pattern formation. Some authors [133, 134, 162–164] paid their attention to the dependence of interferograms contrast on optical system aperture. In single works [165], phase difference between interfering waves is considered as an aperture function. Though in many cases, this change is small (about $\lambda/4$), with the appearance of the possibility for high-accuracy measurements, the necessity appears to take into account the sources of even small mistakes. Moreover, the image phase depends on the used microscopic objective, especially when the sizes of the objects are close to the resolution limit as is shown in [166]. Phase distribution in small phase asperity image is calculated there.

In the work [167], the dependence of interference pattern contrast on the magnification is being discussed, and in the work [168], the attempt of considerable increase in the sensitivity by using optical system longitudinal magnification is grounded. The defocusing problem, connected with the use of the objective is the following. If the object is shifted along the optical axis as the result of deformation, then it transcends the focusing plane. Three-dimensional shift vector $\vec{d}$ describing deformation is usually taken into account only in the expression for the reflecting wave phase, and the change of pulse response for the analysis simplification is neglected even in such generalized studies as [162]. And with it, in the works [169–171], it is clearly shown that during defocusing the image itself considerably changes, and as a result, the interference pattern where there is a defocused image for transparency, phase and diffusive objects—should be also changed. In native and foreign literatures, there is almost no quantitative analysis of such studies. At the best, it is noticed that in the result of deformation, the object should not transcend objective contrast. It is quite obvious that during one and the same shift, defocusing will be different for different objectives, in particular, for microobjectives with small focusing distance, because even small shift leads to considerable defocusing. In such conditions, it is quite problematic to study strongly defocused or excited structures, for example, muscle fibers.

Let us consider the scheme of holographic interference microscope shown in Fig. 2.5a in detail. In its initial state, the object is in the plane $x_0 0 y_0$ described by two-dimensional vector $\vec{x}_0 = (x_0, y_0)$ at a distance d_0 from the objective entrance pupil O_b, which forms the image in the plane $x_i 0 y_i$, $\vec{x}_i = (x_i, y_i)$.

If $U_1(\vec{x}_o)$ is the field distribution behind the object O, and $h(\vec{x}_i - M\vec{x}_0, d_z)$ is the pulse response of isoplanatic objective depending in general on the defocusing value d_z, then in the image plane, the field distribution will be defined by (2.1)

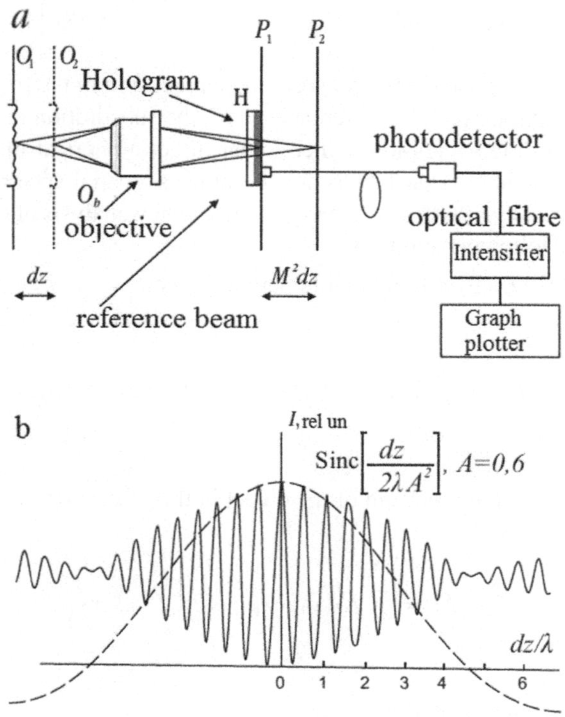

Fig. 2.5 Scheme of holographic interferograms producing **a** at defocusing and **b** the dependence of the contrast (dotted line) and interference fringes intensity (solid line) on the shift value of the reflective diffusive object along the optical axis. Reprinted from [136] with permission

$$V_1(\vec{x}_i) = \int_{-\infty}^{\infty} h(\vec{x}_i - M\vec{x}_0, 0)U_1(\vec{x}_0)d^2\vec{x}_0, \tag{2.1}$$

where $d^2\vec{x}_0$ means the integration using variables x_0 and y_0. This image is recorded, and then, it is reconstructed from a hologram H.

During the experiment, the observing plane $x_i 0_{y_i}$ does not change even after object shift described by the vector $\vec{d}$. Its projection on the optical axis is d_z, and on the plane $x_0 0y$ is $\vec{d}_x$, respectively. In the observing plane, the wave is formed

$$V_2(\vec{x}_i) = \int_{-\infty}^{\infty} h(\vec{x}_1 - M\vec{x}_0, d_z)U_2(\vec{x}_0)d^2\vec{x}_0, \tag{2.2}$$

where $U_2(\vec{x}_0)$ is the field distribution directly behind the deformed object. Intensity distribution in the interference pattern will be defined by the following expression

$$I(\vec{x}_i) = (V_1 + V_2)(V_1 + V_2)^* = |V_1|^2 + |V_2|^2 + 2\mathrm{Re}\{V_1 V_2^*\}. \quad (2.3)$$

The first and the second components are incoherent sum of the focused and defocused images, and the third is interference term. The substitution of (2.1) and (2.2) into (2.3) leads to the expression, which under direct numerical integration needs less computer timetable. That is why for the objects, spatial spectra of which are expressed analytically, it is easier to carry out frequency analysis of the scheme. Let us introduce the following symbols:

$H(\vec{f}, d_z)$ is the transfer function of coherent system:

$$H(\vec{f}, d_z) = \int_{-\infty}^{\infty} h(\vec{x}_i d_z) \exp(-i2\pi \vec{f}\vec{x}_i) d^2\vec{x}_i = F\{h(\vec{x}_i d_z)\}; \quad (2.4)$$

$G_{ol}(\vec{f})$ is the spatial spectrum of the object in the plane $\vec{x}_0 (l = 1, 2)$:

$$G_{ol}(\vec{f}) = \int_{-\infty}^{\infty} U_l(\vec{x}_0) \exp\left[-i2\pi \vec{f}\vec{x}_0 d^2\vec{x}_0\right];$$

$G_l(\vec{f})$ is the image spatial spectrum:

$$G_l(\vec{f}) = \int_{-\infty}^{\infty} V_l(\vec{x}_i) \exp(-i2\pi \vec{f}\vec{x}_i) d^2\vec{x}_i. \quad (2.5)$$

The quantities $H(\vec{f}, d_z)$, $G_{ol}(\vec{f})$ and $G_l(\vec{f})$ are connected by the relation

$$G_l(\vec{f}) = H(\vec{f}) G_{0l}(M\vec{f}), \quad (2.6)$$

where M is the objective magnification.

The intensity distribution in the interference pattern

$$I(\vec{x}_i) = \left|F^{-1}\left\{G_1(\vec{f}) + G_2(\vec{f})\right\}\right|^2, \quad (2.7)$$

where F^{-1} is the reverse Fourier transformation.

Transfer function of diffraction-limited defocused coherent-optical system can be expressed in the following way [150, 151]:

$$H(\vec{f}, d_z) = \lambda^2 \rho_i^2 k P(\lambda \rho_i \vec{f}) \exp\left(\frac{ik\lambda^2 d_z}{2}\right) M^2 \left|\vec{f}\right|^2, \quad (2.8)$$

where P is the pupil function, ρ_i is the distance from the output pupil to the image plane, k is the constant.

Equation (2.8) shows that during the formation of a defocused image, frequency components transmitted by the objective (the first factor) are added with some phase shift proportional to d_z and frequency square.

Let us show the influence of defocusing with several typical examples.

Plane mirror. For example, as such an object we can use a mirror of the hetero-junction laser during the study of its thermal deformation [56, 57, 65] if its size is higher than the microobjective resolution limit. When the plane wave falls on the mirror parallel to the optical axis, reflected plane wave will have the spectrum $G_{01}\left(\vec{f}\right) = \delta\left(\vec{f}\right)$, and the spectrum $U_{02} = U_{01}\exp(-i2kd_z)$ will be different only in phase factor $G_{02}\left(\vec{f}\right) = \delta\left(\vec{f}\right)\exp(-i2kd_z)$, where $\delta\left(\vec{f}\right)$ is the two-dimensional delta function. Then, interference part included in (2.3) and being of utmost interest for us will be equal to:

$$\mathrm{Re}\left\{V_1V_2^*\right\} = 2\mathrm{Re}\left\{F^{-1}\left[H(\vec{f},0)\delta(M\vec{f})\right]\right.$$
$$\left.F^{-1*}\left[H(\vec{f},d_z)\delta(M\vec{f})\exp(-i2kd_z)\right]\right\} = \frac{2\lambda^4\rho_i^4}{M^4}\cos 2kd_z. \tag{2.9}$$

Essentially, the described system is the holographic variant of the Michelson interferometer tuned for the infinite width lines, and in the image plane, there will be cyclic blackening and lightening of all field of view as d_z increases. The defocusing does not influence the interference pattern form because plane wave is localized neither in the object domain nor in the image area.

Diffusely reflected surface. If light is scattered by rough surface, then the image of this surface $V_l(\vec{x}_i)$, as it was noticed, will have a speckle-structure. In the result of coherent addition of two random fields $V_1(\vec{x}_i)$ and $V_2(\vec{x}_i)$, there will be a new speckle-field

$$V(\vec{x}_i) = V_1(\vec{x}_i) + V_2(\vec{x}_i) = \int \{U_{01}(\vec{x}_0)\rho_1(\vec{x}_0)h(\vec{x}_i - M\vec{x}_0, 0)$$
$$+U_{02}(\vec{x}_0)\rho_2(\vec{x}_0)h(\vec{x}_i - M\vec{x}_0, d_z)\}d^2\vec{x}_0, \tag{2.10}$$

where U_{ol} describes the wave reflected from macrorelief (smooth surface), and $\rho_l(\vec{x}_0)$ describes phase chance variation made by the object microrelief in the initial ($l = 1$) and deformed ($l = 2$) states.

In order to detect field macromodulation or interference pattern, it is necessary to eliminate speckle-structure, i.e., to average diffuser ensemble ρ, which further will be marked as $\langle\ldots\rangle$. Let us analyze interference part $\langle V_1(\vec{x}_i)V_2^*(\vec{x}_i)\rangle$ one more time considering that $\langle\rho_1(\vec{x}_0)\rho_1(\vec{x}_0')\rangle = \delta(\vec{x}_0 - \vec{x}_0')$ and assuming that in the result of deformation $U_{02}\left(\vec{x}_0 + \vec{d}\right) = U_{01}(\vec{x}_0)\exp\left[i\phi(\vec{x}_o)\right]$, where $\vec{d}$ is the two-dimensional shift vector in the plane $\vec{x}_0$; $\phi(\vec{x}_0) = \Delta\vec{k}\vec{d}$; $\Delta\vec{k} = \vec{k}_2 - \vec{k}_1$ is the change of wave vector in the result of scattering, $\rho_2\left(\vec{x}_o + \vec{d}\right) = \rho_1(\vec{x}_0)$, and also assuming that $U_{01}(\vec{x}_0)$ is the slowly varying function. Then,

$$\begin{aligned}\langle V_1(\vec{x}_i)V_2^*(\vec{x}_i)\rangle &= \iint \left\{U_{01}(\vec{x}_0)U_{02}^*(\vec{x}'_0)\langle \rho_1(\vec{x}_0)\rho_2(\vec{x}'_0)\rangle h(\vec{x}_i - M\vec{x}_0, 0)\right.\\ &\quad \times h^*(\vec{x}_i - M\vec{x}_0, d_z)\}d^2\vec{x}_0 d^2\vec{x}'_0\\ &= \left|U_{01}\left(\frac{\vec{x}_i}{M}\right)\right|^2 \exp\left[i\phi\left(\frac{\vec{x}_i}{M}\right)\right]c(\vec{d}_0, d_z), \end{aligned} \tag{2.11}$$

where

$$c(\vec{d}_0, d_z) = \frac{1}{M^2}\int h(\vec{x}_i, 0)h^*(\vec{x}_i - M\vec{d}_0, d_z)d^2\vec{x}_i. \tag{2.12}$$

Let us consider when (2.12) is distinct from zero. In particular, it can be calculated in the following way

$$\begin{aligned}F\left\{c(\vec{d}_0, d_z)\right\} &= F\left\{\frac{1}{M^2}h(M\vec{d}_0, 0) \oplus h(M\vec{d}_0, d_z)\right\}\\ &= \frac{1}{M^2}H(\vec{f}, 0)H(\vec{f}, d_z)\\ &= \frac{1}{M^2}\left[k\lambda^2 d_i^2 P(\lambda\rho_i P(\lambda\rho_i \vec{f})\right]^2 \exp\\ &\quad (ik\frac{d_z}{2}\lambda^2 M^2\left|\vec{f}\right|^2 = \frac{k\lambda^2 d_i^2}{M^2}H(\vec{f}, d_z), \end{aligned} \tag{2.13}$$

where $\oplus$ indicates the convolving.

Having found reverse Fourier transformation (2.13), we have

$$c(\vec{d}_0, d_z) = \frac{k\lambda^2 d_i^2}{M^2}h(M\vec{d}_0, d_z). \tag{2.14}$$

Pulse response $h(\vec{x}_i, d_z)$ corresponds to the field distribution within the focus of converging spherical wave at a distance ρ_i from the output pupil. This distribution is described in [170]. It can be also shown that interference pattern contrast is expressed in the following way

$$\gamma = \left[\frac{c(\vec{d}_0 d_z)}{c(0, 0)}\right] = \left[\frac{h(M\vec{d}_0 d_z)}{h(0, 0)}\right] \tag{2.15}$$

In a particular case if the object is influenced only by transverse shift, i.e., $d_z = 0$, then contrast change for diffraction-limited system is described by Bessel function. Interference pattern can be formed until the shift does not limit system resolution that corresponds to the result in [125]. In spite of the results in [125], (2.16) shows that in general the contrast depends on the object transverse shift as well as on the defocusing value. Let us consider the private case of (2.15) when $d_z \neq 0$ and $\vec{d}_x = 0$. The function $h(0, d_z)$ describes the change of the field amplitude in the observing plane center during the shift of the point source along the axis at a distance d_z. In this case,

$$\gamma = \left| \sin c\left(\frac{2\pi q^2 d_z M^2}{\lambda \rho_i^2} \right) \right|, \tag{2.16}$$

where q is the output pupil radius, and interference pattern contrast becomes zero under

$$\left| d_z M^2 \right| = \frac{\lambda}{2} \frac{\rho_i^2}{q^2}. \tag{2.17}$$

The result gives the possibility to explain it simply. As the value

$$\delta = \lambda \frac{p_i^2}{q^2} \tag{2.18}$$

determines the speckle average size along the observing axis, and $d_z M^2$ is the separation of the image planes in the result of the object shift, then for the interference pattern disappearance, it is enough that the distance between two speckle-fields V_1 and V_2 will be more than the speckle size, and correlation between them will disappear.

Passing in (2.17) from the value q/ρ_i to the object input aperture A, it is possible to write interference pattern condition in the following way

$$|d_z| < \frac{\lambda}{2A^2} \tag{2.19}$$

Equation (2.19) shows the general regularity: The increase of the objective resolving power leads to the decrease of measurement limits. But, for example, for $A = 0.2$, it is possible to produce about 30 fringes with the contrast of 0.5 that is quite sufficient in many supplements during the studies of microobjects, for example, thermal deformations of semiconductor elements [55–68]. The character of change of interference pattern with the increase of d_z is shown in Fig. 2.5b, where the dotted line is the visibility dependence and the solid line is the normalized interference pattern rate (ensemble averaged) for $A = 0.6$. Negative visibility values indicate the inversion of interference fringe contrast. It should be noticed that it is necessary to be more careful if A is increased by (2.16). It is connected with the fact that if diffusivity character does not change $\langle \rho_1(\vec{x}_0)\rho_1(\vec{x}_0')\rangle = \delta(\vec{x}_0 - \vec{x}_0')$ (i.e., the number of scatterers, which are situated at a distance equal to objective resolving power remains high), and the allowed distance itself approaches to wavelength, then the diffusive object becomes a mirror object. In other words, with the decrease of the heterogeneity size in comparison with λ, more light is reflected. So, for high A in (2.11), it is necessary to use the correlation function $\langle \rho_1(\vec{x}_0)\rho_1(\vec{x}'_0)\rangle$ different from delta function and dependent on surface particular properties.

Thus, the above-mentioned results confirm the limits of possible shift in holographic microinterferometry schemes. The increase of the object longitudinal shift range can be reached by the decrease of the objective aperture and by the loose of the resolution. During interference studies of the objects, which change light beam amplitude distribution, it is necessary to consider the fact that defocusing distorts the image itself as well as phase difference between the images. It is also obvious that the use of microobjective in the schemes of holographic interference microscopy does not allow considerable increasing sensitivity to longitudinal shift measurements, but it improves spatial resolution. It is shown that the use of 10×–40× magnification objectives allows measuring shifts to tens of micron. This fact along with the simple identification of interferograms, observed in real time, makes the methods of holographic interference microscopy with high spatial resolution quite useful for the studies of diffusive objects [59, 68].

2.1.4 Interference Pattern Localization at Homogeneous Radial Change of Cylinder Object

Some biological objects (nerve, nerve fiber and others) have cylindrical symmetry. Under the influence of different physical agents (for example, magnetic field, pressure), the changes corresponding to three-dimensional compression or object broadening can be observed. In the series of experiments under the influence of pulsed magnetic field on isolated nerve, it is impossible to observe any interference fringes. One of the reasons for this could be the disagreement of the observing plane and localization plane. In connection with this, there was made a preliminary theoretical analysis of double-exposed interferograms under homogeneous radial broadening of the cylinder object in the scheme shown in Fig. 2.6a.

The interference pattern localization plane is assumed as a maximal contrast plane. Let us consider that the object has R_0 radius and is under homogeneous radial deformation ΔR, $\beta = \Delta R/R_0$. Reflection index for all object points is equal to 1, out of it—it is equal to zero. The object is projected by the lens 4 with the focal distance f and the aperture q which is at a distance d from the object, on the plane P, the defocusing parameter (see Fig. 2.6a)

$$\varepsilon = \frac{1}{d} + \frac{1}{p} - \frac{1}{f}. \tag{2.20}$$

Localization plane can be detected by the search of maximum value of module of a complex number:

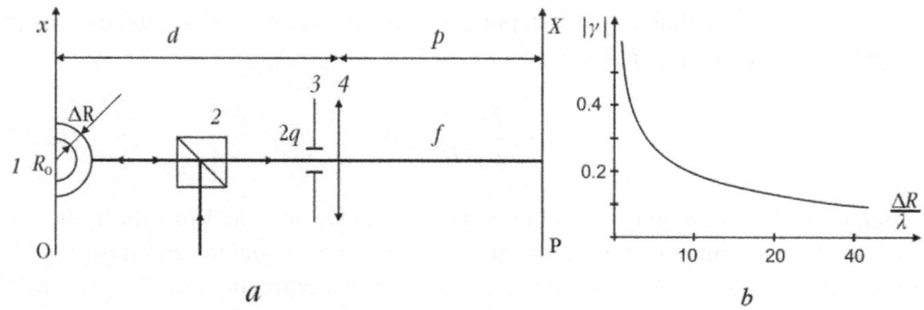

Fig. 2.6 **a** The scheme of interference pattern recording at radial expansion of the cylinder object: 1—the object; 2—the divider; 3—the diaphragm; 4—the lens; P—the recording plane. **b** Dependence of interference pattern contrast on deformation value. Reprinted from [136] with permission

$$
\begin{aligned}
|\gamma| = \int\limits_{-R_0}^{R_0} \exp\Bigg[& i\frac{4\pi}{\lambda}\sqrt{R_0^2 - x^2} + i\frac{4\pi}{\lambda}(1+\beta)\sqrt{R_0^2 - x^2} \\
& + i\frac{\pi}{\lambda}\left(\frac{x^2}{d} + \frac{X^2}{p}\right) - \frac{\left(\frac{x}{d} - \frac{X}{p}\right)^2}{\frac{\lambda^2}{\pi q^2} - i\frac{\lambda}{\pi}\varepsilon} - \frac{\left(\frac{(1+\beta)x}{d} - \frac{X}{p}\right)^2}{\frac{\lambda^2}{\pi^2 q^2} + i\frac{\lambda}{\pi}\varepsilon} \\
& - i\frac{\pi}{\lambda}\left(\frac{(1+\beta)^2 x^2}{d} + \frac{X^2}{p}\right)\Bigg] \mathrm{d}x
\end{aligned}
\tag{2.21}
$$

Let us simplify, considering that $\sqrt{R_0^2 - x^2} \approx R_0 - x^2/2R_0$, and substitute the reflection function for $\exp(-x^2/R_0^2)$. But stated assumptions are reasonable for cylinder radial expansion in the range $|x| \le R_0/4$.

Then, transformed integral is taken accurately within infinite limits, and under normalization $\beta = 0$, the module γ will be written in the following way:

$$
|\gamma| = \left[1 + \left(\frac{k\beta R_o A}{2B}\right)^2\right]^{-\frac{1}{4}} \exp\left\{ -\frac{k^6 q^8 \beta^2 R_0^2 X^2 \varepsilon^2}{8d^2\left(4 + k^2\varepsilon^2 q^4\right) M^2 B} \frac{\left[1 - \frac{2R_0 A}{4q^2 B}\right]^2}{\left[1 + \left(\frac{k\beta R_0 A}{2B}\right)^2\right]} \right\},
\tag{2.22}
$$

where

$$
\begin{aligned}
A &= d^2\left(4 + k^2\varepsilon^2 q^4\right) + R_0 k^2 \varepsilon q^4; \\
B &= d^2\left(4 + k^2\varepsilon^2 q^4\right) + R_0^2 k^2 q^2,
\end{aligned}
$$

where $M = p/d$.

It is easy to notice that in (2.22) exponent tends to zero at $\varepsilon \to \infty$, and moreover, it is equal to zero at the condition

$$1 - \frac{2R_0 A}{\varepsilon q^2 B} = 0. \tag{2.23}$$

Equation (2.23) amounts to cubic equation relative to ε and has the root $\varepsilon = 4R_0/3q^2$. Considering that the exponent is approximately equal to zero, it is possible to show that at $|\varepsilon| > 5R_0/dq$ the exponent is almost equal to zero. As, as a rule, $5R_0/dq < 4R_0/3q^2$, then the case of $|\varepsilon| > 5R_0/dq$ is further considered and at that the contrast is transformed into:

$$\gamma = \left[1 + \left(\frac{\pi \Delta R}{\lambda}\right)^2\right]^{-\frac{1}{4}}. \tag{2.24}$$

If $\gamma \geq 0.1$, then $\Delta R/\lambda < 100/\pi$. Thus, in all planes, which satisfy the condition of $|\varepsilon| > 5R_0/dq$, for object areas corresponding to $|x| < R_0/4$ and at the deformation of $\Delta R/\lambda < 100/\pi$, the interference pattern will be observed with the contrast more than 0.1, and this contrast does not depend on the coordinates in the recording plane and reveals as the function of only deformation value ΔR (Fig. 2.6b). Phase correlation at $\varepsilon = -5R_0/dq \ldots$ will lead to the formation of dark and light fringes:

$$\frac{2\left(\frac{X}{M}\right)^2}{625\pi R_0^2 \frac{\Delta R}{\lambda}}\left[1 - \frac{25\pi^2}{4}\left(\frac{\Delta R}{\lambda}\right)^2\left(\frac{q}{d}\right)^2\right] - 4\pi\frac{\Delta R}{\lambda} = \begin{cases} 2\pi n & \text{—light fringes} \\ (2n+1)\pi & \text{—dark fringes} \end{cases}$$

For example, n-th fringe will be in the point position with the coordinate

$$X_n = 25\pi R_0 M \sqrt{\frac{n + 2\frac{\Delta R}{\lambda}}{1 - \frac{25\pi^2}{4}\left(\frac{\Delta R}{\lambda}\right)^2\left(\frac{q}{d}\right)^2}\frac{\Delta R}{\lambda}}. \tag{2.25}$$

Thus, the calculation shows the limits of holographic recording of cylinder object radial deformation and allows calculating its value by the interference pattern [71].

2.2 Holographic Study of Structural and Functional Characteristics of Phase Microobjects: Nerve Fibers and Lymphocytes

At the initial stage of the studies, we produced holograms and holographic interferograms of onion epidermis cell (Fig. 2.7a–d), frog blood red cells (Fig. 2.8a, b), cavy spleen lymphocytes (Fig. 2.8c), drosophila salivary gland, nerve cells of the crop of sensitive back radicular rat ganglion, pond snails giant neurons, frog sciatic nerve myelinated nerve fiber. During the search of the subject of inquiry, we decided to choose excited microobjects such as nerve cells, nerve fibers, nerve (Fig. 2.8d), muscle fiber, muscle (Fig. 2.8e, f) and also lymphoid cells (Fig. 2.8c). A little bit later it has become known about the studies on nerve fibers carried out in Delaware University (the USA) [172].

2.2.1 About Nerve Cell, Nerve Fiber as the Object of Physical Studies

A nerve cell or neuron is the basic component of the nerve system responsible for performing such functions as the reception of ambient and internal environment conditions, conduction of excitement, signal processing and finally pulsing to effector organs (Fig. 2.9). Morphologically, "adult" neuron corresponds to an example of specialized cell, which lost the ability to division. Nerve cell sizes are varying in wide limits from 5 to 10 μm (mammal cerebellum seed cells) up to 500 μm (invertebrate

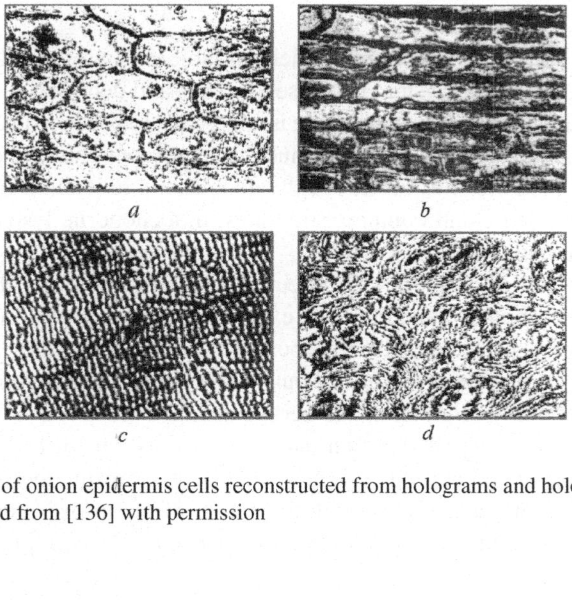

Fig. 2.7 Images of onion epidermis cells reconstructed from holograms and holographic interferograms. Reprinted from [136] with permission

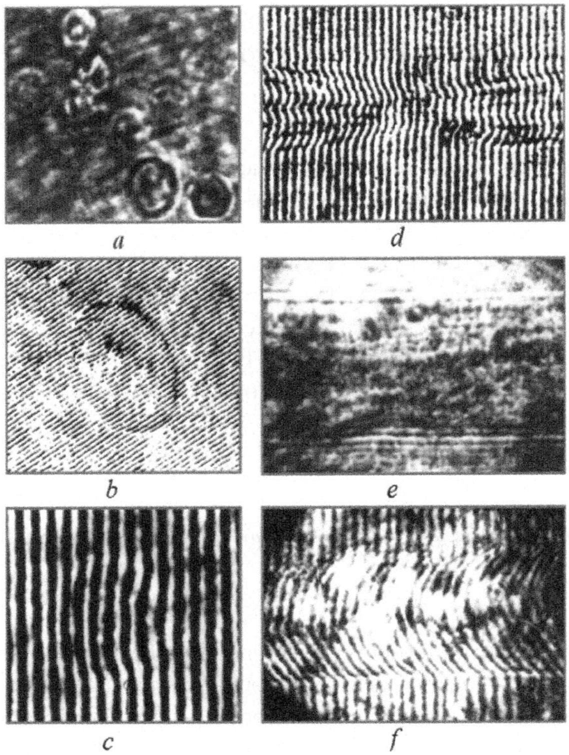

Fig. 2.8 The images reconstructed from holograms and holographic interferograms: **a**, **b** frog blood red cells, **c** cavy spleen lymphocyte, **d** nerve fiber, **e**, **f** muscle fiber. Reprinted from [136] with permission

giant neurons). Nerve cell shape is extremely variable, and it is typically the presence of more or less developed system processes—axons and dendrites [173–175].

In the central part of a nerve cell, there is a large nucleus where usually the nucleolus is situated. In cytoplasm around the nucleus, there are tighter dyed substances—Nissl bodies, which are sites of the rough endoplasmic reticulum. In the cell cytoplasm, there are also Golgi complex structures, mitochondria, lysosomes and also neurofibrils and neurotubules.

Outside the cell is covered with three-layered tunic with the thickness of 8–10 nm. Cell membranes functions are extremely varied. Nerve cell membranes besides mechanical protection also transport different ion molecules. During neuron specialization, the oscillations of membrane permeability have become the basis for generation excitability origination and nerve pulsing.

The problem of nerve pulse has rich and at times even dramatic history. In the XVII century, it was assumed that through nerve fibers, the information was transported from the brain to muscles. Concerning the nature of the information carrier, there were different assumptions, which were corresponded to certain scientific notations

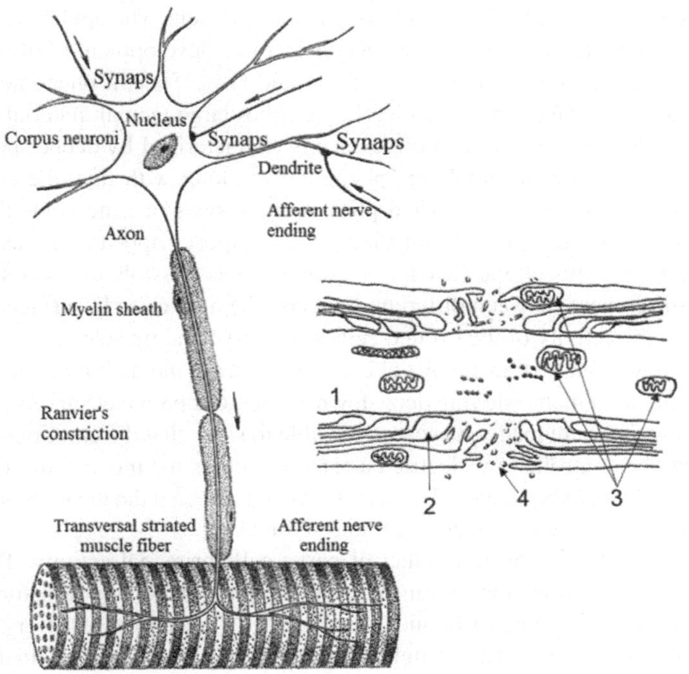

Fig. 2.9 Scheme of nerve fiber structure (frog spinal motor neuron): 1—the axon, 2—the Schwann's cytoplasm, 3—the mitochondria, 4—the connective tissue. Reprinted from [136] with permission

of each epoch. As an example, let us see a very interesting statement of Dekart: "Animal spirits, reminding quite subtle liquid or even extremely pure and active flame, which constantly appears in the heart, are transported to the brain like in special reservoir. Then they are come to nerves, reach the muscles caused contraction and relaxation depending on these spirits quality" [159].

Physiologists using microscopes leaned toward the view that nerve fiber is the tube, through which "nerve fluid" is flowing. I. Newton in "Principia" wrote about elastic wave spread along the fiber. It was quite natural for the mechanic-age. But sooner this hypothesis was disapproved, and electrical concepts took its place.

In the XVIII century, the scientists thought that there was a certain connection between "nerve forces" and external electric field. But, the question about "nerve fluid" remained open. And in 1791, L. Galvani published his "Treatise about electric forces during muscle contraction". Contacting two different metals with frog neuromuscular specimen, L. Galvani observed muscle contraction. He considered that muscle contraction was caused by bioelectricity, and nerve fiber was used as a conductor, which in connection with metal electrodes completed a circuit and assisted in muscle discharge equivalent to the Leyden jar [176–178].

Physical representations of bioelectricity origin were developed by N. A. Bernstein and were based on the works of J. U. Gibbs, G. L. F. Helmholtz, F. V. G. Nernst and V. F. Oswald on thermal dynamics of galvanic element. The application of more perfect experimental devices provided a considerable development of the Bernstein representations. Developed by M. Hodzhkin and O. L. Haksli, phenomenological theory of bioelectrical phenomena clearly describes large factual material [179].

New development in this area of science is characterized by deeper penetration into molecular nature of membrane phenomena. Along with this, the continuous expansion of the scope of biological phenomena takes place, new possibilities of their use in medicine especially for diagnostics purposes appear. And as a rule, it is referred to as extracellular fields. For example, it is possible to measure electric potential differences between different points on the human body surface. It can be done due to the activity of the excited cells set—nerve and muscle.

It was shown that extracellular fields are very weak, the activity of milliards of cells is summed in them. Moving deep down, closer to separate electric brain center, great discoveries were made. It became possible to show that the emotions of higher animals can be controlled by electric current, as well as the movement of frog legs during the Galvani experiments. The scientists even detected the location of different centers such as satisfaction, fear, anger and so on [180].

Extracellular fields are by-product of nerve cells principal activity. The whole information received by the organism from the environment is transformed into nerve impulses having electrochemical character with the help of receptors. Through nerve fibers, which are used as strengthening cable, they are transmitted to the central nervous system. There the information is processed, the decisions are made, and then, the instructions are given to muscles in the form of nerve impulses. One of the most remarkable characteristics of nerve impulses is that they have constant amplitude and form and stimulation character influences only their frequency.

Moreover, myelinated nerve fibers have an additional layer consisted of repetitive spiral wound around the axon layers of satellite Schwann's cells [181]. A nerve cell is also surrounded by glia cells, which differ by the absence of action potential. Glia cells actively participate in neuron functional activity maintenance. More convincing results on the metabolic correlation of neuron and glia (oligodendroglia) concern protein and nucleic acid syntheses [182].

The participation of glia cells (astrosphere) in transportation from capillaries to neuron is obvious [181]. Nerve cells in the tissue are in contact. Such a correlation is expressed in the terms of electric and chemical contacts. Being functionally more mature, chemical contacts or synapsis accurately transfers nerve impulse. In a common form, synapsis is the neuron process thickening—synaptic button—tightly adjoining to another neuron. There is a synaptic gap with the width of 15–25 nm between them. The synaptic button contains a set of vesicles, which contain the resources of transmitter substances. The extreme effectiveness of pulsing is shown electrophysiologically. Chemical synapse providing one-way impulse transmission is also used as a functional regulator of nerve cell electric activity due to the influence of different factors on the mechanism of mediator extraction and adsorption [175].

Present experimental material shows considerable development of all cellular structures of a neuron. Moreover, because of its specialization, nerve cell has typical structures—electroexcitable membrane, neurofibrils, neurotubules, synapse, medullary sheath and fibers. Nerve cell complex structure should also be shown in the variety of morphofunctional shifts during the excitation. Some characteristics of the nerve tissue activity are detected using optical methods.

During the experiments on the cell size and its nucleus measurements at various types of stimulation, true changes of the volume of the body and of the nucleus in comparison with the control were detected statistically. So, [183, 184] show the increase of neuron body sizes of mammal triangular kernel at horizontal rotation. Electric stimulation (ortho- and antidromic) [185] also resulted in the increase of spinal motoneuron body weight. Light flicker strengthened, strengthened motion activity resulted in the increase of the body and nucleus of ganglionic cells of frog and chicken retina [186, 187]. At the same time, there is information about the decrease of function neuron sizes under the influence of excessive motion activity [188], flickering light [187] and electrostimulation [189]. It should also be noted that some authors [190, 191] did not get the true changes in the body and nucleus sizes of active nerve cells.

Analyzing such diverse information, Yu. Geinisman [182] showed the difference of the methods used by the authors for nerve cells stimulation as well as inadequate conditions of preparation histological processing and object geometry measurements. He noted that it was quite difficult to determine the volume using the histological section. And this problem has not still solved.

Some authors [192, 193] consider that the changes in a nerve cell body and nucleus volume are based on the shifts in the content of liquid intracellular components; others consider that it happens because of the changes of dense matter amount, essentially of proteins [188, 194]. But regardless what mechanism prevails in changes of brain volume, the changes themselves can serve as a neurons functional state. Yu. Geinisman [182] showed that the decrease of cell sizes accompanied by continuous stimulation can serve as an indicator of action potential generation and inhibitory postsynaptic potential generation.

In spite of its fragmentation, the above-mentioned information shows the reasonability of carrying out the studies on the measurements of cell geometry of functional nerve tissue.

The study of RNA (ribonucleic acid) in the functioning brain can be made by traditional biochemical method as well as cytospectrophotometry method. The latter allows detecting the localization and qualitative content of RNA in separate cells. But, during simultaneous experiments, considerable shifts of RNA concentration could be detected using the method of cytophotometry, whereas they could not be detected biochemically (in the result of averaging) [195].

Some authors [102, 189, 196] came to the conclusion that the increase of the nerve system functional activity is accompanied with the growth of neuronal RNA content. Simultaneously, with the increase of its quantity in activated neurons, it can also be observed the decrease of RNA content in glia cells surrounded these neurons [190, 196]. Strong neuron functioning can cause the decrease of RNA in them, which is considered as the result of nervous system fatigue and emaciation [192].

The number of RNA varies depending on the cell activity level: biosynthesis level, ATP content, possibly on the level of axoplasmic transport. RNA content shifts in function neurons can serve as an indicator of their activity changes.

Cytospectrophotometry methods [197], which use light absorption by cell substance, are of the utmost interest for lifetime studies. But, absorption band of main cell components—proteins and nucleic acids—is outside of visible region (260–290 nm). The spectra of separate substances can be blocked what complicates their identification. Interesting works [198] on lifetime spectrometry of nerve cells have appeared recently. Using absorption spectra of some pigments, the dynamic of some redox reactions of huge mollusks neurons was studied.

To visualize different substances in the nerve tissue preparations, they are subject to color using the reactions of specific bonding of some dyes. Using histochemical methods, nucleic acids, proteins, fats and ferments distribution were studied [199]. In this respect, more interesting was the use of lifetime dyes, for example, methylene blue, acridine orange, neutral red, Janus green, which under limited disturbing action allow seeing separate structures and its dynamic in living cells.

The majority of nerve cells under lifetime estimation appear to be devoided of dyed structures, but they can differ by refractive index distribution. It became the basis for a wide use of phase-contrast and interference microscopy in nerve tissue lifetime studies [200].

Cell structures, for example, membrane structures, being ordered possess the quality of anisotropy that makes it possible to use polarization devices during the analysis of their structure [201]. These methods by strengthening the image contrast allow receiving the information about lifetime cell structure. Using interference microscopy, quantitative analysis was carried out of integral content of dry substance—proteins, cytoplasm fats and DNA of the nucleus [202].

Till now, we talked about rather slow structural changes of nerve tissue, which show only the general process of metabolism in function neurons. At the same time, there appeared considerable number of the works [184–189] dedicated to the direct analysis of optical changes in excited systems, in particular, in nerve fibers. The changes recorded graphically in these works are extremely small ($\Delta F/F \sim 10^{-5}$–10^{-6}) that unfortunately excludes their direct observation in an object image. So, G. N. Berestovskii [203] showed the true change in huge squid axon of light scattering, optical transmission and birefringence. For example, the authors observed high decrease of birefringence by $0.005 \pm 0.002\%$ during the first millisecond after the stimulation and then slow return to the initial level. The increase of light scattering was recorded [204] in huge squid axon under the angle of 90° relative to the direction of transmitted light during action potential. During the record of small-angle light scattering (10°–30°), converse effect of light scattering decrease is observed ($\Delta F/F \sim 10^{-6}$). Detailed experimental analysis of this firstly detected non-electric process during the soldering [205] allowed determining that light scattering changes after the excluding of axoplasm as well as under axon perfusion that testifies the localization of the observed changes in membrane and near membrane region. Evidently, depolarization causes the amplification of protein hydration and increase of their radial orientation. Membrane and near membrane protein structures thickness increases under this condition. Functional value of these changes can be quite high for the transport process as well as for activity control of membrane ferments.

The studies of membrane structural changes using optical methods considerably complete the use of vital fluorescent dyes. Fluorescence (to be more exact its quantum output, spectrum and polarization) is a quite sensitive test to viscosity change, pH, surrounding polarity of dye molecules. I. Tasaki and co-authors [206] detected that fluorescence intensity of ANS acid dye (1—anilinonaftalin-8—sulfonate), which colored squid axon, increases during action potential. At that the change is about 10^{-4}. They described the form of action potential almost precisely. I. Tasaki [207], later L. B. Koen [205] tested a few scores of different fluorescent dyes, for which they detected potential-dependent fluorescence. It turned out that some dyes, especially merocyanine, detect considerable fluorescence value shift of about 10^{-3} that allows avoiding storage during the recording.

For huge crab axons, there was received the information about the change of some dyes optical density (turquoise—straight fast to light M), sorbed by the envelope under threshold rhythmical stimulation and under single soldering ($\Delta F/F \sim 10^{-6}$) [204, 208].

It follows that nerve tissue is characterized by the set of changes of optical features under the stimulation. They can be detected by direct measuring and under the conditions of preliminary object dying. The data received by optical methods cover a wide set of cell structures from morphofunctional restructuring (the change of body volume, cell nucleus, synthesis or protein decay, RNA) to thin restructuring in nerve fibers membranes. It is important to note that the record of insignificant changes of light scattering, light absorption, birefringence in fluorescence unlike similar in shape information of electrophysiology corresponds to the reflection of molecular stimulation mechanism.

Thus, the basis of vital activity of higher organisms is the functioning of electrically excited cells—muscle and nerve. These cells have much in common, but to be more concrete, we will consider nerve fibers in detail. In general terms, this is morpho-physiological abstract of a nerve cell. The study of optical characteristics of function nerve fiber makes it possible to characterize its structural changes in excitation.

2.2.2 To the Question About Neural Holography and Brain Characteristics as 3D Dynamic Hologram

The brain is the most complicated human organ. But brain tissue consists of cells. And it consists of special nerve cells, or neurons. Exactly with them, all variety of our thoughts, feelings, actions are connected, namely they regulate all vital activity processes of an organism. In the first part of the review, we talked that the main structural element of the nervous system of higher organisms was a neuron.

But neuron has its own characteristic masses of nerve cells concerning directly its function. The main function of a neuron is to get information, process it and hand it further. For this purpose, neuron has a lot of dendrites (Fig. 2.9), which are used for handing the information to the cell and one axon, the length of which can be several meters: Through it, the information leaves the neuron and hands further over the nervous system. At some distance from the cell body, axon branches send its processes to other nerve cells and to the dendrites. Each process ends with special thickness—synaptic patch filled with vesicles, which preserve different chemicals—mediators. They provide communication between neurons.

Nerve impulses, which leave neurons and transmit through axon, are specific electric signals. Axon itself can be compared with electric conductor, the central part of which is formed by nerve fibers and covered with insulation—medullary sheath. It provides a high speed of electric pulse through nerve fiber isolating it from electrochemical influence of other nerve fibers.

Electric pulse moving through the axon reaches the synaptic patch and starts there chemical reactions. As a result, mediators are released and thrown into the synaptic gap (microspace which dividing two membranes into synaptic and postsynaptic). Mediator molecules interact with receptors built-in in the postsynaptic membrane, and due to this, channels for potassium and sodium ions are opened in the cell. Occurred intensive flow of ions excites the cell and generates there electric pulse, which is passed to the following neuron, etc.

But this process is not infinite. If the excitation began to pass through the whole channels of interneural connection, such a chain reaction would result in the disorganization of the brain and even could cause death. It does not happen because along with the excitation, there exists deceleration. The specialists are persistently trying to understand the essence of deceleration, as the role of retarding pulses in the functioning of the brain is as important as excitation pulses. When deceleration processes are damaged and neurons begin "to talk" simultaneously and continuously, it causes the development of one and other mental insanities.

By studying complex mechanisms of nerve pulsing, the specialists determined that the number of branches of neuron processes changes through life due to what the brain is growing and developing. After all, the mature nerve cell is not capable of division and cannot reproduce. Those 10–14 milliards of neurons (as different authors say), which are formed by the birth of a child, does not increase anymore. But the number of dendrites as well as the number of axon branches is constantly changing. The most intensive growth of these elements takes place during the first 5–7 years. Accordingly, the number of neurons synaptic connections is also growing; according to the studies of the specialists, 80% of nerve cell surface can be covered with synapsis.

During last years, the scientists found out a lot of new information about the organization of interneural connection. In particular, they detected that the number of synapsis testifying to the number of neuron connections varies in different nerve cells.

Not long ago, it has been considered that synaptic connection exists only between axon of one neuron and its branches, and the body or dendrites of the other. Using electron microscope, the studies detected contacts between axons of two neurons and even between their bodies. They also detected the dynamics of synaptic connections: Some of them can disappear, others—appear. In this process, the functional load, which neurons receive or do not receive, is very important.

If healthy experimental animals after their birth were kept in utter darkness, then synaptic connections of those neurons of brain visual centers, which perceive and process only the information about light (so-called monosensory neurons), were not developing. In the result, the animals became blind though they had all other visual organ elements: Pupil, nervous tunic of eyeball and moreover all nervous conduction paths were saved. The fewer the animals were kept in darkness, the easier it was to restore the function of monosensory neurons and to restore their sight. Such experiments were carried out with the neurons of the auditory center, and the results were similar.

The experiments prove that neurons of all brain centers—visual, auditory, motor and others—for its healthy growth need information gain and adequate functional gain. Only in this case, multilateral interneuronic connections are formed, which to a considerable degree detect the reliability and plasticity of all mechanisms of the central nervous system including adaptation mechanism, training mechanism and memorization mechanism. Brain language is the language of electrical pulses, as well as chemical pulses [208] and possibly optical pulses (see Sect. 2.2.7).

Now let us describe structural and functional characteristics of a nerve fiber. Nerve fiber consists of axon and layers. Axon in comparison with nerve cell body has other features. For example, impulses in it are transferred with the velocity up to 100 m/s, while in brain gray substance, it is 1–2 m/s. Myelin sheath appears at a distance of 50–100 μm from the cell body and disappears in the nerve ending zone. Myelinated nerve fibers have the diameter from 2 up to 20 μm. The relation of axon diameter and the diameter of the whole fiber is used to be constant and equal to 0.51:1. Myelin sheath because of its high refraction is used as a lens, that is why the axon parts under it seems to be thicker in comparison with the parts in the area of Ranvie interceptions, where the myelin sheath is absent (Fig. 2.9).

As a rule, axons are surrounded by thick fat (myelin) sheath, which is periodically (in 1–2 mm) broken by Ranvie interceptions (1 μm). Myelin segments are used as insulating joint. Nerve fiber on these parts is similar to passive communication. Only the part of cellulated surface in Ranvie interceptions is electrically active. It is very convenient to study unmyelinated squid axons, which sometimes have the diameter of 1 mm. In such fibers, which are also called smooth grain, the whole surface is electrically active.

Axon can be represented as a hollow tube filled with an electrolytic solution, which was dipped into extracellular fluid. This tube wall—axon membrane—consists of lipids and proteins. Membrane thickness is ~7 nm. Electric membrane rest resistance is quite high, it is about 10^3 Ω/cm^2, and its capacity is about 1 mcF/cm^2. Axon membrane separates the internal filling solution from the external one having another composition. So, inside, in axoplasm, in comparison with the environment (K—10 mmol/l, Na—460 mmol/l, Cl—540 mmol/l), the K ions concentration is quite high (400 mmol/l), and Na and Cl ions concentration is quite low (50 mmol/l). Rest cell interior is negatively charged concerning the environment, and the potential difference on the membrane is about 60 mV. The appearance of "rest potential" in the Bernstein theory is explained in the following way. It was assumed that membrane was permeable only for K ions. It turned out that rest membrane is permeable for Na and Cl ion also. Really, Na ions, which are in plenty in the environment, come to the cell under the influence of diffusive and electric forces. Thus, it was assumed that membrane has special mechanism—sodium pump, which provides so-called active transport, i.e., takes Na ions out, against electrochemical potential gradient consuming metabolism energy.

It is postulated in the Hodzhkin–Haksli model that ion channel, for example, potassium, can be in one of the fifths conformational states, and only one of them is conductive. This scheme is not the only possible if we proceed from the conditions of coincidence with experimental data on the fixation of the material.

Not considering the question of membrane theory in detail it is possible to suppose that future experiments in the creation of supersensitive measuring system, which enables to divide K and Na ions contribution in the refraction, will be successful. These ions cross the membrane of living nerve fibers using the wavelength-tuned source in the researched range as it was done by the author together with the co-authors on resonance atomic moving of K and Na in the plasm (see Chap. 1, Sect. 1.3).

At the same time, the researchers were interested in the problems of nerve impulse velocity calculations in homogeneous fiber and also complete excitation propagation mode in active media, for example, in cardiac muscle. It is known that in myocardium there may appear flutter phenomenon, or fibrillation, connected with spontaneous electric activity of a medium. Famous mathematician N. Viner and A. Rozenblut studied fibrillation. They introduced the notion of formal medium excitation where the nature of excitation is insignificant and basic results were got.

Another group of works is connected with impulse velocity calculation based on simple representation about physics of the process. Usually, it was amounted to the fact that when some critical conditions were achieved, membrane capacity discharge began or membrane generator with tailor-made-properties switched on. Such physical modeling allows receiving analytical results and finding out physical regularities.

The most intriguing question in the physics of nerve impulse is the action mechanism of ion channels of excited membranes operated by electric field. The change of channel conductivity is either connected with the noticeable tuning of system molecular geometry or induced by the change of electrostatic component of ion energy.

It seems natural that in the first case the channel will work discretely on the principle of "all or nothing". In the second case, its conductivity will be changing continuously. The measurement of current fluctuation spectrum makes it possible to choose between these two possibilities. Thus, fluctuation analysis of ion current of biological membranes becomes of special significance. But till the present moment, there is still no clarity about the essence of electrical quantity characterizing the channel state. So far, there is no any other method that would make it possible to detect whether transfer or rotation of some excited groups at long distances take place or there is cooperative reorientation of dipoles ensemble of atomic scale. Today, the most actual task for nerve impulse biophysics is to clarify this issue.

The theory of nerve impulses distribution on separate fibers of different geometry is quite detailed. The next step is to consider electrically connected nerve fibers, the so-called excitable medium.

The development of excitable medium theory is connected with considerable difficulties, as besides random parameters distribution it is necessary to take into consideration nonlinear character and memory of source function. Therefore, it is not a surprise that suitable formal apparatus has not been developed till the present moment.

During the study with the use of electron microscope in axon, neuroplasm and neurofibrils were distinguished. In the present, it is considered that the conduction of excitement is the function of interaction between neurofibrils and axoplasm. Myelin sheath has a layered structure. Myelin layers are interleaved by thin protein interlayers. Myelin consists of the molecules of cholesterol, phospholipids and cerebrosides. During the study in polarized light, myelin detects single-layer positive birefringence [209]. It is assumed that myelin is necessary for isolated pulsing. It is easier to excite nerve fiber in the area of Ranvie interceptions, i.e., in the area where there is no myelin.

Nerve fibers connect neurons with each other or with the cells of working organs. The impulses appeared in nerve cell or receptor are distributed through nerve fiber. The fiber can be excited by temperature rising, mechanically and chemically. The highest excitability nerve fiber has concerning electric current [158]. The majority of experiments carried out using electrophysiological methods. And now it is generally accepted that nerve impulse distribution is carried out in electric way. But it is acceptable that mechanic and electric conceptions, which characterize nerve impulse distribution along the fiber, will be replaced by waveguide mechanism of distribution along it (nerve fiber) of electromagnetic radiation like autowave oscillations such as solitons. It is also acceptable that in living neuromuscular tissue, there is some self-induced system (such "substations" concerning nerve fiber are, for example, the area of Ranvie interceptions), which fulfills the most important nerve impulse features, such as the ability to spread along the fiber with the constant speed and without attenuation [210, 211].

The studies in the area of neuron neurophysiology [212–227] are also of the utmost interest. To some extent, they explained the mechanisms of impulse distribution in the neurons. But, the final understanding of fundamental mentation as the dynamic of many impulses in neural net has not been reached. The neural net indicates disorder in the interconnection of the majority of neurons up to 10^{10}. In spite of this feature, there is perfection in mentation function systematization. If any systemized function is in disorder of the element set, then it has to have statistic character. This fact was marked by N. Viner (1948) [216] in its cybernetics theory. On the other hand, in 1971 K. Pribram proposed heuristic and metaphoric statement of the mentation [217]. It is called neural holography, and it provides stable model of memory processes. It was produced on the basis of deep mathematical consideration and has absolutely similar to some optical hologram properties. Memory processes scatter the saved information in neural net dilated domain. Moreover, the saved information is unlimitedly reserved unlimitedly in the limited space of neural net. This fact assumes the validity of neural holography theory. Experimental and theoretical check of the processes of neural holography has not been received yet.

Is there statistic theory of neural net which describes fundamental mentation? Is it possible to detect the processes of neural holography in such a statistic theory?

In this connection, one work on neurophysiology neurons should be marked. This is the article "Stochastic dynamics of the impulse: mathematical basis of neural holography" by Kunio Wasye, Mari Dzhiby, Tetsua Misava [223–227]. They proposed stochastic model, which forms mathematical base for neural holography. It was detected that linear wave equation for complex wave function describes impulse motion in neural net. Neural holography is really statistic effect such as electron holography.

The present work shows statistic theory of impulse collective motion in neural net and stochastic model of neural net with absolutely new theoretical point of view. Fundamental mentation is identified with statistic collective impulse motion. It was detected that wave equation will describe such a collective motion. It has the same form as a wave equation in quantum theory. Neural holography will have real meaning of statistic phenomena similar to electron holography.

It is also important to notice that light holography is physical phenomenon provided by light wave properties. Possibly there is also similarity with wave phenomena in neural net. And, the authors of the present monograph, L. V. Tanin and A. L. Tanin, detected that brain properties and processes are in general similar to the properties and processes of three-dimensional dynamic hologram. The first feature that coincides with the basis of holographic principles, i.e., the brain ability to record, save and read (reconstruct) complete information about the object. In holography, it is happened due to the interference of reference wave and reflected from the object wave: the amplitude, phase, wavelength and the polarization of the reflected light [78, 228– 238]. The second feature is the following: Large amount of information is recorded, and its density can be 10^{20} bit/cm^2. The third feature is that record and reading velocity of optical information from a dynamic hologram, the properties of which (if assume) are based in the brain, is considerably higher than the velocity of electric and chemical processing in a human and animal organisms. It allows more easily understanding and explaining the mechanisms of low-frequency laser radiation influence on human and animal organisms with probabilities of all three models of nerve impulses: electric, chemical, optical and their interconversion [239].

Large amount of information about nerve fiber structure was received with the help of electron microscope. But the electron microscope requires specially prepared drugs and is not used for lifetime studies. Optical, in this case holographic, methods have a set of advantages in comparison with electrophysiological and electron microscopes. First of all, the reason for this is that being contactless these methods do not distort the characteristics of the objects under consideration and on the other hand they make it possible to highly visualize the structural changes. And holographic interference methods allow detecting optical path difference, which is considerably lower than the wavelength of the used radiation. They also allow detecting small changes of structural and functional characteristics.

Taking into consideration the results of the analysis, made in Sects. 2.2.1 and 2.2.2, there appeared the interest in the study of optical characteristics of nerve fibers by highly sensitive noninvasive contactless method of holographic interference microscopy. Such method did not exist before, it had to be developed and adapted to the medical and biological studies by Leonid Tanin and his colleagues.

2.2.3 Development of the Method of Laser Acupuncture and Intravenous Blood Irradiation for Lumbar Osteochondrosis Neurological Manifestation Treatment

Depending on the character and peculiarities of the disease, there were used different methods with different localization of exposure: locally (on supposed lesion focus), on the corresponding reflexogenic zone and acupuncture points. The most effective method, according to our studies, was the method of laser acupuncture.

The development and use of absolutely new coherent-optical methods and devices for medical diagnostics of the peripheral nervous system diseases required the detailed study of the influence of laser, magnetic fields, pressure from 10^{-4} to hundreds of atmosphere on separate nerve cells, fibers, nerve fibers in nerve trunk, living muscle fibers. In particular, according to literary data, low-intensive laser irradiation is successfully used as one of the phototherapy types, in biology and medicine. Stimulating influence of low-power laser irradiation on biologically different objects during the experiment resulted in its wide use for the treatment of different diseases [240–243]. In the works [244, 245], we studied the influence of laser irradiation on animals during treatments simulation—a wound, inflammatory processes, burns and fractures.

The analysis of the works [246–248] showed that low-energy laser irradiation under certain irradiation conditions activates physiological functions in the organism, improves metabolism, takes vasorelaxant action, stimulates erythrogenesis, glycogen synthesis, anagenesis and so on.

Theoretical arguments about the mechanisms of laser influence on biological tissues are quite different. There are a lot of hypotheses, which explain light influence and are based on the study of the characteristics of biological membranes, enzyme system, endogenous and opioid systems and other [249–251]. The majority of the studies has only one common thing: Radiation action of all low-power, low-intensive laser is based on the effect of selective absorption of light energy by definite cell structures, and as a result, biosynthesis processes are stimulated [240, 241, 252, 253].

The method of laser acupuncture for the treatment of some diseases becomes widely spread in clinical practice. The methods of reflex therapy are used for the treatment of allergic diseases, sharp and chronic inflammatory processes, and also for the increase of body resistance under the influence of infections. There is a number of clinical and experimental data, which prove and expand the idea of the immune system role in the realization of laser acupuncture therapeutic effects. The changes of immunoresponsiveness appeared under reflex action should be considered not only during infectious and allergic diseases. Evidently, they highly influence the increase of general non-specific organism resistance, which can be revealed as side positive effect of laser acupuncture. Neuroreflectory mechanism of laser radiation action of low energy level has been proved recently.

Low-intensity laser irradiation is widely used in clinical practice of neurologic diseases [254]. Positive effects of using low-energy laser radiation during the treatment of radiculitis, neuralgia of trifacial and glossopharyngeal nerves, facial neuritis have been obtained [255–258].

In the Laboratory of Physiotherapy and Balneology of Belarusian Research Institute of Neurology, Neurosurgery and Physiotherapy of the Ministry of Health of the Republic of Belarus, the studies on the therapeutic effectiveness of laser acupuncture during the treatment of the peripheral nervous system (PNS) diseases took place. The influence was made by "open" laser beam or through the light guide with the density power from 10 to 20 mW/cm^2. Per procedure there were used from 4 to 8 points, duration is 10–30 s. The main task during separate process design and during the design of the whole laser acupuncture treatment was the correct choice of the points, their combinations and the sequence of actions. Meanwhile, it is necessary to follow the principles of typical manual acupuncture, taking into account present ideas of osteochondrosis pathogenesis and its neurologic manifestations. During the choice of the points, sequence and time of their irritation, it was very important to estimate in a right way the general state of patient, the intensity and prescription of pain syndrome.

The procedure should also include the points of local segmental actions (LSA)—local, "painful", segmental (considering pain topography, pathological center localization); and also the points of so-called "general" action, which determine the development of the common reaction and located in flanged area, in the area of hands and legs.

The point combination for laser acupuncture, sequence and time of influence on each of them was determined first of all depending on the intensity and prescription of pain syndrome. If the intensity of pain syndrome is moderate and lasts more than 1 month, the influence began from the points of "general" action located in the area of hands, legs and flanged area. Then, step-by-step, the points of local segment action were connected to the area of the back, legs. During each further procedure, the number of points of local segment actions was increased from 2 to 8 and time of irritation for each of them from 10 to 30 s.

In a case of expressed pain syndrome and duration less than 1 month, laser acupuncture was started from the points of local segment action by irritating each of them during the first procedure with maximum power and duration. When the pain syndrome decreases, the points of "general" action were wider connected up moving away from the area of the strongest pain and decreasing the number of local segment points.

If the pain intensified, we changed the point number, their stimulating time and also laser radiation power.

The dominance of reflex-tonic disturbances in clinical finding required wider inclusion of the flanged area, hands.

More effective laser acupuncture was during radicular and especially radicular-vascular manifestations of lumbar osteochondrosis. During radicular manifestations, it was necessary to use more localized influence with primary irritation of local segment points.

In the results of examinations in 1981, 340 patients with lumbar osteochondrosis neurological manifestations were treated using laser acupuncture.

The effectiveness of laser acupuncture of 120 patients with radicular (72 patients) and reflex (48 patients) stages of lumbar osteochondrosis at the age of 20–60 years has been studied. The disease period was from 20 to 60 years. There were 22 women and 98 men.

Using laser acupuncture, the patients were treated at the beginning of exacerbation and after ineffective drug and physiotherapy treatment.

Laser therapy was carried out using He–Ne lasers $\lambda = 632.8$ nm (LH-38) with light guides of different diameter and power flux density (PFD) from 3.5 to 16 mW/cm^2 and from LH-56 and LH-72 by the "open" laser beam PFD from 2.5 to 8 mW/cm^2, and also, it was carried out using He–Cd laser $\lambda = 441$ nm (LH-61).

The effectiveness of laser puncture was estimated by the clinical and paraclinical survey techniques. After laser acupuncture treatment, 90 (75%) patients felt amelioration, 18 (15%) patients—small amelioration and 12 (10%) patients were without changes.

During clinical surveys, there were detected the absence or decrease of the degree of manifestation of pain (79%) and reflex-tonic (77%) syndromes at complete or partial recovery of the function of affected roots.

55 patients with different clinical manifestations of lumbar osteochondrosis were treated by pulsed sequential action by low-intensity laser irradiation with blue (441 nm) and red (632 nm) wavelengths on 4–6 acupuncture points with the frequency from 5 to 50 Hz. During the course of treatment, there were 10–12 manipulations. 21 of 55 patients had a reflex syndrome, 34—radicular syndrome of different intensities. In the result of the treatment, 65% of the patients felt amelioration, 15%—small amelioration and 20% stayed without changes.

But only in a quite limited number of studies, the study of laser influence on health and pathological tissues and organs of animals accompanied by the analysis of the changes in peripheral blood and immune-competent organs [255, 256].

In recent years, the method of organism laser biostimulation, namely, the method of intravascular laser blood irradiation (ILBI) when almost the whole circulating blood is irradiated, has been used [16, 18, 21, 26–35, 37, 39–45, 48, 52, 53]. Taking into consideration the fact that blood is polyfunctional system, which also functions as an integrated medium, its irradiation influences the whole organism and tissues and reflects the state of the whole organism in general. By the present moment, the effectiveness of this method is shown using it as the prevention and treatment of many diseases (myocardial infarction, atherosclerosis obliterans of the vessels of lower extremity, peritonitis, coronary disease, disseminated sclerosis and others) [259–262] and has a real basis in a form of accumulated clinical observations and the works on the study of its medical and biological influence.

In particular, it has been established that ILBI intensively influences rheological properties of blood—reduces its viscosity by decreasing the degree and speed of gluing of platelets, reduces the sludging of red blood cells with an increase in fibrinolytic activity and others [263–265] what improves microcirculation and tissue trophism, affects the oxidant system (catalases activity, glutamateperoxidases and others) on the stage of lipids peroxidation, which is the important part of pathogenesis which detects different structural and cellular membrane failures.

In the response of blood cells to laser irradiation, the important role of lymphocytes, which are the carriers of immunological memory and which provide organism self-protection system, should be noted. ILBI actively influences immunological system (the indicators of humoral immunity; functional activity of T-lymphocytes, increase resetting, the level of blast-cell transformation). The above-mentioned factors improve proliferative processes and suppress anaerobic reactions developed in pathologically changed tissues.

In recent years, some stages of pathogenesis of some diseases have been identified. They are based on tissue damages by the antigen-antibody complexes. If any antigens appear, the organism begins to produce antibodies. Simultaneous presence of complementary antigens and antibodies in the organism results in appearance of immune complexes (IC), which should be removed by the process of phagocytosis. If the immunoregulatory mechanisms are impaired, the number of ICs may increase, that, as a rule, causes a chronic relapsing course of PNS-diseases. The determination of IC content in combination with other tests is used in evaluating the effectiveness of treatment.

The ILBI method having a systemic influence on the body is the most effective way of the normalization of immune status, microgemocirculation, transcapillary exchange and tissue metabolism during the damages of their innervation and blood supply [266]. Some authors studied the interaction of laser radiation and blood. The results of these studies showed that this interaction is the most effective in red and infrared range with the wavelength of 600–1000 nm. He–Ne (632.8 nm) and semiconductor (850 nm) laser radiate in these ranges.

Susceptibility of biological structures to laser radiation is provided by the complex of specific and non-specific photoacceptors, which absorb energy of this radiation and provide it transformation during biochemical and biophysical processes [267, 268].

Experimental studies revealed changes in lymphoid tissue that occur at the influence of low-intensity laser irradiation (LILI). Many authors associate the immunomodulatory effect of laser radiation with the influence on the receptor apparatus of immunocompetent cells [269, 270].

Earlier we have experimentally detected [28, 77, 271] that under traumatic and ischemic damages of peripheral nerves ILBI-therapy normalizes prooxidant reactions, enzymatic and hormonal damages, the increase of immunoactive proteins and the activity of the antioxidant system, the improvement of the functional state of neuromotor apparatus and microgemocirculation that strengthens compensatory-adaptive organism reactions in the conditions of model pathology. Consequently, ILBI influences circulating blood and has biostimulating effect, normalizes the state of the important functional systems and organs that allows using this method of non-drug therapy in treating different pathologies including the diseases of the nervous system: disseminated sclerosis, cerebrovascular pathology, vertebragenous diseases of the peripheral nervous system (PNS) [272, 273].

In neurosurgery, ILBI used in treating craniocerebral injury, tumors, vascular diseases and led to the disappearance of blood metabolic acidosis, normalized blood osmolality, microcirculation improvement [274]. The regression of clinic symptomatology was characterized by the decrease of cerebral damage (the degree of consciousness disorder, the intensity of psychomotor damages, the intensity of headache). Brain edema, which was typical to the neurosurgery patients in 3–5 days after the operation, stopped to progress or was ill-defined. The regression of nidal neurologic symptomatology, mainly of cortical genesis, was marked [275]. It allows concluding that neurosurgery patients had slighter postoperative period after ILBI-therapy.

In the works [29, 34, 35, 37, 39–41, 43, 45, 48, 53, 77], there are the results of the studies on the clinic use of ILBI under traumatic damages of peripheral nerves directly after open treatment—neuroraffia.

It is considered promising to take microhemocirculation indicators as criteria for severity and prediction of outcomes of various pathological conditions. [276]. However, the question of objective evaluation of microhemocirculation and the effect of laser radiation on it has not been fully resolved due to the lack of reliable noninvasive methods of study.

The development of unique possibilities of contactless noninvasive speckle-optical diagnostics of cutaneous blood flow changes under the damages of peripheral nerves of different genesis will allow quantitative assessment of the effectiveness of ILBI use on early stages of restoring-regenerating process.

Using the method of speckle-optical myography, it is possible to get objective criteria of miotonus damages and the damages of the contractive activity of denervated muscles and also to assess their function restoration degree under the ILBI influence [277–279].

For the diagnostics of hypoxic tissue changes by damaged peripheral nerves, it is very important to have the information about the level of cutaneous chromophores, hemoglobin and oxyhemoglobin in cutaneous blood flow [279].

Spectral analysis is the most effective method for receiving this type of information [20, 22–24, 25]. The studies of absorption spectrum of animal blood preparation using three-wavelength spectrophotometric method under ILBI allow estimating the influence of LILI ($\lambda = 0.6328$ μm and 0.85 μm) on such important index of the blood oxygen transport function as the degree of oxygenation and use this criterion as a basis for the objectification of the damages of blood oxygen transport function under different peripheral nerves damages (traumatic damages of nerve trunks, partial ischemia of animal hind limb with the development of ischemia neuropathy of sciatic nerve) and also study the possibility of this criterion correction by ILBI.

It is known that some diseases of peripheral nervous system, in particular vertebragenous pathology, are accompanied by a long motion activity limitation, marked pain syndrome, micro and hemodynamics damages, organism immune system. Considering the positive influence of ILBI on the system of hematosis, the technique was developed of clinical use of intravascular blood irradiation of the patients with some pathologies of PNS [16, 18, 21, 28, 32, 41, 42, 46].

It also should be noted that starting from 1978 the employees of the team "Coherent-optical research methods of medico-biological systems" along with the specialists in the area of laser technology of the Optical Holography Laboratory and the Laboratory of Atomic Spectral Analysis of the Institute of Physics of the Academy of Sciences of the BSSR assisted to the first Neurologic Department and the Department of Physiotherapy and Balneology of the Institute of Neurology, Neurosurgery and Physiotherapy of the Ministry of Health of the BSSR in the development and delivery of some definite laser setups for their use in clinical conditions for the purpose of development and implementation of the methods of laser therapy of peripheral nervous system diseases, the diseases of loco-motor system and peripheral vessels.

In the Institute of Neurology, Neurosurgery and Physiotherapy of the Ministry of Health of the BSSR at the participation of L. V. Tanin, there was established one of the first in the Republic of Belarus cabinet of laser puncture, the assistance was provided in the exploitation of laser setups, and the medical personnel was learnt to work with this equipment. The methods of application of low-intensive laser irradiation at diseases of peripheral nervous system were developed, particularly, there was developed and implemented the method for treatment of patients with radicular-vascular manifestations of osteochondrosis [4], the results of clinical use of laser radiation for treatment of patients with this disease were jointly discussed.

It should be noted that for many years the Laboratory of Laser Systems and Devices (V. A. Mostovnikov, G. R. Mostovnikova, V. Yu. Plavsky and others) of the Institute of Physics of the NAS of Belarus develops, improves and delivers new laser setups for the purpose of their use for medical purposes in medical institutions of the Republic of Belarus [280–284]. The obtained clinical effect requires its further active study and also comprehensive study of the peculiarities of this radiation influence on separate living nerve cells, living fibers, living muscle fibers and on the whole human neuromuscular apparatus by highly sensitive, contactless, noninvasive, coherent-optical methods for the purpose of the development of new methods of medical diagnostics and treatment [1–53, 77, 271, 273, 277–279, 285–291].

The clinical effect gained after the use of low-intensity He–Ne laser irradiation, caused interest in further studies. It is shown in the monograph of L. V. Tanin, N. I. Nechipurenko, L. A. Vasilevskaya, G. K. Nedzved, S. E. Rovdo, A. L. Tanin "Laser hemotherapy in peripheral nervous system treatment". Edited by N. I. Nechipurenko, L. V. Tanin, Minsk: Magic Book, 2004).

2.2.4 Isolated Preparations—Adequate Experimental Model of the Study of the Influence of Laser Magnetic Fields, Hyperbary on the Excitability of Nerve and Muscular Tissues

Recently, we have observed the increase of scientific interest and importance of the model studies using isolated preparations. The results of such studies enable to differentially assess the influence of laser and magnetic fields, hyperbary on the definite vital activity links.

The object of the study of the influence of these factors on nerve and muscular tissues were isolated preparations of single nerve fibers, nerve fibers composed of nerve trunk, isolated single muscle fibers, muscles. The preparation of isolated nerve was produced by separation of frog sciatic nerve. Preparation method was typical, the nerve was cut above and below of preliminary ligations, which were used, if needed, for fixing and transporting of the nerve to special hyperbaric chambers. Single isolated nerve fibers were prepared from the frog sciatic nerve diameter of 8–10 μm at the length of the prepared area of 5–7 mm). All preparation processes of single isolated nerve fiber were made according to the method described by I. Tasaki [292]. Before putting the preparation into hyperbaric chambers, electrophysiological control of nerve tissue vital activity was made. Nerve fiber excitability, composed of the isolated nerve trunk, was estimated by its amplitude, action potential time and the quantity of threshold electrostimulation, which causes the potential generation in the preparation. For the graph construction of the relations time-force, potential generation threshold was defined in response to electric stimulation by pulses with time of 50, 100, 200 and 500 μs.

The stimulation and recording of action potential were made through bipolar metal electrode hermetically built in hyperbaric chambers. Electrostimulation, strengthening and recording of action potential were made using electromyography "Medikor".

For separation of single muscle fiber, we prepared biceps of frog hip and studied it under the microscope MBS-9. Muscle fibers were separated by electrolytically edged steel needles under microscope control. The rest part of the muscle was removed. Separated fiber was placed in Ringer's solution for cold-blooded animals.

In the experiment, only intact fibers were used, contracting in response to electrostimulation. Two types of chambers were designed and manufactured for the effect of increased pressure on the nerve and muscle tissue under study: one to maintain gas pressure in the range of 0–5 atm., the other to create hydrostatic pressures in the range of 0–200 atm.

In the works [1–3, 5], the possibility of applying the methods of holographic interference microscopy for studying the nerve fibers in in vivo state and lymphocytes was first demonstrated.

These studies allowed obtaining holographic images of the main structural formations (myelin sheath, Schwann's cell perikaryon, Ranvie interception structure details). In addition, a significant difference in phase incursions was observed,

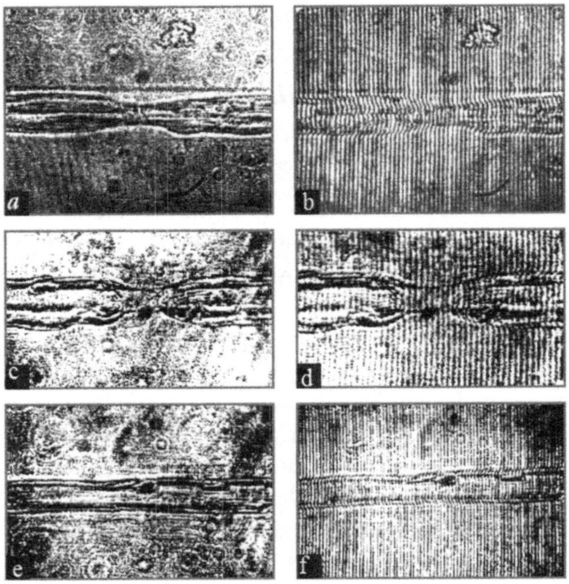

Fig. 2.10 Reconstructed holographic images of the nerve fiber parts (**a**, **c**, **e**) and corresponding holographic interferograms (**b**, **d**, **f**). Reprinted from [136] with permission

between the areas of the interception Ranvie bulb and the area of its nodal section, i.e., the area, where the excitable membrane is located, which is responsible for conducting the nerve impulse along the fiber.

In in vivo state, the nerve fiber preparation is a very labile formation, in which continuous morphological transformations occur. At the same time, the study is significantly limited by the duration of their lives. Holographic registration of the nerve fiber, cells with a posteriori study of the reconstructed image allowed us to avoid this difficulty. As an example of obtaining such information in Figs. 2.10 and 2.11, various regions of the nerve fiber recovered from the hologram (Fig. 2.10a, c, e and 2.11a, c, e, g, i) and the interferogram of the same areas (Fig. 2.11b, d, f, h, j) are illustrated.

The phase shifts obtained from the interferogram show the structure of myelin fiber and changes in it. Both photos clearly illustrate myelin fiber and variations of its form. In the central area of the fiber, there is situated perikaryon Schwann's cell including the nucleus. The interferogram illustrates that the most shift of fringes is observed in the axial region of the fiber and myelin envelope. This shift of fringes is conditioned by the cylindrical fiber shape and essentially larger in comparison with surrounding solution refraction index of lipoprotein structure of myelin envelope.

On the reconstructed from a hologram image (Fig. 2.11e, g, i), there is a slot of Ranvie interception, interception bulbs and their longitudinal wrinkles. Fringes shifts corresponding to interferograms in Fig. 2.11f, h, j are connected to the mentioned-above interception structures. Thus, axon, which does not have a dense structure

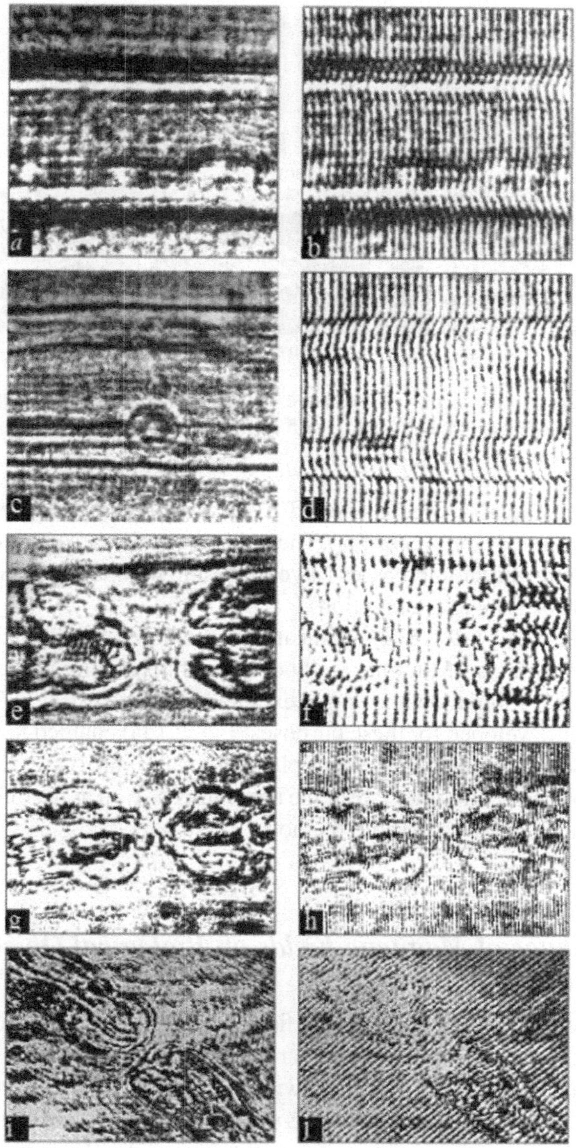

Fig. 2.11 Reconstructed holographic images of different parts of nerve fiber (**a**, **c**, **e**, **g**, **i**) and corresponding interferograms (**b**, **d**, **f**, **h**, **j**). Reprinted from [136] with permission

in the area of the slot, corresponds to minimal disturbance of interference fringes, whereas comparatively dense myelin sheaths of bulb interceptions and their wrinkles are clearly seen on the interferogram.

Figure 2.12a–e shows different on section depth 3D images of the nerve fiber produced by consecutive overfocusing on the stage of hologram reconstruction.

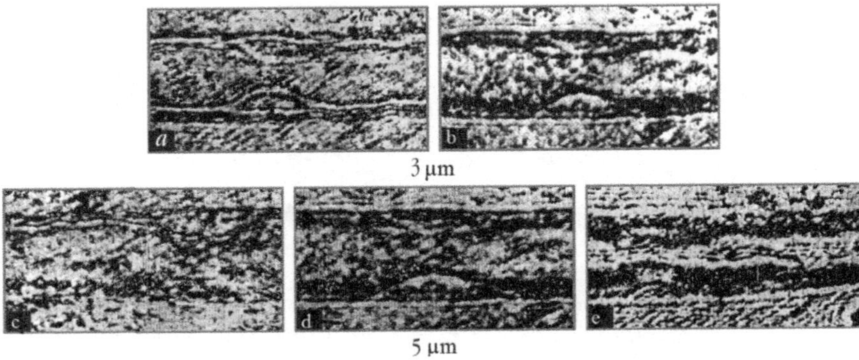

Fig. 2.12 Sections of 3D images of the nerve fiber at overfocusing, reconstructed from one hologram and varying on depth: **a, b** 3 μm, **c–e** 5 μm. Reprinted from [136] with permission

Depth of field (~5 μm) of reconstructed from a hologram image exceeds in 3–5 times the one at photography recording. These studies enabled to produce holographic images of the main structural masses (myelin sheath, perikaryon of Schwann's cell, structure details of Ranvie interception).

The works [1–3, 5, 7–9, 15] are dedicated to the development and improvement of holographic methods of lifetime study of nerve fibers using the regimes of their rhythmical electrostimulation and single excitability action.

The chamber developed for these purposes (Fig. 2.13a) enabled to carry out holographic studies with simultaneous electrophysiological control (Fig. 2.13b) that made it possible to follow the lifetime state of the nerve fiber as well as to control the process of removal synchronization by both methods.

2.2.5 Influence of Magnetic Fields on Biological Objects

More often in clinical practice, magnetic fields are used, and moreover, there can be used magnetic fields with constant and variable magnets in intensity broad band.

Physiotherapy cabinets are equipped with pilot setups (such as "Polus"). But the mechanism of magnetic influence on the whole organism as well as on a separate cell still remains unexplained that restrains the further development and use of the methods of magnetotherapy.

The literary data show two possible interaction mechanisms of the magnetic fields and biological objects: its detection by an organism connected to electromagnetic induction phenomena, and the interaction of the field with object magnetic material [293]. Both mechanisms cause different changes in tissues and even their deformations, which can be recorded by coherent-optical methods. As it is shown, electrical points induced by magnetic field influence, primarily, the excitable structures (nerve and muscle tissues). In the work [294], it was mentioned the functioning of the first

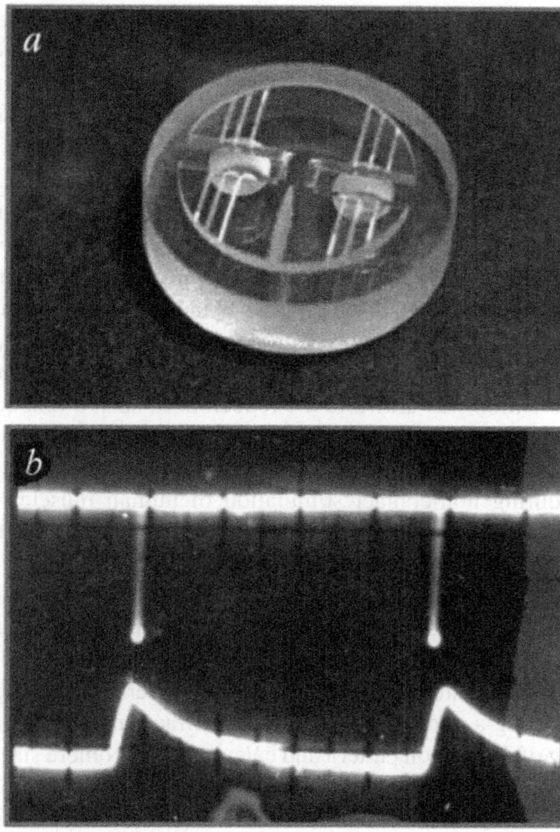

Fig. 2.13 Optical camera for holographic study of nerve fiber (**a**) with simultaneous electrophysiological control (**b**). Reprinted from [136] with permission

interaction mechanism. In this work under overlapping of magnetic field, longitudinal current appeared in frog sciatic nerve, which had undamaged anatomic connection with the central nervous system and peripheral organs. An interesting fact of the changing of electric activity of shell spontaneously active and silent neurons was observed in pulsed magnetic fields [295–299]. But these works do not specify the interaction mechanism of the field and the object. On the other hand, during active potential transport (in the absence of the field) through squid huge axons [300, 301], there was detected the surface shift of about 2–20 nm. Comparing the changes of electrical activity under the influence of the field with deformations under action potential in nerve tissue, it can be assumed that the field detected by the object will also cause surface shift.

The literature also describes the interaction of magnetic field and bioobject magnetic materials. Bending of prepared frog sciatic nerve [302] hanged up toward permanent magnet with the field of 58 mT indicates magnetic properties of the tissue. But other authors note that the results of these experiments require a detailed check

[303]. There are sources, which show that under the influence of permanent magnetic field with the induction of 0.4–1.3 T, there is the change of the capacity of double layer of lipid membranes from egg lecithin in dextran correlated with the membrane area changes, and these changes take place under interaction time about several minutes [304, 305].

Muscular tissue is also sensitive to magnetic fields. In the work [306], morpho-histochemical changes in cat skeletal muscles were registered. They appeared under the influence of pulsed magnetic field of 89 kA during 6 h per day over 20 days or under single influence [307]. On the fifth day after the manipulation in many muscle fibers, contraction nodes appear, in which some of the fibers broaden in a flask-shaped way and the fringes of cross-striation approach. In the detected part, there is sharp myofibril multiplexing, which forms clear longitudinal striation in the muscle fiber.

In addition, attention was drawn to the reduction of such a functional parameter as the threshold of excitability in the neuromuscular preparation when applying a field [308]. Some researchers have noted a contractile response caused by a pulsed magnetic field during non-contact stimulation of human muscles [309], muscle contraction of a frog and a rabbit [310]. When a frog was poisoned, the contraction was absent (the method of poisoning was not specified).

It should be noted that the mentioned structural and functional changes in muscle tissue were observed in some time after field influence. First of all, it can be explained by the fact that for the structural changes, accumulation time is needed, and secondly, by the complexity of the studies directly under field overlapping. The last difficulty can be obviated using a contactless method of diffractometry. This method allows remote recording of diffraction pattern and calculating sarcomere sizes [311–313] of muscle fiber and their changing during the moment of the field action. The method allows detecting the thickness of myofibril cluster [314], the function of sarcomere length distribution [315] and A-disks dynamics [316].

In the review [317], results of studies and perspectives of the action mechanisms of constant and low-frequency magnetic fields are shown. These fields do not influence some membrane and tissue structures. The authors of the work [297] studied the influence of single and set of pulses of small amplitude magnetic field on electric activity of shell neurons. For single pulse of triangle magnetic field, there is detected the dependence of neuron reaction on the magnetic field change velocity. For the set of pulses [318] with the amplitude of 0.1 mV and time of 10 s, there was observed activity depression of snail neurons. The possible mechanism of low-level electromagnetic radiation influence on living organisms, which is based on the assumption of electromagnetic self-oscillations of cellular construction (for example, membrane parts) and qualitative state of living cells, is discussed in the work [319]. The method of the measurement of the latent period and pulsing velocity in isolated nerve after the processing of variable magnetic field is proposed in the work [320]. The author of the work [321, 322] studied therapeutic influence of magnetic field on neuromuscular apparatus and assumed that magnetic field influenced the nervous system peripheral departments. There was also carried out the set of experiments on the influence of variable magnetic field of neuromuscular apparatus where we could observe plus phase amplitude incidence of electromyographic response and its negative component increasing trend.

The mentioned-above publication review allows concluding that the influence of magnetic fields reveals as a weak effect, for the recording of which highly sensitive methods and often long time of observation are needed.

2.2.6 *Study of the Influence of Powerful Pulsed Magnetic Field on the State of Isolated Nerve*

The study of the influence of powerful pulsed magnetic field [323] on the state of the isolated nerve was carried out in the work [14]. Isolated sciatic nerves of brown frog Rana temporaria were the object of electrophysiological inquiry of the influence of powerful pulsed magnetic field. The nerve trunk with an average length of 5 cm was prepared by a common method and placed in Ringer's solution for cold-blooded animals for 1–2 h for demarcation potential decrease. Then, the preparation was transported to specially made moist chamber for electrophysiological studies.

The nerve was stimulated by electric current rectangular pulse with the voltage of 5–10 V and time of 50 μs. Action potential (AP) was recorded using a hanging silver electrode at a distance of 5 mm between them. During the experiment, the chamber was shut tight with the cover with an optically transparent area that during the work allows controlling the conditions of bioelectrical signals retractions and avoiding the preparation drying up. AP was being recorded over a long period of time (up to 48 h).

Experimental setup consisted of an electrostimulator of EW-2 type, a bioelectric-potential amplifier with the bandwidth from 10 Hz to 20 kHz and 10 megaohm input resistance, an oscilloscope and a digital time-domain voltmeter of V4-17-type.

To achieve the goals of the present part of the work, we measured maximum amplitude of the nerve action potential before, directly after and in some time after the influence of magnetic field. AP amplitude has been chosen as the criterion of preparation functional state as it is the integral index of nerve cells participation, which forms nerve trunk, in the conduction of excitement. Moreover, AP amplitude is a stably registrable index, which can be objectively detected by pulsed voltmeter.

The measurement mode was based on the fact that the pulsed magnetic field causes strong framings in electronic circuits, which distort and reject desired signal and, in some cases, disable highly sensitive electromeasuring apparatus. At the moment of pulse magnet influence, the chamber with the preparation was switched off the stimulator and amplifier. Besides it excluded the nerve stimulation by pulsed electric framing on the cyclic conductor–electrode–wire–device input resistance–wire–electrode–nerve.

During the experiment, electric current pulses of increasing amplitude caused peak response of the nerve. Then, stimulus amplitude was decreased until AP appeared, the value of which was half-peaked. By this, some part of nerve fibers in the trunk remained spare. By this, the possibility of the recording "stimulating" and "inhibited" action of pulsed magnetic field on the nerve state was obtained.

After this, the nerve was stimulated by 1 Hz-frequency electric current. Meanwhile, response amplitude value was fixed. The information was continuously recorded during 1–3 h except for the periods of the influence of powerful pulsed magnetic field. The parameters were the following: Field intensity was 25 kE, pulse duration was 0.5 ms, pulse repetition frequency was 1 Hz, and there were 50 pulses.

The obtained results were processed statistically. For the study of the reliability of average difference, we calculated average value (x) with confidence interval (d) on 5% significance level using the following formula: $x \pm d$, where $d = t_{0.005} m_{\bar{x}}$, and $m_{\bar{x}} = \sigma / \sqrt{n-1}$.

In the case of confidence interval overlapping, we considered average value differences doubtful. The average value was calculated every 5 s by 10–15 AP amplitude values and also directly after the influence of pulsed magnetic field. The results of the processing are shown in Table 2.1.

At all, 23 frog sciatic nerves were studied. Each preparation was used for 10–15 measurements of the influence of powerful pulsed magnetic field. The experimental data allow detecting true amplitude changes of nerve AP under the influence of a magnetic field. The results of the second experiment (Table 2.1) reveal the most character pattern of AP amplitude changes under the influence of the pulsed magnetic field. This reaction has a complex time-multi-directional structure.

It is seen from Table 2.1 that the first two influences of pulsed magnetic field lead to the true AP amplitude increase, the third one does not influence the nerve functional state index under consideration, and the following influences (5 tests) cause true decrease of amplitude, which in 2–3 min recovers reference quantity to the value before "inhibitory" action.

One more interesting peculiarity of the influence of a powerful pulsed magnetic field on the indexes under study should be mentioned. In experiment No. 14, we observed spontaneous AP amplitude incidence caused by the damage of the nerve trunk in the process of separation. Along with the progressive decrease of AP value, there was noticed strongly expressed "stimulating" influence of pulsed magnetic field. The essence of this is that the measuring parameter truly increases by 25–40% directly after the influence.

Thus, the following peculiarities can be marked of the influence of powerful pulsed magnetic field on the functional state of frog isolated sciatic nerve:

1. The influence of pulsed magnetic field leads to the true increase of AP amplitude caused by electric stimulation.
2. Time-consecutive stimulations by pulsed magnetic field cause multi-directional changes of AP value: At first, we observed "stimulating" action changed into "inhibitory" action (it was observed in 89% experiments). One of the experiments (No. 14) allows predetermining the modeling influence of the nerve functional state on the value of the response.

Thus, it should be noticed that using pulsed magnetic field in clinical practice, there must be considered the detected dose-dependent effect of the changes in the

Table 2.1 Average value and confidence interval of AP amplitude after the amplification in volts before (a), directly after the influence of pulsed magnetic field (b) and in some time after this influence (c)

Experiment no.	Influence no.	Time after experiment beginning (min)	Average value	Confidence interval	
				Maximum value	Minimum value
2	1	a 35	1.702	1.713	1.690
		b 40	1.780	1.787	1.773
		c 45	1.815	1.827	1.803
	2	a 85	2.117	2.122	1.112
		b 888	1.147	2.155	2.139
		c 90	2.139	2.154	2.124
	3	a 95	2.156	2.162	2.150
		b 98	2.178	2.186	2.171
		c 100	2.172	2.191	2.153
	4	a 105	2.185	2.200	2.170
		b 107	2.190	2.204	2.176
		c 110	2.214	2.225	2.203
	5	a 120	2.221	2.229	2.213
		b 122	2.202	2.206	2.197
		c 125	2.232	2.239	2.225
	6	a 130	2.235	2.248	2.222
		b 134	2.184	2.193	2.174
		c 135	2.193	2.201	2.185
	7	a 140	2.223	2.232	2.214
		b 143	2.151	2.155	2.147
		c 135	2.231	2.235	2.227
	8	a 150	2.251	2.261	2.241
		b 152	2.158	2.162	2.154
		c 156	2.265	2.227	2.254
	9	a 160	2.286	2.297	2.276
		b 162	2.245	2.251	2.238
		c 165	2.296	2.301	2.139
14	1	a 28	0.776	0.795	0.757
		b 31	0.723	0.741	0.705
		c 33	0.671	0.681	0.660
	2	a 40	0.385	0.419	0.351
		b 43	0.469	0.508	0.429
		c 46	0.263	0.300	0.226

(continued)

Table 2.1 (continued)

Experiment no.	Influence no.	Time after experiment beginning (min)	Average value	Confidence interval	
				Maximum value	Minimum value
	3	a 50	0.145	0.176	0.114
		b 53	0.282	0.319	0.245
		c 55	0.130	0.159	0.101

studied structure-functional state as well as their initial functional state. Moreover, we should remember about the influence of the inductive framing on the input stage of diagnostic equipment.

2.2.7 Holographic Interference Microscopy in the Study of Refraction Characteristics of Nerve Fibers (Nerve Fiber Is an Optical Waveguard)

Our scientific interest to the study of optical characteristics of living tissue, in particular neuromuscular, was attracted by the necessity of understanding low-intensity laser irradiation action mechanisms influence on acupuncture points during the treatment of the patients by the method of laser acupuncture with the purpose of the development of modern approaches and methods of treatment and diagnostics of the peripheral nervous system diseases. In connection with this, main efforts were aimed at experiments on the study of optical characteristics of nerve and muscle fibers, nerve fibers in the nerve trunk, muscles, lymphocytes and so on. More appropriate for the stated purposes were the works [324–327]. So, in particular, in the work [324], there were studied optical characteristics using huge nerve fibers of squids in polarized light: squids Liligo vulgaris with the diameter from 250 to 700 μm and Ommatostrephes sagittatus and cuttlefish Sepia officinalis with the diameter of 150–350 μm.

Time delay system allows connecting the observed optical changes in the axon with development phases in action potential time. It is founded that the axon is an optically anisotropic body with the main direction, which coincides with its axis. During the rotation by 90° in the plane perpendicular to the light beam, we can observe four extinction angles and four the brightest angles. A birefringence phenomenon in the huge axon was detected long ago [328], and during the excitation, there were no changes detected in birefringence pattern.

In the result of experiments in the work [324], the author showed that birefringence presents in the axon even after the loss of its excitation and disappears in 20–40 min (the axon becomes black in the field of crossed prisms), apparently, under biological molecule denaturation, which was a part of axoplasm or during the damage of its specific structure. Under axon, rough mechanic damage birefringence in the point of fault disappears, and it becomes dark. The front of the transition from anisotropic state into isotropic moves from the point of fault at a velocity of 20–40 μ/min that could be seen in the field of crossed nicoles as the darkening distribution. On undamaged functioning axons if observed them at the angle close to the brightest, there was seen central dark longitudinal fringe along the whole length of the axon. Huge axon in the nerve trunk is also clearly seen in the field of crossed prisms, and the central dark fringe is seen, and its presence indicates good state of the axon.

At the angles close to the extinction angles (±5°–7°), there were observed frequent narrow dark transverse fringes with the width of 15–30 μm divided by light space of 20–40 μm. The bigger the axon diameter is, the narrower are the fringes and the less is the distance between them. If the axon is damaged, they immediately disappear including its conductivity.

Also, the work [325] should be mentioned, which is dedicated to the optical studies of the changes in nerve membrane structure during pulsing. Earlier, in the work [325] using microinterferometer, there was observed volume-elastic wave in the nerve fiber, which was synchronous to stimulation pulses, but the attempt to simulate the process failed [329, 330].

Later, the true changes in polarized light of optical characteristics of the nerve fiber during action potential have been received [330]. It also became known about the experiments on huge squid axons [331], and due to this, we could see the changes directly in the membrane of the nerve fiber.

The results of the experiments in the work [325] using huge lobster axon with the diameter of 60 μm ($\Delta W/W = 1.3 \times 10^{-4}$, W is the luminous flux and ΔW is its changes) give birefringence increase by 1.5×10^{-4}; the results received using huge squid axon with the diameter of 1 mm ($\Delta W/W = 7 \times 10^{-6}$) [331] give birefringence increase by 1.4×10^{-4}. Reference quantity of birefringence according to the information of different authors is about 10^{-2}–10^{-3}. Fractional increase of membrane birefringence in all cases is about 1.5–15%.

Optical studies of the nerve fiber allow concluding that during potential action generation the change of phase difference between ordinary and extraordinary rays transmitted through nerve fiber of polarized light occurs. Such effect is provided by the change of membrane optical characteristics of the nerve fiber. With a high degree of confidence, it should be noted that it is the result of birefringence increase in membrane (in soldering peak by the value about 1.5×10^{-4}). From the experiments on phospholipidic membranes, the birefringence changes in the living membrane can be explained by the change of its structure during the transmission under electric current of liposoluble ions but not by direct influence of the field on lipid or protein membrane parts.

An important place in organism vital activity is held by optical radiation. Being an integral factor of the environment, light for living organisms is the energy source (photosynthesis) of remote information about the environment, one of exchange reaction product (bioluminescence). In the work [327, 332], biological significance of this last radiation of remote intercellular interactions is shown. And with it, the mechanism of optical radiation distribution in biological tissue (in this case—it is nerve tissue) remains unexplained at a cellular level.

In the present work, the holographic study of refraction structure of the living nerve fibers is carried out. In particular, there were measured radial distribution of Schwann's sheath refractive index, the axon of myelinated nerve fiber and also myelinated nerve fibers.

Aging preparations of single myelinated nerve fibers (the diameter is 8–10 μm at the length λ of the dissected area of 5–7 mm) taken from the brown frog sciatic nerve and nerve cells appendices grown in the sympathetic ganglion dissociated culture served as the object of the study [333]. The laser-holographic block-scheme of the setup for the study of structure-functional characteristics of neurobiological objects is presented in Fig. 2.14. The setup designed for these purposes consists of two parts: a holographic one and an electrophysiological one.

The holographic part includes the source of radiation—continuous wave laser LG-38 (50 mW) in many cases containing a stroboscope, an adjustment laser LG-56, holographic microscopes operating for transmission and in the reflected light (Fig. 2.15), with the help of which studies of phase and diffusing microobjects altered their state were conducted, as well as a real-time hologram processing equipment. This equipment allows obtaining holograms and holographic interferograms in different actuation phases of nerve fibers using the time-elapsed method of real-time holographic interferogram recording, including differential holographic interferometry, and the stroboscopic one.

1. The double-exposed method of microobjects holographic interferogram recording consists in consecutive recording of two microobject states on one detecting medium: the initial and the modified one, for example, with stimulation electric or laser influence.
2. Term "real time interferometry" means that the interferometric pattern alters simultaneously with the object state alteration. In this case, the hologram is exposed one time, is developed and fixed on the same place, where it was during the survey or is processed directly on the detecting point.
3. Stroboholography is holographic repeated processes research methods when holograms are exposed in the light of following light pulses, which are synchronized with a certain process phase.

The holographic interference microscope has the following characteristics:

Size of registered objects	From 1 μm up to 3 mm

Hologram registration time:

Working with a continuous wave laser LG-38	5×10^{-2}:30 s
Working with a pulse ruby laser	3×10^{-8} c

Real-time hologram detection time using strobe

Pulses with off duty ratio 100	1/100 s

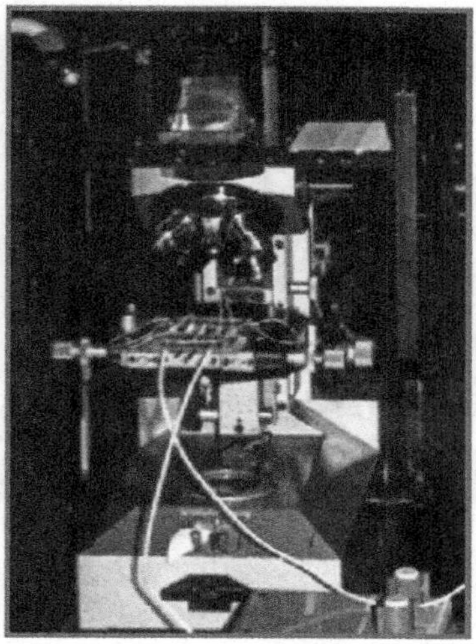

a

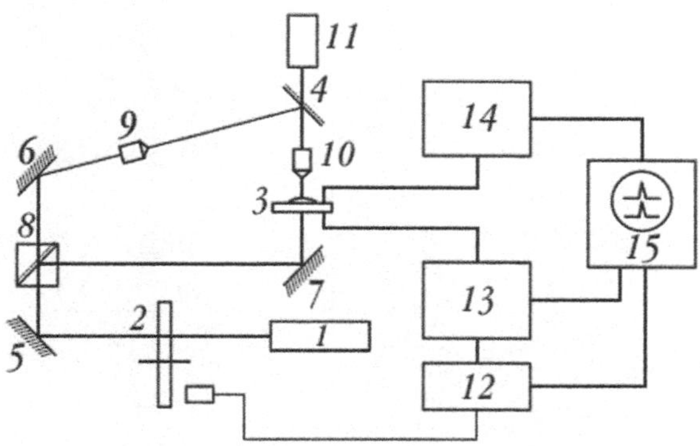

b

Fig. 2.14 Photo (**a**) and the installation diagram (**b**) for holographic study of the nerve fiber in the rhythmical stimulation mode: 1—the laser; 2—the mechanical stroboscope; 3—the nerve fiber chamber; 4—the photoplate; 5, 6, 7—the deflecting mirrors; 8—the beam splitter; 9, 10—the microscope objectives; 11—the eyepiece; 12—the delay unit; 13—the electrostimulator; 14—the bioelectric-potential amplifier; 15—the biradiated oscillograph. Reprinted from [136] with permission

Maximum strobe frequency	5 kHz
Maximum magnification	1800×

Minimum time interval between exposures

In double-exposed interferometry 1.8×10^{-4} s

Wavelengths used

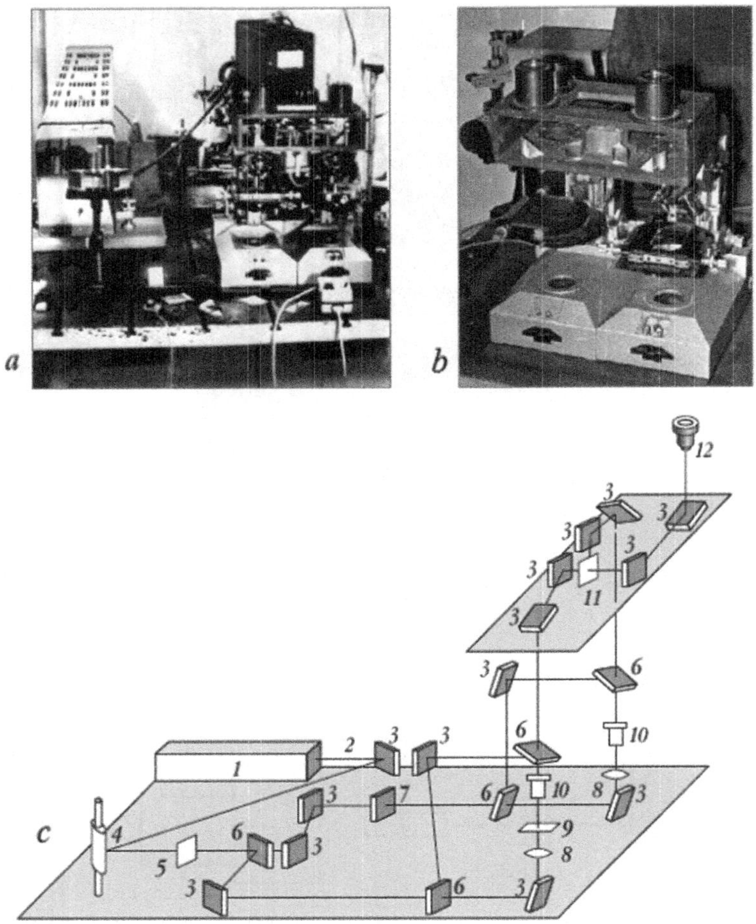

Fig. 2.15 Photo (**a**, **b**) and the diagram (**c**) of a holographic interference microscope for the work in transmitted and incident light: *1*—the laser; *2*—the laser ray; *3*—the non-transmitting mirror; *4*—the electromechanical modulator; *5*—the diaphragm; *6*—the semireflecting mirror; *7*—the quarter-wave plate; *8*—the condenser; *9*—the object stage; *10*—the objective; *11*—the hologram; *12*—the ocular. Reprinted from [136] with permission

LG-38-He–Ne laser	632.8 nm
"Igla 4M"—argon laser	514.5 nm or 472.2 nm
Ruby laser	694.3 nm

Output power:

LG-38	40 mW
Ruby laser radiant energy	0.4–1.5 J
LG-38 single-mode operation coherence length	20 cm

Table 2.2 Photographic plates characteristics

Photographic plate	Sensitivity (erg/cm^2)	Resolution (lines/mm)	Contrast rate attenuation	Emulsion layer thickness (mm)
PE-2	2000	5000	4.5	12–15
LOI-2	4000	5000	–	17

For hologram recording, photographic plates PE-2 and LOI-2 are used. Their characteristics are presented in Table 2.2.

A special attachment was designed for the microscope. It allows implementing the following methods in the study: the colliding beams method [228–230], method with inclined reference beam [231–234], the focused image method [235–238], etc. In this case, holographing of the image, not of the microobject, takes place. In the microscope optical scheme, standard microlenses of $10\times, 40\times, 90\times$ (oil immersion) and eyepieces of $5\times, 7\times, 15\times, 20\times$ were used that allowed modifying magnification in the range of $50\times$–$1800\times$. Hologram exposure time depended on the magnification of the lenses used on the recording stage and was for lenses $10\times$–0.5 s, $40\times$ –6 s, $90\times$–20 s. The illumination of the object with a focused laser beam allowed essentially enhancing object beam brightness. Meanwhile, hologram exposure time was 3–5 s (lens $90\times$). Interfering beams optimum ratio of 1:1 with different magnifications was reached by reference beam attenuation using neutral wedge filters and polaroids. Hologram exposure time was controlled by the electromechanical shutter. Photographic plates LOI-2, PE-2 were used for hologram recording. After exposure, the photographic plates were processed with a developer GP-2 at the temperature of 29 ± 0.5 °C, washed in distilled water and fixed. While conducting real-time study, the photographic plates were processed on the recording place with a special cell that provided stability of their initial state. Developing time is 8 min. The interferograms were registered on the film KN-2. The holographic setup was mounted on a plate that allows eliminating vibrations influence with an air bag.

The electrophysiological method [292, 334] based on electric potentials registration is one of traditional methods for studying such an excitable structure as nerve tissue (a fiber, a cell, etc.). Being the activity index of a nerve fiber or a cell, these potentials bear information only about ion fluxes alteration and do not give data about the structure of fibers and cells. In these experiments, the electrophysiological part is dedicated to nerve fiber activity control and contains an electrical stimulator ESU-2, a special camera to fix the fiber, control oscillographs, a biopotential amplifier.

The camera (Fig. 2.16a) is a device, which provides fixation of the object under study, supply of perfusion solution to it, and has an optical surface, through which laser sensing of nerve fiber is operated. The camera construction presupposed nerve fiber stimulation and potential action leads (Fig. 2.16b) according to Tasaki air bridges method. For scanning a piezoceramic element, device was produced. It is placed on the microscope sample stage, which allowed shifting the preparation at the distance up to 30 μm with controllable pitch (from 0.5 μm and more).

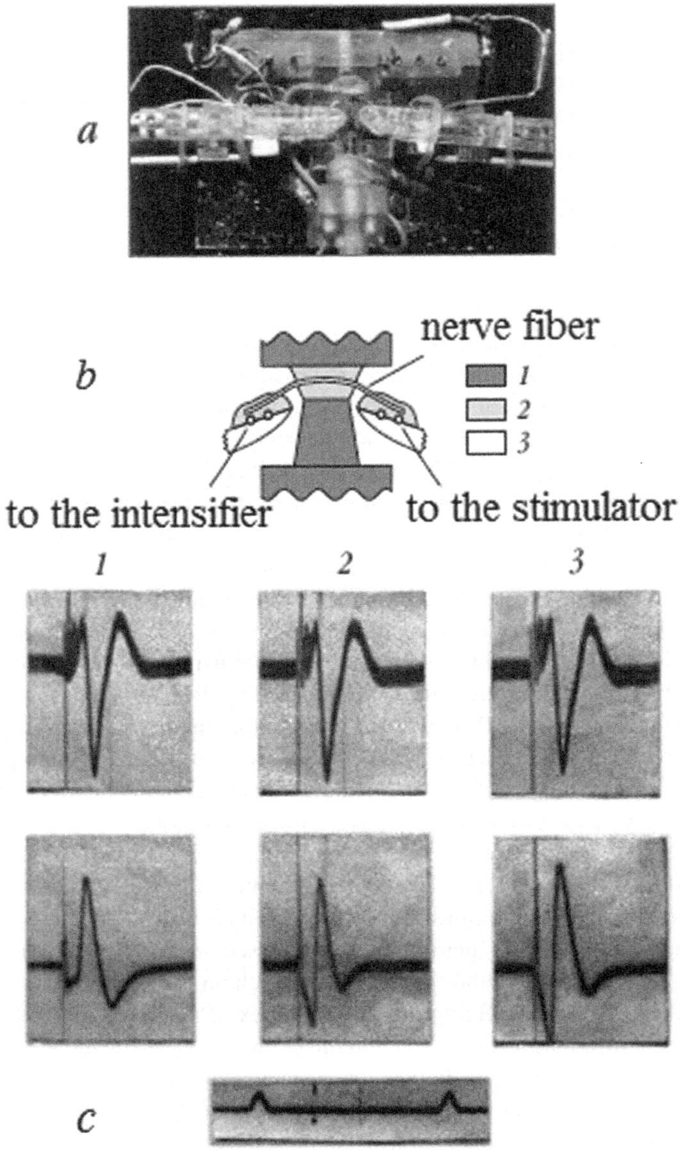

Fig. 2.16 Camera providing the record of holograms and holographic interferograms of nerve fiber with electrophysiological control. Reprinted from [136] with permission

Surviving preparation of a myelinated nerve fiber was made by splitting of a frog sciatic nerve with tungsten needles. Depending on the experiment requirements, the fiber was put into Ringer physiological solution ($n = 1.3337$), silicon ($n = 1.4057$) and vaseline ($n = 1.4805$) oil. The functional state of the nerve fiber, used in the experiments, was estimated according to its ability of stable potential generation in response to electric irritation produced by right-angled current pulses with the duration of 0.1 ms and frequency of 1 Hz. When double-exposed holographic interferograms in bands with infinite width were ready, the first exposure corresponded to preparation addition into the object beam. During the second exposure, the object was absent. Interferograms in infinite width bands were formed by glass wedge injection of 3° and 5° into the object beam (band period was 0.4–0.2 μm). For differential interferogram recording, two stages of the same object separated by some time interval were recorded on one hologram.

To study structure nerve fiber alterations in rhythmic stimulation regime, the stroboholographic method was used (see Fig. 2.14b). In this case, a stroboscope with a system of synchronization was placed between the continuous wave laser and the beam splitter. The character of hologram recording is accumulative. For microobjects like nerve fiber, the properties of which are reversible for excitation frequencies of 100 Hz, it is possible to record holograms with a set of laser synchronized pulses of low intensiveness with a certain excitation phase.

To register holographic interferograms in different excitation phases, a stroboscope with a synchronizing system was used, which included a delay unit. Laser radiation after the stroboscope had the duration of 1 ms with the frequency of 100 Hz. A delay line was launched simultaneously with lightning the object through a photodiode. The form and the duration of the beam pulse were controlled by photodiode FA-27A and double-beam oscillograph 15 (see Fig. 2.14c). The delayed pulse controlled the electrical stimulator in such a way that the action potential took place in the microscope visual field simultaneously with the hologram recording beam. The electrical stimulator occasionally sent a stimulating pike-shaped pulse onto the nerve fiber.

Delay time alterations allow scanning excited nerve fiber phases. To control synchronization, the nerve pulse with the help of microelectrodes and the amplifier as well as the photodiode signal was shifted to the oscillograph. Nevertheless, holographic interferograms for detection of certain nerve fiber excitation phases did not allow detecting obvious structure-functional alterations in it (the desired signal was within the measurement margin of error). Experiment conditions optimization and interferogram recording according to the maximal phase incursion on the nerve fiber can be implemented using an automated system, which can help to reduce laboriousness of interferogram processing and deciphering, enhances accuracy and gaging speed, specifically nerve fiber refractive index distribution, nerve fiber deformation during excitation pulse propagation, neuron pulsation, etc. Increase of interference bands quality (with contrast and possibility of band center detection) can be reached using coherent noises averaging device.

Real-time hologram recoding was the most effective during this stage of the study (see Fig. 2.15). Lens 90× (oil immersion) was used for image recording of the object beam nerve fiber. In this case, initially on a hologram, a comparison wave was recorded which went through all the elements of the optical scheme without the object. Then, the nerve fiber was placed and was lighted up with a plane wave, which transformed into a spherical one behind the lens. As a result of interference of the recovered reconstructed spherical wave and wave having phase perturbation caused by the object, an interference fringes system occurs. To produce finite width bands, the incidence angle of the reconstructed wave on the hologram was modified. Taking into account the fact that the distance between the centers of the spherical waves is much smaller than their curvature radius in the interference region, then the bands can be considered as almost straight.

If coordinate system is placed in such a way that z is an optical axis of the holographic interference microscope, and y is the object symmetry axis, then a hologram turning around x-axis by a small angle shifts the center of the reconstructed spherical wave along y-axis.

In this case, the dark band equation is the following [335]

$$\frac{\delta(x, y)}{2\pi} = \frac{y - y_m}{T}, \tag{2.26}$$

where $\delta(x, y)$ is the phase difference between the object wave and the comparison wave; T is the interference pattern period; $y - y_m$ is the m-order interference fringe deviation.

The phase perturbation $\delta(x, y)$ in the plane directly behind the fiber is the following

$$\delta(x, y) = \delta(x) = 2k \int\limits_{0}^{\sqrt{1-x^2}} \Delta n(x, z) \mathrm{d}z, \tag{2.27}$$

where $k = 2\pi/\lambda$, λ is the wavelength.

The object is supposed to have axial symmetry and be lightened up by the plane wave. A nerve fiber with a good degree of accuracy can be estimated as an axially symmetric object.

On the interferogram, it is very easy to identify the fiber edge and its other areas using the coordinates of its image for correspondence determination. In any plane $z = z_p$

$$\delta(x_p) = \delta\left(-\frac{f}{z_p}x\right).$$

The multiplier $-\frac{f}{z_p}$ is the scale change in the observation plane $z = z_p$ concerning the object plane. Normalizing the coordinates in the observation field in such a way that the image radius was equal to 1, and making the replacement of the variable $x = r\cos\varphi$, $z = r\sin\varphi$ in (2.27) and considering the fact that $\Delta n = \Delta n(r)$ and $\Delta n(r) = n(r) - n_0$ (where $n(r)$ is the nerve refraction index; n_0 is the physiological solution refraction index), (2.26) can be written as follows:

$$\frac{2R_0}{\lambda}\int_x^1 \frac{\Delta n(r) r \mathrm{d}r}{\sqrt{r^2 - x^2}} = \frac{y - y_m}{T} \tag{2.28}$$

The equation is known as the Abel equation (R_0 is the fiber radius).
Formally, the equation converse is

$$\Delta n(r) = -\frac{\lambda}{\pi T R_0}\int_r^1 \frac{(y - y_m)'_x \mathrm{d}x}{\sqrt{x^2 - r^2}} \tag{2.29}$$

In real conditions, $y - y_m$ deviation is measured with an error, which can be both systematic and occasional. Then, derivative $(y - y_m)'_x$ may not exist. In practice, the solution calculated according to (2.29) has a strongly oscillating character. Nevertheless, a regularized decision of this incorrectly set task can be found, if smoothing cubic splines are applied for experiment data approximation. The experiment data approximation task $(y - y_m)$ consists in making function $S(x)$, which approximates function $y - y_m$ according to the initial data. The smoothing cubic spline $S_{n,\alpha}(x)$ is defined as the solution of variational task $\Phi_\alpha\left[S_{n,\alpha}(x)\right] = \inf \Phi_\alpha[S(x)]$, $S(x) \in C^2_{[0,1]}$, where

$$\Phi_\alpha = \alpha\int_0^1 \left[S''(x)\right]^2 \mathrm{d}x + \sum_{i=1}^n p_i^{-1}\left[\left(\widetilde{y - y_m}\right) - S(x_i)\right]^2 \tag{2.30}$$

and depends on the smoothing parameter α.

To select the smoothing parameter, the criterion of optical approximation of experimental data was used obtained in the work [147]. The derivative $S_{n,\alpha}{}'$ is continuous and in this case calculations

$$\int_{x_i}^{x_{i+1}} (x - x_i)^l \left(x_i^2 - r^2\right)^{-1/2} \mathrm{d}x, \tag{2.31}$$

where x_i is the deviation coordinate $\left(\widetilde{y - y_m}\right)_i$, $i = 0, \ldots, p - 1$ (p is the number of measurements conducted); $l = 0, 1, 2$.

Such a scheme does not use quadrature formulae, which bring an additional error at the Abel equation converse [336, 337]. We implemented such an algorithm as a subprogram package written in the Fortran-IV language. The interferograms obtained were measured photometrically with "Leitz" microphotometer MPV-2 and processed with BESM-6 computer. Model function reconstruction is made with the accuracy up to 10^{-3}.

To estimate the error, there was conducted the study of refraction index distribution of a glass capillary, filled with vaseline oil ($R_0 = 75$ μm). The data obtained correspond to the known ones with the accuracy up to 10^{-3}, but it is worth pointing out that the calculation of the same interference pattern for objects with the radius of 5 μm will reduce accuracy up to ~1.5×10^{-2}. To enhance the sensitivity of measurements using the real-time method, a device has been developed containing a photomultiplier with an analog integrator on precision containers, which, when located in the recording plane of holographic interferograms, leads to the accumulation of a useful signal on a set of synchronized with a certain phase excitation pulses. The studies of the refractive index of nerve fibers in in vivo state were carried out for the first time.

The interferograms obtained in the rhythmic stimulation mode of the internodal segments of nerve fibers and the results of their interpretation are presented in Fig. 2.17. Recovered images of nerve fibers were observed with magnification up to 1800×. The refraction index radial distribution study of both myelinated and non-myelinated nerve fiber in the intravital state was conducted. The results are obtained from different myelinated fibers and presented in Fig. 2.18.

The character of the dependence $n(r)$ of the myelinated nerve fiber internodal area has analogy to tubular optical waveguide profile [338] that allows assuming waveguide characteristics of the myelin sheath and probably of the axon itself. The data obtained can become the presupposition for detecting remote intercell interaction mechanism through radiation. In connection with this, it is important to refer to the work [339], which shows that light is conducted through the stretched cells from some parts of a plant to another, and the information transmitted in such a way is of great importance in the development and controlling of all vital functions of a plant. In our case, an experimentally obtained myelin sheath refraction index that is rather high for biological structures (1.44 ± 0.01) matches to the morphological data of a protein-lipoid membrane dense packing of this structure. As is seen in Fig. 2.19, the axoplasm refraction index on a fiber axis is higher, then in the areas directly under the myelin sheath. Thus, using the holographic interferometry method, the myelinated nerve fiber optical characteristics study was conducted. The fiber interferograms with a diameter of 10 μm with the spatial expansion of more than 1 μm were obtained. Nerve fiber refraction index profiles were made (the average axon refraction index is 1.35, the myelin sheath is 1.44). The latter points out the presence of waveguide

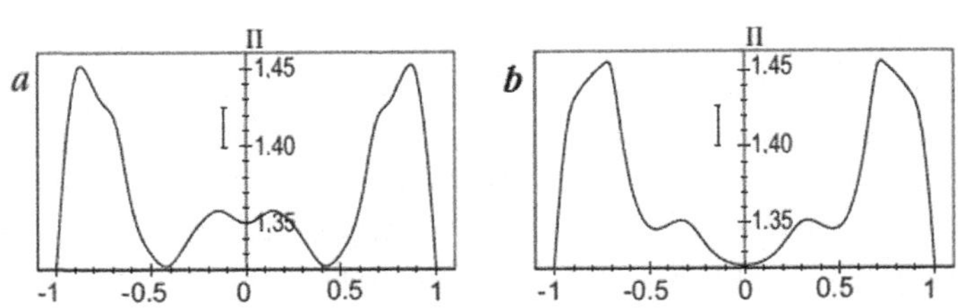

Fig. 2.17 Profiles of refraction index of myelinated nerve fibers, defined by the methods of holographic interferometry. Reprinted from [136] with permission

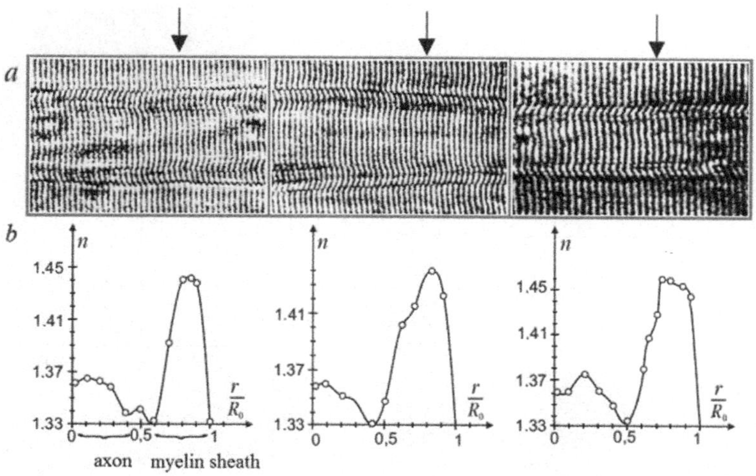

Fig. 2.18 Interferograms (**a**) and calculated on them refraction indices of myelinated nerve fibers (**b**). *Arrow signs show* sections for which measurement was conducted. Reprinted from [136] with permission

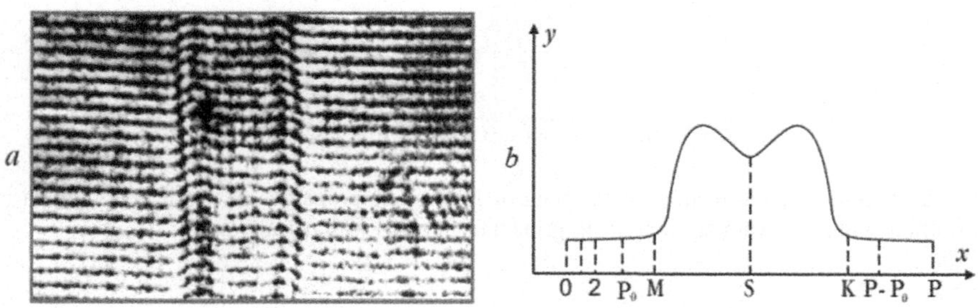

Fig. 2.19 Interferogram of nerve fiber (**a**), separate interference fringe for the detection of radial allocation of nerve fiber refraction indicator (**b**). Reprinted from [136] with permission

characteristics of the myelin sheath and probably of the axis cylinder (the axon) [5, 8–10]. The profile of the myelinated nerve fiber refraction index is determined as well.

To compare the obtained results, the research was conducted using the focusing method [340], applied for light guide studying. Nerve cells appendices grown in the sympathetic ganglion dissociated culture were the object of research [10, 333].

The adaptation of the focusing method to such a unique object is connected with a set of difficulties: small sizes of the fiber (1–5 μm), the necessity immersion selection. The immersion liquid had a refraction index equal to the refraction index at the edge of the fiber. The immersion selection was fulfilled by changing physiological solutions with n_2 = 1.334–1.380 in a simple camera (Fig. 2.20a) where the preparation was situated. The preliminary investigations showed the absence of substantial nerve fiber

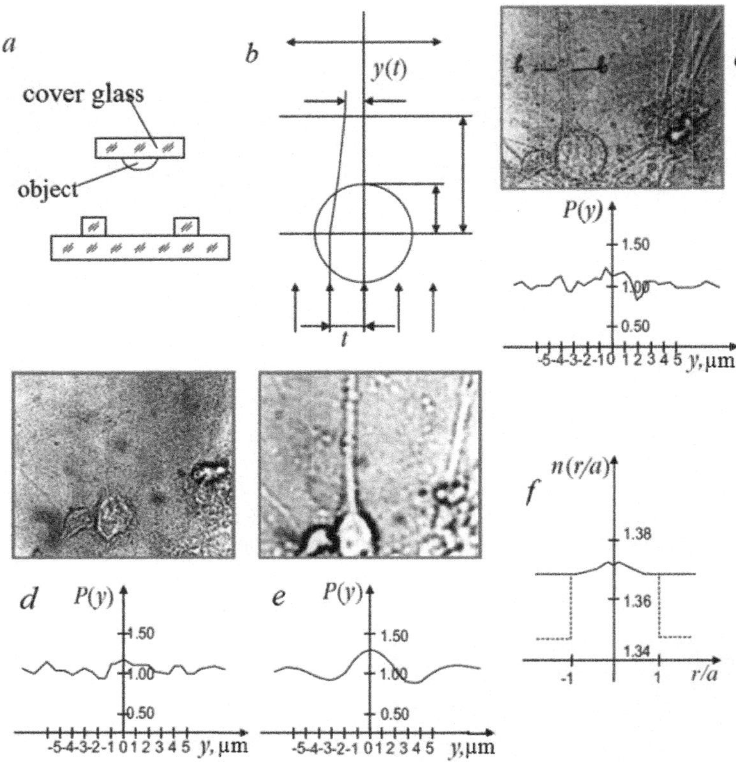

Fig. 2.20 Study by the method of focusing of light guide properties of nerve fibers grown in distortion culture of synthetic ganglia. Reprinted from [136] with permission

absorption (less than 6%). If the refraction index has radial dependence $n = n(r)$, it will work as a cylindrical lens. By measuring intensity distribution $P(y)$ in the plane behind the fiber (Fig. 2.20c), refraction index profile can be calculated as

$$n(r) - n_2 = \frac{n_2}{\pi L} \int_0^{\infty} \frac{t - y(t)}{\sqrt{t^2 - r^2}} \mathrm{d}t. \tag{2.32}$$

Function $y(t)$ is found as the inverse to the function

$$t(y) = \int_o^y P(y')\mathrm{d}y'. \tag{2.33}$$

The immersion selection accuracy is 5×10^{-3}.

The photo of the cell in the physiological solution with the refraction index of 1.334 is shown in Fig. 2.20c, and in Fig. 2.20d—in the medium $n_2 = 1.370$, where the fiber is less visible. Observations in plane $L = 20\ \mu m$ have shown that the nerve fiber appendix (Fig. 2.20c) has focusing characteristics. The intensity distribution along the line b-b' is shown in the lower part of Fig. 2.20c–e. The corresponding refraction index distribution is shown in Fig. 2.20f.

For estimating medium refraction characteristics surrounding the fiber rat serum, refraction index $n = 1.3484$ was measured as a possible intercellular fluid analog. Having insignificant absorption and a quasi-stepped refraction index, the myelinated fiber can be presented as a waveguide characterized by a dimensionless parameter

$$V = \frac{2\pi}{\lambda} a \sqrt{n_1^2 - n_2^2} \approx 6.3, \tag{2.34}$$

where n_1 is the refraction index on the fiber *axis*, n_2 is the immersion refraction index, λ is the wavelength of the radiation used.

Theoretical and experimental researches such as study of nerve fiber intravital states at rest and during stimulation (1978) [1–3] and later research of refraction characteristics of myelinated and non-myelinated nerve fibers (1982) [5, 8–10] aimed at searching for light guide characteristics of neuromuscular tissue were conducted on animals by L. V. Tanin and his colleagues. During these studies, the conclusion was made that the character of refraction index dependence on the radius of internodal part of the meddulated nerve fiber is analogous to the refraction index profile change of separate optical waveguides.

It is interesting to point out that regardless of our studies in 1983 after a series of experiments, biologist D. Mandoli, a scientist in the botanic laboratory of Carnegie Institute (Stanford, California), has established the fact that plants have optical waveguide characteristics, i.e., the presence of a waveguide mechanism of light spreading in plant tissues [339, 341]. But that concerns only flora.

What do a golden gram shoots and telephone fiber-optic cable have in common?

It appeared that the shoot transmits and deflects light in the same way as multi-fiber optical beams, used in international communications.

In the work [341], it was found out that when a narrow-band beam of laser radiation was sent to the one end of the grain arched segment (oat or golden gram), then the light went at least for an inch inside to the other end (Fig. 2.21a, b). It is based on a simple phenomenon known as the total internal reflection: The light going through the plant cell at a certain angle is reflected from the one end to the other going through the whole length of the cell in a zigzag way like in the optical fiber. Despite the fact that plants are not so effective during light transmission as the optical fiber, D. Mandoli points out that this characteristic can play an important role in growth and development of a plant.

During another experiment, she has found out that when light was directed to one side of the coleoptile's apex (first germinal blade), more light went to the dark side not to the illuminated one. Reacting to red illumination of separated areas along the undamaged etiolated oat germ indicates that illumination of the area around a coleoptiles node leads to coleoptile maximum growth stimulation and mesocotyle growth suppression. The quantitative definition of fiber optical characteristics of

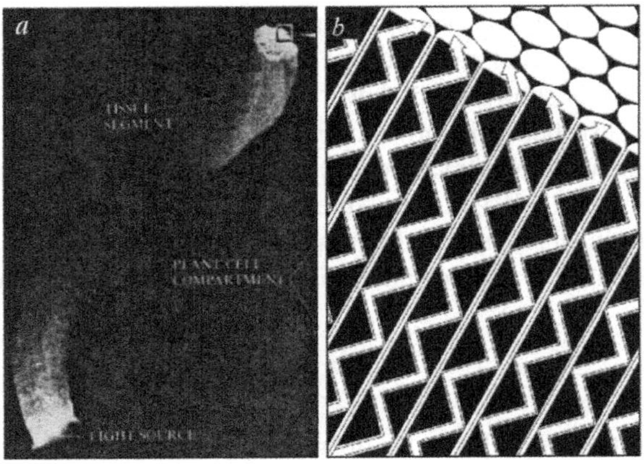

Fig. 2.21 To the question of optical properties of plant fiber (**a**, **b**). Reprinted from [136] with permission

these etiolated tissues has showed that the quantity of paraxial transmitted light is in logarithmic-linear dependence on the distance both for the mesocycle and for the coleoptiles (plus the primary leaf). D. Mandoli assumed that the lighted cells illuminate light from inside directing it around the coleoptile into photosensitive areas on the dark side. The optical characteristics of the germ fiber can potentially allow the germ to intensify significantly the effective light signal received by the photosensitive area. This in turn stimulates faster cell growth on the dark side and turns the plant to the light in such a way. The defined properties of a plant as an optical waveguide influenced greatly the conducted experiments and will influence the future ones, as what was traditionally considered as internal glow in a shoot cannot be localized, but go in and out the plant or spread in the plant as a whole. As a result, the conclusion can be made that plants have a complex optical system, probably not less complex than a human does [341]. The waveguide mechanism of light propagation is found in photoreceptors [342].

2.2.8 To the Question of Studying the Processes of Muscle Contraction (The Muscle Fiber as a High-Performance Diffraction Grating)

Interest to the question of contraction, which is one of the forms of biological movement, is first of all caused by the fact that it is not exactly known, which way muscles really contract. It has been ascertained that the mechanisms implementing different forms of movement consist of tissue proteins myosin and actin, and ATP is the energy source for all these processes. Meanwhile, there is a number of ideas, including the theoretical assumption of academician A. S. Davydov (Institute of Theoretical Physics, Kiev) about the existence of solitons that allows understanding and explaining the process of muscle contraction. The soliton mechanism was demonstrated very convincing in work (A. S. Davydov "Solitons, Bioenergetics, and the Mechanism of Muscle Contraction", Institute of Theoretical Physics, Academy of Sciences of the Ukrainian S. S. R., Kiev, USSR. Int. J. of Quantum Chemistry. 1979. Vol. 16. P. 5–17).

However, let us consider the main known principles of functioning of the contractile apparatus. The nerve fiber contractile apparatus, as a rule, consists of ~10^4 consequently connected identical elements—sarcomeres, each of them contains 10^6 thick and thin filaments. The force developed by the sarcomeres at a constant length is proportional to the number of bridges in the overlap area of the thick and thin filaments.

Closed bridges are the source of the muscle contractile force. From discreteness of acceptors and their filaments incompressibility, it follows that while contracting every bridge must work cyclically: close, develop moving force at a certain interval of the muscle relative movement and then open, but closing and opening of the myosin bridge cannot be considered as elementary chemical acts. It can be assumed that in an excited muscle each of these processes is characterized by the effective rate constant, which depends on the relative position of myosin and actin active centers. This allows applying the kinetic approach to describe muscle contractions [343]. The force recorded at both ends of the muscle fiber is equal to the force developed in any half of any sarcomere. The rate of fiber shortening $v = 2\,N\upsilon$, where N is the number of sarcomeres, and υ is the rate of thick and thin filaments relative movement. The myosin bridges function independently. That means that bridge opening is subjected to monomolecular kinetics. Under steric limitations, only a free active center can be situated near a free bridge, so bridge closing is also subjected to the monomolecular kinetics. Mathematical formulations of muscle contraction are presented in the work [344].

One of the important concepts in researching biological mobility problems is also cross-striated muscle structure studying.

Using optical microscopy for this purpose appeared to be impossible because focusing conditions alter under large deformation amplitudes during contraction. Moreover, the light microscope spatial resolution is not enough, as the size of the structural elements of this object is commensurable to the length of the light wave.

A muscle fiber is a cylinder with the diameter of about 50–100 μm. Most of its volume is filled with myofibrillas which are longitudinal structures with the diameter of 1–2 μm. In the myofibrillas from the adjacent Z plates, two systems of thin filaments move toward each other. Periodical recurrence of such a structure provides fiber cross-striation. In this case, the muscle fiber can be considered as a diffraction grating with the period equal to the length of the sarcomere. The general absorption of the muscle in the visible part of the spectrum does not exceed 1.4×10^{-4} for 1 μm of the fiber thickness, and differences in structural elements absorption will be lower than 10 that makes it possible to neglect the contribution of the proper absorption to diffraction grating creation. Nevertheless, the muscle fiber cannot be considered

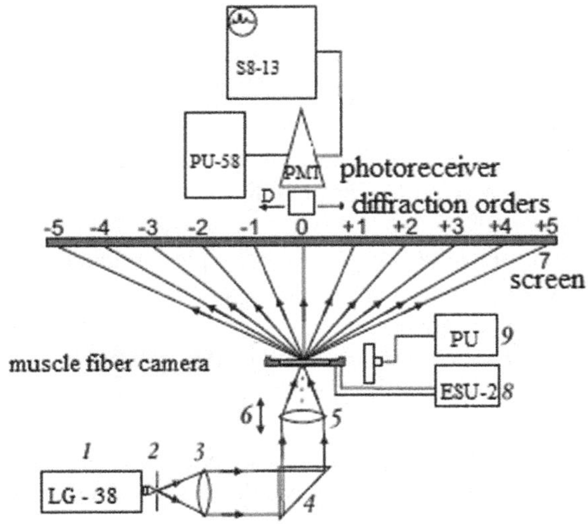

Fig. 2.22 Fundamental scheme of experimental setup for muscle fiber study by laser diffractometry method. Reprinted from [136] with permission

as a pure phase diffraction grating, as its intensity reduces when the parallel light beam passes. The reason of the weakening is the background small-angle scattering on myofibrils, which does not influence the regulated diffraction image.

Optical diffraction methods have considerable advantages connected with contactlessness, non-inertness, required spatial resolution. The main idea of these methods is that focused on the muscle fiber radiation from the laser source gives the diffraction pattern (Figs. 2.22 and 2.24). It is observed due to the periodical phase modulation of the light wave, which goes through different fiber areas. Having high directionality, spectral radiance and radiation monochromatism, the laser sources make it possible to obtain stable, clear, contrast diffraction patterns. Diffraction orders are situated to the left and to the right of the zero one. Intensity distribution in the highest diffraction orders (except the zero one) is stretched in an arched way, which is accounted for the simultaneous muscle fiber motion as a cylindrical lens. The measurement of the sarcomere length at rest and measurement of the sarcomere length during contraction by the angular distance between the zero and ± 1 and other diffraction orders allowed establishing an analogy between the muscle fiber and the diffraction grating (Fig. 2.23).

Isolated solitary muscle fibers of a frog with circular cross section of 70–100 μm at rest were the subject of the study. A camera was made of optically transparent glass that gave the fiber the possibility to be examined with the laser beam. Electrodes were built in the camera to stimulate the fiber with square-wave current pulses. Depending on the object and conditions of the experiment, the stimulating pulse amplitude was selected, which was 5–20 W on the average and the pulse duration was 1 s. The electrophysical part of the setup contained a standard electrostimulator (ESU-2) 8. He–Ne laser (LG-38) 1 with the wavelength of 623.8 nm was used as a radiation

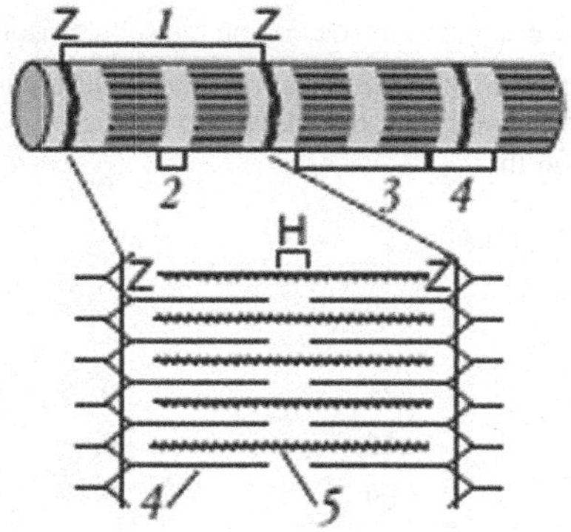

Fig. 2.23 Sarcomere structure: 1—sarcomere; 2—H-disk; 3—A-band; 4—I-band; 5—thin filaments. Reprinted from [136] with permission

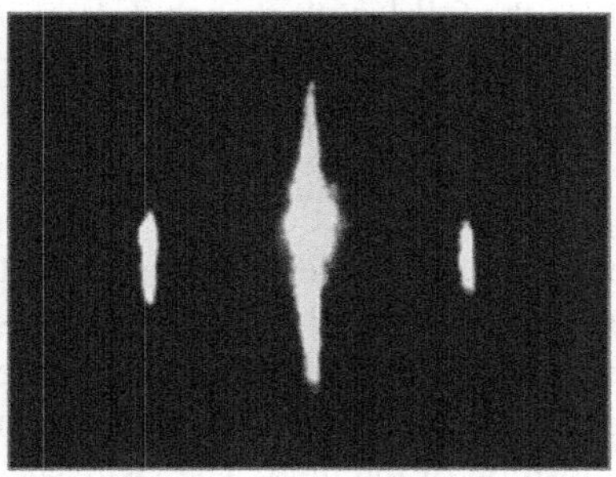

Fig. 2.24 Muscle fiber diffraction image. Reprinted from [136] with permission

source. Coherent laser radiation going through the diaphragm 2 and the rotary prism 4 with the help of the lens 5 focuses on the muscle fiber. As a result of the diffraction of radiation, on the periodic structure of the muscle fiber there occured up to 5 diffraction orders to the right and to the left of the zero one, i.e., +1, ±2, ±3 and so on. The distance between the camera with the fiber and the display was 80 mm.

The distance between the diffraction orders was enhancing in the course of the diffraction maximum number increase. Asymmetry of +1 and −1, +2 and −2 and so on diffraction orders was observed relative to the zero one. The diffraction efficiency of the muscle fiber is rather high that is seen in Table 2.3, which shows relation efficiency changes in a diffraction order according to its increase. Calculated according to the angular distance between the zero and the first diffraction orders grating period is equal to 2.53 μm (see Table 2.3) that corresponds to the sarcomere' length, defined with other methods.

During muscle fiber contraction under stimulation pulse, the distance between the zero and the first diffraction order increases by 5 mm that corresponds to the decrease of sarcomere length by 0.49 μm. In this case, the zero order was not displaced. During the contraction, the number of orders was reduced from 5 to 3 with simultaneous

Table 2.3 Parameters of laser radiation diffraction on the solitary isolated muscle fiber

Diffraction order	Intensity ratio of the first, second, third, fourth orders to the intensity of the zero one	Distance from the zero order (mm)	Diffraction angle sin φ	Diffraction grating period d μm
0	1	–	–	–
1	0.08	21	0.25	2.53
2	0.03	46	0.49	2.54
3	0.01	89	0.47	2.57
4	0.0017	392	0.97	2.58

intensity change. This can apparently be explained by the fact that different sarcomere groups differentiating along the full-length change Z-disks inclination angle relative to the incident ray and thus change the diffractive efficiency of the muscle fiber.

The study was conducted concerning the influence of a powerful pulse magnetic field of ≈25 kOe with pulse duration of 100 ms on the muscle nerve contraction mechanism. On the change of the diffraction image, the moment of contraction can be observed with high sensitivity. After a series of experiments, it was established that structural and functional changes were observed in the muscle tissue in some time after the field effect. First of all, it can be explained by the fact that it takes some time for accumulation of structural changes and secondly by complexity of conducting a study directly by the imposition of the field. The last issue can be settled by using non-contact method of diffractometry that makes it possible to record remotely the diffraction pattern and due to it calculate sizes of the muscle fiber sarcomeres [311–313] and their alteration in the moment of field action. The method also allows defining the thickness of myofibrillars clusters [314], the function of sarcomere distribution along the length [315] and dynamics of A-disks [316].

2.2.9 Study of Structural and Functional State of Lymphocytes Using the Holographic Interference Microscopy Method

In the work [7], the holographic interference microscopy was used for studying the functional state of lymphocytes in the process of their cultivation under the influence of antigens.

It is known that during the immunization process elements getting into contact with the antigen are accompanied by some biological alterations (accumulation of antigen connecting receptors on the cytoplasmic membrane, protein and nucleic acid synthesis increase), the surface potential of a cell, its electropheric mobility

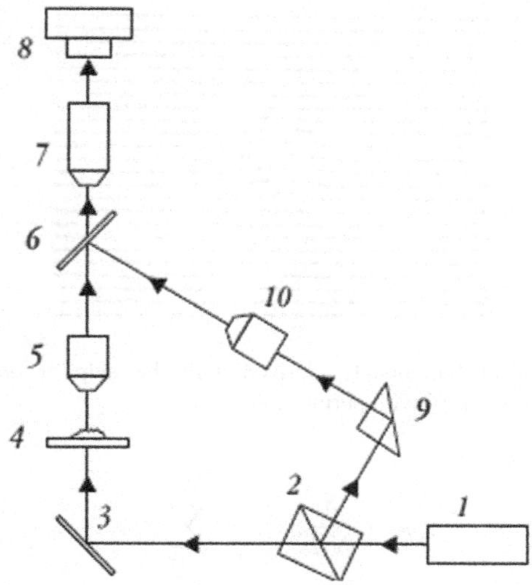

Fig. 2.25 Holographic interference microscope operating in the transmitted light: 1—He–Na laser; 2—the beam splitting cube; 3—the deflecting mirror; 4—the preperation; 5, 10—the lenses; 6—the hologram, 7—the eyepiece, 8—the camera; 9—the prism. Reprinted from [136] with permission

are changing. There is reason to believe that such alterations must be followed by changes in optical characteristics of the cell.

The studies were conducted using the holographic interference microscopy method in real time. The scheme of the corresponding holographic interference microscope operating in the transmitted light is given in Fig. 2.25.

During the work, the general increase was 1350 (lens 90×, eyepiece 15×). For hologram recording, a 50 mW laser was used. The hologram of the object beam was recorded without the preparation, after processing it was placed on the initial place between the lens and the eyepiece and was lighted with the reference beam. The interference of the wave recovered from the hologram and the wave distributing from the preparation under study lead to the formation of an interference pattern in the form of parallel bands curving in the cell area (Fig. 2.26).

Period of interference fringes and their direction were regulated by hologram inclination. The obtained interferogram of a lymphocyte was recorded on film "Mikrat-300". Viable lymphocytes at the height of immune reaction (on the fifth day after immunization) secreted from the spleens of control and experimental animals of white rats of the Wister line served as the object of the study. During the experiments, lymphoid cells of one size (their diameter was 10 μm) were selected with the help of eyepiece micrometer.

During the processing of interferograms of the lymphocytes, interference fringes inclination in the cell area was measured. In our experiments, the magnitude φ was defined for 98 normal and 113 immune cells, selected in the following way:

Fig. 2.26 Interferogram of lymphocyte, derived with the help of holographic interference microscope. Reprinted from [136] with permission

$$\varphi = \frac{S}{Nd}, \quad S = \sum_{i=1}^{N} x_i, \tag{2.35}$$

where S is the average phase incursion, which is carried into the object wave by a lymphocyte; x_i is the inclination of the interference wave in the i-point; N is the number of measurements (the number of points, in which the inclination of the fringe was measured); d is the interference pattern period.

The distributions of normal and immune lymphocytes according to the φ-parameter are presented in Fig. 2.27. Certain population heterogeneity emerges due to this criterion. Thus, range φ is 0.029 when the most probable meanings are 0.31–0.41 for lymphocytes with the diameter of 10 μm (normal and immune). An average quadratic deviation is equal in both cases to $\delta = 0.05$. If the limits of effective range of the phase incursion carried into the object wave are the same (i.e., if the variations

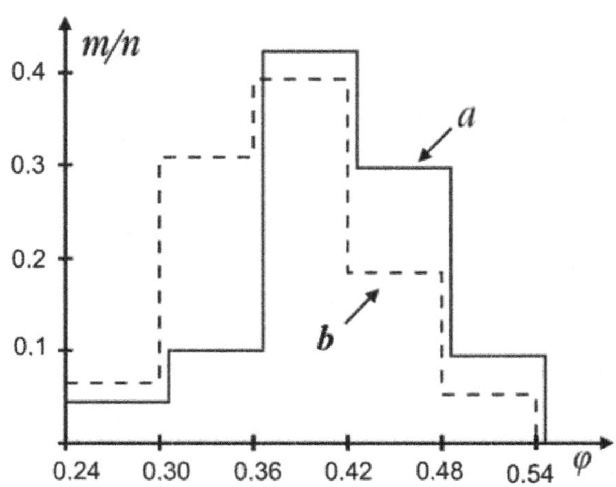

Fig. 2.27 Distribution of regular (**a**) and immune (**b**) lymphocytes by parameter φ. Reprinted from [136] with permission

of transmission density are the same), the sequences of normal and immune lymphocytes differ by the distribution of separate variants. It is seen from the histogram that possibility of finding a cell with φ-magnitude exceeding the magnitude of the modal class is about two times higher for the lymphocytes in the control than for the immune lymphoid cells.

The arithmetic average values $\overline{\varphi}$ for the control and the immune lymphocytes are $\bar{\varphi}_k = 0.40$ and $\bar{\varphi}_i = 0.38$, respectively, when the average measurement error is defined by $\bar{\varphi}$:

$S\bar{\varphi} = 0.005$. To estimate the accuracy of difference between the arithmetic values, the normalized deviation t was calculated [345]:

$$t = \frac{\overline{\varphi_k} - \overline{\varphi_i}}{S_{(\overline{\varphi_k} - \overline{\varphi_i})}}, \tag{2.36}$$

where the root-sum-square uncertainty of the difference is in the denominator.

Calculation of the value by the experiment data gives $t = 2.28$ that corresponds to the high probability of the difference accuracy ($p = 0.9952$). Thus, the shift of densities to small values is observed in the group of immune lymphocytes compared to the normal ones that can serve as a criterion for their distinction.

It is known that in the process of immunization characterized by high nuclear-cytoplasmic relationship and a nucleus with densely aggregated chromatin, small lymphocytes with the diameter of 5–8 μm transform into middle ones with the diameter of 8–12 μm with less dense nuclear chromatin and into big ones (diameter of 12–20 μm), the so-called blasts, which have a nuclear with the loose structure of chromatin. During the immunization there were noted the changes of the average incursion introduced by the cell. Apparently, these changes and changes of density connected with them characterize the process of nuclei loosening, which precedes the growth of lymphoid elements. The obtained data can be of great interest for applied medicine at the development of new laboratory diagnostic methods. They give the opportunity to establish a criterion, which makes it possible to carry out the differentiation of immune and non-immune lymphocytes that can be of interest to contemporary immunology.

2.3 Holographic Study with Electrophysiological Control of Hyperbary Impact on Isolated Preparations of Solitary Nerve Fibers, Nerve Fibers as a Part of the Nerve Trunk and Solitary Muscle Fibers

2.3.1 Approaches to the Study of Mechanisms of Hyperbary Impact in Humans

According to literature data, hyperbary impact on living organisms indicates the complex character of the impact of the given physical factor. During the application of high pressure, the functional state of different systems of an organism (cardiovascular, neuromuscular, respiratory, hematopoietic, endocrine, immune systems) is breaking. Along with the integral reaction of an organism to the pressure, structural and functional changes develop in separate organs and tissues as well. The studies of some isolated preparations allow assuming that long-term exposure of high pressure modifies the structure of membrane scaffold protein, influences membrane processes. A very important issue is the direct mechanical effect of the pressure that leads to injury or destruction of tissues. Biochemical and structural changes supposed and experimentally proved in a number of cases by the impact of high pressures on a live tissue give reasons for conducting such studies using optical methods.

A number of physical factors have a great impact on the human organism. Their study has great applied value. Hydrostatic pressure is one of the most important. The need for deepwater diving and work has significantly grown recently because of oil and gas field development on a continental shelf with enhancing the use of ocean resources. Nowadays, there are methods and equipment helping people to work at a depth of 350–400 m. Improvement of technical means and training of divers give the possibility to claim that in the nearest future deepwater works become possible at a depth up to 500 m [346]. Maximal pressure reached by a human during simulation diving using nitrogenated helium-oxygen mixture is 69.4 kg-force/cm^2 that corresponds to the depth of 68.4 m. During saturation diving, a person is at a depth of 501 m. Diving of small animals was simulated under pressure of 270 kg-force/cm^2. The limits of pressure endured by a human are not established [347].

Along with the necessity of equipment improvement for deep diving, the search of measures, which prevent the negative impact of hydrostatic pressure on a human organism, is of great importance. The latest studies prove early findings and point to disorders in functioning of excitable tissues, first of all of the whole nervous system. By the investigation of 33 professional divers and corresponding to the people from the control group, the following facts were ascertained. Psychomotor coordination impairment and suppressed disorders of central nervous system (CNS) functions are typical for the consequences of the decompression sickness [348].

It was found out that neurological symptomatology is accompanied not only by decompression, but also by an increase in diving depth. Even before the decompression, there was observed the progress in neurological changes, which is expressed by tremor, myoclonic twitchings and spasms [349]. Very soon the researchers who studied impact mechanisms of hydrostatic pressure understood that the problem of providing divers with adequate diving gas is not the only one in providing diving security. They faced with such a phenomenon that is traditionally named hyperbaric neural syndrome of high pressures [350]. It is observed during diving at a depth of more than 200 m. Meanwhile besides the stated neurological symptoms divers mention nausea, vertigo and distinctive changes in electroencephalogram like delta-waves dominating and occurrence of paroxysmal discharges of reflected waves [347].

Recently, reasons for high pressures syndrome have been searched for. And if this problem is to be solved by selecting an adequate normaxic diving gas, then now it is clear that study of mechanisms of high pressures neural syndrome should be developed on the basis of different approaches [351], especially because hyperbary is a multi-component factor that is why it is not always possible to differentiate the impact of the compression and connected with its toxic impacts of partial pressures from stress reactions [352]. Most intensive studies are conducted on the issue of hyperbary impact on the biochemical level [349, 353, 354]. The study of non-oxygen factors of hyperbaric impact on human and animal organisms is significantly developed by the cellular studies. The next important step in the methodical improvement of researches is to study excitable tissues of isolated preparations functioning under hydrostatic pressure [355–357].

Obtained from isolated preparations of excitable tissues of animals heart, experimental data are of prime importance for interpretation of pathological states observed in divers organism during diving. Problems with excitement conduction are the cause for observed hyperbaric changes of human blood circulation, in particular bradycardia [355].

The facts of hyperbaric impact on the cross-striped tissue were found as well both by animals [358] and by humans [359]. By divers under test during the simulation diving, the researchers studied maximal force and time of gastrocnemius muscle contraction during development under the influence of Hoffman reflex electrostimulation as well as of the direct motor reaction stimulation. While diving maximal contraction force increased by 40–60% relative to the control measurements. The reasons of this phenomenon are not fully clear; moreover, there were facts of injurious impact of hydrostatic pressure on the muscle [360]. Three-dimensional compression effect plays an important role in the influence of high pressures on the human and on animals [361].

Analysis of biochemical adaptation by living organisms to high pressures is evidence of the fact that many processes important for life react to hyperbary with slowdown or acceleration of product reaction generation. Meanwhile, the ratio of reagents volume and reaction products plays an important role [362]. These phenomena are more important on the forming stage and on the stage of energy provision of an organism. Moreover, knowledge about the hyperbary impact directly on the functions of the excitable tissues such as excitability, processes of its provision, distribution and transmission are of great importance. Namely, the efforts of the researchers who seek reasons of emergence of so-called high pressures neural syndrome [363] are directed to the study of these processes. The conducted studies showed a complex multi-parametric character of changes in metabolism processes of different substances that provide and modulate the synoptic excitability transmission. It is connected, for example, with alterations of force and duration of convulsive contractions of human muscles under hyperbary [364], isometric muscle contraction of intact animals under high pressure [365].

While studying hyperbary impact on excitable tissues and analyzing the results obtained from the whole organism, the complex character of this physical factor should be taken into account, which influences functioning of different systems of an organism: immune, hematopoietic, respiratory, cardiovascular, endocrine [366]. Thereby studies of hyperbary impact on isolated preparations of excitable tissues gain importance. Though such an approach strictly constrains the interpretation of obtained data, there emerge difficulties while trying to transfer the ascertained regularities to the whole organism, but there occurs a possibility to observe the process under study on a simplified model directly during the hyperbary impact.

In a number of works, the preparations of Purkinje fibers [367], auricle muscle [368] and a portal vein [369] were investigated. In these experiments, it was stated that high hydrostatic pressure decelerates excitability distribution and reduces force and frequency of contraction of a cross-striped muscle tissue as well as of a smooth one. The data coincide with the results of studies of heart work by divers, which was analyzed with ECG. The authors of the study interpret the obtained material as an evidence of atrioventricular conduction decrease [370]. Observation of isolated preparation of the right atrial auricle made it possible for the authors to assume that long effect of helium-oxygen medium under high pressure modifies the primary structure of slow sodium-potassium membrane canals and potassium canals that are responsible for anomalous or inhibited rectification effect [371]. But it concerns only long stays (32 days) of an organism under hyperbary. Under the impact of high pressure on preparation of the excitable tissue, the true reaction mechanisms are not clarified. It is possible that hydrostatic pressure not only modifies but also directly influences the membrane processes, for example, potassium conductance [372]. Also, it was fairly mentioned in the literature that under pressure there increases the impact of the following factors on the isolated preparations: temperature [357] and preparation survival medium [373]. In the works [15, 17], hyperbary impact on neural and muscle fiber, lymphocytes, nerve and muscles was studied.

Thus, the analysis of the literature shows that processes occurring in tissues under high pressure are not studied properly. However, it is theoretically possible that a number of structure-functional changes can emerge, which are directly caused by informational protein changes (framework and contractive cell proteins). Also, changes in enzymatic systems of cell maintenance are assumed and, in some cases, experimentally proved.

2.3.2 Holographic and Electrophysiological Study of Gas Pressure Impact in the Range of 0:5 Atm. on the Isolated Solitary Nerve Fiber

To influence a nerve, a solitary nerve or muscle fiber with high gas pressure in the range of 0:5 atm. was placed in turn in a special camera, the schematic image of which and its photo are shown in Fig. 2.28a, b. The cameras were made of stainless steel in the form of two metal disks with a ring gasket made of vacuum rubber between them. In the central part of both disks, optical windows were pasted with epoxy resin. The windows were made of fused silica, the window thickness was 4 mm with the diameter of 10 mm, and the deviation from the flatness does not exceed $10''$.

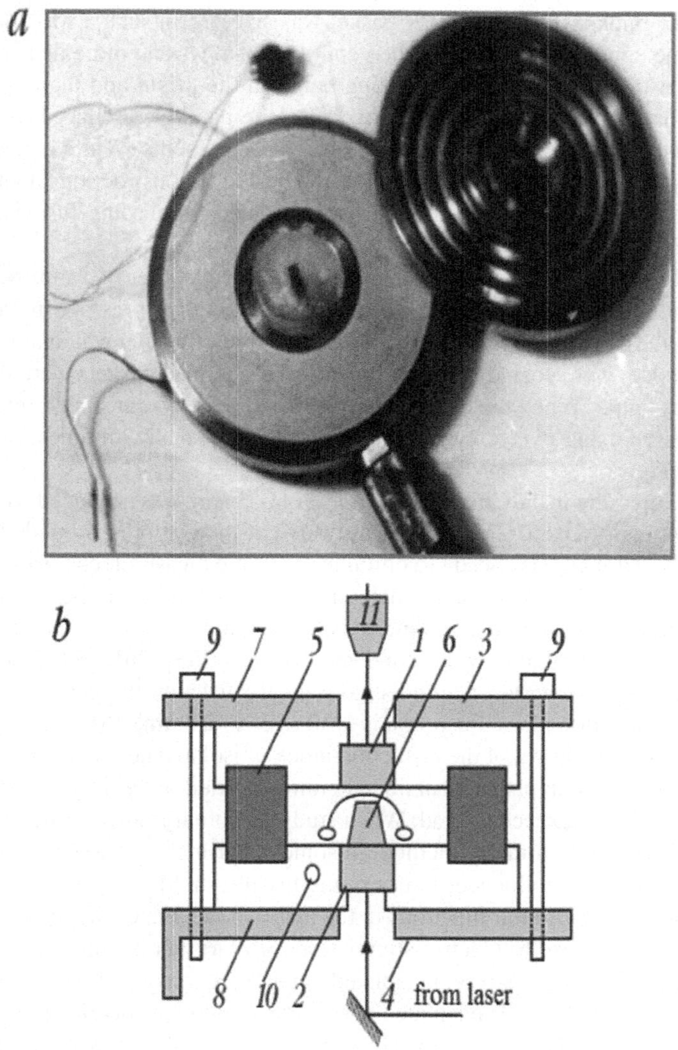

Fig. 2.28 Photo (**a**) and the scheme (**b**) of camera for creation of gas pressure of 0–5 atm.: 1, 2—the optical windows; 3, 4—the metal disks, 5—the gasket made of vacuum rubber; 6—the window prism; 7, 8—the metal plates; 9—the screws; 10—the brunch pipe for supplying gas under pressure; 11—the microscope objective. Reprinted from [136] with permission

Apart from creating pressure, the camera construction made it possible to conduct studies of isolated preparations using the method of holographic interference microscopy and the electrophysiological method. To provide action potential derivation of nerve fiber effect using Tasaki air bridges method [292, 374], a glass prism was glued to the lower optical window of the camera with special optical glue. On the prism in a drop of the physiological solution, the preparation under study was placed. The thickness of the rubber gasket was selected in such a way that the gap between the upper optical camera glass and the prism was no more than 1–1.5 mm.

Then, in the closed camera spreading between the prism and the upper optical window, the physiological solution formed a plane-parallel capillary layer that was a necessary condition to conduct interference measurements. While conducting the electrophysiological studies of the nerve fiber such a construction of the camera made it possible to stimulate and derivate the action potential using Tasaki air bridges method [292, 374] (Fig. 2.28b).

In the lower camera disk, three channels with the diameter of 1 mm were turned to supply stimulating and pickup electrodes and making gas pressure in the working volume of the camera. To seal the the camera the holes, through which the electrodes were extended, were coated with epoxy adhesive, and lead gaskets were used in the pressure gas pipe. When the camera in a working mode was installed into the unit, it was pressed using two screws between two metal plates with holes, which open optical windows.

The gas pressure in the camera in the range of 0–5 atm. was produced with the help of the compressor UK40-2M. The release valve of the compressor made it possible to adjust pressure and its speed. To eliminate possible measurement errors using the real-time holographic interference method, there was conducted the study of active volume deformations in the gas camera under pressure.

A series of experiments were conducted on recording differential holographic interferograms of the nerve fiber under gas pressure of 4 atm. When the measurement accuracy of insertion phase incursion is $\lambda/30$ ($\lambda = 632.8$ nm), the optical characteristics stability was shown of the refraction index of isolated nerve and muscle fibers. It was ascertained that phase incursion alteration carried in by the nerve fiber under hyperbary does not exceed 0.2 rad. When studying solitary muscle fiber contraction using the laser diffractometry method, in some experiments, it was found that high gas pressure (4 atm.) influences fiber excitability threshold.

As a main index of vital functions in the nerve tissue, the excitability parameters were used in response to electric stimulation. Solitary nerve and muscle fibers as well as nerve fibers as a part of an isolated nerve were the object of the study.

Preparation methods. The isolated nerve preparation was obtained by extraction of the sciatic nerve of a frog cutting it higher and lower of the applied ligatures which were used in a case of necessity for fixing and transferring it to special cameras. On the extraction stage and in further studies, the preparation was moistened with a physiological solution (0.9% NaCl) or with Ringer solution (NaCl—110.5; KCl—2.5; $CaCl_2$ 1.8; Tris—5.0 mmol, pH—7.3) to avoid extinction. The preparations were placed in the same medium during the whole period of registration.

The solitary isolated nerve fibers were dissected from the sciatic nerve of a frog. For this purposes, the isolated nerve was placed on the stage of binocular microscope MBS-9 in a drop of the physiological solution. The object was examined for transparency. Optical fibered illuminator OVS-I was used for this purpose. That allowed eliminating overheating of the object during lightning up and decelerating evaporation of water from the physiological solution. Then, all the preparation operations with the solitary nerve fiber were conducted according to the methods, described by I. Tasaki [292, 374].

Methods of estimating the excitability. The excitability of the isolated solitary nerve fiber was estimated according to the presence of action potential, the disappearance of which under continuous stimulation was considered as an irreversible injury of the nerve fiber. The excitability of the nerve fiber as a part of isolated nerve trunk was estimated according to the duration amplitude of the total action potential. In a series of cases, such characteristics were recorded as excitability of the isolated nerve, threshold value of electrical stimulation voltage with different duration of an electric pulse (50, 100, 200 μs), to construct a graph duration-force. The excitability of an isolated muscle fiber was estimated according to the presence of contraction in response to electrostimulation and threshold (minimal) quantity of the voltage which caused the contraction. An electromyograph produced by firm "Medicor" was used for objects stimulation, amplification and observation of the action potentials.

To record the action potentials and isolated solitary nerve fibers stimulation, metal electrodes were used which were placed inside the camera to investigate in such a way that a simultaneous electrophysiological control is carried out during optical measurements (see Fig. 2.14).

Results of the study. During electrophysiological control of vital functions of the solitary nerve fibers, the attention was drawn to the irreversible suppression of generation of action potential in the first seconds of high air pressure application. An electrophysiological study of the isolated nerve was conducted to make sure whether hyperbary is the reason of suppression of the ability of the nerve fiber to react with action potential in response to electrical stimulation or whether it is a consequence of mechanical injuries during preparing of solitary nerve fibers which often accompany it.

Isolated sciatic nerves of a brown frog were object to the study. The total action potential of the nerve fibers was analyzed in response to electrostimulation at the frequency of 50 Hz during and after the high air pressure application (1 or 2 atm.).

The graphs illustrating statistically correct electrical stimulation voltage increase needed for generation of action potential by most sensitive nerve fibers are presented in Fig. 2.29 and show the initial state of the object and the state under hyperbary impact. That means that analysis of the date from Fig. 2.29 can prove only that the excitability of the fibers under hyperbary has decreased. Nevertheless, it should be taken into consideration that the nerve fiber excitability decrease of the isolated nerve occurs with the course of time and without an impact of hyperbary or other external factors. Moreover, it can emerge spontaneously because of the preparation isolation and impossibility to maintain metabolism. This peculiarity of the object under study is shown in the fact that the value of the excitability parameters is constantly changing being practically the time function of preparation aging of the nerve tissue outside a living organism. Apparently, this can explain the constant decrease of action amplitude of the isolated nerve starting from a particular time moment (Fig. 2.30). At the same time, hyperbary impact accelerates the amplitude contraction of the action potential that means that the quantity of fibers capable to generate action potential under the initial electrical stimulation has decreased. Such alteration of excitability

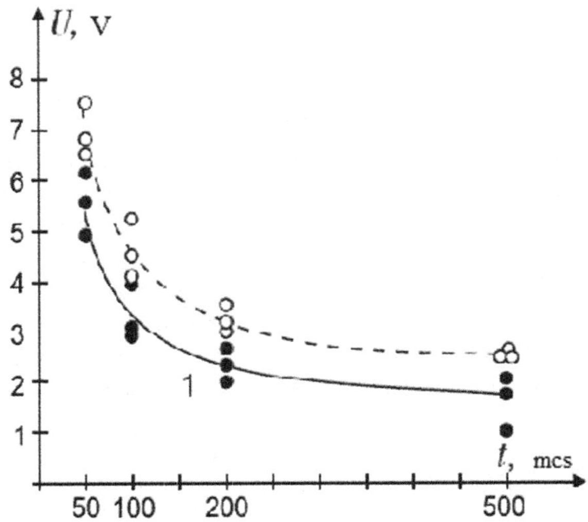

Fig. 2.29 Change in the force-duration dependence of an isolated nerve before (•) and after (||) single exposure of hyperbary (1 atm. within one minute). Along ordinate axis is voltage of threshold stimulations (volts); along abscissa axis is duration of electric pulse. Reprinted from [136] with permission

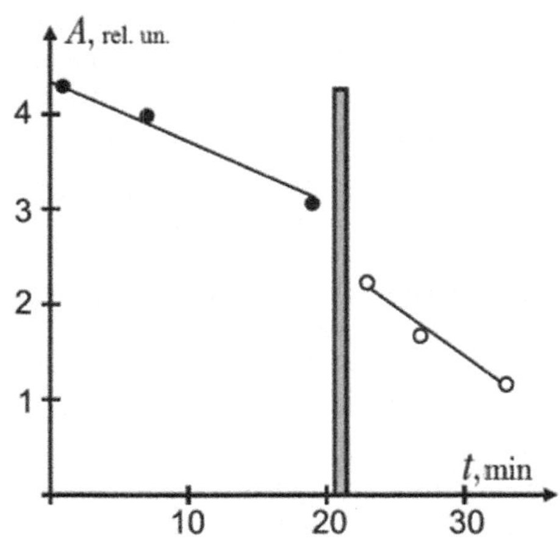

Fig. 2.30 Change in the amplitude of the action potential of an isolated nerve before (•) and after (||) exposure of hyperbary (shaded area—1 atm. within 1 min), along ordinate axis is amplitude of the action potential; along abscissa axis is time pulse. Reprinted from [136] with permission

parameters was most typical for the majority of the examined nerves. A higher degree of hyperbary (up to 2 atm.) and longer in time can lead to excitability suppression till the full disappearance of the action potential (Fig. 2.31). Meanwhile as the data presented in the figure show, the action potential amplitude of a fresh isolated nerve may not undergo substantial changes before as well as during hyperbary and

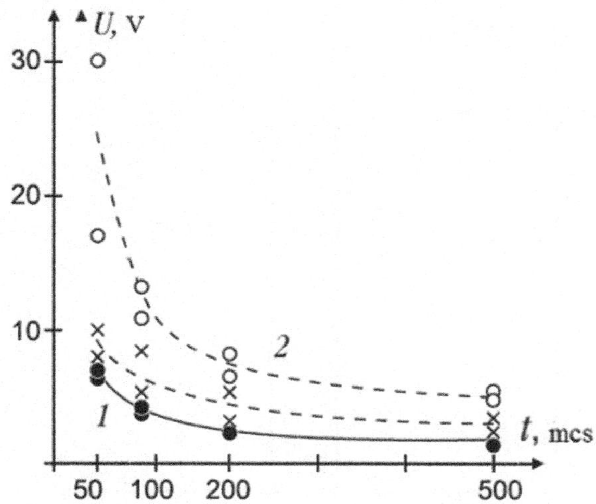

Fig. 2.31 Change in the force-duration dependence of an isolated nerve before (•) during (×) and after (||) single exposure of hyperbary (2 atm. within 15 min). Along ordinate axis—voltage of threshold stimulations; along abscissa axis—duration of electric pulse. Reprinted from [136] with permission

dramatically diminishes after decompression. But observations of threshold quantities of electrical stimulations have shown that excitability decrease already begins during hyperbary impact (Fig. 2.32) that is reflected in growing increase of threshold stimulation especially by the shortest pulse duration.

Analyzing the obtained data on the changes of threshold quantities stimulation and of the action potential parameters under the impact of hyperbary in the air, it can be concluded that the suppressive hyperbary impact on the excitability of isolated nerve fibers really occurs and is most probably caused by damaging factors aggravation of nerve tissue preparation aging (absence of oxygen and energy), that as a result reduces the time during which the nerve fiber is able to generate and conduct the action potential in response to electrical stimulation.

Statistical analysis. It is often impossible to estimate the dependence of experimental data on some factor by means of the single-factor dispersion analyses either because of the failure to implement suppositions about the normal state of distribution, or because of the fact that the observations themselves do not undergo quantitative measurements. In our case, the absence of normality in distribution of voltage threshold values is evident. This voltage causes the action potential during isolated nerve electrostimulation that makes impossible to use parametric statistics. If the observations are placed in the order of increase or decrease, then rank methods of non-parametric statistics can be applied, specifically Crascal–Wallis [Mardia, Zemrot] (distribution table, 1983). For this purpose, all the observations of the samples of different experimental situations (before, during and after hyperbary) are ranged beginning from the lesser with rank $r_{ij} = 1$ and ending with the greater with rank $r_{ij} = n$ or vice versa.

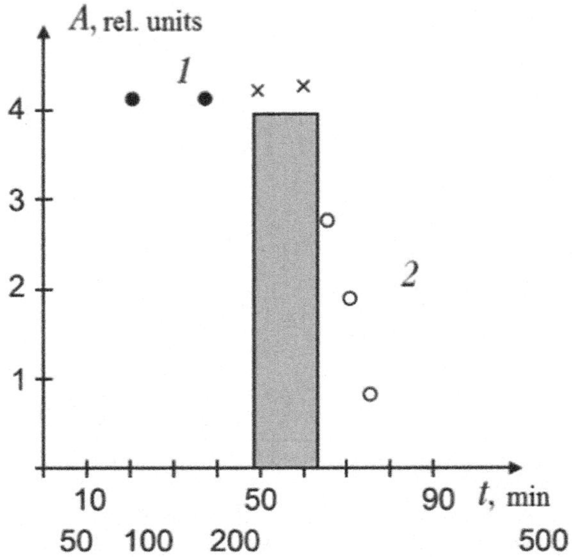

Fig. 2.32 Change in the amplitude of the action potential of an isolated nerve before (•), during (×) and after (‖) exposure of hyperbary (shaded area—2 atm. within 15 min). Along ordinate axis—amplitude of the action potential; along abscissa axis—time. Reprinted from [136] with permission

Supposing that

$$r_i = \sum_{j=1}^{n_i} r_{ij}, \quad \overline{r_i} = \frac{r_i}{n_i}, \quad r_{11} = \frac{1}{n}\sum_{i=1}^{k}\sum_{j=1}^{n_i} r_{ij} = \frac{n+1}{2}, \tag{2.37}$$

statistics of H of Crascal–Wallis's criterion is estimated by the formula

$$H = \left(\frac{12}{n(n+1)}\sum_{i=1}^{k}\frac{r_{i1}^2}{n_i}\right) - 3(n+1). \tag{2.38}$$

According to this criterion, hypothesis H_0 on equality of average random variables is declined if H exceeds some critical value. For its estimation, it is better to use approximation which was suggested by Crascal and Wallis.

The following designations are used

$$M = \frac{n^3 - \sum_i n_i^3}{n(n+1)},$$

$$V = 2(k-1) - \frac{2\left[3k^2 + 6k + n\left(2k^2 - 6k + 1\right)\right]}{5n(n+1)} - \frac{6}{5}\sum_{i=1}^{k}\frac{1}{n_i}. \tag{2.39}$$

Table 2.4 Results of statistical analysis of the electrophysiological study concerning the hyperbary impact on the isolated nerve

#No.	Duration of electrical stimulation (mcsec)	Analysis data according to the criteria of Crascal–Wallis, arbit. unit						The level of criterion significance
		H	M	F	V	V_1	V_2	
11	50	3.86	3.86	1.4×10^7	1.37	0.56	1.60	<0.0001
	100	3.05	3.86	1.1×10^7	1.37	0.56	1.60	<0.1
	200	3.86	3.86	1.4×10^7	1.37	0.56	1.60	<0.0001
	500	3.86	3.86	1.4×10^7	1.37	0.56	1.60	<0.0001
22	50	4.57	4.57	1.2×10^7	2.06	1.31	1.69	<0.0001
	100	3.71	4.57	5.60	2.06	1.31	1.69	<0.2
	200	3.71	4.57	5.60	2.06	1.31	1.69	<0.2
	500	3.71	4.57	5.60	2.06	1.31	1.69	<0.2

Then, the statistics

$$F = \frac{H(M - k + 1)}{(k - 1)(m - H)}, \tag{2.40}$$

obeys F-distribution with degrees of freedom

$$\nu_1 = (k - 1)\frac{(k - 1)(M - k + 1) - V}{\left(\frac{1}{2}\right)MV}, \quad \nu_2 = \frac{M - k + 1}{k - 1}\nu_1. \tag{2.41}$$

Using such a statistical analysis, the researchers estimated the significance level of different threshold values of isolated nerve stimulation within four electric pulse durations before, during and after hyperbary impact. The results of the statistical analysis are shown in Table 2.4. They testify (on the generated significance level <0.2) the reliable decrease of nerve fibers excitability in the isolated nerve as a result of hyperbary impact.

2.3.3 Study of Hyperbary Impact on the Nerve Fiber in the Range of Hydrostatic Pressure of 0–200 Atm.

This study was conducted using a camera, the scheme and the photo of which are presented in Fig. 2.33.

The camera body *1* was made of stainless steel. For optical measurements, there were the plane-parallel optical windows *2, 3* made of sintered quartz. The thickness of the windows was 8 mm, and the diameter was 10 mm. The lower camera window *2* was firmly fixed in its body with the circular section seal ring 4. The Teflon gaskets *5* are located between the camera body and the window in the place of their contact.

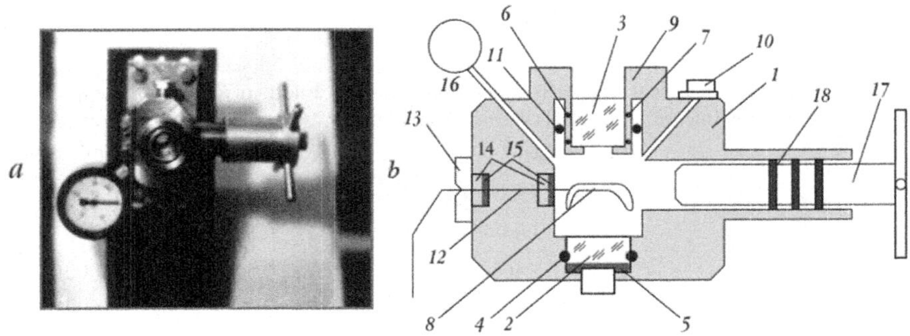

Fig. 2.33 Photo (**a**) and the scheme (**b**) of the camera for creation of hydrostatic pressure 0–200 atm. Reprinted from [136] with permission

The upper optical window *3* was fixed in the metal bushing *7* with the seal rings *6*. The object under study *8* was placed into liquid in the camera active volume then the bushing sank and was fixed with the threaded nut *9*. Excess liquid was removed through the valve *10*. The bushing *7* was hermetically fixed in the body by the packing disk *11*. Metal isolated electrodes were extended through the orifice *12* and were pressed out to the hermetical state in the rubber and Teflon gaskets (*14*, *15*) using a threaded bushing *13*. In the camera body, there were threaded orifices for the manometer *16* as well. In the optical experiments conducted with the simultaneous electrophysical control, the object (nerve trunk) was fixed on the electrodes in vaseline oil. The hydrostatic pressure in the camera was created by compressing the oil with a translational motion of the piston *17*. For hermeticity, the piston was sealed in the camera body with three circular sectioned rubber rings *18*. The studies were conducted under hydrostatic pressure of 0–200 atm. The maximal pressure, for which the construction of the camera was calculated, was 250 atm.

During the experiment, the holographic interferometry method was worked out for studying an isolated nerve under high hydrostatic pressure in reflected light.

Complication of structural and functional nerve organization in comparison with a separate nerve fiber leads to diffuse scattering of light in it and to impossibility of classical interference measurements. Besides a nerve isolated from the organism and aging in an artificial environment (physiological solution vaseline oil) changes its volume in the course of time under osmotic pressure. This leads to changes in the microstructure of the nerve surface and as a result to speckle-fields decorrelation of the radiation scattered on the nerve. Thus, the interference study of the nerve using holographic methods becomes a problem as well.

In our experiments, it was shown that the stabilization of the speckle-pattern in the image of the nerve surface can be reached. For these purposes, the nerve was put into a leak-proof camera in humid atmosphere and was held there for an hour.

The speckle-pattern was controlled with the help of a microscope. The interferogram of the nerve is presented in Fig. 2.34.

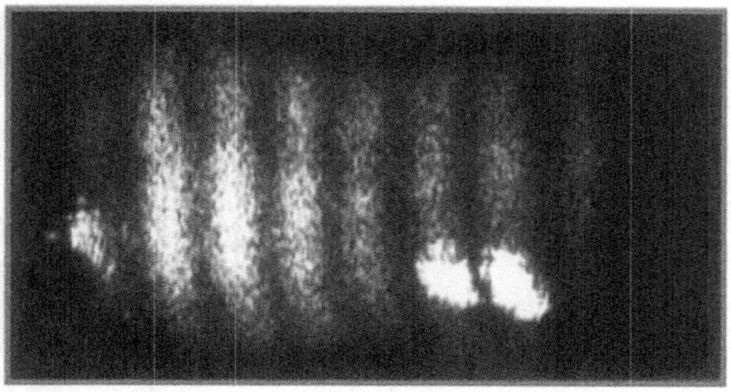

Fig. 2.34 Interferogram of a nerve in the reflected light. Reprinted from [136] with permission

It was obtained with the help of continuous radiation of He–Na laser according to the method of two exposures in reflected light. Between the exposures, the camera with the nerve was inclined at a small angle. The time interval between two exposures was ~1 min. Formation of regular interference pattern is an evidence of the fact that within this time the speckle-fields correlation of the nerve scattered radiation was not lost. This method, namely, holding the nerve in the humid atmosphere until the speckle-pattern stabilizes, can be used, for example, to study elastic characteristics of the nerve while stretching it. But when the external conditions change (alteration of the pH-medium, the influence of chemical substances, effect of technical loads including high pressures) speckle-field decorrelation can occur again.

To avoid this difficulty on the nerve surface, a reflecting mark was fixed in our experiments concerning the study of the isolated nerve compression ratio under high hydrostatic pressure. This was a piece of aluminum foil of 0.5×0.5 mm^2, which was firmly held under capillary forces on the nerve surface. Such a method allowed conducting interference measurements in reflected light according to the scheme presented in Fig. 2.35a, b.

The isolated nerve with the mark was fixed on a reflecting substrate, and then, it was put with the substrate in the chamber with hydrostatic pressure. The nerve was lighted with a narrow laser beam reflecting from the mirrors *3*, *4*, *7*, *8*. The light reflected from the mark on the nerve and from the substrate area adjacent to it was directed to the hologram 14, the lens 12 with focal distance $f = 13$ cm formed the image of the nerve in the photodetector plane. Hologram exposure was made under the pressure in the camera $p = 0$. Photographic plates PFG-03 were used. The exposure time was ~30 s with the laser output power of 55 mW.

The photos of the nerve surface with the reflecting mark and the adequate holographic interferogram obtained in the bands of finite width when changing the angle of the reference beam are shown in Fig. 2.35c. The interference pattern was observed in real time.

The hydrostatic pressure application leads to the motion of interference fringes through the increase of camera active volume because of the shift of its optical windows along the optical axis. The phase incursion in interferogram areas, which correspond to the nerve surface and the reflecting substrate, is the following:

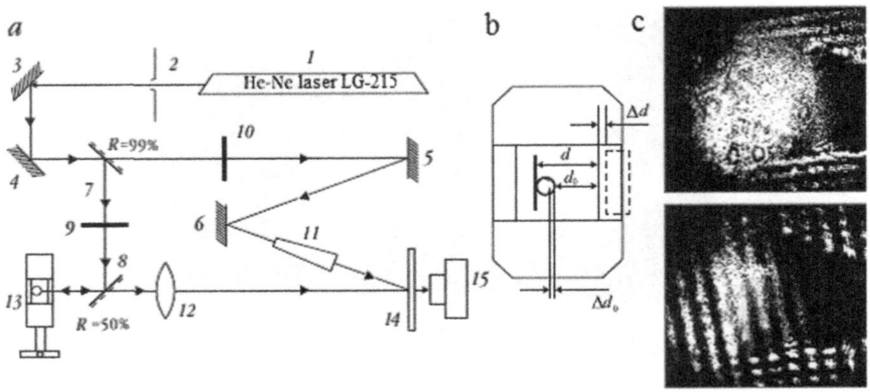

Fig. 2.35 Optical scheme for studying nerve compression when exposed to hydrostatic pressure by holographic interferometry in reflected light: 1—the laser; 2—the diaphragm; 3–6—the deaf mirrors; 7, 8—the beam splitters ($R = 99\%$ and $R = 50\%$); 9, 10—the light filters; 11—the telescopic attachment; 12—the lens; 13—the camera with an object; 14—the hologram; 15—the camera. Reprinted from [136] with permission

$$\begin{aligned} \Delta\varphi &= \frac{2\pi}{\lambda}\Delta L = \frac{2\pi}{\lambda}\{[\Delta d(n_c - n) + \Delta R n_c] - \Delta d(n_c - n)\} \\ &= \frac{2\pi}{\lambda} n_c \Delta R \end{aligned} \tag{2.42}$$

where Δd is the shift of camera window; ΔR is the alteration of the nerve radius; n is the air refraction index; n_c is the liquid refraction index where the nerve is placed.

Thus, the nerve compression under high pressure must lead to the "failure" of interference fringes corresponding to the additional phase incursion $\Delta\varphi$.

The methods described above were tested in three experiments. If the pressure is in the rate of 0–180 atm. and recording at the accuracy of the optical path difference is $\lambda/10$, such a failure in the nerve interferogram was not found that corresponds to changes of its radius at the value smaller than 0.02 μm (i.e., $\frac{\Delta R}{R} \leq 10^{-5}$).

This result testifies of the fact, that in processes, which are the basis of nerve function alterations during high pressure application, failures of metabolism processes play probably the main part, not the mechanical injurious action.

The holographic interferometry method was improved to study a nerve in transmitted light. During hyperbary research of an isolated nerve in transmitted light, namely, during studying the simultaneous changes in its indices of refraction and shape, the main problem is still speckle-field decorrelation.

To solve this problem, we developed an interferometry research method of diffused scattered objects with a slowly changing structure and cylindrical symmetry. The method is based on optical spatial filtering (OSF) [375] of rays passed through the object. It is known that applying the OSF-methods at the stage of hologram recording or during object wave field reconstruction makes it possible to extract most informative component of their spatial frequency spectra.

At decrease of frequency bands, the average speckle size in diffused scattered object field increases, and therefore, the spatial region, in which identical spectra collide, widens that leads to interference pattern localization field expansion. This allows obtaining holographic interferograms at small measurements of one of the interfering fields. The optical scheme of the holographic setup, in which the OSF-method is implemented, is presented in Fig. 2.36. The main principles of this method are based on the following. Let the semitransparent diffused scattering object be lighted by a parallel beam (Fig. 2.36b). The output complex amplitude of the light wave can be presented as two components: $A = A_1 + A_2$, where A_1 is the complex amplitude of the straightly transmitted wave corresponding to transmission of a pure phase object; A_2 is the field of scattered radiation having a chaotic speckle-structure. In the approximation of the cylindrical symmetry of the object:

$$A_1(x_1 y_1) = a_0 \exp\left[-ik_o 2\left(Rn_c + \Delta n\sqrt{R^2 - x_1^2}\right)\right],$$

$$k_0 = \frac{2\pi}{\lambda}; \quad a_0 \equiv 1; \quad \Delta n = n - n_c, \tag{2.43}$$

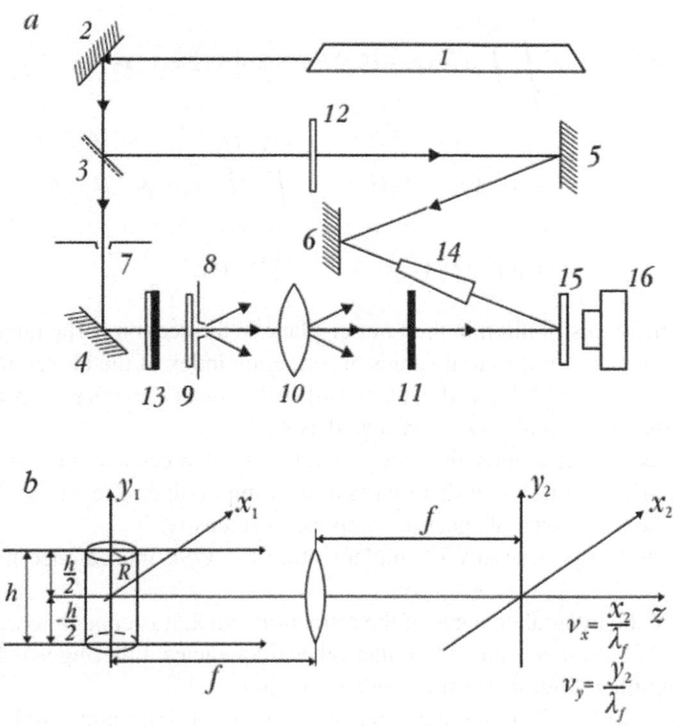

Fig. 2.36 Diagram of a holographic setup for obtaining interferograms of diffusely scattering objects with cylindrical symmetry (**a**): 1—He–Ne laser ($W_{out} = 55$ mW); 2, 4, 5, 6—the deaf mirrors; 3—the transmissive mirror ($R \sim 1\%$); 7, 8—the diaphragm; 9—the camera with an object (nerve); 10—the lens ($f = 130$ mm); 11—the spatial filter; 12, 13—the light filters; 14—the telescopic attachment; 15—the hologram; 16—the camera and circuit in which the method of optical spatial filtering is implemented (**b**). Reprinted from [136] with permission

where R is the radius of the object, n_c is the refraction index of the environment; n is the refraction index of the object.

Fourier-spectrum of this component in the focal plane $(x_2 y_2)$ of the lens 10, which builds image of the object, is the following:

$$A_1(\nu_x \nu_y) = \int\limits_{-R}^{R} \int\limits_{-\frac{h}{2}}^{h/2} \int\limits_{-\frac{h}{2}}^{h/2} a_0 \exp\left[-ik_o 2\left(Rn_c + \Delta n\sqrt{R^2 - x_1^2}\right)\right]$$

$$\times \exp\left[-i2\pi\left(\nu_x x_1 + \nu_y y_1\right)\right] dx_1 dy_1. \tag{2.44}$$

and is oriented perpendicularly to the object axis.

Placing a spatial filter extracting this spectrum to the Fourier-plane of the lens 10 leads to suppression of the speckle-structure in the image of the object under study.

Initially, when conducting model experiments, the cylindrical object was put directly on the way of a laser beam. In this case, the spatial frequencies spectrum of the object wave not registering the diffused component can be described by the following formula:

$$A(\nu_x \nu_y) = \int\limits_{R}^{h/2} \int\limits_{-\frac{h}{2}}^{h/2} a_0 \exp(-ik_o 2Rn_c) \exp\left[-i2\pi\left(\nu_x x_1 + \nu_y y_1\right)\right]$$

$$\times dx_1 dy_1 + A_1(\nu_x, \nu_y) \int\limits_{-h/2}^{-R} \int\limits_{-h/2}^{h/2} a_0 \exp(-ik_o 2Rn_c)$$

$$\exp\left[-i2\pi\left(\nu_x x_1 + \nu_y y_1\right)\right] dx_1 dy_1 \tag{2.45}$$

The intensity distribution in the Fourier-plane in the section perpendicular to the cylinder generatrix for different values of refraction index of the object under study is presented in Fig. 2.37. Here, the intensity distribution at $n = 1$ is presented as well, i.e., when the object is absent, is presented as well.

As it is seen in the graphs, the spectrum intensity changes very little if n changes even in the first decimal digit that makes almost impossible to record such changes due to the measurements of spectral components intensity.

This leads to the necessity of implementing the OSF interference methods of study.

To define the integral changes of the refraction index, it is enough to transmit the zeroth order of the spectrum and to filter other frequencies. Filtering was conducted with an amplitude filter in the shape of a round hole.

According to Fig. 2.36, the diameter of the filter was 0.06 mm. Implementation of such filtering allows obtaining rather contrasting interferograms (Fig. 2.38). But when transmitting to strongly scattering objects, such as a nerve, intensity of the wave passed through the object was significantly less than of the wave traveling outside the

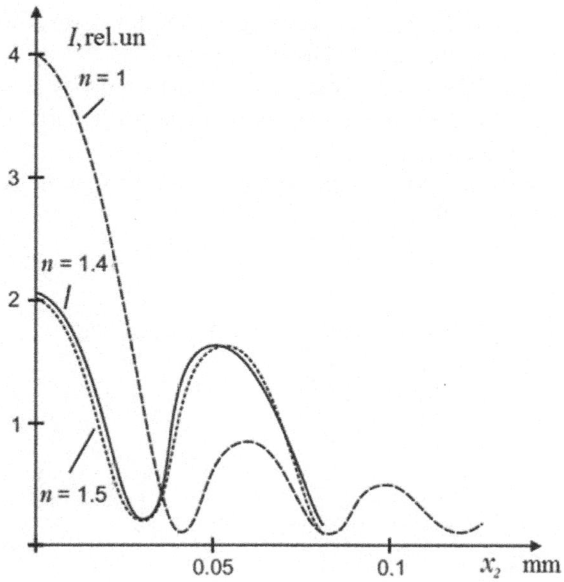

Fig. 2.37 Spatial spectrum of emission in Fourier-plane of a lens. Reprinted from [136] with permission

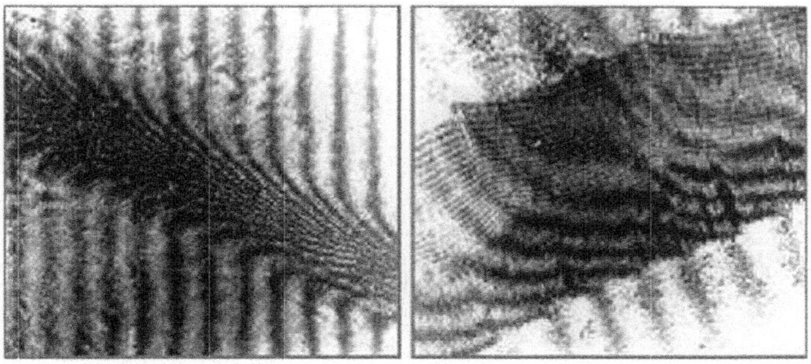

Fig. 2.38 Interferograms of a nerve, found in real time in the transmitted light: **a** after placing the camera with nerve out of the object beam; **b** at camera shift with nerve. Reprinted from [136] with permission

object and scattered on the object that made impossible to form the interferogram required. To extract only the component passed through the object, it was placed between the legs of the slot, the diameter of which does not exceed the diameter of the object. And the spectrum of spatial frequencies of the object wave was changing and was described by (2.45).

The width of the zero-order spectrum practically doubled, and the intensity magnitude nearly quadrupled. Thus, in this case for filtering the zero-order spectrum, a filter twice as big in the diameter can be used. But the energy of the filtered light wave was not enough to form the interference pattern good for observation that is why a narrow slot placed along the axis was used as a filter.

This technique can be applied to study optical characteristics of an isolated nerve in the transmitted light aging in altering environment particularly under the impact of high hydrostatic pressure.

Moreover, the possibility of hologram recording of the focused nerve image is of the great interest (Fig. 2.39).

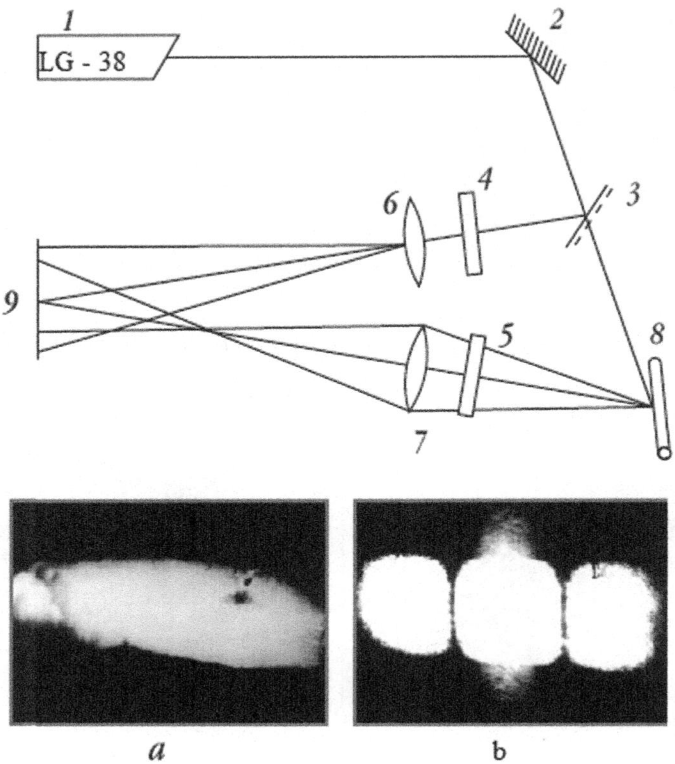

Fig. 2.39 Optical recording scheme of the focused image: 1—the laser; 2—the non-transmitting mirror; 3—the beam splitter; 4, 5—the polaroids; 6, 7—the lenses; 8—the object; 9—the hologram. The holographic image (**a**) and interferogram when rotating nerve of a frog. Reprinted from [136] with permission

2.3.4 *Study of Singe Isolated Muscle Fibers Under Gaseous Hyperbary*

An experimental setup was developed to study the changes of the refraction index and the shape of single muscle fibers under gas pressure. Its scheme is presented in Fig. 2.40.

The setup is a holographic microinterferometer with an inclined reference beam and the preliminary magnification of the object under study at the hologram recording stage. He–Na laser LG-215 with the power of 25 mW was used as a source of radiation. With the mirror 3 with the dielectric coating (reflectance $R = 30\%$ for $\lambda = 638.2$ nm at the incidence angle of 45°), the laser beam was divided into two channels: an object one (I) and a reference one (II). The camera 16 with the muscle fiber under study was placed in the horizontal plane. The object beam probing the muscle fiber was directed into the camera between the mirrors 7 and 8.

The real image of the muscle fiber was formed in the photographic film plane 17 with the consequently located microlens 12 (10×) and the eyepiece 13 (14×). Application of such a consequent magnification was dictated by the necessity of using a microobject with the clearance distance exceeding the thickness of optical camera window (4 mm).

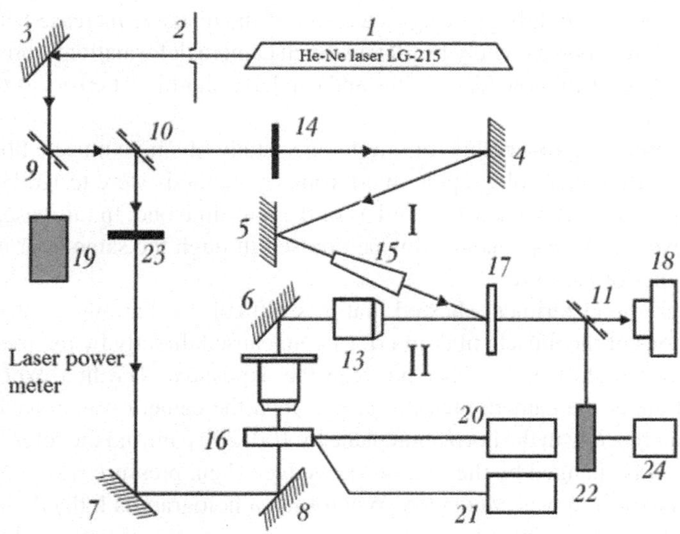

Fig. 2.40 Scheme of the device for the study of single muscle fibers by the method of holographic microinterferometry: I (II)-the object (bearing) laser beam; 1—the laser; 2—the diaphragm; 3–8—the non-transmitting mirrors; 9–11—the beam splitters; 12, 13—the microscope objectives; 14, 23—the light filters, 15—the telescopic attachment; 16—the camera with object; 17—the hologram; 18—the photocamera; 19—the laser power meter; 20—the electromyograph; 21—the gas compressor; 22—the photomultiplier tube; 24—the power unit. Reprinted from [136] with permission

For expansion and collimation of the reference beam, a telescopic attachment was used with magnification of 20×. Balancing of propagation difference between the reference and the object beams was conducted with the delay line made by the mirrors 4 and 5. The intensities of the beams were regulated by the neutral filters 14 and 23 and correlated as 1:1 during recording.

The convergence angle of the reference and the object beams was ~30° that corresponds to spatial frequencies of the bearing interference pattern in the hologram plane of ~600 mm^{-1}. For hologram recording, the photolayers PE-2 and LOI-2 were used with further processing in the developers FMG and GP-2, respectively, with washing in distilled water and drying in ethanol. Hologram fixing was not performed. The best diffraction efficiency was reached at the exposure time of ~15 s.

Holographic interferograms of the muscle fibers under study were recorded with the photocamera 18 (on the film of "Micrat 300" type) or photomultiplier tube (PMT) 22. The camera deformation character was studied beforehand in which the gaseous pressure was created. For this purpose, wave front was recorded on the hologram, which passed through the camera, under pressure $p = 0$. Then, the hologram was reconstructed by the reference beam, and the interference pattern was observed in real time in the finite width bands under the summation of wave fronts reconstructed from the hologram and passing through the camera the pressure of which was increasing.

The experiments showed that "membrane" deformations were not observed for optical windows in the pressure interval $p = 0$–4 atm. Meanwhile, the active volume of the camera was slightly growing as a result of the distance increase between the optical windows approximately by 10 μm. Such camera deformations correspond to additional phase incursion $\Delta\varphi = \text{const}$ and can have significant errors as a result of measurements.

While processing the method of interference study of single muscle fibers under hyperbary, differential holographic interferometry methods were tested both of the double-exposed interferometry as well as of the real-time one. In this case, comparison of wave fronts was conducted which passed through the same fiber under two different types of pressure.

However, the experiment showed that it is difficult to share the contribution to the alterations of the muscle fiber interferogram caused directly by the pressure and by uncontrolled shift of the fiber between the exposures. The first exposure was conducted under the camera pressure $p = 0$, then the camera was moved with the help of a microscrew in the horizontal plane by 100–200 μm, and the reference beam inclination was changed by the turn of the wedge. Then, pressure was pumped, and the second exposure was conducted. When such a hologram is lighted with such a beam, the microobject interferogram is formed in the bands of finite width, and the first image corresponds to the absence of pressure, and the second one—to microobject under hyperbary. For simplification of interferogram processing procedure, the reference bands were oriented perpendicular to the axis of the muscle fiber.

The series of holographic interferograms of single muscle fibers obtained in the bands of finite and infinite width is presented in Fig. 2.41. The local inclination of the interference band $\Delta y(x)$ at point x in the interferogram makes it possible to estimate phase incursion $\Delta\varphi(x)$ carried in by the fiber

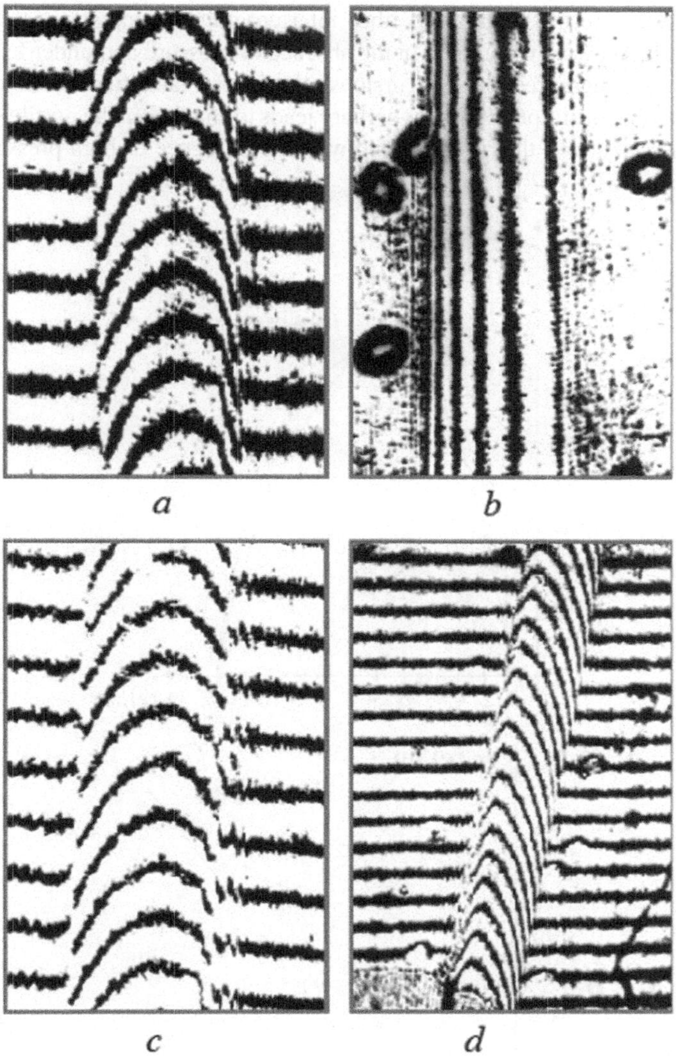

Fig. 2.41 Holographic interferograms of single muscle fiber, derived in the bands of final (**a**, **c**, **d**) and endless (**b**) widths. Reprinted from [136] with permission

$$\Delta\varphi = \frac{2\pi}{\lambda}\Delta n 2\sqrt{R^2 - x^2} = \frac{2\pi\,\Delta y(x)}{D}, \tag{2.46}$$

$$\Delta n = n - n_p,$$

where n is the average refraction index of the muscle fiber; n_p is the refraction index of the surrounding solution; R is the muscle fiber radius; D is the period of interference reference bands.

Comparative results of muscle fiber interferograms processing are shown in Fig. 2.42a, b. They were obtained in the absence of pressure and under hyperbary $p = 2.5$ atm. The measured phase incursions relate to the same fiber section.

It is clear that the impact of gas hyperbary leads to changes in the profile symmetry of the phase incursion that is connected with the influence of high pressure on the

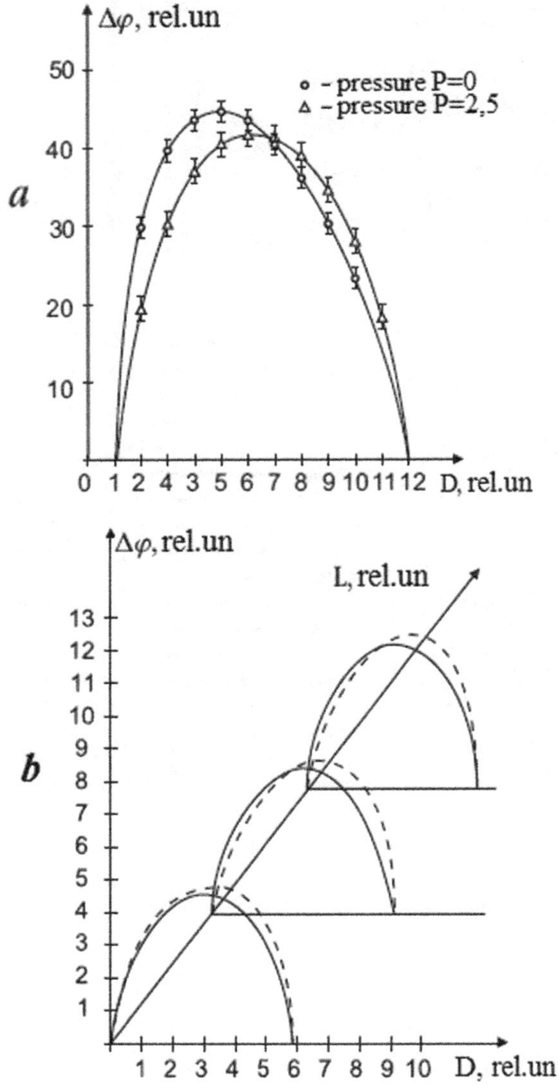

Fig. 2.42 Profile of phase incursion $\Delta\varphi$, created by muscle fiber. The profile is measured on interferograms, received at pressure $P = 0$ and $P = 2.5$ atm.; D-count on interferogram along muscle fiber diameter (**a**, **b**). Reprinted from [136] with permission

refraction index and the shape of muscle fibers. However, it is necessary to gain a certain amount of statistical data for getting information about what contributes more to the phase incursion—the refraction index or the form of the muscle fibers.

Implementation of the holographic methods adapted to the studies of isolated muscle fibers under gaseous hyperbary made it possible to get alteration of phase incursion profiles in the muscle fiber from the interference measurements at pressure application that points out the structural changes in muscle fibers.

2.4 Development and Improvement of the Holographic Interference Microscopy Method for Studying Deformations Occurred Under Thermal Heating of Mirror and Diffusely Scattering Microobjects

An isolated nerve fiber, unlike compounding it microobjects such as an isolated nerve fiber, a cell and muscle fiber in their lifetime, has a reflecting or a diffusing scattering surface that significantly hampers and sometimes makes impossible its studies by the traditional methods of optical microscopy.

It requires search and development of new methodological and technical ways, which allow conducting studies with interferometric accuracy, for example, pulsation amplitudes of a certain area of a muscle, nerve, shift value of their surface, changes of their sizes during their vital functionality or under external factors. As such factors, it is natural to observe temperature, environmental pressure and study their influence on the neuromuscular fiber. Thus, it is known that under normal conditions of nerve cell functioning lipids are in liquid-crystalline, partially ordered state and passes into another crystal phase under decrease of temperature.

Phase transition under temperature change influences greatly biological processes in a cell: Intensity and character of ion transport change sharply, kinetic characteristics of lipids change uppermost because of their viscosity that immediately affects the protein diffusion coefficient. The study of a wide range temperature impact (from low to high temperatures) is of great practical interest for the development of medical physiotherapy methods as well as for cryobiology (preservation of organs). Thus, for example, in the work [111] using holographic interference microscopy, optimal velocities of temperature decrease were ascertained, under which erythrocytes preservation is provided and so on.

The holographic interferometry methods were approved during the study of tympanic membrane vibrations. Alterations of vibrations frequency up to 5000 Hz made it possible to detect imperfections in its structure. The data show the possibility and the appropriateness of conducting the study connected with the development and improvement of the holographic interference microscopy method to analyze the neuromuscular tissue deformations, which occur under thermal heating. But during the search of the microobject, which is optimal for methodological decisions processing at the stage of method development, a model system, namely, the laser diode-heat remover, was chosen. The size of the mirror face is 150 × 259 μm^2, of the diffusing scattered face is 250 × 500 μm ^{2}. Partly the choice depends on the fact that in biology there are methods measuring amplitude of the microobject surface deflection from equilibrium position. They are based on fixing a diminutive mirror on the biological microobject under study. Methodically, it is adequate to shift measurements of the laser diode mirror face during pumping current pulse percolation. It was shown in the experiments that this method has prospects in studying the impact of the electric current, temperature and installation methods of semiconductor structures on the service life of their work.

The research of the laser diodes also attracted attention because of studying the waveguide characteristics of myelinated and non-myelinated nerve fibers. One of the suppositions about information spreading contains the idea of its posteriori transmission by transformation of electrical energy into the light one, of the light energy into the chemical one, of the chemical one into the electrical one and so on, like it happens in the fourth generation computer, where the electrical signal is transformed into the optical one with laser diodes in order to enhance the density significantly. It is interesting to point out that in Ranvie constriction area there is a substance having characteristics of semiconductor structures.

The results of author's studies connected with development of the holographic interference microscopy and the holographic microscope in the reflected light were processed and further were used in a series of collaborate researches of mechanic voltage and laser diodes deformations, which were conducted by the staff of the Semiconductor Optics Laboratory in the Institute of Physics of the NAS of Belarus. These results have an independent scientific and applied interest [55–69].

The holographic interferometry in studying semiconductor laser diodes. In this case, the holographic method is adapted to deformation changes in the laser diode-cold conduit system. The distinctive feature of this method is the possibility of real-time study of deformation with high spatial shift resolution of the mirror as well as of the diffusing scattered faces of a diode during pulse and continuous pumping currents.

Heating and mechanical stress in injection heterolasers significantly affect their power characteristics and service life [376, 377]. The analyses of heating processes in pulse laser diodes were conducted in the works [378, 379] on the basis of resonator deformation measurements with the Mickelson interferometer, semiconductor crystal face served as one of the mirrors. The thermal expansion of the heat removal was not considered. In this case, the holographic interference microscopy method in the real scale was implemented for quantitative estimation of deformation occurring in semiconductor laser diodes under continuous pumping current.

Injection lasers with two-side heterostructure in GaAS-Al_x $Ga_{1-x}As$ system were under study. The samples were soldered with indium to a copper heat sink, one side of which had a polished surface (Fig. 2.43a, b), and worked continuously under the room temperature.

One part of the diodes was fixed from the side of the p-layer, the second one from the side of the n-layer. The heterolaser-heat sink system was firmly fixed in the setup and was insulted from the holder.

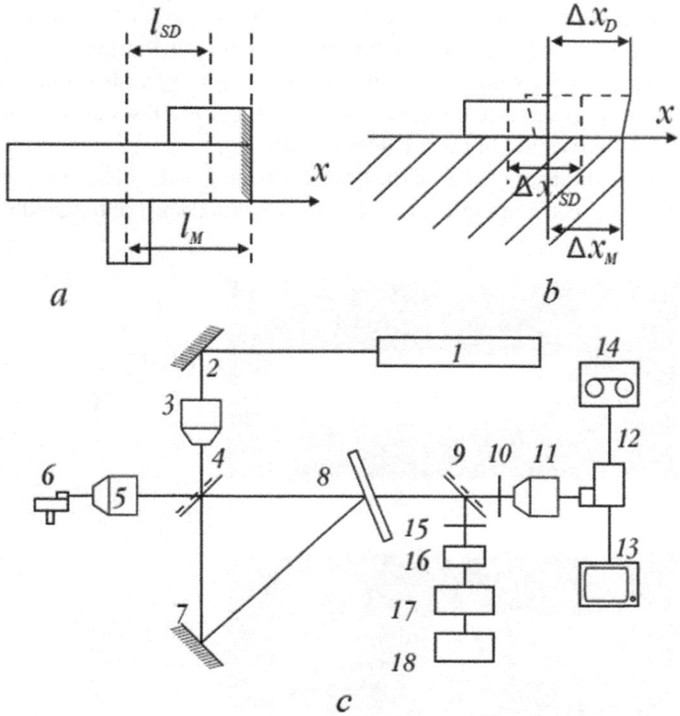

Fig. 2.43 Scheme of the experimental device: 1—He–Ne laser; 2, 7—the mirrors; 3, 5—the microscope objective; 4—the semitransparent separating plate with antireflection layer ($\lambda = 633$ nm); 6—the test specimen; 8—the hologram; 9—the semitransparent plate; 10, 15—the filters absorbing the emission of semiconductor laser; 11—the ocular; 12—the telecamera; 13—TV; 14—the video tape recorder; 16—the photodiode system; 17—the intensifier; 18—the recorder. Reprinted from [136] with permission

For temperature control of the heat sink in its upper part, joints of five thermal couples were installed.

The experimental setup (see Fig. 2.43c) made it possible to record movements of the mirror faces of a laser diode as well as of its heat sink with the accuracy of 7:10 nm. The hologram recording of the microobjects under study was conducted on a holographic interference microscope the optical scheme of which was developed with regard to specific samples [55]. The necessary ratio of the reference and the object beams intensities was reached by selecting a reflection coefficient of a semitransparent separatory plate that minimized light losses in the microscope. The surface was coated with antireflecting layer to eliminate the interference structure affected by re-reflection of the coherent radiation in the separatory element and to improve the diode image quality on one of the divider surfaces.

The holograms were recorded on the photoplates PE-2 which were exposed under the absence of current through the diode and were processed on the point. The interference image which characterizing system deformations was formed under wave front summation spreading directly from the surfaces of the heterolaser and pumping current of the heat sink. Current intensity distribution in the points of the interferogram under study and its modification were recorded with a system of recording photodetectors. Real-time interferograms of the laser diode mirror under the current I are presented in Figs. 2.44 and 2.45.

Heating of the diode as well as of the heat sink occurs under continuous generation of the laser diode. As a consequence of this, the recorded shift of the diode facet Δx_d

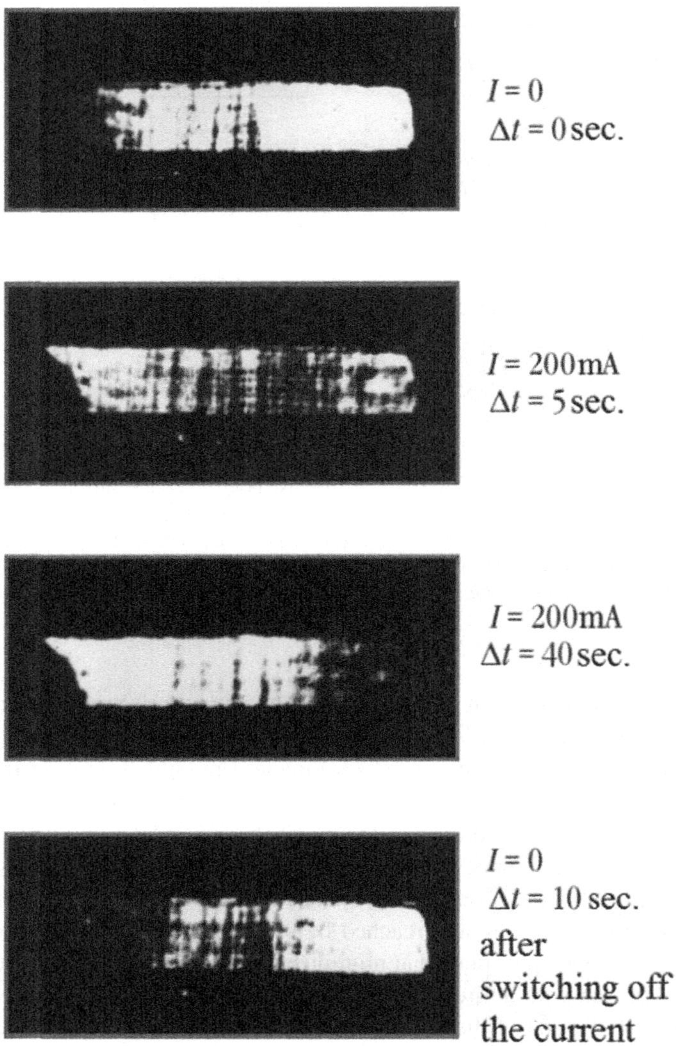

Fig. 2.44 Interferograms of a mirror of laser diode in real time at supply of pumping current I; Δt—period of time from the moment of switching on the pumping current. Reprinted from [136] with permission

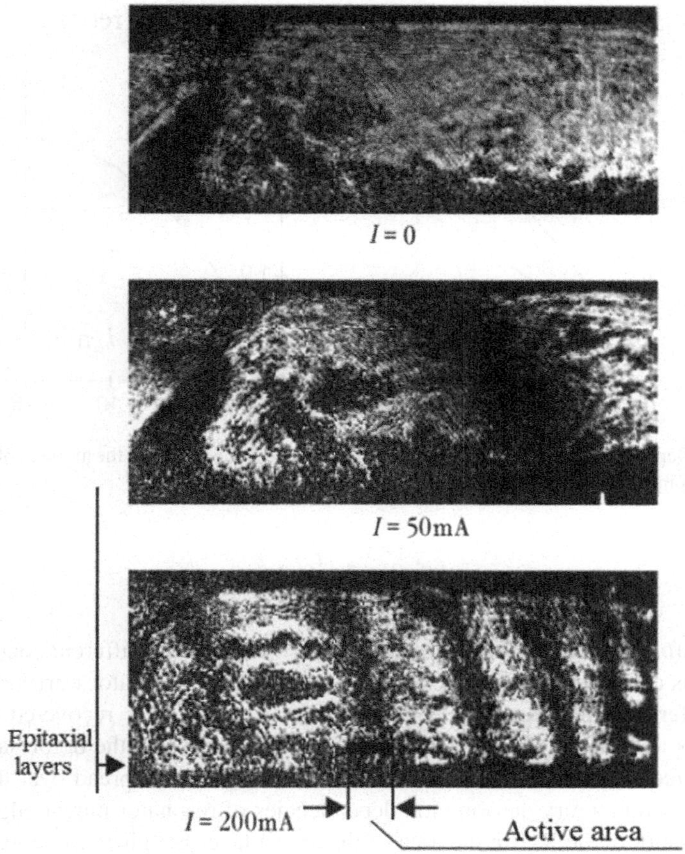

Fig. 2.45 Interferograms of resonator mirrors, derived at different levels of heterolaser excitation. Reprinted from [136] with permission

is affected by the inherent deformations δ of the semiconductor structure as well as by the transfer of the whole diode with an expanding heat sink Δx_{cd}

$$\Delta x_d = \delta + \Delta x_{sd} \tag{2.47}$$

The diode center shift value is defined by the following formula:

$$\Delta x_{sd} = \frac{l_{sd}}{l_m} \Delta x_m \tag{2.48}$$

where Δx_m is the recorded expansion of the copper heat sink; l_m is the distance between the center of the system fixing and the polished side of the heat sink.

Taking this into account, the rating formula for δ is the following

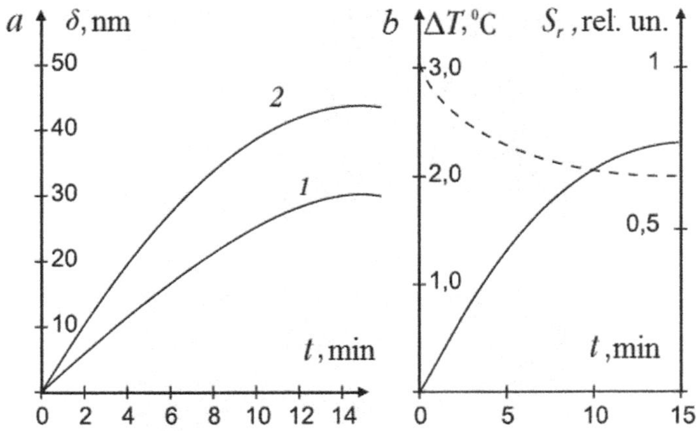

Fig. 2.46 Dependence of deformation quantity δ on the period of time from the moment of switching on the pumping current. Reprinted from [136] with permission

$$\delta = \Delta x_d + \frac{l_{sd}}{l_m} \Delta x_m \tag{2.49}$$

The deformation value of the injection heterolasers under different conditions of continuous current pump was measured. The images of resonator mirrors obtained under different excitation levels of the heterolaser [55] were recovered from the holograms and the holographic interferograms. As it is seen, the deformations are not localized only in the boundaries of the active layer and spread over the whole diode edge. Temporary deformation dependencies of resonator mirror edges of the semiconductor diode R-7 in the area of the active layer are given as an example in Fig. 2.46. After switching on the current, the deformation value δ begins to increase and reaches the maximum in the moment of steady-state thermal conditions. For studying the diode and the kind of the heat sink applied, the output in the steady-state conditions took 10–15 min that corresponds to the alterations of the power generation S_r and the temperature of the copper holder ΔT over this period of time (see Fig. 2.46b).

The data make it possible to estimate the mechanical stress occurring during the work of the diode with different holders and levels of continuous current pump. The dependence of power characteristics was studied as well as the service life of the heterolasers on the resonator deformations values. The suggested methods of deformation definition of semiconductor crystals can be used for resource prognostication of injection heterolasers.

Two-channel holographic interferometry in the study of semiconductor laser diodes [67]. In the works [58, 63], the holographic interference microscopy method was implemented in real time to find thermoelastic deformations and connected with them mechanical stresses in the semiconductor crystals of the injection lasers. While conducting such measurements, it is important to consider carefully the mechanical characteristics of the solder and the thermal deformations of the cooling vessel. One of the ways to solve this task is the simultaneous registration of the front edge shift of the semiconductor crystal and the polished surface of the cooling vessel [57, 59]. The results obtained in such a way depend on the geometry of the diode fastening and the degree of homogeneity of the temperature distribution in the cooling vessel [60]. In this work, the methods for the study of the thermoelastic deformations of the injection heterolasers are suggested. It is based on the control of two opposite edge shifts of the semiconductor crystal.

Such an approach makes it possible to minimize the influence of the cooling vessel and to expand the sphere of implementation of the holographic interferometry method for studying thermal characteristics of laser diodes.

The optical scheme of the setup is presented in Fig. 2.47. The holographic microinterferometer has two object channels I and II where thermoelastic shifts of two edges of the diode resonator are registered. With two totally reflecting mirrors 2–9, the beam splitters 10–12 and the microlenses 13 and 14, the image of the resonator edges is projected in the holography plane 19 onto two spatially separated areas, which are located within the boundaries of the reference beam incidence area. The latter is formed by the mirrors 8 and 9 and the lenses 16, 18. The hologram was recorded on the plates PE-2. The plate was processed on the place of exposure. The holographic image was recovered with the initial reference beam. In the area of the resonator edge image, an interference pattern in infinite bandwidth was observed, which corresponded to the superposition of the object wave recovered from the hologram and of the wave spreading from the diode in the moment of its excitation. The interferogram illumination was measured in real time. The calculation of the shift value in the point of the heterolaser facet along the direction of observation can be found in the works [57, 62].

The scheme of fastening the laser diode on the copper cooling vessel and the corresponding notations used during calculation of the temperature expansion of the semiconductor crystal in the diode-cooling vessel system are shown in Fig. 2.47b.

The pump current transmission through the laser diode leads to heating and thermal expansion of the semiconductor crystal as well as of the cooling vessel.

If the shifts of the two opposite edges are signed as Δx_1 and Δx_2 recorded in the interferometer channels I and II, respectively, then

$$\Delta x_1 = \delta_1 + \Delta x_{xl}; \quad \Delta x_2 = \delta_2 - \Delta x_{xl} \tag{2.50}$$

where δ_1, δ_2 are the inherent thermoelastic diode deformations; Δx_{cd} is the diode shifts as the whole along the observation axis by the expanding cooling vessel.

Thus, the full thermoelastic elongation of the semiconductor crystal is the following:

$$\Delta l = \delta_1 + \delta_2 = \Delta x_1 + \Delta x_2, (\Delta x_1 > 0). \tag{2.51}$$

Designation Δx_2 is defined by the calibrated modification of the optical path in object channel II. During calculation of Δl, the value compensation of thermal expansion of the cooling vessel occurs in channels I and II. This makes it possible

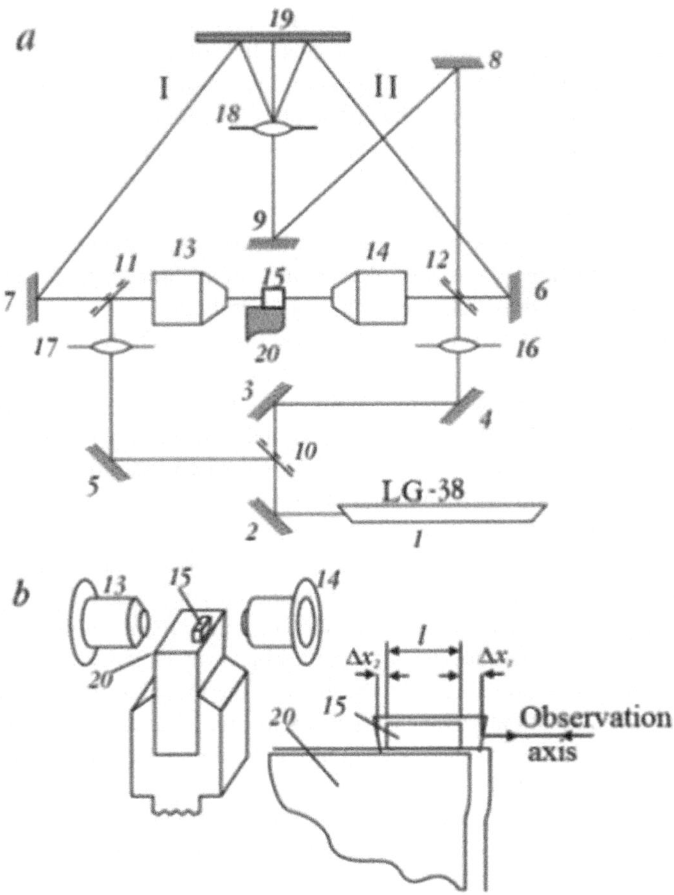

Fig. 2.47 Optical scheme of the device for the study of thermoelastic deformations of injective heterolasers (**a**): 1—He–Ne laser; 2, 9—the non-transmitting mirror; 10–12—the splitters; 13, 14—the microobjectives; 15—the object of study, 16–18—the lense; 19—the hologram) and the scheme of attachment of the laser diode 15 on the copper cooling vessel 20 (**b**): 13, 14—the microobjective. Reprinted from [136] with permission

to remove a number of restrictions during measurements constrained by the cooling vessel. In particular, the temperature distribution along the cooling vessel can have a heterogeneous character.

The methods described above were tested while defining the heating temperature of the laser diodes operating in the continuous mode of excitation. In the experiments, there were used GaAs/AlGaAs injection lasers with strip geometry of the contact fixed with the indium solder on the copper cooling vessel.

One part of the samples was mounted on a temperature compensating gasket made of pseudoalloy. Heating of the laser crystal active area relative to environmental temperature was calculated according to the formula:

$$\Delta T = \frac{(1 + a^* t_{av})\Delta l}{a^* l} \tag{2.52}$$

where l is the device resonator length under the temperature of t_{av}.

The influence on the Δl value of the resonator thermal elongation of the semiconductor crystal constraint with the cooling vessel was considered by the appropriate choice of the coefficient a^* [69]. During the experiment, the magnitude a^* was estimated due to the results of the holographic measurements relative to the edge shifts of the diode and the cooling vessel [59].

The dependencies of values ΔT on the pumping current of laser diodes gathered on the plate made of pseudoalloy and on the copper cooling vessel are shown in Fig. 2.48. The values of the superheating temperatures of the active layer in the devices under study correspond well to the data obtained with the help of other methods.

For one of the laser diodes, the comparative estimations of the structure superheating were conducted due to the measured thermoelastic deformation δ_1 of just one crystal edge without taking into account the stiffness of their connection to the cooling vessel ($\Delta T'$) (it was supposed that resonator deformation Δl of the heterolaser is equal to the doubled value δ_1) and due to Δl defined by the two-channel deformation measurement of the both mirrors, but without taking into account the stiffness of the connection in the diode-cooling vessel system ($\Delta T''$).

The results obtained are reflected in Table 2.5 and are indicative of the fact that heterogeneity of the cooling vessel heating as well as consideration of the mechanical characteristics of the connection can significantly correct the design quantity of the active layer superheating of the laser diode. The table shows the value of the active layer superheating of the GaAs/AlGaAS heterolaser, defined by the thermoelastic shifts of the device resonator shifts under different pump currents.

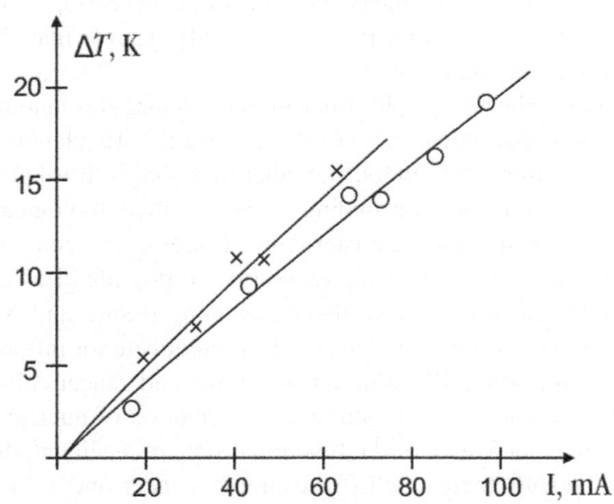

Fig. 2.48 Dependence of overheating of active area of laser crystal on the pumping current. Reprinted from [136] with permission

Table 2.5 Superheating value ΔT of the active layer of the GaAs/AlGaAs heterolaser under different pump currents I

I (mA)	20	40	60	80	100
AT′	31	74	123	139	204
AT″	12	23	35	46	58
AT	4	8	12	15	19

The results of the tests conducted are indicative of the fact that the two-channel holographic microinterferometry based on sounding of the laser diode resonator from its two opposite edges makes it possible to significantly increase the measurement accuracy of thermaelastic deformations of the injection heterolasers and thus judge more objectively about its thermal characteristics.

2.4.1 Possible Perspectives of Holographic Microscopy Development

Today, a wide practical use of holographic methods in microscopy edges a number of difficulties connected with the development of (1) methods of coherent noises removal, made by the laser systems; (2) coherent light sources with smooth wavelength broad band tuning (especially in the UV-area); (3) optical methods of contrast and luminance enhancing of certain microobject structures in the lifetime condition; (4) high-resolution photosensitive recording media; (5) automated systems of optical image processing aimed at optimization of experiment conditions during lifetime study of short-living biological microobjects; (6) storage devices, multi-pass schemes and others. That is why the development of coherent-optical, highly sensitive, contactless methods of lifetime microobject study, in particular, the holographic microscopy, is still an urgent problem.

The possibilities of the holographic microscopy enhance substantially with implementation of laser source radiation convertible along the wavelength (for example, dye lasers, F-center lasers, converters, and other [380–382]). It is known that pathologic lesions differ from the surrounding tissues in their biological composition and therefore in their spectral characteristics. Thereby, the consistent choice of the optimal wavelength of the probing radiation can provide clear structure isolation of the pathological lesions from the surrounding tissues and will influence a certain contrast of the interference pattern. The prerequisite for infrared holography is different absorption of the IR-radiation by normal and cancer cells. The infrared microscopy can be implemented to study the distribution of nucleic acids in cells. Such a distribution is indicative of the functional state of a cell and differs from the normal state in the pathological cell (for example, cancer ones). Moreover, as the radiation of the coherent sources moved to the short-wave area, then the resolution of the holographic microscopy method is advancing, and if such lasers appear, recording and studying 3D-images of macromolecules can be possible. The improvement of the resonance interferometry methods concerning a separate cell can give the possibility to study its vital functions, in particular to define the spatial distribution of the K and Na ion concentration and the dependence of the distribution alteration connected with the membrane permeability. Also, the development of the dynamic hologram recording method can be promising, where a separate cell or a whole set of cells (erythrocytes, lymphocytes or nerve cells) can be the recording medium that gives possibility to define in their volume a number of the so-called fundamental parameters, such as the coefficients of thermal conduction, diffusion and so on.

The papers are also known on electrical activity measurements of objects at the level of a separate cell of blood and a nerve tissue while using voltage-sensitive dyes, i.e., the dyes, which light absorption or fluorescence intensity changes depending on the electrical potential value. Thus, for example, the alteration of the dye fluorescence-merocyanine 540, injected into the nerve fiber during stimulation with light, the wavelength of which is 570 nm, accurately reproduces modifications of the membrane potential with the time lag of only 40 μs. The optical effect magnitude depends on the reconfigurable over the laser radiation frequency and on the potential efficiency of sensitive dyes. Development of the selective excitation methods using the switching laser radiation concerning biological preparations can favor the emergence of new "optical dyeing" methods (enhancing of the brightness contrast), which make it possible to visualize a certain structure lesion in a living cell.

2.5 Results and Conclusions

1. Preliminary research was conducted, during which different static microobjects were studied. In particular, the possibility to obtain contoured maps of microreliefed surfaces was shown using simultaneous generation of four-long-wave switching radiation of the dye pulse laser "Raduga-3M". The model object of the holographic recording was a metal relief with the height no more than 1 mm. The forming conditions of the holographic interferograms of the relief were defined at time order of 10^{-7} s, where the level difference between the neighboring bands was about 100 μm. As a result, principally the possibility of microscopic objects contouring was shown in perspective and biologically in order to clarify the peculiarities of their shape.

A number of coherent-optical (including holographic and speckle-optical) methods, setups and devices were developed that make it possible to conduct a study of the neuromuscular tissue with simultaneous electrophysiological control. In particular, a holographic setup for analyses of the morphological and functional states of excitable microobjects (a nerve fiber, a muscle fiber, a nerve) was developed and assembled. The setup allowed recording holograms and holographic interferograms of nerve fibers using different methods—the real-time one, the stroboscopic holographic one and the differential holographic interferometry method.

Specially for these purposes, there was developed and manufactured the camera for electrostimulation and biopotential derivation of the nerve fiber, which contains an optical part that enabled to conduct studies simultaneously using two independent methods: the electrophysiological one and the interference holographic one. New methods of preparation and electrophysiological control of neuromuscular tissue preparation viability (nerve, muscle fiber, nerve fiber) are mastered. The system of synchronization provided firing of the laser pumping block with a sync pulse from the electrical stimulator. The delayed relative to the firing one stimulation pulse excited the nerve fiber. In this case, the lightening of the nerve fiber with the laser pulse coincided with passing through its action potential. The recording of holographic interferograms of the nerve fiber was conducted in different excitation phases, provided by the delay line and the synchronization system.

The methods of holographic microscopy and interferometry were developed for studying intravital states of the neuromuscular tissue [1, 2, 3, 5, 7, 8, 9, 10, 11, 15]. The defocusing impact on the formation of holographic interferograms (changes of band shape, contrast), which is caused by the longitudinal object shift, was studied. It is shown that the implementation of lenses in the schemes of holographic interferometry gives no possibility to enhance sensitivity to the longitudinal shift measurement but improves the spatial resolution. The contrast of the interference pattern emerged under the shift of the diffuse object is defined by the function, which is a three-dimensional pulse response of the lens with the argument equal to the shift vector. There were examined cases for plane mirror and the diffuse object shift. For the latter case, it is shown that the range of the shifts measured is limited by the depth of focus of the lens. The conducted study confirms in terms of quantity the qualitative conclusions about the limits of possible shifts in the holographic interferometry schemes. The range increase of the object longitudinal shifts can be reached by the loss of resolution by aperture decrease. During the interference study of objects, which change amplitude distribution of the illuminating beam, it is important to consider the fact that defocusing distorts the image itself as well as the phase difference between the images.

The experimental samples of holographic microscopes were developed, which operate on transmission and in the reflected light and applicable for the study of phase, specular reflected and diffusing microobjects that change their state with time. For the first time, the possibility to study the structures of nerve fibers, muscle fibers, lymphocytes was shown using the holographic microscopy and interferometry methods. The conditions of nerve fiber hologram and holographic interferogram forming were studied, which made it possible to single out and visualize its main structural formations.

There was shown the absence of noticeable damaging effect of the laser radiation in doses necessary for hologram and holographic interferogram recording on the structure and the function (action potential transmitting) of the nerve fiber. Also, optimal conditions for hologram and holographic interferogram recording of a nerve, a muscle fiber, lymphocytes were studied in the bands of finite and infinite width. The method of single nerve cells cultivating was perfected (dissociated cultures). Recording of bioelectrical activity of a single nerve cell was mastered. The possibility of recording a hologram or a holographic interferogram of a nerve trunk in the transmitted as well as in the reflected light was experimentally shown. The obtained holograms make it possible to judge about structural and phase peculiarities of these objects.

For the first time, the refraction structure (of the refractive index profile) of intravital myelinated nerve fibers was studied on the created experimental sample of the transflective holographic interference microscope. There were obtained interferograms of the fibers with the diameter of ~10 μm and spatial resolution of more than 1 μm, and the interference fringe frequency, which is one order higher than the fringe frequency in a classic interference microscope.

On the nerve fiber interferograms, the myelin sheath in different shape variations could be easily observed. In the center of the nerve fiber area, the perikaryon of Schwann's cell with the nucleus is situated. In the interferograms, it is seen that the most significant band movement is observed in the axial area of the fiber and the myelin sheath. This band movement is conditioned by the cylindrical shape of the fiber and a significantly higher refractive index of lipoprotein structures of the myelin sheath in comparison with the environment.

The software was created to calculate the radial distribution of the refractive index (in this case the nerve fiber) of axially symmetric phase objects. There was developed and implemented as a software one of the most prospective deciphering algorithms of holographic interferograms, namely, the Abel inversion algorithm based on the approximation of experimental data by cubic spline-functions with boundary conditions corresponding to interference fringe alterations. Radial distribution of nerve fiber refractive index was calculated.

A fundamental result is experimentally measured for the first time the refractive index distribution of intravital myelinated nerve fibers with the electrophysiological control. The result was obtained using the developed by the author and the co-workers holographic interference microscopy method and the transflective holographic microscope with the spatial frequency of the interference fringes, which is one order higher than the fringe frequency in a classic interference microscope.

The radial distribution of the refractive index of Schwann's sheath and the myelinated nerve fiber axon were studied. The obtained values of the refractive index of myelin (1.44) and axoplasm (1.36) correspond to the native state of these structures in the fiber. The experimentally measured rather high for biological structures value of the refractive index of the nerve fiber myelin sheath (1.44 ± 0.01) matches morphological data about the thick package of the protein-lipoid membranes of this structure. The changes of the axoplasm on the fiber axis are much bigger than on the areas directly under the myelin sheath. It is ascertained that the dependence character of the refractive index on the radius of the internodal area of the myelinated nerve fiber is similar to the profile changes of the refractive index of separate optical waveguides.

The refractive index distribution of Schwann's sheath and the axoplasm of the myelinated nerve fiber is analogous to the refractive index distribution in the tubular optical fiber, and the muscle fiber is a perfect diffractive grating, a splitter and a light energy re-distributor as it occurs in optical waveguide communication lines. This fact gives us the reason to hypothesize about the optical mechanism of information transmission in the neuromuscular system (probably of the infrared radiation or radiation of other ranges). It is interesting that Ranvie constriction area is a light semiconductor in its essence (a kind of heterolaser) and this is also included in this hypothesis as a self-sustained system (substation) that provides the nerve pulse signal from attenuation, like the soliton spreading along the neuromuscular system. In addition to that, the refractive characteristics of the myelinated nerve fibers are investigated. The data also indicate the presence of the waveguide characteristics of the nerve fibers and as a result of the nerve tissue. The further studies in this direction make it possible to understand better the mechanism of action of laser radiation while treating PNS-diseases, in particular, using the laser acupuncture method. It is important to point out that the laser stimulating effect enhances if several biologically active points, which are responsible for the disease, are simultaneously irradiated (the interference effect). Under the experimental conditions, there were no structural alterations in the sciatic nerve myelinated fiber of a frog to observe. In addition to that, stimulating impact of the continuous laser radiation on the nerve fiber was observed that is presented in a number of cases in a form of slight increase of action potential. It was found that the laser radiation positively influences the regeneration of the nerve system.

Considering the results of our research connected with detection of a waveguide mechanism of electromagnetic radiation distribution of waves along the nerve fibers and the results of clinical study (using the laser acupuncture method and the intravascular laser blood irradiation methods), we can assume that along with an electrical, chemical model, which explains the information transmission mechanisms in a human organism via nerve formations, there exists an optical waveguide model as well, which functions in the infrared radiation range as well as in radiations of other ranges with the possibility of interconversions and interactions of electrical, chemical, optical (waveguide) pulses with each other: the chemical one into the optical one, the electrical one into the optical one, the chemical one into the electrical one and so on [223–227, 383–438].

2. The urgency of expanding the methodological base of the studies of hyperbary impact mechanisms on the human organism follows from the analysis of literature. Being complex in its essence, this impact requires complex methodological support.

The result of the conducted study is the development of several coherent-optical (holographic and speckle-optical) methods for intravital research of isolated preparations of nerve and muscle fiber under normal and high pressure with simultaneous electrophysical control.

The study conducted made it possible to make a number of conclusions.

It was the first time when there were shown perspectives of implementation of the holographic interference method as a contactless noninvasive way for quantitative recording of structural alterations of nerve and muscular fibers under hyperbary.

There were developed and tested two types of optical cameras for holographic study of structural and functional changes in the aging nerve and nerve fiber preparations under gaseous and hydrostatic pressure that make it possible to implement high pressure modes in broad limits from 0 to 200 atm.

To estimate the alterations of the refractive index and shape of the isolated nerve and muscle tissue preparations under high gaseous pressure, the holographic method was developed that makes it possible to obtain interferograms of the microobjects under study with magnification of 140× in the pressure range of 0:5 atm. There were obtained finite band holographic interferograms of the nerve and muscle fibers under normal conditions and under high pressure of 4 atm. with the accuracy of path difference measurement of interfering beams $\lambda/30$. The implementation of this method showed in a number of cases the profile changes of phase incursions in a single muscle fiber under pressure application that is indicative of its possible structural changes.

The adaptation of the electrophysiologically controlled laser diffractometry method for the study of function alteration of muscle fiber under aging hyperbary proved that pressure application (~3.5 atm.) can lead to spasmodic irreversible excitability oppression. On the whole, the fibers under the gaseous pressure were characterized by stronger marked excitability fluctuations and had statically lesser time of preparation aging than in the control.

Using the holographic interference microscopy method adapted to studying the degree of compression of isolated nerve preparation under high hyperbary (up to 200 atm.), the absence of mechanical damaging effect of the high hydrostatic pressure ($\Delta R/R \leq 10^{-5}$) was ascertained that is indicative of the fact that failures of membrane processes are likely to underlie changes of nerve functions under these conditions.

In order to study the changes in the refractive index of the isolated nerve under hydrostatic pressure, there was developed the method of obtaining hologram of semitransparent diffusely scattering objects with a varying microstructure and having cylindrical symmetry. The method is based on the spatial filtration of Fourier-spectrum of the radiation passed through the object.

During the electrophysiological study of excitability of isolated nerve tissue preparations, it was found that the hyperbary impact (air up to 3.5 atm.) is oppressed by electrogenesis that provides generation and conduction of action potential caused by electrical stimulation of isolated nerve fibers as well as nerve fibers as a part of the isolated sciatic nerve of a brown frog. This manifests in action potential attenuation of an isolated nerve fiber in response to the application with pressure. The impact of the gaseous pressure (3.5 atm.) leads to irreversible electrogenesis loss of single fibers. The alteration of the nerve fiber compound action potential of the single nerve in response to the hyperbary impact has a more complex character: The first applications on the fresh-separated nerve can influence slightly the electrogenesis that however intensifies as the time of preparation aging increases. This leads to action potential amplitude contraction, an increase of dispersion of latent period and length, growth of stimulation threshold values that finally leads to the loss of ability to generate the action potential in response to the electrical stimulation. This occurs significantly earlier than in conditions of aging without any impact on the isolated nerve. In the case of the isolated nerve, the oppressive action of hyperbary increased by the following decompression. Under hydrostatic pressure, the decrease of nerve excitability was observed beginning from 160 atm. and had a deo-dependent character. This effect had a reversible character: after full decompression, there was observed a temporary recovery of excitability of the isolated nerve was observed.

The method of lymphoid cells cultivation was developed to obtain viable suspensions. There were conducted experiments in studying the lymphocytes functional activity under antigens impact with the holographic interference microscopy method. It was ascertained that the antigens impact leads to decrease of optical density of the immune (pathologic) lymphocytes in comparison with the control (normal) lymphocytes. With the holographic interference microscopy, the results were obtained, which gave possibility to determine the criterion for differentiation of the immune and non-immune lymphocytes that can be of great interest for the applied medicine for laboratory diagnostics improvement.

The hyperbary impact was tested for proliferative activity of animals lymphocytes as well. It was ascertained that the proliferative activity in the spleen cells as well as in the thymus ones increases under mitogens that testifies the intensification of the organism response under high pressure.

3. There was developed and tested the holographic interference microscopy method for thermal process study in laser diodes. In particular with its help, there were studied in real time the resonator edge deformations of two-sided heterostructured injection lasers in the system under the pump current perlocation (100–200 mA) and the cooling vessel heating in the range of temperature from 289 to 423 K. The measurement accuracy of laser diode edge deformation in the direction of observation for interference fringe recording was 15–20 nm when the absolute values of deformation were 1–2 μm. The application of the developed method made it possible to study the dependence of the relative velocity of the heterolaser degradation on the thermoelastic voltage values. The holographic research of edge deformations of the laser diode and the cooling edge under heating of the latter made it possible to study the mechanical characteristics of the connection of the system injection laser–indium solder–cooling vessel in the temperature interval of 289–423 K.

To study the thermal process dynamics in pulse heterolasers (pump currents are 50–200 mA), a holographic method is developed, which makes it possible to conduct measurements considering oscillation in the time of the zeroth band of the interference pattern. The holographic way of measurement with the accuracy of 2% of thermal deformations and linear expansion coefficients of thin (5–10 μm) epitaxial heterolaser layers was developed based on the multiple reflection of the object wave from the surface of the layer under study and introduction of the calibrated phase delay in the reference wave.

The two-channel holographic microinterferometry method was developed based on the resonator sounding of the laser diode from the side of its opposite edges that made it possible to increase significantly the measurement accuracy of the thermoelastic deformations of the injection heterolasers by minimizing the influence of the cooling vessel and thus more objectively judge their thermal characteristics.

The conducted experimental studies of laser diode deformations make it possible to conclude that these holographic interference microscopy methods have good perspective for studying thermal and mechanical characteristics of heterostructures.

4. Using the methods of laser diffractometry and electrophysiology, the impact of a powerful pulse magnetic field on the neuromuscular preparation state (nerve, muscle fiber) was studied. The direct influence of the powerful pulse magnetic field on the contractile reaction of the muscle fiber was not observed. It is shown that in the moment of the magnetic field impact (~25 kOe) the diffraction pattern of the muscle fiber is stable. The functional state of the nerve under magnetic field estimated by the action potential amplitude was changing. This reaction has a complex structure multi-directional in time. It is ascertained that the character of changing the action potential amplitude is affected by the initial functional state of the nerve conductors.
5. The results of the study on the impact of the pressure and the pulse magnetic field on the nerve and muscle fibers, lymphocytes, a nerve were used in the Institute of Medical and Biological Problems (Moscow), Pavlov Institute of Physiology of the Academy of Sciences of the USSR (Leningrad), the Institute of Neurology, Neurosurgery and Physiotherapy of the Ministry of Health of the BSSR, the Institute of the Rigid Body and Semiconductors of the Academy of Sciences of the BSSR, the Institute of Physics of the Academy of Sciences of the BSSR (Minsk). The developed methods of acupuncture and intravascular blood irradiation for treatment of PNS-diseases were widely implemented in the Institute of Neurology, Neurosurgery and Physiotherapy of the Ministry of Health of the BSSR and in the Institute of Neurology (Moscow, Russia). With the help of coherent-optical methods and devices, in particular the holographic interference method, developed with the active participation of L. V. Tanin, the results were obtained, which became the basis of a number of theses.

References

1. L.E. Alikevich, A.V. Goroshkov, A.S. Rubanov, L.V. Tanin, et al., *Abstracts of III All-Union Conference on Holography*, Ulyanovsk, the USSR, 1978, p. 105. (in Russian)
2. L.E. Alikevich, A.V. Goroshkov, A.S. Rubanov, L.V. Tanin, et al., Proc. AS BSSR, Biol. Ser. **6**, 118 (1979). (in Russian)
3. L.V. Tanin, A.V. Goroshkov, G.K. Nedzved, Peripher. Nerv. Syst. **4**, 76 (1981). (in Russian)
4. K.P. Antonov, G.K. Nedzved, V.I. Kozyro, N.V. Solovey, M.I. Kalyuta, L.V. Tanin, Copyright Certificate No. 1183100 (USSR), 08 June 1985. (in Russian)
5. A.V. Goroshkov, I.V. Makrhvida, A.S. Rubanov, L.V. Tanin, V.N. Kalyunov, *Abstracts of IV All-Union Conference on Holography*, Yerevan, the ASSR, 1982, Part 2, p. 842. (in Russian)

6. I.P. Antonov, G.K. Nedzved, N.V. Solovey, V.V. Panteleyev, L.V. Tanin, *Proceedings of the Conference "Problems of Development and Application in the Economy of Laser and Optoelectronic Technology"*, Minsk, 1982, Part 2, p. 115. (in Russian)
7. L.E. Batai, L.V. Tanin, G.V. Abramchik, R.M. Tanina, Proc. AS BSSR Biol. Ser. **4**, 90 (1982). (in Russian)
8. I.P. Antonov, A.V. Goroshkov, V.N. Kalyunov, I.V. Markhvida, A.S. Rubanov, L.V. Tanin, Proc. SPIE **370**, 133 (1982)
9. I.P. Antonov, A.V. Goroshkov, V.N. Kalyunov, I.V. Markhvida, A.S. Rubanov, L.V. Tanin, Zh. Priklandnoi Spektroskopii **39**(1), 103 (1983). (in Russian)
10. I.V. Markhvida, A.V. Goroshkov, L.V. Tanin, Proc. AS BSSR Biol. Ser. **4**, 104 (1983). (in Russian)
11. N.V. Solovey, G.K. Nedzved, L.V. Tanin, *Abstracts of the Conference "Application of Lasers in Medicine in the Copper"*, Krasnoyarsk, the USSR, 1984, p. 134. (in Russian)
12. G.K. Nedzved, V.V. Panteleyev, N.V. Solovey, L.V. Tanin, *Proceedings of All-Union Conference "The Use of Lasers in the National Economy*", Zvenigorod, the USSR, 17–20 May 1985, p. 191. (in Russian)
13. A.V. Goroshkov, L.V. Tanin, B.J. Wilner, Proc. AS BSSR Biol. Ser. **1**, 115 (1987). (in Russian)
14. A.S. Rubanov, L.V. Tanin, V.V. Dubovik, Proc. AS BSSR Biol. Ser. **6**, 102 (1988). (in Russian)
15. L.V. Tanin, L.E. Bataille, V.V. Dubovik, Peripher. Nerv. Syst. **12**, 41 (1989). (in Russian)
16. V.A. Lapina, R.M. Tanina, *Abstracts of the Conference "The Action of Low-Intensity Laser Radiation on Blood"*, Kiev, the UkSSR, 1989, p. 47. (in Russian)
17. R.M. Tanina, V.J. Latysheva, *Abstracts of I All-Union Congress of Immunology*, Sochi, the USSR, 1989. (in Russian)
18. V.A. Lapina, R.M. Tanina, *Abstracts of the Scientific and Practical Conference "Low Intensity Laser Radiation in Medical Practice"*, Khabarovsk, the USSR, 1990. (in Russian)
19. L.V. Tanin, L.A. Vasilevskaya, G.K. Nedzved, *Technical Digest of ICO International Topical Meeting on Optical Computing*, Minsk, 1992, p. 309
20. L.V. Tanin, V.A. Lapina, S.K. Dik, S.A. Aleksandrov, R.M. Tanina, Proc. SPIE **1887**, 158 (1993)
21. V.A. Lapina, R.M. Tanina, Proc. SPIE **2086**, 7 (1993)
22. L.V. Tanin, S.K. Dik, S.A. Aleksandrov, R.M. Tanina, *Optic for Protection of Man and Environment Against Natural and Technological Disasters*, 239 (1993)
23. S.A. Aleksandrov, S.K. Dik, L.V. Tanin, V.A. Lapina, R.M. Tanina, Peripher. Nerv. Syst. **16**, 13 (1993). (in Russian)
24. S.A. Aleksandrov, S.K. Dik, M.M., Loyko, L.V., Tanin, G.G. Petrovsky, Patent of Russian Federation 2061514, 10 Jun 1996. (in Russian)
25. S.A. Aleksandrov, S.K. Dik, M.M. Loyko, L.V. Tanin, G.G. Petrovsky, R.M. Tanina, Patent application 93028292, 01 Jun 1993. (in Russian)
26. R.M. Tanina, V.A. Lapina, et al., *Optical Methods in Biomedical and Environmental Sciences: Selected Contributions of the 3rd International Conference on Optics within Life Sciences OWLS III*, Waseda University, Tokyo, Japan, 10–14 Apr 1994, p. 199
27. L.V. Tanin, G.G. Petrovsky, R.M. Tanina, *Abstract Book of European Biomedical Optics Week Bios Europe'96*, vol. 2922(75), p. 96 (1996)
28. N.I. Nechipurenko, A.R. Gavrilova, R.M. Tanina, G.G. Petrovsky, L.V. Tanin, *Abstracts of the Third Congress of Belarusian. Society Islands Photobiology and Biophysics*, Minsk, 21–23 Oct 1998, p. 185. (in Russian)
29. A.F. Smeyanovich, N.I. Nechipurenko, A.L. Tanin, Patent of the Republic of Belarus BY4289, 23 Apr 1998. (in Russian)
30. I.P. Antonov, G.K. Nedzved, N.I. Nechipurenko, L.V. Tanin, et al., *Proceedings of the Congress of Neurologists and Neurosurgeons of the Republic of Belarus*, Minsk, Belarus, 2003. (in Russian)
31. V.A. Lapina, L.I. Matusevich, R.M. Tanina, *Abstracts of the International Conference "Laser optics'95"*, St.-Petersburg, Russia, 1995. (in Russian)

32. I.P. Antonov, L.V. Tanin, L.A. Vasilevskaya, S.K. Dik, R.M. Tanina, Final Report on R&D No. 1996381, Minsk, 28 Mar 1996. Minsk, 1998. (in Russian)
33. I.P. Antonov, L.V. Tanin, N.I. Nechipurenko et al., Adv. Med. Sci. Belarus **3**, 5 (1999). (in Russian)
34. A.L. Tanin, Actual problems of neurology and neurosurgery A, 134 (1999). (in Russian)
35. A.L. Tanin, L.A. Vasilevskaya, A.F. Smeyanovich, Actual Problems Neurol. Neurosurg. **1**, 138 (1999). (in Russian)
36. L.A. Vasilevskaya, N.I. Nechipurenko, L.V. Tanin, Application of low intensity radiation of helium-neon laser for diagnostics and treatment of ischemic damage to peripheral nerves, Paper presented at I Caucasian Symposium on Life Sciences, Tbilisi, 1999. (in Russian)
37. N.I. Nechipurenko, L.A. Vasilevskaya, A.L. Tanin, Pathological physiology and experimental therapy, 13 (1999). (in Russian)
38. N.I. Nechipurenko, L.A. Vasilevskaya, R.M. Tanina, Theory and practice of medicine: a review. Sci. Pract. Yearbook **2**, 88 (2000). (in Russian)
39. A.L. Tanin, N.I. Nechipurenko, A.F. Smeyanovich, Med. News **7**, 55 (2000). (in Russian)
40. A.F. Smeyanovich, A.L. Tanin, N. Nechipurenko et al., Eur. J. Neurol. **7**(3), 1079 (2000)
41. A.L. Tanin, Actual Problems Neurol. Neurosurg. **2**, 171 (2000). (in Russian)
42. R.M. Tanina, YuI Musienko, Actual Problems Neurol. Neurosurg. **3**, 157 (2001). (in Russian)
43. A.L. Tanin, Actual Problems Neurol. Neurosurg. **3**, 152 (2001). (in Russian)
44. N.I. Nechipurenko, L.A. Vasilevskaya, L.V. Tanin, *Abstracts of the International Conference on Optics within Life Sci.*, Lucerne, 2002
45. A.L. Tanin, V.I. Hodylev, Actual Problems Neurol. Neurosurg. **4**, 138 (2002). (in Russian)
46. Yu. I Musienko, R.M. Tanina, A.S. Mastykin, T.I. Cherpuhina, Actual Problems Neurol. Neurosurg. **4**, 128 (2002). (in Russian)
47. L.A. Vasilevskaya, N.I. Nechipurenko, L.V. Tanin, *Proceedings of III Third International Scientific and Technical Conference "Means of Medical Electronics and the New Medical Technologies,"* Minsk, 2002, p. 262. (in Russian)
48. A.L. Tanin, N.I. Nechipurenko, L.A. Vasilevskaya, *Proceedings of III Congress of Neurosurgeons of Russia, St. Petersburg*, 2002, p. 538. (in Russian)
49. N.I. Nechipurenko, L.A. Vasilevskaya, L.V. Tanin, Y.I. Musienko, The assessment of efficiency of intravenous laser irradiation of blood in experimental cerebral ischemia, paper presented at VII International Conference on Optics Within Life Sciences, Luzern, Switzerland, 1–5 Sept 2002
50. L.V. Tanin, L.A. Vasilevskaya, Paper presented at the International conference" Laser 2003," St. Petersburg, 2003. (in Russian)
51. L.A. Vasilevskaya, L.V. Tanin, Paper presented at the International conference "Laser physics and laser applications," Minsk, 2003. (in Russian)
52. N.I. Nechipurenko, L.V. Tanin, L.A. Vasilevskaya, *Proceedings of the International Conference "Laser and Optical Technologies of Biology and Medicine"* 1, 85 (2004). (in Russian)
53. L.V. Tanin, N.I. Nechipurenko, L.A. Vasilevskaya, G.K. Nedzved, S.E. Rovdo, A.L. Tanin, *Laser Hemotherapy in the Treatment of Diseases of the Peripheral Nervous System* (Magic Book, Minsk, 2004) 148 p. (in Russian)
54. L.V. Tanin, *Proceedings of Russian Seminar "Yuri Denisyuk—The Founder of the National Holography*, St. Petersburg, 2007, p. 216. (in Russian)
55. L.E. Batai, L.N. Guzov, V.F. Voronin, G.T. Pak, G.I. Ryabtsev, L.V. Tanin, I.V. Yashumov, *Abstracts of All-Union Conference "The Use of Lasers in Science and Technique,"* Leningrad, 1980, p. 114. (in Russian)
56. L.E. Batai, V.F. Voronin, G.I. Ryabtsev, L.V. Tanin, *Proceedings of the Conference "Problems of Development and Application in the Economy of Laser and Optoelectronic Technology"*, Minsk, 1982, p. 46. (in Russian)
57. L.E. Batai, V.F. Voronin, A.N. Kuzmin, G.T. Pak, G.I. Ryabtsev, L.V. Tanin, *Proceedings of the IV All-Union Conference of Holography*, Erevan, 1982, p. 759. (in Russian)

58. L.E. Batai, V.F. Voronin, A.N. Kuzmin, G.T. Pak, G.I. Ryabtsev, L.V. Tanin, *Proceedings of III All-Union Conference on Physical Processes in Semiconductor Heterostructures*, Odessa, 1982, p. 130. (in Russian)
59. L.E. Batai, Y.L. Bessonov, V.F. Voronin, A.N. Kuzmin, G.T. Pak, G.I. Ryabcev, L.V. Tanin, S.M. Sapozhnikov, Tech. Phys. Lett. **9**(1), 6 (1983). (in Russian)
60. A.N. Kuzmin, G.I. Ryabtsev, L.V. Tanin, in *Lasers and Optical Nonlinearity*, ed. by P.A. Apanasevich, A.S. Rubanova (Institute of Physics, Minsk, 1984) p. 185. (in Russian)
61. L.E. Batai, I.V. Markhvida, A.N. Kuzmin, L.V. Tanin, Optics. Commun. **57**(5), 307 (1986)
62. L.E. Batai, A.N. Kuzmin, G.I. Ryabtsev, L.V. Tanin, *Proceedings of the Conference "Holography in Industry and Scientific Researches"*, Grodno, 1986, p. 15. (in Russian)
63. A.N. Kuzmin, S.S. Kurlenkov, G.I. Ryabtsev, S.M. Sapozhnikov, L.V. Tanin, Proc. AS BSSR Phys. Ser. **5**, 76 (1986). (in Russian)
64. L.E. Batai, L.V. Tanin, *Proceedings of VIII Resp. Conference of Young Researchers in Physics*, Minsk, 1986, p. 134. (in Russian)
65. L.E. Batai, A.N. Kuzmin, G.I. Ryabtsev, L.V. Tanin, *Abstracts of All-Union Conference "The Use of Modern Physical Methods in the Non-destructive Research and Control,"* Khabarovsk, 1987, p. 213. (in Russian)
66. L.E. Batai, A.N. Kuzmin, G.I. Ryabtsev, M.I. Sverdlov, S.A. Sosnovsky, L.V. Tanin, *Proceedings of VII Inter-republic Scientific Seminar "Lasers and Optical Nonlinearity"*, Minsk, 1986, p. 153. (in Russian)
67. L.E. Batai, A.N. Kuzmin, G.I. Ryabtsev, L.V. Tanin, *Proceedings of VIII Inter-republic Scientific Seminar "Lasers and Optical Nonlinearity"*, Vilnius, 1987, p. 236. (in Russian)
68. A.N. Kuzmin, G.I. Ryabtsev, V.P. Duraev, L.V. Tanin, Proc. AS BSSR Phys. Ser. **3**, 67 (1988). (in Russian)
69. A.N. Kuzmin, G.I. Ryabtsev, et al., Proc. AS BSSR Phys. Ser. **4**, 63 (1985). (in Russian)
70. Yu.I. Ostrovsky, L.V. Tanin, ZhTF **45**(8), 1756 (1975). (in Russian)
71. L.I. Rachkovsky, L.V. Tanin, *Proceedings of the 19th All-Union School on Coherent Optics and Holography*, Grodno, 1988. (in Russian)
72. Yu.I., Ostrovsky, L.V. Tanin, *Proceedings of II All-Union Conference "Lasers Based on Complex Organic Compounds and Their Usage"*, Minsk, 1977, p. 331. (in Russian)
73. I.V. Markhvida, Correlational properties of dynamic speckle-fields and their use for measuring the velocities and displacements of diffuse objects, Abstract of the Dissertation, Belarusian State University, 1989. (in Russian)
74. S.E. Rovdo, Clinical evaluation of the effectiveness and the state of peroxide oxidation of lipids in patients with neurological manifestations of osteochondrosis in the treatment of intravenous laser blood irradiation and photoblocks, Abstract of the Dissertation, Belarusian State Institute for Advanced Training of Doctors, 1998. (in Russian)
75. L.A. Vasilevskaya, Functional disorders of neuromotor apparatus, their speckle-optical diagnostics and correction by low intensity laser radiation in certain diseases of peripheral nervous system, Abstract of the Dissertation, Minsk, 1999. (in Russian)
76. S.K. Dik, Methods and means of speckle-optics for the control of the movements and vibrations of diffuse objects, Abstract of the Dissertation, Minsk, 1999. (in Russian)
77. A.L. Tanin, Recovery of function of peripheral nerves after neurography under the influence of laser hemotherapy, Abstract of the Dissertation, Minsk, 2003. (in Russian)
78. D. Gabor, Nature **161**(4098), 777 (1948)
79. D. Gabor, Proc. R. Soc. **197**, 454 (1949)
80. D. Gabor, Proc. Phys. Soc. B **64**, 449 (1951)
81. D. Gabor, Research **4**, 107 (1951)
82. D. Gabor, *Proceedings of the Congres de Microscopie Electronique* (1952)
83. G. Appelt, *Introduction to the Methods of Microscopic Examination*, 3rd edn. (Medgiz, Moscow, 1959) 426 p. (in Russian)
84. G.M. Frank, *Modern Methods and Techniques of Morphological Logical Studies* (Nauka, Leningrad, 1978). (in Russian)
85. E.J. Feleppa, Phys. Today **22**(7), 25 (969)

86. Yu.I. Ostrovsky, M.M., Butusov, G.V. Ostrovskaya, *Holographic interferometry* (Nauka, Moscow, 1977) 335 p. (in Russian)
87. M. Zussman, *Developmental Biology* (Mir, Moscow, 1977). (in Russian)
88. V.M. Ginsburg, B.M. Stepanov, et al., Proc. VNIIOFI. Hologr. Ser. B **2** (1972). (in Russian)
89. R.F. van Lighten, H. Osterberg, Hologr. Microsc. Nat. **211**, 282 (1966)
90. G. Ellis, Science **154**, 1195 (1966)
91. G. Knox, Science **153**, 989 (1966)
92. K. Snow, R. Vanderwarker, Appl. Opt. **7**(3), 549 (1968)
93. V.M. Ginsburg, B.M. Stepanov, DAN BSSR **302**(2), 313 (1972). (in Russian)
94. M.E. Gurary, A.A. Magomedova, G.I. Rukshan, V.K. Sakharov, DAN BSSR **208**(1), 233 (1973). (in Russian)
95. V.M. Ginsburg, Instrum. Exp. Tech. **3**, 21 (1975). (in Russian)
96. Heflindger, G.L. Stewart, Appl. Opt. **17**(6), 944 (1978)
97. Heflindger, G.L. Stewart, Appl. Opt. **17**(6), 951 (1978)
98. G. von Bally, Laser + Electro Opt. **2**(32), 1977 (1977)
99. G. von Bally, *Proceedings of the Conference "Applications of Holography and Optical Data Processing,"* Jerusalem, 1977, p. 593
100. H.M. Presby, Appl. Phys. Lett. **21**(1), 31 (1972)
101. T. Uyemura, H. Matsumoto, Bull. Jap. Soc. Press **10**(3), 101 (1976)
102. V.L. Koulis, *Proceedings of Works of Postgraduate Students of LITMO*, Leningrad (1974). (in Russian)
103. N. Barakat, A.M. Hamed, Optic **76**(3), 102 (1987)
104. P.A. Dorokhin, et al., ZhTF **54**(3), 511 (1984). (in Russian)
105. A.M. Karger, G.M. Holeman, Appl. Opt. **11**(7), 1646 (1971)
106. V.M. Ginsburg, E.G. Semenov, S.N. Smirnova, B.M. Stepanov, DAS BSSR **302**, 313 (1972). (in Russian)
107. E.J. Feleppa, Phys. Today **22**(7), 25 (1969)
108. Z. Baranowski, Cytobiologie **13**, 118 (1976)
109. S. Himeno, et al., *Proceedings 14 Congress International Communication Optics*, Quebec, 24–28 Aug 1987, p. 155
110. N.K. Sharova (ed.), *Application of Holography in Biology and Medicine* (Nauka, Leningrad, 1977), p. 35. (in Russian)
111. V.M. Ginsburg, Y.A. Utkin. E.N. Lekhtsier, Y.P. Presnyakov, N.S. Pushkar, E.G. Semenov, B.M. Stepanov, *Application of Holography in Biology and Medicine*, ed. by N.K. Sharova (Nauka, Leningrad, 1977), p. 35. (in Russian)
112. M.E. Glicksmana, R.J. Schaefera, J.A. Blodgettb, J. Cryst. Growth **13–14**, 68 (1972)
113. F. Bedarida, J. Cryst. Growth **79**(1–3), 43 (1986)
114. K.-E. Follstrom, K. Olofsson, H.O. Saldner, S. Schedin, Opt. Lasers Eng. **24**(5–6), 429 (1996)
115. K.A. Jacob, V. Dayal, B. Ranganayakamma, Exp. Mech. **23**, 49 (1983)
116. K.D. Vandergriff, J.S. Olson, J. Biol. Chem. **259**(20), 12619 (1984)
117. E.A. Evans, Y.-C. Fung, Microvasc. Res. (4), 335 (1972)
118. E.A. Evans, JOSA **61**(8), 991 (1971)
119. E.A. Evans, P.F. Lelond, Biology **10**, 393 (1973)
120. T.V. Gudim, et al., Laboratory Case (10), 761 (1987). (in Russian)
121. J.S. Beck, Theor. Biol. **75**, 487 (1978)
122. A.M. Chernukh, P.N. Aleksandrov, O.V. Alekseev, *Microcirculation* (Meditsina, Moscow, 1984), p. 38. (in Russian)
123. A.N. Metelkin, V.N. Monin, A.G. Rumyantsev, Pediatrics **5**, 43 (1980). (in Russian)
124. V.I. Shepotinovsky, Pathol. Physiol. Exp. Ther. **2**, 70 (1984). (in Russian)
125. R. Danliker, Prog. Opt. **17**, 1 (1980)
126. M.P. Duncan, J. Reintjes, T.J. Manuccia, Opt. Lett. **7**(8), 350 (1982)
127. J.C. Earnshaw (ed.), *The Application of Laser Light Scattering to the Study of Biological Motion* (Springer, New York, 1983)
128. S. Einav, H.J. Berman, J. Biomed. Eng. **10**, 393 (1988)

129. G. Merritt, E. Monson, E. Betzig, R. Kopelman, Ultramicroscopy **71**(1–4), 183 (1998)
130. D.K. Hamilton, C.I.R. Sheppard, Opt. Acta **29**(12), 1573 (1982)
131. R.N. Smartt, W.H. Steel, JOSA **69**(10), 1455 (1979)
132. K.I. Zemskov, A.A. Isaev, M.A. Kazarian, G.T. Petrash, Quantum Electron. **3**(1), 35 (1976). (in Russian)
133. F.V. Bunkin, K.I. Zemskov, M.A. Kazarian, DAS USSR **243**(6), 1568 (1978). (in Russian)
134. E.N. Leith, J. Upatnieks, J. Opt. Soc. Am. **55**, 569 (1965)
135. E.N. Leith, J. Upatnieks, K.A. Haines, J. Opt. Soc. Am. **55**, 981 (1965)
136. L.V. Tanin, A.L. Tanin, *Biomedical and Resonance Optics: Theory and Practice* (Belaruskaya navuka, Minsk, 2011), 687 p. (in Russian)
137. M.E. Cox, Microscope **19**, 137 (1971)
138. N. Royer, Opt. Commun. **20**(1), 73 (1977)
139. L. Toth, S.A. Collins, Appl. Phys. Hett. **13**, 7 (1968)
140. R.W. Smith, T.N. Williams, Optik **39**(2), 150 (1972)
141. R. Collier, K. Berhart, L. Lin, *Opt. Hologr.* (Mir, Moscow, 1973). (in Russian)
142. C. West, *Holographic Interferometry* (Mir, Moscow, 1982), 504 p. (in Russian)
143. R. Pawluczyk, Z. Kraska, Appl. Opt. **24**, 3072 (1985)
144. Improvement of image quality in holographic microscopy. JOSA **66**(10), 1061 (1976)
145. I.S. Klimenko, *Holography of Focused Images and Speckle Interferometry* (Nauka, Moscow, 1985) 222 p. (in Russian)
146. R.F. van Ligten, Appl. Opt. **12**(2), 255 (1973)
147. H.J. Caulfield, Opt. Commun. **26**(3), 322 (1978)
148. I.S. Klimenko, E.G. Matinyan, G.I. Ruklan, Opt. Spectrosc. **29**, 160 (1970). (in Russian)
149. I.S. Klimenko, V.P. Ryabukha, ZhTF **53**(5), 888 (1983). (in Russian)
150. I.S. Zeylyukovich, N.M. Spornik, Opt. Spectrosc. **62**(3), 659 (1987). (in Russian)
151. V.M. Ginsburg, V.M. Kurbatov, G.N. Pavlygin, Holographic methods and equipment, application in the physical studies, 82 (1974). (in Russian)
152. I.S. Klimenko, E.G. Matinyan, Opt. spectrosc. **27**, 367 (27). (in Russian)
153. R. Pavlyuchek, *Proceedings of V Proceedings School of Holography*, Leningrad, 1973, p. 314. (in Russian)
154. R. Pawluczyk, Opt. Commun. **7**(4) (1973)
155. G.S. Safronov, T.V. Tishko, A.M. Garagataya, Ukr. Phys. J. **28**(10), 1472 (1983). (in Russian)
156. H.R. Worthington, JOSA **56**, 1987 (1966)
157. F. Heinigen, F. Laeri, T. Tsudo, Appl. Opt. **29**(19), 3378 (1981)
158. P.J. Maggill, A.P. Wilson, J. Appl. Phys. **19**, 471 (1968)
159. V.C. Sainov, S.C. Sainov, S.N. Ossikovska, Rep. Bulg. Ac. Sc. **36**, 1383 (1983)
160. R. Danliher, Prog. Opt. 17 (1980)
161. K. Iwata, R. Nagata, Opt. Acta **25**, 19 (1978)
162. K. Iwata, R. Nagata, Opt. Acta **25**(1), 19 (1973)
163. W.T. Welford, Opt. Commun. **1**, 123 (1969)
164. W.H. Steel, Opt. Acta **17**, 873 (1970)
165. J. De Jong, *Proceedings of the Conference "Holographic Data Nondestructive Testing,"* Yugoslavia, 1982
166. A.V. Sokolov, Coherent optical methods for studying biological microobjects, Abstract of the Dissertation, Leningrad, 1981. (in Russian)
167. I.S. Klimenko, E.I. Kucheryavenko, E.G. Matinyan, Opt. Spectrosc. **34**(2), 360 (1973). (in Russian)
168. R. Chislead, P. Sucin, I. Cucurezeann et al., Opt. Commun. **30**(6), 351 (1981)
169. J. Goodman, *Introduction to Fourier Optics* (Mir, Moscow, 1970), 364 p. (in Russian)
170. M. Born, E. Wolf, *Principles of Optics* (Nauka, Moscow, 1964), 856 p. (in Russian)
171. A. Marechal, M. Franson, *The Structure of the Optical Image* (Mir, Moscow, 1964), 295 p. (in Russian)
172. M. Scharnoff, N.J. Rumer, B. Coher et al., Biol. Bull. **155**(2), 465 (1973)
173. S. Oks, *Fundamentals of Neurophysiology* (Mir, Moscow, 1969). (in Russian)

174. P.K. Anokhin, Success Physiol. Sci. **5**(2), 5 (1974). (in Russian)
175. *In Vivo Microscopy of the Neuron. Methods for Physiological Studies* (Nauka, Leningrad, 1978). (in Russian)
176. A. Lebedinsky, *Selected Works of Animal Electricity* (Biomedgiz, Leningrad, 1937). (in Russian)
177. L. Galvani, A. Volta, *Selected Works on Animal Electricity* (Biomedgiz, Leningrad, 1937), 430 p. (in Russian)
178. M. Liotstsi, *History of Physics* (Mir, Moscow, 1970), 460 p. (in Russian)
179. M. Hodgkin, *Nerve Impulse* (Mir, Moscow, 1965), 125 p. (in Russian)
180. P.V. Simonov, *What Is an Emotion?* (Nauka, Moscow, 1966), 94 p. (in Russian)
181. J. Sade, D. Ford, *Fundamentals of Neuroscience* (Mir, Moscow, 1976), 350 p. (in Russian)
182. Y.Y. Geynisman, *Structural and Metabolic Manifestations of Neuron Function* (Nauka, Moscow, 1974), 208 p. (in Russian)
183. E.N. Kosmarskaya, *Problems of Development of the Brain and the Effect of Negative Factors* (Medgiz, Moscow, 1960), p. 209. (in Russian)
184. V.Y. Brodsky, L.A. Danukyan, Cytology **2**, 1528 (1969). (in Russian)
185. C.L. Haydarliu, Cytology **9**, 644 (1967). (in Russian)
186. P.B. Kazakova, N.I. Nezlina, Neurophysiology **4**, 41 (1972). (in Russian)
187. M.E. Niverova, V.Y. Brodsky, Vestnik Moscow Univ. Biol. Soil Sci. Ser. **26**(4), 21 (1971). (in Russian)
188. V.A. Brumberg, DAS USSR **184**, 1231 (1969). (in Russian)
189. L.S. Zaguskin, Bull. Sci. **3**, 30 (1965). (in Russian)
190. L.Z. Pevzner, *Mechanisms of the Activity of Central Neuron* (Nauka, Moscow, 1966) p. 7. (in Russian)
191. G.G. Arakelov, E.N. Sokolov, V.Y. Brodsky, Cytology **2**, 865 (1969). (in Russian)
192. D.Z. Pevzner, S.H. Haydarliu, Cytology **9**, 840 (1967). (in Russian)
193. J. Roth Bradley, John P. Wikswo, Math. Biosci. **76**(1), 37 (1985)
194. H. Hyden, in *The Cell*, ed. by J. Brachet, A.E. Mirsky (Academician Press, New York, 1960), p. 215
195. L.Z. Pevzner, *Functional Biochemistry of the Neuroglia* (Nauka, Leningrad, 1972), 200 p. (in Russian)
196. H. Hyden, A. Pigon, J. Neurochem. **6**, 57 (1960)
197. L.G. Agroskin, G.V. Papayan, *Cytophotometry* (Nauka, Leningrad, 1977), 295 p. (in Russian)
198. V.N. Karnaukhov, Biophys. Live Cells **4**(1), 19 (1973). (in Russian)
199. M.E. Neverova, *Functional Histochemistry of Some Parts of the Central Nervous System: in the Ontogenesis* (Moscow, 1960). (in Russian)
200. V.Y. Brodsky, A.F. Kuznetsova, Cytology **1**(3), 89 (1962). (in Russian)
201. H. Appelt, *Introduction to the Methods of Microscopic Examination* (Medgiz, Moscow, 1959), 426 p. (in Russian)
202. G. Veid (ed.), *Introduction to Quantitative Cytochemistry* (Mir, Moscow, 1969). (in Russian)
203. G.N. Berestovskii, E.A. Lieberman, V.Z. Lunevsky, G.M. Frank, Biophysics **15**, 62 (1970). (in Russian)
204. L.B. Cohen et al., J. Physiol. **224**, 701 (1972)
205. L.B. Cohen et al., J. Physiol. **224**, 727 (1972)
206. J. Tasaki, et. al., Proc. Nat. Acad. Sci. USA 885 (1961)
207. J. Tasaki, et. al., Proc. Nat. Acad. Sci. USA 1362 (1964)
208. S.V. Levin, *Structural Changes in Cell Membrane* (Nauka, Leningrad, 1976). (in Russian)
209. *Great Medical Encyclopedia* **20**, 674 (1961). (in Russian)
210. V.A. Bahtagozin, J.P. Buhay et al., The Biomed. **16**, 66 (1976). (in Russian)
211. A.C. Scott, Phys. Multiparticle Syst. **12**, 37 (1987)
212. T. Kato, *Perturbation Theory for Linear Operators* (Springer, Berlin, 1966), p. 592
213. M. Nagasava, J. Math. Biol. **9**, 213 (1980)
214. M. Nagasava, *Proceedings of 1st Internernational Ascona Como Meeting, Stochastic Processing, Classical and Quantum System*, Ascona, Switzerland, 1986

215. E. Nelson, Phys. A: Stat. Mech. Appl. **124**(1–3), 509 (1984)
216. N. Wiener, *Cybernetics* (The Technology Press, Cambridge, 1948)
217. K. Pribram, *The Languages of the Brain* (Englewood Cliffs, New Jersey, 1971), p. 432
218. K. Yasue, J. Math. Phys. **22**, 1010 (1981)
219. K. Yasue, J. Funct. Anal. **51**, 133 (1983)
220. J.C. Zambrini, Int. J. Theor. Phys. **24**, 277 (1985)
221. J.C. Zambrini, Phys. Rev. **33**, 1532 (1986)
222. J.C. Zambrini, J. Math. Phys. **27**, 2307 (1986)
223. K. Uase, M. Jiby, T. Misawa, Stochastic dynamics of impulse: mathematical foundations of neural holography. (in Russian)
224. P.J. van Heerden, Appl. Opt. **2**, 393 (1963)
225. P.J. van Heerden, Nature **227**, 410 (1970)
226. P.J. van Heerden, Nature **225**, 177 (1970)
227. P.J. van Heerden, Phil. Phenomenol. Res. **29**(4), 608 (1969)
228. Y.N. Denisyuk, DAS BSSR **144**(6), 1275 (1962). (in Russian)
229. Y.N. Denisyuk, Opt. Spectrosc. **15**(4), 522 (1963). (in Russian)
230. Y.N. Denisyuk, Opt. Spectrosc. **18**(2), 275 (1965). (in Russian)
231. A. Lohmann, Opt. Acta **3**, 97 (1956)
232. E.N. Leith, J. Upatnieks, J. Opt. Soc. Am. **53**, 1377 (1963)
233. E.N. Leith, J. Upatnieks, J. Opt. Soc. Am. **54**, 1295 (1964)
234. E. Leith, J. Upatniexs, Phys. Today **18**, 26 (1965)
235. J.T. Winthrop, C.R. Worthington, Phys. Lett. **15**, 124 (1965)
236. G.W. Stroke, R.C. Restrick, Appl. Phys. Lett. **7**, 229 (1965)
237. L. Rosen, Appl. Phys. Lett. **9**, 337 (1966). (in Russian)
238. G.W. Stroke, Phys. Lett. **23**, 325 (1966). (in Russian)
239. P.J. van Heerden, Appl. Opt. **2**, 387 (1963)
240. N.F. Gamaleya, *Lasers in Experiments and Clinics* (Meditsina, Moscow, 1972), 231 p. (in Russian)
241. V.M. Inyushin, Histophysiological study of the effect of monochromatic red light (lasers and other slight devices on the animal organism, Abstract of the Dissertation, Lvov, 1972. (in Russian)
242. S.D. Pletnev, N.D. Deviatnikov, V.P. Belyaev et al., *Gas Lasers in the Experimental and Clinical Oncology* (Meditsina, Moscow, 1979). (in Russian)
243. A.P. Rakcheev, Lasers Clin. Med. **225** (1981). (in Russian)
244. T.G. Polanji, H.C. Bredemeier, T.W. Davis, Med. Biol. Eng. **8**(6), 547 (1970)
245. S.H. Tomson, Ann. N.V. Acad. Sci. **2**, 191 (1976)
246. V.M. Inyushin, P.R. Chekurov, *Biostimulation by a Laser Beam and the Bioplasm* (Alma-Ata, Kazakhstan, 1975), 119 p. (in Russian)
247. P.R. Chekurov, A.V. Eremin, Problems of bioenergy of the organism stimulation by laser beam, 114 (1976). (in Russian)
248. V.V. Shur, N.S. Makeeva, *Proceedings of the Conference "Some Questions of Biodynamics and Bioenergy of a Body in Health and Disease Condition, Laser Biostimulation by Radiation,"* Alma-Ata, 1972, p. 47. (in Russian)
249. L.I. Usachenok, A.P. Tulikova, L.F. Podmarenkova, New Laser Med. Surg. **4**(2), 115 (1990). (in Russian)
250. I.M. Korochkin, A.V. Kartemishev, G.M. Kapustina, et al., *Proceedings of the International Conference "The Use of Lasers in Clinics and Experiments,"* Moscow, 1987, p. 105. (in Russian)
251. N. Zarkovic, H. Manev, D. Pericic et al., Lasers Surg. Med. **9**(1), 63 (1989)
252. L.A. Komarova, T.F. Markova, Occup. Health **2**, 8 (1976). (in Russian)
253. S.D. Pletnev, O.M. Karpenko, M.S. Abdurazakov, Lasers Clin. Med. **201** (1981). (in Russian)
254. F.E. Gorbunov, A.V. Kochetov, A.A. Minenkova, et al., New Laser Med. Surg. **4**(2), 103 (1990). (in Russian)
255. T.M. Shakirova, Z.V. Burnakova, Bull. Biol. Med. Exp. **1**, 95 (1970). (in Russian)

256. B.V. Ognev, A.A. Wisniewskiy, I.K. Troickiy, Bull. Exp. Biol. Med. **6**, 20 (1972). (in Russian)
257. I.P. Antonov, G.K. Nedzved, V.N. Kozyro, N.V. Solovey, M.I. Kalyuta, L.V. Tanin, Copyright Certificate No. 1183100 (the USSR), 1985. (in Russian)
258. B.A. Kudryashov, V.E. Pastorova, G.G. Bazazyan, Mechanisms of damage to the resistance, adaptation and compensation 1, 377 (1976). (in Russian)
259. G.G. Shvalb, A.L. Zakharchenko, A.A. Sigaev, M.I. Kataev, Sov. Med. 3 (1990). (in Russian)
260. U.N. Sirenko, S.Y. Savitsky, S.S. Krasnitsky, A.V. Shabilynov, L.I. Popova, Sov. Med. 3 (1990). (in Russian)
261. N.M. Korochkin, E.V. Babenko, Sov. Med. 3 (1990). (in Russian)
262. S.V. Dmitrienko, E.E. Korobkov, V.V. Osipov, K.P. Vorobyeva, Modern methods of laser therapy (1989). (in Russian)
263. I.M. Korochkin, M.V. Poslavsky, Clin. Med. **3**, 102 (1986). (in Russian)
264. V.A. Voronina, et al., The use of lasers in medicine, 16 (1985). (in Russian)
265. A.D. Gyscha, E.A. Stroev, S.V. Tarasenko, V.A. Yudin, *Proceedings of International Symposium on the Application of Lasers in Surgery and Medicine* 1, 510 (1989), p. 510. (in Russian)
266. B.N. Zhukov, N.A. Lysov, E.A. Vasilieva, *Proceedings of the 1-st International Congress "Laser and Health '97"*, Limassol, Cypris, 1997, p. 13. (in Russian)
267. N.D. Devyatkov, S.M. Zubkov, I.B. Laprun, Present-day success. Biology **103**(1), 31 (1987). (in Russian)
268. A.A. Minenkov, T.V. Konchugova, D.B. Kulchitskaya, Issues. kurortology **2**, 11 (1992). (in Russian)
269. O.V. Milanov, A.R. Evstigneev, Immunology **4**, 88 (1988). (in Russian)
270. A.D. Meyers, J.J. Joyce, Med. **6**(6), 540 (1987)
271. L.A. Vasilevskaya, Peripheral. nerve. Syst. **21**, 6 (1998). (in Russian)
272. V.V. Skupchenko, T.G. Makhovskaya, G.N. Handurina, *Proceedings of the Conference "Low-Intensity Laser Radiation in Medical Practice,"* Khabarovsk, 1990, p. 148. (in Russian)
273. S.E. Rovdo, A.P. Gavrilova, N.Y. Chumakova et al., Peripheral. nerve system **21**, 39 (1998). (in Russian)
274. V.L. Stadnik, A.G. Fedorchuk, L.M. Skivka, E.A. Karasevskaya, *Proceedings of the Conference "The Effect of Low-Intensity Laser Blood Irradiation,"* Kiev, 1989, p. 160. (in Russian)
275. J.A. Zozulya, P.V. Spasichenko, G.M. Oleinik, O.F. Ponomareva, G.M. Yakhnenko, Z.M. Rudykh, *Proceedings of the Conference "The Effect of Low-Intensity Laser Blood Irradiation,"* Kiev, 1989, p. 92 (in Russian)
276. B.J. Saligeperova, Formation of roundabout routes of blood flow in the pelvic limbs under laser irradiation, Abstract of the Dissertation, Kharkov, 1991. (in Russian)
277. L.V. Tanin, S.A. Aleksandrov, A.A. Kumeysha, M.M. Loiko, I.V. Markhvida, S.K. Dik, L.A. Vasilevskaya, Peripher. Nerv. Syst. **17**, 50 (1994). (in Russian)
278. L.V. Tanin, A.A. Kumeysha, I.V. Markhvida, M.M. Loiko, S.A. Aleksandrov, S.K. Dik, L.A. Vasilevskaya, Proc. SPIE **2083**, 280 (1993)
279. L.V. Tanin, S.K. Dik, S.A. Aleksandrov, M.M. Loiko, A.A. Kumeisha, I.V. Markhvida, L.A. Vasilevskaya, *Proceedings of Laser Optics ICONO'95* 1, 304 (1995)
280. A.S. Kruck, V.A. Mostovnikov, I.V. Hohlov, N.S. Serdyuchenko, *Therapeutic Efficiency of Low-Intensity Laser* (Science and Technology, Minsk, 1986), 328 p. (in Russian)
281. B.I. Stepanov, V.A. Mostovnikov, A.N. Rubinov, I.V. Hohlov, DAS USSR **236**(4), 1007 (1977). (in Russian)
282. V.A. Mostovnikov, G.R. Mostovnikova, V.U. Plavsky, S.A. Tretyakov, Proc. AS USSR. Phys. Ser. **54**(8), 1636 (1990). (in Russian)
283. V.Y. Plavsky, V.A. Mostovnikov, A.B. Ryabtsev, G.R. Mostovnikova, L.G. Plavskaya, N.K. Nikeenko, I.A. Leusenko, V.V. Ginevich, A.V. Mostovnikov, V.S. Ulashchik, N.S. Serdyuchenko, P.S. Rusakevich, A.V. Volotovskaya, I.A. Rybin, Opt. J. **74**(4), 27 (2007). (in Russian)

284. V.A. Mostovnikov, G.R. Mostovnikova, V.Y. Plavski, L.G. Plavskaja, R.P. Morozova, Proc. SPIE **2370**, 541 (1994)
285. F.V. Oleshkevich, A.V. Shamkalovich, A.S. Fedulov, I.G. Liandres, A.P. Shkadorevich, I.I. Sakovich, Laser Med. **12**(3), 18 (2008). (in Russian)
286. L.V. Tanin, Lasers Biomed. **73** (2002). (in Russian)
287. N.L. Nechipurenko, L.A. Vasilevskaya, L.V. Tanin, J.I. Musienko, *Proceedings of International Conference on Optics Within Life Science*, Lucerne, Switzerland, 2002, p. 72
288. L.V. Tanin, *Abstracts of IX International Conference on Lasers Applications in Life Science*, Vilnius, 2002, p. 73
289. N.I. Nechipurenko, L.V. Tanin, L.A. Vasilevskaya, *Abstracts of the International Conference "Laser and Optical Technologies in Biology and Medicine,"* Minsk, 2004, p. 31. (in Russian)
290. I.G. Lyandres, S.I. Leonovich, A.P. Shkadarevich, *Lasers in Clinical Surgery* (Minsk, 1997), 227 p. (in Russian)
291. A.V. Volotovskaya, N.M. Kozlova, A.N. Antonovich, E.I. Slobozhanina, V.S. Ulaschik, Proc. Natl. Acad. Sci. Belarus. Biol. Ser. **3**, 44 (2001). (in Russian)
292. I. Tasaki, *Implementation of the Nerve Impulse* (Izdatelstvo inostrannoy literatury, Moscow, 1955) 186 p. (in Russian)
293. R. Frankel, Annu. Rev. Biophys. Bioeng. **13**, 85 (1984)
294. A.S. Osenniy, N.K. Kurendina, Biophysics **20**(1), 78 (1955). (in Russian)
295. K.K. Saudauskas, J.P. Schuranova, Biophysics **29**(4), 681 (1984). (in Russian)
296. K.K. Saudauskas, J.P. Schuranova, *Proceedings of All-Union Symposium "The Biological Effect of Electromagnetic Fields,"* Pushkin, 1982, 58. (in Russian)
297. V.I. Danilov, V.V. Parshintsev, V.V. Turkin, Biophysics **29**(1), 109 (1984). (in Russian)
298. V.I. Danilov, V.V. Parshintsev, V.V. Turkin, The influence of a series of pulses of the magnetic field of small amplitude electrical activity of the neurons **29**(2), 320 (1984). (in Russian)
299. V.I. Danilov, V.V. Parshintsev, V.V. Turkin, Biophysics **31**(5), 838 (1986). (in Russian)
300. S.V. Levin, DAS USSR **289**(1), 208 (1986). (in Russian)
301. J. Tasaki, K. Jwasa, Jap. J. Physiol. **32**, 69 (1982)
302. P. Kolta, Acta Physiol. Acad. Sci. **3**(1), 89 (1973)
303. Y.A. Kholodov, H.A. Shishlo, *Electromagnetic Fields in Neurophysiology* (Nauka, Moscow, 1979). (in Russian)
304. A.N. Simonov, T.V. Vyshenskaya, V.A. Livshits, *Proceedings of All-Union Symposium "The Biological Effect of Electromagnetic Fields,"* Pushkin, p. 72, 1982. (in Russian)
305. A.N. Simonov, V.A. Livshits, A.N. Kuznetsov, Biophysics **31**(5), 777 (1986). (in Russian)
306. A.M. Merkulova, V.V. Amosova, Proc. Nauchn. Tp. Kirghiz. Med. Inst. **100**, 91 (1974). (in Russian)
307. A.M. Merkulova, G.A. Panfilova, Macro-and microstructure of the tissue in normal, pathological and experimental condition **3**, 49 (1976). (in Russian)
308. P.P. Lazarev, Medicine **13**(11), 1583 (1935). (in Russian)
309. S.P. Kozlovsky, G.A. Govor, L.E. Kozlovskaya, Modern Problems Med. Equipm. **79** (1983). (in Russian)
310. R. Bickford, B. Flemming, *Modern Electronics in Physiology and Medicine* (Kiev, 1967), p. 31. (in Russian)
311. V.B. Emelyanov, I.S. Fedorov, V.A. Pechatnikov, Biophys. Biochem. Muscul. Reduct. **227** (1976). (in Russian)
312. V.B. Emelyanov, Mol. Cell. Biophys. **288** (1977). (in Russian)
313. J.D. Altringham, G. H. Pollack, *Proceedings of 2nd International Symposium on Cross-Bridge Mechanical Muscle Lan tract. Wash*, Leningrad, 1984, p. 473
314. Kevin Burtain, R.J. Baskin, Pflugers Arch. **406**(4), 409 (1986)
315. M.M. Judy, V. Summerour, Biophys. J. **37**(2), 475 (1982)
316. J. Yen, B.G. Pinsky, Biophys. J. **42**(1), 83 (1983)
317. L.A. Piruzyan, A.N. Kuznetsov, Proc. AS USSR. Biol. Ser. **6**, 805 (1983). (in Russian)
318. V.I. Danilov, V.V. Parshintsev, V.V. Turkin, Biophysics **29**(2), 320 (1984). (in Russian)
319. N.D. Devyatkov, M.B. Golant, A.S. Tager, Biophysics **28**(5), 895 (1983). (in Russian)

320. V.I. Lavrinenko, Proc. Kirghiz. Med. Inst. **100**, 34 (1974). (in Russian)
321. V.N. Loshchilov, *Proceedings of Scientific-Practical Conference "Clinical Application of Magnetic Fields,"* Perm, 1977, p. 17. (in Russian)
322. V.V. Padalko, *Proceedings of the Conference "Using the Ultrasound and New Types of Energy in the Diagnostics, Therapy and Surgery,"* Moscow, 1982, p. 75. (in Russian)
323. G.A. Govor, V.M. Dubrovenskiy, R.G. Kuzmich, et al., Patent 3839979/28-14 (USSR), 10 Jan 985. (in Russian)
324. I.P. Shmelev, G.M. Frank, DAS USSR **166**(6) (1966). (in Russian)
325. G.N. Berestovsky, E.A. Lieberman, V.Z. Lunevsky, G.M. Frank, Biophysics **15**(A), 62 (1970). (in Russian)
326. S.V. Revenko, O.S. Sotnikov, B.I. Hodorov, Neurophysiology **10**(4) (1978). (in Russian)
327. V.P. Kaznacheev, L.P. Mikhailova, *Ultraweak Radiation in Intracells Interactions* (Novosibirsk, 1981) 260 p. (in Russian)
328. F.O. Schmitt, O.H. Schmitt, J. Physiol. **98**, 26 (1940)
329. R. Sandlin, I. Lerman, W. Barry, I. Tasaki, Nature **217**, 575 (1963)
330. G.N. Berestovsky, V.Z. Lunevsky, V.D. Razhin, *Proceedings of Symposium*, Moscow, 1968. (in Russian)
331. L.B. Cohen, R.D. Keynes, B. Hille, Nature **218**, 438 (1968)
332. L.P. Kaushin, R.G. Ludkovskaya, Proc. Inst. Biol. Phys. Acad. Sci. USSR **1**, 40 (1955). (in Russian)
333. B.Y. Vilner, N.I. Lushchytskaya, Proc. AS BSSR. Biol. Ser. **3**, 71 (1982). (in Belarusian)
334. A.I. Lakomkin, I.F. Myagkov, *Electrophysiology* (High School, Moscow, 1977). (in Russian)
335. V.M. Ginzburg, B.M. Stepanova (ed.), *Optical Holography. Practical Applications* (Sovetskoe Radio, Moscow, 1978), 237 p. (in Russian)
336. Y.E. Voskoboinikov, *Abel Inversion and Its Generalizations* (Novosibirsk, 1978), 208 p. (in Russian)
337. Y.E. Voskoboinikov, *Processing Algorithms and Means of Automation of Thermophysical Experiment* (Novosibirsk, 1978), 154 p. (in Russian)
338. H.G. Unger, *Planar Optical Waveguides and Fiber* (Mir, Moscow, 1980), 656 p. (in Russian)
339. D.F. Mandolli, W.R. Briggs, World Sci. **10**, 66 (1984). (in Russian)
340. D. Marcouse, Appl. Opt. **18**(1), 9 (1979)
341. D.F. Mandolli, U.R. Briggs, Sci. Dig. **91**(2), 19 (1983)
342. F.G. Gribakin, *Mechanisms of Insect Photoreception* (Leningrad, 1981). (in Russian)
343. V.I Descherevsky, *Mathematical Models of Muscle Contraction* (Nauka, Moscow, 1977), 160 p. (in Russian)
344. H.E. Huxley, W. Brown, J. Mol. Biol. **30**(2), 383 (1967)
345. P.F. Rokitsky, *Biological Statistics* (High School, Minsk, 1967), 328 p. (in Russian)
346. J.P. Imbert, Traraux **618**, 28 (1987)
347. A.I. Selivra, *Circulatory Physiology, the Regulation of Blood Circulation* (Nauka, Leningrad, 1986), p. 480. (in Russian)
348. A.M. Williamson et al., Brit. J. Int. Med. **44**, 459 (1987)
349. S. Gilman, J. Colton, J. Hallenbeck, Appl. Physiol. **61**, 2067 (1986)
350. J. Naguet, Arch. Ital. Biol. **120**, 201 (1982)
351. S.A. Gular, V.N. Ilyin, Fiziol. J. **33**, 86 (1987). (in Russian)
352. J. Tamura et al., Environ. Med. **29**, 23 (1985)
353. H.J. Marriage, W. Paton, Undersea Biomed. Res. **13**, 369 (1986)
354. H.J. Little, D.L. Thomas, J. Physiol. **374**, 387 (1986)
355. T.J. Doubl, Undersea Biomed. Res. **9**, 241 (1982)
356. N.C. Ornhagen, Undersea Biomed. Res. **8**, 113 (1981)
357. J.T. Doubl, P.M. Hogan, Undersea Biomed. Res. **3**, 211 (1980)
358. D.D. Makkenzi, K.R. Dandas, Proceedings of 7 International Congress 2, 57 (1983). (in Russian)
359. D.J. Harris, P.B. Bernett, J. Appl. Physiol. **54**, 1209 (1983)
360. N.N. Arrout, M.F. Konstantinova, Cytology **2**, 147 (1969). (in Russian)

361. G.L. Salzman, Human Physiol. **10**(4), 659 (1984). (in Russian)
362. P. Hochachka, J. Somero, *Biochemical Adaptation Strategy* (Moscow, 1977). (in Russian)
363. R. Naquet, Arch. Ital. Biol. **120**(1–3), 201 (1982)
364. D. Harris, J. Bernett, B. Petter, J. Appl. Physiol. **54**(5), 1209 (1983)
365. J.D. Makkenzi, K.R. Dandas, Proceedings of 7 International Congress 2, 43 (1983). (in Russian)
366. E.M. Kreps, G.L. Zaltsman (ed.), *The Organism in Conditions of Prolonged Hyperbary* (Nauka, Leningrad, 1977), 192 p. (in Russian)
367. T.J. Doubl, P.M. Hogen, Undersea Biomed. Res. 9 (1982)
368. NCh. Ornhagen, Undersea Biomed. Res. **8**(2), 113 (1981)
369. S.B. Sigurdsson, Undersea Biomed. Res. **7**(3), 171 (1980)
370. G.N. Evstronova, Bull. Experts Biol. Med. **84**(9), 259 (1977). (in Russian)
371. V.V. Sysoev, in *The Organism in Conditions of Prolonged Hyperbary*, ed. by E.M. Kreps, G.L. Zaltsman (Nauka, Leningrad, 1977), p. 150. (in Russian)
372. A.I. Selivra, *Circulatory Physiology, the Regulation of Circulatory Absorption* (Nauka, Leningrad, 1986), p. 480. (in Russian)
373. C. Armon, D. Kerem, Undersea Biomed. Res. **2**, 121 (1981)
374. O.S. Sotnikov, *Dynamics of Structure of Living Neuron* (Leningrad, 1985). (in Russian)
375. D. Keysensent (ed.), *Optical Information Processing* (Mir, Moscow, 1980), 349 p. (in Russian)
376. Y.I. Kruzhilin, V.I. Shveikin, N.V. Antonov, Y.I. Koloskov, Electronics **12**(7), 692 (1969). (in Russian)
377. P.G. Eliseev, A.V. Khaidarov, Quantum Electron. **2**(1), 127 (1975). (in Russian)
378. M.P. Tychinskaya, R.A. Vanetsian, V.P. Zakharov, I.T. Rassokhin, Quantum Electron. **10**(4), 101 (1972). (in Russian)
379. R. Keller, R. Salathe, T. Tschudi, IEEE J. Quantum Electron. QE **8**, 783 (1972)
380. L.F. Mollenauer, *Tunable Solid-State Lasers for Remote Sensing. Proceedings of the Conference*, Stanford, CA, 1–3 Oct 1984, p. 60
381. R. Beigang, *Group VIII Advanced Materials and Technologies* 2, 141 (2008)
382. E.M. Dianov, I.A. Bufetov, V.M. Mashinsky, V.B. Neustruev, O.I. Medvedev, A.V. Shubin, M.A. Melkumov, A.N. Guryanov, V.F. Khopin, M.V. Yashkov, Quant. Electron. **34**(8), 695 (2004). (in Russian)
383. W. Barret Terence, Math. Bio-sci. **9**, 49 (1970)
384. Yu.N. Denisyuk, Nature **2**, 2 (1971). (in Russian)
385. B. Vaitovich, Electron. techn. obliczen. Nowosci. **14**(3), 48 (1975). (in Russian)
386. S.M. Zubkova, Z.A. Sokolova, Problems Med. Chem. **24**(3), 326 (1978). (in Russian)
387. O. Ersoy, Appl. Opt. **18**(15), 2543 (1979)
388. J. Cadusseau, G. Galand, Exp. Brain. Res. **40**(3), 339 (1980). (in Russian)
389. E. Robinson Charles, Theor. Biol. **9**(1), 231 (1981). (in Russian)
390. J. Dempsher, Acta biotheor. **30**(2), 121 (1981). (in Russian)
391. V.P. Kaznacheev, L.P. Mikhailova, *Ultraweak Light Emission in the Inter-Cellular Interactions* (Nauka, Novosibirsk, 1981). (in Russian)
392. G. Brown, J. Walken, *Liquid Crystals and Biological Structures* (Mir, Moscow, 1982). (in Russian)
393. P.S. Gukosyan, N.V. Tabiryan, Tech. Phys. Lett. **9**(4), 238 (1983) (in Russian)
394. L.O. Svaasund, R. Ollingsen, Photochem. and Photobiol. **38**(3), 293 (1983). (in Russian)
395. V.A. Lapina, A.I. Filipovich, *Methodological processing of medical and scientific information*, Leningrad, 1983, p, 153. (in Russian)
396. F. Iannotti, J.T. Hoff, G.P. Schielke, J. Neurosurg. **62**(1), 83 (1985)
397. D.F. Mandoli, W.R. Briggs, Sci. Am. **251**(2), 80 (1984). (in Russian)
398. V.A. Welling, Phusiol. J. USSR **72**(5), 694 (1986). (in Russian)
399. F. Bahz, Akupunktur **3**, 3 (1986). (in Russian)
400. P.D. Wade, J. Taylor, P. Siekevitz, Proc. Nat. Acad. Sci. USA **85**(23), 9322 (1988). (in Russian)
401. A.A. Vedenov, *Modeling of Thinking Elements* (Nauka, Moscow, 1988) 160 p. (in Russian)
402. M. Ueda, T. Sato, JOSA **61**, 418 (1971)

403. Y.V. Gulyaev, E.E. Godik, A.V. Petrov, et al., DAS BSSR **277**(6), 1486 (1984). (in Russian)
404. E. Skat, *Waves in Active and Nonlinear Media Attached to the Electronics* (Sov. Radio, Moscow, 1977). (in Russian)
405. T.V. Holsh, *Nonlinear Spectroscopy* (Mir, Moscow, 1979), p. 41. (in Russian)
406. V.A. Teplov, S. Beylish, *Autowave Processes in Systems with Diffusion* (Gorkyi, 1981), p. 190. (in Russian)
407. J.R. Meyer-Arendt, AdVol. Opt. Electron. Microsc. **8**, 1 (1982). (in Russian)
408. E.M. Diyakov, A.M. Prokhorov, V.N. Serkin, DAS BSSR **273**(5), 1112 (1983). (in Russian)
409. J.G. Walker, Opt. Acta **30**(9), 1197 (1983). (in Russian)
410. MRZ **4**(9), 1985. (in Russian)
411. K.J. Blow, Ross. J. Phys. 1986-7L1092 (1986)
412. N.P. Ryabukha, V.D. Jareniya, *Proceedings of Materials of Science and Technical Conference "The Use of Lasers in Science and Technology,"* Leningrad, 1969, pp. 24–27. (in Russian)
413. K. Pribram, *Languages of the Brain* (Progress Publishers, Moscow, 1975). (in Russian)
414. U.L. Gogolitsin, U.D. Kropotov, Met. Problems. Medicine, 117 (1975) (in Russian)
415. V.A. Bahtigozin, J.P. Bugay, U.A. Kumanin, V.G. Chervov, Problems Biol. **16** (1976). (in Russian)
416. B.I. Belyi, Human Physiol. **5**(6), 998 (1979). (in Russian)
417. G.Y. Hvoles, B.V. Bohutsky, A.P. Konko, *Proceedings of Researcher Conference on an Experimental Balneology and Physiotherapy*, Moscow, 1962, p. 305. (in Russian)
418. A.A. Bildyukevich, H.S. Khamitov, E.G. Ulumbekov, et al., *Proceedings of the Conference "The Influence of Artificial Magnetic Fields on Living Organisms,"* Baku, 1972, p. 14. (in Russian)
419. V.I. Bankov, Bull. Exp. Biol. Med. **9**, 14 (1972). (in Russian)
420. V. Scruggs, D. Landowne, Biol. Bull. **159**(2), 491 (1980). (in Russian)
421. I. Tasaki, K. Iwasa, Jap. J. Physiol. **30**(6), 897 (1980). (in Russian)
422. V.I. Lavrinenko, Proc. Sci. Pap. Kyrgyz Med. Inst. **100**, 34 (1974). (in Russian)
423. L. Nagy, Acta biochim et biophis. Act. Sci. hung. **18**(1–2), 39 (1983)
424. J.C. van Gemert Martin, Laser Surg. Med. **5**(3), 235 (1985). (in Russian)
425. K. Veber, E. Osborn, World Sci. **12**, 52 (1985). (in Russian)
426. B.C. Wilson, H.S. Patterson, Photochem. Photobiol. **46**(5), 601 (1987). (in Russian)
427. S.V. Yagodin, V.M. Slutsky, et al., DAS USSR **287**(6) (1986). (in Russian)
428. M.L. Calvo, M. Chevalier, C. Carreras, Appl. Opt. **23**(2), 324 (1989)
429. L.K. Srebnitkaya, G.N. Vishnyakov, S.A. Neiman, Z.E. Rozhdestvenskaya, O.A. Andreev, G.G. Levin, Biophysics **46**(3), 518 (2001). (in Russian)
430. C.L. Curl, C.J. Bellair, P.J. Harris, B.E. Allman, K.A. Nugent, L.M. Delbridge, Exp. Pharmacol. Physiol. **31**(12), 896 (2005)
431. L.A. Erokhova, S.M. Novikov, G.L. Lazarev, T.A. Kazakova, D.A. Orlov, K.V. Indukaev, G.V. Maksimov, Bull. Exp. Biol. Med. **140**(2) (2005)
432. S.L. Jacques, Photochem. Photobiol. **45**(44) (1987). (in Russian)
433. G. Eiring, *Molecular Biology* (Moscow, 1963), pp. 92–101. (in Russian)
434. D. Mandoli, W. Briggs, Plant Cell Environ. **5**, 137 (1982)
435. P.J. van Heerden, Appl. Opt. **2**(4), 393 (1963)
436. R.V. Paddington, *Optical Spectroscopy of Bias and Correlation of the Photons* (Mir, Moscow, 1987). (in Russian)
437. N. Arinchin, *Peripheral "Heart" of Man* (Nauka i technika, Minsk: Science and Technology, 1980), 80 p. (in Russian)
438. J.M. Nadel, *Waveguide Link* (Znanie, Moscow, 1975), 63 p. (in Russian)
439. P.S. Magill, A.D. Wilson, J. Appl. Phys. **39**(10), 4717 (1968)
440. R.F. Leighten, Opt. Technol. **28**, 71 (1969)
441. B.J. Thompson, J.H. Ward, W. Zynky, Appl. Opt. **6**, 519 (1967)
442. J.W. Goodman, W.H. Huntley, D.W. Jackson, M. Lehmann, Appl. Phys. Lett. **8**, 311 (1967). (in Russian)
443. E. Spits, Compt. Rend. **264** B, 1449. (in Russian)

444. E.B. Champagne, N.G. Massey, Appl. Opt. **8**, 1879 (1969). (in Russian)
445. K.A. Stetson, Appl. Phys. Lett. **11**, 225 (1967). (in Russian)
446. K.A. Stetson, Appl. Phys. Lett. **12**, 362 (1968). (in Russian)
447. H. Nassenstein, Optik **29**, 597 (1969). (in Russian)
448. O. Bryngdahl, J. Opt. Soc. Am. **59**, 1645 (1969). (in Russian)
449. E.J. Feleppa, J. Cell Biol. **40**, 838 (1969)
450. G. Nomarski, JOSA **60**(11), 1575 (1970)
451. M.E. Cox, Laser Focus **7**(2), 41 (1971)
452. G. Pierratini, Opt. Commun. **5**(1), 41 (1972)
453. M.L. Mendelsohn, J. Biophys. Biochem. Cytol. **11**, 509 (1961)
454. M. Berus, Biophys. J. **16**(8), 913 (1976)
455. M.O. Breitmeyer et al., J. Assoc. Adv. Med. **6**, 365 (1972)
456. W.E. Martin, Appl. Opt. **13**(9), 2112 (1974)
457. M. Opas, J. Protokol. **22**, 74 (1975)
458. E. Handler et al., Laser-Elek-Opt. **6**(4), 46 (1974)
459. P. Hariharan, et al., Laser-Elek.-Opt. **6**(4), 46 (1974). (in Russian)
460. K. Scrimannarayaka, Pad. Venkategware, Opt. Appl. **7**(1), 19 (1977)
461. V.P. Nayyar, N.K. Verma, Appl. Opt. **16**(9), 2461 (1977)
462. M.E. Cox, K.J. Vanala, Appl. Opt. **17**(9), 1455 (1978)
463. A.V. Sokolov, V.N. Bykov, Proc. LPI of M.I. Kalinin, **124** (1978). (in Russian)
464. Y. Shono, Inuzuka, Appl. Phys. Lett. **33**(1), 111 (1978)
465. E.N. Lekhtsier, Meas. Tech. **9**, 22 (1978). (in Russian)
466. V.F. Klimkin, V.V. Pickalov, J. Appl. Mech. Tech. Phys. **3**, 14 (1979). (in Russian)
467. D.L. Lessor, J.S. Hartman, R. Gordan, JOSA **69**(2), 357 (1979)
468. P. Dunn, I.M. Walls, Appl. Opt. **18**(3), 263 (1979)
469. D. Courjon, J. Bulafois, J. Opt. **10**(3), 125 (1979)
470. A.N. Metelkin, Holographic interferometry and laser microscopy of red blood cells, Abstract of the Dissertation, Moscow, 1981. (in Russian)
471. E.N. Lekhtsier, A.N. Metelkin, Proceedings of VHIIOFI: Hologr. methods and equipment used in the physical studies, 49 (1982). (in Russian)
472. J.G. Solem, G.C. Baldwin, Science **218**(4569), 229 (1982). (in Russian)
473. A.P. Yakimovich, Opt. Spectrosc. **58**(3), 490 (1983). (in Russian)
474. J.J. Clair, A.M. Hamed, Optic **64**(2), 133 (1983)
475. E.N. Lekhtsier, B.M. Stepanov, Instrum. Exp. **6**, 152 (1983). (in Russian)
476. E.N. Lekhtsier, PTE **11**, 41. (in Russian)
477. F.F.S. Yu, F.K. Hsu, Opt. Commun. **52**(6), 195 (1984)
478. C. Ozkul, Opt. Acta **31**(3), 337
479. A.U. Vikarov, V.V. Kafarov, V.A. Bykov, et al., DAS BSSR **284**(4), 986 (1985). (in Russian)
480. M. Sharnoff, JOSA **2**(10), 1619 (1985)
481. J.H. McGeher, R.F. Wuerker, Proc. SPIE Int. Soc. Opt. Eng. **523**, 70 (1985)
482. S.V. Abuladze, N.P. Kutikov, A.V. Lukin, L.T. Mustafina, R.A. Nigmadzyanov, A.N. Chugunov, in *V Proceedings of the Conference for Holographic Physics*, vol. 2, pp. 37 (1985). (in Russian)
483. R.F. Wuerker et al., Proc. SPIE Int. Soc. Opt. Eng. **523**, 60 (1985)
484. A.C. Testa, U.P. Wild, J. Phys. Chem. **90**(18), 4302 (1986)
485. Z. Kunxian, Chin. J. Lasers **14**(11), 667 (1987)
486. R.A. Goncharov, T.D. Corlov, et al., Preprint **4**, 12 (1987). (in Russian)
487. N. Ohyama, Opt. Commun. **61**(2), 96 (1987)
488. G.S. Safronov, T.V. Tishko, PTE **2**, 249 (1987). (in Russian)
489. S.S. Jong, D. Christensen, J. Zhang, C.H. Wang, J. Phys. Chem. **91**(17), 4504 (1987)
490. Yu. Xiling, Wei Aijian, J. Cryst. Growth **83**(1), 11 (1987)
491. H. Royer, Proc. SPIE Int. Soc. Opt. Eng. **491**, 970 (1985)
492. V.B. Konstantinov, V.A. Babenko, A.F. Maliy, ZhTF **77**(12), 92 (2007). (in Russian)
493. V.A. Babenko, V.B. Konstantinov, A.F. Maliy, ZhTF **33**(8), 77 (2007). (in Russian)

494. L.A. Piruzyan, O.V. Kolosov, R.G. Maev, V.M. Lyavin, T.A. Senyushkin, DAS BSSR **280**(5), 1115 (1985). (in Russian)
495. L.S. Agroskin, N.V. Korolev, Opt. Spectrosc. **6**, 832 (1959). (in Russian)
496. L.A. Ferin, L.S. Agroskin, Biophysics **4**(4), 476 (1959). (in Russian)
497. L.S. Agroskin, N.V. Korolev, Biophysics **6**(1), 478 (1961). (in Russian)
498. E.M. Brumberg, I.Y. Barsky, J. Gener. Biol. **22**, 459 (1961). (in Russian)
499. I.L. Barsky, N.I. Polyakov, V.A. Yakubenas, *Contact Microscopy* (Meditsina, Moscow, 1976). (in Russian)
500. B.K. Vainstein, A.M. Mikhailov, I.A. Andriashvili, DAS USSR **234**(3), 699 (1977). (in Russian)
501. O.V. Bogdanovich, et al., Meas. Technol. (12), 30 (1978). (in Russian)
502. R.N. Smart, W.H. Steek, JOSA **69**(10), 1455 (1978)
503. R.N. Yungerman, P.C.D. Hobbs, G.S. Kino, Appl. Phys. Lett. **45**(8), 846 (1984)
504. J.E.H. Pruijs, Grant. Annal. And Hed. Photogram. Proc. Int. Conf., Oxford, 1987, p. 68
505. E. Salmon, P. Tran, Metjods Cell Biol. **56**, 153 (1998)
506. G.A. Dunn, Proc. R. Microsc. Soc. **33**, 189 (1998)
507. U. Schemmert, J. Kruger, J. Weitkamp, Schemmert Du Microporous and Mesoporous Materials **32**(1–2), 101 (1999)
508. D. Murphy, *Fundamentals of Light Microscopy and Digital Imaging* (Wiley-Liss, New York, 2001), p. 153
509. O. Geier, S. Vasenkov, E. Lehmann, J. Karger, R.A. Rakoczy, J. Weitkamp, Stud. Surf. Sci. Catal. **135**, 154 (2001)
510. G.N. Vishnyakov, G.G. Levin, V.L. Minaev, V.V. Pickalov, A.V. Likhachev, Microsc. Anal. **87**, 19 (2004)
511. G.G. Levin, A.A. Kovalev, V.L. Minaev, K.A. Sukhorukov, Meas. Technol. **4**, 62 (2004). (in Russian)
512. A. Bosseboeuf, J.P. Grandchamp, C. Breluzeau, S. Lani, J. Palomo, D. Bouville, Microsyst. Technol. **12**(10–11), 1063 (2006)
513. J. Kärger, L. Heinke, P. Kortunov, S. Vasenkov, Stud. Surface Sci. Catal. **170**(1), 739 (2007)
514. D. Tzoulaki, W. Schmidt, U. Wilczok, J. Karger, Microporous Mesoporous Mater. **110**(1), 72 (2008)
515. L. Hung-Wing, M. McCloskey, Y. He Edward, S. Yeung, Anal. Bioanal. Chem. **387**(1), 63 (2007)
516. C.L. Curl, C.J. Bellair, P.J. Harris, B.E. Allman, K.A. Nugent, L.M. Delbridge, Clin. Exp. Pharmacol. **31**(12), 896 (2004)
517. L.K. Srebnitskaya, G.N. Vishnyakov, S.A. Neumann, Z.E. Rozhdestvenskaya, O.A. Andreev, G.G. Levin, Biophysics **46**(3), 518 (2001). (in Russian)
518. N.G. Vlasov, ZHNPF **4**(5), 67 (1999). (in Russian)
519. S. Noda, T. Funami, M. Nakauma, I. Asai, R. Takahashi, S. Al-Assaf, S. Ikeda, K. Nishinari, G.O. Phillips, Food Hydrocolloids **22**(6), 1148 (2008)
520. H.N. Pishkenari, M. Behzad, A. Meghdari, Chaos, Solitons & Fractals **37**(3), 748 (2008)
521. D.A. Sidorov-Biryukov, A.N. Naumov, S.O. Conorov, A.B. Fedotov, A.M. Jeltikov, Quantum Electron. **30**, 1080 (2000). (in Russian)
522. L. Heinke, C. Chmelik, P. Kortunov, D.B. Shah, S. Brandani, D.M. Ruthven, J. Karger, Microporous Mesoporous Mater. **104**(1–3), 18 (2007)
523. K. Habib, K. Muhana, Opt. Lasers Eng. **38**(6), 577 (2002)
524. J. Kärger, C. Chmelik, E. Lehmann, S. Vasenkov, Stud. Surf. Sci. Catal. **154**(2), 1791 (2004)
525. K. Habib, K. Al-Muhana, A. Habib, Opt. Lasers Eng. **46**(2), 149 (2008)
526. L. Wang, J. Ke, Opt. Lasers Eng. **9**(2), 111 (1988)
527. F. Corvaro, M. Paroncini, Appl. Therm. Eng. **28**(1), 25 (2008)
528. J.P. Hohimer, P.J. Hargis, Appl. Phys. Lett. **30**(7), 344 (1977). (in Russian)
529. A. Angelica, Laser Focus **13**(2), 38 (1977). (in Russian)
530. H.-W. Li, M. McCloskey, Y. He, E.S. Yeung, Anal. Bioanal. Chem. **387**(1), 63 (2007)

531. V.V. Tuchin, in *Proceedings of the V International Conference "Laser Technology '95"*, Russia, 24–26 June 1995. (in Russian)
532. D.G. Veller, F.F. Usikov, E.D. Romanova, et al., *The Use of Lasers in Medicine* (Kiev, 1985). (in Russian)
533. V.A. Voronina, et al., *The Application of Lasers in Medicine* (Moscow, 1985), p. 16. (in Russian)
534. U.N. Sirenko, S.U. Savitsky, S.S. Krasnitsky, A.V. Shabilynov, L.I. Popova, Sov. Med. **3** (1990). (in Russian)
535. A.R. Rahishev, V.P. Coi, in *Proceedings of the 43rd Conference of Alma-Ata,* Alma-ata, 1973, p. 94. (in Russian)
536. A.R. Rahishev, *Tools and Methods of Electronics in Medicine* (Saratov, 1976), p. 155. (in Russian)
537. M.F. Barnothy, Plenum Press **2**(4), 11 (1969)
538. A.M. Merkulova, V.V. Amosova, Proc. Sci. Pap. Kirghiz Med. Inst. **100**, 91 (1974). (in Russian)
539. R.O. Becker, Plenum Press **2**(4), 207 (1969)
540. J.K. Woolsey, J. Roth Bradley, J.P. Wikswo, Math. Biosci. **76**(1), 1 (1985)
541. L.N. Prokhorova, Holographyic methods and apparatus, applicable in physical studies. Proc. VNIIOFI **5**, 4 (1974). (in Russian)
542. V.R. Kuchta, V.V. Lopatin, P.G. Petrov, Opt. Spectrogr. **56**(1), 178 (1984). (in Russian)
543. V.F. Klimkin, O.V. Pokalev, Opt. Spectrogr. **64**(A), 159 (1988). (in Russian)
544. W.H. Carter, A.A. Dougall, J. Opt. Soc. Am. **56**(12), 1754 (1966)
545. M. De, L. Sevigny, Appl. Phys. Lett. **10**(3), 78 (1967)
546. S.V. Levin, K.A. Golfand, A.S. Troshin, DAS BSSR **252**(4), 1014 (1980). (in Russian)
547. B.C.R. Ewan, Appl. Opt. **18**(18), 3156 (1979)
548. R.R. Anderson, I.A. Parrish, J. Invest. Dermal. **77**(1), 13 (1981)
549. N.F. Gamaleya, E.D. Shishko, Y.V. Yanin, DAS BSSR **273**(1), 224 (1983). (in Russian)
550. N.Y. Talinskaya, V.N. Kazanov, M.G. Zvyarintsev, in *Proceedings of the Conference "Magnetic Field in the Theory and Practice of Medicine,"* Kuibyshev, 1984, pp. 17–21. (in Russian)
551. E.M. Krens, in *The Organism in the Conditions of Hyperbary* (Nauka, Leningrd, 1984), p. 13. (in Russian)
552. V.N. Chernigovsky, I.A. Salov, *The Organism in the Conditions of Hyperbary* (Nauka, Leningrd, 1984), 159 p. (in Russian)
553. P. Greguss, Opt. Laser Technol. **17**(3), 151 (1985)
554. Sh.S. Saakian, M.G. Asratyan, A.K. Dadivanian, Sci. Rep. High School Biol. Sci. **9**, 23 (1986). (in Russian)
555. A. Ashkin, J.M. Dziendzie, T. Xamane, Nature **330**(6150), 769 (1987)
556. A.U. Vinarov, V.V. Kafarov, V.A. Bykov, E.S. Shitikov, S.P. Karlov, DAS BSSR **284**(4) (1985). (in Russian)
557. M.E. Cox, R.G. Buckles, D. Whitlow, Appl. Opt. **10**(1), 128 (1971)
558. V.S. Markin, Y.A. Chizmodzhev, *Physics of Nerve Impulse* (Znanie, Moscow, 1977), 63 p. (in Russian)
559. G.N. Berestovsky, E.A. Liberman, V.Z. Lunevsky, G.M. Frank, Biofizika **15**(1), 62 (1970). (in Russian)
560. V.B. Emelyanov, I.S. Fedorova, V.A. Pechatnikov, *Biophysics of Biochemistry Muscular Reductions* (Moscow, 1976), p. 227. (in Russian)
561. V.B. Emelyanov, et al. in *Molecular and Cellular Biophysics* (Mir, Moscow, 1977), p. 288. (in Russian)
562. G.Y. Anishchenko, Z.M. Polyanskaya, I.G. Dallakyan, *Laser Puncture in Neuropathy: Methodological Recommendations* (The Ministry of Health of the USSR Chap. Exercise. organization of med. Assistance, Moscow, 1991), 19 p. (in Russian)
563. G.N. Vishnyakov, V.P. Minaev, in *Proceedings of the VII International Conference "Holography EXPO-2010,"* Moscow, 2010, p. 88. (in Russian)

564. G.N. Vishnyakov, G.G. Levin, V.L. Minaev, A.G. Lomakin, in *Proceedings of the VII International Conference "Holography EXPO-2010,"* Moscow, 2010, p. 237. (in Russian)
565. V.V. Orlov, in *Proceedings of the VII International Conference "Holography EXPO-2010,"* Moscow, 2010, p. 330. (in Russian)
566. V.V. Orlov, Opt. J. (9), 83 (2006). (in Russian)
567. V.V. Orlov, Opt. Spectrosc. **92**(5), 862 (2002). (in Russian)
568. V.V. Orlov, in *Proceedings VI International Conference "Holography EXPO-2009,"* Kiev, 2009, p. 99. (in Russian)
569. Z.S. Bekyasheva, A.V. Pavlov, in *Proceedings of the VII International Conference "Holography EXPO-2010,"* Moscow, 2010, p. 334. (in Russian)
570. A.V. Pavlov, in *Proceedings of the IV International-Practical Conference*, Kolomna, May 28–30, 2007, p. 282. (in Russian)
571. A.M. Alekseev, A.V. Pavlov, Opt. Spectrosc. **107**(6), 1038 (2009). (in Russian)
572. V.V. Chernaya, V.N. Borovitskii, in *Proceedings of VI International. Conference "Holography EXPO-2009,"* Kiev, 2009, p. 214. (in Russian)
573. A.V. Pavlov, A.M. Alekseev, in *Proceedings of the VII International Conference "Go-loekspo-2010,"* Moscow, 2010, p. 136. (in Russian)
574. O.P. Kuznetsov, News Artif. Intell. 2 (1998). (in Russian)
575. A.V. Pavlov, Artif. Intell. Decis. Making **1**, 13 (2010). (in Russian)
576. A.V. Pavlov, Opt. J. **77**, 26 (2010). (in Russian)
577. E.F. Pen, R.M. Rusakov, in *Proceedings of the VII International Conference "Holography EXPO-2010"*, Moscow, 2010. (in Russian)
578. E.D. Lantukh, S.N. Pashkevich, S.N. Letuta, E.K. Alidzhanov, A.A. Kulsarina, in *Proceedings of the VII International Conference "Holography EXPO-2010"*, Moscow, 2010. (in Russian)

Chapter 3
Holographic Interferometry for Studying Time-Varying States of the Human Surface Circulatory System

The Development of Scientific Approaches to Implementation of the Holographic Interferometry in Medicine.

The holographic interferometry methods, which are a powerful means of nonvolatile control, make it possible to investigate with high resolution defined by the light wave fraction as opposed to the classical interferometry not only phase and mirror objects, but also diffuse ones [1–3], as well as the processes occured with them in statics and dynamics. They are implemented to:

- analyze elasticity and fatigue, residual voltages, material aging [4, 5],
- analyze homogeneity distortions of mixtures and compound materials [6–8],
- optimize constructions through vibration mode analyses (compressor blades and buckets, motor details, reservoirs, etc.) [9–12],
- control pressure homogeneity of welded, bolted and rivet joints of manifold details, especially under conditions requiring high reliability [6, 13],
- study flows and thermal gradients [14–17],
- study processes of microcrack occurrence [4, 18],
- study growth of vegetables, crystals, microparticle tracks, characteristics of gases and liquid [19–22],
- analyze the sizes and spatial distribution of particles [23, 24],
- measure velocity and size of micro- and macroparticles [25–27],
- study relief of static and dynamic surfaces of objects [22, 28–58].

Using the holographic interferometry interference, the comparison of rough surfaces is conducted that makes it possible to use it for detection of material defects or technologies [59, 60].

The holographic interferometry is efficiently used for plasma diagnostics [61–65] and study of non-stationary processes, for example, of gas-dynamic flows [66, 67] that is necessary for aerodynamic research during development of the optimal form of flying vehicles, etc.

L. Tanin et al., *Biomedical and Resonance Optics*, Bioanalysis 11,
https://doi.org/10.1007/978-3-030-60773-9_3

The possibilities and spheres of application of the holographic interferometry methods have significantly broadened as a result of powerful pulse solid lasers implementation in holographic systems. The studies and the development made in a number of foreign companies (Rottencolber Holosystem (FRG), Apolo Laser Corp. (USA), etc.) have shown good prospects of this school. Most significant results were obtained during the implementation of specialized lasers in holographic systems. The lasers operate in the generation mode of two giant pulses with regulated pulse-repetition intervals and high spatial and time coherence of the radiation.

The peculiarity of the studies conducted in this chapter is the usage of unique possibilities of pulsed dye lasers and ruby lasers that enabled to develop and implement the methods of two- and multi-long-wave holographic interferometry with regulated spectral range and of two- and multi-exposure holographic interferometry with regulated temporal interval between generation pulses, as well as the methods of holographic contouring of surface relief using resonance, absorbing and optically active media.

The holographic contouring methods are easy and suitable for obtaining contoured maps of arbitrarily shaped objects. There exist different possibilities to form such maps: The multi-long-wave method of recording a hologram of an object is conducted in laser radiation containing several wavelengths [68, 69]; the immersion one—the hologram is recorded when the object is placed in the camera [70] where the refractive index of liquid or gas, which fill the camera, is changing between two exposures; the several sources method—two sources spatially separated are used during recording a hologram of an object [68], etc.

The method of holographic study of an object shape using several wavelengths was most developed. In the work [69], it is shown that in the case of two point sources emitting different wavelengths the interference fringes are formed when the object crosses the family of paraboloids of revolution. It is pointed out that while using collimated beams it is possible to obtain contour maps with plane equidistant secant surfaces. It is ascertained [28–31] that the most perspective light source for sensitivity regulating in this method is a dye laser. It allows smoothly converting the wavelength in a wide range and implementing simultaneous generation of several wavelengths. In the works [58, 71], the theoretical analysis of phase ratios under holographic interferogram recording in several wavelengths is conducted.

Conditions of contour maps obtaining for Fresnel, quasi-Fourier and holograms of a focused image are examined in the work [72], and for topograms of real and virtual image—in the work [73].

Despite great possibilities of the holographic contouring method, it has its limitations and disadvantages. Two-long-wave interferometers require a rather complicated adjustment procedure, though a number of decisions are found, which simplify it.

It is known that for multi-long-wave contouring ruby [74], pulse xenon [75] and other ion gaseous lasers were implemented, which are characterized by high radiation power and coherence length that enabled to get contours on real objects. Implementation of convertible along the wavelength radiation of dye lasers with continuous pumping [71] and pulse pumping [28, 76] is most advantageous when there is need of smooth regulating interval between broad bands. That is why dye lasers are of great interest where it is possible to record surface relief changing in time by regulating the interval between radiation pulses. In pulse radiation, the contoured maps in sphere motion are obtained [77]. Real-time contouring was conducted with krypton ion laser and electro optical crystal with minimal mentioned in literature depth interval equal to 2.67 μm [78]. The contoured maps of retina were obtained [79], wear of artificial bones was studied, tests of mirrors [80] and aspherical surfaces [81, 82] were conducted, and reflective index of acrylic rubber was defined [59]. Thus, the analysis of the literature shows that the main advantage of the two-long-wave contouring method is the possibility to study shape of objects in motion or changing their shape with time. This advantage comparing to other contouring methods occurs due to laser equipment improvement and, in particular, development of modes, in which lasers simultaneously generate several wavelengths.

Almost at the same time with the two-long-wave contouring method the immersion method appeared [70]. This method is double exposed in its essence and is implemented in the following way. The object under study is placed in the immersion chamber with a flat window. Before the first exposure, the chamber is filled with liquid or gas with the refractive index n_1, and before the second exposure with the refractive index n_2. After recording the double-exposed hologram contour, bands were observed on the object surface.

The immersion method was used to obtain contoured maps of a punch for turbine bucket producing [83], and it was used for the study of wear [84].

It is easy to see that if to produce a cell, transparent window of which has internal surface with a relief opposite to the relief of the standard, then the layer thickness of the liquid will characterize the difference relief [85]. The contoured map of the relief gradient can be obtained if the surface under study moves in the transverse direction relative to its transparent replica serving as the cell window [86].

Another contouring method is the so-called two sources method [69]. In the holographic version of this method, the hologram is exposed twice by movement of illuminating source. The depth interval between the bands in this case depends on changing the object entrance angle. But as while hologram recording side illumination of the object must be used, then its surface appears to be partially black-out. The solution of this situation can be found in illuminating the object from different sides [87] or in implementing hologram recording in colliding beams [88]. Along with this at solution of each particular task, the selection of the method of surface relief research must be conducted after detailed study of advantages and disadvantages of these methods, as they nearly differ according to their metrological characteristics (the depth interval obtained). Moreover, from the analysis of the literature it is seen that the holographic contouring methods make it possible to solve several types of tasks: to detect the surface relief, difference of the two reliefs, to examine the surface relief gradient and the medium refractive index. In all approaches, the assumption was used that the sequence order of planes cutting the object is known; i.e., the shape of the object is known to be regular. But for objects with a surface shape, which is beforehand unknown, with random alternate convexities and concavities or while comparing shapes of two surfaces such an assumption leads to errors, and that is why in holographic contouring as well as in the whole holographic interferometry, there exists a problem of absolute sequence order detecting of interference fringes.

Besides it must be pointed out that holographic contouring appeared as a sensitivity roughening, which is characterized by an efficient wavelength reaching several centimeters. The hologram recording and its observation are conducted in ordinary light waves. The appeared discontinuity in sensitivity of holographic contouring and holographic interferometry was compensated in two ways. By lowering the efficient wavelength to several millimeters, it became possible to diminish significantly the depth interval between the contour bands. In its turn, narrowing of interference maxima allowed enhancing the accuracy of their location detection on the surface under study. Some aspects of absolute relief detection will be described in this chapter.

The generation mode of pulses with regulated repetition interval and high spatial and temporal radiation coherence is of great interest for the contemporary medicine. One of its important and urgent tasks is to develop and implement coherent-optical (holographic and speckle-optical) methods and to create on their basis new equipment of new generation for remote noninvasive disease diagnosing, especially during mass examination of population: of laser-holographic complex (laser cardiograph) and laser specklometer.

Nowadays, the contemporary medicine is in great need of creation and introduction of new contactless distant noninvasive methods of study of physiological state of a human organism. A perspective tendency for development of such methods is implementation of the latest achievements in the sphere of holography and holographic interferometry.

Their implementation in the applied medicine will promote early diagnostics of pathological anomalies in cardiovascular and nervous systems. This gives the possibility to provide timely treatment of diseases, which are most widespread and hardly defined on early stages that will decrease the number of people with chronic diseases and reduce the duration of temporary disability.

The laser-holographic complex is an equipment of a new class, which helps to study the activity of cardiovascular system, respiratory organs, muscular systems, etc. The complex principle of operation is based on the detection and analysis of shift of points of human chest surface, which are caused by the activity of mechanical waves that appear under cardiac contraction.

The examination of functional state of human cardiovascular system occurs via oscillation recording of skin in a holographic interferogram with its further automated computer processing.

Usage of double-pulse mode of ruby laser radiation generation and its synchronization with the process of chest hologram recording in different phases of the cardiologic cycle gives the possibility to form relevant holographic interferograms of different states of human chest surface.

The formed graphic pattern of the field shifts in different phases of the cardiologic cycle makes it possible to get information about the interference fringe distribution in the image of chest surface, which corresponds to the region of surface of the movement points. As a result of comprehensive analysis of holographic interferograms, it will be possible to get the idea of cardiac murmurs and of the functional state of valves and cardiac muscle region, as well as information about the whole or partial hemodynamics on the whole surface of the human chest.

Usage of the laser-holographic complex (holographic cardiograph) in the sphere of cardiology opens ample opportunities to conduct contactless noninvasive study with cardiovascular pathologies.

As these methods of getting information about the state of human health are contactless, efficient and safe it improves work of cardiological centers and makes it more resultant during professional examination and selection. It also becomes determinative while solving expert questions in disability level estimation, provides possibility of mass diagnostic study of population during radiation damage.

According to medical specialists of the Institute of Neurology, Neurosurgery and Physiotherapy and the Institute of Cardiology of the Ministry of Health of the Republic of Belarus who participated in this study, the laser-holographic complex (holographic cardiograph) will give the possibility to conduct search and study of factors that enable to investigate deeper the regulating mechanism of nervous, endocrine, cardiovascular and muscular systems, on its basis the ways of influence on these mechanisms will be developed and it will be implemented in clinical practice.

The application of the highly efficient holographic interferometry methods is rather perspective for the study and diagnostics of vertebragenous diseases of peripheral nervous system, in pathogenesis of which the vascular factor plays an important role. Detection of hemodynamic malfunctions and establishment of correlations of system hemodynamics changes with regional blood flow indices in patients with osteochondrosis neurologic manifestations will make it possible to find out the pathogenesis of a disease by a particular person that gives the possibility to conduct therapy of this pathology differentially.

Along with this the laser-holographic complex can help to get data about functional state of skeletal muscles and intensity of anatomical changes of the spine that is very important while solving the question concerning the appropriateness of developing indices to kinesitherapy including manual therapy. The complex can be implemented in the research of muscle biomechanics and micro-hemodynamics while studying osseous–articular apparatus, in otolaryngology, in dentistry and traumatic surgery for detection of different bone tissue changes in statics and in dynamics, in bioengineering while transplant creating and artificial organs, etc.

In many cases, implementation of the holographic complex will give the possibility to eliminate the necessity of traditional examination methods, for example, roentgenology, that is very important for the republic to ease the consequences of the accident at the Chernobyl nuclear power station.

Along with this the complex can help to shorten time of examination and therefore of purposeful treatment of patients.

The author of this book is the initiator, the research supervisor, the principle investigator and just the executor of the project "Development and assimilation of the laser-holographic complex for detection of the state of human circulatory system." This study was conducted by the scientific group of "Coherent-optical research of medical–biological systems" founded by the author in the Research Institute of Neurology, Neurosurgery and Physiotherapy of the Ministry of Health of the Republic of Belarus. In the experimental sample development of the laser-holographic complex (holographic cardiograph), there took part a lot of scientists and specialists from different establishments of the NAS of Belarus and of the Ministry of Health of the Republic of Belarus organized by the author of this project into a temporary working group. These are the Institute of Cardiology of the Ministry of Health of the Republic of Belarus, the Institute of Neurology, Neurosurgery and Physiotherapy of the Ministry of Health of the Republic of Belarus, the Institute of Electronics of the NAS of Belarus, the Institute of Applied Physics of the NAS of Belarus, the Institute of Physics of the NAS of Belarus and many others.

Great support and help in implementation of this project was provided by P.A. Apanasevich, Academician, Professor, Doctor of physical and mathematical sciences, Director of the Institute of Physics of the NAS of Belarus; A.S. Rubanov, Academician, Professor, Doctor of physical and mathematical sciences, Head of the Optical Holography Laboratory of this establishment; I.P. Antonov, Academician, Professor, Doctor of medical sciences, Director of the Institute of Neurology, Neurosurgery and Physiotherapy of the Ministry of Health of the Republic of Belarus; G.I. Sidorenko, Academician, Professor, Doctor of medical sciences, Director of Institute of Cardiology of the Institute of Neurology, Neurosurgery and Physiotherapy of the Ministry of Health of the Republic of Belarus; A.V. Frolov, Doctor of biological sciences; V.V. Mironchik, Candidate of medical sciences; A.A. Kovalev, Professor, Doctor of physical and mathematical sciences, Deputy Director of the Institute of Electronics of the NAS of Belarus, Corresponding Member; B.N. Tyushkevich, Doctor of physical and mathematical sciences; Y.F. Morgun, Candidate of physical and mathematical sciences; F.G. Drik, research worker of the Institute of Applied Physics of the NAS of Belarus, Candidate of physical and mathematical sciences. The author is very grateful to everyone who helped in realization of this project.

As a basis there were assumed author's evidences in the sphere of holographic interferometry, as well as the results of the research published in a number of articles of native and foreign magazines, for example [89].

It is also important to point out that the author of this project would not have begun this project if in the Institute of Electronics of the NAS of Belarus there had been no groundwork in developing a ruby laser with a double radiation monopulses and convertible period of their sequence. This laser exceeds on its parameters the best developments of foreign companies (Rottencolber holosystem (FRG), Apolo Laser Corp. (USA), etc.).

The author's idea connected with the creation (development and production) of the holographic cardiograph was realized within the framework of the program "Beloptics" as an experimental sample. But the increased in 1994 rates of inflation, the sudden growth of equipment costs, as well as increase of expenses for remuneration of labor, which was not indemnified due to indexation required correction of the respective parts of the theme. This reason and the necessity of development of expensive hard- and software for controlling the working regimes of the complex with inputting and qualified analysis of holographic interferograms on computers required the extension of terms for fulfillment of this task and better financing of the work. That is why the decision was taken to create only an experimental sample of the holographic cardiograph. At the same time to implement the experimental sample of this device, there was just one step left.

Later L.V. Tanin, the author, managed to realize another idea in an experimental sample of the laser specklometer, which is successfully used today as a device for diagnostics in neurological practice (it is described more detailed in Chap. 5).

3.1 The Holographic Method of Contouring of Static and Time-Changing Surfaces Using Multi-long-Wave Dye Laser Radiation with Laser Pump and Resonance, Absorbing and Optically Active Media

3.1.1 The Holographic Method of Surface Relief Contouring Two- and Four-Long-Wave Dye Laser Generation Mode with Laser Pump and Regulated Spectral Interval Between Them

Let us consider the possibility of using rhodamine 6G laser with laser pump reconstructed over the radiation spectrum in the study of surface relief with the help of colliding beam holography [90, 91].

The surface hologram obtained in the two-long-wave radiation regime from two independent lasers through wave front reconstruction makes surface relief contour [69]. The same hologram obtained in multi-long-wave radiation regime gives narrower bands as it is shown in the work [71] as they appear due to multi-beam light interference. But this method has a disadvantage: Dispersion of the hologram obtained with an inclined reference beam leads to transverse shift of the reconstructed images corresponding to different wavelengths. This leads not only to contour distortion but also to band localization at a significant distance from the surface [92]. To eliminate hologram dispersion and these undesired effects, the scheme of hologram recording in colliding beams was used [22, 28–31]. Thus, the colliding beam holography method using the one- and multi-long-wave radiation generation regime of dye lasers makes it possible to manage the spatial resolution and sensitivity of the holographic method of contouring of static and time-changing surfaces depending on the step between the spectrum components and the number of these spectral components.

Increasing sensitivity of holographic contouring methods occurs at the decrease of the depth interval between the bands that is best provided while using a laser with gradually changing wavelength with regulated spectral interval. But with the growth of this interval between two wavelengths the shift between corresponding points on the object increases as well that leads to decorrelation and worsening of band luminous efficiency. The accuracy of calculation of interference fringe center can be enhanced through the half-width decrease.

The ability of the holographic method to preserve and further reconstruct the wave front recorded by the hologram makes it possible to realize two generation regimes of dye lasers radiation in the studies of surface reliefs [28, 29, 31].

The holographic relief surface contouring in generation regime of one reconstructed wavelength of dye laser radiation.

During the experiment, it was ascertained that the quality of the hologram recorded separately in dye laser radiation with the wavelengths of 590, 596.5 and 585 nm stayed almost the same. In its turn, it made it possible using the double-exposure method to record two wave fronts with different wavelengths, the interval between which can be smoothly changed in the limits of 0.01–30 nm on one type of dye, in radiation of one laser consequently on one hologram. While reconstructing the wave front according to such a hologram, we will obtain the relief surface contours according to the work [69]. And as the distance between the neighboring contour lines is definitely connected with interval between wavelengths used in hologram recording, then the change of this generation interval from 0.01 to 30 nm leads to the depth distance change from 1 cm to 6 μm. It is obvious that quick change of the dye will allow extending the limits of wavelength generation reconstruction and, thus, increasing the minimal resolution along the depth between the neighboring contour relief lines. In this case, the possibility of holographic recording will be limited by the spectral sensitivity of the photomaterial to wavelengths of the radiation recording the hologram and its resolution. Such a hologram recording has few differences from recording using the immersion method [93]. The difference is only that in one case in the interval between the exposures the wavelength changes, and in another one—the refractive index of liquid or gas. A common disadvantage of both types of recording is that they are suitable only while studying the relief of static surfaces.

The holographic contouring of surface relief in multi-long-wave converting regime of dye laser radiation generation [22, 30, 31].

Making holograms of surface relief in colliding beams using multifrequency regime of 6G rhodamine laser generation was conducted on the set-up, the optical scheme of which is presented in Fig. 3.1.

As it was shown in Sect. 3.1.1, introduction of 6G-rhodamine laser with laser pump of Fabry–Perot interferometer, particularly, with the base of 100 μm into the resonator gave the possibility to obtain simultaneous generation of four wavelengths.

Thus, by changing the free dispersion region of the interferometer the multi-long-wave generation regime can be implemented. This regime is especially important during holographic study of relief of a surface changing in time. In this case, the width of the lines determines narrowing (definition) of the surface relief contours, and the interval between the generation lines specifies the distance due to the depth, which corresponds to the neighboring contour lines. The duration of the generation pulse regulates the dynamics of fast processes. The most typical surfaces: a sphere, a cylinder, a wedge and some surface areas of turbine bucket model, were the objects of the study. The hologram recording was conducted in 6G rhodamine laser radiation, the generation spectrum of which consisted of four lines splitted by the interval of 1.8 nm. In this case in the plane of the hologram during the exposure, four interference structures with spatial frequency of about 5000 lines/mm emerge, which appear as

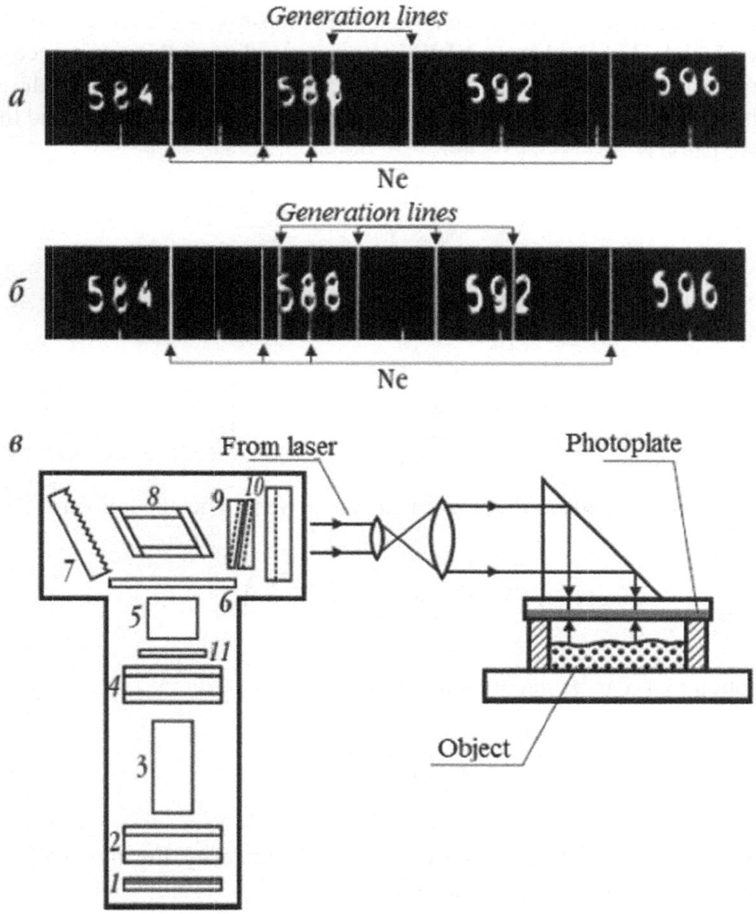

Fig. 3.1 Holographic scheme of an object in colliding beams (**a**) using two (**b**) and four-wavelength generation modes of 6G-rhodamine laser with laser pump. Reprinted from [94] with permission

a result of interference between four pairs of object and reference waves. The phase difference φ_n between the nth object and the nth reference waves in the point is

$$\varphi_n = 4\pi \frac{z}{\lambda_n} \tag{3.1}$$

where z is the distance between the hologram and the corresponding point of the object; n is the index of the wavelength in its ascending order.

On the reconstruction stage, the reference beam is replaced by a light beam starting from the continuous spectrum source. In this case, the reconstruction process can formally be described in such a way that the hologram structure in point R reconstructs N partial waves, amplitudes of which are approximately proportional to amplitudes of object waves, and their phases relative to the reconstructing beam are equal to the phases which had object beams relative to corresponding to them reference beams during the recording process. Therefore, the phase difference between the waves reconstructed by the hologram structure, which was formed by the nth and the $n+1$th wave pairs, is

$$\delta\varphi = |\varphi_n - \varphi_{n+1}| \approx \frac{4\pi\Delta\lambda}{\lambda^2} \tag{3.2}$$

where λ is the central wavelength of the dye laser spectrum and $\Delta\lambda = \lambda_n - \lambda_{n-1}$.

Then, in each point, we have interference of four waves:

$$E = E_0 \sum_{n=0}^{3} \exp\left(i\frac{4\pi z\Delta\lambda n}{\lambda^2}\right), \tag{3.3}$$

For full light intensity in the point of an image, we will get the following formula:

$$I = I_0 \left|\sum_{n=0}^{3} \exp\left(i\frac{4\pi z\Delta\lambda n}{\lambda^2}\right)\right|^2. \tag{3.4}$$

Equation (3.4) can be rewritten in a simpler way:

$$I = I_0 \frac{\sin^2\left(4\frac{\delta\varphi}{2}\right)}{\sin^2\left(\frac{\delta\varphi}{2}\right)} \tag{3.5}$$

Provided that all the generation lines had the same intensity, the maximal intensity in the point of the image is reached when:

$$\delta\varphi = 2\pi k, \tag{3.6}$$

where k is the integral number.

These image points are connected by the contours, which are the intersection of object surfaces with the family of surfaces specified by:

$$\delta\varphi = 2\pi k,\ z = \frac{k\lambda^2}{2\Delta\lambda} \tag{3.7}$$

In such a case, the distance along the normal between the neighboring contour lines is

$$l = \frac{\lambda^2}{2\Delta\lambda} \tag{3.8}$$

It also follows from (3.4) that the use of several beams leads to interference beam narrowing: half-width of the interference maximum is in inverse proportion to the quantity of interfering beams. In a case with four beams, which occurs in our studies, the bandwidth is twice as little as the double-beam one.

The results of the experiment are in good agreement with calculation due to (3.8). The full relief depth of the surfaces under study was measured with micrometer from the base to the envelope of the object and is 1 mm for the sphere, 1.5 mm—for the cylinder and 0.9 mm—for the wedge.

It is worth mentioning that in the conducted earlier works on holographic relief detection with radiation containing several equidistant spectral lines [71], the scheme with an inclined reference beam was used that led to hologram dispersion, i.e., to in-plane shear of the reconstructed surfaces of the object (see, for example [92]). In our work, for this purpose, the scheme with colliding beams was used for the first time, which is known to form holograms without transverse dispersion [92]. To reduce exposure time, the objects were covered with paint, which has a high diffusion reflection coefficient that made it possible to eliminate points preventing image reconstruction. The photographs of the surface relief contours reconstructed by the hologram are presented in Fig. 3.2, where it is seen that 10 contour lines were reconstructed for the sphere, 15—for the cylinder, 9—for the wedge. Moreover, the relief contours spread over the whole surface of the objects under study.

3.1.2 To the Question of the Possibility of Increasing the Spacial Resolution of the Holographic Surface Relief Contouring Method While Using the Resonance Media

On the basis of a two-long-wave method, there were developed a rather great number of holographic contouring modifications. Several works were dedicated to the development of sensitivity enhancing principles [30, 71].

A new method is offered, with the help of which spatial resolution can be regulated. Resonance medium is applied for its realization. Such media have already been used during the study of plasma [95].

The scheme of holographic interferogram recording is presented in Fig. 3.3. The object was placed into a transparent chamber with a plane window containing the resonance medium. The medium can be gas or liquid with absorption line x (e.g., hemoglobin or cyanine solutions, alkali vapor of such metals as sodium, cesium, potassium, etc.). The object is lighted by plane waves 1 and 2 from the side of observation. Implementation of plane waves allows obtaining intersections of the object surface by the family of planes that significantly simplifies the interpretation of the contour map. The reflected light interferes on the hologram with two reference

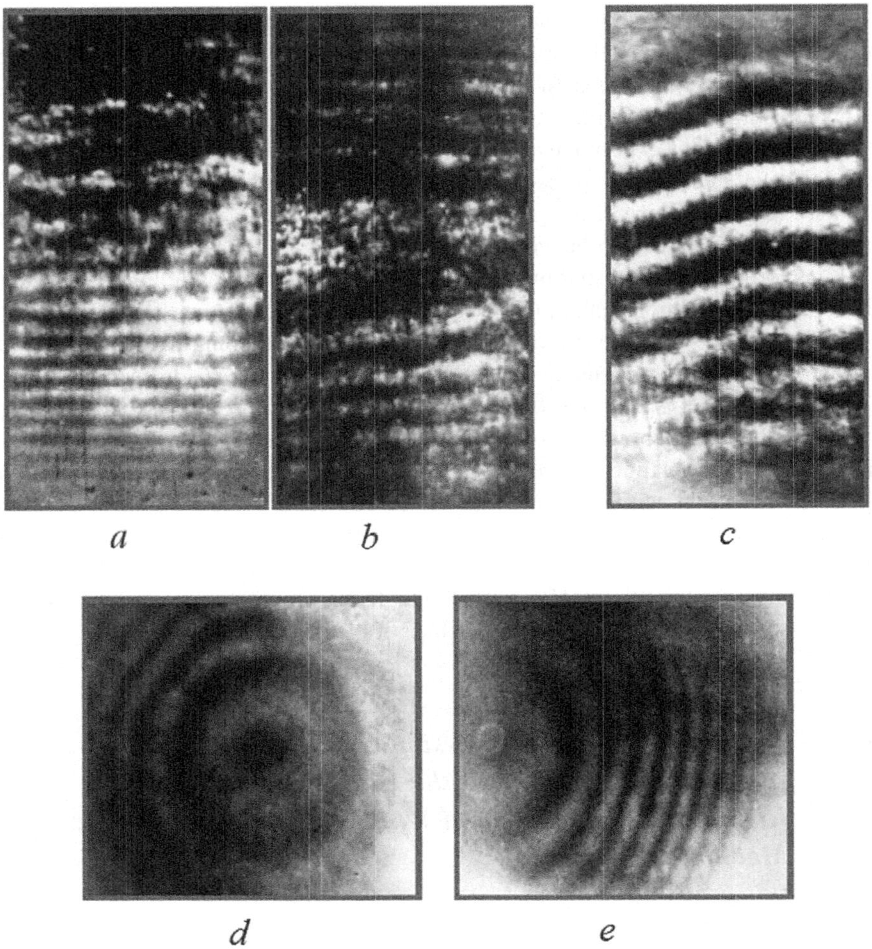

Fig. 3.2 Image of the contours of the relief of surface: **a**, **b**—of a cylinder, **c**—of a wedge; **d**, **e**—of spheres, restored by a hologram, recorded in the oncoming beam. Reprinted from [94] with permission

beams 1′ and 2′ directed into the hologram under the angles α_1 and α_2. To eliminate the image shift, the following condition should be met in different wavelengths

$$\frac{\alpha_1}{\lambda_1} = \frac{\alpha_2}{\lambda_2}. \tag{3.9}$$

This condition is automatically fulfilled during hologram recording in colliding beams. Phase difference for two wavelengths for the corresponding point of the object is set by the expression

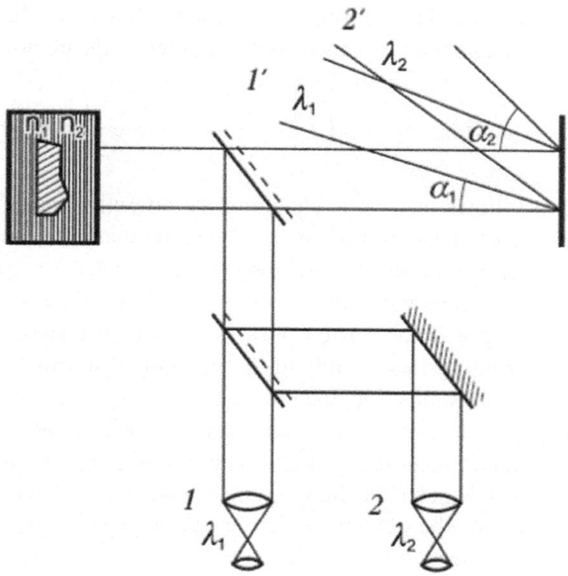

Fig. 3.3 Scheme of the registration of interferograms with holographic resonance contouring. Reprinted from [94] with permission

$$\Delta\varphi = 2\pi m\left(\frac{z_1 + z}{\lambda_1} - \frac{z_2 + z}{\lambda_2}\right) + 2\pi m\left(\frac{z_p n_1}{\lambda_1} - \frac{2 z_p n_2}{\lambda_2}\right), \tag{3.10}$$

where z_1, z_2 is the distance between the sources λ_1 and λ_2 to the wall of the chamber, respectively; n_1 is the refractive index of the object under study; n_2 is the refractive index of the resonance medium; z is the distance between the chamber wall and the hologram; z_p is the distance between the chamber wall to the point of the object.

It is worth mentioning that the first summand in (3.10) does not influence the value of the depth interval. Under changes of phase difference in 2 πm where $m = \pm 1, 2, 3\ldots$, on the objects there can be observed light bands with the depth interval calculated according to the formula

$$\Delta z_p = \frac{\frac{\lambda_1}{n_1}\frac{\lambda_2}{n_2}}{2\left|\frac{\lambda_1}{n_1} - \frac{\lambda_2}{n_2}\right|}. \tag{3.11}$$

It is known that the depth interval for the two-long-wave method is set by the formula

$$\Delta z_{\lambda_1\lambda_2} = \frac{\lambda_1\lambda_2}{|2n(\lambda_1 - \lambda_2)|}. \tag{3.12}$$

After some mathematical transformations, it can be shown that the depth intervals in the two-long-wave and the resonance contouring methods are bounded

$$\Delta z_{\lambda_1,\lambda_2} = \Delta z_p\left(1 + \frac{n_2 - n_1}{n_1}\frac{\lambda_1}{\lambda_1 - \lambda_2}\right). \tag{3.13}$$

From (3.13), it follows that while changing the refractive index of the medium (e.g., at the expense of concentration, pressure and temperature) and adjusting the wavelength of the laser radiation to the absorption line of the resonance medium the value of the depth interval can be smoothly regulated in the resonance method relative to the two-long-wave one. Thus, in case if the first wavelength λ_1 is adjusted to the half-height of the resonance line of the medium absorption, and the second one λ_2 is situated far from this line, then the maximal increase of spatial resolution is reached, which according to the calculation amounts to the order value of 30%.

The offered resonance contouring method can also be applied for the study of the shape of phase objects, which have the symmetry plane perpendicular to the incident radiation, as well as for the objects, which are homogeneous, and the resonance medium is inside the object.

An example of such an object can be a blood cell—an erythrocyte containing hemoglobin. An erythrocyte in native state takes a shape of a discocyte, a spherocyte or an ellipsoid of rotation, which have corresponding symmetry planes.

For the phase object, the depth interval between the bands transforms to the following form

$$\Delta z_p = \frac{\frac{\lambda_1}{(n_1-n)}\frac{\lambda_2}{(n_2-n)}}{2\left|\frac{\lambda_1}{n_1-n} - \frac{\lambda_2}{n_2-n}\right|}. \tag{3.14}$$

It follows from (3.14) that the value of the depth interval is determined by the wavelengths as well as by the value of the refractive index corresponding to them, which are determined by the resonance value. After transformation the validity of the following correlation can be shown

$$\frac{1}{\Delta z_p} = \frac{1}{\Delta z_{\lambda_1,\lambda_2}} + \frac{1}{\Delta z_{\text{im}}}, \tag{3.15}$$

where the depth interval during resonance contouring can be set by (3.11), and the depth intervals $\Delta z_{\lambda 1,\lambda 2}$ and Δz_{im} are estimated by the formula

$$\Delta z_{\lambda_1,\lambda_2} = \frac{\lambda_1\lambda_2}{2n_2(\lambda_2 - \lambda_1)}, \tag{3.16}$$

$$\Delta z_{\text{im}} = \frac{\lambda_1}{2(n_1 - n_2)} \tag{3.17}$$

and correspond to the two-long-wave and immersion contouring methods. Equation (3.15) shows the connection of depth distances between contours for resonance, two-long-wave and immersion methods of surface relief detection. It allows estimates the contribution to the depth interval for the resonance contouring method introduced by the interval between the used wavelengths as well as by the difference of refractive indices of the resonance medium corresponding to them.

3.1.3 Holographic Two-Long-Wave Method of Absolute Surface Relief Determination Based on the Application of Absorbing Media

The resonance medium gives the possibility to implement only effects connected with dispersion dependence of the refractive index close to the absorption line and, thus, to determine how the depth interval depends on the used wavelengths and their adjustments to the absorption line.

Along with this the influence of the absorption coefficient on the interference pattern formation is also of great interest. Let us consider the case of hologram recording in colliding beams (Fig. 3.4).

The object is placed into the immersion medium with the refractive index n_m and the absorption coefficient $\varkappa_m$ and lighted with a collimated beam with the wavelength λ_m, where $m = 1.2$. Let us consider that the recording of a double-exposure hologram is conducted in the linear regime. In such a case, the hologram reconstructs waves, the amplitudes of which are proportional:

$$A_m = R_m a_m \exp\left[-4\pi i \frac{z}{\lambda_m}(n_m - i\varkappa_m)\right], \tag{3.18}$$

where R_m is the coefficient of surface reflection; a_m is the amplitude of the wave incident on the absorption medium for the object illumination; z is the distance between the surface of the absorbing medium and the point of the object.

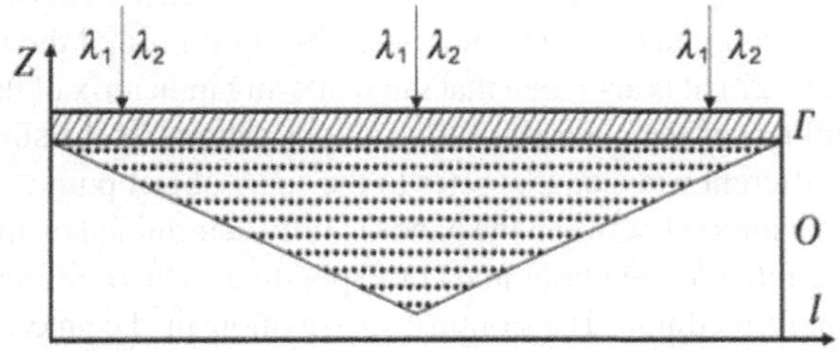

Fig. 3.4 Scheme of holograms registration by two-long-wave method of defining absolute relief of the surface. Reprinted from [94] with permission

As a result of interference of these reconstructed waves, we can get the intensity distribution

$$I = B_1^2 + B_2^2 + 2B_1 B_2 \cos\left[4\pi z\left(\frac{n_1}{\lambda_1} - \frac{n_2}{\lambda_2}\right)\right], \tag{3.19}$$

where

$$B_m = R_m a_m \exp(-4\pi z\frac{\varkappa_m}{\lambda_m}).$$

The depth distance Δz between the bands is calculated from (3.19) containing the cosine and is equal to

$$\Delta z = \frac{\frac{\lambda_1}{n_1}\frac{\lambda_2}{n_2}}{2\left|\frac{\lambda_1}{n_1} - \frac{\lambda_2}{n_2}\right|}. \tag{3.20}$$

Due to the absorbing medium, the intensity of the interference fringes in maxima and minima appears to be modulated. Let us calculate the contrast P of the interference pattern using

$$P = \frac{I_{\max} - I_{\min}}{I_{\max} + I_{\min}}. \tag{3.21}$$

After a number of mathematical transformations, we obtain that

$$P = \left\{\mathrm{ch}\left[4\pi z\left(\frac{\varkappa_1}{\lambda_1} - \frac{\varkappa_2}{\lambda_2}\right) + \ln\frac{R_2 a_2}{R_1 a_1}\right]\right\}^{-1} \tag{3.22}$$

If $\frac{R_1 a_1}{R_2 a_2} = 1$ the contrast is monotonously falling under the depth increase of the absorbing layer z. This makes it possible while registering the change of the contrast from band to band to determine their absolute depth location relative to each other. The laser radiation is partially absorbing while going through the medium that leads to an alteration of intensity in the maxima of the interference pattern. The more the density of the absorbing layer is, the more the absorption is and the more the value decreases. From (3.22), it is also seen that the scattering indicatrix of the object under study weakly influences the contrast of the contour pattern, as the surface reflection coefficients on different wavelengths refer to the same object point.

Thus, the depth interval between the bands is fully determined by the wavelengths of the radiation emitted by the laser and corresponding to them refractive indices of the used absorbing medium. The contrast distribution in the interference pattern depends only on wavelengths and connected with them coefficients of medium absorption.

While comparing the differences between wavelengths and the width of the spectral line of the absorbing medium two extreme cases can be observed.

1. If the absorption line is rather narrow and the absorption coefficient as the function of the wavelength changes rather fast, then the absorption on one of the used wavelengths can be much more intensive than on the other wavelength. Then, the formula for the contrast is transformed to the following form

$$P = \left[\operatorname{ch}\left(4\pi z \frac{\varkappa_1}{\lambda_1} + \ln \frac{R_2 a_2}{R_1 a_1}\right)\right]^{-1} \tag{3.23}$$

From (3.23), it follows that the value of the contrast is determined only by absorption and wavelength corresponding to one of the exposures. Realization of this case is possible if gaseous media are used as immersion.

2. If the refraction coefficient remains almost constant under changes of wavelengths between the exposures, then $\varkappa_1 = \varkappa_2 = \varkappa$. Such spectral lines are typical for liquid media and the refractive index for two wavelengths can also be considered as identical. Then taking into account (3.20), (3.21) simplifies to

$$P = \left[\operatorname{ch}\left(\frac{2\pi z \varkappa}{\Delta z} + \ln \frac{R_2 a_2}{R_1 a_1}\right)\right]^{-1} \tag{3.24}$$

Comparison of (3.23) and (3.24) shows that for the obtained identical values of the depth distance and the thickness of the absorbing layer, on which the contrast incidents to the minimal recorded value, gaseous media are preferable as the value of the absorption index from (3.23) will be significantly lower than the corresponding value in (3.24). This leads to the fact that the intensity of the interference patter is greater due to the absolute value if the object is placed in gaseous immersion.

To illustrate the possibilities of the method of absolute surface relief estimation let us show the normalized intensity distribution (see Fig. 3.5a) and the dependence of contrast (Fig. 3.5b) on the thickness of the absorbing layer between the object (Fig. 3.4) and the wall of the chamber calculated for a wedge-shaped surface. The parameters corresponding to the sodium vapor were chosen for the calculation:

$$\lambda_1 = 589.5 \text{ nm}, x_1 = 0.004, n_1 = 1.004,$$
$$\lambda_2 = 620 \text{ nm}, x_2 = 5.4 \times 10^{-7}, n_2 = 1.000,$$
$$R_1 a_1 = R_2 a_2.$$

To simplify the calculation the intensity distribution of the beams illuminating the object was supposed to be homogeneous. As to define the absolute surface relief the contrast of the interference pattern is used, this distribution does not play an important role.

Thus, application of the absorbing medium in the two-long-wave contouring method makes it possible to determine the absolute surface relief of the objects which have an arbitrary shape. Owing to the fact that different sources exist, which emit two wavelengths simultaneously, this method makes it possible to register the shape of moving objects or time-changing objects

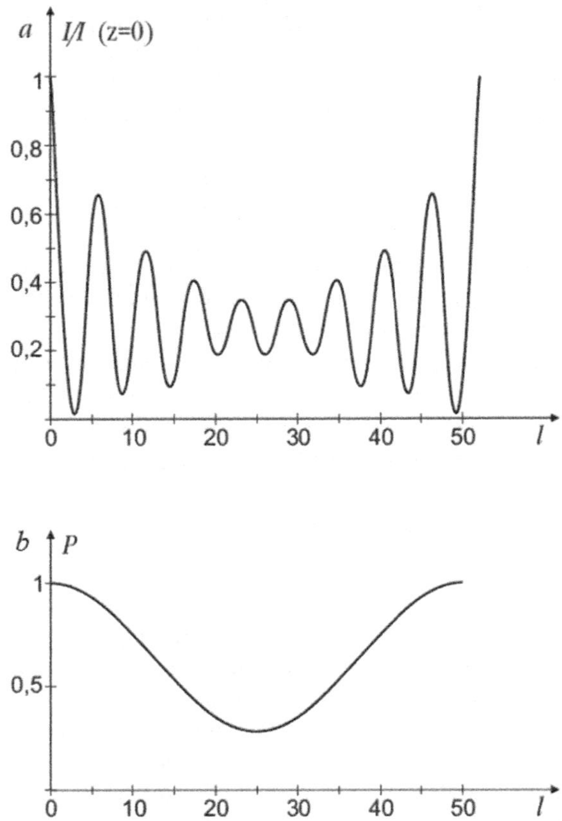

Fig. 3.5 Dependence of the normalized distribution of intensity (**a**) and contrast (**b**) for a wedge-shaped surface. Reprinted from [94] with permission

3.1.4 Study of the Absolute Surface Relief Through the Immersion Method

Another solution for determining the absolute surface relief is to use two immersion media. This method is a double-exposure one, and to obtain the contour map, the refractive index of the immersion medium changes. The phase difference emerging due to the change of the refractive index is proportional to the difference between the walls of the immersion chamber and the object. This made it possible to form interference fringes characterizing the surface relief. In order the contrast of the interference pattern to change from band to band along with the increase of the thickness of the immersion medium layer as it occurs in the two-long-wave absorptive method, at least one of the two applied media must be a narrow absorption line.

Using the approach implemented in the two-long-wave contouring method, it can be considered that two-wave fronts interfere after reconstruction with the amplitudes, which are proportional to

$$A_m = \alpha_m R \exp\left[-i\frac{4\pi z}{\lambda}(n_m - i\varkappa_m)\right] \tag{3.25}$$

After some mathematical transformations, we obtain that the depth distance between the contours is estimated due to the correlation

$$\Delta z = \frac{\lambda_1}{2|n_2 - n_1|}. \tag{3.26}$$

It is worth mentioning that in the most common case of absorbing media the refractive index is estimated by the Kramers–Kronig relation:

$$n = n_o + \frac{2}{\pi\lambda}\int_{\lambda_H}^{\lambda_K}\frac{\lambda_0^2\varkappa(\lambda_0)}{\lambda^2 - \lambda_0^2}\,d\lambda_0, \tag{3.27}$$

where λ_0 is the integration parameter; λ_H, λ_K are the wavelengths determining the integration interval; n_0 is the refractive index determined by the spectrum area located outside the integration interval.

The contrast of the interference pattern changes according to the thickness of the absorbing layer and is calculated according to the formula

$$P = \left[\mathrm{ch}\left(\frac{4\pi z}{\lambda}(\varkappa_1 - \varkappa_2) + \ln\frac{a_2}{a_1}\right)\right]^{-1} \tag{3.28}$$

This correlation (3.28) is valid for the case when holograms are recorded under normal incidence of a collimated reference beam on the interface: The absorbing medium is the window of the immersion chamber.

The comparison of the possibilities of the two-long-wave and the immersion methods of study of absolute surface relief is of interest. To make this it should be mentioned that in both methods two values are used for calculation of the absolute surface relief: the depth interval between the bands and the contrast. Namely, due to estimating the contrast changes from band to band, the possibility emerges to determine the order of sequence of separate bands.

But it is known that if the contrast of the interference pattern incidents lower than a certain level α (usually being a value of the order of 0.05), then the contour bands cannot be seen with an eye. Thus, the third value, which characterizes the both ways, is the depth, on which the contrast decays lower than the registered level α. This value is described by the expression

$$z_\alpha = \frac{\ln\left(\frac{1+\sqrt{1-\alpha^2}}{\alpha}\right)}{4\pi\left|\frac{\varkappa_1}{\lambda_1} - \frac{\varkappa_2}{\lambda_2}\right|} \tag{3.29}$$

for the two-long-wave method, and for the immersion method of estimation of the absolute surface relief, there is a formula

$$z_\alpha = \frac{\ln\left(\frac{1+\sqrt{1-\alpha^2}}{\alpha}\right)}{\frac{4\pi}{\lambda}|\varkappa_1-\varkappa_2|} \quad (3.30)$$

Now let us compare (3.29) and (3.30) for depth intervals and depths of contrast decay to the level α for both methods. It is worth mentioning that coefficient $\varkappa_i$, which is a part of (3.29), is the function of wavelength, and the depth interval between the bands in the two-long-wave method depends on the wavelengths used as well as on refractive indices corresponding to them $n_i = n(\varkappa_i, \lambda_i)$. That means that in the two-long-wave method for this absorption line x different but connected values Δz and z_α will correspond to different wavelengths combinations. Therefore, it is impossible, for example, to hold the value z_α as a constant and simultaneously change Δz; i.e., it is impossible to provide independent regulation of the contrast and of the depth distance.

Using different immersion absorbing media in the immersion, the contouring method gives the possibility to make the values z_α and Δz independent of optical constants of the absorbing medium during the second exposure. If for example during the second exposure the non-absorbing substance $\varkappa_2 = 0$ was studied, then the parameter z_α is determined only by $\varkappa_1$. Then holding the constant depth of contrast decay to the level α, it is possible to independently regulate the depth distance Δz through replacement of the first immersion with the second one with an independent refractive index and therefore to conduct an independent regulation of contrast and of the depth interval.

It is worth mentioning that the immersion method has positive differences from the long-wave one because it helps to easily obtain arbitrary correlations of the depth interval between contours and the depth where the contrast decays to the value α.

The two methods are almost identical in the possibility of obtaining equal values of the depth interval that corresponds well to the literary data for ordinary contouring methods.

But it should be pointed out that the long-wave method has certain advantages over the immersion method of absolute surface relief estimation. In particular, as the immersion method is time-elapsed, it does not provide any possibility to study the shape of dynamic objects.

The immersion method of absolute surface relief estimating was experimentally implemented. In the conducted experiment, the hologram of aluminum wedge surface was registered according to the scheme in colliding beams in He–Na laser radiation ($\lambda = 632.8$ nm). During the first exposure, an oxasin-9-perchlorat dye solution was poured into the chamber. The absorbing coefficient of the solution was 1.671×10^{-7}, and its refractive index was 1.3333 on the wavelength of the laser radiation. The absorption coefficient was determined by absorption spectra measured with a double-beam spectrophotometer "SPECORD."

The refractive index was controlled with a refractometer and corresponded well with a value calculated using to the Kramers–Kronig relation. After exposure, the chamber was cleaned of the rests of the solution and was dried for 10 min to eliminate the residual drops of the liquid, which could lead to some dilution of the second solution used. As the second immersion, a mixture of distilled water and ethyl alcohol was used. The refractive index of the solution was 1.3336. These two liquids provide a depth interval between the bands of 1.05 mm. The densitogram of the obtained interferogram of the wedge-shaped surface is presented in Fig. 3.6. The envelope, which passes through the band maximum, is set by the shape of the object, by the medium absorption and by Gaussian intensity distribution in the illuminating beam. But the contrast of the interference pattern shown in Fig. 3.7 does not depend on the latter factor; i.e., the type of intensity distribution does not influence the determination of the absolute surface relief due to the change of the contrast from band to band.

The inclination angle of the wedge is calculated according to the distance between the contour bands in the interferogram based on the depth interval found earlier. The contrast decay from great values on the left to small values on the right points to the corresponding direction of the wedge inclination that is the determination of the absolute surface relief.

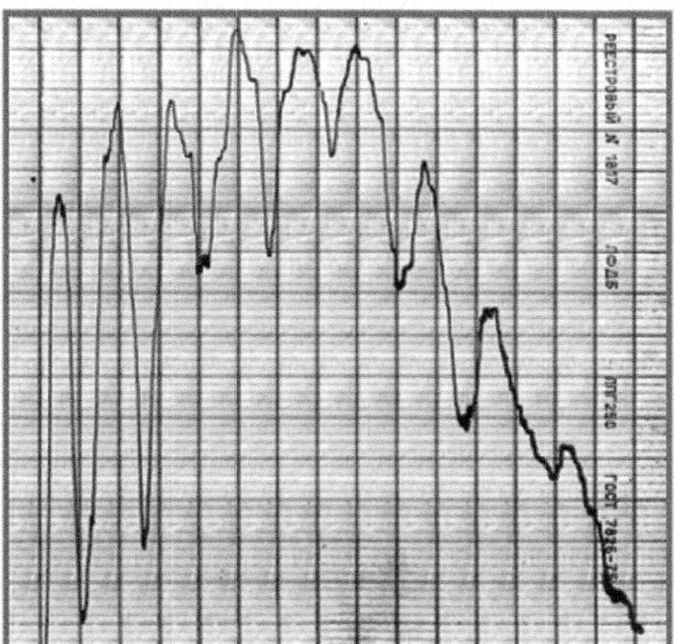

Fig. 3.6 Densitogram of contour chart of the wedge-shaped surface. Reprinted from [94] with permission

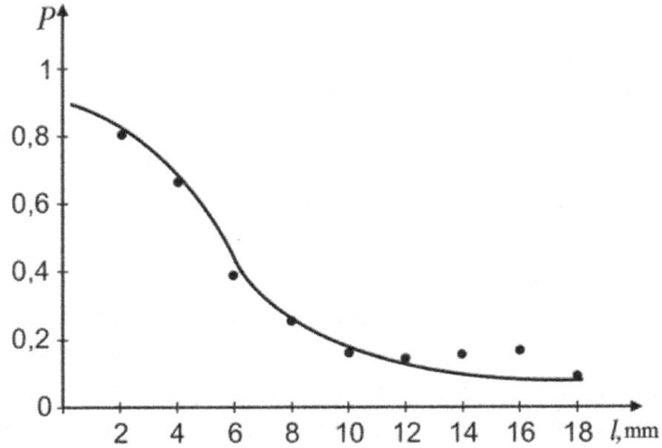

Fig. 3.7 Allocation of contrast along counter chart of the wedge-shaped surface. Reprinted from [94] with permission

3.1.5 Study of Possibility of Expanding the Class of the Researched Objects in the Holographic Methods of Absolute Surface Relief Estimation

As it has already been mentioned, in the two-long-wave and immersion methods of absolute surface relief estimating the shape of the object is reconstructed due to the distance between the bands in the interferogram from the set depth interval and due to the change of the contrast from band to band. But the decay of the contrast lower than a certain registered level α restricts the possibilities of the methods; i.e., the part of the object where the contrast decayed lower than the α-level is hardly registered and therefore its shape is not estimated. From the analysis of (3.22) and (3.28) for the contrast in the two-long-wave and immersion methods, it follows that the contrast distribution in the interference pattern also depends on the amplitudes of the illuminating beams used during both exposures. It appeared that at corresponding change of the ratio of these two amplitudes, the point on the object with the maximal contrast can be shifted along the thickness of the absorbing layer. Meanwhile two different values Δz will correspond to the same value of contrast; i.e., the contrast stops being a value, which definitely characterizes the sequence order of interference fringes. This difficulty can be overcome if illuminating beams are used with different intensity distribution along the beam section during the registration and reconstruction of the hologram.

The methods of absolute surface relief estimation are slightly modified in this case. The object is placed in the immersion chamber. The dependence of the absorption coefficient and the refractive index of the medium on the wavelength are calculated beforehand. Collimated beams with homogeneous intensity distribution are used for object illumination. Any of the well-known methods is used to obtain such a homogeneous profile, for example, based on heterogeneous absorbing filter. To enhance the registered depth of the contoured surface, the relation of intensity in the illuminating beam under different exposures is changed in a way to set the greater intensity of the illuminating beam for a greater relation $\varkappa/\lambda$. The accurate correlation between the intensities is selected according to exact values $\varkappa/\lambda$ and the required growth of the registered depth. Collimated beams with homogeneous intensity distribution are also implemented on the reconstruction stage. The relative relief is estimated due to the distribution between the bands of the image and the known depth interval.

Information about absolute surface relief is obtained through the comparison of intensities in the interference fringe maxima. Greater maxima intensity corresponds to the points on the object, which are situated nearer to the window of the immersion chamber. In this case the intensity in the maxima is set by the formula:

$$I_{\max} = \left[R_1 a_1 \exp\left(-\frac{2\pi\varkappa_1 z}{\lambda_1}\right) + R_2 a_2 \exp\left(-\frac{2\pi\varkappa_2 z}{\lambda_2}\right)\right]^2 \tag{3.31}$$

and incidents under the growth of the coordinate z, that allows using $I_{\max}$ for absolute surface relief estimation.

Let us show in detail that at such a determination of the absolute relief of the surface the increase of the registered depth takes place. The depth z_α, on which the contract decays up to the value α, in a general case is set by the equation

$$z_\alpha = \frac{\ln\left(\frac{1\pm\sqrt{1-\alpha^2}}{\alpha}\right) - \ln\left(\frac{R_2 a_2}{R_1 a_1}\right)}{4\pi\left(\frac{\varkappa_1}{\lambda_1} - \frac{\varkappa_2}{\lambda_2}\right)} \tag{3.32}$$

where indices 1 and 2 correspond to two different exposures (wavelength is the same in the immersion method).

At $R_1\alpha_1 = R_2\alpha_2$ the maximal contrast corresponds to the points, which lie closely to the chamber window, and the formula for z_α transforms to the form

$$z_\alpha = \frac{\ln\left(\frac{1+\sqrt{1-\alpha^2}}{\alpha}\right)}{4\pi\left|\frac{\varkappa_1}{\lambda_1} - \frac{\varkappa_2}{\lambda_2}\right|} \tag{3.33}$$

From (3.33), it is seen that the object has to be placed as close as possible to the interface of the chamber window and the absorbing medium. Otherwise, the contrast of the bands in the interferogram will change not from one to zero, but from the value less than one to zero that will limit the possibilities of the methods in absolute surface relief estimation.

In this case, there is a univocal connection between the depth and the value of the contrast P as it was shown beforehand that makes it possible to use the contrast as a parameter for absolute surface relief estimation. It is obvious that under the decrease of the relation $R_1\alpha_1/R_2\alpha_2$ there will occur the increase of z_α if $\varkappa_1/\lambda_1 > \varkappa_2/\lambda_2$. In the formula

$$\ln\left(\frac{1+\sqrt{1-\alpha^2}}{\alpha}\right) = \left|\ln\left(\frac{R_2 a_2}{R_1 a_1}\right)\right| \tag{3.34}$$

the depth z_α doubles.

In this case, the absolute surface relief can also be estimated using the values $I_{\max}$ as a parameter.

It is obvious that the increase enhancing of the regulated depth of the object occurs due to the transition from the contrast to intensity values in maxima of the interference pattern during measurements, i.e., due to the transition from relative measurements to absolute ones. This slightly enhances the requirements to the accuracy of intensity measurements and to time stability of the illuminating sources.

Let us observe the calculated intensity distribution I (for convenience of presentation the values I normalized for value 1 when $z = 0$ are shown in Fig. 3.8) and contrast P depending on distance l for an ordinary wedge-shaped object placed in an absorbing medium (e.g., sodium vapor).

It is supposed that the method is applied with two wavelengths, and the following parameters were chosen for the calculation:

$$\lambda_1=589.5 \text{ nm}, \varkappa_1=0.004, n_1=1.004, \lambda_2=620 \text{ nm}, \varkappa_2=5.4\cdot 10^{-7}, n_2=1.000.$$

The dependencies $I(z)$ (curve 1) and $P(z)$ (curve 2) for cases $R_1\alpha_1 = R_2\alpha_2$ and $R_1\alpha_1 = eR_2\alpha_2$ are presented in Fig. 3.8a,b respectively. It follows from Fig. 3.8, a that the contrast decays to the level of 0.2 at the depth of approximately 27.97 μm. The contrast lower than the level $\alpha = 0.2$ in Fig. 3.8b corresponds to the depths up to 39.16 μm, meanwhile the decrease of intensity values in maxima from the left to the right points to the corresponding inclination of the wedge. For the case presented in the example, the registered depth of the contoured surface increases 1.4 times.

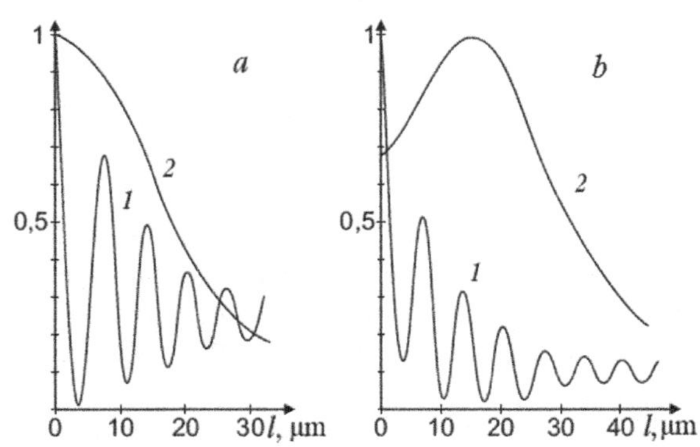

Fig. 3.8 Dependance of allocation of intensity I (curve 1) and contrast P (curve 2) on the thickness of absorbing layer in case when ratio $R_1\ \alpha_{1-} = R_2\ \alpha_2$ equals to: **a**—1, **b**—*e*. Reprinted from [94] with permission

3.1.6 Multi-angle Method of Surface Relief Contouring

The increase of accuracy in holographic contouring can be reached in two ways: due to decrease of the depth interval between the bands or due to narrowing of interference maxima. In the latter case, the location of contours on the surface of the object under study can be more accurately calculated.

As it was previously stipulated, the multi-long-wave method [22, 28–31] and the multi-immersion medium [96] method are applied for obtaining contour maps with narrower interference lines. But both interference methods are rather difficult to implement in experiments, and that is why there was a task to develop an easier and flexible method for obtaining narrow interference contours. As it is known the suggested in the work [68, 69] contouring method with two spaced sources is the simplest of the holographic methods of surface-shape detection nowadays. In the work [88], it is shown that the depth interval between the bands on the reconstructed image can be decreased at changing the observation angle. This makes it possible to conduct a detailed study of the shape of the object. Moreover, in the two-source method, the depth interval can be reduced through the increase of the incidence angle between two waves illuminating the object under study. But the side illumination relative to the observer leads to darkening of almost half of the studied surface. To eliminate this disadvantage, several methods are suggested [88, 97, 98]. Two-sided lateral illumination of the object and image filtration on the reconstruction stage are used during holographic moire contouring [97]. In case when the object is registered in colliding beams under surface illumination from the side of the observer, the number of shaded areas significantly decreases [88].

Based on the two-source method with registration in colliding beams, a technique of contour maps forming with narrower interference fringes is suggested. The object is illuminated with a collimated beam through the hologram, and the light reflected from the object forms an object beam. The exposed hologram is m-fold registered and either the object-hologram (G) system or the illuminating beam turns through angle $\Delta\alpha$ between the exposures. Hologram recording can also be implemented under the object illumination simultaneously by m beams evenly spaced along the angle. The scheme of the experiment is presented in Fig. 3.9. As a result of the interference of m waves with amplitudes, which are proportional to

$$a_m = \exp\left[i\frac{2\pi z}{\lambda}\left(\cos\alpha_m - \frac{1}{\cos\beta} + tg\beta\sin\alpha_m\right)\right], \tag{3.35}$$

where λ is the wavelength; z is the distance between the hologram and the object; $\alpha_m = \alpha + m\Delta\alpha$ is the inclination angle of mth illuminating wave; and β is the observation angle.

The object crosses the system of interference fringes characterizing by the intensity distribution.

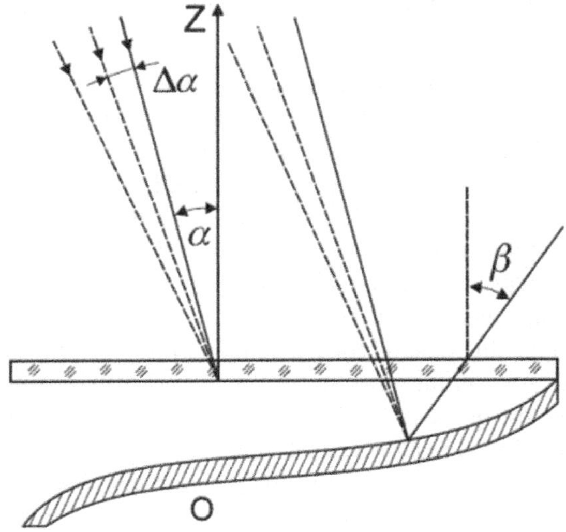

Fig. 3.9 Scheme of multi-angle method of contouring. Reprinted from [94] with permission

$$I = \left\{ \frac{\sin\left[\frac{\pi}{\lambda} z \Delta\alpha m (tg\beta \cos\alpha - \sin\alpha\right]}{\sin\left[\frac{\pi}{\lambda} z \Delta\alpha (tg\beta \cos\alpha - \sin\alpha\right]} \right\}^2. \tag{3.36}$$

The main maxima on the half-height is *m*/2 times narrower than under traditional beam contouring that leads correspondingly to accuracy increase during detection of band location.

This interference pattern has a period between the main maxima set by the depth interval calculated as

$$\Delta z = \frac{\lambda \cos\beta}{\Delta\alpha \sin(\alpha - \beta)}. \tag{3.37}$$

Due to the wave phase in (3.35) is proportional not to the angle α_m itself, phase mismatch can occur, and the interference pattern will loose regularity and can get blurred under certain conditions. Due to this reason, the maximal rotation angle must satisfy the condition

$$\Delta\alpha \ll \frac{2tg\alpha}{m}. \tag{3.38}$$

The interferograms of the model object (bolt) for cases of twofold and sevenfold interference correspondingly demonstrating the narrowing of the bands and increasing determination accuracy of their location in 3.5 times are presented in Fig. 3.10a, b. But in these experiments, it was proved on the model objects that the increase of exposures of the same hologram leads to reduction of its diffractive efficiency, which is expressed in intensity decrease in the maxima of the interference pattern. As it is known, this effect is caused by two factors: time of one exposure

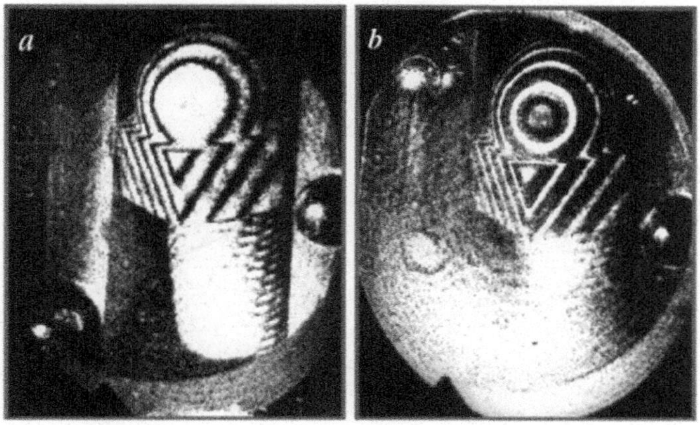

Fig. 3.10 Contour charts of a bolt in a case of double (**a**) and sevenfold (**b**) interference. Reprinted from [94] with permission

and the number of exposures. The dependence of relative intensity of interference pattern maxima for two-, three, four-, fivefold exposed holograms on time of one exposure is presented in Fig. 3.11. It is obvious that all the curves have their best value. To specify the best conditions of interferogram recording a series of additional tests were conducted. The data are presented in Fig. 3.12. The total time needed for the exposures was held at the level of 10 s. Theoretical analysis shows that intensity in maxima of such interferograms must be the same (the dotted straight line) (see

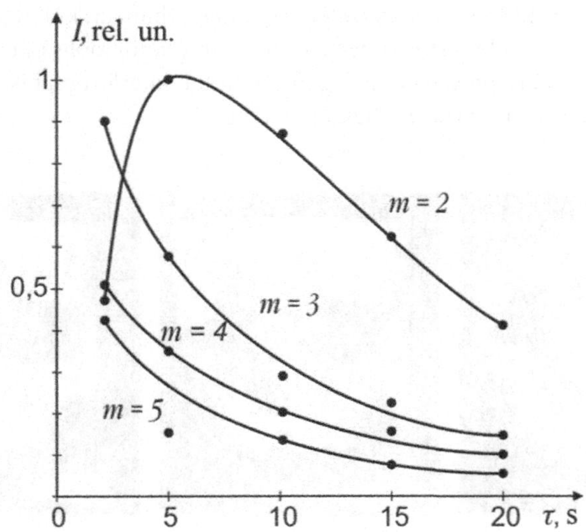

Fig. 3.11 Intensity of the maximum of the interference pattern depending on the time vof one exposure and the number of exposures. Reprinted from [94] with permission

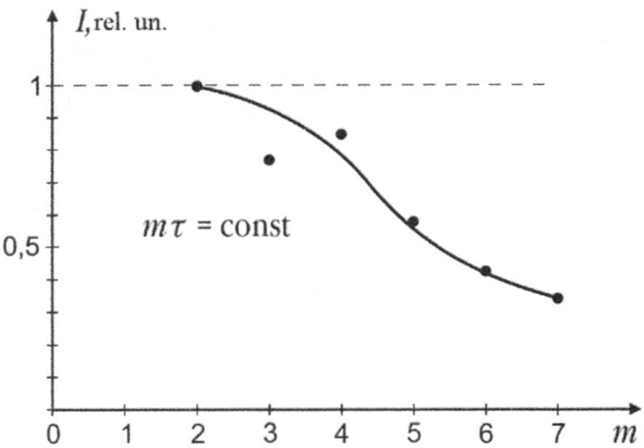

Fig. 3.12 Dependence of the intensity at the maxima of the interference pattern on the number of exposures. Reprinted from [94] with permission

Fig. 3.12). But it is supposed that hologram registration is conducted on the linear section of the characteristic curve. In reality, it is not like this: A part of the exposure does not get in this area. It is likely to be the key factor, which leads to deviation of the experimental curve from the theoretical one. But despite this deviation the preservation of the number of exposures as a constant during one exposure in the experiment improves the diffractive effectiveness.

These holographic contouring methods were tested in the experiment on medico-biological objects (anatomical bone structures). We have shown the possibility of their application for detection of spatial characteristics of objects with complex surface topography. In particular, the interferograms of human skull were obtained for cases of two-, three- and fourfold interferences. The photographs of contoured maps of the human skull are presented in Fig. 3.13. In the interferograms, the contoured maps characterizing the surface relief are obtained.

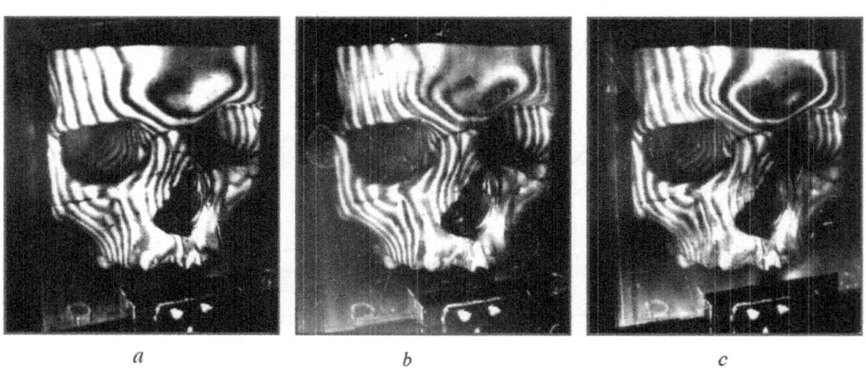

Fig. 3.13 Contour charts of a human skull for the cases of two- (**a**), three- (**b**) and fourfold (**c**) interference. Reprinted from [94] with permission

The band brightness on the obtained image decreases as the points of the object move away from the plane of the hologram. This indicates the possibility to use these methods for absolute relief estimation of static structures as well as of highly labile biological tissues.

Thus, the developed multi-angle contouring method gives possibility to enhance the accuracy of detection of contour maxima on the object in *m*/2 times comparing with the traditional methods of surface relief detection.

Though it should be pointed out that the study of surface peculiarities of the human body and their changes during treatment is connected with certain difficulties and requires further developments in the sphere of creation of diagnostic equipment, which uses coherent-optical waves. Transfer to holographic contouring of living tissues can be accomplished through pulse lasers.

Implementation of these methods for surface relief detection will make it possible to get a huge amount of information, by which one can judge both about the state of soft surface tissues and indirectly about the state of the muscular system and tuberculosis structures.

3.1.7 Briefly About Moire Surface Relief Contouring

As one of the examples of practical application of coherent-optical methods, there can be mentioned the implementation of moire methods for estimation of manual therapy efficiency while detecting functional blocks and the level of deblocking of spinal motional segments.

In a number of cases, the moire contouring methods are simpler in their realization [99]. Their essence is that while illuminating the object under study with two periodic systems of light bands a third structure is formed, which bears information about the shape of the surface of the objects under study. Several different methods of moire pattern forming are known. The most widespread method is based on using periodic grating that can be produced with the method suggested in the works [100, 101] and through putting dark lines on the glass surface. To reach the maximal contrast of the moiré pattern, the width of a dark line must be equal to the width of the light space between them. The principle scheme of implementation of this method is shown in Fig. 3.14. Located behind the grating *D* with the period d the surface of the object O under study is illuminated with the light source 1.

As a result of this, dark and light alternate lines are projected onto the surface under study. While observing the studied surface through the grating from the direction, which differs from the illumination direction, the projection of the strokes of the surface of the object with the stroke of the grating occurs, and thus, the moiré pattern is formed. The surface under study appears to be a crossed near the alternating planes, the distance between which is approximately the same and is equal to *h*. The value *h* is known as the depth interval. While using plane-parallel light beams $h_1 = h_2 = h$ (Fig. 3.14), the observed pattern of moiré bands is a contoured map (topogram) of the surface under study. The resolution of the moiré topogram is estimated by the depth

Fig. 3.14 Scheme of projection moire method of contouring. Reprinted from [94] with permission

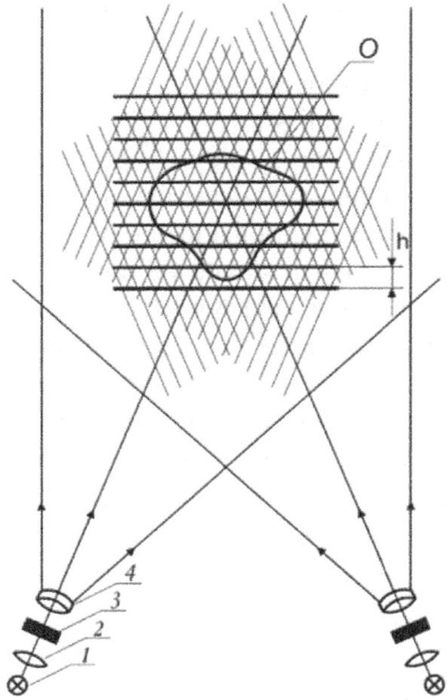

interval *h*. The obtained topograms of the objects which have spherical, conical and cylindrical surfaces are presented in Fig. 3.15a. The depth interval is 7 mm in this observation.

Figure 3.15b illustrates moiré topograms of the same objects, but with the depth interval of 3 mm and the resolution increased more than in two times. According to the topograms obtained, we can make a profile of the studied surface in any section that is interesting for a researcher. For example, a moiré topogram of cylindrical surface (Fig. 3.16a, b) is not strictly cylindrical as its generatrix has some curvature. Maximal deflection value of the cylinder geneatrix estimated due to the moiré topogram is 6 mm.

Moreover, the possibility of implementation of the moiré contouring method is shown for getting a 3D image of vertebral deformation. There were obtained moiré

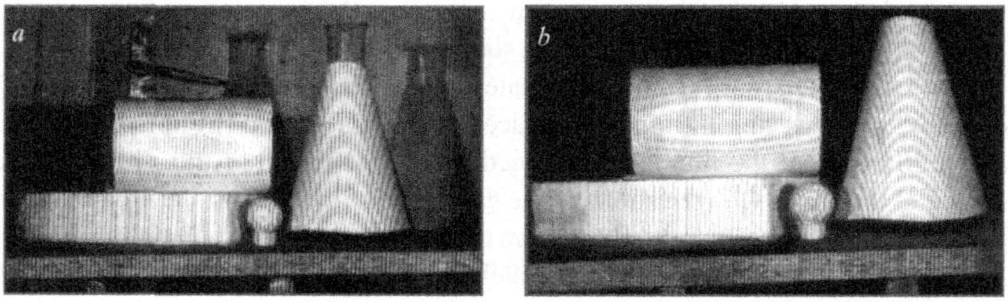

Fig. 3.15 Moire basis surface chart of model objects, deep interval: ~7 mm (**a**), ~3 mm (**b**). Reprinted from [94] with permission

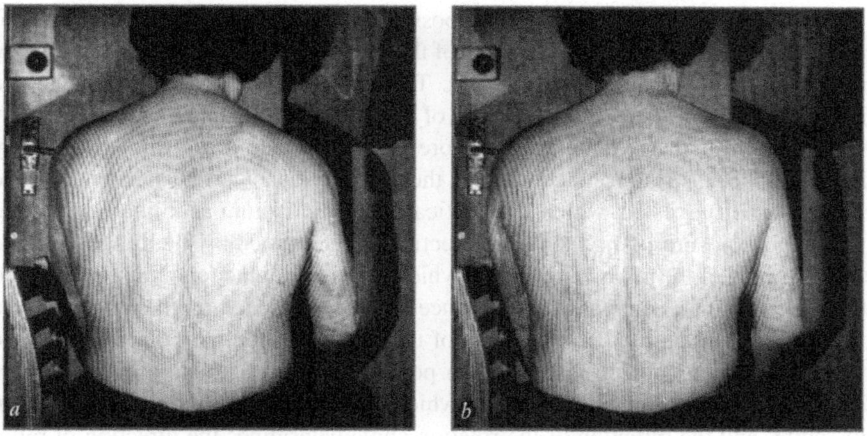

Fig. 3.16 Example of a moire picture of dorsal human body surface. Reprinted from [94] with permission

topograms of dorsal body surface of healthy people and of patients with vertebral deformations. The moiré topogram is presented in Fig. 3.16. It is seen that before manual therapy treatment the patient had deformation perpendicular to the plane of the photo-image. Externally, these two photos are almost indistinguishable but the moiré pattern in Fig. 3.16 has high frequency along the pectoral and the jugular areas of the spine that is the indicative of the high gradient of changing of the surface shape.

This method makes it possible to detect minor vertebral deformations accompanied by local increase of muscle tone and can be used for detection of extremity deformations. The main advantage of this method along with the simplicity of implementation, high informativity and harmlessness is the possibility of its implementation for detection of dynamics of pathological process and the efficiency of the used therapeutic complexes [102].

3.1.8 Holographic Method of Surface Relief Contouring Based on the Change of Polarization State of Objective Waves

In holographic interferometry, the information about the studied object parameters, for example, about the surface relief, is contained in phase difference of two object waves. As two-phase differences identical in the absolute value but opposite in the sign of function, random under superposition of two coherent waves form two totally identical interferency patterns, then information about the sign of phase difference disappears. Thus, the detection of relief change direction and convexity–concavity of the surface under study is a complex and in some cases unsolvable task.

As it has been shown, one of the possible problem solutions of detection of relief height sign is the implementation of the immersion method, and the absorbing medium is used as immersion [34, 38]. The implementation of absorbing media made it possible to detect the direction of relief change. Meanwhile, the accuracy of estimation of the relief height in more remote surface areas was significantly decreasing because the band contrast on these areas was lowering. The holographic contouring is suggested, which uses optically active medium as immersion during one of the exposures. And the relief direction is determined by the direction of the rotation angle of the polarizer, through which the holographic topogram is observed when the maximal contrast of interference fringes is reached on the analyzed area of the surface under study. As a result of using this method the polarization plane of total (resultant) object beam in each point of the reconstructed image appears to be turned at the angle, the value of which is proportional to the relief height in this point, and the direction of the rotation angle determines the direction of relief change. This method gives the possibility to detect not only the value but also the sign of the relief height, as well as to enhance the resolution and the measurement accuracy. Generally, the phase difference of two reconstructed forms, which form contour lines of constant relief height, can be connected with the height of the relief surface $\Delta\varphi(x, y) = \varphi_2(x, y) - \varphi_1(x, y)$ using the expression

$$\Delta\varphi(x, y) = \varkappa z(x, y), \tag{3.39}$$

where $\varkappa$ is the value determined by the topogram forming.

Let us express x from (3.39)

$$\Delta\varphi(x, y, z + h) - \Delta\varphi(x, y, z) = 2\pi, \tag{3.40}$$

where h is the depth interval of the topogram, i.e., the distance between the neighboring bands in the equidistant band system. As a result of their intersection with the studied surface, a holographic topogram is formed.

Using (3.39), we get

$$\varkappa[z(x, y) + h] - \varkappa z(x, y) = 2\pi, \tag{3.41}$$

whence

$$\varkappa = \frac{2\pi}{h} \tag{3.42}$$

For example, for the immersion method

$$\varkappa = \frac{4\pi}{\lambda}|n_2 - n_1| \tag{3.43}$$

where n_1, n_2 are the refractive indices of the immersion media, in which the object under study was placed during the first and the second exposures.

Let us substitute (3.42) into (3.39) we will get

$$\Delta\varphi(x, y) = \frac{2\pi}{h} z(x, y). \tag{3.44}$$

Equation (3.44) describes the intensity distribution in the polarization topogram.

Each hologram of the surface under study will be recorded and reconstructed with its reference waves, which are spaced and orthogonally polarized. If the polarization planes are located diagonally to the coordinate axes and the amplitudes are equal, the reconstructed waves can be written in the following form

$$\begin{aligned} \vec{\mathcal{E}}_1(x, y, z) &= \mathcal{E}_0\big[\exp(-i\varphi_1(x, y, z))\vec{e}_1 + \exp(-i\varphi_1(x, y, z))\vec{e}_2\big], \\ \vec{\mathcal{E}}_2(x, y, z) &= \mathcal{E}_0\big[\exp(-i\varphi_2(x, y, z))\vec{e}_1 + \exp(-i\varphi_2(x, y, z))\vec{e}_2\big], \end{aligned} \tag{3.45}$$

where $\vec{e}_1$ ($\vec{e}_2$) are the unit vectors along the x(y)-axis.

Let the waves run through a diagonal crystal phase plate $\lambda/4$ and then through the polarizer turned at the angle α relative to the axis of the coordinate system (Fig. 3.17). In this case, the total complex amplitude of waves passing through the polarizer will be equal to

$$\begin{aligned} \mathcal{E}_p &= \mathcal{E}_{1p} + \mathcal{E}'_{1p} + \mathcal{E}_{2p} + \mathcal{E}'_{2p} \\ &= \mathcal{E}_0\left\{\exp\left[-i\left(\varphi_1 - \alpha + \frac{\pi}{2}\right)\right] - \exp\left[-i\left(\varphi_2 + \alpha - \frac{\pi}{2}\right)\right]\right\}. \end{aligned} \tag{3.46}$$

The corresponding intensity distribution has the form

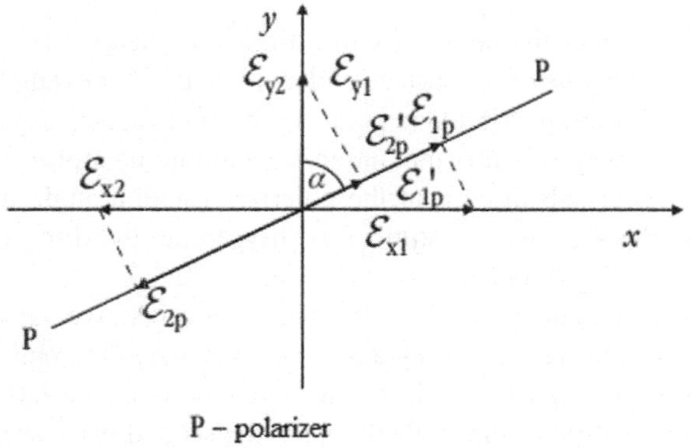

Fig. 3.17 Scheme of rotation of polarizer P relative to the vertical axis of the x, y coordinate system. Reprinted from [94] with permission

$$I = 4\mathcal{E}_0^2 \cos^2\left(\frac{\Delta\varphi}{2} + \alpha\right) \quad (3.47)$$

The intensity distribution in (3.47) is a topographic map of the surface under study. The lines of constant intensity, which correspond to the lines of constant relief height, are observed in those points where the rotation angle of plane polarization of the total object wave differs in 180°.

Using (3.47) and (3.44), the relief height as the function of intensity value and polarization angle $\alpha(x, y)$ is expressed as

$$z(x, y) = \frac{h}{\pi}\left\{\arccos\left[\sqrt{\frac{I(x, y)}{2\mathcal{E}_0}}\right] - \alpha(x, y)\right\}. \quad (3.48)$$

In two neighboring points of the surface with coordinates x_1, y_1 and x_2, y_2, the relief height will differ on the value

$$\Delta z(x, y) = \frac{h}{\pi}\left\{\arccos\left[\sqrt{\frac{I(x_2, y_2)}{2\mathcal{E}_0}}\right] - \arccos\left[\sqrt{\frac{I(x_1, y_1)}{2\mathcal{E}_0}}\right] - \Delta\alpha(x, y)\right\}, \quad (3.49)$$

where $\Delta\alpha(x, y) = \alpha(x_2, y_2) - \alpha(x_1, y_1)$. If $\Delta\alpha(x,y)$ is so that $I(x_2, y_2) = I(x_1, y_1)$, then

$$\Delta z(x, y) = -\frac{h}{\pi}\Delta\alpha(x, y) \quad (3.50)$$

Equation (3.50) gives the possibility to estimate the height and the direction of relief change for all points of the surface under study. Let, for example, the intensity maximum (or minimum) be located in point A with coordinates x_1, y_1. The polarizer is turned in such a way that this maximum (or minimum) moved to the analyzed point $B(x_2, y_2)$. After substitution of the polarizer rotation angle into (3.50) with consideration of the sign, let us estimate the height and the direction of the relief changes in point $B(x_2, y_2)$ relative to the axis $A(x_1, y_1)$.

Let us consider the change of the relief height as a positive value in the case if the analyzed point is moved for a longer distance than the initial one. If the polarizer turns clockwise angle $\Delta\alpha > 0$, and if it turns anticlockwise, $\Delta\alpha < 0$. Let also $\Delta\varphi > 0$, i.e., under illumination of the studied object, the second wave spread in a denser medium or had a smaller wavelength, etc. Then if the polarizer should be turned clockwise for dark (light) band shift from the initial point $A(x_1, y_1)$ to the analyzed point $B(x_2, y_2)$, i.e., $\Delta\alpha > 0$, then according to (3.50), $\Delta z < 0$; i.e., the point $B(x_2, y_2)$ is closer to the observer than the point $A(x_1, y_1)$.

If the polarizer is to be turned anticlockwise, i.e., $\Delta\alpha < 0$, then $\Delta z > 0$ and the point $B(x_2, y_2)$ are more remote from the observer than the point $A(x_1, y_1)$. The measurements of the value and direction of the relief can be conducted in each point of the image through shifting the topographical lines on the surface of the object under study by turning the polarizer that increases resolution. Meanwhile, the value of the relief height is estimated by the change of the rotation angle of the polarizer and by calculations according to (3.50). Thus, the suggested method makes it possible to move from linear measurements to the angular ones that increase accuracy of the measurements.

One of the possible versions of the optical scheme of the experimental set-up for implementation of this method is presented in Fig. 3.18. The immersion method was used. The shape of the surface of the plastic ball was determined. Firstly, the chamber with the object 17 was filled with optically neutral medium.

During the first exposure, the hologram was recorded on the photoplate 11 with a reference beam formed by the beam splitter 6 and the mirror 5.

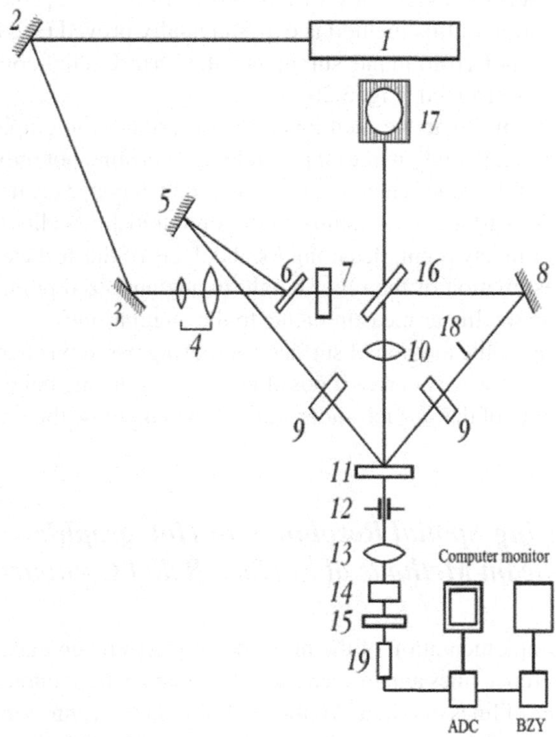

Fig. 3.18 Optical scheme of the set-up of surface relief contouring based on the change of polarization condition of object waves: 1—the laser; 2, 3, 5, 8—the mirrors; 4—the telescopic system; 6, 16—the light splitters; 7,14—the crystal wafers; 9—the light filters; 10,13—the lenses; 11—the hologram; 12—the spacial filter; 15—the polarizer; 17—the camera with prototype system; 18—the shutters; 19—CCD matrix. Reprinted from [94] with permission

Main directions of the crystal phase plates λ/2 are oriented at the angles of 0 and 90° to the direction of oscillation of linearly polarized laser beam, and the second reference beam is overlapped. Then, the optically neutral medium was changed for an optically active one in the chamber 17, and the second hologram was recorded on the same photoplate. During the second exposure, the hologram was recorded with the help of the reference beam formed by the mirror 8, and the first reference beam was overlapped. The plate 7 was placed diagonally.

During the reconstruction, the holographic plate 11 was illuminated simultaneously by two reference waves, and the plate 7 was in a diagonal state.

The contoured map of the surface was observed behind the diagonally situated crystal phase plate 14 (λ/4) and the polarizer 15; in the image plane formed by the telescopic system 10, 13, 14, the contoured map was visually observed with the microscope 19. Polarization of the reconstructed waves was turned into one plane with the analyzer 15. While turning the polarizer 16 clockwise the rings on the contoured map were subtended to the center, i.e., to more approximate points of the surface. This shows that the surface is convex. Thus, using this method convexes and concavities of the surface under study are easy to determine. The principle possibility of practical realization of this method is experimentally proved (see Fig. 3.19). Due to the results of the polarization measurements, the surface relief contoured maps in several sections were formed (Fig. 3.20).

As it follows from the above mentioned the described holographic polarization method of surface relief study makes it possible to determine not only the height but also the direction of the relief change, i.e., to distinguish convexes from concavities.

Scanning of the surface under study with contoured lines allows changing the height of the relief in any point, detecting local defects of the surface and increasing the accuracy of estimation of the relief height in any analyzed point at the expense of movement from the linear measurements to the angular one.

Thus, this holographic method of surface contouring based on changing the polarization state of object waves makes it possible to determine the height as well as the direction of the relief of the surface under study, and to increase the spatial resolution.

3.1.9 Increasing Spatial Resolution of Holographic Multi-beam Methods of Surface Relief Contouring

It is known that implementation of the multi-beam interference makes it possible to narrow the interference lines and in such a way to enhance the accuracy of determination of coordinates of lines maxima. Applied to holographic contouring methods, this leads to the increase of the accuracy of relief detection [28, 103]. For example, in the work [28], it is suggested to use dye laser generating radiation with different wavelengths for multi-beam topograms forming. In the work [103], the multi-immersion method is used for the same reason; i.e., the object under study is placed in a chamber with the immersion medium, and before each hologram registration, the refractive

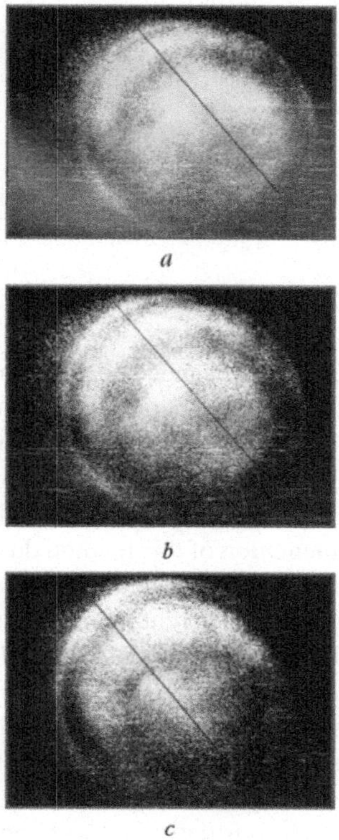

Fig. 3.19 Contour chart of the surface relief of a plastic ball, derived by immersion method with the usage of optically active medium Reprinted from [94] with permission

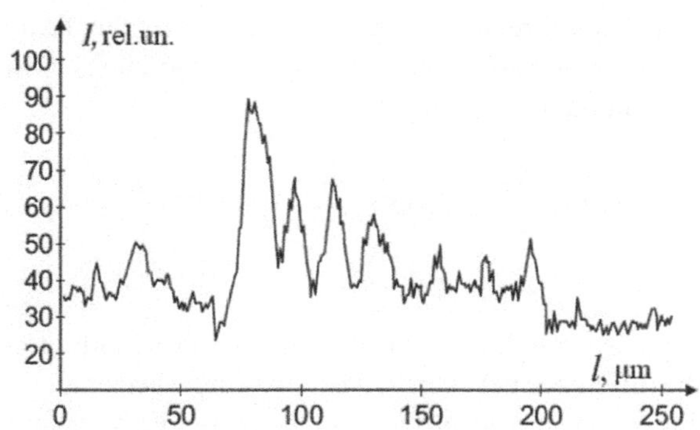

Fig. 3.20 Surface relief contour of the same object in ozne of the cross sections based on polarization measurements. Reprinted from [94] with permission

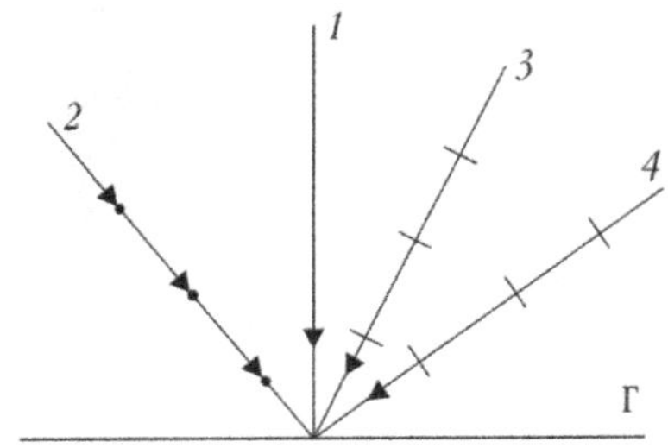

Fig. 3.21 Principle scheme of the method realization. Reprinted from [94] with permission

index of the immersion medium is changed for the same value Δn, for example, by changing the composition of the immersion medium. In this work, a method is suggested, which makes it possible to enhance the resolution of multi-beam holographic contouring methods. The method consists in forming and superposition of two multi-beam topograms with a set shift of the one relative to the other.

Let us observe the implementation of this method during the realization of multi-long-wave contouring. The object surface under study is illuminated by collimated beams of coherent radiation with m different wavelengths, and $m \geq 3$ (Fig. 3.21). On the light-sensitive layer 5 of the holographic plate m holograms of wave 1 scattered by the surface under study are registered using, for example, the reference beam 2. In addition to that m holograms are registered using other reference light beams 3 and 4, which are spatially separated between one another, and the beam used for the record of initial holograms.

And at least one of m holograms is registered with the reference beam 4, others—with the reference beam 3. Reconstruction of all the holograms is conducted with the same three reference beams used during the record, but they are linearly polarized. The polarization plane of the light beam 2, which reconstructs m initial holograms is to be turned at 90° relative to two other reconstructing beams, for example, with a diagonally placed crystal plate λ/2. If $m = 4$, the intensity distribution after reconstruction of m initial holograms is

$$I_1 = 2a^2\left(2 + 2\cos 4\pi z\frac{\Delta\lambda}{\lambda^2} + \cos 8\pi z\frac{\Delta\lambda}{\lambda^2} + \cos 12\pi z\frac{\Delta\lambda}{\lambda^2} + \cos 16\pi z\frac{\Delta\lambda}{\lambda^2}\right), \tag{3.51}$$

where a is the real wave amplitude; λ is the central wavelength of the used spectral interval; Δλ is the difference between two neighboring wavelengths; z is the distance between the point on the studied surface and the holographic plate; $z = z(x, y)$.

The obtained intensity distribution corresponds to the distribution while using the well-known multi-long-wave contouring method and is shown in Fig. 3.22 (curve 1). In the reconstructing beam 4, regulated phase delay is put, for example, with a compensator. In this case, if the delay is equal to π, then the intensity distribution during reconstruction of additional holograms is

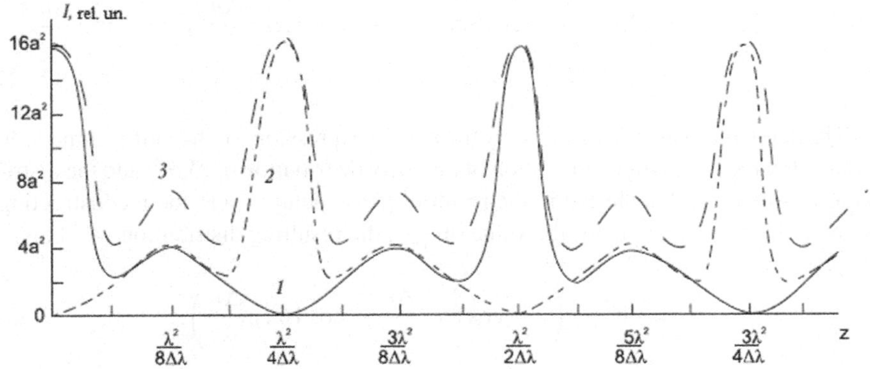

Fig. 3.22 Intensity distribution in the formed hologram in the direction of observation. Reprinted from [94] with permission

$$I_2 = 2a^2\left(2 - 2\cos 4\pi z\frac{\Delta\lambda}{\lambda^2} + \cos 8\pi z\frac{\Delta\lambda}{\lambda^2} - 2\cos 12\pi z\frac{\Delta\lambda}{\lambda^2} + \cos 16\pi z\frac{\Delta\lambda}{\lambda^2}\right) \tag{3.53}$$

The curve 2 in Fig. 3.22 corresponds to the intensity distribution. This curve is the same as the curve 1, but it is shifted for half a period along 0 z-axis. As planes of wave polarization reconstructed from m initial holograms are perpendicular to the planes of wave polarization reconstructed from the additional holograms, then the total intensity distribution in the formed holographic contoured map will be equal to

$$I = I_1 + I_2 = 4a^2\left(2 + \cos 8\pi z\frac{\Delta\lambda}{\lambda^2} + \cos 16\pi z\frac{\Delta\lambda}{\lambda^2}\right) \tag{3.53}$$

This intensity distribution corresponds to the curve 3 in Fig. 3.22. The corresponding depth interval is

$$h = \frac{\lambda^2}{4\Delta\lambda} \tag{3.54}$$

The depth interval described by (3.54) has a two times smaller value than the depth interval of the contoured map formed through traditional multi-long-wave method with the same number of waves.

A similar result can be obtained, for example, by implementing this approach during the realization of the known traditional multi-immersion contouring method [103]. In this case, the intensity distribution during reconstruction of m initial holograms for $m = 4$ is

$$I_{1n} = 2a^2\left(2 + 2\cos 4\pi z\frac{\Delta n}{\lambda} + \cos 8\pi z\frac{\Delta n}{\lambda} + 2\cos 12\pi z\frac{\Delta n}{\lambda} + \cos 16\pi z\frac{\Delta n}{\lambda}\right). \tag{3.55}$$

The depth interval is determined by the known expression for the multi-immersion contouring method. After summation of intensity distribution of (3.55) and the distribution similar to it but shifted in the result of phase delay π into the reconstructing beam 4 along the 0 z-axis for the value of $\frac{\lambda}{4\Delta n}$, the resulting distribution will be

$$I_n = 4a^2 = \left(2 + \cos 8\pi z\frac{\Delta n}{\lambda} + \cos 16\pi z\frac{\Delta n}{\lambda}\right). \tag{3.56}$$

And the depth interval is

$$h_n = \frac{\lambda}{4\Delta n}. \tag{3.57}$$

So with the same number of immersion liquids, this method also made it possible to reduce to a half the depth interval value in comparison to the known method.

Thus, implementation of the suggested polarization approach to the holographic multi-beam methods of surface relief contouring allows increasing their spatial distribution.

3.2 Holographic Interferometry Using Generation Regime of Double Monopulses of a Ruby Laser with the Regulated Time Interval Between Them

3.2.1 Holographic Study of Deformations of Human Lower Jaw in Radiation of Continuous He–Ne Laser

The research is based on the direction closely connected with the development and implementation of the laser-holographic unit for studying cardiologic problems; thus, in future, totally new noninvasive express methods are planned to be introduced into the clinical practice.

However, before creating the experimental sample of the laser-holographic unit (holographic cardiograph) as a device for human cardiovascular system diagnostics, it was necessary to estimate experimentally absolutely new conditions for hologram recording of living objects, and then to adapt and test holographic interferometry methods using ruby- and He–Ne pulsed laser radiation for studying such medical and biological (diffusely reflective) objects as tissues of hands, vertebras with onsets of lumbar osteochondrosis, human jaw for solving prosthetics problems, etc. This would give the possibility to detect and broaden possibilities and limits of application of this method. Moreover, the research of deformations, which occur through functional overuse of human upper and lower jaws, was influenced by the necessity of development of this holographic methodical approach to the solution of the mentioned problem for specialists of Surgical Dentistry Department of the Belarusian State University of advanced medical studies headed by A.S. Artyushkevich, Doctor of medical sciences, and of Orthopedic Dentistry Department of Belarusian State Medical University headed by Professor S.A. Naumovich, Doctor of medical sciences. The knowledge of bone overload distribution is of great-applied importance for surgeons and orthopedic dentists while selecting and planning treatment of patients with injuries and abnormalities of dentition.

Holographic interferometry is one of the most informative methods of physical agent deformation detection [104]. This method was developed in collaboration with dentists of the Ministry of Health of the Republic of Belarus and is successfully implemented today in dentistry for studying tooth-substituting objects, plastic models of the lower jaw and skull bones.

In our earlier studies, the experiments with overload were conducted without taking into account the whole variety of mechanical forces caused by muscle contraction, which lift and drop the lower jaw [105–108]. Later, spatial deformations of the lower jaw during bite and deglutition were studied. The results of the research are of great value for the development and planning of methods of treatment in surgical and orthopedic dentistry. In particular a special loading apparatus was developed [109–111], which gives the possibility to imitate pressure of all front and back mastication muscles with high accuracy, i.e., direction and absolute force of each group of muscles. The holographic unit was used in the experiment, which gives the possibility to record holographic interferograms with fastening of holographic plates directly on the object, and that is why "parasitic fringes" on the interferogram were avoided during loading of the object (Fig. 3.23). On the interferogram, it is seen that the lines present only on the base where the object is fixed, and on the object itself not observed shear lines are compensated. As in this case, the object was not deformed, then there were no lines on it. As a light source, there was used radiation of He–Ne laser LGN-215, which generates single-mode radiation with the length of the coherency length of 20–25 cm and the wavelength of 632.8 nm. Interferogram registration was conducted on high-resolution holographic plates PFG-03. They were fixed in the lower jaw near a canine with glue "Cyacrine," which provided accurate and immovable plate fixing.

As a result, deformation fields were obtained and investigated in horizontal and front planes on a level with alveolar process, sinciput and lower side of the jaw (Fig. 3.24).

The qualitative estimation of holographic interferograms of the lower jaw under different functional overloads has shown that the object experiences different deformations in the horizontal and front planes. The estimation of deformations of lower jawbone was conducted according to the spatial distribution of the interference fringes and their directions. The density of interference fringes depended on the level of object deformation.

a

b

c

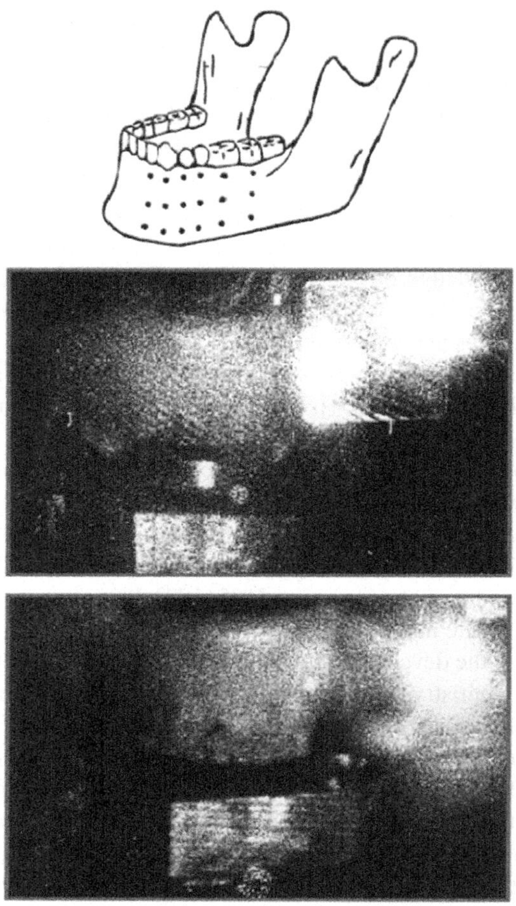

Fig. 3.23 Holographic interferograms of the surface of a lower jaw (**a**) wherein functional deformation, interferogram of a shifted object (**b**), interferogram of an object with shift compensation (**c**) were detected. Reprinted from [94] with permission

For more accurate deformation detection, qualitative estimation of interferograms was conducted. In this case, the technique of absolute band arrangement was used. As in the first approximate calculation the object is presented as a flat body, then interference scheme with symmetrical observation direction was used. The bands were calculated beginning from the zeroth level, which was the place of fixing the holographic plate. The deformation value of the object was estimated for three coordinates according to the equations of sensitivity after using the interferograms in every detected point on the lower jawbone surface from different directions of observation.

Based on the character of interference fringe distribution, we can conclude that maximal deformations are observed in the area of teeth and the alveolar process,

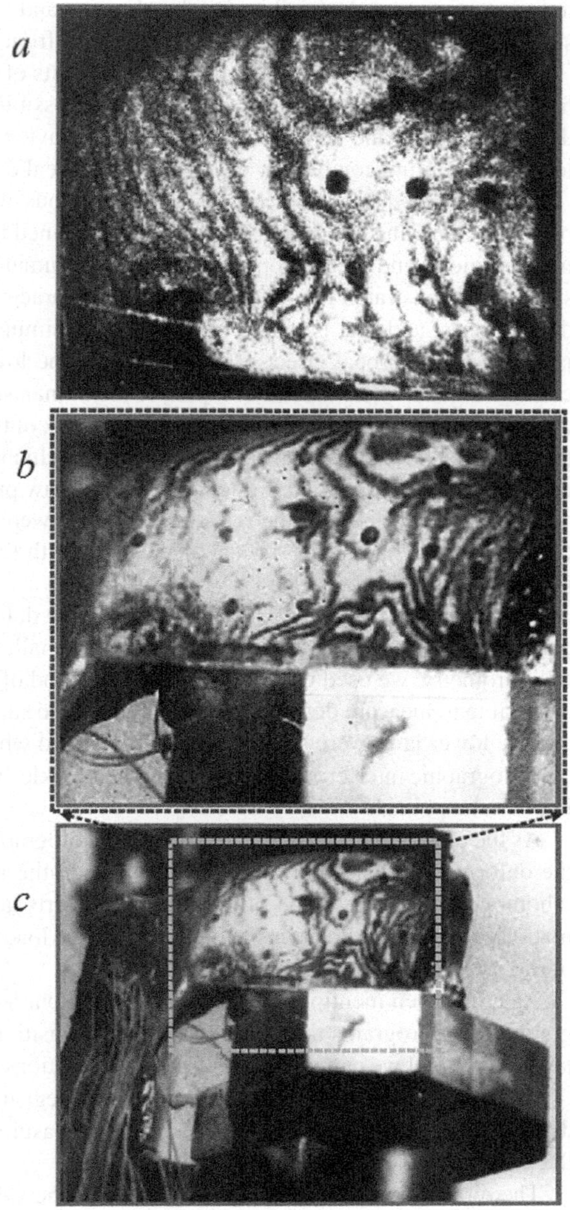

Fig. 3.24 Interferograms of lower jaw at different loads: **a**—temporal muscle, **b**, **c**—mylohyoid muscle. Reprinted from [94] with permission

lower part of the lower side of the jaw. In the chin area, the deformation is less expressed than in the area of the body and the corner of the jaw.

The quantitative processing of holographic interferograms made it possible to establish how deformations are distributed in the area of the alveolar process, the border of the corner and the symphysis in horizontal and front planes. Maximum deformations were observed in the alveolar area and the jaw joint in both planes. A significant diminution of deformation was in the front area of the jaw.

The obtained qualitative and quantitative results of the study of this object using holographic methods of measurement give the possibility to estimate measurements during the loads and further to observe the behavior of the object under different loads and limiting conditions, which simulate a real dental equipment.

In this study [109–119], based on the previous works, deformations of facial part of the human skull under loads were determined using methods of holographic interferometry and speckle-optics. The study of human skull bones isolated from the skull itself constrains to a certain extent the character of deformation distribution during loads, and that is why a task was set to simulate muscle load on the upper jaw as well reciprocal effect of the upper and the lower jaws during their joining considering the whole muscle complex of the human skull. A loading apparatus was created, which includes the whole muscle complex of the human skull. The model of the muscle load included all the main muscles, which influence during joining and opening of the jaws and their interaction during jaw pressing.

The two-exposure and the real-time methods were used while investigating the deformations of the facial surface of the skull with the holographic interferometry method.

As the sensitivity of these measurements to the deformations, which are directed parallel to the surface under study, is extremely small, in addition to the holographic interferometry, we used the speckle-optical method of measurements, which makes it possible to measure deformations along the bone surface. Holograms of the upper and the lower jaws were simultaneously registered while measuring with the use of the holographic interferometry methods. The speckle-interferograms were registered in the same points.

As the qualitative analysis of the obtained holographic interferograms has shown the deformation distribution depends greatly on the load. While pressing the jaw, inhomogeneous distribution of the points of efforts takes place. We observe deformations of different directions with increasing the load because of the sliding contact of the jaw teeth caused by the functional bite.

As it has been mentioned before while using our scheme of registration of holographic interferograms the vector of maximal sensitivity is directed normally to the jaw surface and we can measure normal deformations as well.

During two-exposure speckle-interferogram registration with the same load, the deformations of the object under study were measured which were directed along the plane of the object.

The direction of the deformation vector and the value were measured according to the period of the interference fringes.

During the registration of the speckle-interferogram, there was used a lens to form a speckle-field in the focal area and the object image modulated by the speckle-structure.

As a result of analysis of the experimental data, the distribution of the deformation fields and the shift of the object under study were obtained. Thus, if the load of the lower jaw is increasing, the gradual asymmetric deformation extension is observed

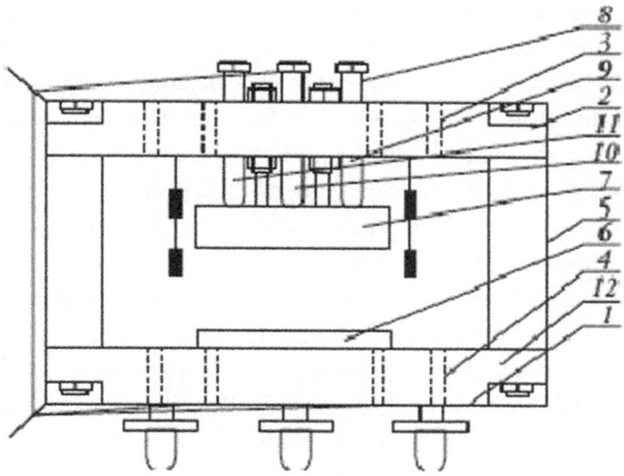

Fig. 3.25 Scheme of the device for simulating functional skull load. Reprinted from [94] with permission

connected with the functional bite. The rupture of the interference fringes occurs on the upper jaw under certain load that is connected with heterogeneous bone structure in this place. Measurements of deformations and shifts with the speckle-interferometry method show the shift of the bone surface in different directions.

According to the registered interferograms, the deformation field of the facial surface of the lower jaw under a certain load of approximately 15 kg was estimated.

The experimental study using unique capabilities of the holographic interferometry method (two-exposure or real-time) gave the possibility to understand and substantiate stress dependence mechanisms of the dentofacial area under bridge prosthetics.

While simulating the physiological conditions of dentofacial area load, a device for simulating functional muscle loads of the facial skull area was applied [109] (Fig. 3.25). This device contains the base in the form of two plates (the upper one 1 and the lower one 2) with symmetrical row of the slit-like 3 and round holes for muscle load simulating connected with each other by the fixed racks 5; a supporting and fixing system being two removable forms 6 and 7 for skull fixing, and the form 7 can be regulated by the check screws 8; the system of modeling thrust blocks 9–11 and the mechanism of lower plate regulating in the form of three regulated screws 12. While conducting the experiments, two types of load were applied, which corresponded to muscle contraction while biting and the force of the contraction while chewing the food.

For detection of deformations and shifts in different points of the dentofacial area a program was designed, which gave the possibility to present visually the occurring changes and measure the needed parameters according to the inputted coordinates of the points. Different prosthetics variations were considered for investigating biomechanical processes in the system tooth-periodontal-bone under bridge prosthetics. The development of the equipment and methods for studying the state of strain in the dentoalveolar system, the suggestion of bridge prosthetics constructions with experimental–clinical research and justification of their usage for lowering the level of the states of strain in the dentofacial area related to the tasks of the study.

The result of such research was the creation of a device for functional chewing load simulation in the whole dentofacial area that gives the possibility to reproduce with maximal accuracy the conditions of teeth load for studying with optical methods. With the help of the developed holographic interferometry methods, the regularities of the states of strain were for the first time studied, which occur in the intact dentoalveolar area under conditions, which maximally simulate the clinical conditions, and the device was developed for simulation of chewing load in real bridge dentures. The methods of load estimation were developed, and the character of load transfer in the whole dentofacial area was concretely defined while applying different constructions of bridge dentures. Based on the conducted study, more specific information was obtained concerning indications for orthopedic treatment of bounded edentulous spaces with different constructions of bridge dentures. The techniques of optimization of the states of strain during the prosthetics of patients with bounded edentulous spaces were developed.

Thus, a complex experimental–clinical study was conducted, as a result of which the construction of a device was designed, which provides simulating of functional muscle loads of the skull facial part, which are maximally approximate to the physiological ones that made possible using high-precision optical methods to study the impact of real orthopedic constructions on the whole dentofacial area [109, 108]. Investigation of the states of strain in the dentofacial area using the optical holography methods (non-destructive methods) gave the possibility to obtain qualitative and quantitative data of the objects and to reliably reproduce current changes under real applied loads using different orthopedic constructions including bridge dentures [110, 111, 114–116]. The results of experimental research have shown certain regularities of deformations occurring in intact dentofacial area under two types of load. The conducted quantitative analysis of deformations allowed concluding that the applied load is inadequate along the direction but has its ways of distribution. The anatomic peculiarities of the jaws, the form of dentitions and contact points between teeth support the dentofacial area in the state of strained balance. Under chewing load, each tooth has it turn center [112, 113, 117–119]. (Patent No. 3336 of the Republic of Belarus for the invention "Device for functional muscle load simulation of the facial part of the skull" was obtained).

It is pleasant to mention that this joint groundwork was started years ago thanks to competent, highly professional attitude, sensitivity to novelties of Professor S.A. Naumovich, Doctor of medical sciences, Head of the Orthopedic Dentistry Department of the State Educational Establishment "Belarusian State Medical University." Today, it turned into a whole specialized scientific school with its well-equipped modern basis and specialists that allows developing and introducing his ideas into clinical practice (see monograph "Holographic methods of study in dentistry" edited by S.A. Naumovich, (Minsk 2009) [120].

3.2.2 Holographic Recording of Muscle Stress of Hand in Radiation of Pulsed Ruby Laser with Double Monopulse

The study of living objects using holographic methods is impossible in continuous gas laser radiation due to high mobility of the objects under study. In this case, pulsed laser gives great possibilities for holographic research of shifts and deformations of human skin, which makes it possible to record an interferogram in short periods of time. It also helped to eliminate firm fixing of the object under study.

The two-pulsed holographic interferometry is one of the most effective methods to obtain information about the dynamics of fast processes, deformations of movable diffuse objects. The implementation of paired pulse laser sources with the pulse duration of 10–100 ns and the regulated interval between them from 100 to 700 μs enables to increase significantly the number of tasks solved with the holography methods. The regulation of the interval of radiation pulse sequence is mostly determined by the optical shutter used [121–152].

In those cases when accurate synchronization (microsecond unit order) is needed, step connection of laser Q-code is necessary. In the works [128, 129], stress on the electrooptical shutter (EOS) of the laser was measured twice to obtain paired pulses of the laser. Each pulse was generated after corresponding stress release. Approximate equality of amplitudes of generated laser pulses can be reached through selecting the step amplitude. Time interval between the steps determines the interval between the monopulses. The main disadvantage of this method is that the laser resonator transforms stepwise into supraliminal state, and that is why time of development of a giant pulse is little ($\approx$100 ns). Besides, highly qualitative sorting elements should be used in the resonator to obtain sufficient radiation coherence.

One of the most effective ways of generation of giant pulses with identical monofrequent spectrum and regulated interval of the sequence is the injection of monofrequent laser with quasi-stationary pulse of millisecond duration generation. In this case, the pulse transformation interval is determined by the duration of inoculating radiation and naturally by the pump pulse of the driven laser [132, 133, 142–150]. But the technical implementation of this method is rather difficult, and the laser itself becomes rather cumbersome.

A combined shutter is usually used in the laser to obtain the regime of generation of paired pulses generation when OES stabilizes the moment of appearance of the giant pulse, and the antireflective filter provides a narrow generation spectrum. In the work [133], the control circuit of OES was suggested and investigated, the principle of operation of which is analogous to that of the combined shutter. In this case, the scheme of positive electrooptical feedback (PEOF) functions as an antireflective filter. Thereafter, the inclusion mode providing the effective generation spectrum narrowing and stabilization of moments of giant pulse appearance are fulfilled with the same OES. In our case, the combined shutter is used in the laser. The stress is gradually released from the OES to generate the paired pulses.

The optical scheme of the resonator of the optical laser is presented in Fig. 3.26.

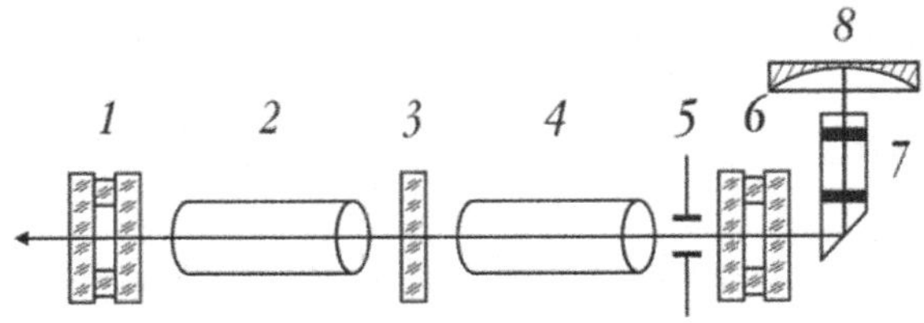

Fig. 3.26 Optical scheme of resonator of the laser of paired pulses 1, 6—the resonant reflector; 2, 4—the active elements from synthetic corundum RA2B with 8120/180 mm; 3—the intracavity selector; 5—the diaphragm with 2 mm diameter for basic mode defining; 6—the cuvette with alcoholic bleaching dye solution 1044, initial transmission $T_0 = 55\%$; 7—the electrooptic shutter on the basis of LEZ-1 element; 8—the non-transmitting mirror with radius of 1.5 m. Reprinted from [94] with permission

In the laser, the active elements with the chamfered Brewster ends are used. The active elements are placed into illuminators, and pumping of each element was from of opal glass MS-20 with one pulsed lamp ISP-5000. The duration of pulse pumping is 1.5 ms; the form is similar to rectangular.

The spatial distribution of radiation intensity on the laser output is presented in Fig. 3.26. For spectrum registration, the Fabry–Perot interferometer was used with the base of 3 cm and a resolution of 300 MHz (Fig. 3.27). The total energy of two giant pulses is up to 0.5 J (Fig. 3.28b). The regulated interval is from 100 to 700 μs. The duration of each pulse of laser radiation is 20–120 ns. The width of the spectral line of laser radiation is not more than 100 MHz. The oscillograms of paired pulses of generation are presented in Fig. 3.29. A passive antireflective shutter [151–165] was used. Generation of two pulses with the duration of about 30 ms and the sequence interval of 100 μs could be reached through the selection of transmission of the passive filter in the range of 40–60%. The total energy of pulses reached 1 J (with an amplifier). Regimes of hologram recording and processing were optimized.

Fig. 3.27 Spatial distribution of radiation intensity at the laser output. Reprinted from [94] with permission

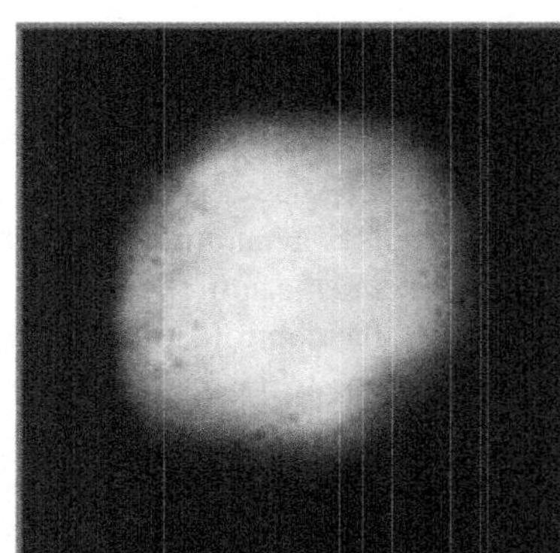

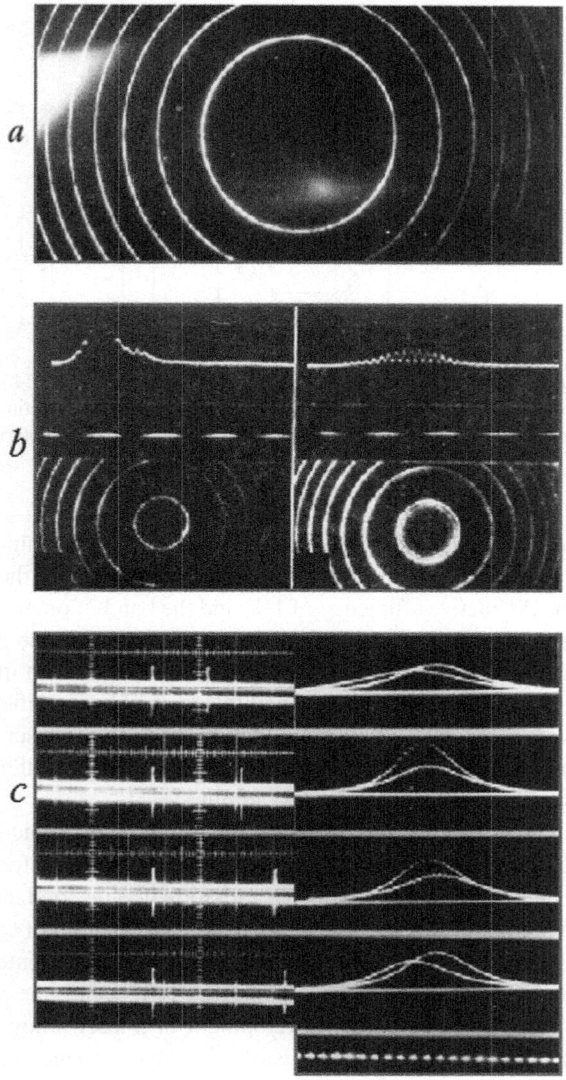

Fig. 3.28 Interferogram of laser radiation spectrum ($T_0 = 50\%$, interferometer basis −3 cm) (**a**); oscillograms of giant pulses and corresponding interferograms of radiation spectra, signs are 50 ns (**b**); oscillograms of paired pulses of deflection generation are 200 μs divisions and 250 μs divisions, correspondingly, label −10 ns (**c**). Reprinted from [94] with permission

A double pulse ruby laser unit and registration systems with thermoplastic carrier (see Fig. 3.29) were conducted to investigate shifts and deformations of living objects using the holographic interferometry method.

A registration cell contains a charger, a corona discharger, a pattern of the registering medium and a temperature-sensitive element. A pattern of the registering

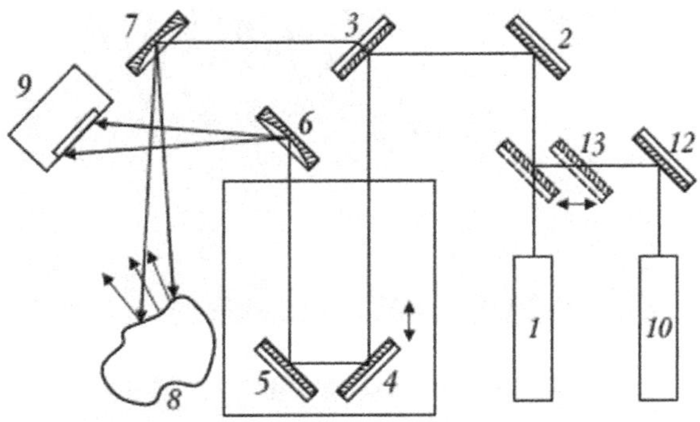

Fig. 3.29 Optical scheme of holographic device: 1—the ruby laser; 2, 4, 5, 12—the non-transmitting mirrors; 3, 13—the semi-reflecting mirror; 6, 7—the non-transmitting spherical mirror; 8—the holographic object; 9—the registration compartment on the phototermoplastic carrier; 10—He–Ne laser LGN-215. Reprinted from [94] with permission

medium is a polymer film applied on a glass substrate of 40–50 mm with a conductive SnO layer. The sensitivity of the used material is 10 J/cm^2 on the wavelength of 632.8 nm and the diffractive efficiency of 1%, and the band of operating frequencies is 500–1500 lines/mm. The number of re-recording cycles on one plate is not less than 300, and maximal diffraction efficiency is 20%. The hidden image turns into the mechanical surface relief. The developed image can be intensified if needed. To do this, the system is re-started without switching on the ruby laser. Erasure of the image is conducted similar to the development, but in this case, the control circuit emits current pulse of greater duration and amplitude.

A human hand was taken as the object for recording holographic interferograms. He–Ne laser LGN-215 with the wavelength of 632.8 nm was used for reconstruction of the recorded interferogram. To do this, its radiation was inserted into the circuit of the ruby laser with the relocatable mirror 13. The interferogram of strained muscles of human hand with an interval between pulses of 100 μs is presented in Fig. 3.30.

The conducted preliminary experiments helped L.V. Tanin, the author, to come close to holographic research on the development of a laser-holographic complex for studying the functional state of the cardiovascular system. The complex was named a "holographic cardiograph," and its detailed description was given further in Sect. 3.3.3.

It is also important to mention that the author of this project would not have begun its implementation if there had been no great groundwork on designing a ruby laser with double radiation monopulse and tunable period of their sequence [133–141, 144, 150–152, 164, 166] in the Institute of Electronics of the Academy of Sciences of the BSSR. This laser exceeds best developments of foreign companies (Rottencolber, Holosystem (FRG), Apolo Laser Corp. (USA), etc.) in its parameters.

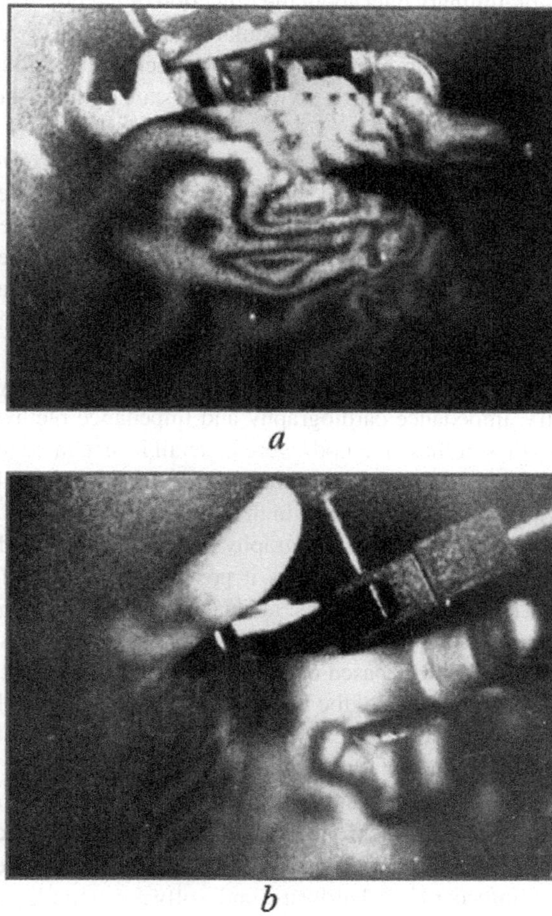

Fig. 3.30 Interferogram of a hand. Palmar surface. Reprinted from [94] with permission

3.3 Laser-Holographic Complex (Holographic Cardiograph) for Investigation of the State of Human Cardiovascular System

Laser-holographic complex (LHC) is a holographic cardiograph aimed at investigating the functional state of human organism through registration of holographic interferograms of human skin oscillations caused by the state of one's cardiovascular system and other systems and organs with further automatic computer processing.

The holographic cardiograph is based on the hypothesis about the possible impact of mechanical oscillations of muscular and cutaneous coverings of the chest on its whole area on the formation of a holographic interferogram of this part of the body. Obtaining holographic pattern of real-time changes can be connected with skin deformation speed, i.e., with process of change in skin position. The analysis of preliminary data about the reflection of the laser beam from skin presupposes that the oscillation speed of muscular and cutaneous coverings will be to a great extent connected with hemodynamic and mechanical shifts. They are based on pumping and mechanical heart activity and the state of the vascular or arterial system of blood circulation.

There was conducted an experimental analysis of research methods of mechanic, electrical, acoustic and pumping heart activity as well as research methods of the vascular system, measurements of pressure and time of blood circulation in order to select the methods of study of blood circulation needed for comparison with the holography pattern of the chest. Methodical informative essence of the following methods was analyzed: apex-, kineto-, ballisto-, dynamo- and phonocardiography, phase analysis of cardiac cycle, method of cardiac contraction speed detection, echocardiography, echoangiography, sphygmography, oscillography and oscillometry, impedance cardiography and impedance plethysmography. Methodical peculiarities of these methods were in detail investigated in collaboration with specialists of the Institute of Cardiology of the Ministry of Health of the Republic of Belarus who directly participated in this study.

The impedance cardiography method was selected from the most promising and informative ways. It makes it possible to obtain information about pumping and mechanic heart functions, arterial pressure and peripheral vascular resistance very fast. This method allows singling out hemodynamic syndromes, which are detected on the one hand based on high, normal and low heart release and on the other hand based on the state of the vascular resistance. Namely, these parameters can determine to a large extent the real-time hemodynamic shifts and that underlies the oscillations of human muscular and cutaneous coverings.

Traditional medical equipment must be adapted to provide cardiovascular system control and to synchronize the moment of hologram recording with the needed phases of cardiac contractions during the study using the laser-holographic complex.

"Impecard," a hardware and software complex [167–170], was selected and mastered following the conducted analysis of technical–medical characteristics of the produced equipment. The complex has a cardiologic output good for sync pulse formation and gives possibility to conduct syndrome diagnostics of the state of blood circulation without covering the areas of human chest, which are studied with the laser-holographic complex. Functional interrelations of the hardware and software complex "Impecard" with the main parts of the laser-holographic complex were determined.

The suggested optical scheme of the complex gives possibility to record interferograms of the surface of human chest. As a result of muscle contractions shift of points of human chest surface occurs. The contained in the inteferogram information about distribution of these shifts along the surface of the human chest, about the direction value and the speed of movements leads to the conclusion about the state of the human blood circulation system.

This laser-holographic complex ensures changes in movement of point of human chest surface within the time interval from 100 to 700 μs and the determination of cardiac cycle phase to an accuracy of not more than 0.5 μm. The main point of the suggested method is the use of the two-pulse holographic interferometry for the detection of shift and deformations of points of the human body surface and processes in the human organism. The mechanic waves occurring during heart contraction spread in the tissues, which surround the body and preserve the form of oscillation of the source generating them. When they reach the chest, these waves cause shift of its surface and bear information about the state of the source itself as well as of the medium, in which they are spread.

Corresponding holographic interferograms can be registered using the two-laser regime of the ruby laser and synchronizing the process of hologram recording in the chest area in different phases of the cardiac cycle. The distribution of the interference fringe on the surface of the chest image corresponding to the shift field of the surface points gives the possibility to get the idea about the nature of cardiac murmurs and functional state of the corresponding valve or of a cardio muscle area, as well as information about regional and local dynamics on the whole chest area.

The suggested development is based on scientific achievements of the Institute of Physics, Electronics and Applied Physics of the Academy of Sciences of the BSSR, the Institute of Neurology, Neurosurgery and Physiotherapy of the Ministry of Health of the Republic of Belarus and the Institute of Cardiology of the Ministry of Health of the Republic of Belarus. The development of the holographic cardiograph and the study with its help were conducted under the leadership and direct participation of L.V. Tanin and in close collaboration with the specialists of the above-mentioned institutes. There is no analogous instrument to the developed laser-holographic complex.

In the created laser-holographic complex, a two-pulse ruby laser with tunable pulse interval was used, which exceeds the well-known domestic and foreign analogs (Apollo Laser, Quanta Ray, Quantronics, Rottencolber) in their spectral energetic characteristics and in radiation coherence. These parameters are crucial and give the possibility to investigate a large area of the human body surface in a broad frequency range at the same time and, thus, to obtain full and trustworthy information about the state of one's functional systems.

3.3.1 Structure of the Laser and Holographic Complex for Determining the State of Human Circulatory System

The block diagram of the laser-holographic complex is presented in Figs. 3.31, 3.32 and 3.33.

It includes:

- the ruby laser with a high-voltage power unit and cooling unit;
- the high-voltage control unit with electrooptical shutter;
- the optical system for illuminating beam formation;
- the optical system for the formation of recording and restoration beams with an electrooptical system of their spatial separation;

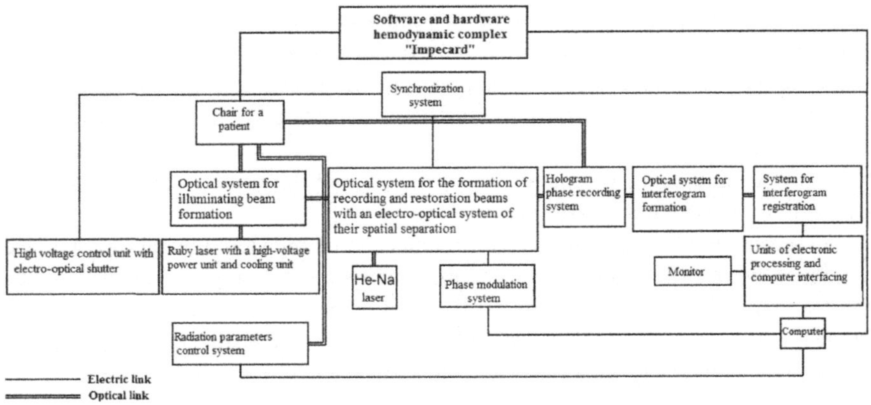

Fig. 3.31 Block diagram of laser-holographic complex for determining the state of circulatory system. Reprinted from [94] with permission

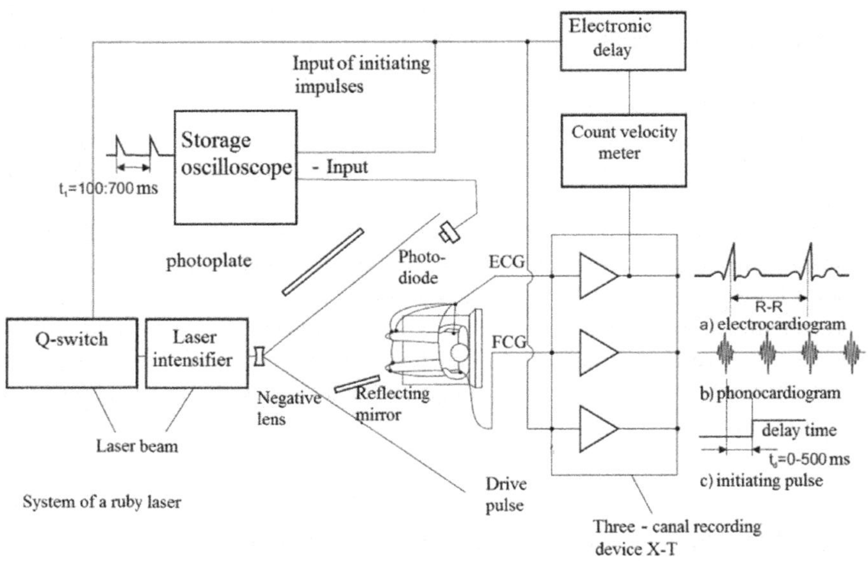

Fig. 3.32 The block diagram of the laser-holographic complex synchronization system to define state of human circulatory system. Reprinted from [94] with permission

- the phase modulation system of regenerative light beam;
- the hologram phase recording system;
- He–Ne laser;
- the optical system for interferogram formation;
- the system for interferogram registration;
- the units of electronic processing and computer interfacing;
- the monitor, the computer;

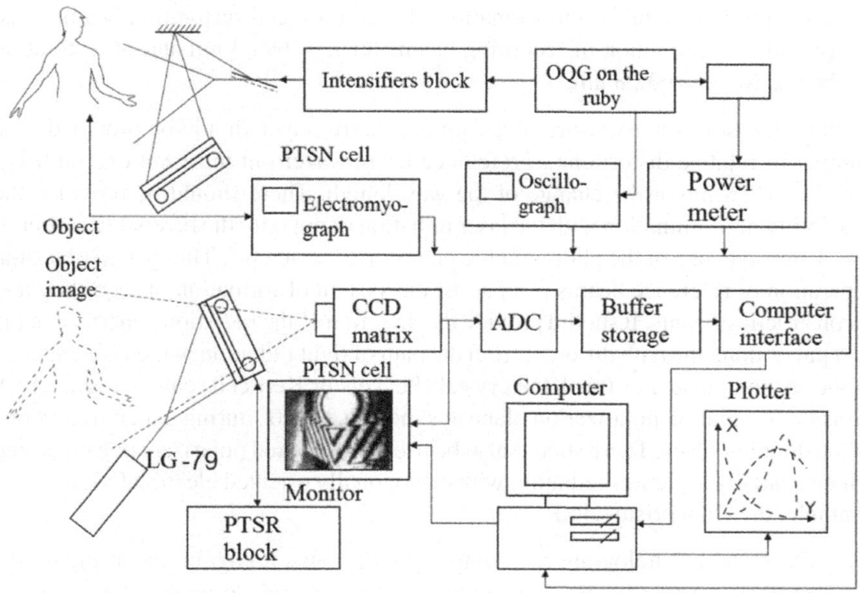

Fig. 3.33 The block diagram of the holographic cardiograph registration system. Reprinted from [94] with permission

- the chair for a patient;
- the illuminating radiation parameters control system;
- the synchronization system;
- the software and hardware hemodynamic complex "Impecard".

Common requirements to the main parts of the complex

1. The high-voltage control unit of the electrooptical shutter must provide safe laser functioning under sync pulse.
2. The ruby laser, the cooling and power units must meet the requirements developed by the Institute of Electronics of the National Academy of Sciences of Belarus.
3. The optical system for illuminating beam formation should provide even illumination of the object with the size of $50 \times 60\ cm^2$. And the energy of the illuminating radiation must be sufficient for a reliable hologram registration.
4. The object under study—a person—must sit in a special chair in a position providing steady illumination of its chest and placement of sensors to control the state of one's heart activity.
5. The control system of the radiation parameters should provide prompt and obvious control of the shape, duration and energy of each pulse of the laser radiation and the time interval between the pulses (a photodiode or a photomultiplier tube can be used which is placed in the field of the illuminating beam near the object). Computer processing of the obtained electrical signal is desirable.

6. The optical system for the formation of recording and restoration beams must provide the formation of recording beams for $\lambda = 694.3$ nm and of restoration beams for $\lambda = 632.8$ nm.

In both cases even exposure of the light-sensitive layer should be provided. The number of relative distortions of reference beam wavefront must not exceed 0.1 λ for $\lambda = 632.8$ nm at the change of the wavelength. There should be provided the possibility of illumination with the laser radiation of not only the reversible carrier of $4 \times 4\ \text{cm}^2$, but also of the photographic plates of $30 \times 40\ \text{cm}^2$. The system of spatial separation of reference beams is a part of the system of formation of recording and reconstructing beams. It should provide the direction of the radiation generated in the first pulse along the path different from the path of light radiation, which is generated in the second pulse. For this KDP crystal (Pockels or Kerr cell) can be used, which turns laser radiation polarization plane at sync pulse at 90° during generation of one of the double-pulses. There should also be used two crossed polarizers or a polarizer, which spatially separates radiation with orthogonally oriented electrical vectors (for example, the Glan prism, etc.).

7. The system of hologram recording should provide hologram recording on the reversible carrier for prompt formation of reconstructed images. The requirements for the reversible carrier for hologram recording are described separately. The recording system should be capable of providing recording on holographic plate of $9 \times 12\ \text{cm}^2$ as well. There should also be provided the possibility of fast removal of the plate for photoprocessing and putting it back into the same place with high accuracy.
8. The system of interferogram formation should provide obtaining qualitative contrast interferograms, which are the image of the human chest covered with interference fringes. The size of the image should correspond to the size of the receiving area of the registration system. There should be a possibility of interferogram formation for several observation directions in order to determine the spatial shifts (in case of utilization of large holographic plates).
9. The system of phase modulation should provide periodic changes with the set phase frequency of one of the reconstructing beams for implementation of the phase regime of interferogram processing. The phase change can be fulfilled with oscillation of a piezoelectric flat mirror (the mirror 8 on the optical scheme in Fig. 3.34).
10. The registration system should provide reliable registration of two-dimension intensity distribution in set moments of time. A photodiode matrix, a vidicon, a CCD matrix can be used.
11. The units of electronic processing of joining with a computer should provide extraction of a desired signal of the set frequency, noise filtering, taking information in set moments of time and transmitting it to the computer, and displaying the obtained image in real time.
12. The display must provide visualization of the holographic interference pattern during the process of reconstruction of the two-exposure hologram.

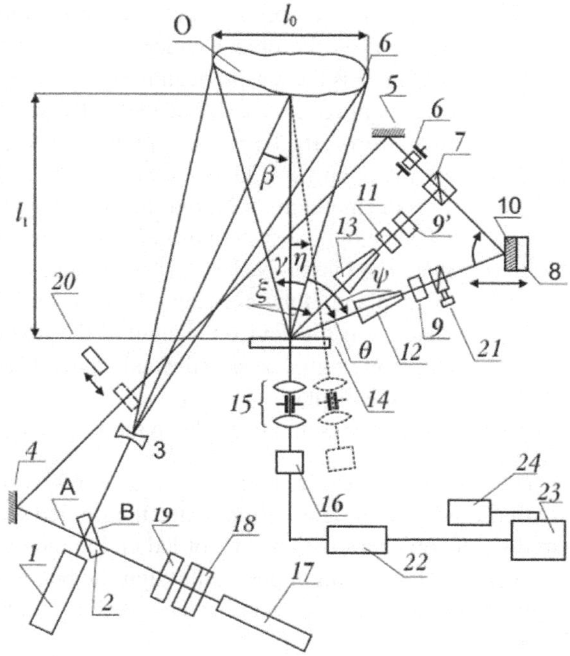

Fig. 3.34 The principal optical scheme of a laser-holographic complex. Reprinted from [94] with permission

13. The computer should provide processing of two-dimension data arrays and giving information about the size and direction of the shifts, speeds and state of the patient's circulatory system in a most convenient way. Besides, the computer should process data of the soft- and hardware hemodynamic complex "Impecard" and give the corresponding parameters, as well as set the moments of time in certain phases of the heart activity to synchronize them with the moment of operation of the ruby laser launch block. It is also desirable to use the computer for analyzing the parameters of the illuminating laser radiation.
14. He–Ne laser should provide hologram reconstruction and formation of qualitative interferograms. It is recommended to use LGN-215 and later modifications (LGN-220) of the "Dendrary" type.
15. The soft- and hardware hemodynamic complex "Impecrd" should control the state of a person by giving information, which serves for interpretation of the shift field parameters determined with the laser-holographic complex in order to estimate the cardiovascular activity of a patient, as well as for determination of the moments of time in the needed phases.
16. The synchronization system should provide synchronization of the launching moment of ruby laser generation pulses with the needed phase of patient's heart contraction and synchronization of two separate pulses of laser radiation in the channel of the reference beams in two directions.
1. Technical description of the optical scheme of the laser-holographic complex:

1.1. The principle optical scheme of the complex is presented in Fig. 3.31 where O is the object under study; 1 is the ruby laser; 2 is the beam splitter; 3 is the lens (a negative one); 4, 5, 10 are the flat mirrors; 6 is the electrooptical device (based on KDP crystal, Pockels or Kerr cell); 7 is the two-beam-refractive prism of Glan; 8 is the flat piezoelectric mirror; 9, 9' are the polarizers; 11 is the $\lambda/2$ plate; 12, 13 are the telescopic systems; 14 is the medium for hologram recording, including the reversible carrier; 15 is the optical system of interferogram formation; 16 is the registration system; 17 is He–Ne laser; 18 is the plate for laser beam shift perpendicular to the plane of the drawing; 19 is the $\lambda/4$ plate; 20 is the semitransparent plate; 21 is the device for input of the set phase shift; 22 is the device for electronic signal processing and input into a computer; 23 is the computer; 24 is the display.

1.2. The optical scheme works in the following way.

The ruby pulse laser radiation with $\lambda_1 = 694.3$ nm is reflected from the beam splitter 2 and with the lens 3 illuminated the studied surface of the object O. The scattered by the object radiation is recorded on the medium for the hologram recording 14. A part of the radiation passed through the beam splitter 2 with the mirrors 4, 5 is directed to the electrooptical device 6, which can be in the form of KDP crystal, Pockels or Kerr cell.

When one of the pulses of the two-pulse laser radiation passes to the device 6, electrical radiation is applied, as a result, the polarizer plane of the passing linear-polarized radiation turns into 90°. The beam splitter 7 splits the radiation into two beams, one of which passes through the polarizer 9′, the $\lambda/2$ plate 11, broadens with the telescopic system 13 and illuminates the light-sensitive layer of the holographic medium 14. The second beam is reflected from the mirror 10, which gives the possibility to oscillate with the help of the piezoelectric element 8, passes through the polarizer 9, which is set in the crossed state in relation to the polarizer 9′, broadens with the telescopic system 12 and also illuminates the medium for hologram recording 14. As the voltage onto the device 6 is supplied at passing of one of the pulses, the pulses of the laser radiation will be reciprocally orthogonally polarized and thus each of the polarizers 9 and 9′ will pass only one of the pulses. As a result of this each of the holograms formed between two pulses will be recorded with its reference waves, which are spatially separated. Reconstruction of the holograms is conducted with the continuous He–Ne laser 17, $\lambda = 632.8$ nm. The laser radiation passes through the plane-parallel plate 18, which can turn in the vertical plane and serves for adjustment shifts of the laser beam in the plane, which is perpendicular to the plane of the drawing. Then the radiation becomes circularly polarized after having passed through the diagonal plate $\lambda/4$ and using the mirrors 4, 5 is directed onto the beam splitter 7. The beam splitter 7 splits the radiation into reconstructing beams, which spread in the same directions as the reference ones do. One of the beams passes through the polarizer 9′, then through the plate $\lambda/2$, which turns its polarization plane at 90°, and then it broadens with the telescopic system 13 and illuminates the hologram 14. The second beam is reflected from the mirror 10, passes through the polarizer 9, broadens with the telescopic system 12 and illuminates the hologram 14 as well. The reconstructed holographic image of the object, which is covered with the interference fringes is transferred onto the light-sensitive area of the registration system 16 with the help of the optical system of interferogram formation.

The mirror 10 should be capable of harmonic oscillations with the help of the piezoelectric element 8 in the direction, which is perpendicular to its reflecting surface to provide interferogram processing in the phase regime.

Ellipticity can appear under reflection of circularly polarized radiation from the beam splitters and the mirrors that will lead to intensity inequality of the reconstructed waves. To compensate this, a reverse set radiation ellipticity can be introduced with the photoplate $\lambda/4$ 19. It is obvious that chromatic aberrations can appear if λ_1 is replaced by λ_2 during reconstruction. Moreover, the aberrations will appear because of diameter difference of laser beams and different divergence. It is required to eliminate them, for example, by filtering the radiation. But if the values of such aberrations are equal for both reconstructing beams it should not lead to distortions of the interference pattern.

The plane-parallel semitransparent plate 20 serves for adjustment and matching of the lasers.

As the localization surface of the interference fringes cannot coincide with the surface under study and be located at essential distance from it, then it is necessary to use the small aperture optical system to simultaneously observe the image of the surface and of the interference fringes. But the size of the speckles increases and that considerably diminishes the quality of the interference pattern. To find a compromise it is important to have the possibility of changing the aperture of the interferogram forming system. As the field of view is narrowing if the aperture decreases, then the system turns around the axis, which passes through the central hologram point, should be considered to provide the possibility of analysis of the whole human chest surface under study.

1.3. Technical requirements to the elements of the optical scheme.

1.3.1. Ruby laser

a. the wavelength of the laser radiation should be 694.3 nm,
b. the width of the spectral line of each pulse of the laser radiation should be not more than 300 MHz;
c. the operation mode: single-mode generation of pair monopulses with the regulated limits of 100–700 μs with intervals between the pulses.
d. the energy of each of the two radiation pulses of the driving generator should reach 0.01 J, of the two-stage amplifier—0.25 J, of the total one—0.5 J;
e. the duration of each pulse of the laser radiation should reach 20–120 ns;
f. the radiation is polarized in the vertical plane.

The electrical vector oscillations of the generated radiation should be in the plane perpendicular to the incidence plane on the beam splitter 2.

1.3.2. The beam splitter 2 serves for splitting radiation from the laser 1 into illuminating and reference beams, for input of reconstructing beams from the continuous He–Ne laser 17, and for adjustment of schemes with the continuous laser 17. The transmission coefficients of the beam splitter should be about 95% and the reflection coefficients—of about 5%. Interference pattern should not be observed at the wavelengths $\lambda_1 = 694.3$ nm and $\lambda_2 = 632.8$ nm in the transmitted and reflected beams. It is recommended to produce a beam splitter in the form of a plane-parallel plate, on the edge of which an antireflective coating should be put (for $\lambda = 694.3$ nm). The coating should be laser-resistant. And the wavefront distortions in the reflected and the transmitted beams for λ_1 and λ_2 at the diameter of 10 mm should not exceed 0.1λ for $\lambda = 632.8$ nm. The beam splitter 2 should have two smooth adjustment spins (in the plane of the drawing and in the perpendicular plane), light sensitivity—$2''$ and also parallel shift in the drawing plane along the direction of the beam generated by the laser 17, the sensitivity is 0.01 mm.

1.3.3. The lens 3 should provide even illumination of the surface of the object under study of 50×60 cm^2. Laser damage resistance should be observed. A negative lens should be used as an objective.

1.3.4. The mirrors 4, 5, 10 should provide the required direction of the recording and the reconstructing beams. The reflecting surface should be frontal, and the reflection coefficient should be maximal for λ_1 and λ_2. The mirrors 4, 5, 10 should be capable of angle adjustments in two mutually perpendicular planes.

1.3.5. The electrooptical device 6 should provide the rotation at 90° of the polarization plane of the transmitted linear-polarized radiation with $\lambda_1 = 694.8$ nm under voltage and not change the polarization state in its absence. KDP crystal, Pockels or Kerr cell can be used as such a device.

1.3.6. The beam splitter 7 should provide formation of two spatially separated reference beams. The ratio between transmission and reflection coefficients of the beam splitter should be chosen according to the conditions of equality of intensities of the reference beams in the plane of the light-sensitive layer of the reversible carrier 14. There should be no interference structures in the reference beams. The covering should be laser-resistant. The beam splitter 7 should have the possibility of angle adjustments in two mutually perpendicular planes.

1.3.7. The polarizers 9 and 9′ are to be in crossed state during hologram recording and in parallel (or remote) during reconstruction. There should be no shifts and inclinations, which pass through the beam polarizers when λ_1 is replaced by λ_2. When the polarizers turn or move away, change of distortions of the wave front should not exceed 0.05 λ. If it is impossible, then the plate $\lambda/2$ 11 and $\lambda/4$ 19 should be used and the polarizers are to be crossed.

1.3.8. The mirror 10 with the piezoelectric element 8 should provide phase modulation of the reference beam. To do this the piezoelectric element 8 should provide harmonic oscillations of the mirror 10 in the direction, which is perpendicular to its surface with the amplitude of 2 μm and the frequency providing the effective operation of the electronic system of signal processing. Meanwhile, the mirror turns should be no more than 2″. The phase modulation should be conducted at the stage of hologram reconstruction.

1.3.9. The device for the input of the set phase shift 21 should provide smooth change of the optical path length of the transmitting light wave from λ_2 to $\pm$ 2 μm without oscillations of more than 0.05 λ.

For example, a compensator can be used as such a device. It is consisted of two wedges put together in such a way that a plane-parallel plate is received, as it is shown in the scheme. The thickness of the plate changes if one of the wedges is shifted. In case the mirror 10 is slowly shifted in a set direction with the piezoelectric element 8 that leads to visual detection of the shift direction of the interference fringes, then the device 21 is not necessary.

1.3.10. The plate λ/2 11 should provide the rotation of the polarization plane of the transmitted radiation at 90°.

1.3.11. The telescopic systems 12, 13 should provide formation of plane fronts of recording and reconstructing beams and even illumination of the light-sensitive surface of the reversible carrier 14 of the size 40 × 40 mm^2. Operative restructuring of the telescopic systems for wavelengths λ_1 and λ_2 is acceptable. The difference of changing wavefront distortion passed through the telescoping systems 12, 13, if λ_1 is replaced by λ_2 along the whole front, should not exceed 0.05 λ.

The alternative variant can be the use of one illuminating lens achromatized for λ_1 and λ_2 along the field of view. But the requirements to wave front distortions remain the same over the whole active field of view.

1.3.12. The medium for hologram recording should provide reliable hologram recording. As such a medium, it is suggested to use a reversible carrier, the requirements to which are described additionally.

There should be a possibility to use holographic plates with the size of 9 × 12 cm^2 instead of the reversible carrier. In this case, instead of the telescopic systems 12, 13 lenses should be installed, which form spherical waves for even illumination of large photoplates. There should be a possibility of fast removal of the plate for photoprocessing and placing it into the same place with an accuracy of 0.005 mm.

1.3.13. The optical scheme for formation of the interferogram 15 should provide construction of a qualitative image of the human chest (50 × 60 cm^2) on the light-sensitive area of the registration system 16. The optical system 15 should have a possibility to change the aperture angle γ, at which the object under study is observed within the limits from 1° to 2 $\mathrm{arctg}(l_0/2l_1)$, where l_0 is the size of the object; ($l_0 = 60$ cm), l_1 is the distance between the object and the hologram. This can be implemented through changing the diaphragm size situated in the focus of the first lens. The same diaphragm can be used for filtering of phantom images, which are reconstructed by each of the reconstructing waves.

The distortion of the interference fringes under image transmitting should not exceed 0.1 λ.

There should be a possibility of smooth turns of the optical system 15 with the registration system 16 in the horizontal and the vertical planes in the range $\pm\gamma/2$ around the axes, which pass through the central point of the reversible carrier 14 ($\gamma = 2\,\mathrm{arctg}(l_0/2l_1)$).

There should be a possibility of smooth turns of the optical system 15 with the registration system 16 in the horizontal and the vertical planes about axes, which pass through the central point of the object. The length of the zones of the angle turn $\pm\,\mathrm{arctg}(l_2/2l_1)$ where l_2 is the size of the hologram.

1.3.14. The registration system 16 should have parameters (sensitivity, dynamic range, inertia, etc.), which make it possible to safely register the oscillating interference pattern in definite moments of time. A photodiode matrix, a PEA-matrix or a vidicon can be used in the system.

1.3.15. As a source of continuous radiation 17, it is recommended to use He–Ne laser LGN-215 or its later modifications. The oscillation plane of the electrical vector of the generated radiation should be perpendicular to the plane of the drawing.

1.3.16. The plane-parallel plate 18 should provide parallel shift of the laser beam of He–Ne laser 17 in the plane perpendicular to the plane of the drawing at $h \pm 2$ mm. For this, it should have smooth inclinations in the plane perpendicular to the plane of the drawing. The value of the inclination angle can be calculated according to the formula

$$n = \mathrm{d}\sin\alpha(1 - \cos\alpha/\sqrt{n^2 - \sin^2\alpha})$$

where d is the thickness of the plate; n is the refractive index.

1.3.17. The plate λ/4 19 is intended to obtain circularly polarized radiation of He–Ne laser. This plate should be used in the case of the turn of one of the polarizers 9 or 9′ into 90° or deleting of both polarizers will lead to changes of wavefront distortions more than 0.05 λ. In this case, the polarizers can be left crossed.

1.3.18. The semitransparent plane-parallel plate 20 should have a possibility to be put in and out the ray path for adjustment of the ruby laser 1. It should have adjustment inclinations, which make it possible to join the reflected beam with the incident one. The sensitivity of the angle adjustments should not be worse than 2″.

1.3.19. All the elements of the optical scheme should be firmly fixed in the set states. There should be no cutting of light beams. There should be boots protecting the optical surfaces from dust and diaphragms for the elimination of spurious glares.

1.3.20. The optical elements 7–14 should be located as compact as possible and be firmly fixed to the common base. It is desirable to fulfill them with one block. The distances between the beam splitter 7 and the mirror 10 and between the beam splitter 7, the mirror 10 and the reversible carrier 14 should be as minimal as possible.

1.3.21. The distance l_1 between the object and the hologram should be as minimal as possible.

1.3.22. The angles ξ and ψ (see Fig. 3.34) should be in the range from 35° to 45° ($l_1 = 1$ m). At smaller distance, the range is also smaller, because it is determined by the resolution of the reversible carrier (see requirements to the reversible carrier). The angles θ and β should be as minimal as possible.

1.3.23. All the elements of the optical scheme should be placed on one massive base and be firmly fixed to it and the base should be vibroinsulated.

3.3.2 *Principle of Operations of the Complex*

The principle of operation of the complex can be described in the following way (Figs. 3.35 and 3.36). As a result of cardiac contraction, shift of points of human chest occurs. Information about the distribution of these shifts along the human chest surface, about the value, the direction and the speed of these shifts can help to make a conclusion about the state of human cardiovascular system.

It is known that registration and reconstruction of a double-exposure hologram, one exposure of which corresponds to the initial state of the object under study, and the other one—to the altered state (for example, deformed) give possibility to form

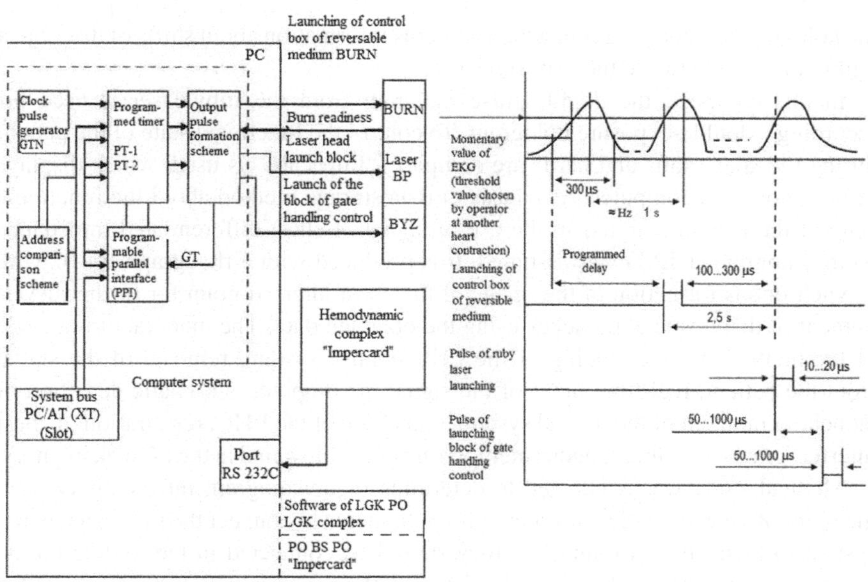

Fig. 3.35 Temporal diagram of the complex operation. Reprinted from [94] with permission

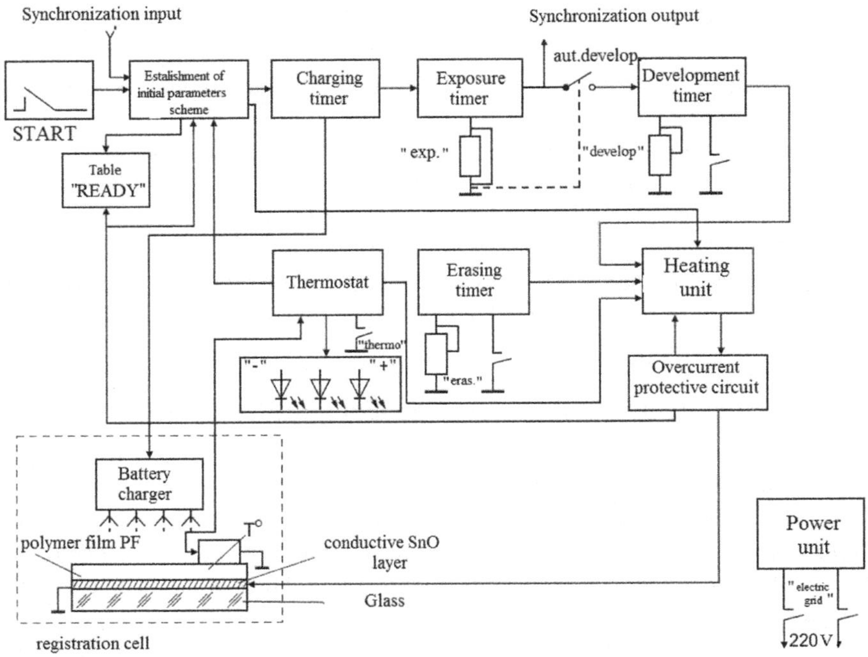

Fig. 3.36 Structural scheme of the control unit for recording holographic interferograms on a photothermoplastic carrier. Reprinted from [94] with permission

the holographic interferogram, which contains information about shifts of this object within the time between the exposures [1].

In this complex, the double-pulse radiation from the ruby laser is used for recording a double-exposure hologram. To control the functional state of the patient during the study soft- and hardware complex, "Impecard" is used, which displays information on a computer. As a basis of data storage method about the functional state of the patient's organism the use of signal of chest differentiated impedance plethysmogram (DIPG) was assumed. It is produced with a rheograph R4-02 with a synchronous formation of the image of the chest interferogram for further development of diagnostic approaches using the obtained data. The manufacturing cycle of the method includes tuning of the LHC control system, removal of the signal from the patient, real-time input of the signal in computer, automatic detection of launching moment of the optical system, launching of the LHC, registration of chest interferograms and simultaneous memorizing of the biosignal in the PC disk memory.

Medical study was conducted to determine plethysmogram informativity from the point of view of choice of optimal time moments to connect the synchronization system to them. It was found that they should be connected in the middle on the ascending or the descending branch of the interferogram curve, as namely here the greatest informativity relative to the amplitude of mechanical shifts of oscillation zone of chest cutaneous covering is reached.

The sync pulse is formed with the synchronization system in required phases of the cardiac cycle in moments of time set by an operator. This sync pulse launches the ruby laser. The patient sits in a special chair and his chest is evenly illuminated with the optical system of light beam formation. The scattered radiation is recorded on the hologram with two reference beams, which were obtained with the help of system for formation of recording and reconstructing beams.

Reconstruction of the double-exposure hologram was conducted with the continuous He–Ne laser.

The image of the chest with the interference pattern is formed on the receiving area of the registration system with the optical system of interferogram formation. With the electronic processing unit and conjugation with the computer analog-digital processing of the image, its input in the computer and parallel displaying for visual estimation of the interference pattern occur.

The values of shifts and speeds of points of the object surface are computer-calculated.

Time diagram of operation of the complex is shown in Fig. 3.35. The thermoplastic interferogram recording unit, the pulse ruby laser, the shutter control unit are consecutively switched on by the synchronization system when the selected by an operator value of the curve on the computer cardiogram is reached. Two consequent laser radiation pulses defined by the operator as delay get into the optical scheme between them.

A beam, which illuminates the person, and two reference beams are formed in the scheme. These beams correspond to the first and the second illuminating pulses of the ruby laser.

The control unit of holographic interferogram (HI) recording on the photothermoplastic carrier (PTPC) was designed taking into account the possibility of using any type of reversible carriers produced (Fig. 3.37) [171–183]. Universality is provided, on the one hand, constructively through mounting changeable cassettes, and on the other hand, schematically through setting all the parameters of recording on a temporary diagram of the working cycle.

The hologram registration device is aimed at multiple recording and erasing of optical holograms on the films of reversible photothermoplastic semiconductors (PTS) with real-time processing of the carriers during registration (Fig. 3.38) [166, 184–200].

Multiple uses of the carriers and rapid registration process give the possibility to promptly correct the optical scheme and registration conditions, which provide obtaining the set data. The functional scheme of the reversible hologram registration device is presented in Fig. 3.36. The control unit includes a charging timer, an exposure timer, processor and erasing timers, a scheme of setting initial parameters, a thermostat, an indicator of sample temperature, a heating unit, an overcurrent protective circuit, a power unit. Registration cell includes a battery charger, a corona-forming gap, a sample of the registering medium, a temperature-sensitive element.

The sample of the registering medium is a polymer film (PF) plotted on a glass plate with a conductive SnO layer (Fig. 3.38).

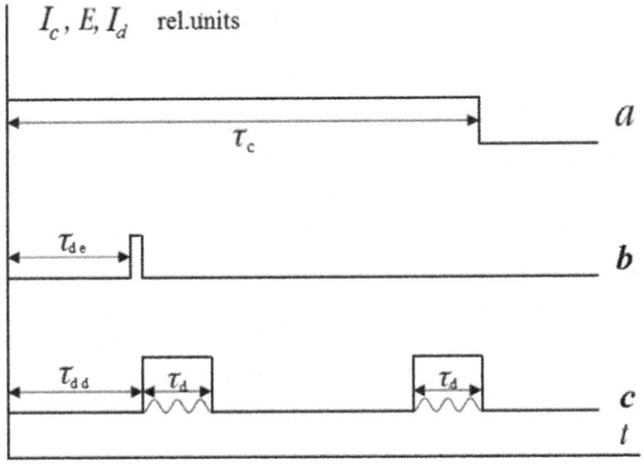

Fig. 3.37 Temporal diagram of recording modes of two-exposure holographic interferograms on PTS material: τ_c—corona discharge duration; $\tau_{\rm de}$—exposure delay time relative to corona discharge; $\tau_{\rm dd}$—development delay time relative to corona discharge; τ_d—pulse development duration. Reprinted from [94] with permission

The signal from the exposure timer is directed to the sync input of the pulse laser if the pulse laser is used as a radiation source. After the exposure pulse, if the switcher of the automatic processor is turned on, which is joined with the "ex" regulator, then "The development timer" is automatically launched, the duration of which is set by the regulator "PR". The signal from the processor timer gets into the heating block input, and as a result of this, the heating voltage of the "heating block" gets into conductive sublayer of the sample of the registration medium through the "overcurrent protection circuit". This leads to heating of the sublayer, which heats PF.

Electrical forces deform PF and the latent image turns into surface geometry. After registration of the interferogram, it should be processed; i.e., the field of shifts and deformations of the object (in this case the human chest) should be made up on the image of fringes on the interferogram. For this purpose, the continuous He–Ne laser and the phase modulation system of one of the reference beams are used. The block diagram of one of the experimental units is presented in Fig. 3.39. With its help, the regime of the recording of holograms and holographic interferograms on the extensive reversible photothermoplastic semiconductor were processed.

3.3.3 Equations for Interference Fringes Interpretation

Let the studied area of the chest surface AB is illuminated by a point source placed in the point O (Fig. 3.40).

The light scattered by the surface under study goes through the hologram into the observation point P. The double-exposure hologram is registered with a doubled

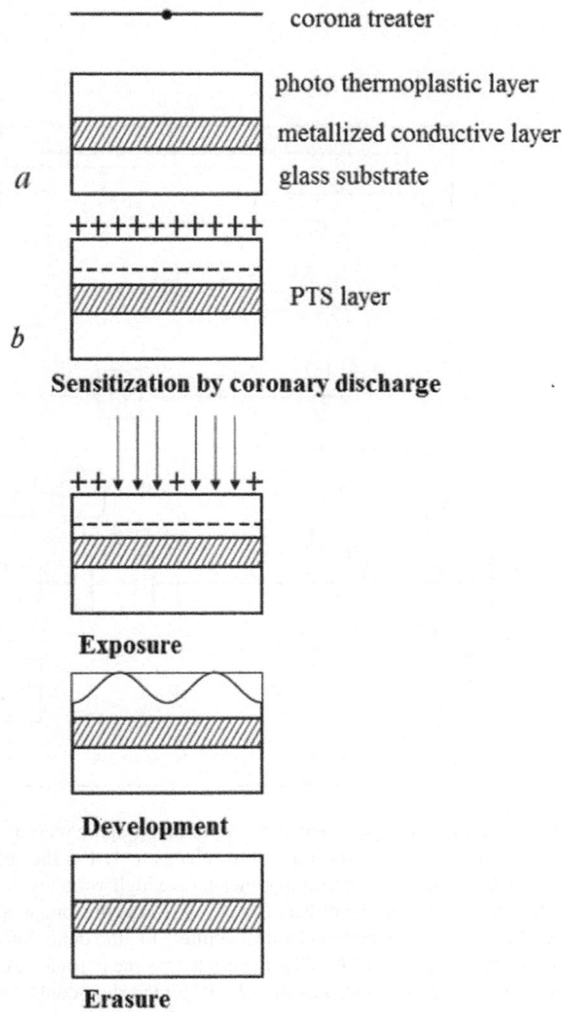

Fig. 3.38 Stages of hologram recording onto reversible photothermoplastic semiconductor. Reprinted from [94] with permission

pulse of the ruby laser, and the shift of the studied point A into position A' between the exposures is determined by vector $\vec{L}$.

As the role of the hologram is to reproduce light scattered by the object in the initial as well as in the shifted state, it can be not taken into account in this consideration.

Phases of light beams coming into the point of observation from shifted and non-shifted points of the chest surface have the following value:

$$\varphi_1 = \vec{k}_1 \vec{r}_{\text{ob1}} + \vec{k}_{\text{ob1}} \left(\vec{r}_{\wedge} - \vec{r}_{\text{ob1}}\right) + \varphi_r \tag{3.58}$$

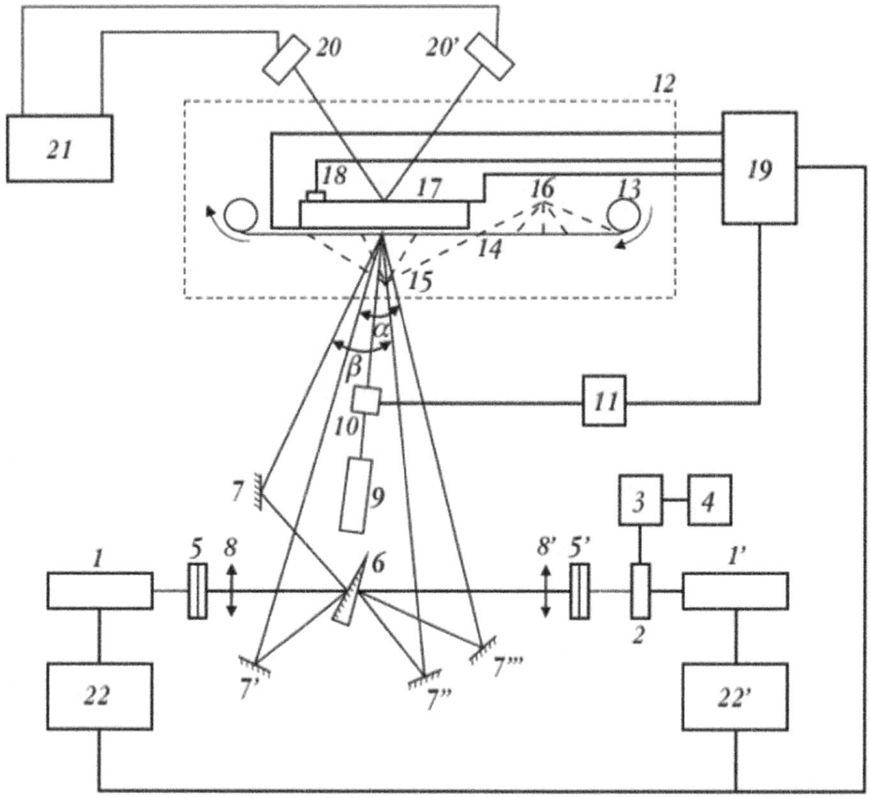

Fig. 3.39 One of the variants of the experimental set-up for holographic recording onto reversible photothermoplastic semiconductor with film-traction mechanism: 1, 1′—the ruby lasers; 2—the electrooptic shutter; 3—the discharger with forming line; 4—the high-voltage power supply source, 5, 5′—the neutral light filter; 6—the beam splitter; 7, 7′, 7″, 7‴—the mirrors; 8, 8′—the expanding lens, 9—the probing laser; 10—the electromechanical shutter; 11- the delay line; 12—PTS registration; 13- the film-traction mechanism;14—PTS material; 15—the corona treater; 16—the electrizer; 17—the heater; 18—the temperature sensor; 19—PTS recording control unit; 20, 20′—the photodiodes; 21—the oscilloscope; 22, 22′—the power units. Reprinted from [94] with permission

$$\varphi_2 = \vec{k}_2\vec{r}_{ob2} + \vec{k}_{ob2}\left(\vec{r}_\wedge - \vec{r}_{ob2}\right) + \varphi_r \tag{3.59}$$

where φ_r is the initial wave phase; $\vec{k}_1$ and $\vec{k}_2$ are the wave vectors of the illuminating waves; $\vec{k}_{ob1}$ and $\vec{k}_{ob2}$ are the wave vectors of the scattered object waves, which come to observation point P.

By denoting minor changes of wave vectors $\Delta\vec{k}_1$ and $\Delta\vec{k}_2$ we will get

$$\vec{k}_2 = \vec{k}_1 + \Delta\vec{k}_1 \; \vec{k}_{ob2} = \vec{k}_{ob1} + \Delta\vec{k}_2 \tag{3.60}$$

Then, the phase difference in this point will be equal to

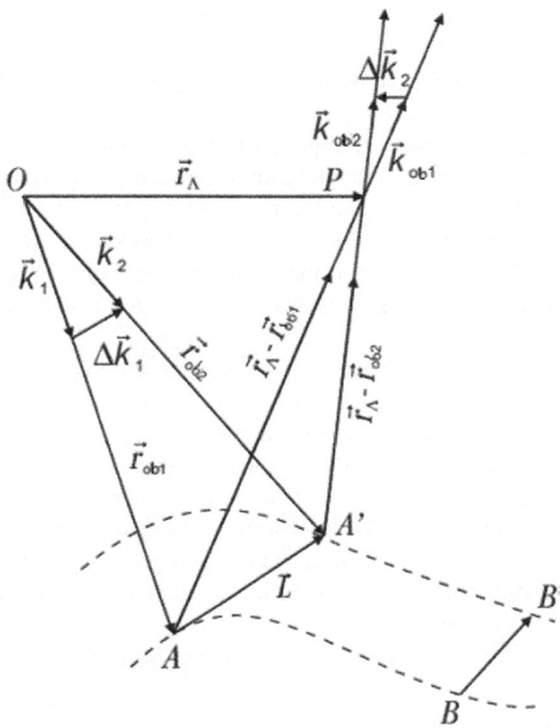

Fig. 3.40 Scheme of illumination and observing of the surface under study. Reprinted from [94] with permission

$$\Delta\varphi = \left(\vec{k}_{ob1} - \vec{k}_1\right)(\vec{r}_{ob2} - \vec{r}_{ob1}) + \Delta\vec{k}_1\vec{r}_{ob2} + \Delta\vec{k}_2(\vec{r}_\wedge - \vec{r}_{ob2}) \tag{3.61}$$

In the real experiment $|\vec{r}_{ob1}| \gg \left|\vec{L}\right|$ and $|\vec{r}_{ob2}| \gg \vec{L}$, where $\vec{L} = \vec{r}_{ob2} - \vec{r}_{ob1}$; that is why we can consider that $\Delta\vec{k}_1 \perp \vec{r}_{ob2}$; and $\Delta\vec{k}_2 \perp (\vec{r}_\wedge - \vec{r}_{ob2})$; and $\Delta\vec{k}_1\vec{r}_{ob2} \to 0$; $\Delta\vec{k}_2(\vec{r}_\wedge - \vec{r}_{ob2}) \to 0$

Then (3.61) will be rewritten in the following form

$$\Delta\varphi = \left(\vec{k}_{ob1} - \vec{k}_1\right)\vec{L} \tag{3.62}$$

The sensitivity vector $\vec{k} = \vec{k}_{ob1} - \vec{k}_1$ coincides with the bisectrix of the angle between directions of illumination and observation. Equation (3.62) was firstly obtained in the work [89] and determines connection between directions of illumination of the studied object, the observation direction, the shift vector and the phase difference. Using (3.62), the projection of the shift vector on to the bisectrix of the angle between the illumination and observation directions can be estimated for this observation direction. Thus, having the system of three equations like (3.62) three-dimensional shift vector L can be found. For example, for parallel shifts of the object under study as a whole (3.62) can be transformed to the following one

$$\Delta\varphi = \vec{g}\vec{r}_{\mathrm{ob1}} - g_1 \tag{3.63}$$

where g is the constant vector; $\vec{r}_{\mathrm{ob1}}$ is the solitary vector of the scattered wave; $\vec{k}_{\mathrm{ob1}} = 2\pi\vec{r}_{\mathrm{ob1}}/\lambda$, $g_1 = \mathrm{const}$.

It is necessary to equate to zero variations of phase difference under changes of the observation angle within the receiver aperture to determine the surfaces of localization of the interference fringes.

After the scalar product from (3.63) is decomposed along the axes of the Cartesian coordinate system and the result of differentiating the phase difference function in the observation angle is equaled to zero we obtain the localization condition

$$\alpha_2 = \mathrm{const} \tag{3.64}$$

where α_2 is the observation angle.

To satisfy this condition the final aperture receiver should be situated in infinity; i.e., in the remote field where the angular surface sizes visible from the receiver are negligible. Thus, to observe contrast interference fringes it is necessary to use the objective and register the interference pattern in its focal plane as the system of parallel interference fringes localized in the infinity corresponds to the gradual shift of the object.

Under a small turn of the object under study, (3.62) is the following

$$\Delta\varphi = -k_0\theta x + (\sin\alpha_1 + \sin\alpha_2) \tag{3.65}$$

where θ is the angle of the turn; α_1 is the angle of illumination of the object under study.

Unlike the case of (3.63) the phase difference at the turn of the object depends on the x-coordinate of points on the investigated surface. The point *P* is supposed to be on the localization surface at a distance *h* from the surface of the object under study. Then having differentiated (3.65), having equated the differential to zero, having solved the equation $\delta(\Delta\varphi) = 0$ relative to *h* and after using the ratio

$$\mathrm{d}x = \frac{h d\alpha_2}{\sin^2\alpha_2}, \tag{3.66}$$

we get

$$h = \frac{-x\cos\alpha_2\sin^2\alpha_2}{(\sin\alpha_1 + \sin\alpha_2)}. \tag{3.67}$$

If the direction of observation is normal to the surface, i.e., $\lambda_2 = \pi/2$, then $h = 0$ and the fringes are localized on the surface under study. If $\lambda_2 \neq \pi/2$, the localization surface will lie either in front of the surface or behind it, but it always crosses the surface under study along the axis of rotation $x = 0$.

According to the conducted analysis in common case the localization surface of the interference pattern does not coincide with the surface of the object under study; and in the case of gradual shifts of (3.64), it is placed in the infinity. But it is supposed that mainly deformation shifts will present during the study of the point shifts of human chest; i.e., the localization surface will be near the object under study.

3.3.4 *Methods of Estimating the Shift Value of Points on the Surface of the Object Under Study Considering the Direction of Shifts*

Let us suggest several methods for estimating the direction of shifts. For implementation of these methods it is necessary to use two spatially separated light beams for recording and reconstructing of holograms. Separating of reference beams can be conducted in different ways, for example, using KDP crystal (Pockels or Kerr cell) and the beam splitter, and in each of the reference beams the polarizers are installed, which are crossed between each other. In this case, holograms will be recorded and reconstructed by two mutually orthogonal polarized waves.

If each hologram is recorded and reconstructed with two reference waves, which are spatially separated and mutually orthogonally polarized, then the intensity of the sum wave scattered by the object in this observation direction can be written as follows [201]:

$$
\begin{aligned}
I = {} & \alpha_1^2 A_1^2 + b_1^2 B_1^2 + a_2^2 A_2^2 + b_2^2 B_2^2 \\
& + 2\alpha_1 \alpha_2 A_1 A_2 \cos(\varphi_{2x} - \varphi_{1x}) \\
& + 2 b_1 b_2 B_1 B_2 \cos(\varphi_{2y} - \varphi_{1y}),
\end{aligned} \tag{3.68}
$$

where

$$\varphi_{1x} = \vec{k}_1 \vec{r}_{\text{ob1}} + (\vec{r}_\wedge - \vec{r}_{\text{ob1}})\vec{k}_{\text{ob1}} + \vec{r}_r \vec{k}_{\text{ref}x1} + \delta_{x\text{ob1}} + \delta_{x\text{ref1}};$$

$$\varphi_{2x} = \vec{k}_2 \vec{r}_{\text{ob2}} + (\vec{r}_\wedge - \vec{r}_{\text{ob2}})\vec{k}_{\text{ob2}} + \vec{r}_r \vec{k}_{\text{ref}x2} + \delta_{x\text{ob2}} + \delta_{x\text{ref2}};$$

$$\varphi_{1y} = \vec{k}_1 \vec{r}_{\text{ob1}} + (\vec{r}_\wedge - \vec{r}_{\text{ob1}})\vec{k}_{\text{ob1}} + \vec{r}_r \vec{k}_{\text{refy1}} + \delta_{y\text{ob1}} + \delta_{y\text{ref1}};$$

$$\varphi_{2y} = \vec{k}_2 \vec{r}_{\text{ob2}} + (\vec{r}_\wedge - \vec{r}_{\text{ob2}})\vec{k}_{\text{ob2}} + \vec{r}_r \vec{k}_{\text{refy1}} + \delta_{y\text{ob2}} + \delta_{y\text{ref2}};$$

A_1, B_1, a_1, b_1 are the amplitudes of E_x and E_y the components of object and reference waves; k_1, k_{ob1}, $\vec{k}_{\text{ref}x1}$, k_{refy1} are the wave vectors of illuminating, object and reference waves; $\vec{r}_{\text{ob1}}$ and $\vec{r}_{\text{ob2}}$ are the position vectors of the object surface before and after changing of its state; $\vec{r}_\wedge$ is the position vector of the interference pattern localization surface; $\vec{r}_r$ is the position vector on the hologram surface; δ_{xob1}, δ_{yob1}, δ_{xref1}, δ_{yref1} are the initial phases of E_x and E_y components of object and reference waves.

Equation (3.68) contain full information about the state of object wave polarization, and using this method of hologram recording and reconstruction, the shift of the interference fringes is proportional to the quantity of their polarization parameter changing. And the phase differences $\Delta\varphi_1 = \varphi_{2x} - \varphi_{1x}$ and $\Delta\varphi_2 = \varphi_{2y} - \varphi_{1y}$ occurring between object waves are conditioned by the change of not only illumination direction, wavelength and coordinates of object points, but also by the change of polarization parameters $tg\varphi = A_1/B_1$ and $\delta_{o\text{б}1} = \delta_{xo\text{б}1} - \delta_{yo\text{б}1}$. Using this fact on the changes in polarization parameters of each pair of reconstructed waves, it is possible to determine either the direction of change of the surface shape under its contouring or shift direction of surface points during investigation of its deformations and shifts.

From (3.68), the formulae follow, which connect the quantity and direction of the relief surface as well the quantity and the direction of shifts of the surface points with polarization characteristics of the resulting object wave. For example, while investigating the shape of the surface, it is necessary to set parameter change of the illuminating wave $\vec{k}_1$ and while investigating the shift points—change of coordinates of object surface points $\vec{r}_{ob1}$. It is possible to implement combined methods to increase sensitivity and to broaden the measurement range.

Another method for detecting the quantity and direction of relative shift is a method based on introduction of set phase shifts in one of the reference waves. For its implementation, it is also necessary to use two spatially separated reference waves during recording and reconstruction. An optical path-altering device is inserted in one of the waves, for example, a compensator. Alteration of the length of the compensator leads to interference fringes shift in the observed interference pattern. By increasing or decreasing the phase difference at a certain quantity, the direction of relative shifts of points on the surface under study can be determined according to the shift direction of interference fringes. For example, let $\Delta\varphi > 0$ and the phase shift δ be inserted into the wave φ_2. The intensity value in the initial 1 and the analyzed 2 points will be equal to:

$$I_1 = 4\mathcal{E}_0^2\cos^2\Delta\varphi_1/2,\ I_2 = 4\mathcal{E}_0^2\cos^2\Delta\varphi_2/2 \tag{3.69}$$

Let us change the phase difference in such a way that $I_1 = I_2$. If it required increasing φ_2, i.e., $\Delta\varphi_2 = \Delta\varphi_1 + \delta$, then it means that the point 2 moved away from the registration plane relative to point 1. If $\Delta\varphi_2 = \Delta\varphi_1 - \delta$, then vice versa, the point 2 approached the observation plane relative to point 1.

3.3.5 Methods of Direct Measurement of the Function of Phase Difference

Spatial separation of reference beams during recording and reconstruction of holographic images is necessary to increase the accuracy of measurements directly connected with the measurement of phase difference of the interfering waves in the needed points of the image [202]. For this, a device responsible for frequency shift should be inserted into one of the reference beams, for example, a spinning diffraction grating. Then, the reconstructed waves will have the following form

$$\vec{\mathcal{E}}_1 = \mathcal{E}_0 \cos(\omega_1 t + \varphi_2)\vec{i} + \mathcal{E}_0 \cos(\omega_1 t + \varphi_1)\vec{j};$$
$$\vec{\mathcal{E}}_2 = \mathcal{E}_0 \cos(\omega_2 t + \varphi_2)\vec{i} + \mathcal{E}_0 \cos(\omega_2 t + \varphi_2)\vec{j}. \tag{3.70}$$

Equation (3.70) consider superposition of oscillation planes of the reference waves after their separation with a crystal (or two crossed polarizers).

The intensity distribution in the plane of the image will be equal to

$$I = 4\mathcal{E}_0^2 \cos^2[(\Delta\omega t + \Delta\varphi)/2] \tag{3.71}$$

where $\Delta\omega = \omega_2 - \omega_1$.

According to (3.71), the interference fringes will be shifted in the formed interferogram. The photoreceiver in each point of the image will fix the variable signal with the frequency $\Delta\omega$.

In all needed spots of the image, there can be measured the phase of this signal; i.e., the phase difference $\Delta\varphi(x, y)$ can be determined, using special phase-measuring equipment. Apart from phase difference measuring, the direction of relative shifts can be detected with this method. Let, for example, the shift of the frequency be conducted for the second object wave, which is scattered by the shifted surface, and the phase of the registered signal is greater in the analyzed point 2 than in the initial point 1.

Then in a certain moment of time t_1

$$\Delta\omega t_1 + \Delta\varphi_1 < \Delta\omega t_1 + \Delta\varphi_2 \tag{3.72}$$

The same value of the phase in the analyzed point will be as in the initial one in a certain moment of time Δt, i.e.

$$\Delta\omega t_1 + \Delta\varphi_1 = \Delta\omega(t_1 + \Delta t) + \Delta\varphi_2 \tag{3.73}$$

and $I_1 = I_2$.

So during time Δt, the intensity value I_1, which was in the initial point 1 in the moment of time t_1, will shift into the analyzed point 2. Thus, the interference fringes will shift from more remote from the observer points to less remote ones; i.e., the direction of movement of the interference fringes determines the relative proximity or remoteness of points of the deformed surface and the shift directions. For example, if there are annular structures in the interference pattern, then for this case, contraction of the rings to the center will prove the presence of convexity, and their recession from the center—the concavity of the surface.

3.3.6 *Ways of Estimating the Shift Points of the Object Under Study with the Interference Holographic Method*

It is known that registration and reconstruction of a double-exposure hologram, one exposure of which corresponds to the initial state of the object under study P and the other one—to the altered one P_1 (for example a deformed one), give the possibility to form a holographic interferogram containing the information about shifts of this object in the period of time between the exposures (Fig. 3.35).

And the phase difference of the obtained interferogram is estimated by the expression

$$\Delta\varphi = \Delta\vec{K}_1 \cdot \vec{L}, \tag{3.74}$$

where $\Delta\vec{K}_1 = \vec{K}_2 - \vec{K}_1$ is the vector of sensitivity; $\vec{L}$ is the shift vector; $\vec{K}_1$ is the wave vector of the illuminating wave; $\vec{K}_2$ is the wave vector of the object wave.

As it follows from (3.74), it is only possible to determine the projection of the shift vector $\vec{L}$ to the bisectrix of the angle between the directions of illumination and observation for these fixed directions of illumination and observation. To find the space vector $\vec{L}$ of shift points of the object, it is necessary to have at least three equations like (3.74) and solve the system of equations.

To do this, either the illumination direction should be altered or the interference pattern should be observed from different directions.

The average speed of shifts of this object point can be estimated by the formula

$$\vec{V} = \vec{L}/\Delta t \tag{3.75}$$

where Δt is the time interval between pulses of the laser radiation.

It is suggested that holograms are recorded and reconstructed with two spatially separated reference beams to determine the direction of relative shifts of points on the surface under study. If the optical path length of one of the reconstructed beams is gradually changed through the known method at the reconstruction stage, then it is possible to determine the direction of the shift vector of points on the surface under study according to the shift direction of interference fringes (Fig. 3.41).

For example, let the phase difference $\Delta\varphi = \varphi_2 - \varphi_1 > 0$, and the gradual phase shift δ be inserted into the wave φ_2. It is necessary to determine the shift direction of the analyzed point 2 relative to the initial point 1.

The intensity quantity will be in these points

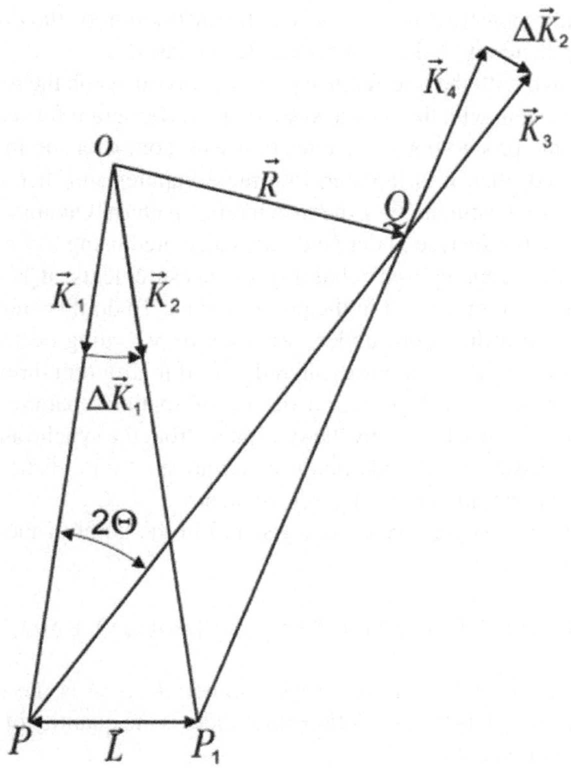

Fig. 3.41 To the calculation of relocation of the point of the experimental object. Reprinted from [94] with permission

$$I_1 = 2\mathcal{E}_0^2(1 + \cos\Delta\varphi_1);$$
$$I_2 = 2\mathcal{E}_0^2(1 + \cos\Delta\varphi_2). \tag{3.76}$$

Let us change the phase difference so that $I_1 = I_2$. If the phase shift had to be increased to do this, i.e., $\delta > 0$, then the point 2 was mere remote after the shift than point 1.

The double-pulse ruby laser radiation is used in this complex for recording a double-exposure hologram. To control the functional state of the patient during the investigation, "Impecard" software and hardware complex is used, which displays the information on the computer.

A sync pulse is formed with the synchronization system in the needed phases of the cardiac cycle in the moments of time set by an operator. The sync pulse launches the ruby laser.

The chest of a patient who sits in the special chair is evenly illuminated with the optical system of illuminating beam formation. The scattered radiation is recorded on the hologram with two reference beams obtained using the system of formation of recording and reconstructing beams. The reconstruction of the double-exposure hologram is conducted with the continuous He–Ne laser.

The chest image with the interference pattern is produced on the receiving area of the registration system with the optical system of interferogram formation. Using the block of electronic processing and connection with computer the image is analog-digitally processed, then it is inputted into the computer and then outputted onto the display for visual estimation of the interference pattern. Quantities of shifts and speeds of point on the surface under study are calculated using a computer.

To increase the accuracy and reliability of measurements, it is suggested that the interferograms are processed in the phase regime. To do this and to provide the determination of shift directions under formation of recording beams, the first and the second ruby laser radiation pulses should spread in different directions.

This can be reached with a special device of spatial separation of recording beams, which should be activated by the sync pulse from the synchronization system. Besides, it is necessary to provide phase modulation of one of the reconstructing beams due to the harmonic law with a set frequency.

Then the electrical signal, which is registered in the point of the image, can be written as follows

$$I(x,t) = I_0(x,y)\{1 + K(x,y)\cos[\Delta\varphi(x,y) + \delta(t)]\} \tag{3.77}$$

where $t = \varkappa_0\ A_0\cos 2\pi\nu$, $\varkappa_0$ is the wave number; $K(x, y)$ is the contrast of the interference pattern; A0 is the oscillation amplitude; is the quantity of the phase shift between the measurements.

It is essential to single out the desired signal with frequency μ during processing that will help to eliminate noises caused by vibration, change of environmental conditions, etc. The measurements should be conducted at least in three moments of time and within one period of oscillations. For example, it can be written for measurements after equal time intervals

$$\begin{aligned} I_1(x,y) &= I_0(x,y)\{1 + K(x,y)\cos\Delta\varphi(x,y)\}, \\ I_2(x,y) &= I_0(x,y)\{K(x,y)\cos[\Delta\varphi(x,y) + \delta]\}, \\ I_3(x,y) &= I_0(x,y)\{1 + K(x,y)\cos[\Delta\varphi(x,y) - \delta]\}. \end{aligned} \tag{3.78}$$

Having estimated the distribution $\Delta\varphi(x, y)$ along the surface of the object then with the known formulae, we can estimate the fields of shifts, deformations, speeds, etc. Taking into account the shift direction of the interference fringes gives the possibility to determine the direction of shift under phase shift δ as it has been shown earlier

$$\Delta\varphi(x,y) = \operatorname{arctg}\left[\sqrt{3}(I_3 - I_2)/(2I_1 - I_3 - I_2)\right] \tag{3.79}$$

In practice, this algorithm is implemented in the following way. The intensity is read in frames three times in each point of the image and its values $I_1(x, y)$, $I_2(x, y)$, $I_3(x, y)$ are written into the memory. CCD matrix is the most convenient form of a

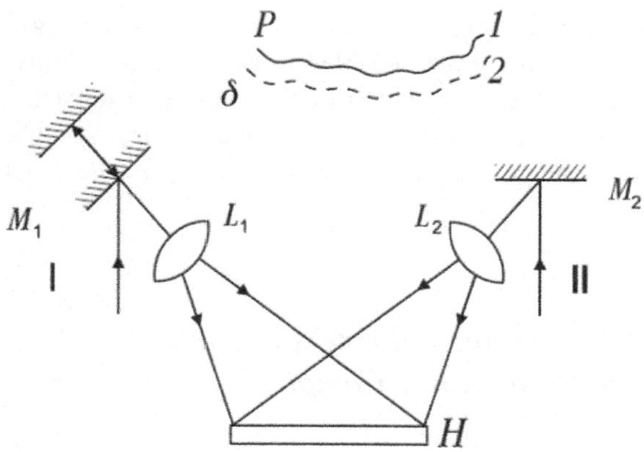

Fig. 3.42 Scheme of eliminating the influence of dispersion during recording and restoration of a holographic image with different wavelengths. Reprinted from [94] with permission

photoreceiver. Then, the computer calculates the phase in each point of the image according to (3.79). It should be mentioned that coordinates of each cell can be accurately estimated in (3.79) with CCD matrix. And as the intensity is normalized, then the spread of the cells according to the sensitivity does not influence the accuracy of measurements. This method gives the possibility to measure shifts to 0.01 μm.

During the implementation of the method some difficulties appeared connected with different wavelengths during recording $\lambda = 694.3$ nm (ruby laser) and reconstructing of the image $\lambda = 632.8$ nm (He–Ne laser).

If the image *P* of the object was in position 1 (a solid line) during recording with the ruby laser, then during radiation reconstructing in He–Ne laser the image reconstructed by the reference beam I was shifted to the position 2, and the image reconstructed by the reference beam II was shifted to the position 1. The influence of dispersion can be eliminated with the scheme (Fig. 3.42). The left reference beam I is to be put in the same position as the right one, then the carrier spatial frequency where the first and the second holograms are recorded will be almost the same, and the images reconstructed with each of the beams will shift horizontally. To remove overwhelming superposition of the images appeared during diffraction of the reference beam II in the hologram recorded by beam I and vice versa, the lens *L* is inserted into the object beam between the object *P* and the hologram *H*. This lens produces an image on the surface of the registering material of the hologram *H*.

Experimental check of elimination of the influence of the dispersions was conducted according to the scheme in Fig. 3.36. Green (514 nm) and blue (457 nm) fringes of the argon laser were used. Positive results were obtained during recording the interferograms on photoemulsion PFG-03. The fringes were preserved and shifted under voltage on the piezoelectric ceramics.

Thus, the conducted work made it possible to optimize the choice of the method of the designed unit through using two reference beams and phase shift for phase measurement at the accuracy, which corresponds to the analogs of the interference holographic systems. It also enabled to show the possibility of reading and processing of images after elimination of the dispersion impact. Continuation of works presupposes the study of the record of focused images on photoplastic materials and improvement of circuitry on principle optical scheme of the developed set-up.

3.3.7 *One of the Variants of Optical Scheme of Laser-Holographic Complex*

On the basis of previous analysis, the optical scheme presented in Fig. 3.43 can be optimal for solving the assigned task. In this figure, *O* is the object under study;

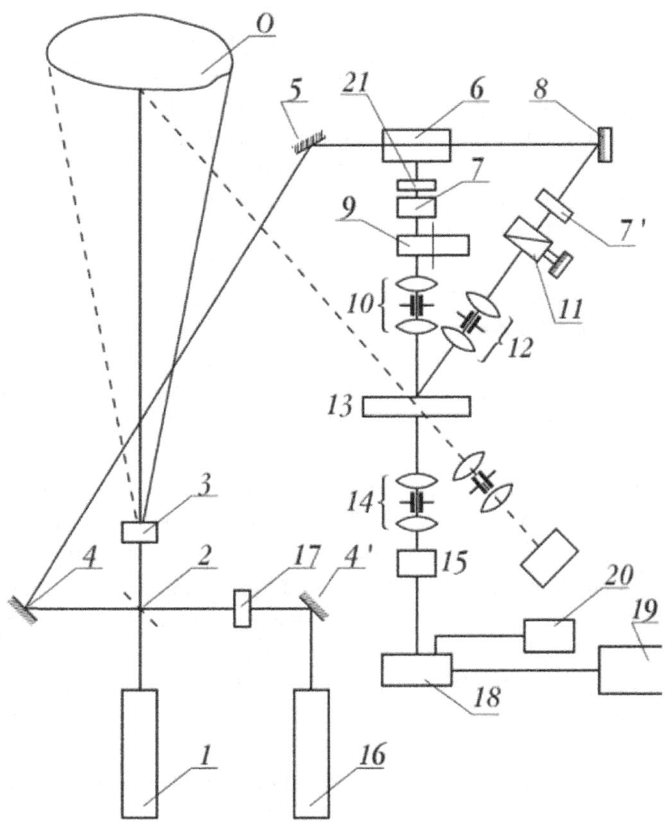

Fig. 3.43 Optical system for holographic interferograms formation and registration. Reprinted from [94] with permission

1 is the ruby laser; 2 is the beam splitter; 3 is the illuminating lens (negative); 4, 5 are the flat mirrors; 6 is the device for reference beams separating; 7, 7′ are the polarizers; 8 is the piezoelectric flat mirror; 9 is the device of frequency shift; 10, 12 are the telescopic systems; 11 is the device for input of phase shift; 13 is the reversible carrier; 14 is the optical scheme for interferogram formation; 15 is the registration system; 16 is the He–Ne laser; 17 is the $\lambda/4$ plate; 18 is the device for electronic processing of signal and input them into computer; 19 is the computer; 20 is the display; 21 is the $\lambda/2$ plate.

Holograms are recorded with a pulse ruby laser. Spatial separation of reference beams is conducted with the device 6. The $\lambda/2$ plate 21 and $\lambda/4$ plate 17 serve for matching reference beam polarizations if KDP crystal with a doubly refracting prism (or with two crossed polarizers) is used as the device 6.

This scheme provides possibility to implement the method of detecting relative quantities, speeds and shift directions. For this purpose, the device of frequency shift is inserted into the reference beams. This device can be in a form of a rotating diffractive grating and a device for inputting the set phase shift 11, in particular, an optical compensator. The piezoceramic oscillating mirror 8 can be used instead of the grating 9. Both a reversible thermoplastic carrier and high-resolution photographic plates can be used as the registering medium. In a case with the plates, the size of the holograms can be rather big—$30 \times 40\ \text{cm}^2$, and that is why observations from three directions can be conducted and the spatial shift vector can be determined. In this case, the system of expressions like (3.62) is composed and solved. For recording and reconstructing such a hologram, it is reasonable to use spherical reference waves, and that is why instead of the telescopic systems 10, 12, it is required to establish microdiaphragm lenses. It is possible to use one lens, through which both reference and reconstructing waves will pass at a small angle. Hologram reconstruction is conducted with the continuous He–Ne laser 16 and the radiation should be spread from it along the same paths as from it 1. The $\lambda/4$ crystal plate 17 is necessary for efficient spatial separation of reference beams with the device 6.

The obtained electrical signals are processed with the device 18, digitalized and inputted using the computer 19. The interference pattern is simultaneously displayed in real time on the screen 20 for visual observation and control. The computer processes the obtained data, determines the shift field and displays it in a way convenient for interpretation. It is worth mentioning that as the object is illuminated by the spherical wave, the illumination direction along the surface of the object changes and that should be taken into account while determining the shifts according to (3.62).

Using a two-pulse laser gives the possibility to estimate relative shift speed of points on the surface under study, which are calculated according to formula $\mu = 0.5\, N\lambda/t$, where t is the quantity of time interval between pulses; N is the number of interference fringes between the analyzed and the initial points. If the phase method of hologram processing is used when there is no counting of interference fringes and shift quantities and the determination is conducted through a change of phase shifts, the relative shift speeds can be calculated according to formula $\mu = L_z/t$ where L_z is the longitudinal shift component.

3.3.8 Optical Scheme of Laser-Holographic Complex (Holographic Cardiograph)

Practical realization of the analyzed variants of optical schemes of the laser-holographic complex (Figs. 3.34 and 3.43) appeared to be impossible, because of the absence of the necessary elementary base: an electric shutter of the Glan prism, half-wave plates, etc. Thereby, the necessity of principle scheme processing appeared taking into account concrete possibilities. The optical scheme of the laser-holographic complex (holographic cardiograph) implemented in our experiments is shown in Fig. 3.44, where 1 is the ruby laser; 2 is the He–Ne laser; 3 is the shutter; 4, 10 are the beam splitters; 5–9 are the mirrors of the compensator of path differences between the reference and the object beams; 11, 16, 17 are the mirrors; 12, 13 are the electromechanical shutters; 14, 15 are the reference beam expanders; 18 is the hologram; 19 is the object beam expander; 20 is the object; 21 is the focus rendering lens; 22 is the video camera.

Process of holographic interferogram recording occurs in the following way. Radiation of the ruby laser 1 gets onto the beam splitter 4 after that the majority of radiation (to 90%) passes through the expander 19, which forms the beam for illuminating the object 20. The radiation scattered by the object 20 passes through the lens 21, which makes the focused image of the object 20 in the hologram 18 plane. The hologram is recorded on a photoplastic carrier or on photoemulsion.

The reflected by the beam splitter 4 smaller part of the ruby laser radiation (10%) goes for formation of the reference beams. The system of 100%—mirrors 5–9 is a

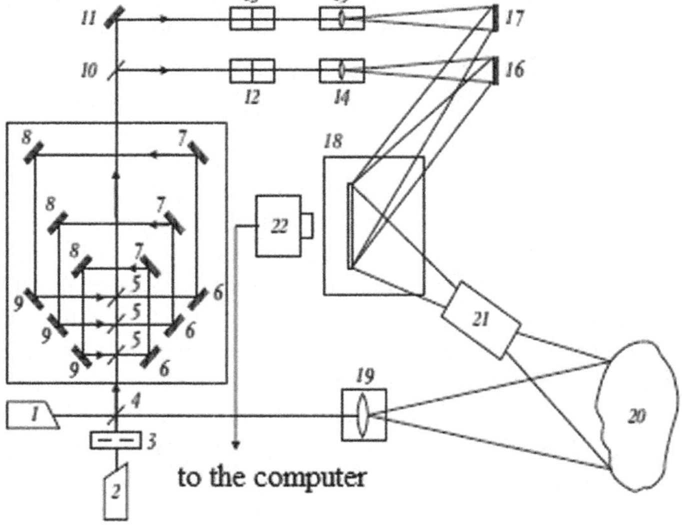

Fig. 3.44 Optical scheme of laser-holographic complex (holographic cardiograph). Reprinted from [94] with permission

line of mirrors for equalizing waves of reference and object beams. The mirror 5 has a two-sided reflecting covering (6, 9). After the beam splitter 4 the laser radiation gets onto the mirror 5, reflects from it and goes along the path 5–10 coinciding in direction with arrows in Fig. 3.44. The beam splitter 10 directs half of the radiation to the expander of the reference beam 14, after which the light is reflected from the mirror 16 and gets to the hologram 18. The radiation passed through the beam splitter is reflected by the mirror 11 mounted on the piezoceramics for phase shift formations, then it passes through the expander 15, is reflected by the mirror 17 and gets onto the hologram 18. The shutters 12, 13 work in such a way that during interferogram recording only one of the channels of the reference beams is opened by the ruby laser radiation. The first pulse goes along the path 10–11–13–15–17–18. The shutter 12 is closed. After the first pulse passed the shutter 13 closes, the shutter 12 opens and the radiation goes along the path 10–12–14–16–18. Thus, the first hologram on a double-exposure interferogram is recorded with the reference beam of the mirror 17, and the second one—with the reference beam from the mirror 16. While reading the interferogram with radiation of the He–Ne laser 2, which goes along the path 2–4–5–6–7–8–9–10 and further, both shutters open and two reference beams from the mirrors reconstruct the interferogram of the object with "living" fringes. Further processing of the interferogram is conducted according to the method described in Sect. 3.3.6.

Intensity is measured in each point of the image, and $I_1(x, y)$ for the moment when the mirror 11 is in position 0, I_2 corresponds to the mirror 11 in position $B(+\delta)$, $I_3(x, y)$—to the mirror in position $A(-\delta)$. Further, the phase distribution is calculated according to (3.79).

The conducted experiments proved functionality of the scheme. In the image, the "living" interference fringes were observed, which displaced during the shift of the mirror 11. The scheme offered in Fig. 3.44 is much easier, cheaper and more reliable in work than the scheme in Fig. 3.42.

Moreover, if there is no need for accurate calculation of shifts and the quantitative estimation of the fringe image character is enough as, for example, in a case of non-destructive testing, then after removing the beam splitter 10, we will get the scheme of a classical interferometer with one reference beam 11, 13, 15, 17, 18. A scheme of compensation node of path difference between the reference and the object beams was developed. It is based on using the mirror optical loop (Fig. 3.38). There was shown the possibility of determining not only the quantities, but also relative shift directions of points on the surface under study while using the scheme of hologram recording and reconstruction with two spatially separated reference beams. Thus, the conducted work gave the possibility to optimize the choice of methods of the developed equipment, i.e., using two reference beams the phase shift for measurements of the phase with the accuracy, which corresponds to analogs of interference holographic systems, and to demonstrate reading and processing of images having eliminated the impact of dispersion. The study of the conditions of recording the focused images on photothermoplastic materials gave the possibility to improve the principle optical scheme of this device.

3.3.9 Recording on Photothermoplastic Carriers

The principle of operation of the laser-holographic complex is presented in Sect. 3.3.2. The registering medium should satisfy hologram recording and meet the following requirements:

1. provide recording of high-quality holograms with a low signal-noise coefficient and high contrast that is necessary for further automatic data processing;
2. this hologram shout come back to the place of the image at the interferometry accuracy;
3. the registering materials should give static repetition at least within the limits of one series.

Photothermoplastic (PTP) carriers are relatively new registering material, which allows recording holograms and interferograms with processing intervals of 2–3 s. And they do not need "wet" processing that is typical for argentiferous materials. We will remind how the PTP-material works. Before the exposure the PTP-material is sensitized giving a high voltage of about 5–15 kV on corona-forming electrode located before the PTP-layer. As a result of ionization, the PTP-layer is discharged. During the exposure, the PTP-layer is discharged under the light depending on the illumination.

In order to develop a latent image, it is necessary to warm up the thermoplastic to the fluid state when the surface deformation of the thermoplastic layer occurs under electrostatic forces, which is determined by the charge remained on it. To fix the image, it is enough to cool the PTP-layer 5–10 °C. If the PTP-layer is warmed up to the temperature, which is higher than the fluidity temperature, the surface relief will become even and the image will be erased. There are PTP-materials on solid transparent substrate (quartz, glass) or on Mylar. The glass PTP endures up to hundred recording-erasing cycles, but with some quality losses. Both film- and matrix PTP are processed in the place of exposure, and that is why the problem of hologram return with interference accuracy is absent. In the experiments matrix, PTP-material on glass support was used.

The possibility of hologram and interferogram registration with the pulse laser was tested on a photothermoplastic carrier as well as on high-resolution photoplates. In the last case, the hologram size can be rather large—$30 \times 40\,\text{cm}^2$. Due to this fact, it is possible to observe from three directions, and to determine the space shift vector, in this case, a system of (3.62) is composed and solved. It is advisable to use spherical reference waves for recording and reconstruction of such a hologram. To fulfill this, microdiaphragm lenses should be used instead of the telescopic systems 12, 13. It is possible to use one objective, through which both reference and reconstructing beams will pass under a small angle.

The holograms were recorded in convergent beams in carrier frequency of about 1000 lines/mm that corresponded to the angle between the reference and the object beams of about 40°. The monopulse laser energy was 0.5 J. The pulse duration was 40 ns, and the pulse interval was 400 μs. During the study, the holographic research method of human chest was developed, the structure of the laser-holographic complex as well as its principle of operation for estimating the state of human circulatory system were realized. There was developed a functional block scheme of the complex, including both main units of the complex and also functional interactions between them. The obtained results became the base for the formulated medical-technical requirements to the laser-holographic complex. A principle optical scheme was designed for implementation of these methods and includes the systems the of reference and reconstructing beams, which illuminate the object under study, as well as a phase modulation system and a system of interferogram formation.

It was shown the possibility of determining not only quantities but also relative shift directions of the points on the surface under study using the scheme of recording and reconstruction of holograms with two spatially separated reference beams.

As a result of the study on the experimental sample, it was established that the PTP-medium is in principle more sensitive and convenient in a real experiment. There is no need in wet processing, in gear for returning the hologram to the exposure place. The cycle "recording-reconstruction" lasts for a couple of seconds. The PTP-matrix by "Newport Report Corporation" (USA) would completely solve the problem. To record a hologram on photoemulsion, the pulse laser energy should be increased at least 10 times, and for this purpose, ruby amplifier stage is needed. Works on hologram recording on a thermoplastic carrier were conducted in collaboration with the Institute of Applied Physics of the Academy of Sciences of the BSSR.

A glass PTP-matrix was used in the experiments. Interferograms of human chest were obtained.

He–Ne laser LGN-215 with the wavelength of 632.8 nm was used to reconstruct the recorded interferogram. To do this, the radiation was inserted in the path of the ruby laser with the relocatable mirror 23.

The results of the experiments on interferogram recording of a human body are shown in photos (Fig. 3.45). In the figure, holographic interferograms of human chest are shown. Time interval between the pulses is 300 and 400 μs. As it is seen from the holographic interferograms, the radiation parameters of the used laser and parameters of hologram recording system provide survey of the human chest. In Fig. 3.45, the holographic interferogram of human body from the neck to the waist is presented. In figures, oscillation sites are clearly seen in the collarbone area (1), in the larynx area (2), and in the diaphragm area (3). Figure 3.45 illustrates a part of the left shoulder and forearm is seen.

The interferograms give the possibility to determine two areas of concentration of fringes, which correspond to maximal shifts of the object surface. One concentration area is located near the larynx (upwardly), and the other is in the area of the diaphragm (below). It is easily seen that the fringes are distributed on the body more evenly, though the fringes are located more densely in the areas, which correspond to the maximum of shifts on the first interferogram. Thus, the surface shift is greater in these places than in other parts of the object under study, but their value is considerably less in the second case than in the first one. The obtained time-shifted holographic interferograms illustrate differences in distribution of pulsation centers of activity on the surface of a human body. Analyses of results of the medical study of circulatory

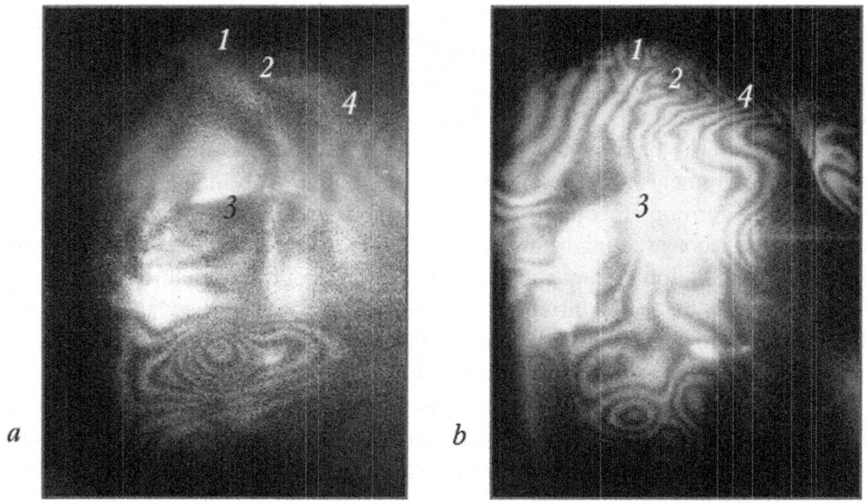

Fig. 3.45 Holographic interferograms of human chest, derived in dual generation regime of a ruby laser: **a**—the interval between pulses is 300 μs; **b**—the interval between pulses is 400 μs. Reprinted from [94] with permission

system have shown that the change of holographic interferogram of human chest surface under heart contraction obtained with LHC is characterized by the intensity level of interference fringe crowding.

In the centers of chest activity (zones of the right and left ventricles, aortic and pulmonary outflow tracts, etc.) at the increase of the tension of the vessel walls in the phase of ventricular systole and further period of emptying the greatest deformation of the tissues is registered that finds its reflection in the interferogram in the form of intensifying interference fringe concentration. This coincides with anacrotic inclination of a rheographic curve registered with "IMPECARD-3" complex. During the diastole phase, the holographic pattern changes (see Fig. 3.45).

It is obvious that the registered holographic interferogram (concentration of interference fringes, their width, contrast, occurring of additional centers of activity, modification of the previous ones, etc.) will considerably differ from the interferogram of a healthy person if there is pathology of the cardiovascular system. It causes abnormalities of pulse oscillations of blood filling that depends on stroke volume, speed of the blood flow, which is mostly defined by the size of the vessel lumen, i.e., by compliance and tonic contraction of a vessel wall. To establish a stricter correspondence between the state of a patient and an interferogram, it is necessary to conduct numerical simulation of the interferogram according to the methods described above, to get statistical data and their analysis.

This complex can be of wide application in the sphere of dentistry and traumatology for detection of different changes in bone tissues in statics and dynamics, as well as in biotechnology while conducting different transplants and artificial organs.

3.4 Holographic Research Methods in Biology and Medicine

Visual information plays a great role in medical diagnosis. But a human eye cannot penetrate different hard-to-reach cavities of a human organism. That is why it is extremely important to obtain information hidden from an eye and in the fullest form.

With the holographic method, a set of tasks can be determined, the solution of which is difficult or principally not possible using traditional methods. Possibility to obtain a 3D-image, recording and storage of great amount of information, and convenience of its processing predetermine usage of these methods in medicine and biology [203, 204]. The peculiarity of the holographic method is that not an image is recorded in the light-sensitive high-resolution emulsion layer but the wave field of the object, or rather phase wave correlations reflected from the object and forming the interference pattern. If the hologram was formed after photochemical processing and during reconstruction the light incidents on its diffractive structure, then it is reflected at such an angle that the beams form a three-dimensional image. These beams concentrate in space or on the area, which was previously occupied by the object itself. If we continue to record the object on the same holographic plate, we can obtain a form of kinematic reconstruction of the object in motion.

One of the most important tasks in medicine is an early diagnostics of diseases. To form clinical diagnosis information about the exact location of the pathological process, its degree of manifestation, histological structure of the invaded tissues and other factors are necessary, which cannot be obtained with an ordinary medical examination and the widespread methods of investigation (radiography, roentgenoscopy, endoscopy, ophthalmoscopy).

Comparing to the widespread methods, the main advantages of the optical holography methods in medicine are qualitatively new possibility of registration and reconstruction of three-dimensional images, increase of informativity of the obtained images, considerable improvement of resolution capacity and possibility of accurate measurement of the spatial location of the structures under study [205].

In the sphere of medical diagnosis, direct observation plays a great role. That is why the holographic reconstruction method gives the possibility to realize large-scale and integral survey of the object under study. Two main applications are of great interest for medicine and biology, namely observation of internals and externals and optical processing of information. The objects of the study can be divided into three main groups. The first group contains open objects, i.e., external organs, which are accessible for laser illumination (e.g., face, chest, arteries, extremities, etc.) and separate organs after postmortem examinations. The second group contains internals, i.e., cavities accessible for optical light-transporting elements (e.g., the oral cavity, bronchi, the esophagus, the rectum, the uterine cavity). The third group contains cavities, which are not accessible for inserting light-transporting elements, which are an eye and its contents—the cornea, the crystalline lens, the eye-ground, etc. The fourth group includes moving objects, i.e., objects, which have the aim of holographic registration to investigate measurements of the object states in a certain period of time. From the physical point of view, all the biological objects are divided further into three categories: fully reflective, including diffuse, fully transparent and semitransparent (which can partially absorb, transmit and reflect light).

Observation of medical and biological objects has two main goals: Firstly, it is recording of 3D-images including ones with the help of holograms; secondly, it is registration and storage of great amounts of information. About ten thousands of biological states can be written on a holographic plate of scaled-down sizes. For instance, using such a method archives of medical pathologies can be made. To control each of these states, which can be in particular cardiographic records of states of heart and other organs, it is enough to change the inclination angle of the laser beam in such a way that different points of the photoemulsion are illuminated. Thus, great amounts of micro- and macro-preparations can be classified. At the same time, this method is implemented not only for preventive analysis, but also for control during therapy. In particular on the photofilm, there were recorded electrocardiograms of consequent stages of different myocardium situations. All this was checked and classified by an expert committee (of specialists), and it gave the possibility of comparison with every new electrocardiogram. In fact, automatic processing of an image, which was analogous to a new one, was conducted that allowed finding out the probable pathology automatically. Therefore, holography can be a very accurate means for preventive examination of infraction processes.

Holography can also be used in the sphere of dentistry while studying materials for dentures in order to have an idea, for example, about the impact of food temperature on deformation of materials used for filling and protection of teeth or for creating a model that gives the possibility to investigate the distribution of tension inside the oral cavity (see Chap. 3, Sect. 3.2.1 [105–107, 109–120]).

Principally, the holography can be implemented for waves of any band. A hologram can be recorded in X-ray coherent radiation as well as in gamma rays. But today, it is possible only in theory as holograms need coherent wave sources. Despite all the world studies, X-ray laser was not created yet.

Another important sphere, which uses the holographic method, is cryobiology. This is a science, which investigates perfect conditions for preservation of organs for transplantation. Of course, it is connected with the problem of changes, which can influence tissues during cooling and thermal reconstruction. Nowadays, cells of blood, brain, cells for reproduction and cells of skin of internals are studied. As they consist of a great amount of water, it is extremely important that the process must not create deformations and damages remediless for the tissue. This study helped to determine the perfect cooling speed -30° per minute. Water crystallization does not cause damages in cells at such a speed of temperature lowering. Besides, the deep study of the object gives the possibility, for example, to follow such phenomena as cell separation or phagocytosis.

One of the main advantages of this technique is the possibility to record the needed data very accurately without direct contact with the object under study.

Actually, there are myographic or electrical ways of analysis, which require direct contact. Even simple pressure measuring or an electrocardiogram should be made through a contact. This is often connected with psychophysical state of the object under study, in the result of which in some cases the data can be incorrect and unreliable. And vice versa while using the coherent-optical methods, a person can even not know when the recording is conducted.

One of real applications of the method mentioned above is obtaining holographic portraits of a living object that is valuable while investigating rare as well as typical forms of diseases [206]. Taking into account dynamics of the aspect of disease and changing or disappearing of the form (e.g., death of a patient), it is important to preserve the shape of the object under study. This method can also be implemented for taking pictures of preparations, models, phantoms, etc. Thus, holographic portrait can be a means of medical diagnosis and serve as demonstration material at lectures or in museums of pathologic anatomy for students training and so on.

A step forward in terms of informativity improvement can be circular holographic registration. The works [207–210] are dedicated to the circular holographic recording. In the work [209], the circular holographic plate can be used instead of plates, which limit the number of angles during reconstruction. An optical scheme and a device for circular recording in colliding beams are presented in the articles [211, 212].

Unlike the flat carrier registration when the image is fixed from one angle, during circular holographic recording the object is fully covered, i.e., under an angle of 360° [207, 208, 210, 213]. The application of the circular holography method in medicine can serve for a more qualitative estimation of the object state under traumatic operations, malignant neoplasms, transplanting of living tissues or during cosmetic surgeries.

Holographic survey of different human internal cavities, which are easy to reach, can be technically conducted with further posteriori processing of 3-D data. This estimation of the pathological state is important for oncology tasks and allows solving different diagnosing tasks as well as choosing the method of treatment. High resolution and accuracy of measurements are most essential in this case, as it is stipulated [214, 215] that a tumor goes through 40 periods until it reaches the critical state and the diameter of ~40 μm corresponds to the initial period. The contemporary endoscopic study gives the possibility to register the minimal tumor diameter of 150 μm that corresponds to approximately 12 periods of its existence (or 30% of all the time of development). Holographic registration can give information about the color and relief of the mucous membrane, about the shape and the size of lesions. Such a test in combination with biopsy will give the possibility to get information about histologic structure of the damaged tissues, to estimate the necessity and the volume of surgery. The repeated study makes it possible to observe the dynamics of the tumor development in some cases, which has an essential meaning during estimation of the results of treatment, to control healing of ulcers and to detect their malignization on time. Using complex research, different pretumor changes can be detected in some organs, the possibilities for determination of more accurate changes of mucous membrane can be created, which precede the development of tumors [214, 215].

In the works of von Bally [216–218] and Gregush [219], there are a lot of examples of how holographic interferometry can be applied in medicine. These are the study of artificial cardiac valve deformation during its creation, the investigation of the middle ear membrane during transmission of sounds using holography with time averaging and successful implementation of double-pulse holographic interferometers in dentistry, possibilities of the "dark reconstruction" method for the study of tremor, breathing, etc.

In particular, the work [218] concerns the development of a non-destructive noncontact highly sensitive method using holographic interferometry. The valve is placed in the test cell with holes for visual observation. Cusp deformations under pressure are recorded with the holographic interferometry. The finite interferogram gives the possibility to establish some defects or structural abnormalities, which can also present in the material, of which the valve is made. Meanwhile, in the work [218], three types of changes are described, which can be determined with screening-tests.

The suggested method is expected to become an effective means of detecting hidden defects of heart valves for replacement. It has already been considered as a promising means of quality control, especially while producing artificial heart valves that will help to detect first signs of calcinosis. Thus, the application of holographic test without destroying the sample can significantly improve the quality and lifetime of artificial heart valves.

In the works [216, 219], oscillations of the ear tympanic membrane were investigated with holographic interferometry that gave the possibility to determine pathologies in this area. It is difficult to study, for example, inner part of the ear cavity using this method. But, it is possible to determine behavior of the membrane in its depth where the pathology can be localized. Ultrasonic holography can be used to study inner organs, parts which are situated behind the nontransparent membrane.

During the experiments [220], a new method of estimation of holographic interferograms with circular shift was developed and tested. The dispersed phase distribution index was calculated, and brightness modulation of interference fringes was determined. Besides the framing, mask was produced during analyses of histograms and determination of the local minimum of brightness modulation distribution a binary. The mask is used for valid and non-valid dispersed phase distribution areas during further phase unwrapping and interpolation as well as its path. Estimation of the obtained holographic interferograms with circular shift shows that this method can be successfully applied for processing of a complex interference pattern even if there are heterogeneous interference fringes and dark areas.

The article [221] shows that to measure deformations in medical and technical spheres with highly sensitive methods of holographic interferometry the methods of research of real-time dynamic processes, for example, during analyses of the environmental impact on biological and other objects, are of special interest. Unfortunately, most traditional materials used for holographic interferometry measurements have a number of disadvantages: long processing time (halogenide-silver materials) or bad resolution of lines. An important problem emerges in the case of implementation of real-time classical inteferometry: After processing of the object, its position and surface should not be changed in the boundaries of the interferometry series. This requirement is rather difficult to meet especially with biological objects or in real production. A possible solution to this problem is to create a recording of interferogram sequence. Photorefractive crystals are especially useful for these purposes as except for good line resolution they do not require much time for processing and are optically erasable. The main purpose is to develop a double-exposure holographic interferometer with high repetition frequency based on photorefractive crystals; that is why the argon-ion laser has been widely used. The obtained samples of the interferograms should be computer-estimated to provide the best effectiveness.

The work [222] points out that miniaturization of the devices becomes an urgent problem in production of contemporary optical equipment, especially in the sphere of endoscopy. Implementation of this technique together with holography provides a number of advantages, for example, 3D-documentation, and a display with high zooming capacity, as well as metrology in the form of holographic interferometry of objects, which are difficult to be optically reached. Basically, implementation of holographic metrology in the sphere of endoscopy is a step forward to creation of the metrological base for quantitative diagnostics in body cavities that opens new possibilities in medical diagnosis and environmental research. Moreover, miniature holographic endoscopic equipment requires a miniature illumination system and an image forming system as well as a small recording device. Usage of photorefractive crystals provides the possibility of development of such a holographic memory unit with high repetition frequency and line resolution.

Recently, the concept of virtual reality has appeared, and the study was conducted in the sphere of "vision" and "motion" of two most important human functions, which underlie such technologies [223]. This study includes: (1) perception model, which explains how a person mentally reproduces three-dimensional forms from a two-dimensional image projected on the retinal; (2) research of tight interconnection between different types of perception, for example, between the auditory and visual perception, visual information and stimuli of muscle motion. An illustrative example of practical application of this technology is measurement of eye motion in order to detect the Alzheimer's disease on the first stages. Also, main problems are discussed, which occur during the creation of flat 3D-displays.

The work [224] shows the possibility of using holographic methods to determine biomechanical characteristics of vessels and their prostheses for the purpose of comparative evaluation.

Medical and biological practice is often in need of motion pictures of fast processes. It is connected with great informativity and possibility to observe the development of different processes in dynamics.

Registration of objects in motion (holographic movie) plays a special role among holographic methods. The works [225–229] are dedicated to this issue. One of the variants of the creation of a 3D movie is described in the article [227]. Motion is realized in the following way: Consequent positions of the object are recorded on one hologram at different angles. During reconstruction, the hologram is illuminated by the same sources and under the same angles. Seven images were recorded on one hologram using the overlap method. During reconstruction, the plate was gradually turning and repeating the angles of the reference beam, and it created an impression of motion. The work [228] describes the obtaining of consequent holographic frames of living objects. In particular, holograms of biological objects—swimming fish were obtained with a pulse ruby laser. The usage of the cineholography method will make it possible to accurately estimate 3D images of the object [225, 226, 230, 231].

With the help of 3D film images, it is possible to document and conduct a comparative qualitative analysis at different stages of such diseases as Parkinson's disease, Basedow's disease, paralyses of facials parts, etc.

Ophthalmology takes a special place in the sphere of application of optical holography methods in medical diagnostics. The eye is an organ, for holographic registration of which it is enough to illuminate it from the outside, because the eye structures are transparent for radiation of the optical range [232, 233]. Registration and study of states of the eye bulb and the eye ground is very important for diagnostics of pathological changes of the visual analyzer as well as for estimation of systemic diseases: tumor, essential hypertension, atherosclerosis, blood and kidney diseases, etc. [234, 235]. For example, information recorded in one hologram of eye-bulb contents can eliminate the necessity of obtaining several hundreds of photographs taken at different angles. High resolution of a hologram gives the possibility to observe retina vessels with a diameter of ~10 μm.. And the theoretical resolution of Zeiss's cameras used for studying of the eye-ground is 19 μm.

The first holographic portrait of a living object was obtained in 1967 [236]. Then, this experiment was also repeated in other laboratories [237–239] where mainly powerful multistage amplification lasers were used.

The works [240–244] are dedicated to the application of the optical holography method to tasks of endoscopic diagnostics.

Maximum permissible levels of radiation impose certain restrictions during holographic registration of the external and internal structures of a living object. Due to this reason, during experimental survey special attention should be paid to tolerances of the laser light impact on the living tissue [245–247].

Specific character of the living tissue sets a certain requirement to illuminating units, which is predetermined by the maximum permissible level (MPL) of radiation [236–239, 248, 249]. It consists of the following: The light directed to the registered area should be diffused in a number of cases.

It is known that a lot of pathological formations differ from the surrounding tissues by their biological structure, and therefore by their spectral characteristics [250]. During experimental study, it is important to conduct a concerted choice of the wavelength that will provide accurate separation of the structure of the pathological formations from the surrounding tissue and obtaining maximal contrast of the interference pattern.

Before using laser radiation for diagnostics, it is necessary to ascertain tolerances of the laser radiation impact on the organ of sight, skin and the mucous membrane [245–248]. The works [251–253] are dedicated to the impact of laser radiation on the skin. Usually, value of energetic illuminance of the object is taken as quantitative measures of radiation impact on the medium. The energetic illuminance is expressed in joules per square centimeter or in watts per square centimeter. MPL is established for direct specular reflected laser radiation as well as for diffusely reflected laser light. MPL for diffusely reflected radiation is determined through maximum allowed radiances (Watt $\cdots$ cm^{-2} sterad^{-1}) of diffusely reflecting objects, onto which the direct laser radiation gets. The works [254, 255] contain the most detailed information about MPL, which was obtained in domestic laboratories.

There are two ways of image registration to solve the problem of intracavitary holographing [243, 244, 256–258]: at the proximal end and the distal one. Based on theoretical and experimental results, which were conducted on the samples, the work [258] shows the advantage of holographic endoscopy in comparison to usual endoscopy, which consists in the possibility of resolution increase.

Hologram copying is a very important issue for medical–biological objects [259–261]. One should decipher direct copying [261], which is a process close to contact printing, from getting a secondary hologram when a wave reconstructed by the original hologram interferes with an independently directed reference wave. The obtained copy is a negative of the initial hologram. But the received image is positive and identical to the image reconstructed by the original [262–264].

3.4.1 Holographic Recording of Anatomical Preparations of Vertebrae with Manifestation of Lumbar Osteochondrosis and Corrosion Preparations of Blood Vessels of Human Liver

There was shown the possibility of registration of large reflecting holograms of 280 $\times$ 406, 400 $\times$ 600 mm^2 of some anatomical objects [41, 54, 265]. For these purposes, there was produced an experimental set-up with the protection from vibrations for holographing unique medical objects in colliding beams. The necessary accessory for fixing of different objects was created, which provided their placement both in horizontal and in vertical planes. 30 mWatt single-mode He–Ne laser LG-38 was used for hologram recording and photoplates PFG-02 and PFG-03 were used as the registering medium.

Anatomical objects for hologram recording were human lumbar and jugular vertebrae with such pathologies as osteochondrous calcifications and corrosion preparations of liver blood vessels. The objects were made through common methods of boring, fixation, cutting in different projections, digestion, preservation, injection of the vascular system with different self-harding compositions with dyes. To eliminate defects connected with hologram drying, they were placed into baths with 50-, 70-, 80% alcohol solution. Optimal modes were determined, which should be used for recording and processing holograms and holographic interferograms of human vertebrae. 3D holographic images give the possibility to examine human vertebrae in the range of 180° (Fig. 3.46).

Besides, circular holographic recording of human liver corrosion preparation blood vessels was conducted (Fig. 3.47). As we performed holographic experiments on obtaining highly informative holographic images of living tissue, there appeared a necessity to form a rather huge amount of beams for illuminating the

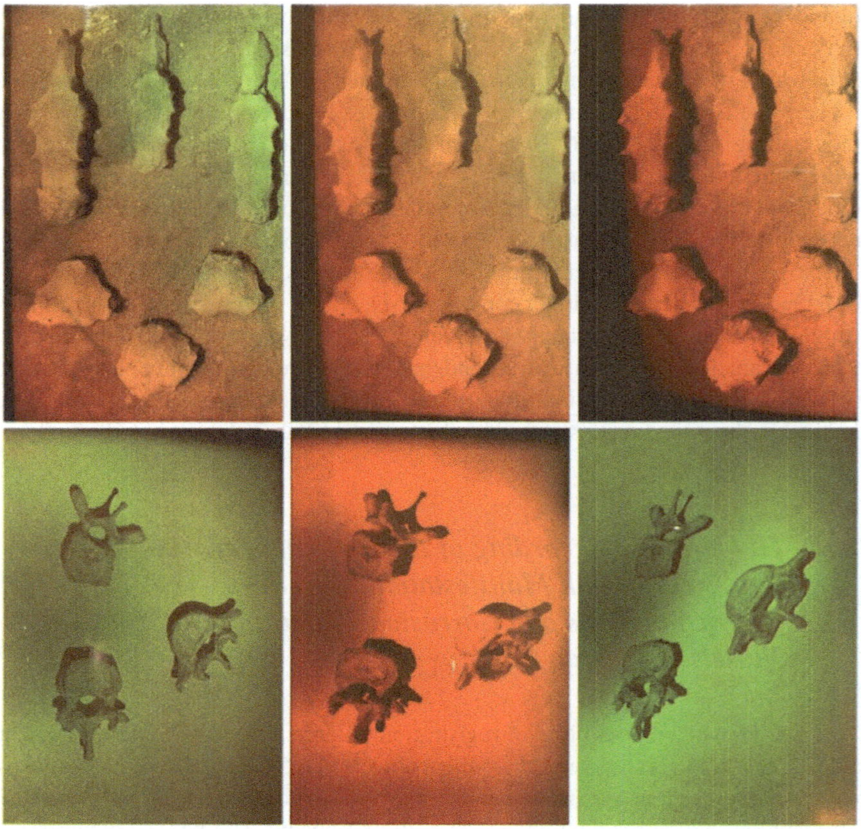

Fig. 3.46 Holographic images of anatomical devices of the human vertebrae with the manifestation of lumbar osteochondrosis (hologram recording was performed by L.V. Tanin in 1982). Reprinted from [94] with permission

holographed object from different sides. The same situation appeared while holographing of prolonged scenes, and the main difficulty is that the existing lasers have a limited length of coherent radiation and that leads to total disappearing of interference pattern on the light-sensitive layer. But it is known [16] if the optical path difference is smaller than the coherence length of laser radiation in relation to the reference beam, this will lead to a stable, contrast interference pattern. Besides, the image was fully seen on one photoplate through recording the image of a rotating object using the slot method [266–268]. While observing such an image it is enough to move the plate relative to the observer in the horizontal (vertical) plane in order to create an illusion of object rotation. Using this method, holograms of human skull were recorded with a small angle of view of the reconstructed image of ~40° (Fig. 3.48).

As the characteristics of biological objects reflection, for example, of skin, as well as any other reflecting objects depend to a great extent on the radiation wavelength, it

Fig. 3.47 Holographic images of corrosive preparations of human liver blood vessels restored from the hologram from different perspectives (the recording of the hologram was carried out by L.V. Tanin in 1982). Reprinted from [94] with permission

is supposed that using monochromatic laser radiation will give the possibility to detect more accurately pathological formations, areas of inflammation, etc. [269–272].

Another difficulty was connected with providing immobility and necessary contrast of the interference pattern during registration of living movable tissue, and it was eliminated through reducing the exposure time [273–275]. Thus, for example, there are methods of holographic recording of a living object, i.e., fast processes, which are based on deliberate frequency change in the object and the reference wave [273]. These methods give the possibility to register the interference pattern of movable objects without reducing the duration of exposure. Another possibility of hologram recording is the local reference wave method [274, 275], which consists in that the reference wave is formed as a part of the object one. This leads to an automatic phase modulation of the reference wave. In the case when the holographic object is in motion as a whole, all or any part of the focused reference wave can

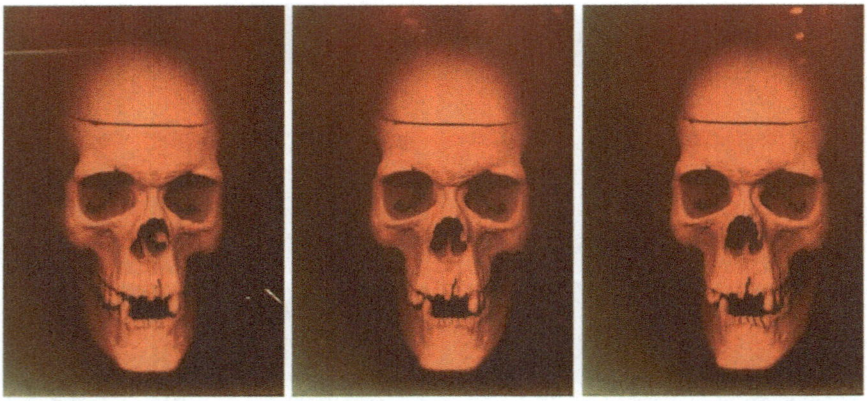

Fig. 3.48 Volumetric image of human skull restored by the hologram (the recording of the hologram was carried out by L.V. Tanin in 1982). Reprinted from [94] with permission

be used for hologram recording. Great importance is given to even illumination of the holographic scene during holographic registration of the object. In our case, the use of diffused illumination through placing between the source and the object of the diffuser, for example, opal glass, leads to a significant widening of spatial frequency spectrum of the hologram. The diffuser formed a noise-type signal with a spectrum of spatial frequencies, which is often much wider than the spectrum of the object. As a result, each spatial frequency component of noise accumulated side bands of desired signal frequencies, which bear information about the object [276]. The diffuser gave the possibility to equalize the exposure along the whole hologram surface that simplified the correct choice of the operating point on the characteristic curve of the photomaterial. It is worth mentioning that the quality of the reconstructed image depended to a certain extent on the properties of the diffuser.

The task of the study of living objects does not give the opportunity to use the continuous laser because of the great mobility of the objects under research. The study of shifts and deformations of human skin requires implementation of a pulse laser that would give an opportunity to record an interferogram of an object in short periods of time and to escape firm fixing of the object under research. For this purpose, we used a pulse ruby laser, which generates in double monopulse mode. It emitted two monopulses with the order duration of 100 ns with the sequence interval of up to 100 μs. AQ passive clearing shutter was used in the laser. The power of the pulses reached 1 J. Interferograms of tense muscles of a human hand were recorded. Modes of recording and processing of holograms were optimized [277] (Fig. 3.30a, b).

A laser with an active shutter was used for a more detailed study of deformations and shifts of living objects (see Sect. 3.2.2). It gave the possibility to accurately synchronize the laser generation pulses with electrical stimulation and registration of electrophysiological parameters of the object [128–131, 277].

The given holograms evoked great interest at exhibitions in our country and abroad (Austria, Germany, Great Britain, Japan, Finland, Australia, India, Czechoslovakia, etc.) in 1985–1995.

3.4.2 *Methods of Formation of Combined Images*

Methods of formation of images were created, which are in obtaining of combined complementary three-dimensional and plane images, which can be used for visualization of structural changes of medical–biological objects as well. The possibility was studied for formation of coherent and non-coherent images. The ways of image forming enhance the effect of perception of optical information. Enhancement of the perception effect is reached at the expense of obtaining complementary combined plane and three-dimensional images. This method gives the possibility to form and reconstruct combined images in color, motion and bulk. It provides continuous visible information with simultaneous volume regulation. The continuous information is provided by the plane image of the object and the holographic image contains the additional information about the object.

The holographic image is formed through hologram recording in colliding beams (Fig. 3.49). The laser beam 1 incidents on the object 3 going through the photoplate 2 and scatters backwards by the object. In the emulsion layer of the photoplate 2, there occurs interference of direct and scattered radiation, and, thus, a hologram is formed.

For example, let us examine the recording of the holographic image of a cone split by plane A into two parts 4 and 5 (Fig. 3.50). On the photoplate 2_1, a reverse image of the upper part of the cone is recorded (Fig. 3.51) and an orthoscopic cone image in front of the splitting plane is created. The second hologram on the photoplate 2_2 is the rest part of the cone 5 and is recorded in the usual way (Fig. 3.52). Then, the first hologram is combined with the second one in such a way that the image 4_1 coincides

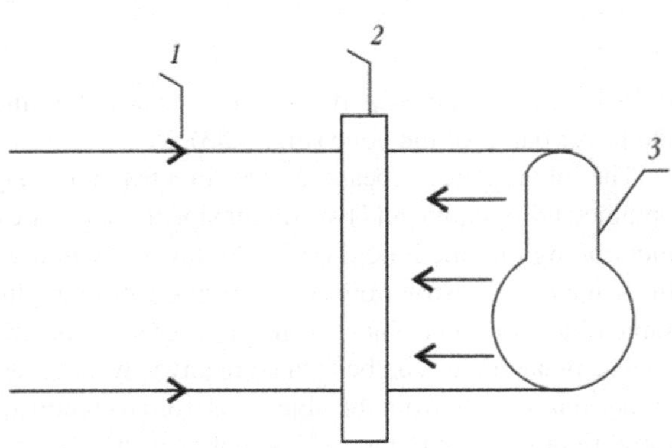

Fig. 3.49 Optical system of holographic interferograms formation and registration. Reprinted from [94] with permission

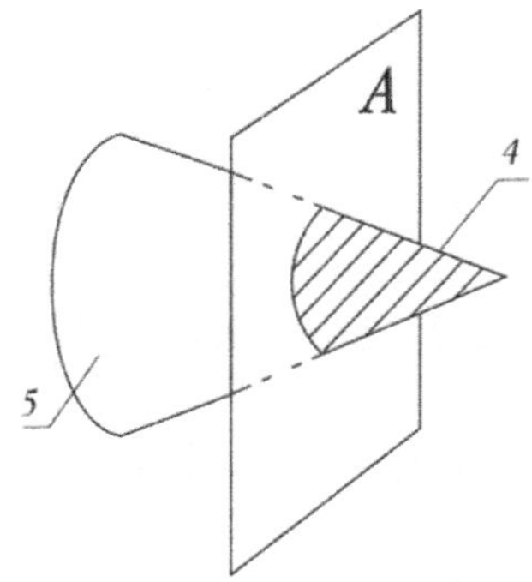

Fig. 3.50 Object conventionally divided by plane. Reprinted from [94] with permission

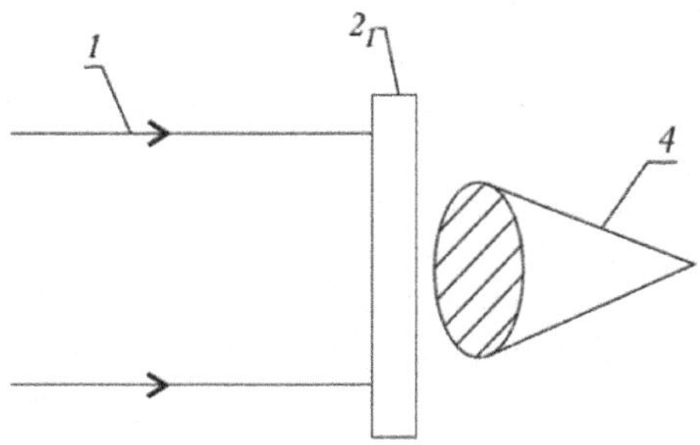

Fig. 3.51 Recording scheme of a holographic image of a part of an object in front of a separating plane. Reprinted from [94] with permission

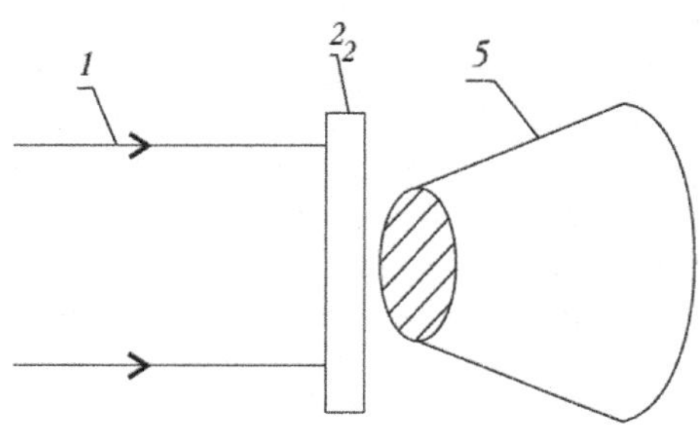

Fig. 3.52 Recording scheme of a holographic image of a part of an object behind a separating plane. Reprinted from [94] with permission

in its base of the upper part 4 of the cone with the image 5_1 of the upper surface of the lower part 5 of the cone (Fig. 3.53).

The photoplate 6 bears a two-dimensional image of the same object. The obtained holographic and two-dimensional images are combined. Besides, it is worth mentioning that the hologram is illuminated with a variable intensity light beam at the stage of reconstruction. Apart from the holographic image, a plane image of the same object or other objects (one or several) is formed on its own substrate.

The plane image can be formed in any way, for example, in a photographic one. It comes out directly from the objects or from its reconstructed holographic image with further copying using the methods of projection or contact printing. Or it is formed with other methods on different data carriers (substrates). If photoemulsions and

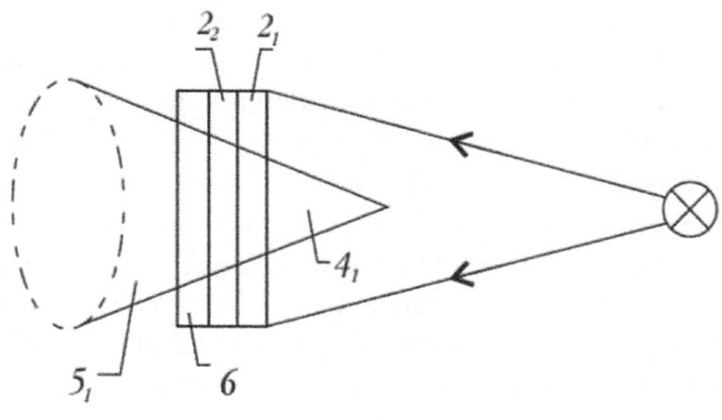

Fig. 3.53 Scheme of a combined image reproduction. Reprinted from [94] with permission

photoresists (photoresistive material) are applied for image reflection improvement the substrate consists of two layers, and the light-sensitive layer is covered with a reflective layer of silver, aluminum, titanium, chromium or their compounds with further photochemical processing after the exposure.

In this case, the plane image is reproduced at the expense of reflection of the incident radiation of the substrate. Depending on distinctive features of the object under recording (size, color of most marked details of the image) layers of argentiferous emulsions, nonsilver emulsions, negative and positive photoresists are used as a light-sensitive layer. Any of widely known emulsions is used as an argentiferous emulsion, and chrome-plated gelatin or shellac can be used as nonsilver emulsions. Besides, the plane image is formed on a substrate, and its material can reproduce a plane image at the expense of intrinsic emission, electrochemical, electroluminescent, semiconductor indicators, on which information about an image is programed and altered under electrical and thermal impact, and when the substrate for the plane image is an indicator of transient data.

The obtained effect gives the possibility to observe transfer of one image into the other one, i.e., transfer of the two-dimension image into the three-dimensional (holographic) one and vice versa and that makes it possible to implement the combination of 3D and plane images, which are complementary to each other. Any effect of visual perception can be obtained with the help of this transfer. In this case, the hologram is the main component of the image formation method that helps obtaining combined images. Meanwhile, the hologram is constantly present and forms a multilayer structure with a substrate with a plane image. Arbitrary orientation of substrates with a plane and a 3D image is possible. Observation of a combination (in space and time) of 3D and plane images is a complex interconversion, constant image formation and transfer of the image into the plane one and vice versa and, thus, realizing continuous obtaining of visual data. Simultaneous regulation of the volume of the incoming data is mainly determined by the intensity value, the level of spatial coherence, spectral and polarization characteristics of the light beam (Fig. 3.53). This is reached in particular by the fact that in the image forming method, which consists in recording of holograms in colliding beams and in reconstructing of holograms by a variable intensity beam and orientation of a two-dimension projection of a reconstructed image with a focused image of the same object several holograms of the objects are formed on one or several light-sensitive layers recorded on different wavelengths, and for image reconstruction a light beam with a variable spectral composition is used.

A number of sources of different lengths can be used (or a tunable laser, for example, Raduga-3 M, Raduga-6) [26, 39, 61, Chap. 1]. And while obtaining a combined image, hologram of the same object (and possibly of different objects during creation of combined composition scenes) is recorded several times on separate emulsion layers and the wavelength of the laser source radiation is tuned each time. For example, a three-color object is recorded on the first hologram at a wavelength $\lambda_1 \sim 640$ nm (red color), on the second one—the same object at a wavelength $\lambda_2 \sim 550$ nm (green color) and at a wavelength $\lambda_3 \sim 450$ nm (blue color) (Figs. 3.49, 3.51 and 3.52). Sources with tunable spectral characteristics are used to control reconstruction process (e.g., with changeable light filters), as a result the recorded images are reconstructed in turn in different wavelengths due to high spectral selectivity of a 3D hologram. If the spectral radiation composition is changed very fast, the effect of colorful image perception is reached, which is analogous to one during reconstruction of the three initial holograms (each was recorded at different wavelengths) with a source with a broad spectral composition (Fig. 3.53). The similar operation of image control can be conducted through illumination of a colorful hologram, which was recorded on one emulsion in three wavelengths in the method mentioned above. But qualitative recording of colorful images on one hologram is inefficient due to limitations in spectral sensitivity and low diffraction efficiency of the registering media.

If a plane image of an object is presented in a form of a hologram, then it will be added with three-dimensional components. Meanwhile, the substrates are mutually located in such a way that during reconstruction contours of the holographic and the plane images coincide with each other. Light with variable spectral composition is used during reconstruction (see Fig. 3.53). The suggested sequence of actions gives the possibility to change the intensity of any spectral component of the reconstructing radiation and as a result to correct the color distribution in the image.

Another possibility is when holograms of at least one object are formed in its different positions on different light-sensitive polarization layers, and to reconstruct the formed image a light beam with variable directions of polarization vector is used. In this case, enhancement of visual efficiency of this method is reached through registration of several holograms of the object on one substrate with a set orientation of the object. The holograms were recorded under change of polarization vector direction in compliance with the object orientation. And for reconstruction of the image, both substrates (with holograms and plane images) are illuminated by a light beam with changeable direction of polarization vector. And the substrate with the plane image can present changeable data. The polarization filter 7 (see Fig. 3.54) is placed in front of the image of the object, the hologram of which is recorded on the photoplate 2, and the formed image is reconstructed with a light source with changeable polarization vector direction. In this case, any light-sensitive material can serve as a substrate: photographic emulsion, chromed gelatin, shellac, liquid crystals, photoresists, etc. A light source of a certain spectral composition and with changeable polarization vector direction is used. If the polarization vector direction is changed

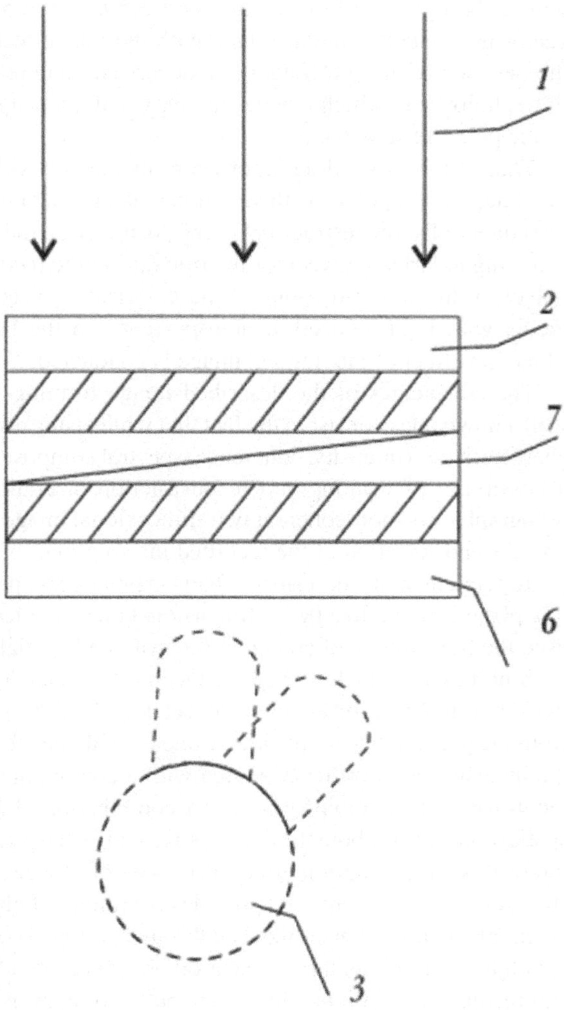

Fig. 3.54 Scheme of image recording using a light beam with a changing direction of the polarization vector. Reprinted from [94] with permission

with the image, which is recorded on the first hologram, there appears an image reconstructed from the second hologram. Maximal depth (intensity) of the image is reached under optimal transmissivity of the polarization filter. If holograms are recorded at different wavelength, the intensity of the images, which are reproduced from the holograms behind the polarization filter, does not depend on the intensity of the first hologram. If all the holograms are recorded at the same wavelength, the intensity of the reconstructed from the holograms image behind the polarization filter is in inverse proportion to the diffraction efficiency of the first hologram. Thus, image reproduction can be controlled through the change of light source polarization.

Holograms of different states of an object or of different objects are recorded on emulsion layers from different sides of the substrate to form changeable images without hologram shifts (see Figs. 3.49, 3.50, 3.51, 3.52 and 3.54). If the polarization vector is changed simultaneously with the image recorded on the holograms from the one side of the substrate, there occurs an image reconstructed on the other side of the hologram, which reaches the maximal intensity under optimal transmissivity of the polaroid substrate.

Thus, during recording on ordinary media, a possibility occurs to smoothly turn one image into another with simultaneous visualization of all images. It is possible to sequentially reconstruct different positions of the holographed objects through changing polarization vector direction during the record of holograms of the shifted object in different directions of the polarization vector while using light-sensitive media with light-induced anisotropy (e.g., on the base of the Weigehrt's effect). Thus, the effect of moving 3D image is conducted.

The advantages of the described image forming method in comparison to the well-known ones consist in the fact that while using a source with regulated radiation characteristics (intensity, coherence, spectral composition, polarization) there occurs a possibility of forming and reconstruction of combined holographic, as well as holographic and non-coherent two-dimensional images, control of the reconstruction process and selection of the recorded information.

To implement the described effects in practice, in particular the effect of transition of a plane image into a three-dimensional one and vice versa, L.V. Tanin, the author used the light source of changing intensity and spatial coherence.

A hologram, which was created using the image forming method, bears qualitatively new information about the object. It realizes the possibility of gradual transition from the plane image to the three-dimensional one. To do this, it is enough to change the intensity of an ordinary source with a continuous spectrum of radiation, which reconstructs the combined image. A combination of 3D and plane image enhances spatial conception about the object presented on the plane. The image forming method intensifies the perception effect of images of objects, scenes, compositions, reconstructed from holograms and provides continuity of obtaining and the increase of the taken information. This is reached through getting a combined plane and 3D images, which are complimentary to each other. This method gives the possibility to form and reconstruct combined images in bulk, color, motion.

On the advice of USSR State committee on inventions and discoveries (letter No. 18–25/2, 872, 202 dated 26.03.80) materials on the application for a patent No. 2872202/25 dated 06.02.80, "The image forming methods" was filed in the USSR and abroad, in the USA, France, FRG, the ChSSR, the PRP, and for them, patents were acquired [278–286]. In 1993, "Samsung Electronics Corporation, Ltd." addressed a request to L.V. Tanin, the main author of the patent and one of the authors of this monography, for transfer of rights for the patented in the USA invention "The image forming methods" (patent No. 4420218 dated 13.12.83).

This invention was realized in the first holographic industrial model in the USSR (certificate of authorship No. 10534 dated May 20, 1980) "The Olympic holographic sign." On the occasion of XXII Olympic Games in Moscow on the basis of the described methods, a technology of holographic signs was designed, and serial production was organized at Minsk Mechanical Plant named after S.I. Vavilov (about

40,000 signs were produced). The main idea was that a photolithographic image with the Olympic symbols was put on a plane substrate.

A three-dimensional image of the Olympic symbols was recorded on the holographic plate. Then the hologram was put on the substrate with the photolithographic image so that the contours of these images matched each other. Then, the emulsion layer was brought for protection from holographic damages. And the observer was located from the side of the light source relative to this multilayer structure and observed the 3D and the plane image or both simultaneously depending on the light conditions. The observer identified the distance between the light source and the angle of turn of the multilayer structure himself by turning it till most bright and accurate images appear.

Another example of practical implementation of this method is a biochromed gelatin watch. The perception effect is intensified through the usage of a hologram with the help of which memorable dates, events can be depicted. In this sample, it is the date of the 25th anniversary from the day of launching the first spaceship with the man on board. In the same time, the hologram served as protective glass. In the watch the effect is used, which can be described by changing the perceived color of the face under different illumination conditions. The face is seen under normal illumination. Under direct light from a slide projector or the sun, a 3D image of a spaceship with bás-relief of Y. Gagarin appears on the face. The developed image forming method is practically feasible and was implemented into serial production by L.V. Tanin. In 1986, Minsk Watch Plant for the first time introduced several thousands of mini alarm-clocks, which realize the image forming methods using combined 3D and plane images. The clocks are very reliable, compact, serve as alarm, have small weight and can be used in road conditions.

Especially for serial production of these clocks S.B. Shevchenko, M.K. Shevtsov and other workers of the State Optical Institute named after S.I. Vavilov developed a technology of obtaining high-performance biochrome gelatin layers, which were then used to cover glass substrates.

Besides, a polymer-registering medium is used to enhance efficiency of holographic images recording [287]. But it should be mentioned that the author pays special attention to studying the possibilities of different schemes of hologram recording and copying while implementing the idea stipulated in patents "The image forming methods" [278–287]. Among them, the work by Sh.D. Kakichashvili can be pointed out, which uses conic glass body of revolution with polished faces as a source of reference beams during circular holographing [288, 289] (see Fig. 3.55). Besides, the interference copying method should be mentioned, which was suggested by V.A. Vanin. It is used during copying of any type of a hologram, but for the Lippman–Denisyuk's holograms it is the only possible method. Interference copies have all the characteristics as the original one. In this case, the original hologram is made according to the scheme with an inclined reference beam. The scheme of recording is presented in Fig. 3.56. It should be mentioned that recording was conducted with four-sided illumination [290, 291], and the scheme of hologram copies obtaining is presented in Fig. 3.57 [290–295].

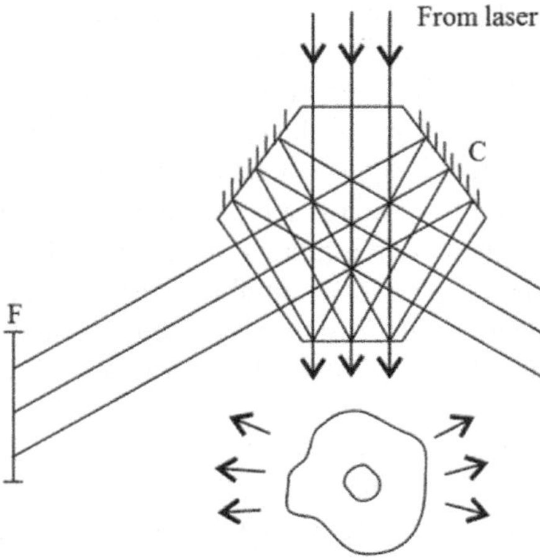

Fig. 3.55 Recording of a full image: *C*—the conical solid of rotation with polished butts; *O*—the object; *F*—the cylindrical film. Reprinted from [94] with permission

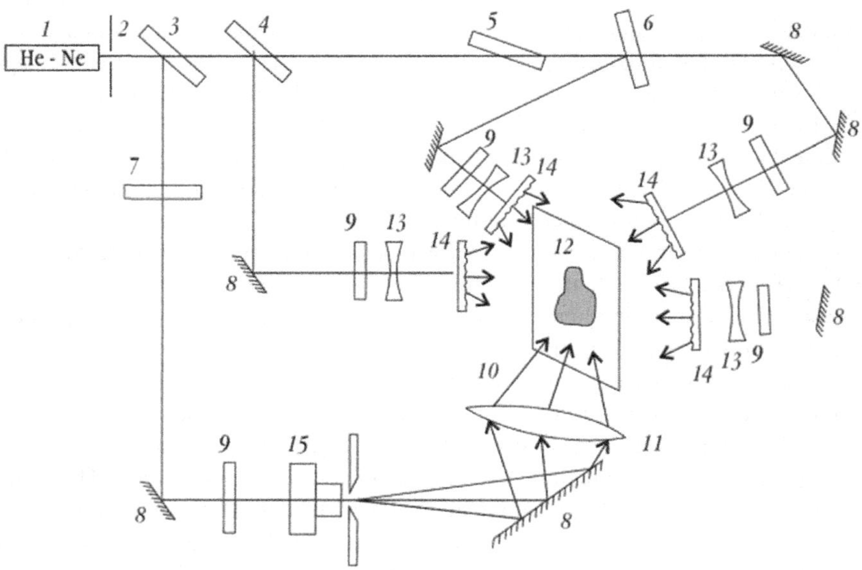

Fig. 3.56 Scheme of the recording of the holograms—originals of transmission type: 1—the laser LG-38; 2—the shutter; 3–6—the beam splitters with variable coefficient of reflecting; 7—the plane-parallel glass plate; 8—the aluminum mirrors; 9—the quarter-wave planes; 10—the photoplane; 11—the lens; 12—the object; 13—the short-focus lenses; 14—the diffusers; 15—the microscope objective with spatial filter. Reprinted from [94] with permission

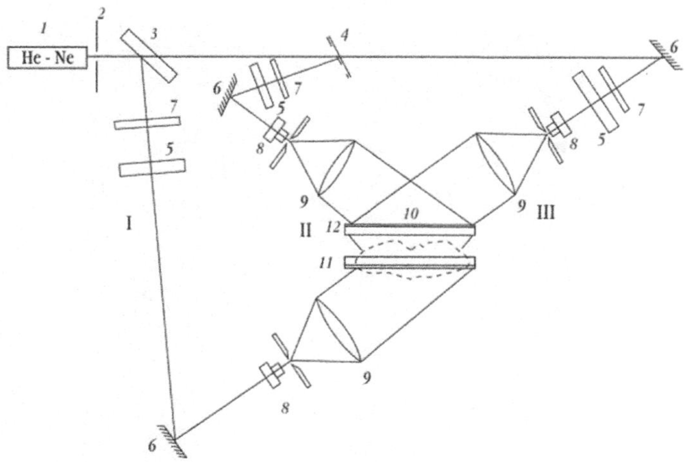

Fig. 3.57 Scheme of the recording of hologram copies- originals of transmission type: 1—the laser LG-38; 2—the shutter; 3,4—the beam splitters; 5—the plane-parallel plates; 6—the mirrors; 7—the half-wave plates; 8—the microscope objectives with spatial filter; 9—the lenses; 10—the intermediate hologram; 11—the photographic plate; 12—the real image; I—the reference beam, II, III—the reconstructing beams. Reprinted from [94] with permission

Patent search and analysis of materials, which were published in our country and abroad, of schemes and constructions suitable for serial production of picture holograms, have shown [296] that the colliding beams scheme suggested by Yu.N. Denisyuk is the most simple and the best in displaying optical information [91, 92].

Recently, for further improvement of the idea relating to the image forming methods, the work was continued, which is connected with the development of technology of obtaining of highly qualitative colliding beams holograms [278–287]. The main advantage and practical value of this colliding beam scheme by Yu.N. Denisyuk (certificate No.88, certificate No. 276270) consists in the fact that it allows giving maximal information (color, extensionality, luminance range transmission, etc.) about an object through the image reconstructed by an ordinary white light source.

It is also necessary to find the intensity ratio of the beams scattered by the object and the optical path difference between the object and the reference beams that is conditioned by the reflective characteristics of the object and the scene depth, and the laser radiation power necessary for the recording of spatial and time coherence. For more complex objects (scenes, compositions), it is recommended to implement the hologram registration scheme with two or more side lights (Fig. 3.58).

Today, this idea is successfully implemented in a new project, which is connected with recording of a large-size reflective hologram of the Cross of Saint Euphrosyne, Princess of Polotsk. It was the first time in the world when the holographic image of the Cross of Saint Euphrosyne, Princess of Polotsk, one of the greatest heritages of the Belarusian people and all the Christians, was reproduced (the hologram with the

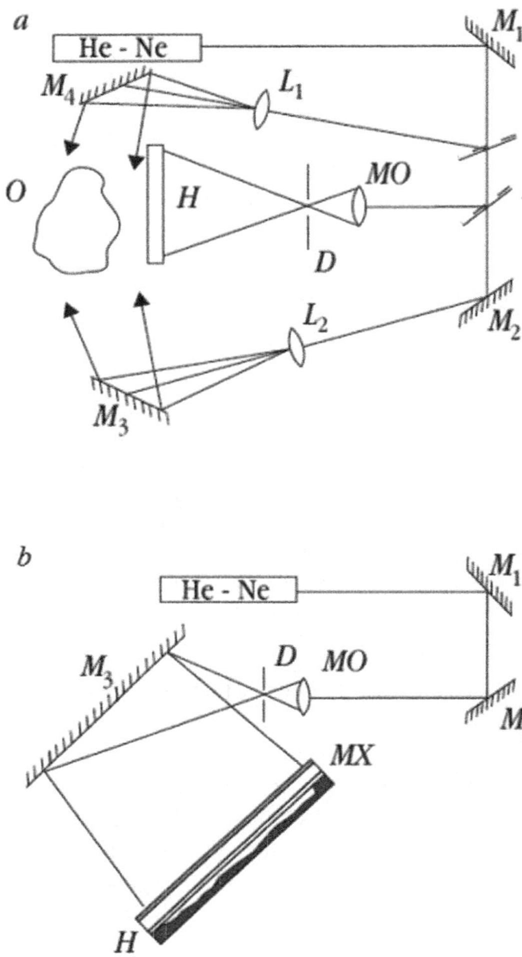

Fig. 3.58 Schemes of the recording of reflecting hologram-original with two side lights (**a**) and receiving the copy of reflecting hologram-original (matrix) (**b**): M_2–M_4—the mirrors; BS—the translucent plates; MO—the microscope objective; L_1, L_2—the lenses; H—the photographic plate; O—the object; D—the pinhole aperture; H—the photographic plate; MX—the hologram-original (matrix). Reprinted from [94] with permission

size of $79 \times 90\,\text{cm}^2$ was recorded by L.V. Tanin and S.N. Ginak) (see Fig. 3.59). The hologram was created by joint efforts of scientists and specialists of "Light Magic Ltd.", one of the founders of which is L.V. Tanin, the author of the book, with support of the Orthodox Church in collaboration with the National Academy of Sciences of Belarus.

This work is the continuation of the efforts to renew the shrine that is respected in the whole orthodox world and specially patronizes White Rus. According to the

Fig. 3.59 Image restored by the hologram "Cross of Saint Euphrosyne of Polotsk" presented to His Holiness Patriach of Moscow and All Russia Alexy II during his visit to the Republic of Belarus on the occasion of the 1020th anniversary of Christening of Russia (October 23, 2008). Reprinted from [94] with permission

Synod of Belarusian Orthodox Church, "It represents our penitence, our faith, hope and love to God and Motherland."

Lost during the Great Patriotic War, The Cross of Saint Euphrosyne of Polotsk as well as the famous Amber Room is one of the ten most precious disappeared objects of the world's art. In 1997, on the eve of the Holy Cross Day the reconstructed

Cross of Saint Euphrosyne was given to the Saviour Cathedral of Polotsk. The Cross keeps a piece of the True Cross with a drop of the Blood of Christ, a piece of the Holy Sepulcher and the Sepulcher of the Most Holy Mother of God, a drop of the blood of Dmitry of Solun, great martyr, a piece of relics of Saint Stephan and Saint Panteleimon.

This picture hologram "The Cross of Saint Euphrosyne of Polotsk" is an achievement in the world practice of large-size hologram recording and gives the possibility to preserve 3D image of the Cross protecting it from unexpected circumstances and to see the image of the shrine to a greater number of people.

Experiments on recording and forming of combined 3D and plane image of the reproduced image of the Cross of Saint Euphrosyne of Polotsk are continued [297].

3.5 Results and Conclusions

1. A number of methods were theoretically developed to determine absolute surface relief, which have a prior character. It was established that implementation of absorbing media in holographic contouring methods gives the possibility to modulate the contrast of the interference pattern. The maximal contrast fringe is interpreted as a zeroth order fringe; other contours are numerated in the order of their visibility decrease.

There were developed the resonance contouring method, which enhances the spatial resolution in 1.3 times, and the holographic method of surface relief study, which gives the possibility to register objects of a double depth through matching light amplitudes in reference waves with absorption coefficients of the immersion medium.

A multi-angle method was theoretically proved and experimentally tested, which consisted in obtaining of contoured maps while illuminating it with M beams simultaneously, which are located at an equal distance along the angle. It allows narrowing the interferences maxima in 3.5 times at the expense of multi-beam interferometry and thus enhancing the accuracy of identifying their location on the surface of the object. As a result of the conducted optimization of interferogram registration conditions, it was shown that if in the experiment the product of time of one exposition and their number stays the same, this significantly enhances the diffraction efficiency of multi-exposure holograms. This gave the possibility to obtain contoured maps of the human skull with half as narrow interference contours.

There were conducted theoretical studies of holographic interferograms forming processes taking into account the polarization characteristics of the interfering waves. A possibility was shown to control the holographic topograms contrast through impact on corresponding polarization components of the object waves. On the grounds of the conducted study, the holographic contouring method was developed, which is based on partition of polarization state of object beams by using optically active media. The contouring method was developed, which does not require usage of holographic recording of object waves. These methods give the possibility to estimate not only the height, but also the relief direction of the surface under study.

Holographic multi-beam surface relief contouring methods were developed, which are based on the analysis and usage of the polarization characteristics of reference and object light waves. One of such methods requires implementation of the source, which generates at several wavelengths, and the other one—several immersions with different refraction indices, which differ at the same value. Implementation of these methods gives the possibility to double the spatial resolution, i.e., the sensitivity of the measurements.

In the result of theoretical and experimental research during studying vertebral pathology, the projection double-raster moiré method of surface relief contouring was developed and was suggested for usage. The experimental processing of this method was conducted on the model objects and a human. In particular, the experiments on the usage of moiré contouring methods were conducted to get the full 3D image of physiological curvatures of human spine. There were given examples of moiré topographic maps of the dorsal surface of human body.

The obtained results of construction of 3D surface form according to the maps prove the possibility of estimating the value of spine curvature. In the result, it was established that the moiré method under corresponding adaptation and improvement can be used for determination of spine curvatures and study of pathological processes, which are connected with such curvatures.

In one of the dye laser modes, the multifrequency one, there were obtained holographic contoured maps of the relief of 3D objects according to the colliding beams scheme. It was shown that while using this holographing scheme (unlike the inclined reference beam scheme) the defects are eliminated, which are connected with the hologram dispersion, for example, contours distortion, localization of bands at a significant distance from the surface under study.

2. For the first time in collaboration with dentists, a complex experimental–clinical study was conducted concerning deformations, which occur under functional overloads of the upper and the lower human jaws. The study gives the possibility to make a firm conclusion about good prospects of usage of holographic and speckle-interferometry methods in dentistry.

A special interest we, experimenters, pay to the study of movable living objects. For these purposes, the pulse ruby laser was developed and produced, which generates double monopulses with tunable time interval between them. The recording method of living object holographic interferogram was perfected. Using the double-pulse holographic interferometry method, the interferogram of strained hand muscles with interval between pulses of 100 μs was obtained. These preliminary holographic experiments gave the possibility to come close to the creation of the holographic cardiograph for studying the functional state of the cardiovascular system.

3. Holographic methods of study of human chest shifts as well as optical schemes for their realization were developed. These holographic methods allow obtaining information about shifts of surface points of the human chest, which are caused by mechanical waves appeared during heart contraction and blood flow pulsation. They give the possibility to qualitatively estimate the value and the direction of the

spatial vector of relative shifts of surface points, which are connected with the state of the circulatory system. Optical schemes of systems for formation of illuminating and recording light beams provide possibility to obtain holographic interferograms of the human chest, which contain information about its shifts. Conditions were determined concerning the process of synchronization of human chest holograms recording with different phases of heart contractions. A test sample of the laser-holographic complex (holographic cardiograph) was developed and produced to study the state of the human circulatory system, which was developed according to the technical requirements; it was in working conditions and gave the possibility to conduct physical and medical study. The compatibility of the complex apparatus according to the signals parameters with the following characteristics, such as interference immunity, vibroprotection, mutual synchronization, including those with timing in any phase of heart contractions, was provided. The test sample consisted of a holographic table, a ruby laser radiator, He–Ne laser, an optical unit of a holographic interferograms registration block on reversible PTPC, "Impecard" complex, a computer with a display and a synchronization block.

The test sample was adjusted that provided the possibility of autonomous operation of different LHC blocks and in the complex with synchronization providing the possibility of adjustment to certain phases of heart contractions. It was established that LHC synchronization adjustment should be conducted in the middle of the ascending or descending branch of the impedance plethysmogram curve. The choice of the time moments is conducted by the computer operator due to the ECG curve in the "Operator's menu" mode.

A test sample of the laser radiator was assembled and adjusted. The laser radiator wavelength is 694.3 nm. The ruby laser radiator operated in the double monopulses generation mode with the regulated interval between pulses in the single-mode regime. Energy of each of the two pulses of the generator radiation is 0.01 J, of the amplifier is 0.025 J, and of the total one is 0.5 J. The width of the laser radiation spectral band is not more than 100 MHz. The laser radiation pulse duration is 20–120 ns. The time interval between the radiation pulses (regulated in the interval) is 100–700 μs.

Due to the functional requirements, the synchronization block formed the necessary pulse drivers of:

- the power supply of the ruby laserRuby laser with the regulated period (10–20 μs with the accuracy of +1 μs);
- the shutter control unit (50–1000 μs);
- the reversible carrier control unit (100–400 μs).

The unit of hologram recording on reversible photothermoplastic carriers was developed and produced.

The temperature control of the carrier is provided in the preparation mode within the limits of 40–90°. The range of the energy regulation of the developing pulse is 1–4 J/cm along the thermoelement area. In the charge mode, the corona charge voltage is regulated within the limits of 3–10 kV, and the duration of discharge is within the limits of 0.1–10 s.

In the exposure mode, the pulse duration of the shutter control can be regulated within the limits of 0.1–10 s.

For the soft RC, the "frame transfer" mode provides the RC tape advancement with the regulated frame length within the limits of 5–40 mm, and continuous broach.

Recording modes of holographic interferograms on low- and high-resistance types of reversible photothermoplastic carriers (PTPC) were perfected. For low-resistance PTPC, a recording method is used, during which operation of charge, exposure and development of a hologram is realized simultaneously. Before registration of the second hologram, the first one is partially erased and then reconstructed simultaneously while developing the second hologram. Charging was conducted by the voltage on the corona filament within the limits of 4–5 kV, and development—of 65–70°, the exposure was continuous.

For high-resistance PTPC, the separate holographic method was used, during which the mentioned operations can be separately realized. Charging is conducted under the voltage on the corona filament mentioned above, the temperature of thermostating is 40 °C, the power of the developing pulse is 2–3 J/cm^2, and the exposure is within the limits of 0.01–0.2 s.

Technical tests were conducted concerning the test sample of LGK on soft and hard reversible carriers with different values of surface resistance of the working Table (0.1–0.2 kohm). It was established that recording and erasing of holograms is conducted on both types of carriers. But reliability and longevity of the conductive layer of the thermoelement in both cases and for the available samples were on the level, which does not exclude a failure. That means that under the order development of 1–2 J/cm^2, which have to be supplied to the carrier for 5–15 ms, the conductive layer of the thermoelement can be destructed. The possibility of this fact is greater if the hard reversible carrier is used. The main reason for destruction is high speed of temperature increase in the conductive layer, which reaches the value of 2–510 °C/s. This reason is partially eliminated through smoothing of the duration of the rise-up portion of the developing pulse up to 1–2 ms. But as far as we are concerned, the decisive factors for enhancing reliability and longevity of the thermoelement are technology and the structure of the evaporation layer, on which the strength of adhesion and the layer itself depend.

Recording modes of holographic interferograms of human chest on reversible carriers (RC) were studied using the test sample of the LHC. The main factor that characterizes the specific character of recording of the above-mentioned interferograms consists in uncertainty of launching moment of the ruby laser, i.e., by the choice of the exposure relative to charging operation of RC and its development. This uncertainty is determined by the necessity of synchronization of the LGC work with the rate of heart contractions and can reach several seconds depending on the synchronization algorithm. During this time, dark decrease of the potential level occurs on the surface of the thermostatic layer; that is why the contrast of a latent electrostatic image formed during exposure and consequently the contrast of the interferogram decay.

This fact was established to significantly weaken the interferogram contrast up to 3–5 s of uncertainty while using multilayer RCs with an underlayer, which has surface resistance of 3–4 kohm/cm^2. But for single-layer RCs relaxation of the static surface discharge at time up to 10 s does not influence the quality of the interferogram.

It was ascertained that the interferogram quality, which is provided by the photothermoplastic RC, is reached if all the firm requirements to the stability of the recording parameters are met. Thus, the evenness of development temperature along the frame surface should be not lower than 0.2:0.4 relative to the optimum, which is selected for every type of RC and lies within the limits of 60–80 °C.

The schemes of obtaining "living" interference fringes using two reference beams were studied. As a basic variant, a scheme was selected with reference beams situated in a vertical plane and with a focused image. This scheme provides the mode of "living" fringes while using radiation of different length, recording interferograms of the ruby laser radiation $\lambda = 694.3$ nm and its reconstruction with a He–Ne laser radiation $\lambda = 632.8$ nm. The conducted study gave the possibility to select the optimal optical scheme of the complex, which gives accuracy during estimation, a scheme of the interferometer with two reference beams and regulated phase shift.

An optical block was developed where the scheme of the interferometer with two reference beams was realized. The reference beams are regulated by the phase shift.

The study was conducted connected with elimination of hologram dispersion impact on the images shift, which appears as a result of different wavelengths of radiation during recording ($\lambda = 694.9$ nm) and reproduction ($\lambda = 632.8$ nm) of interferograms, which give the possibility to use He–Ne laser radiation for interferogram sensing, but preserving the advantages of the two reference beam scheme. The scheme of compensation node of path difference between reference and object beams was developed, which is based on the usage of mirror optical loop.

Preliminary medical study was conducted. In order to select research methods of blood circulation needed for comparison with the holographic interferogram of the chest, there was conducted the analysis of methods to study mechanical, electrical and pumping heart activity, as well as methods to study the vascular system, to measure pressure and time of the blood flow.

The impedance cardiography method was chosen as the most informative way. It allows quickly obtaining information about pumping and mechanical function of the heart, about the state of the arterial pressure and peripheric resistance. This method was realized with the help of "Impecard," a hard- and software complex. The implemented algorithm of automatic estimation of hemodynamics indices gives the possibility to localize all the phase structures of the bioimpedance signal.

Usage of the signal of the chest differentiated impedance plethysmogram was taken as a principle of the method of data storage about the functional state of the cardiovascular system of the examined patient. The plethysmogram is obtained with rheograph R4-02 with synchronous image forming of a chest interferogram for further development of new diagnostic criteria.

It was established that LHC synchronization system should be connected in the middle on the ascending and descending branches of the curve of the impedance plethysmogram. And the greatest informativity is reached relative to the mechanical shifts amplitude of the oscillation zones of the human chest cutaneous covering.

The holographic interferograms of the human chest were obtained (time interval between the pulses was 300–400 μs) that gives possibility to visualize on the interferogram the oscillation zones in the area of the right clavicle, the left shoulder and forearm, as well as in the areas of the larynx and the diaphragm. As it is seen from the given holographic interferograms, the radiation parameters of the created laser and the parameters of the hologram recording device provide filming of holograms of the human chest.

The analysis of the medical research results of the cardiovascular system has shown that alteration of the obtained with LHC holographic interferogram of the human chest surface during heart contraction is characterized by the degree of intensity of interference fringes crowding. In the chest activity centers (zones of the left and right ventricles, tracts of the aortal and pulmonary outflows, etc.), the greatest deformation of the proper tissues is observed at the tension of the vessel walls growth in the ventricular systole phase and further period of quick emptying. In the interferogram, it looks like intensification of the interference fringes concentration and it coincides with the anacrotic inclination of the rheographic curve that is registered with "Impecard-3." But during the diastole period, the holographic image changes. The holographic interferogram reflects pulse oscillations of the blood filling of the area under study. The oscillations depend on the heart stroke volume as well as on the speed of the blood flow, which is mostly determined by the size of the vessels lumen, elasticity and tonic contraction of the vessel wall. It is obvious that if there is pathology of the cardiovascular system, which causes malfunctions of the described functions, then the holographic interferogram (the interference fringes frequency, their width, contrast, occurrence of additional activity centers, modification of the previous ones, etc.) will greatly differ from the interferogram of a healthy person.

Thus, this study gives the possibility to get information about relative intensity of blood filling, the state of the vascular tone in different areas of the human chest. Assembly and adjustment of the experimental sample of the laser radiator were conducted. The radiatior operates in doubled monopulses generation regime with a regulated interval (100–700 μs) between pulses in the single-mode regime.

4. The possibility of obtaining complimentary combined 3D and plane images was studied. The efficiency of visual perception of the images was intensified. The methods of recording and formation of images were developed, which can widely be applied in data storage systems (holographic archives) during the creation of atlases and classifiers of pathology of medical–biological objects. The pathologies, which are registered in the hologram, can serve for medical workers as an objective source of information during medical study as well as during preparation of surgeons for a surgery, and they can be used as a visual tutorial during demonstration for medical students. There were obtained highly qualitative holograms of human vertebrae preparations with lumbar osteochondrosis manifestations, corrosion preparations of the liver blood vessels (see Fig. 3.47). Holographic images give more accurate data about the localization of the process, the degree of its intensity, etc. Transfer to pulse laser sources gave the possibility to conduct holographic

registration of fast processes, to "freeze" the structural states in time. That is why obtaining holograms of a living object is of special interest while studying rare as well as typical diseases.

Combined images can be applied in data storage systems for spatial distribution of images, in indication systems, during construction of artistic panorama, during creation and multiplication of copies of artworks, registration of holographic portraits, in advertisement, and for production of visual and teaching aids, technical stands, souvenirs, etc.

"Image forming methods" were suggested and patented in the USSR, the USA, France, Germany, the People's Republic of Bulgaria, the ChSSR.

For the first time in the Patent Office of the Republic of Belarus, there was registered the transfer of patent rights for the invention "Image forming methods" to foreign firms (in particular, to "Samsung Electronics Corporation, Ltd.") under the number 93-10001.

References

1. Yu.I. Ostrovsky, M.M. Butusov, G.V. Ostrovskaya, *Holographic Interferometry* (Nauka, Moscow, 1977), 400p. (in Russian)
2. N.G., Vlasov, S.G., Galkin. U.P. Preasnyakov, in *Methods and Apparatus of Optical Holography* (05–13 Nov 1983), p. 148. (in Russian)
3. N.G. Vlasov, O.G. Lisin, U.I. Savilova, J. Tech. Phys. **54**(10), 2037 (1984) . (in Russian)
4. W. Schuman, M. Duba, *Analysis of Deformation of Opaque Objects by the Method of Holographic Interferometry* (Mashinostroenie, Leningrad, 1982), 170p. (in Russian)
5. S.A. Dobrolezh, I.S. Klimenko, V.V. Klimovskikh, N.A. Sishenok, Electron. Ind. **10–11**(116–117), 83 (1982). (in Russian)
6. R.K. Erf, *Holographic Non-destructive Study* (Mashinostroenie, Moscow, 1979), 448p. (in Russian)
7. C. West, *Holographic Interferometry* (Mir. Moscow, 1982), 504p. (in Russian)
8. Y.N. Denisyuk, S.I. Soskin, Opt. Spectrosc. **31**(6), 992 (1971). (in Russian)
9. R. Jones, K. Yaips, *Holographic and Speckle-Interferometry* (Mir, Moscow, 1986), 328p. (in Russian)
10. J.P. Watees, J. Opt. Soc. Amer. **62**, 373 (1972)
11. T. Nakagima, H. Saito, Jap. J. Appl. Phys. **14**(1), 277 (1975). (in Russian)
12. M. Murata, Jap. J. Appl. Phys. **14**(1), 271 (1975). (in Russian)
13. L. Kersch, Mater Evaluat. **29**, 125 (1971)
14. J. Vienot, P. Smigilsky, A. Ruays, *Optical Holography: Development and Application* (Mir, Moscow, 1973), 216p. (in Russian)
15. Yu. I. Ostrovsky, *Holography and Its Applications* (Nauka, Leningrad, 1973). (in Russian)
16. N.N. Zakharov, I.S. Zeylikovich, N.V. Karnaukhov et al., Proceedings of CIAM **1026**, 366 (1983). (in Russian)
17. V.C. Sainov, S.C. Sainov, S.N. Ossikovska, Dokl. Bulgaria. Acad. Sci. **36**(11), 1383 (1983). (in Russian)
18. C.M. Vest, Nondestruct. Testing. **3**, 351 (1972)
19. G.V. Ostrovskaya, Yu.I. Ostrovsky, Prog. Opt. **22** (1985)
20. A.K. Beketova, A.F. Belozerov, A.N. Bereziuk, *Holographic Interferometry of Phase Objects* (Nauka, Leningrad, 1979), 232p. (in Russian)
21. M.L. Gurary, A.A. Magomedova, G.I. Rukshan, V.K. Sakharov, DAS BSSR **208**(1), 233 (1973). (in Russian)

22. G.V. Ostovskaya, Yu.I. Ostrovsky, L.V. Tanin, *Proceeding of Symposium on Applied Holography,* (Prague, 1978), p. 229
23. F.R. Reich, B.P. Helibrand, Precipit. Scaveng. **213** (1970). (in Russian)
24. T. Jschudi, G. Herziger, A. Engel, IEEE. J. Quant. Electron. **9**(6), 631 (1973). (in Russian)
25. F. Crane, P. Dunn, P.M. Malyak, B.J. Thompson, *Proceedings of the Society of Photo-Optical Instrumentation Engineers* 1982, 348/2; in 15 *International Congress on High-Speed Photography and Photonics,* San-Diego, 21–27 August 1982, p. 634. (in Russian)
26. O.V. Angelskiy, V.V. Yashchenko, D.I. Derkach, Opt. Spectrography **57**(5), 899 (1984). (in Russian)
27. A.M. Chernykh, V.S. Siniyakov, N.P. Karavankina, et al. in *Proceeding of I All-Union Conference Tools and Methods of Quantum Electronics in Medicine* (Saratov, 1976), p. 205. (in Russian)
28. Yu.I. Ostrovsky, L.V. Tanin, *ZhTF* **45**(8), 1756 (1975). (in Russian)
29. Yu.I. Ostrovsky, L.V. Tanin, *Proceedings of the II All-Union Conference on Holography* (Kiev, 1975). (in Russian)
30. Yu.I. Ostrovsky, L.V. Tanin, *Proceeding of the Conference "Laser Based on the Complex Organic Compounds and their Use* (Dushanbe, 27–30 September 1977), p. 331. (in Russian)
31. L.V. Tanin, *Holographic Methods of Research,* (1978), 194. (in Russian)
32. L.I. Rachkowsky, L.V. Tanin, Copyright Certificate 1522867, (The USSR), (15 Jul 1989). (in Russian)
33. L.I. Rachkowsky, L.V. Tanin, A.S. Rubanov, Copyright Certificate 4680200, (the USSR) (21 Sept 1989). (in Russian)
34. L.I. Rachkowsky, L.V. Tanin, A.S. Rubanov, *Proceedings of VI All-Union Conference* (Vitebsk, 1990). (in Russian)
35. L.I. Rachkowsky, L.V. Tanin, A.S. Rubanov, Copyright Certificate 1629749, (the USSR) (22 Oct 1990). (in Russian)
36. L.I. Rachkowsky, L.V., Tanin, I.E. Drobot, Copyright Certificate 1714352, (the USSR) (22 Oct 1991). (in Russian)
37. L.I. Rachkowsky, L.V. Tanin, A.S Rubanov, V.K. Zabarovsky, Copyright Certificate 1696852 (the USSR) (08 Aug 1991). (in Russian)
38. L.I. Rachkovsky, L.V. Tanin, A.S. Rubanov, Proc. SPIE **1461**, 232 (1991)
39. L.I. Rachkovsky, L.V. Tanin, Copyright Certificate 1707469 (the USSR) (22 Sept 1991). (in Russian)
40. L.I. Rachkovsky, I.L. Drobot, L.V. Tanin, Proc. SPIE **1507**, 450 (1991)
41. V.K. Zabarovsky, L.I. Rachkowsky, L.V. Tanin, Peripheral Nerv. Syst. **15**, 30 (1992). (in Russian)
42. L.V. Tanin, S.A. Aleksandrov, Patent Application 5008033/28 (Russia), positive decision dated 15 Sept 1992. (in Russian)
43. L.V. Tanin, S.A. Aleksandrov, Patent Application 5008034/28 (Russia), Positive Decision Dated 15 Sept 1992. (in Russian)
44. L.V. Tanin, S.A. Aleksandrov, A.S. Rubanov, Patent Application 5060096/25 (Russia), Positive Decision Dated 26 Aug 1992. (in Russian)
45. S.A. Aleksandrov, L.V. Tanin, Patent Application 92005238/28 (11 Nov 1992). (in Russian)
46. L.V. Tanin, L.I. Rachkowsky, ZhTF **62**(12), 34 (1992). (in Russian)
47. L.V. Tanin, S.A. Aleksandrov, Patent Application 92005950/28 (11 Nov 1992). (in Russian)
48. S.A. Aleksandrov, L.V. Tanin, *Technical Digest of ICO International Topical Meeting on Optical Computing* (Minsk, 1992), p. 29E18
49. S.A. Aleksandrov, L.V. Tanin, Proc. SPIE **1806**, 437 (1992)
50. S.A. Aleksandrov, L.V. Tanin, Laser Optics-93: Proceeding. Reports **2**, 499 (1993). (in Russian)
51. S.A. Aleksandrov, L.V. Tanin, Proc. SPIE **2994**, 355 (1993)
52. S.A. Aleksandrov, L.V. Tanin, Proc. SPIE **2051**, 833 (1994)
53. S.A. Aleksandrov, L.V. Tanin, A.S Rubanov, Lett. ZhTF **19**(18), 40 (1993). (in Russian)

54. L.V. Tanin, S.A. Aleksandrov, V.K. Zabarovsky, M.M. Loiko, Peripheral Nerv. Syst. **17**, 44 (1994). (in Russian)
55. L.V. Tanin, *in the 3 International Conference Optical Methods in Bio-medical and Environmental Sciences* (Tokyo, 1994), p. 77
56. S.A. Aleksandrov, L.V. Tanin, A.S. Rubanov, ZhTF **64**(2), 201 (1994). (in Russian)
57. S.A. Aleksandrov, L.V. Tanin, Patent 2085835 of the Russian Federation (27 Jul 1997). (in Russian)
58. T. Uyemura, H. Matsumoto, Bull. Jap. Soc. Precis. Eng. **10**(3), 101 (1976); RZh. phys. **511**, 1100 (1977). (in Russian)
59. J.L. Goldberg, Jap. J. Appl. Phys. **14**(1), 253 (1975); EI IRS **14**(1) (1976). (in Russian)
60. E.G. Semenov, U.I. Filenko, Quantum Electron. **6**(11) (1979). (in Russian)
61. G.V. Ostrovskaya, *The Development of Optical Methods of Plasma Diagnostics and Their Application to Studies of Laser Sparks in Gas*, Abstract of the Dissertation, (Order of Lenin Physical and Technical Institute. A.F. Ioffe, Leningrad, 1979)
62. G.V. Ostrovskaya, ZhTF **78**, 9 (2008). (in Russian)
63. I.I. Komissarova, G.V. Ostrovskaya, V.N. Filippov, E.M. Shedova. ZhTF **53**(2), 251 (1983). (in Russian)
64. H.P. Alum, U.V. Kovalchuk, G.V. Ostrovskaya, ZhTF **54**(5), 896 (1984). (in Russian)
65. I.I. Komissarova, G.V. Ostrovskaya, V.N. Filippov, E.N. Shedova, ZhTF **57**(2), 377 (1987). (in Russian)
66. A.K. Beketova et al., WMD **7**, 24 (1980). (in Russian)
67. N.P. Daukevich, et al., Quantum Electron. **12**(10), 2029 (1985). (in Russian)
68. B.P. Hildebrand, K.A. Haines, Phys. lett. **21**(4), 422 (1966)
69. B.P. Hildebrand, K.A. Haines, JOSA **57**, 157 (1967)
70. T. Tsuruta, N. Shiotake, J. Tsujiuchi, Jap. J. Appl. Phys. **6**(5), 661 (1967)
71. W. Schmidt, A. Vogel, D. Preussler, Appl. Phys. **1**(2), 103 (1973)
72. A.A. Friesem, U. Levy, Appl. Opt. **15**(12), 3009 (1976)
73. S.T. De, V.F. Kim, A.G. Kozachok, et al., Holographic measuring systems, **50** (1987). (in Russian)
74. L.O., Heflinger, R.F. Wuerker, Appl. Phys. Lett. **15**(28) (1969)
75. P.D. Henshaw, S. Ezekiel, Appl. Opt. **12**, 2550 (1973)
76. V. Schmidt, A.P. Perchner, Opt. comm. **3**(5), 363 (1971)
77. S. Blit, U. Levy, Opt. Lett. **2**(3), 77 (1978)
78. P.M. Kuchel, H.J. Jiziani, Opt. Comm. **38**(1), 257 (1981)
79. C. Leonard, J. Varner, P. Calkins, JOSA **60**, 1568 (1970)
80. J.C. Wyant, Appl. Opt. **10**(9), 1269 (1971)
81. J.C. Wyant, S. Oreb, P. Hariharan, Appl. Opt. **23**(22), 4020 (1984)
82. I.M. Nagibina, D.V. Kozak, Instrument **17**(12), 88 (1974). (in Russian)
83. D.M. Bavelsky, B.C. Listovets, Yu.I. Ostrovsky, et al., Energ. Mach. Build. **8**, 21 (1976). (in Russian)
84. D. Groves, M.J. Labor, N. A Cohen, J. Phys. E: Sci. Instrum. **13**, 741 (1980)
85. L.L. Marty, B. Moreno, Yu.I. Ostrovsky et al., Lett. ZhTF **10**, 1395 (1984). (in Russian)
86. L.L. Marty, Yu.I. Ostrovsky, N.O. Reyngand, Copyright Certificate 1404814 (the USSR) (23 Jun 1988). (in Russian)
87. I.M. Nagibina, V.L. Kazak, T.A. Ilinskaya, Opti. Holography Appl. **3** (1974). (in Russian)
88. P.D. Henshaw, S. Ezekiel, Opt. Comm. **12**, 39 (1974)
89. H. Bjelkhagen, B. Hok, K. Nilson. Sweden. imaging of chest motion due to heart action by means of holographic interferometry, in *International Conference on Optical Computing in Research and Development*, Preprints Visegrad, 4–9th October 1977, pp. 39–43
90. Y.N. Denisyuk, DAS USSR **144**(6), 1275 (1962). (in Russian)
91. Y.N. Denisyuk, Opti. Spectrosc. **15**(4), 522 (1963). (in Russian)
92. R. Collier, C. Burkhart, L. Lin, *Optical Holography* (Mir, Moscow, 1973), 698p. (in Russian)
93. K. Matsuda, Jap. Journ. Appl. Phys. **7**, 904 (1968)

94. L.V. Tanin, A.L. Tanin, *Biomedical and Resonance Optics: Theory and Practice* (Belaruskaya Navuka, Minsk, 2011), 687p. (in Russian)
95. G.V. Dreiden, A.N. Seidel, G.V. Ostrovskaya, Yu.I. Ostrovsky, N.A. Pobedonostseva, L.V. Tanin, et al. Plasma Phys. **1**(3), 462 (1975). (in Russian)
96. G.V. Ostrovskaya, Yu.I. Ostrovsky, N.O. Reyngand, Copyright Certificate 1315802 (the USSR). (in Russian)
97. V. Sainov, E. Simova, Comptes rendus de 1'Academic Bulgare des Sci. **41**(7), 39 (1988)
98. E. Menzel, Optik **40**(5), 557 (1974)
99. R.E. Brooks, L.O. Heflinger, Appl. Opt. **8**, 935 (1969)
100. H. Takasaki, Appl. Opt. **9**(6), 1467 (1970)
101. H. Takasaki, Appl. Opt. **12**(4), 845 (1973)
102. S. Willner, Moire. Acta Orthop. Scand. **50**(3), 295 (1979)
103. Yu.I. Ostrovsky, Copyright Certificate 1315802 (the USSR) (1987). (in Russian)
104. I.M. Naghibina, *Interference and Diffraction of Light* (Mashinostroyeniye, Leningrad, 1985), p. 241
105. A.S. Artushkevich, L.V. Tanin, I.L. Drobot, Healthc. Belarus **12**, 19 (1991). (in Russian)
106. A.S. Artushkevich, L.V. Tanin, I.L. Drobot, Healthc. Belarus **9**, 28 (1991). (in Russian)
107. A.S. Artushkevich, L.V. Tanin, I.L. Drobot, Dentistry **3–6**, 12 (1992). (in Russian)
108. S.A. Naumovich A.I. Golovko, Achievements of medical science in Belarus **3**(110) (1998). (in Russian)
109. S.A. Naumovich, L.V. Tanin, I.L. Drobot, A.S. Artushkevich, A.I. Golovko, A.P. Dmitrochenko, Patent BY3336 (Belarus), 30 Jun 2000 (State Patent department of Rep. of Belarus **2**(25), 143 (2000). (in Russian)
110. S.A. Naumovich, L.V. Tanin, I.L. Drobot, A.I. Golovko, *Proceedings of the Third Congress of Dentists of Belarus* (Minsk, 1997), p. 220. (in Russian)
111. S.A. Naumovich, A.I. Golovko, L.V. Tanin, I.L. Drobot, Mod. Dent. **4**, 23 (1998). (in Russian)
112. S.A. Naumovich, L.V., Tanin, I.L., Drobot, A.I. Golovko, Adv Med. Sci. Belarus **2**, 78 (1997). (in Russian)
113. S.A. Naumovich, L.V. Tanin, A.I. Golovko, et al. in *International Conference of New Technologies in the Humanities and Fourth International Conference on Optics Within Life Sciences* (Munster, 9–13 July 1996), p. 22
114. S.A. Naumovich, A.I. Golovko, *Proceedings of the 4th Congress of the Dentists in Belarus* (Vitebsk, 2000), p. 19. (in Russian)
115. S.A. Naumovich, A.I. Golovko, Adv. Med Belarus **4**, 85 (1999). (in Russian)
116. S.A. Naumovich, A.I. Golovko, Achievements of Med Sci. Belarus **3**, 111 (1998). (in Russian)
117. A.I. Golovko, *Work of Young Scientists.: Scientific. Works*, ed. S.L.Kabak (BSMU, Minsk, 2001), p. 42. (in Russian)
118. A.I. Golovko, *Work of Young Scientists.: Scientific Works*, ed. S.L. Kabak (MGMI, Minsk, 1998), p. 293. (in Russian)
119. S.A. Naumovich, A.I. Golovko, *Proceedings of Respiratory Science Conference of Young Scientists and Students Actual Problems of Modern Medicine,* 24–25 April 1997 (Minsk, 1997), p. 18. (in Russian)
120. S.A. Naumovich, F.G. Drik, A.I. Golovko, S.S. Naumovich, *Holographic Methods of Study in the Dentistry* (Belarusian State Medical University, Minsk, 2009), 172p. (in Russian)
121. J. Ellis, K. Forney, TNIER **51**(6), 951 (1963). (in Russian)
122. E.A. Kuznetsova, B.M. Stepanov, Instruments and Experimental Techniques **6** (1972). (in Russian)
123. I.K. Pears, J. Phys. E: Sci. Instrum. **10**(10), 961 (1977)
124. J. Antonik, Techn. Phys. **21**(2), 185 (1990)
125. N. Abramson, Laser Focus **4**(26) (1968)
126. S.N. Rowe, W.T. Welford, Nature **216**, 786 (1967)
127. Yu.I. Ostrovsky, *Holographic Interference Methods of Measuring Deformations* (Nauka, Moscow, 1988), 246p. (in Russian)

128. Y.F. Morgun, A.V. Agashkov, L.A. Lavrovsky, L.V. Tanin, *Abstracts of Representatives Conference "Laser-Holographic System for Scientific, Technical and Biomedical Studies,"* 25–28 September 1995, Grodno. (in Russian)
129. A.S. Rubanov, L.V. Tanin, F.G. Drik, Y.F. Morgun, A.V. Agashkov, L.A. Lavrovsky, V.V. Manikalo, S.K. Dik, *Materials of International Conference "Light and Biological Systems* (Wroclaw, Poland, 3–6 July 1995)
130. L.V. Tanin, A.V. Agashkov, S.A. Aleksandrov, S.K. Dik, F.G. Drik et al., J. Biomed. Opt. **2390**, 54 (1995)
131. L.V. Tanin, A.V. Agashkov, S.A. Aleksandrov, S.K. Dik, F.G. Drik, L.A. Lavrovski, V.V. Manikalo, Y.F. Morgun, Y.S. Tipenko, *Materials of International Symposium SPIE on Biomedical Optics-95* (Opt. Biophys., San Jose, 1995)
132. E.O. Amman, J.M. Tarborough, Appl. Phys. Lett. **20**(3), 117 (1972)
133. A.A. Kovalev, N.I. Kabaev, V.I. Yurevich, et al., ZhPS **30**(4), 639 (1979). (in Russian)
134. A.A. Kovalev, B.N., Tushkevich, V.G. Brovkovich, et al., ZhPS **48**(2), 330 (1988). (in Russian)
135. B.N. Tushkevich, V.A. Okushko, Copyright Certificate 1564582 (the USSR). (in Russian)
136. B.N. Tushkevich, M.N. Nizhnik, *Abstracts of Scientific-Practical Seminar Holography in Industry and Research* (Grodno, 1989), p. 13. (in Russian)
137. B.N. Tushkevich, M.N. Nizhnik, Aircraft industry **4**, 76 (1990). (in Russian)
138. B.N. Tushkevich, M.N. Nizhnik, *Abstracts of Reports of Scientific-Practical Sem* (Togliatti, 1989), p. 52. (in Russian)
139. B.N. Tushkevich, M.N. Nizhnik, *Proceeding of Reports of Scientific and Engineering Conference* (Minsk, 1989), p. 4. (in Russian)
140. M.N. Nizhnik, B.N. Tushkevich, Copyright Certificate 1636686 (the USSR). (in Russian)
141. M.N. Nizhnik, B.N. Tushkevich, *Proceeding. reports of IV Nijnevoljskiy Scientific Sem* (Volgograd, 1991), p. 35. (in Russian)
142. I.G. Zubarev, S.I. Mikhailov, Quantum Electron. **1**(3), 625 (1974). (in Russian)
143. A.N. Oraevsky, Quantum Electron **14**(6), 1255 (1987). (in Russian)
144. A.A. Kovalev, Y.V. Razvin, S.V. Serac, et al., ZhPS **27**(1), 42 (1977). (in Russian)
145. V.M. Baev, K.Y. Boller, A. Weiler, Opt. Comm. **62**(6), 380 (1987)
146. K.H. Braun, M.S. Perlmutter, Patent 4627732 (USA)
147. M.A. Dupertuis, R., Saiomaa, at al., IEEE J. QE. **23**(7), 1217 (1987)
148. J.P. Partanen, M.J. Shaw, Appl. Phys. **43**(4), 231 (1987)
149. V.B. Fedorov, I.V. Fomenkov, Brief Rep. Phys. **1**, 36 (1987). (in Russian)
150. A.V. Agashkov, *Proceeding Reports. Republic VII. Conference Young Researchers in Physics* (Minsk, 1982), p. 89. (in Russian)
151. V.A. Pilipovich, Effect of loss and anisotropy of intensification on the work of optical-quantum generator, Abstract of the Dissertation (Minsk, 1971). (in Russian)
152. V.A. Pilipovich, A.A. Kovalev, Y.V. Razvin et al., PTE **2**, 177 (1973). (in Russian)
153. B.R. Belostotsky, Y.V. Lubavsky, *Basics of Laser Technology* (Sov. Radio, Moscow, 1972). (in Russian)
154. W.R. Sooy, Appl. Phys. Lett. **4**, 36 (1965)
155. D.A. Kotomtseva, A.M. Samson, ZhPS **10**(3), 443 (1969). (in Russian)
156. V.I. Malyshev, A.S. Markin, A.A. Sychev, ZhPS **10**(2), 248 (1969). (in Russian)
157. V. Daneu, G.A. Sacchi, O. Svelto, IEEE JQE **2**(8), 290 (1966)
158. W.W. Buchman, Patent 3997854 (USA)
159. L.G. Zimin, V.P. Gribkovsky, Copyright Certificate 631031 (the USSR) (25 Jul 1979). (in Russian)
160. Patent 3500230 (USA)
161. F.P. Schaeffer (ed.), *Dye Lasers* (Mir, Moscow, 1976), 330p. (in Russian)
162. L.A. Mikaelian, J.G. Turkov, V.G. Saveliev, JETP Lett. **6**(6), 675 (1967). (in Russian)
163. V.A. Alekseev, L.K. Denisov, B.A. Kalachev, et al., *Proceedings of Proceeding Conference Lasers Based on Complex Organic. Compounds* (Minsk, 1975), p. 82. (in Russian)
164. V.A. Pilipovich, A.A. Kovalev, *Optical Quantum Generators with Bleaching Filters* (Nauka i Tehnica, Minsk, 1975) 216p. (in Russian)

165. N.E. Korneev, A.V. Folomeev, ZhPS **8**(Z), 503 (1968). (in Russian)
166. B.N. Tushkevich, V.A. Okushko, M.N. Nizhnik, Proceedings of the National Security Agency. Sor. Sci. Sciences **4**, 67 (1990). (in Russian)
167. A.P. Vorobiev, L.N., Lahanko, L.Z., Polonetsky, A.V. Frolov, Copyright Certificate 1671264 (the USSR) (Bull of Invention 3, 1991). (in Russian)
168. G.I. Sidorenko (ed.), *Instrumental Methods in Cardiology* (Nauka i Technika, Minsk, 1994), 272p. (in Russian)
169. A.P. Vorobiev, L.N. Lahanko, L.Z. Polonetsky, A.V. Frolov, Microprocess. Environ. Syst. Prop. **3**, 2 (1987). (in Russian)
170. G.I. Sidorenko, V.V. Mironchik, A.V. Frolov, et al., Proc. Belarusian Acad. Eng. **1**(3), 10 (1997). (in Russian)
171. V.G. Vorobjev, L.M. Panasyuk, *Non-silver and Unusual Environments for Holography* (Nauka, Leningrad, 1978) pp. 71–81. (in Russian)
172. M.S. Borodkina, I.A. Malakhova, T.V. Cheltsova, Fundam. Bas. Opt. Mem. Environ. **13**, 134 (1982). (in Russian)
173. E.N. Maltseva, F.K. Dzesova, L.I. Bekicheva, A.G. Baratov, *Proceeding of All-Union Conference Single-Layer PTSC for Holographic Data Recording* (Vilnius, May 1980), p. 105. (in Russian)
174. YuP Getmanchuk, N.I. Sokolov, Fundam. Opt. Mem. Med. **14**, 11 (1983). (in Russian)
175. V.I. Barba, O.I. Zhovtanetsky, A.I. Zubrik, et al., *Proceeding of All-Union Conference "Single-Layer PTSC for Holographic Data Recording,"* (Vilnius, May 1980), p. 25. (in Russian)
176. B.D. Eisenberg, A.G. Baratov, Electron. Ind. **10**, 12 (1980). (in Russian)
177. N.G. Nakhodkin, N.G. Kuvshinskii, I.M. Pochernyaev, Methods of Recording on Non-silver Carriers **5**, 3 (1974). (in Russian)
178. V.I. Monastyrskaya, E.S. Klimov, N.I. Ludkin, et al., J. Scient Appl. Photo Cinematogrophy **33**(4), 241 (1988). (in Russian)
179. Y.A. Cherkasov, M.S. Borodnina, E.A. Egorov, et al., *New Recording Media for Holography* (Leningrad, 1983) p. 109. (in Russian)
180. I.A. Akimov, Y.A. Cherkasov, M.I. Cherkashin, *Sensitized Photo Effect* (Nauka, Moscow, 1980), 384p. (in Russian)
181. N.G. Kuvshinskii et al., Prob. Holography **3**, 161 (1973). (in Russian)
182. O.V. Galaburda, V.A. Okushko, B.N. Tushkevich, Autometry **1**, 114 (1991). (in Russian)
183. O.V. Galaburda, V.A. Okushko, B.N. Tushkevich, Autometry **1**, 18 (1991). (in Russian)
184. T.L. Gredell, F.J. Spong, RGA Rev. **33**(3), 206 (1973)
185. J.G. Bellamy, D.B. Ostrowsky, M. Poindron, E. Spitz, Appl. Opt. **10**(6), 1558 (1971)
186. D.B. Ostrowsky, M. Poindron, Revue Technigues **3**(4), 655 (1971)
187. M.Y. Bazhenov, D.I. Staselko, *Abstracts of 3 All-Union Conference "Recording Media, Methods and Apparatus of Holography* (Kishinev, 1980), pp. 96–97. (in Russian)
188. O.I. Zhovtanetsky, A.I. Zyubrik, O.G. Levchenko et al., Optoelectron. **1**, 99 (1983). (in Russian)
189. V.I. Anikin, O.J. Korshak, L.M. Panasyuk, et al., *Properties of Supersensitive Materials and Their Use in Holography* (Nauka, Leningrad, 1987), p. 100. (in Russian)
190. V.A. Komarov, O.V. Zaichenko, J. Scientific Appl. Photogr. Cinematography **26**(3), 202 (1981). (in Russian)
191. V.G. Vorobjev, F.I. Dimov, L.M. Panasuk, A.A. Forsch, *New Recording Media for Holography* (Leningrad, 1983), p. 130. (in Russian)
192. A.I. Zubrik, PTE **5**, 229 (1981). (in Russian)
193. V.P. Verhovoi, O.V. Zaichenko, V.A. Komarov et al., Autometry **3**, 107 (1981). (in Russian)
194. V.V. Krukov, Opt.-Mech. Prom-St **4**, 26 (1980). (in Russian)
195. A.A. Ayazyan, L.K. Mamylia, S.M. Savrovsky et al., Autometry. **2**, 84 (1980). (in Russian)
196. A.A. Akaev, S.A. Alymkulov, K.M. Jumaliev, et al., *Holographic Storage Techniques, Conversion and Processing* (Frunze, 1983), p. 20. (in Russian)
197. M.A. Ahmametiev, *Optic-electronic Systems and Devices* (Novosibirsk, 1983), pp. 71–81. (in Russian)

198. V.A. Komarov, S.I. Soroka, A.A. Kozak, Fundam. Opt. Mem. Med. **9**, 139 (1978). (in Russian)
199. M.U. Bajenov, Fundam. Opt. Mem. Environ. **10**, 3 (1979). (in Russian)
200. O.V. Zaichenko, V.A. Komarov, PTE **1**, 238 (1987). (in Russian)
201. S.A. Aleksandrov, V.L. Kazak, Opt. Spectrosc. **57**(5), 894 (1984). (in Russian)
202. R. Dändliker, B. Jneichen, F. Mottier, Opt. Commun. **9**, 12 (1973)
203. V.A. Diakov, I.V. Shchukin, U.A. Aleksandrov, A.O. Bakrunov, E.P. Lyasnikov, Prob. Holography **2**, 145 (1973). (in Russian)
204. S.A. Kirillov-Postnikov, Problems of Holography **4**, 17 (1974). (in Russian)
205. R.I. Utiamishev, *Application of Holography in Medicine and Biology* (Nauka, Leningrad, 1977), p. 4. (in Russian)
206. S.A. Kirillov-Postnikov, G.A. Sobolev, E.P. Suhman, Opt. Instrum. Holography **8**, 19 (1976). (in Russian)
207. T.H. Jeong, P. Ruddle, A. Kuckett, Opt. Soc. Amer. **65**, 1263 (1966)
208. Sh.D. Kakichashvili, A.I. Kovaleva, V.A. Rukhadze, Opt. Spectrosc. **24**(4), 627 (1968). (in Russian)
209. JSh Vieno, P. Smigilskiy, A. Roiye, *Optical Holography, Development and Application* (Mir, Moscow, 1973). (in Russian)
210. L.N. Tanner, J. Sci. Instrum. **43**(5), 346 (1966)
211. B.P. Jugeli, Sh.D. Kakichashvili, Copyright Certificate 366779, 1972. (in Russian)
212. ShD Kakichashvili, B.P. Jugeli, ZhNiPFiK **3**, 179 (1973). (in Russian)
213. K. Murata, K. Kunugi, Appl. Opt. **7**, 1798 (1977)
214. U.A. Tsepelev, V.N. Sazontova, A.E. Sorokina, Symposium on proceeding of first all-union. Scientific and technical. Opt Instrum. Holography **8**(9) (1976). (in Russian)
215. M.E. Nemirovsky, News of medical instrumentation A, 96 (1973). (in Russian)
216. G. von Bally, Laser + Electro Opt. **2**(32) (1977)
217. G. von Bally, *Applications of Holography and Optical Data Processing*, ed. by E. Marom, A.A. Friesem (Pergamon Press, 1977), p. 593
218. A.M. Zarubin, A.W. Geiger, G. von Bally, H.H. Scheld, The Journal of Heart Valve Disease **2**, 440 (1993)
219. P. Greguss, Opt. and Las. Technol. **8**(4), 153 (1976)
220. X.-Y. Su, A.M. Zarubin, G.von Bally, Opt. Comm. **105**, 379 (1994)
221. D. Dirksen, G. von Bally, *Optics for Protection of Man and Environment Against Natural and Technological Disasters,* ed. by G. von Bally, H.I. Bjelkhagen (Elsevier Sciences, 1999)
222. G. von Bally, D. Dirksen, F. Matthes, *Optical Methods in Biomedical and Environmental Sciences,* ed. by H. Ohzu, S. Komatsu (Elsevier Sciences, 1994)
223. H. Ohzu, K. Habara, Proceedings of the IEEE **84**(5) (1996)
224. O.P. Bolshakov, T.A. Ilinskaya, N.E. Protsenko, *Proceeding of the Conference Achievements of Biomechanics in medicine.* (Riga, 1986), p. 91. (in Russian)
225. V.G. Komar, *Proceedings of the Second All-Union Conference on Holography* (Kiev, 1975), p. 82. (in Russian)
226. V.G. Komar, *Proceedings of the Third All-Union Conference of Holography* (Leningrad, 1978), p. 171. (in Russian)
227. E. Leyt, U. Upatnieks, K. Khoem, N. Massey, Foreign Rad. Electron. **1**, 142 (1967). (in Russian)
228. H. Paques, P. Smigielski, Opt. Acta **12**(4), 359 (1965)
229. B.P. Konstantinov, Suc. Phys. Sci. 2, 185 (1970). (in Russian)
230. Y.N. Denisyuk, Opt.-Meh. Prom-St., **18** (1967). (in Russian)
231. E.N. Leith, D.B. Brumm, S.S. Hsiao, Appl. Opt. **11**(9), 2016 (1972)
232. E.G. Matinyan, E.I. Polyanskaya, N.M. Elagina, Tools and methods of quantum electronics in medicine, **200** (1976). (in Russian)
233. L.G. Kopeiko, K.K. Huhlaev, Proceedings of the first proceeding scientific and engineering, in *Symposium Opting Instrument Making Holography*, vol. **8** (1976), p. 36 . (in Russian)
234. S.D. Leonard, J.L. Calkins, J. Varner, R. Zech, J. Opt. Soc. Amer. **60**, 1668 (1970)

235. R.M. Tamarova, U.A. Tsepelev, A.G. Reznichenko, Opt. Instr. Making Holography **8**, 7 (1976). (in Russian)
236. N.K. Sheridon, Appl. Phys. Lett. **12**, 316 (1968)
237. D.A. Ansley, Appl. Opt. **9**(4), 815 (1970)
238. U.I. Filenko, Problems of Holography **1**, 83 (1973). (in Russian)
239. U.I. Filenko, Opt. Spectrosc. **33**(3), 596 (1972). (in Russian)
240. D.I. Mirovitsky, V.F. Dubrovin, I.F. Budagyan, V.N. Sobolev, V.I. Tsvetayev, A.V. Goncharov, Prob. Holography **5**, 161 (1975). (in Russian)
241. I.F. Budagyan, V.F. Dubrovin, N.N. Evtihiev, D.I. Mirovitsky, V.V. Usatyuk, Prob. Holography **2**, 109 (1973). (in Russian)
242. Holographic device for cavity diagnostics. VDNKh booklet 1974. (in Russian)
243. D.I. Mirovitsky, *Proceedings of VI All-Union School of Holography* (Leningrad, 1974), p. 494. (in Russian)
244. E.G. Matinyan, *Proceedings of the VIII All-Union School of Holography* (Leningrad, 1976), p. 298. (in Russian)
245. I.M. Arefief, A.V. Bibikova, K.K. Huhlaev, *Tools and Methods of Quantum Electronics in Medicine* (Publishing House of Saratov, Saratov, 1976), p. 231. (in Russian)
246. A.I. Semenov, I.N. Ushkova, *Tools and Methods of Quantum Electronics in Medicine* (Publishing House of Saratov, Saratov, 1976), p. 241. (in Russian)
247. S.D. Pletnev, N.D. Deviatkov, V.P. Belyaev, M.Sh. Abdurazakov, *Gas Lasers in Experimental and Clinical Oncology* (Meditsina, Moscow, 1978), p. 36. (in Russian)
248. I.M. Arefief, A.V. Bibikova, *Report on topic 7348. Investigations with the Purpose of Developing Standards and Regulations for Safety in the Use of Lasers in Health Care* (Moscow, 1975). (in Russian)
249. D.I. Staselko, Y.N. Denisyuk, A.G. Smirnov, J. Sci. Appl. Photogr. Cinematography **15**(2) (1970). (in Russian)
250. M.M. Petiaev, *Biophysical Approaches to the Diagnostics of Malignant Tumors* (Meditsina, Moscow, 1972). (in Russian)
251. M.M. Petiaev, M.Sh. Abdurazakov, A.I. Ilyina, *Proceeding of the Conference Occupational Health and Biological Effects of Lasers* (Moscow, 1974), p. 25. (in Russian)
252. R.E. Kavetsky, V.G. Chudakov, N.F. Gamaleya, *Proceedings of the Annual Scientific Conference KNIIEKO* (Kiev, 1966), p. 39. (in Russian)
253. A.A. Vishnevsky Jr., *The Possibility of Using Optical Quantum Generators in Surgery (Clinical and Morphological Studies), Abstract of the Dissertation* (Moscow, 1973). (in Russian)
254. *Temporary Sanitary Rules when Working with Optical Quantum Generators* (Ministry of Health Moscow, 1972). (in Russian)
255. *Interim Guidelines for Working with the Laser Radiation in Medical Establishments* (Ministry of Health Moscow, 1978). (in Russian)
256. Sh.D. Kakichashvili, A.M. Labartkava, B.P. Dzhugeli, *The Use of Holography in Medicine and Biology* (Nauka, Leningrad, 1977), p. 7. (in Russian)
257. V.V. Smirnov, *Materials of the 6th All-Union School of Holography* (Leningrad, 1974), p. 520. (in Russian)
258. D. Handbawnik, Optik **45**(1), 21 (1976)
259. F.S. Harris, G.S. Sherman, B.H. Billing, Appl. Opt. **5**, 665 (1966)
260. M.J. Landry, Appl. Phys. Lett. **9**, 303 (1966)
261. A.D. Jacobson, V. Evtuhov, J.K. Neeland, Appl. Phys. Lett., **120** (1969)
262. L.N. Beinarovich, N.P. Larionov, A.V. Lukin, K.S. Mustafin, Opt. Spectrosc. **30**(6), 345 (1971). (in Russian)
263. V.A. Vanin, *Prospect VDNKh, PNII* (Electronics, Moscow, 1974). (in Russian)
264. V.A. Vanin, L.N. Vagin, *Proceedings of the 7th All-Union School of Holography* (Leningrad, 1975), p. 225. (in Russian)
265. I.P. Antonov, S.D. Bezzubik, S.K. Dik, V.V. Dubovik, I.V. Markhvida, L.V. Tanin, The physical factors and technical means in medicine **32** (1986). (in Russian)

266. Laser Focus **4**(9), 16, 18 (1968)
267. Bell Labs, Rec. **46**(3), 100 (1968)
268. M.C. King, Appl. Opt. **7**(8), 1641 (1968)
269. L.V. Tanin, S.K. Dik, S.A. Aleksandrov, Laser Optics-93: Proc. Reports **2**, 627 (1993). (in Russian)
270. L.V. Tanin, S.K. Dik, S.A. Aleksandrov, Proceeding of Russian academy of sciences. Phys. Ser. **59**(6), 84 (1995). (in Russian)
271. S.A. Aleksandrov, S.K. Dik, L.V. Tanin, V.A. Lapina, R.M. Tanina, Peripheral Nervous System **16**, 13 (1993). (in Russian)
272. I.V. Markhvida, L.V. Tanin, V.V. Dubovik, Arch. Anatomy Histol. Embryol. **5**, 73 (1988). (in Russian)
273. V.M. Ginzburg, B.M. Stepanov (eds.), *Holography, Methods and Devices* (Sov. Radio, Moscow, 1974). (in Russian)
274. H.J. Caulfield, Appl. Phys. Lett. **16**(6), 234 (1967)
275. F.M. Mottier, Appl. Phys. Lett. **15**(2), 44 (1969)
276. G.A. Sobolev, E.R. Tsvetov, Video Film Telev. **2**, 27 (1968). (in Russian)
277. I.L. Drobot, L.V. Tanin, A.S. Rubanov, *Proceedings of VI All-Union Conference on Holography* (Vitebsk, 1990), p. 158. (in Russian)
278. A.S. Rubanov, L.V. Tanin, L.V. Vassilieva, A.N. Kalinin, V.A. Bursky, F.W. Vidmant, Patent N 4420218 (U.S.) (13 Dec 1983). (in Russian)
279. A.S. Rubanov, L.V. Tanin, L.V. Vassilieva, A.N. Kalinin, Copyright Certificate 34301 (NRB) (10 May 1984). (in Russian)
280. A.S. Rubanov, L.V. Tanin, L.V. Vassilieva, A.N. Kalinin, V.A. Bursky, F.W. Vidmant, patent 8102263 (France) (25 Feb 1984). (in Russian)
281. A.S. Rubanov, L.V. Tanin, L.V. Vassilieva, A.N. Kalinin, Copyright Certificate 1099310 (the USSR) (22 Feb 1984). (in Russian)
282. A.S. Rubanov, L.V. Tanin, L.V. Vassilieva, A.N. Kalinin, V.A. Bursky, F.W. Vidmant, Copyright Certificate 1108903 (the USSR) (15 Apr 1984). (in Russian)
283. A.S. Rubanov, L.V. Tanin, L.V. Vassilieva, A.N. Kalinin, V.A. Bursky, F.W. Vidmant, patent 225838 (GDR) (07 Aug 1985). (in Russian)
284. A.S. Rubanov, L.V. Tanin, L.V. Vassilieva, A.N. Kalinin, V.A. Bursky, F.W. Vidmant, patent 3103895 (FRG) (23 March 1987). (in Russian)
285. A.S. Rubanov, L.V. Tanin, L.V. Vassilieva, A.N. Kalinin, Copyright Certificate 253 253 (Czechoslovakia) (04 May 1988). (in Russian)
286. A.S. Rubanov, L.V. Tanin, L.V. Vassilieva, A.N. Kalinin, V.A. Bursky, F.W. Vidmant, Copyright Certificate 253 254 (Czechoslovakia) (05 May 1988). (in Russian)
287. G.I. Borovkov, F.V. Vidmant, A.V. Ermakov, I.P. Zyatkov, A.N. Kalinin, V.V. Mogilny, A.V. Pavlov, A.A. Stavrov, S.N. Sharashkin, L.V. Tanin, A.P. Shkadarevich, Copyright Certificate 1345874 (the USSR) (15 Jun 1987). (in Russian)
288. Sh.D. Kakichashvili, *Proceedings of the II All-Union School of Holography* (Leningrad, 1971). (in Russian)
289. Sh.D. Kakichashvili, B.P. Dzhugeli, J. Appl. Sci. Photogr. Cinematography **18**(3), 179 (1973). (in Russian)
290. V.A. Vanin, *Prospect VDNKh PNIN* (Electronics, Moscow, 1974). (in Russian)
291. V.A. Vanin, L.N. Vagin, *Proceedings of the VIII All-Union School of holography* (Leningrad, 1975), p. 225. (in Russian)
292. V.A. Vanin, Quantum Electron. 5(7), 1413 (1978). (in Russian)
293. V.A. Vanin, L.N. Vagin, Journal of scientific and applied photography and cinematography 20(6), 416 (1975). (in Russian)
294. V.A. Vanin, L.G. Nazarov, Problems of Holography A, **53** (1973). (in Russian)
295. V.A. Vanin, L.N. Volin, V.A. Korolev, *Recording Media for Holography* (Nauka, Leningrad, 1975). (in Russian)
296. I.N. Kotova, S.N. Ginak, A.N. Kalinin, F.V. Vidmant, L.V Tanin, *Proceeding of the Conference. Holography in Industry and Research* (Grodno, 1986), p. 45. (in Russian)
297. L.V. Tanin, *Proceedings of the VII Intern. Conference. Holography EXPO-2010* (Moscow, 2010), p. 8. (in Russian)

Further Reading

298. B.G. Turukhano, N. Turukhano, ZhTF **38**(4), 757–758 (1968). (in Russian)
299. B.G. Turukhano, N. Turukhano, Yu.M. Lavrov, Opt. Laser Technol. **28**(4), 251 (1996)
301. B.G. Turukhano, V.P. Gorelik, S.N. Kovalenko, N. Turukhano, Opt. Laser Technol. **28**(4), 263 (1996)
302. N. Turukhano, *Proceeding Proceedings V International Conference Holoexpo-2008* (St. Petersburg, 2008), p. 95. (in Russian)
303. N.G. Vlasov, Y.P. Presnyakov, E.G. Semenov, Opti. Spectrography **37**(2), 369 (1974). (in Russian)
304. N.M. Spornik, V.I. Yanichkin, Opt. Prom-St **8**, 38 (1971). (in Russian)
305. N.M. Spornik, ZhPS **34**(3), 534 (1981). (in Russian)
306. N.M. Spornik, Opt. Meh.-Prom-St **1**, 74 (1974). (in Russian)
307. N.M. Spornik, Opt.-Meh. Prom-St **2**, 77 (1973). (in Russian)
308. N.M. Spornik, ZhTF **1**(4), 184 (1975). (in Russian)
309. I.S. Zeylikovich, N.M. Spornik, ZhTF **46**(5), 1054 (1976). (in Russian)
310. M.M. Butusov, ZhTF **42**, 409 (1972). (in Russian)
311. A.F. Belozerov, N.M. Spornik, Opt.-Fur. Prom-St **3**, 9 (1971). (in Russian)
312. I.S. Zeylikovich, N.M. Spornik, Opti. Spectrosc. **46**(2), 393 (1979). (in Russian)
313. K.S. Mustafin, V.A. Seleznev, Opt. Spectrosc. **32**(5), 993 (1972). (in Russian)
314. I.S. Zeylikovich, A.M. Lyalikov, N.M. Spornik, Quantum Electron. **13**(7), 1386 (1986). (in Russian)
315. I.S. Zeylikovich, N.M. Spornik, A.J. Smolyak, et al., *Optical Holography and Its Application in Industry* (LDNTP, Leningrad, 1976), p. 96. (in Russian)
316. I.S. Zeylikovich, N.M. Spornik, Opt.-Meh. Prom-St **9**, 77 (1975)
317. J.S. Zeylikovich, N.M. Spornik, Opt. Spectrosc. **53**(5), 863 (1982). (in Russian)
318. A.P. Burmakov, A.A. Labuda, V.M. Lutkovsky, Eng. Phys. J. **49**(3), 499 (1975). (in Russian)
319. A.V. Berdennikov, N.M. Spornik, J. Scient. Applic. Photogr. Cinematography **4**, 295 (1975). (in Russian)
320. B.A. Belgorodsky, M.M. Butusov, U.G. Turkevich, Autometry **1**, 47 (1972). (in Russian)
321. L.A. Borinyak, S.I. Gerasimov, V.A. Zhilkin, Optoelectronics **1**, 17 (1982). (in Russian)
322. M.M. Butusov, Opt. Spectrosc. **37**, 532 (1974). (in Russian)
323. M.M. Butusov, U.G. Turkevich, J. Sci. Appl. Photograph. Cinem. **4**, 303 (1971). (in Russian)
324. N.G. Vlasov, A.E. Shtanko, *Proceedings of the VII All-Union School on Holography* (Leningrad, 1975), p. 212. (in Russian)
325. S.T. De, A.G. Kozachok, A.V. Loginov, *Optical Holography and Its Application* (Len. House of Scientific Technical. Propaganda, Leningrad, 1974), p. 12. (in Russian)
326. V.O. Indisov, V.S. Pisarev, V.P. Schetinov, V.V. Yakovlev, in *Deformation and Destruction of the Materials and Structures of the Nuclear Technology* (Energoatomizdat, Moscow, 1983), p. 45. (in Russian)
327. I.S. Klimenko, V.P Ryabukho, Opt. Spectrosc. **59**(2), 398 (1985). (in Russian)
328. I.S. Klimenko, V.P. Ryabukho, B.V. Feduleev, Opt. Spectrosc. **55**(1), 140 (1983). (in Russian)
329. V.N. Kudreev, U.A. Panibrattsev, G.S. Safronov et al., MicroelectrOnics **8**, 166 (1979). (in Russian)
330. O.G. Lisin, Opt. Spectrosc. **50**(3), 521 (1981). (in Russian)
331. V.S. Listovets, Yu.I. Ostrovsky, ZhTF **44**, 1345 (1974). (in Russian)
332. N.V. Morozov, H.P. Alum, Yu.I. Ostrovsky, ZhTF **51**, 355 (1981). (in Russian)
333. I.M. Nagibina, V.V. Hopov, Proc. Univ. Mach. Build. **26**(2), 80 (1983). (in Russian)
334. M.P. Petrov, S.I. Stepanov, A.V. Homenko, *Photosensitive Electrooptic Media in Holography and Optical Information Processing* (Nauka, Leningrad, 1983), 270p. (in Russian)
335. A.I. Tselikov, B.A. Morozov, O.G. Lisin, V.S. Aistov, Probl. Strength **6**, 106 (1976). (in Russian)
336. V.P. Schepinov, V.V. Yakovlev, ZhTF **49**(5), 1005 (1979). (in Russian)

337. H. Bjelkhagen, N. Abramson, Appl. Opt. **16**, 1727 (1977)
338. H. Bjelkhagen, N. Abramson, Appl. Opt. **17**, 187 (1977)
339. P.M. Boone, R. Verbiest, Opt. Acta **16**(5), 555 (1969)
340. P.M. Boone, Opt. Laser Technol. **4**(4), 162 (1972)
341. P.M. Boone, Opt. Acta **22**, 579 (1975)
342. J.M. Burch, A.E. Ennos, R.J. Wilton, Nature **209**, 1015 (1966)
343. Y. Chen, Trans. ASME **E42**, 499 (1975)
344. R. Dändliker, B. Ineichen, F.M. Mottier, Opt. Comm. **9**, 412 (1973)
345. R. Dändliker, Progr. Opt. **17**, 1 (1980)
346. P. Hariharan, Z.C. Hegedus, Opt. Comm. **9**, 152 (1973)
347. L.O. Heflinger, R.F. Wuerker, Appl. Phys. Lett. **15**, 28 (1969)
348. P.W. King III, Appl. Opt. **13**, 231 (1974)
349. K.A. Stetson, Opt. Technol. **2**(2), 80 (1970)
350. T. Tsuruta, N. Shiotake, Y. Itoh, Jap. J. appl. phys. **6**, 661 (1967)
351. T. Tsuruta, N. Shiotake, J. Tsujiuchi et al., Jap. J. Appl. Phys. **6**, 419 (1967)
352. C.H.F. Velzel, J. Opt. Soc. Amer. **10**, 419 (1970)
353. M. Yonemura, Opt. Acta **27**(10), 1537 (1980)
354. V.I. Evseev et al., Orthopaedics. Traumatol. Prosthet. **7**, 73 (1980). (in Russian)
355. O.J. Burchett, Mater. Eval. **30**, 25 (1972)
356. P. Greguss, Opt. Laser. Technol. **7**(6), 253 (1975)
357. N. Bochner, J. Ripman, Appl. Phys. **9**, 1825 (1976). (in Russian)
358. V.I. Antonov, U.A. Bykovsky A.I. Larkin, Use Holography Med. Biol. **81** (1977). (in Russian)
359. W. Karmowski et al., J. Techn. Phys. **23**(3–4), 239 (1982)
360. K.M. Hagenbuck, S.P. Sarraf, Opt. Lett. **7**(7), 325 (1982)
361. P.M. Saari, R.K. St. Charles, A.K Rebane, Quantum Electron. **12**(4), 672 (1985). (in Russian)
362. G. von Bally, Opt. Sci. **18** (1979)
363. G. von Bally, H.I. Bjelkhagen (Elsevier Science Publishers B.V. 2, 1993)
364. V.I. Adyalov, V.E. Makeev, I.R. Utyamyshev, Tools and methods of quantum electronics in medicine **219** (1976). (in Russian)
365. V.I. Adyalov, V.E. Makeev, I.R. Utyamyshev, *The Use of Holography in Medicine and Biology* (Nauka, gorod, Leningrad, 1977), p. 67. (in Russian)
366. M.N. Kochetkov, N.V. Biryukova, *Holographic Methods of Information Processing* (Moscow, 1978). (in Russian)
367. B.D. Komarov, Yu.A. Bykovsky, A.I. Larkin et al., J. Med. Sci. USSR **11**, 87 (1982). (in Russian)
368. M.L. Pool, H.D. Spangerberg, Laser Unconventional Opt. **61**, 3 (1976). (in Russian)
369. E.D. Pigulevsky, A.F. Ryzhkevich, Ultrasound, **46** (1983). (in Russian)
370. K.A. Stetson, R.L. Powell, JOSA **56**, 1161 (1966)
371. A.E. Ennos, J. Phys. Ser. E: Sci Instrum. **1**, 731 (1968)
372. K. Biedermann, N.E. Molin, J. Phys. Ser. E.: Sci Instrum. **3**, 669 (1970)
373. E. Wolf, JOSA **60**(1), 16 (1970)
374. N.S. Stromilov, Quantum Electron. **5** (1971). (in Russian)
375. R.D. Matulka, D.I. Collins, J. Appl. Phys. **42**(3), 376 (1971)
376. V.A. Afanasyeva, K.S., Mustafin, V.A. Seleznev, Opti. Spectrosc. **33**(5–6), 1183 (1972). (in Russian)
377. R.J. Adams, Proc. Techn. Programme Elect.-Opt. Int. Conf., Brighton, **3**, 337 (1972)
378. I.S. Klimenko, E.I. Kucheryavenko, E.G. Matinyan, Opt. Spectrosc. **34**(2), 360 (1973). (in Russian)
379. R.L. Kurtz, M.P. Laurence, Appl. Opt. **12**(12), 2815 (1973)
380. M.H. Zambuto, A. Hanaty, Opt. Eng. **14**(4), 372 (1975)
381. G.V. Dreiden, E.N. Shedova, *Optic Holography* (Leninrgad, 1975), p. 71. (in Russian)
382. A.S. Bogomolov, A.S. Romashov, V.G. Seleznev, Opt. Spectrosc. **38**(5), 85 (1975). (in Russian)
383. V.P. Chernov, B.B. Gorbatenko, Opt. Spectrosc. **39**(5), 963 (1975). (in Russian)

384. C.M. Vest, Appl. Opt. **14**(7), 1601 (1975)
385. A.N. Titov, D.I. Mirovitskaya, Probl. Holography **6**, 56 (1975). (in Russian)
386. M.Y. Lieberman, J. Ophthalmol. **4**, 77 (1975). (in Russian)
387. G.A. Tyler, B.J. Thompson, Opt. Acta **23**(9), 685 (1976). (in Russian)
388. R. Dändliher, E. Marom, J. Opt. Soc. Amer. **66**(1), 23 (1976). (in Russian)
389. H. Vachutka, Optik **44**(3), 289 (1976). (in Russian)
390. O.B. Serov, O.N. Slobodchikova, O.A. Pletnev, Biol. Sci. **9**, 126 (1976). (in Russian)
391. A.M. Kondratenko, A.N. Skripsky, Optoelectronics **2**, 3 (1977). (in Russian)
392. P.R. Wedendal, H.I. Bjelkhagen, Laser Appl. Med. and Biol. **3**, 221 (1977). (in Russian)
393. W.B. Saunders, Gardner. Appl. Opt. **16**(9), 2368 (1977)
394. V.T., Chernykh, I.N. Zelinsky, WMD **10**, 26 (1977). (in Russian)
395. H. Royer, Opt. Comm. **20**(1), 73 (1977). (in Russian)
396. N. Abramson, Appl. Opt. **16**(9), 2521 (1977)
397. S. Toyooka, Appl. Opt. **16**(4), 1054 (1977). (in Russian)
398. W.T. Armstory, Appl. Opt. **16**(1), 229 (1977). (in Russian)
399. E.M. Birger, Proc. Central Upper-Air Observatory **130**, 73 (1977). (in Russian)
400. W. Schumann, Optik **47**(4), 391 (1977). (in Russian)
401. H. Kohler, Optik **47**(2), 135 (1977); EI KIT 26, **21** (1977). (in Russian)
402. V.V. Smirnov, G.B. Semenov, *Application of Holography in Medicine and Biology* (Nauka, gorod, Leningrad, 1977), p. 12. (in Russian)
403. N. Abramson, H. Bjelkhagen, Appl. Opt. **17**(2), 187 (1978)
404. S. Cha, C.M. Vest, JOSA **69**(10), 1474 (1979)
405. B. Hofstetter, et al., Laser-Electro-Opt. **3**, 30 (1978). (in Russian)
406. T. Kubota, T. Ose, Opt. Comm. **28**(2), 159 (1979)
407. F.T.S. Yu, A. Tai, H. Chen, Appl. Opt. **18**(2), 212 (1979). (in Russian)
408. K. Murata, N. Baba, K. Kunugi, Optic **53**(4), 285 (1979)
409. H.-G. Unger, *Planar Optical Waveguides and Fiber* (Mir, Moscow, 1980). (in Russian)
410. A.A. Rassokha, *13 Proceeding School on the Physical Principles of Holography* (Sochi, 1981), p. 104. (in Russian)
411. Ch. Rady, S. Petru, Opt. Comm. **39**(6), 351 (1981)
412. M.W. Berus, Science **213**(4507), 305 (1981)
413. I.S. Amosov, *Modern Aspects of Medical Roentgenology and Radiology* (Obninsk, 1981), p. 26. (in Russian)
414. E.A. Morozova, A.M. Prokhorov, V.R. Savransky, et al., Reports **261**(6), 1460 (1982). (in Russian)
415. A.K. Beketova, T.V. Silina, WMD **10**, 7 (1982). (in Russian)
416. A.P. Yakimovich, Opt. Spectrosc. **55**(3), 490 (1983). (in Russian)
417. T.D. Dudderar, J.A. Gilbert, Appl. Phys. Lett. **43**(8), 730 (1983); EI KR. **29** (1984). (in Russian)
418. P. Varga, G. Kish, Quantum Electron. **10**(1), 111 (1983). (in Russian)
419. D.M. Rowley, Opt. Laser Technol. **15**(4), 194 (1983)
420. Ph Lam, J.D. Gaskill, C. Wyant, Appl. Opt. **23**(18), 3079 (1984)
421. R.W. Peterson, G.M. Robinson, R.A. Carlsen, Appl. Opt. **23**(10), 1464 (1984). (in Russian)
422. G.S. Safronov, T.V. Tishko, Ukr. Phys. J. **30**(7), 994 (1985). (in Russian)
423. Yu.I. Ostrovsky, G.V. Dreiden, M. Stoiky, M.I. Etinberg, WMD **10**, 27 (1985). (in Russian)
424. P. Stimpfling, Smigielski. Opt. Eng. **24**(5), 821 (1985)
425. R.J. Pryputniewicz, Opt. Eng. **24**(5), 820 (1985)
426. T. Mitsuo, Z. Tuny, J. Opt. **16**(3), 127 (1985)
427. F. Cyimesi, Z. Fuzessy, Opt. Comm. **53**(1) (1985)
428. V.C. Sainov, S.C. Sainov, D.A. Tontchev, Proc. Bul. Acad. Sci. **39**(10), 67 (1986)
429. G. von Bally, E. Brune, W. Mette, Appl. Opt. **25**(19), 3425 (1986)
430. V.S. Siniakov, Bull. experts. biol. and medicine 102(10) 495 (1986). (in Russian)
431. K. Satoshi, Ya. Koichi, N. Ryo, H. Matogoshi, Appl. Opt. **25**(10), 1638 (1986)
432. H. Zhang, S. Wang, C. Bao, Opt. Eng. **5**, 677 (1986)

433. D.B. Neumann, Opt. Eng. **24**(4), 625 (1985). (in Russian)
434. A.A. Gouchaev, P.B. Vankov, B.P. Zakharchenya, ZhTF Lett. **13**(7), 404 (1987). (in Russian)
435. K. Creath, Appl. Opt. **26**(14), 2810 (1987). (in Russian)
436. Lira, Ch.M. Vest, Appl. Opt. **26**(18), 3919 (1987). (in Russian)
437. Y. Gongksin, Laser J. **8**(3), 164 (1987). (in Russian)
438. F. Shufen, J. Chen, Opt. Comm. **67**(6), 417 (1988). (in Russian)
439. A.A. Bugaev, A.B. Vankov, T.U. Dunayeva, Phys. Firm Body **30**(3), 775 (1988). (in Russian)
440. R. Dändliker, R. Thalmann, Opt. Eng. **24**(5), 824 (1985)
441. J. Chaplin, L. Wood, Successes Phys. Sci. **121**, 331 (1977). (in Russian)
442. A.D. Drake, D.C. Leiner, I.E.E.E. Trans, Biomed. Eng. **31**(7), 507 (1984). (in Russian)
443. A.R. Evstigneev, Y.V. Golubenko, Electron. Mater. Proc. **2**, 77 (1985). (in Russian)
444. I.S. Zeylikovich, A.M. Lyalikov, Quantum Electron. 13(7), 1386 (1986). (in Russian)
445. K. Qin, J.W. Chen, R. Wang, Chin. J. Lasers **14**(7), 420 (1987). (in Russian)
446. I.S. Zeylikovich, A.M. Lyalikov, Quantum Electron. **14**(8) (1987). (in Russian)
447. K. Betzler, A. Grone, P. Voigt, Rev. Sci. Instrum. **54**(4), 652 (1988)
448. D. Choudhury, P.N. Puntambekar, A.K. Chakrab, J. Opt. **22**(1), 6 (1993)
449. G. von Bally (ed.), *Holography in Medicine and Biology* (Munster, 1979)
450. G.H. Caulfield (ed.), *Optical Holography* (Mir, Moscow, 1982), 376p. (in Russian)
451. A.W. Lohmann, J. Opt. Soc. Am. **55**(11), 1555 (1965)
452. G.Y. Sirat, J. Opt. Soc. Am. A **9**(1), 70 (1992)
453. G.Y. Sirat, J. Opt. Soc. Am. A **9**(1), 84 (1992)
454. H.R. Wornington, J. Opt. Soc. Am. **56**(10), 1397 (1966)
455. O. Bryngdahl, A.W. Lohmann, J. Opt. Soc. Am. **58**(5), 625 (1968)
456. G. Cohran, J. Opt. Soc. Am. **56**(11), 1513 (1966)
457. S.B. Aleksandrov, A.M. Bonch-Bruevich, ZhTF **37**, 360 (1967). (in Russian)
458. L.B. Likhterman, *Ultrasound Imaging and Thermal Imaging in Neurosurgery* (Moscow, 1983). (in Russian)
459. R. Konig, S. Mory, A. Rosenfeld, J. Phys. E: Sci. Instrum. **20**(2), 200 (1987)
460. A.L. Mikaelian, A.N. Razumov, U.G. Turkov, et al., JETP Lett. **5**(5), 148 (1967). (in Russian)
461. M.M. Asimov, A.G. Varnahovich, A.N. Rubinov, ZhPS **47**, 389 (1987). (in Russian)
462. P. Shajenko, Appl. Phys. **44**(7), 3200 (1973). (in Russian)
463. B.P. Hildebrand, K.A. Haines, Phys. Lett. **21**(4) (1966); JOSA **57** (1967)
464. T. Tsuruta, N. Shiotake, I. Tsujiuchi, Jap. J. Appl. Phys. **6**(5) (1967). (in Russian)
465. L.R. Vitushkin, M.Z. Smirnov, Measuring Equipment **11**, 19 (1984). (in Russian)
466. M.A. Sreseli, O.P. Bolshakov, et al. Arch. Anat. Histol. Embryol. **71**(9), 104 (1976). (in Russian)
467. G. von Bally, H. Bjelkhagen, *Optics Within Life Sciences 2* (Amsterdam, 1993), p. 13
468. H. Ohzu, K. Habara, *Behind the Scenes of Virtual Reality: Vision and Motion, Fellow* (IEEE, 1996)
469. A.D. Gara, Jap. J. Appl. Phys. **14**(1), 217 (1975). (in Russian)
470. N. Vachutka, Optik **43**(4), 309 (1975). (in Russian)
471. M. Chander, M.M. Bindal, A. Kulshreshtha, B.K. Agarwala, Appl. Opt. **15**(12), 2964 (1976)
472. V.D. Zimin, P.G. Frick, Opt. Spectrosc. **40**(3), 578 (1976). (in Russian)
473. J. Ebbeni, Appl. Opt. **16**(9), 2543 (1977). (in Russian)
474. K. Yoshihara, K. Nakashimo, Jap. J. Appl. Phys. **15**(6), 1169 (1976). (in Russian)
475. P. Hariharan, F.S. Hegedus, Appl. Opt. **15**(4), 848 (1976). (in Russian)
476. N. Abramson, Appl. Opt. **15**(4), 1018 (1976). (in Russian)
477. F. Montel, Rev. Franc. Mec. **60**, 73 (1976). (in Russian)
478. K. Asai, Jap. J. Appl. Phys. **16**(10), 1805 (1977). (in Russian)
479. H.S. Shaalak, J.M. Jonathan, Opt. Acta **25**(11), 1025 (1978)
480. R. Jones, G. Wykes, Opt. Acta **25**(6), 449 (1978). (in Russian)
481. N.P. Harihara, B.F. Oreb, Opt. Comm. **51**(3), 142 (1984)
482. P.C. Gupta, A.K. Aggarwal, Appl. Opt. **15**(12), 2961 (1976). (in Russian)
483. K.A. Stetson, JOSA **66**(6), 627 (1976). (in Russian)

484. C.A. Sciammarella, J.A. Gilbert, Appl. Opt. **15**(9), 2176 (1976). (in Russian)
485. S.T. De, A.G. Kazachok, A.V. Loginov, U.N. Solodkin, Quantum Electron. **4**(1), 103 (1977). (in Russian)
486. U.S. Gubkin, I.M. Nagibina, ZhPS **28**(2), 214 (1978). (in Russian)
487. A.A. Voevodin, V.L. Kazak, ZhTF **52**(7), 703 (1982). (in Russian)
488. A.I. Grimanov, S.T. De, A.V. Kukharenko, V.A. Handogin, Optoelectronics **4**, 90 (1986). (in Russian)
489. K.A. Stetson, Appl. Opt. **14**, 272 (1975)
490. K.A. Stetson, Appl. Opt. **14**(9), 2256 (1975)
491. R.J. Pryputniewicz, K.A. Stetson, Appl. Opt. **15**(3), 725 (1976)
492. R.J. Pryputniewicz, Appl. Opt. **17**(22), 3613 (1978)
493. L.M. Panasuk, *New Recording Media for Holography* (Nauka, Leningrad, 1983), p. 122. (in Russian)
494. R. Urbach, Meier. Appl. Opt. **5**(4), 666 (1966)
495. L.I. Bekicheva, A.G. Baratov, A.P. Zhulai, Recording medium, method and apparatus for holography, **41** (1980). (in Russian)
496. U. Killut, D. Terrell, Opt. Acta **24**(4), 441 (1977)
497. G.I. Goglidze, G.N. Manushkevich, G.N. Ryabukha, *Proceeding pf IV All-Union Conference "Non-silver and Extraordinary Photographic Processes"* (Vilnius, 1980), p. 78. (in Russian)
498. V.P. Kollontaev L.M. Panasyuk, E.P. Pokatilov, *The Dielectric Electronics* (FAN, Tashkent, 1974), p. 230. (in Russian)
499. V.G. Vorobiev, L.M Panasyuk, *The new Recording Medium for Holography* (Nauka, Leningrad, 1983), p. 136. (in Russian)
500. L.N. Lin, J. Opt. Soc. Am. **61**(2), 203 (1971)
501. R.J. Farker, Opt. Acta **25**(8), 787 (1978)
502. K. Frusem, Z. Rav-Noy, Shron. Opt. Eng. **19**(5), 659 (1980)
503. T. Saito, S. Oshura, T. Honda, Tsujichi. Opt. Comm. **16**(1), 90 (1976)
504. V.A. Komarov, S.I. Soroka, J. Scient. Appl. Photogr. Cinematography. **22**(2), 136 (1977). (in Russian)
505. N.G. Kuvshinskiy, N.G. Chuprin, A.A. Kostyuk, Fundam. Opt. Memory Medium **10**, 97 (1979). (in Russian)
506. A.M. Belonozhko, N.G. Kuvshinskii, V.A. Pavlov, et al., J. Sci. Appl. Photogr. Cinematography **33**(2), 133 (1988). (in Russian)
507. Friesem, Katzir, Z. Rav-Noy, et al., Opt. Eng. **19**(5), 659 (1980)
508. Z. Hirshfeld, A.A. Friesem, Z. Rav-Noy, Appl. Phys. **52**(2), 605 (1981)
509. M.Y. Bazhenov, N.G. Kuvshinskii, N.G. Nahodkin, Fundam. Opt. Memory Med. **14**, 3 (1983). (in Russian)
510. S.I. Soroka, S.I. Ratnikov, Fundam. Opt. Mem. Med. **2**, 47 (1980). (in Russian)
511. A.A. Airapetov, A.P. Eliseev, B.I. Sedunov, *Proceeding of III Conference "Non-silver and Extraordinary Photographic Processes,"* (Vilnius, May 1980), p. 106. (in Russian)
512. P.Ya. Ganich, M.N. Nizhnik, V.A. Okushko, Proceedings of the national security agency. Sor. Sci. Sci. **1**, 66 (1984). (in Russian)
513. N.G. Nakhodkin, N.G. Kuvshinskii, *Spatial Light Modulators* (Nauka, Leningrad, 1977), p. 53. (in Russian)
514. F.I. Dimov, A.A. Muzalevskii, L.M. Panasyuk, *Proceeding of III Conference "Non-silver and Extraordinary Photographic Processes,"* (Vilnius, May 1980), p. 132. (in Russian)
515. J.G. Urbach, R.W. Meer, Appl. Opt. **8**(11), 2269 (1969)
516. S.A. Zubko, F.G. Drik, et al., Defectoscopy 9 (1981). (in Russian)
517. O.I. Zhovtanetsky, et al., The use of parasitic memory fototermoplastiks in holographic interferometry **1**, 99 (1983). (in Russian)
518. S.B. Gurevich (ed.), *Spatial Light Modulators* (Nauka, Leningrad, 1977), p. 4. (in Russian)
519. B.P. Jugeli, E.G. Matinyan, *Proceedings of the 3rd All-Union Conference of Holography* (Leningrad, 1978), p. 190. (in Russian)
520. Sh.D. Kakichashvili, Video Film Telev **3**, 17 (1971). (in Russian)

521. Sh.D. Kakichashvili, B.P. Jugeli, *Proceedings of the Conference on Introduction Into the Practice of Museums of the Achievements of Modern Optics* (Moscow, 1975), p. 47. (in Russian)
522. Sh.D. Kakichashvili, V.S. Kontridze, J.N. Antelava, B.P. Jugeli, *New in the Treatment and Diagnostics of Eye Diseases* (Metsnieretba, Tbilisi, 1976), p. 10. (in Russian)
523. N.I. Kirillov, N.V. Vasilieva, E.M. Feldsherov, *Proceedings of the 2nd All-Union School on Holography* (Leningrad, 1975), p. 299. (in Russian)
524. I.S. Klimenko, G.V. Skrotskii, Successes Phys. Sci. **109**, 269 (1973). (in Russian)
525. V.G. Komar, *Proceedings of the II Proceeding Conference on Holography* (Kiev, 1975), p. 82. (in Russian)
526. N.L. Kosobokova, V.S. Obraztsov, *Proceedings of the III Proceeding Conference on Holography* (Leningrad, 1978), p. 352. (in Russian)
527. E.G. Matinyan, E.I. Polyanskaya, N.M. Elagina, *Proceedings of I All-Union Conference "Means and Methods of Quantum Electronics in Medicine,"* (Saratov, 1976), p. 200. (in Russian)
528. S.D. Pletnev, M.Sh. Abdurazakov, A.I. Ilyina, *Proceedings of the Conference "Occupational Health and Biological Effects of Lasers,"* (Moscow, 1974), p. 25. (in Russian)
529. I.R. Protas, Y.A. Krakau, V.I. Mikhailova, P.T. Sidorenkova, Probl. Holography **3**, 76 (1973). (in Russian)
530. Y.I. Rovinskaya, N.S. Gafurova, N.A. Prosalova, V.P. Mikheeva, I.A. Pishchulina, A.M. Sagdeeva, *Proceedings of the IX All-Union School on Holography* (Leningrad, 1977), p. 339. (in Russian)
531. Y.E. Usanov, *Recording Media for Holography*, ed. by N.I. Kirilov (Leningrad, 1975), p. 98. (in Russian)
532. D.B. Brumm, Appl. Opt. **5**, 1946 (1966)
533. D.B. Brumm, Appl. Opt. **6**, 588 (1967)
534. A.L. Burton et al., Tex. Rep. Biol. Med. **29**, 535 (1971)
535. M. Mishkin, T. Eppentsener, World Sci. **8**, 30 (1987). (in Russian)
536. P.M. Ostergard, Patent 2045760 (USA) 1960
537. E.N. Leith, J. Upatnieks, J. Opt. Soc. Amer. **52**, 1123 (1962)
538. R.F. van Lighten, B. Grolman, R. Lawton, Amer. J. Optom. **43**, 351 (1966)
539. A.E. Jones, A.J. McCartney, Inv. Ophthalmol. **5**, 474 (1966)
540. M. Hercher, B. Ruff, J. Opt. Soc. Amer. **57**(1), 113 (1967)
541. J.W.C. Gates, J. Sci. Instr. **1**, 989 (1968)
542. H.J. Caulfield, Lu Sun, I.L. Harris, J. Opt. Soc. Amer. **58**(7), 1003 (1968)
543. H. Kogelnik, K.S. Pennington, J. Opt. Soc. Amer. **58**(2), 273 (1968)
544. L.D. Siebert, Proc. IEEE **56**(7), 7 (1968)
545. K. Biedermann, Optic **28**(2), 160 (1969)
546. H. Nassenstein, N. Dedden, H.J. Metz, M.E. Rieck, D. Schultze, Photogr. Sci. Eng. **13**, 194 (1969)
547. C.S., Vikram, R.S. Sirohi, Appl. Opt. **10**(9), 2195 (1971)
548. R.L. Wiggins, K.D. Vauchan, Arch. Ophth. **88**(1), 75 (1972)
549. C.S. Vikram, R.S. Sirohi, Appl. Opt. **11**(12), 293 (1972)
550. J. Govignon, O. Pomerantzeff, Trans. Am Acad. Opthalmology Octolarynhology **76**(5), 1214 (1972)
551. H. Ohzu, *Proceedings of the International Symposium on Holography in Biomedical Sciences* (New-York, 1973), p. 26
552. M.L. Wolbarsht, D.H. Sliney, Laser Appl. Med. Biol. **2**, 309 (1974)
553. F.S. Barnes, Laser Appl. Med. Biol. **2**, 205 (1974)
554. K.D. Vauggar, R.A. Laing, *Laser Application in Medicine and Biology*, ed. by M.L. Wolbarsht (Springer, Boston, 1974), p. 77
555. A.N. Rosen, Opt. Laser Technol. **7**(3), 127 (1975)
556. J. Upatnieks, C.D. Leonard, B.J. Martilla, *International Optical Computing Conference* (IEEE, 1975) 108

557. I.L. Calkins, *A Specular Illuminating Arrangement for Holographically Stress-Testing Corneal Wounds in Post-Operative Patients, paper presented at Photo-optical Instrumentation Symposium* (San Diego, 1977)
558. T. Matsumoto, R. Nagata, M. Saishin, T. Matsuda, S. Nakao, Appl. Opt. **17**(22), 3538 (1978)
559. N.L. Kosobokova, Yu.E. Usanov, Opto-Meh. Prom-St **12**, 57 (1975). (in Russian)
560. O.V. Andreeva, Z.A. Zagorsk, T.I. Kadysh, V.I.Sukhanov, *Optical Holography* (Leningrad, 1972). (in Russian)
561. N.I. Kirillov, *Recording Media for Holography* (Nauka, Leningrad, 1975), p. 5. (in Russian)
562. A.D. Vorzobova, A.A. Leshchev, A.L. Semenov, V.G. Sidorovich, D.I. Staselko, Opt. Spectrosc. **45**(4), 779 (1978). (in Russian)
563. M.P. Gapoyan, E.G. Zemtsova, Opto-Meh. Prom-St **2**, 2 (1976). (in Russian)
564. Yu.E. Usanov, M.P. Ermolaev, Opto-Meh. Prom-St **2**, 39 (1972). (in Russian)
565. O.V. Andreeva, V.I. Sukhanov, Opto-Meh. Prom-St **30**(4), 780 (1971). (in Russian)
566. N.G. Maslenkova, A.S. Petrenko, N.I. Kirillov, *Proceeding of GosNIIHim Photo Project* (1975). (in Russian)
567. P. Hariharan, C.S. Ramanathan, G.S. Kaushik, Appl. Opt. **12**(3), 611 (1973). (in Russian)
568. N.I. Kirillov, *High-Resolution Photographic Materials for Holography and Their Processing* (Nauka, Moscow, 1979). (in Russian)
569. K.S. Pennington, Y.S. Harper F. Lanung, Appl. Phys. Lett. **18**(3), 80 (1971)
570. T.G. Ovechkina, L.P. Vakhtanova, et al., J. Sci. Appl. Photogr. Kinematography **21**(5) (1976). (in Russian)
571. N.G. Maslenkova, A.S. Petrenko, I.F. Tolkacheva, I.E. Gaponenko, N.I. Kirillov, *Registering Media for Holography* (Nauka, Leningrad, 1975), p. 110. (in Russian)
572. J.P. Sosnowski, H. Kogelnik, Appl. Opt. **9**(9), 2188 (1970)
573. N.N. Vsevlodov, V.A. Poltoratsky, ZhTF **55**(10), 2093 (1985). (in Russian)
574. U. Schnars, J. Opt. Soc. Am. A **11**(7), 2011 (1994)
575. G. Pedrini, H.J. Tiziani, Y. Zou, Opt. Laser Eng. **22**, 199 (1997)
576. G. Pedrini, H.J. Tiziani, Y. Zou, J. Mod. Opt. **42** (1995)
577. A.L. Barannikov, N.M. Ganzherli, S.B. Gurevich, V.B. Konstantinov, I.A. Maurer, S.A. Pisarevskaya, B.F. Ryadinsky, A.A. Serebrov, V.N. Sobolev, V.M. Stolovitzky, M.S. Cheberyak, D.F. Chernih, ZhTF Lett. **9**(11), 659 (1983). (in Russian)
578. N.M. Ganzherli, P.V. Maurer, P.V. Granskiy, ZhTF **74**(A), 68 (2004). (in Russian)
579. K. Creath, Appl. Opt. **24**(8), 3035 (1985)
580. S.I. Stepanov, *International Trends in Optics*, ed. by J.W. Goodman (Academic Press, New York, 1991), Ch. 9
581. V.V. Tuchin (ed.), *Handbook of Optical Biomedical Diagnostics*, 2nd edn. (SPIE Press, Bellingham, 2002), 637p
582. L. Sekaric, M. Zalalutdinov, S.W. Turner, A.T. Zehnder, J.M. Parpia, H.G. Craighead, J. Appl. Phys. **80** (2002)
583. I.P. Gurov, Opt. J. **67**, 17 (2000). (in Russian)
584. L.F. Yu, L.L. Cai, J. Opt. Soc. Am. A **18**, 1033 (2001)
585. Zh Liu, G. Centurion, G. Panotopoulos, J. Hong, D. Psaltis, Opt. Lett. **27**, 22 (2002)
586. M. Sato, B.E. Hubbard, L.Q. English, A.J. Sievers, B. Ilic, D.A. Czaplewski, H.G. Craighead, CHAOS **13**(2), 702 (2003)
587. U. Schnars, W. Jueptner, *Digital Holography* (Springer, Berlin, 2005), p. 164
588. A.P. Popov, A.V. Priezzhaev, *The Method of Computing the Effectiveness of the Protective Properties of the Nanoparticles at Irradiation of Materials and Biological Tissues by Light in the UV-A and UV-B range* (Moscow, 2006) 36 p. (in Russian)
589. S. Girolamo, A.A. Kamshilin, R.V. Romashko, Yu. N. Kulchin, J.-C. Launay, Opt. Express **15**(2), 545 (2007)
590. V.P. Tychinskiy, UFN **177**, 535 (2007). (in Russian)
591. V.S. Gurevich, V.I. Redkorechev, A.M. Isaev, V.E. Gaponov, M.E. Gusev, I.V. Alekseenko, V.A. Melnikov, V.A. Kalinin, *Proceedings of V International Conference "Holography EXPO-2008,"* (St. Petersburg, 2008), p. 100. (in Russian)

592. I.V. Alekseenko, M.E. Gusev, *Proceedings of V International Conference "Holography EXPO-2008,"* (St. Petersburg, 2008), p. 109. (in Russian)
593. R.Kh. Makaeva, *Proceedings of V Internat. conference "Holography EXPO-2008,"* (St. Petersburg, 2008), p. 105. (in Russian)
594. G. Pedrini, I. Alekseenko, W. Osten, H.J. Tiziani, Appl. Opt. **42**(29), 5846 (2003)
595. I.V. Alekseenko, M.E. Gusev, Optoelectronics **44**(1) (2008). (in Russian)
596. M.E. Gusev, I.V. Alekseenko, V.S. Gurevich, *Proceedings of V International Conference "Holography EXPO-2008,"* (St. Petersburg, 2008), p. 189. (in Russian)
597. V.S. Gurevich, V.I. Redkorechev, A.M. Isaev, V.E. Gaponov, M.E. Gusev, I.V. Aleseenko, V.A. Melnikov, V.A. Kalinin, *Proceedings of V International Conference "Holography EXPO-2008,"* (St. Petersburg, 2008), p. 100–105. (in Russian)
598. M.E. Gusev, A.A. Voronin I.V. Alekseenko, V.S. Gurevich, *Proceedings of VI International Conference "Holography EXPO-2009,"* (Kiev, 2009), p. 61. (in Russian)
599. V.I. Redkorechev, I.A. Kulagin, V.S. Gurevich, M.E. Gusev, Yu.N. Zakharov, Opt. Spectrosc. **107**, 433 (2009). (in Russian)
600. A.G. Poleschuk, R.K. Nasyrov, A.E. Matochkin, V.V. Cherkashin, *Proceedings of VI International Conference "Holography EXPO-2009,"* (Kiev, 2009), p. 64. (in Russian)
601. G.G. Levin, G.N. Vishnayakov, V.L. Minaev, A.G. Lomakin, *Proceedings of VI International Conference "Holography EXPO-2009,"* (Kiev, 2009), p. 68. (in Russian)
602. U.D. Lantuh, G.A. Kentsle, S.N. Pashkevich, S.N. Letuta, E.K. Alidzhanov, A.A. Kulsarina, *Proceedings of VI International Conference "Toloekspo-2009,"* (Kiev, 2009), p. 94. (in Russian)
603. S.B. Odinokov, N.M. Verenikina, D.S. Lushnikov, V.V. Markin, E.A. Usovich, A.S. Goncharov, A.I. Nikolaev, *Proceedings of VI International Conference "Toloekspo-2009,"* (Kiev, 2009), p. 102. (in Russian)
604. V.P. Koronkevich, A.I. Lohmatov, A.E. Matochkin, *Proceedings of VI International Conference "Toloekspo-2009,"* (Kiev, 2009), p. 156. (in Russian)
605. V.A. Babenko, A.F. Maly, I.U. Fedorov, *Proceedings of VI International Conference. "Holography EXPO-2009,"* (Kiev, 2009), p. 158. (in Russian)
606. V.I. Redkorechev, I.A. Kulagin, Z.T. Azamatov, V. Gurevich, M.E. Gusev, U.N. Zakharov, *Proceedings of VI International Conference. "Holography EXPO-2009,"* (Kiev, 2009), p. 161. (in Russian)
607. G.N. Vishnyakov, E.Y. Levina, V.L. Minaev, N.N. Moiseev, I.U. Tselmina, *Proceedings of VI International Conference. "Holography EXPO-2009,"* (Kiev, 2009), p. 163. (in Russian)
608. T.A. Efimov, R.V. Romashko, *Proceedings of the VII International Conference "Holography EXPO-2010,"* (Moscow, 2010), p. 237. (in Russian)
609. R.V. Romashko, S. Di Girolamo, Y.N. Kulchin, A.A. Kamshilin, J. Opt. Soc. Am. **27**(2), 311 (2010)
610. M.N. Bezruk, R.V. Romashko, *Proceedings of the VII International Conference "Holography EXPO-2010,"* (Moscow, 2010), p. 398. (in Russian)
611. A.A. Kamchilin, R.R. Romashko, YuN Kulchin, J. Appl. Phys. **105**(2), 2 (2010)
612. I.V. Krasnikov, A.U. Seteykin, A.P. Popov, *Proceedings of the VII International Conference "Holography EXPO-2010,"* (Moscow, 2010), p. 381. (in Russian)
613. V.S., Gurevich, A.M., Isaev, V.E. Gaponov, M.E. Gusev, *Proceedings of the VII International Conference "Holography EXPO-2010,"* (Moscow, 2010), p. 83. (in Russian)
614. V.I. Redkorechev, I.A. Kulagin, Z.T. Azamatov, V.S. Gurevich, M.E. Gusev, *Proceedings of the VII International Conference. "Holography EXPO-2010,"* (Moscow, 2010), p. 97. (in Russian)
615. G.N. Vishnyakov, G.G. Levin, V.L. Minaev, A.G. Lomakin, *Proceedings of the VII International Conference. "Holography EXPO-2010,"* (Moscow, 2010), p. 237. (in Russian)

Chapter 4
Speckle-Optical Methods and Devices for Studying Human Skin and Muscle Tissue

Theoretical and Experimental Prerequisites for Usage of Correlation Properties of Dynamic Speckle-Fields in Medicine

Coherent optics, holographic interferometry, correlated spectroscopy, Doppler anemometry, as well as speckle-photography and speckle-interferometry play an important role in scientific study and development of new metrological means. These speckle-optical methods, which are based on the study of correlated properties of optic fields, give the opportunity to determine such characteristics of objects, which cause diffraction or radiation scattering as a degree of roughness, relief, shift, deformations distribution, velocity, time of processes relaxation, etc. Recently, diffusive object interferometry is intensively developing, which was connected from the very beginning with random interference phenomena—speckles. It is known [1] that speckles appear as a result of interference of many elementary waves, spreading from separate surface diffusers. As the number of such diffusers is great and they randomly change the wave phase, then the resulting intensity distribution and the field phase will be random and such a field can only be described with statistic methods. The statistic character of reflection of coherent laser light from the rough surface, passing through phase screens, turbulent medium, multi-mode waveguides, liquid crystals and a number of other phenomena lead to forming a speckle-structure (Fig. 4.1).

If initially speckles were estimated as optical noise, then for about 20 years new metrological methods have been developed on their basis, especially speckle-interferometry. Study of correlated characteristics of dynamic speckle-fields formed by a moving diffusing object became a natural development of speckle-optics. Movement of the diffuser causes complex change of a speckle-field, which bears however not a stochastic character, but as some studies showed [3–6], rather regular one. Theoretical study of spatial and temporal correlations of speckle-fields becomes complicated, because speckle-fields do not meet the condition of mutual spectral frequency [7], and amplitude spatial and temporal correlated function do not decompose into a product of the spatial and the temporal parts. That is why time intensity fluctuations cannot be analyzed independently from the spatial ones, as it occurs in case of the Brownian movement of diffusing particles.

L. Tanin et al., *Biomedical and Resonance Optics*, Bioanalysis 11,
https://doi.org/10.1007/978-3-030-60773-9_4

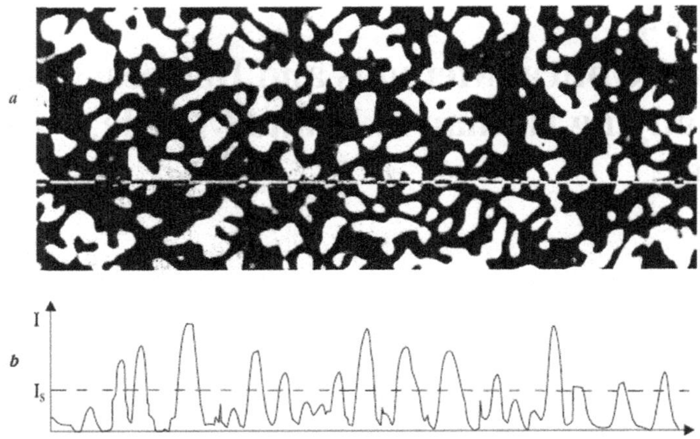

Fig. 4.1 Photo of the speckle-field formed by the reflection of coherent laser light from a diffuse surface (**a**) and the intensity distribution in one of the sections of the speckle-field (**b**). Reprinted from [2] with permission

The random field is most accurately described with joint distribution of density probability of complex amplitude, which gives the possibility to calculate the average values of amplitude fluctuation and intensity fluctuation of field, etc. But in praxis, there were used less informative but much simpler characteristics, for example, correlated functions. The main of them is determined in the following way. The correlated function of amplitude V of field Γ_v and normalized correlated function of amplitude γ_v are

$$\Gamma_v(\vec{x}_1, \vec{x}_2; t_1, t_2) = \langle V(\vec{x}_1, t_1) V^*(\vec{x}_2, t_2) \rangle$$

$$\gamma_v(\vec{x}_1, \vec{x}; t_1, t_2) = \frac{\Gamma_v(\vec{x}_1, \vec{x}_2; t_1, t_2)}{\langle |V(\vec{x}_1, t_1)|^2 \rangle^{1/2} \langle |V(\vec{x}_2, t_2)|^2 \rangle^{1/2}},$$

Correlated function of field intensity Γ_I, normalized correlated function of intensity γ_I and normalized correlated function of intensity fluctuation $\gamma_{\Delta I}$ are

$$\Gamma_I(\vec{x}_1, \vec{x}_2; t_1, t_2) = \langle I(\vec{x}_1, t_1) I(\vec{x}_2, t_2) \rangle$$

$$\gamma_I(\vec{x}_1, \vec{x}_2; t_1, t_2) = \frac{\Gamma_I(\vec{x}_1, \vec{x}_2; t_1, t_2)}{\langle I^2(\vec{x}_1, t_1) \rangle^{1/2} \langle I^2(\vec{x}_2, t_2) \rangle^{1/2}},$$

$$\gamma_{\Delta I}(\vec{x}_1, \vec{x}_2; t_1, t_2) = \frac{\Gamma_I(\vec{x}_1, \vec{x}_2; t_1, t_2) - \langle I(\vec{x}_1, t_1) \rangle \langle I(\vec{x}_2, t_2) \rangle}{\langle I^2(\vec{x}_1, t_1) \rangle^{1/2} \langle I^2(\vec{x}_2, t_2) \rangle^{1/2}}$$

Spatial correlation functions of intensity give the possibility to determine average longitudinal and cross sizes of speckles, contrast of the speckle-field, etc. When the compared fields are diffused in different time moments, then in γ_v and γ_I information about shift and deformation of the object are reflected. If correlated functions of the field amplitude play an important part in holographic interferometry, then in speckle-interferometry intensity correlated functions get main importance. Though process of speckle-interferogram formation can be described with elementary models, when speckles are represented by apertures of the same size on non-transparent display [8], then usage of mathematical apparatus of coherent optics appears to be more convenient, as it gives the opportunity to obtain additional information about decorrelation of speckle-fields [9, 10]. Shift of the peak of correlated function reflects the shift, which in its turn determines orientation and period of fringes [11]. Recently, study of not only spatial but also spatial and temporal correlated functions has become more interesting. These functions can be measured using special electronic device. The latter is more informative and gives the possibility to study the dynamics of the process and in a particular case to describe characteristics of speckle-interferograms. To do this, it is enough to know the duration of exposures and interval between them.

Correlated approach to the research of dynamic speckles, which began in the works [12, 13], was applied to the study of scattering of a laser Gaussian beam on a moving diffusing surface [14–20] and analysis of subjective speckles in the area of image of one- [21, 22] and double-lens [23] optical systems. Main results of these works establish connections between parameters of the object, which moves perpendicular to the optical axis, and characteristics of the speckles: quantity and direction of their shift, size, contrast, degree of decorrelation, lifetime and spatial spectrum of intensity. Two substantially different types of speckle movement were found out: shift and "boiling," and the optical system of observation can cause significant changes in the movement of speckle-structure. Time spectrum of power of diffused radiation was studied by the authors of the works [24, 25]. In other works, it was mentioned that the incidence rate of correlated intensity functions integrated in time and space speckle-structures [26–28].

As a result of analyses of spatial and temporal correlated functions of amplitude and intensity of dynamic speckle-field for two arbitrary space points of images in the work [29], it was established that mechanical trajectory of speckles is three dimensional. Along with lifetime, they are the most important characteristics of dynamic speckles. Depending on the selection of initial point of observation trajectories can be either straight passing through the focus or curved of the second order. Two reasons of decorrelation were quantitatively determined: changing the number of elementary diffusers, which are engaged in formation of a separate speckle and changing phase difference between elementary diffused waves. Influence of these processes on the "boiling" of speckles is significantly different for various areas. The first process prevails in the area of images and the second one—in the area of great defocusing.

Gaussian statistics of speckle-field amplitude is supposed in most experimental studies [1, 30]. Correlated functions of the first-order bear information about such a field, and though they do not transmit significant information about the structure of the diffuser, they almost fully describe its movement. In this case, correlated function of intensity fluctuation is determined by squared absolute value γ_v. A number of speckle-optical methods for object velocity determination are based on this assumption. In a number of cases, this model cannot be applied [20], when elementary waves form incoherently (I); surface roughness is small (II); the total number of elementary waves is small, and application of the central limit theorem is not possible (III). Besides, non-Gaussian statistics emerges in the case of multiple light diffusion [31]. Such a situation occurs in experimental praxis, for example, if diffusing object is illuminated by scattered illumination.

It was noticed that longitudinal shift of the diffuser can significantly influence the change of speckle-field causing decrease of determination accuracy of lateral deformations, vibrations, velocity, up to full inappropriateness of methods. In addition to that knowledge of laws of 3D speckle transformation is of great interest for development of new ways of determination of complex spatial shifts.

As it was mentioned earlier, most widespread description of speckle-fields is an approach based on the analysis of correlated functions, and significant role is played by correlated amplitude functions [32–34] as well as correlated intensity functions of intensity, which give the possibility to describe the measured parameters of speckles, contrast, longitudinal and transversal sizes, shift, degree of decorrelation, lifetime, etc. [35].

As a rule, two main methods of intensity correlation studying are used. In the first one, the correlated intensity function is determined as squared absolute value of amplitude correlated function; in the second one—expression for complex amplitude of specific field realization in the given optical scheme is obtained in a form of integral transformation, then the amplitude correlated function as well as correlated intensity function is calculated through ensemble averaging, as well as correlated intensity function. Ensemble averaging gives the possibility to analyze features of the field, in particular its temporal and spatial stationarity.

If the diffuser velocity is zero, then expressions for parameters of static speckle-field formed by stationary diffuser should follow in a natural way, which was studied in the work [1] for transverse coordinated, and for the longitudinal ones —in the work [36].

In holographic and speckle-interferometry and laser anemometry, diffusive surface is illuminated in a number of cases not with a Gaussian beam, but with diffusive coherent radiation. If laser illumination preliminarily passes through phase chaotic screen, turbulent medium, multi-mode waveguide, then a secondary speckle-structure is formed on the object as a result of scattering. Correlation features of secondary speckles depend significantly on the distribution of field amplitude on the diffusive object, as well as on its spatial and time coherence, and in common case they are not identical to the features of primary speckles [37].

In the work [29], the results of the calculation for phase screen with a right-angled mask are given, which give the possibility to find out main laws of changing correlated functions and judge about the character of speckle movement. It was shown that the secondarily speckle-field will not experience shift, and its changing is accompanied only by "boiling".

Static case of light diffusing with two stationary surfaces due to the speckle-field theory was first examined in the work [38] and was experimentally studied in the work [39]. It was shown that the amplitude probability density is described by K-distribution, which is characterized by great fluctuation value comparing to normal distribution. Along with this, the obtained results can be used for elimination of noises in speckle-interferometry using light guides [40, 41], for increase of image quality in microscopy [42], further development of methods of roughness determination [43].

Further behavior was examined of the speckle-field, which is formed as a result free light diffraction on one or two diffusers, when elementary waves diffused by the whole object surface participate in formation of a separate speckle.

At the same time, presence of optical lens system can significantly change character of forming and motion of speckles. Firstly, intensity in some point of an image is determined by the result of coherent addition of elementary waves scattered in the area around the image point, the size of which is equal to the order of resolution limit. A rather huge number of elementary diffusers must be in this area to form a developed speckle-structure. Secondly, real speckle shift will be determined not only by shift of the object, but also by extension of the optical system. Moreover, correlation of speckle intensity will change depending on aperture, aberrations and objective resolution [21, 44–46]. View of speckle movement can significantly change under different positions of observation plane: from pure movement to “boiling”. These factors lead to the fact that observation of speckle-fields formed by optical systems became an independent part of speckle-optics, and such speckles were named subjective. Now the study of impact of longitudinal movement of the diffuser on changing the structure of the spotted field is almost absent, though the impact of the transverse shift is studied rather fully [23, 47–50].

The great practical value of speckle-photography methods, on the one hand, and their maximal sensitivity to transverse shift, on the other hand, caused rather detailed study of influence of transverse shift of the object on behavior of speckle-fields.

Unlike transverse movement of the diffuser [11, 51, 52] where decorrelation is determined by the fact that some areas of the diffuser leave the illuminated area, and other replace them, during longitudinal movement either new areas are added to the illuminated ones or some of them disappear. That is why $\gamma_{\Delta I}$ can be calculated in different ways. During transverse movement, it can be calculated as a ratio of superposed pupils’ area (the initial one and shifted at the value equal to shift of speckles) to the whole area of the pupil. During longitudinal movement, $\gamma_{\Delta I}$ is determined with some other, symmetrized relative to r_2 and r_1 expression, and decrease as well as increase of r_2 equally leads to decorrelation (where $r_{i=1.2}$ can be interpreted as an effective area of the diffuser, which participates in formation of a separate speckle).

Depending on the illumination conditions and observation “boiling” or shift of the speckle-structure can prevail.

For double-lens optical systems, the dependence of areas of translation and “boiling” of speckles from geometric parameters is considered in the work [22].

Notion “speckle lifetime” τ_0 is introduced for characterizing a dynamic speckle-image [47]. This notion characterizes full decorrelation of compared speckle-images, i.e., disappearance of the peak of correlation function. It was shown above that on the one hand speckle-fields occurring as a result of free diffraction can be considered as a particular case of speckles formed in lens systems, and correlated functions of subjective speckle-fields can be used for their description. On the other hand, the knowledge of properties of optical dynamic speckles significantly helps in the study of changes of subjective speckle-fields. On the other hand, the knowledge of characteristics of objective dynamic speckles significantly helps to study subjective speckle-fields. As analyses showed behavior of speckles formed by the optical system can be studied using two equivalent ways: (a) by examining actual transformation of speckles in image space during longitudinal movement of an object; (b) by formally calculating all the values in the space of objects, and then moving on to real values in image space taking into account changes of phase correlation with a lens, which in fact will come to heterogeneous transformation of object space into image space using a lens, as well as influence of aperture.

If the detection plane is situated behind the focal plane, then speckles "scatter" from the axis when the diffuser approaches the objective and when it moves away they gather. This effect is conditioned by the change of magnification under longitudinal movement.

For the first time, the usage of speckle-field correlation for determining longitudinal shifts was mentioned in the work [53].

The method is based on radial "scattering" of speckles in the image plane, obtaining double-exposure speckle-photographs and forming not traditional equidistant fringes but characteristic system of rings [8]. The upper limit of measurements is restricted by the depth of objective focus used during recording of speckle-structures. This method was further developed in the works [44, 54, 55] where in particular it is suggested that ring diaphragms should be used to increase the quantity and contrast of the observed rings, and manifestation of speckle fine structure is investigated.

But in speckle-interferometry of longitudinal shift, there are a number of problems, which require further study of connected with coherent-optical processing of specklograms, study of optimal conditions of spatial filtration, determination of localization area of speckle-interferograms and extension of measurement range. During the analysis of a specklogram, which reflects the transverse shift of the object, Fourier transformation of specklogram is conducted, and equidistant fringes are observed, which are equivalent to Jung's fringes.

In this case instead of pulse response of the objective, it is necessary to put the expression, which responses to Fourier transformation. This particular case was analyzed in detail in literature, for instance in the works [56–63].

Interferograms of longitudinal shift have the appearance of concentric rings and are observed under other circumstances [53].

Interferograms appearing under optic coherent processing of specklograms, on which pairs of speckles are registered, shifted on the same value are registered, have direct analogy in transverse direction with interference experiments on two pin holes and are named Jung's fringes.

Specklograms of longitudinal shift have another structure [64, 65]. They have symmetry center, each pair of identical spectrums is situated on a radius, and value of separation between them is in direct proportion to distance from the center. Under plane wave diffraction on this specially obtained diffuser, completely different type of interference phenomena appears, which is not well-investigated.

Schemes for obtaining speckle-interferograms using pin hole diaphragms were empirically established [66, 67], and their elementary geometrooptical interpretation was given [8]. However, theoretical attempts to explain the origin of ring interferograms and optimize conditions of their obtaining [68] contradict the accumulated experimental data, in particular dependence of rings period on the radius. Later it was experimentally found that annular speckle-interferograms can be obtained in lens systems as well. In practice, they appeared to be more effective. Now it is clear that they are more high aperture, give the possibility to alter the scale and increase the number of fringes and their contrast. But there are a number of conditions and limitations for location of lens elements and aperture. Their usage requires knowledge of interferograms localization and optimal conditions of their obtaining.

It is worth mentioning that experimental test of speckles trajectories is rather easy to realize during the study of transversal shift of speckles using double-exposure speckle-photography method [69]. But standard methods are inapplicable for studying longitudinal shift.

In the method developed below, three-dimensional character of speckle trajectories was used for estimation of longitudinal components that means that longitudinal shift of speckles is firmly connected to their radial "recession".

Transversal shift of diffuser leads to shift of interference rings from the axis that is why contrast of fringes can be enhanced through the application of annular diaphragms, and speckle-interferograms can be used for determination of shift Δd, which were obtained during reconstruction of specklograms with a thin laser beam [44].

It should also be mentioned that the obtained results can be useful while making speckle-interferometers, which use longitudinal correlation of speckle-fields, including real-time ones [70–74]. Analysis of correlated function of intensity fluctuations gives the possibility to find basic laws of changing subjective speckle-image and obtain correlations, which underlie methods to determine velocity of solids moving perpendicular to the optical axis [8, 75–77], frequency-contrast characteristics of optical systems [8, 78], determination of optical parameters of an eye [8, 79] and measurement of deformations and shifts using speckle-interferometry methods [8, 80–83].

Some information about correlation of speckle-fields is contained in a number of works on speckle-interferometry of longitudinal shift. The articles [36, 47] theoretically examine theoretically dependence of the speckle-field on small defocusing and determines longitudinal sizes of speckles. More fully decorrelation of the speckle-pattern near image plane is examined in the work [84] with the help of correlation functions apparatus.

The works [85–87] are dedicated to theory and practical application of new methods based on speckle-images obtained in coherent light. Such methods open new possibilities concerning measurement of shifts, deformations, vibrations, determination of shape and quality of diffused objects and image processing.

The works [87, 88] analyze the methods of quantitative determination of density fields velocity, temperature of flows of gases, liquids, plasma, including turbulent flows using new experimental technical instruments based on double-exposure speckle-photography and speckle-interferometry. Basic formulae are given, which characterize statistic properties of speckle-fields. Experimental setups and apparatus were described needed for realization of these methods. Methods of amortization, collecting and processing of speckle-interferometric measurements were studied, and examples of developed complexes of apparatus for such automation were given.

In the article [89] after introduction into theory of general probability and random processes, all the main questions of statistical optics are examined: theory of coherence of the first and higher orders, impact of partial coherence in systems forming the image, influence of random heterogeneous media and theory of photoelectric registration of light. The work [90] examines statistical characteristics of coherent radiation scattered on rough surface concerning the task of non-contact determination of frequency and amplitude of body vibrations.

Statistical characteristics of coherent radiation scattered by the rough body, which moves in its plane due to the random law, are analyzed in the work [91]. On the basis of this analysis, the ratio for spatiotemporal correlation function of fluctuation intensity of scattered coherent radiation was obtained, which can serve as the basis for the study of different types of movements of bodies. In the case of oscillatory motion of body on several frequencies, the ratio for fluctuation intensity spectrum was obtained. Experiments conducted with He–Ne laser radiation prove the possibility of measurement of mechanical oscillations spectrum of the rough body through registration of spectrum of intensity fluctuations of scattered radiation.

The authors of the work [92] state the theory of methods of holographic and speckle-interferometry considering specific tasks and practical schemes of a device that gives the possibility to implement these methods. In the work [93], the possibility of determination of plane rough surface vibration parameters is studied according to statistic parameters of the image of dynamic speckles formed by it. In particular, the expression for spatiotemporal correlation function of the generalized Wiener–Khintchin theorem for non-stationary processes was obtained, similarity ratio between spectrum of distribution of the scattered radiation and spectrum of vibration (in approximation of small amplitudes) was established. The result was proved experimentally. The authors of the article [94] established theoretically and experimentally connection between transversal velocity of gaseous flow and frequency spectrum of intensity fluctuation of scattered on microheterogeneities laser radiation. The suggested method is notable for sufficient sensitivity and simplicity of realization.

4.1 Correlation and Spectral Characteristics of Dynamic Speckle-Field Formed by Rotating Diffuser

Speckle movement is a combination of shift and boiling during rotation of diffusing object as well as during uniform motion. At the same time, movement of speckles is periodical as the same diffusers appear in the illuminated zone with every turn of the diffuser.

Mathematic description of movement of the speckle-image from the diffuser moving at a constant velocity was made by many authors [14, 16, 47, 84, 95]. Spatiotemporal autocorrelated intensity function in the normalized form is determined in the work [96] in the following way:

$$C_1(\vec{p}_1, \vec{p}_2, t_1, t_2) = \frac{\langle I(\vec{p}_1, t_1)I(\vec{p}_2, t_2)\rangle - \langle I(\vec{p}_1, t_1)I(\vec{p}_2, t_2)\rangle}{\left[\langle I^2(\vec{p}_1, t_1)\rangle - \langle I(\vec{p}_1, t_1)\rangle^2\right]^{\frac{1}{2}}\left[\langle I^2(\vec{p}_2, t_2)\rangle - \langle I(\vec{p}_2, t_2)\rangle^2\right]^{\frac{1}{2}}}, \tag{4.1}$$

where $I\ (\vec{p},\ t)$ is the optical intensity in point $\vec{p}$ of the observation plane in time moment t.

Let us examine the results of the study and the main parameters of dynamic speckle-field from the rotating diffusing object while illuminating it with laser radiation [96].

A plane disk is rotating at a frequency ω in $\vec{p}$ plane around the axis, which is perpendicular to the plane of the object (Fig. 4.2). As a light source, a laser was used, radiation of which runs parallel to the rotation axis and illuminates the area on the disk, and the center of which is shifted for $\vec{d}$ relative to the rotation axis. Observation is conducted in $\vec{p}$ plane at a distance z from the plane of the disk.

In (4.1) t_1 can be replaced by t and t_2 by $t + \tau$ for the case of dynamic speckle-field.

Taking into account the fact that optic field scattered on the disk will be a complex Gaussian one according to the work [97], it is possible to write the following

$$\begin{aligned}\langle I(\vec{p}_1, t)I(\vec{p}_2, t+\tau)\rangle &= \langle V(\vec{p}_1, t)V^*(\vec{p}_1, t)V(\vec{p}_2, t+\tau)V^*(\vec{p}_2, t+\tau)\rangle \\ &= |\Gamma(\vec{p}_1, \vec{p}_2, \tau)|^2 + \langle I(\vec{p}_1, t)\rangle\langle I(\vec{p}_2, t+\tau)\rangle,\end{aligned} \tag{4.2}$$

where V is the complex optical field amplitude.

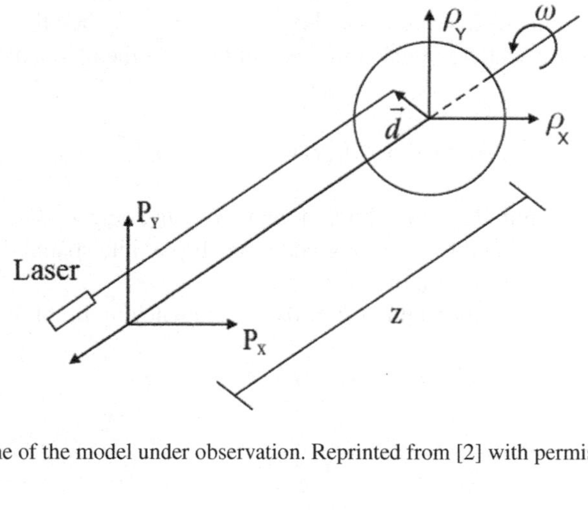

Fig. 4.2 Scheme of the model under observation. Reprinted from [2] with permission

Moment of the second order is determined as

$$\Gamma(\vec{p}_1, \vec{p}_2, \tau) = \langle V(\vec{p}_1, t)V^*(\vec{p}_2, t+\tau)\rangle \quad (4.3)$$

Then (4.1) can be rewritten in the following way:

$$C_1(\vec{p}_1, \vec{p}_2, \tau) = \frac{|\Gamma(\vec{p}_1, \vec{p}_2, \tau)|^2}{\Gamma(\vec{p}_1, \vec{p}_1, 0)\Gamma(\vec{p}_2, \vec{p}_2, 0)} = |\gamma(\vec{p}_1, \vec{p}_2, \tau)|^2, \quad (4.4)$$

where $\gamma\ (\vec{p}_1, \vec{p}_2, \tau)$ is the normalized second moment of the complex amplitude of the field.

Using proximal approximation and the Huygens–Fresnel's principle [98], the field amplitude in observation surface can be expressed through the amplitude in object surface:

$$V(\vec{p}, t) = \frac{k}{2\pi iz}\exp(ikz)\int V(\vec{\rho}, t)\exp\left[\frac{ik}{2z}(\vec{p}-\vec{\rho})^2\right]d^2\rho \quad (4.5)$$

where k is the optical wave number; z is the distance between the observation surface and the object surface.

Then it is possible to assume that within the limits of the Fresnel's diffraction field

$$\Gamma(\vec{p}_1, \vec{p}_2, \tau) = \left(\frac{k}{2\pi z}\right)^2 \iint \Gamma(\vec{\rho}_1, \vec{\rho}_2, \tau)$$
$$\exp\left\{\frac{ik}{2z}\left[(\vec{p}_1-\vec{\rho}_1)^2 - (\vec{p}_2-\vec{\rho}_2)^2\right]\right\}d^2\rho_1 d^2\rho_2 \quad (4.6)$$

We consider that dispersion from different points is not correlated. Let us investigate the center of dispersion within the disk situated in the point with coordinate $\vec{\rho}_1$ at time t. If the same diffuser is situated in $\vec{\rho}_1$ at time $t + \tau$, then the moment of the second order of the scattered field complex amplitude in the object plane is given as in the work [99]:

$$\Gamma(\vec{\rho}_1, \vec{\rho}_2, \tau) = V_1(\vec{\rho}_1, t)V_1^*(\vec{\rho}_2, t+\tau)\delta(\vec{\rho}_2 - \vec{\rho}_1), \quad (4.7)$$

where V_1 is the amplitude of the illuminating field in the object plane; δ is the Dirac delta function and ensemble averaging taken on all possible spatial distributions of point-scattering centers within the disk.

The disk is rotating around point $\vec{\rho} = 0$ at a frequency ω. In this case

$$\vec{\rho}_1^{\,\prime} = \vec{\rho}_1\cos(\omega\tau) + \vec{A}\vec{\rho}_1\sin(\omega\tau), \quad (4.8)$$

where $\vec{A}$ is the $\pi/2$-turn operator.

This operator transforms vector $\vec{\rho}_1$ into vector of the same quantity but directed perpendicular to $\vec{\rho}_1$ and is parallel to the current velocity of the disk in point $\vec{\rho}_1$. Expressing $\vec{\rho}_1$ and $\vec{A}$

$$\vec{\rho}_1 = \begin{pmatrix} x_1 \\ y_1 \end{pmatrix}, \tag{4.9}$$

$$\vec{A} = \begin{pmatrix} 0 & 1 \\ -1 & 0 \end{pmatrix} \tag{4.10}$$

so that

$$\vec{\rho}_1^{\,\prime} = \begin{pmatrix} x_1 \\ y_1 \end{pmatrix} \cos(\omega\tau) + \begin{pmatrix} 0 & 1 \\ -1 & 0 \end{pmatrix}\begin{pmatrix} x_1 \\ y_1 \end{pmatrix} \sin(\omega\tau) = \begin{pmatrix} x_1 \cos(\omega\tau) + y_1 \sin(\omega\tau) \\ y_1 \cos(\omega\tau) - x_1 \sin(\omega\tau) \end{pmatrix}. \tag{4.11}$$

We suppose that the disk is illuminated by a Gaussian beam, the center of which is situated in point $\vec{d}$. Thus, the illuminating field can be written in the following form [100]:

$$V_1(\vec{\rho}_1, t) = V_0 \exp\left[-\left(\frac{1}{W^2} + \frac{ik}{2\mu}\right)\left(\vec{p} - \vec{d}\right)^2\right], \tag{4.12}$$

where V_0 is the field amplitude in the center of illuminated disk area; W is the radius of $1/e^2$ intensity of illuminated region; μ is the curvature radius of the wave front.

Equations (4.6–4.8) and (4.12) give the possibility to obtain

$$\begin{aligned} \Gamma(\vec{p}_1, \vec{p}_2, \tau) = &\left(\frac{k}{2\pi z}\right)^2 |V_o|^2 \iint \exp\left[-\left(\frac{1}{W^2} + \frac{ik}{2\mu}\right)\left(\vec{\rho}_1 - \vec{d}\right)^2\right. \\ &\left. -\left(\frac{1}{W^2} + \frac{ik}{2\mu}\right)\left(\vec{\rho}_2 - \vec{d}\right)^2 + \frac{ik}{2z}(\vec{p}_1 - \vec{\rho}_1)^2 - \frac{ik}{2z}(\vec{p}_2 - \vec{\rho}_2)^2\right] \\ &\times \delta\left(\vec{p}_2 - \vec{\rho}_1 \cos(\omega\tau) - \vec{A}\vec{\rho}_1 \sin(\omega\tau)\right) d^2\rho_1 d^2\rho_2 \end{aligned} \tag{4.13}$$

Calculating the integrals in (4.13) and taking

$$\vec{p} = \vec{p}_2 - \vec{p}_1,\ \vec{P} = (\vec{p}_1 + \vec{p}_2)/2,$$

we obtain

$$\begin{aligned} \gamma\left(\vec{P}, \vec{p}, \tau\right) = \exp&\left\{-\frac{1}{2}\left(\frac{kW}{2z}\right)^2 p^2 - (1 - \cos(\omega\tau))\right. \\ &\left[\frac{\vec{d}^2}{W^2} + \left(\frac{kW}{2z}\right)^2\left(\vec{P} - \frac{Z}{\mu}\vec{d}\right)^2 - \frac{1}{4}\left(\frac{kW}{2z}\right)^2 p^2\right] \end{aligned}$$

$$- \sin(\omega\tau)\left(\frac{kW}{2z}\right)^2\left(\vec{P} - \frac{z}{\mu}\vec{d}\right)\vec{A}\vec{p}$$
$$+\frac{ik}{z}\left[\frac{1}{2}(1+\cos(\omega\tau))\vec{d}\vec{p} - \sin(\omega\tau)\vec{P}\vec{A}\vec{d} - \vec{P}\vec{p}\right]\Bigg\} \quad (4.14)$$

where it is supposed that scalar product of vectors and p means quantity of vector $\vec{p}$.
Equation (4.14) is the following:

$$C_1\left(\vec{P}, \vec{p}, \tau\right) = \exp\left\{-\left(\frac{\mathrm{kW}}{2\mu}\right)^2 p^2 - (1-\cos(\omega\tau))\right.$$
$$\left[\frac{2\vec{d}^2}{W^2} + 2\left(\frac{\mathrm{kW}}{2z}\right)^2\left(\vec{P} - \frac{Z}{\mu}\vec{d}\right)^2 - \frac{1}{2}\left(\frac{\mathrm{kW}}{2z}\right)^2 p^2\right]$$
$$\left.-2\sin(\omega\tau)\left(\frac{\mathrm{kW}}{2z}\right)^2\left(\vec{P} - \frac{z}{\mu}\vec{d}\right)\vec{A}\vec{p}\right\} \quad (4.15)$$

Temporal correlation function can be obtained supposing $\vec{p} = 0$:

$$C_1\left(\vec{P}, 0, \tau\right) = \exp\left\{-2\left[\frac{\vec{d}^2}{W^2} + \left(\frac{\mathrm{kW}}{2z}\right)^2\left(\vec{P} - \frac{z}{\mu}\vec{d}\right)^2\right](1-\cos(\omega\tau))\right\}. \quad (4.16)$$

According to (4.16), the correlation function has a periodic character. Near $\omega\tau = 0$

$$C_1\left(\vec{P}, 0, \tau\right) = \exp\left\{-\left[\frac{\vec{d}^2}{W^2} + \left(\frac{\mathrm{kW}}{2z}\right)^2\left(\vec{P} - \frac{z}{\mu}\vec{d}\right)^2\right]\left(\frac{\omega\tau}{2}\right)^2\right\} \quad (4.17)$$

Both expressions in (4.17) describe two different decorrelation processes. The first one describes "boiling" of speckles, and according to [96] this effect has decorrelation time

$$\tau_{d_1} \approx \frac{W}{\omega d}, \quad (4.18)$$

which is necessary for a new group of diffusers to move into a beam. The second expression describes the rotation of speckle-pattern around point $\vec{P} = (z/\mu)\vec{d}$.
Decorrelation time due to speckle-rotation will be

$$\tau_d \approx \frac{2z}{\mathrm{kW}\omega}\left|\left(\vec{P} - \frac{z}{\mu}\vec{d}\right)\right|^{-1}. \quad (4.19)$$

Statistical characteristics of dynamic speckle-field, which are studied for a model of diffusely reflecting disk, can be applied for a model of deep phase screen, which was examined by the authors of work [16].

In this case, rotating matt diffusely scattering disk can serve as a phase screen. Obtaining for the rotating disk (4.17) can be reduced to the view which corresponds to an object moving at a constant velocity [16]. In this case, temporal correlation function of speckles intensity fluctuation will be

$$\gamma_{\Delta I}(0,\tau) = \exp\left(-\tau^2/\tau_c^2\right), \tag{4.20}$$

where τ_c is the time correlation determined as

$$\tau_c = \frac{1}{|\vec{\upsilon}|}\left(\frac{1}{W^2} + \frac{\sigma^2}{\Delta x^2}\right)^{-\frac{1}{2}}, \tag{4.21}$$

where $|\vec{\upsilon}| = \omega d$ is the value of linear velocity of diffusers shift.

The parameter σ, which determines diffraction conditions, is calculated according to formula

$$\sigma = \frac{R}{\mu} + 1, \tag{4.22}$$

where R is the distance between the object and the observation plane; μ is the radius of wave front curvature.

For this scheme, geometry schemes in (4.17) can be written as

$$\vec{P} - \left(\frac{z}{\mu}\right)\vec{d} = P + \frac{z}{\mu}d = \rho_0\left(1 + \frac{R}{\mu}\right) = \rho_o\sigma, \tag{4.23}$$

as $z = R$, d is the distance between the center of the illuminated zone to the center of disc; P is the transverse distance;

$$k = \frac{2\pi}{\lambda} \text{ is the wave number;} \tag{4.24}$$

$$\Delta x = \lambda R/\pi W \text{ is the average speckle size.} \tag{4.25}$$

The beam width W in the object plane at a distance $R = z$ from waist is calculated according to formula

$$W = W_0\left(1 + \frac{z^2}{d^2}\right)^{\frac{1}{2}}, \tag{4.26}$$

$$\mu = z\left(1 + \frac{a^2}{z^2}\right), \tag{4.27}$$

where

$$a = \frac{\pi W_0^2}{\lambda}, \tag{4.28}$$

a is the parameter, which determines the area of waist of the Gaussian beam; W_0 is the width of the beam in the area of waist.

Spectral density function of intensity fluctuations of dynamic speckle-field can be presented according to the Wiener–Khintchin theorem [16]

$$\Phi_{\Delta I}(\omega) = \exp\left(-\frac{\tau_c^2\omega^2}{4}\right) \tag{4.29}$$

with time correlation in (4.21).

As it is seen from (4.21), the correlation length τ_c of temporal autocorrelation function of dynamic speckles is inversely proportional to the modulus of velocity of a moving object and constant of proportionality is connected with the average speckle size Δx and illuminating conditions with a Gaussian beam through parameters W and μ.

At that spectral width is

$$\Delta\omega = \frac{2}{\tau_c} = 2|\vec{\upsilon}|\left(\frac{1}{W^2} + \frac{\sigma^2}{\Delta x^2}\right)^{\frac{1}{2}}. \tag{4.30}$$

It is seen that it is directly proportional to the velocity of object movement and depends considerably on geometry of illumination W, σ and observation Δx.

Considerable difference in this case from the one studied in the work [16] is periodicity of movement during diffuser rotation. Periodic temporal autocorrelation function gives line power spectrum of intensity fluctuations, and the frequency interval corresponds to frequency of disk rotation and is also directly proportional to linear velocity of shift of separate diffusers during rotation. Despite this difference, (4.29) and (4.30) are applicable for the envelope of spectral harmonics. And the width of spectral curve determines the correlation length τ_c of the temporal autocorrelation function.

4.2 Application of Spectral Characteristics of Dynamic Speckle-Field Intensity Fluctuations for Determining Longitudinal Shift of an Object

While moving on to experimental study, it is better to express rotation frequency in hertz that is why moving on to $\nu = \omega/2\pi$ in (4.29), using (4.22), (4.25)–(4.28) and putting them into (4.29) after some transformations, we obtain expression for amplitude of spectral harmonic on this frequency:

$$A_{\Delta I}(\nu) = \frac{1}{\sqrt{2\pi}} \exp\left[-\pi^2 \nu^2 \upsilon_1^2 \left(\frac{(\lambda R a^2)^2 + (\pi W_0^2((z^2 + a^2) + Rz))^2}{(\lambda R W_0 a)^2 + (z^2 + a^2)}\right)\right]. \tag{4.31}$$

Let us assume that $\upsilon_1 = 2\pi \rho_0 \nu_0$, where ν_0 is the frequency of disk rotation; ρ_0 is the distance between the center of the disk and the center of the illuminated area. Then

$$A_{\Delta I}(\nu) = \frac{1}{\sqrt{2\pi}} \exp\left[-4\pi^4 \nu^2 \rho_0^2 \nu_0^2 \left(\frac{(\lambda R a^2)^2 + (\pi W_0^2((z^2 + a^2) + Rz))^2}{(\lambda R W_0 a)^2 + (z^2 + a^2)}\right)\right]. \tag{4.32}$$

Thus, the following can be written for amplitude of the n-th harmonic taking into account that $\nu = n\nu_0$

$$A_n = \frac{1}{\sqrt{2\pi}} \exp\left[-4\pi^4 n^2 \rho_0^2 \nu_0^4 \left(\frac{(\lambda R a^2)^2 + (\pi W_0^2((z^2 + a^2) + Rz))^2}{(\lambda R W_0 a)^2 + (z^2 + a^2)}\right)\right]. \tag{4.33}$$

Equation (4.33) points out significant dependence of the amplitude of spectral harmonic on the change of geometric parameters of the scheme. Main parameters influencing the amplitude are the width of the illuminated area W, the distance between the center of the illuminated area and the center of disk rotation ρ_0 and the transversal size of speckles Δx in observation plane. Taking into account (4.22), (4.25)–(4.28), we can conclude that as a result spectra of intensity fluctuations of dynamic speckle-field are defined by the distance between the rotating diffuser and location of waist of focusing lens, and by distance R between the diffuser to registration plane. Defining the dependence of spectral harmonics amplitude on these parameters is the main task of this study of spectrum changes of intensity fluctuations under longitudinal shift of elements of the experimental scheme.

There are three independent cases of longitudinal shift:

1. $z_i = \text{const}$, i.e., shift of observation plane at constant z_i;
2. $z_i + R = \text{const}$. This is a case when rotating diffuser shifts in longitudinal direction if the distance between plane of a beam waist and observation plane is constant;
3. $R = \text{const}$, i.e., changes at fixed distance R.

The above-examined common experimental scheme has several degrees of freedom, and different ways of measurement of longitudinal shift can be implemented on its basis (Fig. 4.3).

Diffusely transmitting matt disk was rotating at a constant frequency $\nu = 4.2$ Hz around axis O, which is parallel to the optical axis of the system and locating from it at a distance $\rho_0 = 8$ cm. It was illuminated with He–Ne laser radiation using a

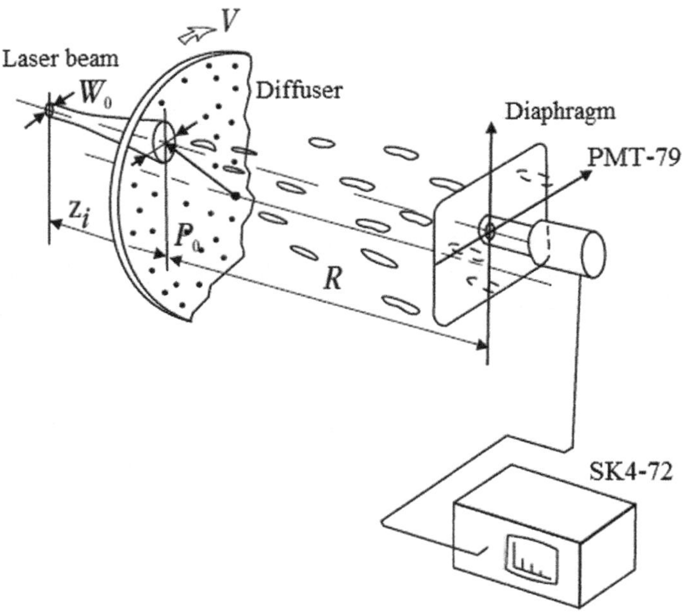

Fig. 4.3 Experimental scheme for the study of intensity fluctuations spectra of dynamic speckle-field during longitudinal shift of the diffuser and alteration of conditions of observation and illumination. Reprinted from [2] with permission

lens with the focal distance $f = 4.5$ cm and the width of waist $W_0 = 20$ μm at a set divergence degree.

Rotation of the disk led to the change of the speckle-field formed by coherent radiation from this diffuser. There were performed registration and transformation of intensity fluctuations of dynamic speckle-field into electrical signal through a field diaphragm using PMT 79 (photomultiplier tube) at a distance R from the diffuser. Then the electrical signal was processed with the spectrum analyzer SK4-72.

1. *Changes of the distance between the diffuser and the photoreceiver* $zi = const$. The distance between the illuminator and the rotating diffusely transmitting disk is registered, and longitudinal shift of the object under study is made as a result of changes of distance *R*. The photoreceiver is fixed to the shifted object. In this case, decorrelation velocity and changes of spectral harmonics value are conditioned by the growth of sizes of speckles according to

$$\Delta x_i(R) = \frac{\lambda R}{\pi W(z_i)}$$

for different values z_i, such as $z_i = 60, 120, 180, 240$ μm, and growth of values of diffraction parameter $\sigma(R)$ calculated from (4.22).

The change of the amplitudes of the first spectral harmonics $A_1(R)$ of intensity fluctuation spectra of dynamic speckle-field at fixed values z_i is presented in Fig. 4.4.

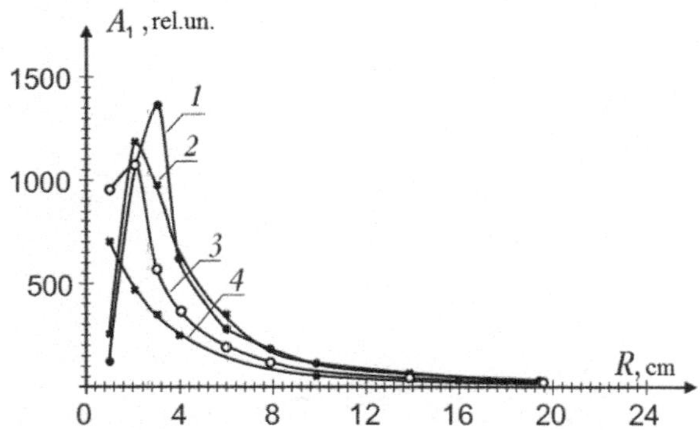

Fig. 4.4 Dependence of amplitude A_1 of the first spectral harmonic during longitudinal changes of distance R between the diffuser and a photoreceiver: 1—$z_i = 60$ mm; 2—$z_i = 120$ mm; 3—$z_i = 180$ mm; 4—$z_i = 240$ mm. Reprinted from [2] with permission

Changes showed areas of monotony in dependence graph $A_1(R)$, the slope of which was determined by z_i value. The more $z_i = \text{const}$ value is, the less the slope of $A_1(R)$ curves is. Thus, longitudinal coordinate (or shift) of the object can be detected while measuring amplitude values (or changes) of the first spectral harmonic of intensity fluctuation spectrum on monotony areas for any chosen $z_i = \text{const}$.

It is worth mentioning that significant weakening of average intensity occurs at the shift of the object under study with a photoreceiver fixed on it along the optical axis. The value of the average intensity changes inversely to a squared distance R^2. This should be taken into account through normalization of the measured values of spectral intensity.

2. *Shift of rotating diffuser in the longitudinal direction at a fixed distance between the position of beam waist and the observation plane* $z_i + r$=const. It is the most complicated situation when all the parameters of the scheme change: width of the illuminated zone $W(z)$, wave front curvature $\mu(z)$ (Fig. 4.5), distance between the diffuser and the observation plane $R(z) = N - z_i$, distance between illuminator and diffuser z_i, which enter into (4.33) for the fluctuation spectrum harmonic and the amplitude of intensity of dynamic speckle-field. Thus, the described scheme is the most sensitive to longitudinal shifts, but it also has the least flexibility.

A new method was developed for determining longitudinal shift of rotating (or oscillating) diffusing object, which velocity or oscillation frequency is unknown [101, 102]. The longitudinal coordinate to the stationary object can be found, but in this case the moving diffuser must be firmly fixed on the object under study. A diffusely transmitting disk was illuminated by a laser beam with a set degree of divergence. The disk was rotating at a velocity v in a plane, which was perpendicular to the axis of beam spreading, and it was situated at a distance z_i from the lens focus. The amplitude $A_1(z)$ of the first spectral harmonic was registered in spectrum of intensity fluctuations of dynamic speckle-field from the rotating diffusely transmitting disk in position characterized by coordinate z_i.

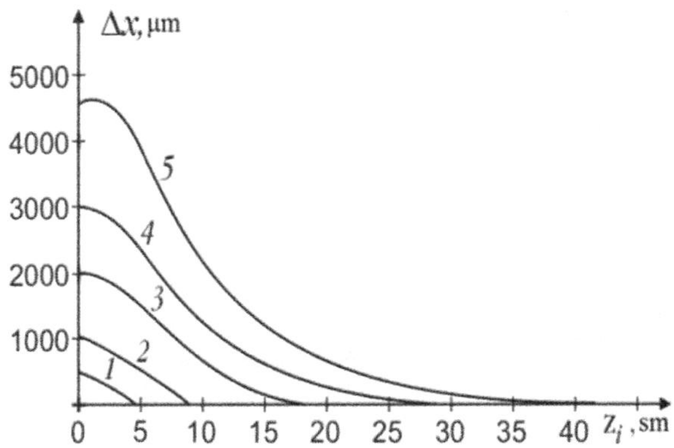

Fig. 4.5 Dependence of the average speckles size Δ in the observation plane during longitudinal shift of the diffuser for $z_i + R = N$ = const: 1—$z_i + R = 50$ mm; 2—$z_i + R = 100$ mm; 3—$z_i + R = 200$ mm; 4—z_i + R=300 mm; 5—$z_i + R = 450$ mm. Reprinted from [2] with permission

The object under study was shifted in direction of spreading the laser beam and the amplitude $A_1(z)$, spectral harmonic of spectrum of intensity fluctuations was registered in position with coordinate z_2, then in the position with coordinates z_3, z_4, etc. Having conducted measurements and plotted values of the amplitudes $A_1(z)$ of spectral harmonics against coordinates z_1, z_2, z_3, etc., which characterize longitudinal shift of the object under study, the dependences have been made for some fixed values $z_i + R = N$ =const. The dependences were calibration curves for this measuring circuit if $z_i + R = 50, 100, 200, 450$ mm (Fig. 4.6). On some areas, the change of the amplitude of spectral harmonic $A_1(z)$ is directly proportional to longitudinal shift of the object, and the straight portion of this curve (for $z_i + R = 100$ mm) is optimal for registration of longitudinal shift (Fig. 4.7). Measurement of spectral harmonic amplitude $A_1(z)$ made it possible to unequivocally determine longitudinal shift of

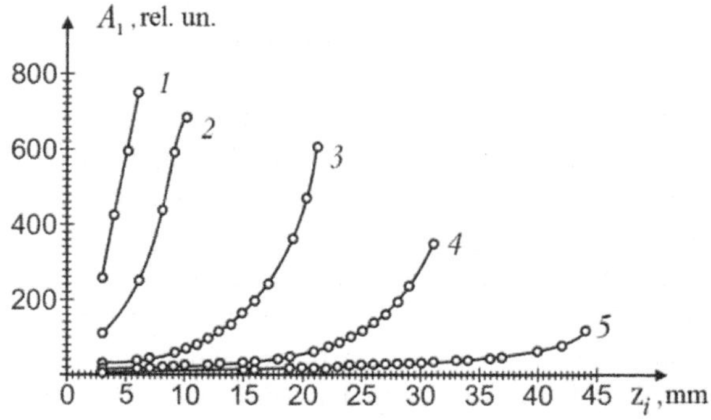

Fig. 4.6 Dependence of the amplitude A_1 of the first spectral harmonic with the longitudinal shift of the rotating diffuser for $z_i + R = N$ = const: 1$-z_i + R = 50$ mm; 2—$z_i + R = 100$ mm; 3—$z_i + R = 200$ mm; 4—$z_i + R = 300$ mm; 5—$z_i + R$=450 mm. Reprinted from [2] with permission

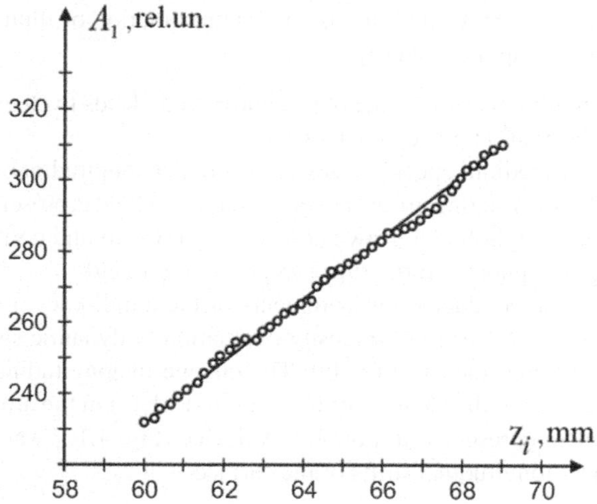

Fig. 4.7 Area of calibration curve of amplitude A_1 of the first spectral harmonic dependence on the value of longitudinal shift of diffuser for $z_i + R = 100$ mm. Reprinted from [2] with permission

diffusely transmitting disk over the range of 60 mm with the accuracy up to 100 μm using the calibration curve. Line spectrum typical for periodic rotary motion gives the possibility to easily determine linear velocity of the object according to formula

$$\upsilon = \Delta \nu \rho_0,$$

where $\Delta \nu$ is the registered interval between adjacent spectral harmonics; ρ_0 is the distance between the rotation axis of the disk and the center of the illuminated zone.

Frequency of diffuser oscillations can be determined according to the distance between spectral harmonics.

1. *Changing of the distance between the illuminator and the diffuser.* This method is based on changing the correlation function and spectra of intensity fluctuation of dynamic speckle-field when distance between the illuminating waist and the diffuser z is changed [103, 104]. The object, in which position and shift should be determined, is firmly connected to the illuminating head, which forms a laser beam and sets the position of the waist. Or it can be connected with a block consisting of a rotating diffusely transmitting disk and a photoelectric sensor, between which distance R does not change. Shift of the object at some distance z along the direction of the illuminating beam leads to alteration of the width of the illuminated area W and the curvature of the wave front μ on the diffuser. For $z << a$ where a is the waist length of the Gaussian beam, which is determined from (4.27) and $W \approx W_0$. Parameter μ has a positive or a negative sign depending on distance z. Curvature μ has the minimal value $2a$ when $|z| = a$ and becomes unlimited for $|z| << a$ and $|z| \gg a$. If $|z| \gg a$ that corresponds to far-field diffraction

region, then $W = \lambda z\pi W_0$ and in this case approximation of illumination with a spherical wave is applicable [16].

In its turn, longitudinal shift of the object under study leads to changes of speckle sizes $\Delta x(z)$ in observation plane due to (4.28).

Dependence of medium speckle sizes $\Delta x_i(z)$ under longitudinal shift for some fixed distances R from rotating diffuser to the observation field is presented in Fig. 4.8. In this figure, the straight line is drawn at level $\delta = 140$ μm and corresponds to the size of diaphragm of photoelectric sensor in observation field.

The results of the conducted measurements of the amplitudes of the first spectral harmonics $A_1(z)$ of spectra of intensity fluctuation of dynamic speckle-field for fixed values R are presented in Fig. 4.9. The change of longitudinal shift can be determined according to the changes of the amplitude $A_1(z)$ in the area of monotony of any dependence represented in Fig. 4.9. A device (Fig. 4.10) was developed for determining linear longitudinal shifts of the surface.

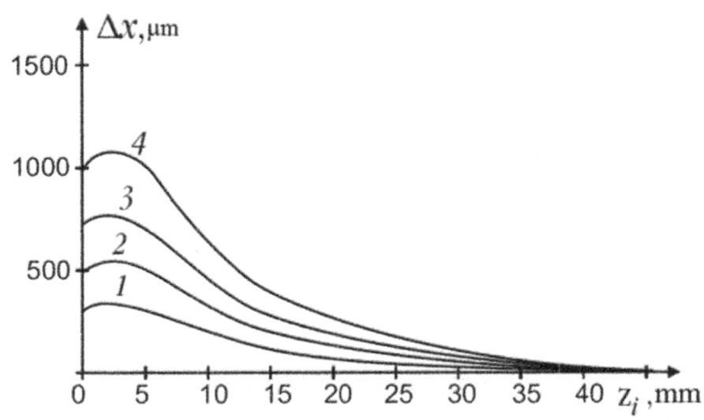

Fig. 4.8 Dependence of average speckle size Δx on the longitudinal shift z_i for R = const: 1—R = 30 mm; 2—R = 50 mm; 3—R = 70 mm; 4—R = 100 mm. Reprinted from [2] with permission

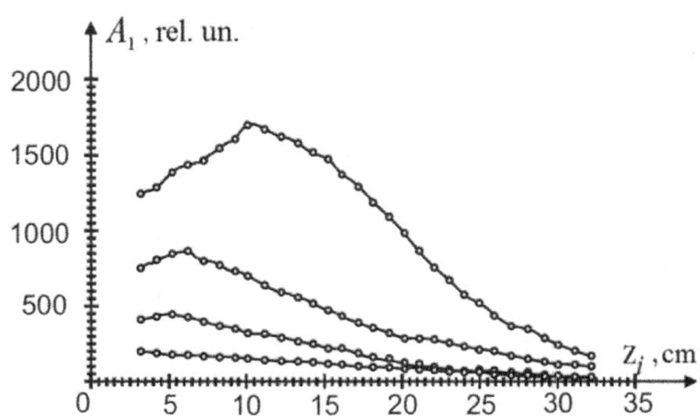

Fig. 4.9 Dependence of the amplitude A_1 of the first spectral harmonic at measuring the distance in the longitudinal direction for fixed R: 1—R = 30 mm; 2—R = 50 mm; 3—R = 70 mm; 4—R = 100 mm. Reprinted from [2] with permission

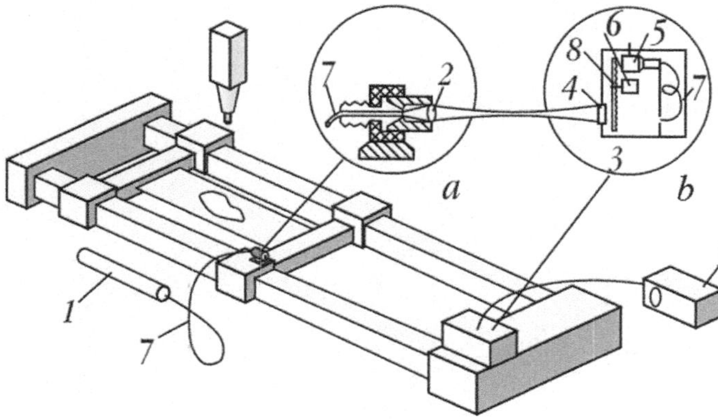

Fig. 4.10 Device for determining linear longitudinal shifts of the surface: a- the forming block of the illuminating beam; b- the registration block; 1—He–Ne laser; 2—the lens; 3—the diaphragm; 4—the monochromatic filter; 5—the photomultiplier tube; 6—the microelectric motor; 7—the light guide; 8—the diffusely transmitting disk; 9—the spectrum analyzer SK4-72. Reprinted from [2] with permission

Radiation of He–Ne laser 1 through the light guide 7 is directed into the block, which forms a beam of set divergence (Fig. 4.10a). The output face of the light guide 7 and the lens 2 is mounted in a small holder of about a couple centimeters. The holder is placed on a magnetic base and is easily fixed on the surface of the illuminated object. The radiation is focused at a distance of about 200 mm from the block.

The second functional block is a registration unit (Fig. 4.10b). It consists of monochromatic filter adjusted to the length of laser radiation $\lambda = 632.8$ nm and cutting flare light. The filter aperture is selected depending on maximal width sizes of the illuminated zone W of the laser beam on the diffuser of the receiver.

It is passed through the filter laser radiation and illuminates the diffuser 8, which is matt diffusely transmitting plate rotated with a microelectric motor. Changes of velocity of movement along the illuminated zone and finite sizes of the beam on the illuminated disk introduce some changes into (4.33) describing the amplitude of harmonics, which are the spectrum of a signal, because the velocity of movement will be averaged to some extent. But it does not change the operating principle and is taken into account only during adjustment and calibration of the sensor.

Behind the rotating disk, there was a receiver—the diaphragm 3 at a distance up to 100 mm, and a light guide was situated behind it. Intensity fluctuations of dynamic speckle-field are registered by means of this light guide with the photomultiplier tube 5, which was mounted in the same unified body of the receiving block. The signal is given from the photomultiplier to the spectrum analyzer, which helps to decompose it into spectrum, and the amplitude of first spectral harmonic $A_1(z)$ of intensity fluctuation spectrum is measured.

The sensor was tested at the following parameters of the scheme: the focal distance of the lens is 45 mm, and the distance from the rotating disk to the receiving aperture

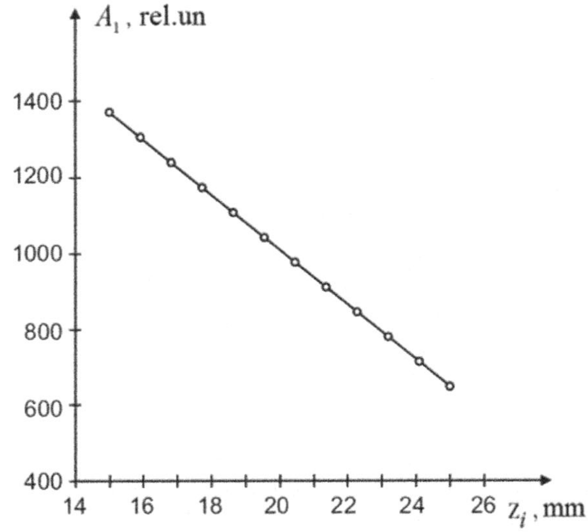

Fig. 4.11 Piece of calibration curve if the distance between the rotating diffuser and the receiving aperture is $R = 30$ mm. Reprinted from [2] with permission

was fixed: $R = 30$, 50, 70, 100 mm. It investigated the possibility of determining the linear longitudinal shift of the traveling bogie with a fixed illuminating head through registration of the change of the amplitude of the first spectral harmonic of the speckle-field, which was formed by laser radiation scattered from the rotating disk. It was registered at rotation frequency during linear shift of the traveling bogie in longitudinal shift. The area of calibration curvature for $R = 30$ mm is shown in Fig. 4.11, on which changes of amplitude of spectral harmonic (frequency is 4.2 Hz) of speckle-field fluctuations from diffusely transmitting disk are in direct proportion to longitudinal shift of the mobile bogie. The range of shifts is 100 mm.

The developed sensor for control of linear longitudinal shift gave the possibility to extend range of measurement of linear longitudinal shift by about two orders in comparison with the sensors described in the works [103, 104] and to enlarge the class of the objects under study at the expense of measurements of shifts regardless of form and optical characteristics of the object [105, 106].

The changes of distance R between the diffuser and the plane of observation give the possibility to alter the inclination of the curvature and, thus, to decrease or increase sensitivity.

4.3 Statistical Properties of Dynamic Speckle-Field Scattered by the Diffuser Oscillating in the Longitudinal Direction

The case of rotation of diffusely scattering object, which was examined in Sect. 4.1, concerned the movement of the diffuser in its plane, i.e., transverse movement. And conditions of object illumination and formation of the speckle-field were constant in time. Oscillations of the diffuser illuminated with coherent radiation along the optical axis lead to changes of medium sizes of speckles of dynamic speckle-field in registration plane, as well as of parameters, which characterize the illuminating beam and the conditions of observation. As a result of this, correlation and spectral properties of dynamic speckle-field considerably differ depending on the movement of the object. But due to considerable difficulties connected with the presentation in analytical form of the results of theoretical calculations of statistical characteristics of intensity fluctuations of dynamic speckles under longitudinal oscillations of diffuser [107], only periodical oscillations and vibrations of diffuse objects in longitudinal direction are rather fully examined [108, 109].

We investigate statistical characteristics of dynamic speckle-field scattered by the diffuser, which oscillates in longitudinal direction using the approach applied while studying longitudinal movement of the object at a constant velocity [110].

We examine the model of oscillating diffusing face with the amplitude h along the z-axis at the frequency ω illuminated by coherent Gaussian beam with the wavelength λ focused by the lens with waist W_0 in the plane of oscillations equilibrium (Fig. 4.12). The observation plane is situated at a distance R_0 from the focus and the position of equilibrium of diffuser oscillation. Speckle-image is formed in the far-field region of diffraction at a distance R_0 from the beam waist. The structure of the speckle-image is altered in time as a result of longitudinal oscillations of the object. Variable

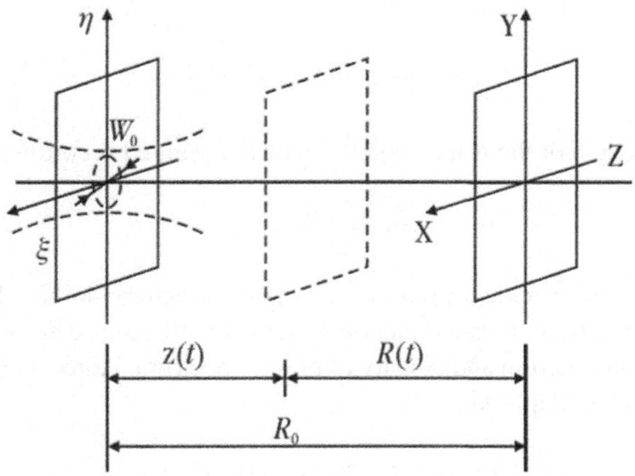

Fig. 4.12 Scheme of the diffuser model under consideration. Reprinted from [2] with permission

amplitude of the speckle in point $\vec{x}=(x, y, R_0)$ in time moment t

$$V(\vec{x}, t) = \int_{\infty}^{+\infty} E_0\left(\vec{\xi}, z\right) \exp\left[i\Phi\left(\vec{\xi}\right)\right] \exp\left(ik\frac{\vec{\xi}\vec{x}}{R(t)}\right) d\vec{\xi}, \tag{4.34}$$

where $\vec{\xi}$ is the vector of point (ξ, η) in the object plane; $k = 2\pi\lambda$ is the wave number; Φ $(\vec{\xi})$ is the variable random phase; E_0 $(\vec{\xi}, z)$ is the distribution of amplitude of the illuminating beam [110], $R(t)$ is the time varying due to longitudinal oscillations distance between the object and the observation plane (x, y); $z(t)$ is the time-varying distance between the beam waist and the object plane

$$R(t) = R_0 - z(t) \tag{4.35}$$

and

$$z(t) = h \sin(\omega t). \tag{4.36}$$

Amplitude distribution in point z is set for the Gaussian beam with width of waist W as

$$E_0\left(\vec{\xi}, z\right) = \frac{W_0}{W(z)} \exp(ikz)\exp, \tag{4.37}$$

where $W(z)$ and $\mu(z)$ are the width of the beam and radius of curvature of the wave front in the object plane, respectively, which can be calculated using the formula

$$W(z) = W_0\sqrt{1 + z^2/a^2} \tag{4.38}$$

and

$$\mu(z) = z\left(1 + a^2/z^2\right) \tag{4.39}$$

where a is the length of the waist, which can be determined according to the formula

$$a = \pi W_0^2/\lambda. \tag{4.40}$$

As in (4.34), the speckle amplitude is a value, which randomly changes in time under object oscillations, then its statistical characteristics can be most fully described with a correlation function and density of power spectrum. Autocorrelation function of dynamic speckle amplitude

$$\Gamma(\vec{x}, t_1, t_2) = \left\langle V(\vec{x}, t_1)V^*(\vec{x}, t_2)\right\rangle, \tag{4.41}$$

where $\langle\ldots\rangle$ is the ensemble averaging.

Using (4.34) in (4.41), we obtain

$$\Gamma(\vec{x}, t_1, t_2) = \left\langle \iint_{-\infty}^{+\infty} E_0[\xi_1, z_1(t_1)] E_0^*[\xi_2, z_2(t_2)] \exp(i[\Phi(\xi_1) - \Phi(\xi_2)]) \times \exp\left[ik\left(\frac{\xi_1}{R(t_1)} - \frac{\xi_2}{R(t_2)}\right)\vec{x}\right] \mathrm{d}\xi_1 \mathrm{d}\xi_2 \right\rangle \quad (4.42)$$

As this random phase function $\Phi(\xi)$ does not depend on z, ensemble averaging is made only for $\exp(i[\Phi(\xi_1) - \Phi(\xi_2)])$ in (4.42), and (4.42) becomes

$$\Gamma(\vec{x}, t_1, t_2) = \iint_{-\infty}^{+\infty} \iint_{-\infty}^{+\infty} E_0[\xi_1, \eta_1, z_1(t_1)] E_0^*[\xi_2, \eta_2, z_2(t_2)] \times \langle \exp(i[\Phi(\xi_1, \eta_1) - \Phi(\xi_2, \eta_2)]) \rangle \exp\left[ik\left(\frac{\xi_1 x + \eta_1 y}{R_0 - z_1(t_1)} - \frac{\xi_2 x + \eta_2 y}{R_0 - z_2(t_2)}\right)\right] \mathrm{d}\xi_1 \mathrm{d}\eta_1 \mathrm{d}\xi_2 \mathrm{d}\eta_2, \quad (4.43)$$

where $\xi_1 = (\xi_1, \eta_1)$, $\xi_2 = (\xi_2, \eta_2)$

If it is assumed that approximation of white noise is applicable for diffused object, which is characterized as deep phase screen, then given in (4.43) ensemble averaging is reduced to

$$\langle \exp(i[\Phi(\xi_1, \eta_1) - \Phi(\xi_2, \eta_2)]) \rangle \approx \Delta S \delta(\xi_1 - \xi_2) \delta(\eta_1 - \eta_2), \quad (4.44)$$

where δ is the delta function; ΔS is the correlation area $\Phi(\xi, \eta)$.

Using (4.44) in (4.43), we obtain

$$|\Gamma(\vec{x}, z_1, z_2)|^2 = \Delta S \iint_{-\infty}^{+\infty} E_0[\xi, \eta, z_1] E_0^*[\xi, \eta, z_2] \times \exp\left[ik\left(\frac{1}{R_0 - z_1} - \frac{1}{R_0 - z_2}\right)(\xi x + \eta y)\right] \mathrm{d}\xi \mathrm{d}\eta \quad (4.45)$$

where z_1 and z_2 are substituted with t_1 and t_2 as $z_1 = h\sin(\omega t_1)$, $z_2 = h\sin(\omega t_2)$.

Expression $\left[E_0[\xi, \eta, z_1] E_0^*[\xi, \eta, z_2]\right]$ in (4.45) can be written using (4.37) in the following form

$$E_0[\xi, \eta, z_1] E_0^*[\xi, \eta, z_2] = \frac{W_0^2}{W(z_1) W(z_2)} \exp[ik(z_1 - z_2)] \exp\left[-\left(\frac{1}{W^2(z_1)} + + \frac{1}{W^2(z_2)}\right)\left(\xi^2 + \eta^2\right)\right] \exp\left[-\frac{ik}{2}\left(\frac{1}{\mu(z_1)} - \frac{1}{\mu(z_2)}\right)\left(\xi^2 + \eta^2\right)\right]. \quad (4.46)$$

Let us assume that time-varying amplitude of speckles obeys Gaussian statistics. In this case for speckle intensity

$$I(\vec{x}, t) = |V(\vec{x}, t)|^2$$

autocorrelation function of speckle intensity fluctuations determined as

$$\Delta I(\vec{x}, t) = I\ (\vec{x}, t) - \langle I(\vec{x}, t)\rangle$$

is expressed through

$$\begin{aligned}\langle \Delta I(\vec{x}, t_1)\Delta I(\vec{x}, t_2)\rangle &= \langle \Delta I(\vec{x}, z_1)\Delta I(\vec{x}, z_2)\rangle \\ &= \langle I(\vec{x}, z_1)\rangle\langle I(\vec{x}, z_2)\rangle |\gamma(\vec{x}, z_1, z_2)|^2 \end{aligned} \tag{4.47}$$

The task of calculation of correlated function of speckle intensity fluctuations is reduced to calculation of the correlated function $\Gamma(\vec{x}, z_1, z_2)$ of changes of speckle amplitude. Main parameters, which characterize dynamic speckle-field under longitudinal oscillations of the diffuse object, are the correlation function and density of power spectrum of speckle intensity fluctuations [107, 110]. It follows from (4.47) that for their determination it is necessary to calculate squared module of the autocorrelation function $\Gamma(\vec{x}, z_1, z_2)$ of time-varying speckles. Squared module of autocorrelation function is obtained through substitution of (4.46) into (4.45):

$$\begin{aligned}|\Gamma(\vec{x}, z_1, z_2)|^2 = &\frac{W_0^4 \Delta S^2}{W^2(z_1) W^2(z_2)} \left| \iint\limits_{-\infty}^{+\infty} \exp\left[-\left(\frac{1}{W^2(z_1)} + \frac{1}{W^2(z_2)}\right)(\xi^2 + \eta^2)\right] \right. \\ &\times \exp\left[-\frac{ik}{2}\left(\frac{1}{\mu(z_1)} - \frac{1}{\mu(z_2)}\right)(\xi^2 + \eta^2)\right] \\ &\times \left. \exp\left[ik\left(\frac{1}{R_0 - z_1} - \frac{1}{R_0 - z_2}\right)(\xi x + \eta y)\right] \mathrm{d}\xi \mathrm{d}\eta \right| \end{aligned} \tag{4.48}$$

Formula [110] should be used for calculation of integral in (4.48):

$$\begin{aligned}&\int\limits_{-\infty}^{+\infty} \exp(i a \xi^2) \exp(-\beta \xi^2) \exp(-i \rho \xi x) \mathrm{d}\xi \\ &= \exp\left(-\frac{\beta \rho^2 x^2}{4(\alpha^2 + \beta^2)}\right) \exp\left\{-i\left[\frac{\alpha \rho^2 x^2}{4(\alpha^2 + \beta^2)} - \frac{1}{2}\tan^{-1}\left(\frac{\alpha}{\beta}\right)\right]\right\},\end{aligned} \tag{4.49}$$

where parameters α, β and ρ correspond to the following values in (4.48):

$$\alpha = -\frac{k}{2}\left(\frac{1}{\mu(z_1)} - \frac{1}{\mu(z_2)}\right); \tag{4.50}$$

$$\beta = \frac{1}{W^2(z_1)} + \frac{1}{W^2(z_2)}, \tag{4.51}$$

$$\rho = k\left(\frac{1}{R_0 - z_1} - \frac{1}{R_0 - z_2}\right). \tag{4.52}$$

Using (4.49) and parameters of (4.50)–(4.52), the set in (4.48) autocorrelation function of speckle intensity fluctuations can be presented as

$$|\Gamma(\vec{x}, z_1, z_2)|^2 = \frac{\pi W_0^4 \Delta S^2}{W^2(z_1)W^2(z_2)(\alpha^2 + \beta^2)} \exp\left[-\frac{\beta\rho^2(x^2 + y^2)}{2(\alpha^2 + \beta^2)}\right]. \tag{4.53}$$

Parameters α, β and *W(x)* are presented in values z_1, z_2 and *a,* using (4.38) and (4.39) in (4.53):

$$\alpha = \frac{\pi(z_2 - z_1)(z_2 z_1 - a^2)}{\lambda(z_1^2 + a^2)(z_2^2 + a^2)}, \tag{4.54}$$

$$\beta = \frac{\pi a(z_1^2 + z_2^2 + 2a^2)}{\lambda(z_1^2 + a^2)(z_2^2 + a^2)}, \tag{4.55}$$

$$\alpha^2 + \beta^2 = \left(\frac{\pi}{\lambda}\right)^2 \frac{(z_1 - z_2)^2 + 4a^2}{(z_1^2 + a^2)(z_2^2 + a^2)}, \tag{4.56}$$

$$W^2(z_1)W^2(z_2) = \left(\frac{W_0^4}{\alpha^4}\right)(z_1^2 + a^2)(z_2^2 + a^2). \tag{4.57}$$

Substituting (4.54)–(4.57) into (4.53), we get

$$|\Gamma(\vec{x}, z_1, z_2)|^2 = \frac{\lambda^2 a^4 \Delta S^2}{(z_1 - z_2)^2 + 4a^2} \exp\left[-\frac{\lambda a(z_1^2 + z_2^2 + 2a^2)\rho^2}{2\pi(z_1^2 - z_2^2) + 4a^2}(x^2 + y^2)\right] \tag{4.58}$$

It follows from this that autocorrelation function of speckle intensity fluctuations is generally non-stationary, as (4.58) is the function of two independent variables z_1 and z_2. Nevertheless, it becomes stationary in the middle of the far-field region, i.e., $\vec{x} = 0$ $(x = y = 0)$ being only interval function $(z_1 - z_2)$ of two points along the optical axis. Alteration of intensity is also stationary near the center in the far-field region of diffraction as the given in (4.52) parameter ρ has a very small value in this case.

Normalized autocorrelation function in the center $\vec{x} = 0$ can be obtained from (4.58):

$$\gamma(0, z_1, z_2) = |\Gamma(0, z_1, z_2)|^2 / |\Gamma(0, 0, 0)|^2. \tag{4.59}$$

Using (4.58) and (4.59), we obtain

$$\gamma(0, z_1, z_2) \equiv \gamma(0, z_1 - z_2) = \frac{4a^2}{(z_1 - z_2)^2 + 4a^2} \tag{4.60}$$

As experimental determination of two-dimensional autocorrelation function is a labor-consuming task [111], the investigation of spectral density of intensity fluctuations of the scattered radiation is of practical interest while studying statistical characteristics of dynamic speckles formed by the diffuser oscillating in the longitudinal direction.

Power spectrum of dynamic speckles fluctuations of the periodic process will be linear with frequency interval ω and according to the generalized Wiener–Khintchin theorem [16, 111] for the amplitude of n-harmonic

$$b_n = \left(\frac{\omega}{2\pi}\right)^2 \iint\limits_0^{\frac{2\pi}{\omega}} \gamma(t_1, t_2) \exp[-in\omega(t_1 + t_2)]\mathrm{d}t_1\mathrm{d}t_2. \tag{4.61}$$

Substituting expression, which describes oscillation in harmonic approximation into (4.60)

$$z_i = h \sin(\omega t_i), \tag{4.62}$$

and substituting into (4.61), we obtain

$$b_n = \left(\frac{\omega}{2\pi}\right)^2 \iint\limits_0^{2\pi/\omega} \frac{4a^2 \exp[-in\omega(t_1 + t_2)]}{h^2[\sin(\omega t_1) - \sin(\omega t_2)]^2 + 4a^2} \mathrm{d}t_1\mathrm{d}t_2. \tag{4.63}$$

The obtained dependence in (4.63) connects relative intensity b on the n-harmonic and oscillation amplitude of diffuser h.

4.4 Application of Spectrum of Dynamic Speckles Intensity Fluctuations for Determining the Amplitude of Object Oscillating in Longitudinal Direction

Power spectrum of intensity fluctuations of dynamic speckles is connected with alteration of the amplitude of diffuser oscillations along the optical axis in transversal direction and depends on conditions of illumination [107, 109]. Broadening of the spectrum can be characterized by the ratio of relative intensity b_2 of the second harmonic to relative intensity of the first harmonic b_1.

But analytical solution of (4.63) is not possible because of difficulties with calculations. Therefore, the assigned task was solved through numerical simulations on PC and presenting graphically dependences of ratio of amplitudes of the first two harmonics b_2/b_1 on the amplitude of oscillations of the object h for different parameters determining conditions of illumination of the diffuser and observation of the speckle-field, which are the width and the length of the waist of a focusing lens, wavelength of laser radiation λ and oscillation frequency of the diffuser ω [112, 113].

Fourier transformation b_n can be presented in the form [107]:

$$|b_n| = \sqrt{b_n b_n^*}. \tag{4.64}$$

Having calculated the module $|b_n|$, the amplitude value of n-th harmonic can be obtained

$$b_n = \left(\frac{\omega}{2\pi}\right)^2 \left[\left(\iint\limits_0^{\frac{2\pi}{\omega}} \frac{\omega \sin[\omega(t_1 + t_2)] \mathrm{d}t_1 \mathrm{d}t_2}{A^2[\sin(\omega t_1) - \sin(\omega t_2)]^2 + 1} \right)^2 + \left(\iint\limits_0^{\frac{2\pi}{\omega}} \frac{\sin[n\omega(t_1 + t_2)] \mathrm{d}t_1 \mathrm{d}t_2}{A^2[\sin(\omega t_1) - \sin(\omega t_2)]^2 + 1} \right)^2 \right]^{\frac{1}{2}} \tag{4.65}$$

where $A = h/2a$.

The calculations were performed for wavelength $\lambda = 0.6328$ μm and oscillation frequency $\omega = 50$ Hz, as well as for values of waist width of the focusing lens W_0, which lie in the range from 10 μm to 10 mm, as further increase is unreasonable for these selected values W_0 because in this case $b_2/b_1 \to 1$.

The results of numeric simulations of dependence of amplitude harmonics b_2/b_1 on alteration of waist width of lens W_0 for different oscillation amplitudes of diffuser are presented in Fig. 4.13. The analysis of the obtained data has showed that increase W_0 of the focusing lens W_0 leads to narrowing of spectrum of intensity fluctuations of the speckle-field and decrease of the ratio of spectral harmonic amplitudes b_2/b_1 for this value of diffuser oscillation amplitude. Ratio b_2/b_1 changes more slowly at the increase of oscillation amplitude if the value of waist W_0 of the focusing lens is larger.

Decorrelation of speckles is determined by the change of wave phases from each elementary diffuser and will be changing the quicker, the narrower the beam waist is. Decrease of waist width leads to the increase of ratio b_2/b_1 for the given value of amplitude of diffuser oscillations h.

Thus, the conducted calculations gave the possibility to develop a method for determining oscillation amplitude of objects through changing the ratio of spectral harmonics amplitudes b_2/b_1.

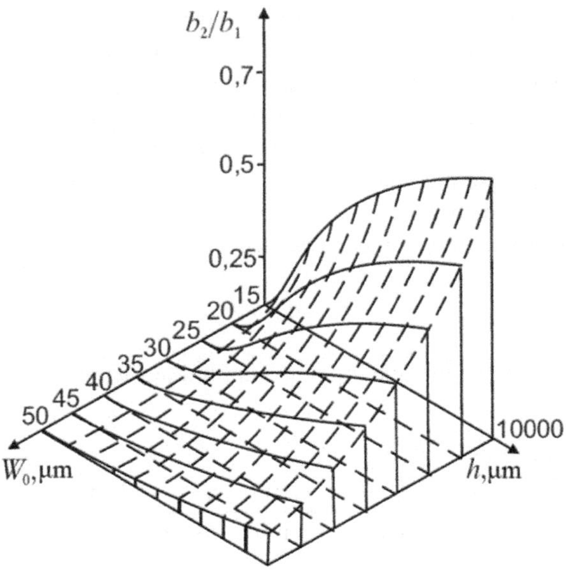

Fig. 4.13 Dependence of the ratio of amplitudes of spectral harmonics b_1/b_2 on the changes of the width of illuminating lens for different diffusor oscillation amplitudes. Reprinted from [2] with permission

Numerical simulations of dependence of spectral harmonics ratio b_2/b_1 on parameter A are presented in Fig. 4.14. Parameter A is determined by ratio of diffuser oscillation amplitude h to the length of the waist of the focusing length. Dependence of the given parameter A on oscillation amplitude h and waist width of lens W_0 are shown in Fig. 4.15. It is seen in the graphs that parameter A determines the range and accuracy of measurements of the object oscillation amplitude using this method

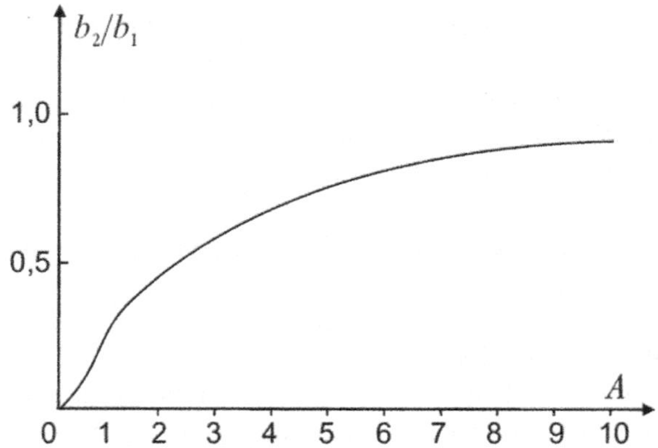

Fig. 4.14 Dependence of amplitude ratio of spectral harmonics b_1/b_2 on parameter A. Reprinted from [2] with permission

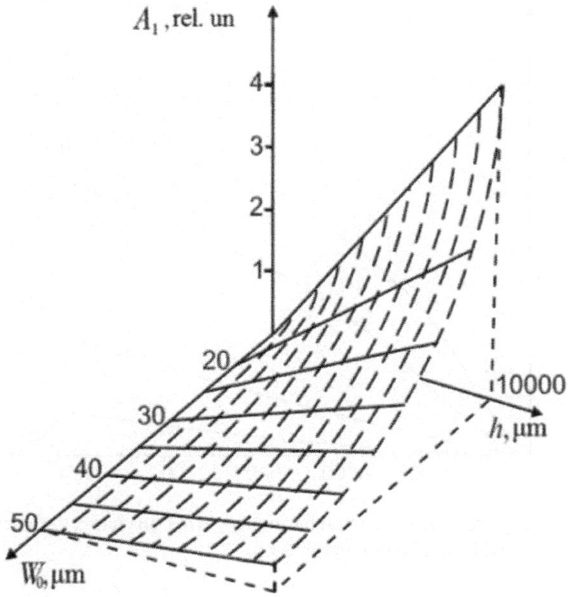

Fig. 4.15 Dependence of parameter A on oscillation amplitude h of the diffuser and width of lens waist W_0. Reprinted from [2] with permission

at the expense of selection of optimal ratio between values h and W_0 because if $A > 5$, then ratio of spectral harmonics is $b_2/b_1 \rightarrow 1$. For experimental proof, He–Ne laser was used, which was illuminated by a focusing lens with $W_0 = 15$ μm, which matted on one side glass plate actuated with the help of a vibrator in direction of spreading of a laser beam with frequency $\omega = 50$ Hz.

Produced by radiation dynamic speckle-field was registered with a photoelectron sensor PMT-79. Electric signal came from its output into the input of a multi-channel spectrum analyzer CK4-72 that gave the possibility to accumulate and repeatedly average spectrum of noise-like signal over the range up to 20 kHz. In speckle-field intensity spectrum, amplitude ratio was fixed for different values of amplitude of matted plate oscillations. The speckle-field intensity spectrum was obtained through time averaging of random signal on 16 realizations of the given spectrum.

The results of numerical simulations and experimental measurements of b_2/b_1 ratio at the change of oscillations amplitude of the scattering surface h are presented in Fig. 4.16.

The obtained experimental data satisfactorily agree with the data of numerical simulation and give the possibility to plot a calibration curve for the given optical scheme ($W_0 = 15$ μm), according to which the value of oscillation amplitude of the object under study can be determined by measuring b_2/b_1 ratio. The linear spectrum, which is typical for periodical oscillating movement of the diffuser, gives the possibility to easily determine the frequency of diffuser oscillations through measuring frequency distance between spectral harmonics.

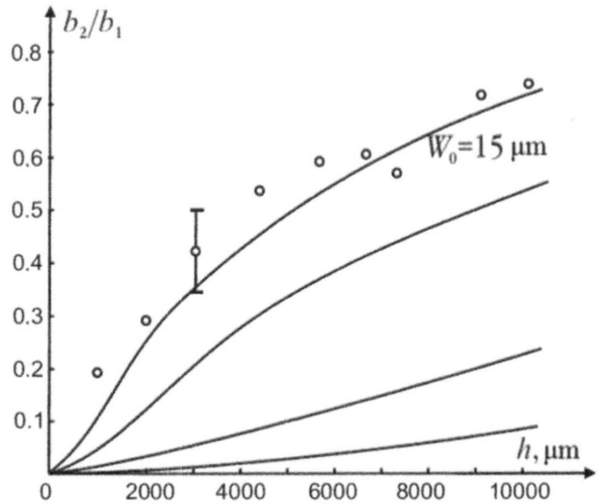

Fig. 4.16 Dependence of the ratio of spectral harmonics $b1/b2$ on changes of oscillation amplitude h of the diffuser. Reprinted from [2] with permission

4.5 Correlation of Speckle-Fields Formed by Diffuse Object Moving Along the Optical Axis

Possibility of non-contact determination of velocities and deformations of scattering objects gives the opportunities to study processes of contractions of muscle tissue, circulation in blood vessels, etc. Determination of transversal component of velocity of diffuse object is based on studying speckle behavior in a plane, which is perpendicular to the optical axis. The creation of measurement methods of longitudinal component of velocity requires knowledge of speckle properties, which depend on alterations of the longitudinal coordinate of diffuser or observation points, which are investigated to a lesser extent. The works [16, 36, 66, 114–116] examine speckle-field in diffraction area directly behind the diffuser. It was shown that on the average speckles are of prolate form [114]. In the work [66], it is mentioned that the speckle-structure, which is formed through illuminating of the diffuser with a point source, changes its scale if the diffuser is shifted along the optical axis. In the work [36], correlation functions of the speckle-field are examined for an analogous scheme. The situation when the diffuser is illuminated with a Gaussian laser beam was in detail examined by N. Takai and others [115, 116]. The speckle-field formed by the objective in the vicinity of the image plane was studied in the work [84]. It is seen from the results of the work [84] that the random field is non-stationary, and the average size of the speckles along the optical axis is determined by the depth of the field of the objective.

In the present case, correlation functions of a field are investigated, which is formed by the diffuse object under its longitudinal shift for any two points of the image space. Let us examine the scheme presented in Fig. 4.17. The optical system

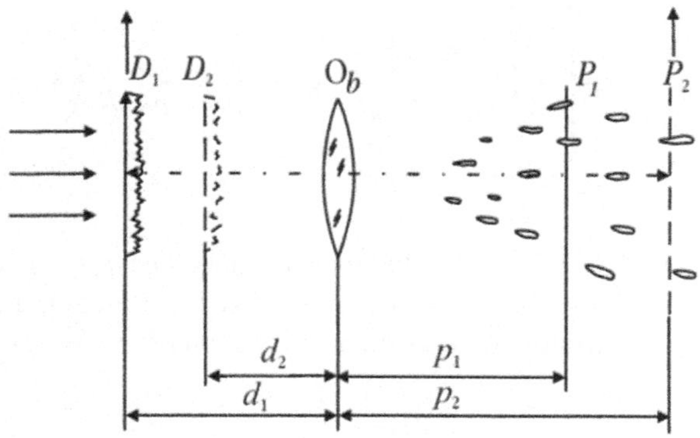

Fig. 4.17 The optical scheme of specklogram registration. Reprinted from [2] with permission

O_b with a pulse response k ($\vec{x}$, $\vec{X}$, $\vec{x}$, d, p) forms a field in the planes P_1 and P_2. The field is scattered by the same diffuse surface D in the planes D_1 and D_2, respectively. Coordinates of the planes P_1 and P_2 are described by two-dimensional vector $\vec{X} = (X, Y)$, and in the planes D_1 and D_2—by vector $\vec{x} = (x, y)$. If the complex transmission (or reflection) coefficient of the diffuser is denoted as $t(\vec{x})$ and distribution of the complex amplitude of the incident wave—u ($\vec{x}$), then the complex amplitude of the field in the observation plane will be

$$V(\vec{\mathrm{X}}, d, p) = \int_{-\infty}^{\infty} u(\vec{x})t(\vec{x})K(\vec{x}, \vec{\mathrm{X}}, d, p)d^2\vec{x} \tag{4.66}$$

Supposing that D is a purely minute-structured phase diffuser, which introduces random phase difference from $-\pi$ to π, we have $\langle t(x)t^*(x')\rangle = \delta(x - x')$, and it is easy to get correlation function of field amplitude:

$$\Gamma_{\upsilon}(\vec{\mathrm{X}}_1, \vec{\mathrm{X}}_2; d_1, d_2; p_1, p_2) = \int_{-\infty}^{\infty} |u(\vec{x})|^2 K(\vec{x}, \vec{\mathrm{X}}, d_1, p_1)K * (\vec{x}, \vec{\mathrm{X}}_2, d_2, p_2)d^2\vec{x}. \tag{4.67}$$

Time-space correlation function can be directly derived from (4.67) assuming that $d_2 = d_1 - \upsilon\tau$ Further we will consider the case when the diffuser is illuminated by a plane wave. Concretizing the view $K(x, X, d, p)$, it is worth mentioning that if the optical system is simulated with a thin lens, then it is very difficult to conduct analytical calculation (4.67). Therewith as it is shown in the works [84, 117], they used for calculations pulse response of the objective with an anodizing screen leads to results well agreed with the experiment. In this case

$$K\left(\vec{x}, \vec{X}, d, p\right) = \exp\left\{ i\frac{k}{2}\left(\frac{|\vec{x}|^2}{d} + \frac{\left|\vec{X}\right|^2}{p}\right)^2 - \left|\frac{\vec{x}}{d} - \frac{\vec{X}}{p}\right|^2 \Big/ \left(4k^2q^2 - i\frac{2}{k}\varepsilon\right)\right\}, \tag{4.68}$$

where $k = \frac{2\pi}{\lambda}$, $\varepsilon = \frac{1}{d} + \frac{1}{p} - \frac{1}{f}$; f is the focal distance O_b; q is the effective radius.

Using (4.67) and (4.68), we define normalized correlation function of the amplitude and its connection to the correlated function of intensities fluctuations.

$$\gamma_{\Delta I} = |\gamma_\nu|^2. \tag{4.69}$$

And after some algebraic transformations, we obtain

$$\gamma_{\Delta I} = 4\alpha^2\left[\left(\alpha^2+1\right)^2 + \beta^2\left(\alpha^2\sigma - 1\right)^2\right]^{-1} \times \exp\left\{-\frac{k^2q^2}{2}\frac{\left(\alpha^2+1\right)}{\left(\alpha^2+1\right)^2}\frac{\left|\frac{\vec{X}_1}{p} - \alpha\frac{\vec{X}_2}{p}\right|^2}{\left(\alpha^2+1\right)^2 + \beta^2\left(\alpha^2\sigma - 1\right)^2}\right\}, \tag{4.70}$$

where $\alpha = d_2/d_1$; $\sigma = \varepsilon_2/\varepsilon_1$; $\beta = \frac{kq^2\varepsilon_1}{2}$.

Equation (4.70) depends on a number of parameters that is why we should consider particular cases to illustrate it. If $\alpha = 1, \sigma = 1$, then the field is examined in some plane behind the objective. Analysis of the exponential factor leads to the known result that the fields stay correlated for the points, the distance between ich is determined by the observation aperture (in particular, it corresponds to the resolution limit for the image plane) and these fields are stationary. If the correlation function alters under small defocusing ($d_2 \approx d_1$) in the same plane $p_2 \approx p_1$ in the same point $\vec{X}_1 = \vec{X}_2$, then we come to the situation, which was considered in the work [84]. It is worth mentioning that in common case, the maximum of the correlation function will be observed for points connected by the correlation.

$$\frac{d_1}{p_1}\vec{X}_1 = \frac{d_2}{p_2}\vec{X}_2 \tag{4.71}$$

When the correlation function is determined for points, which are situated in the image planes P_1 and P_2, then (4.71) means that maximal correlation is observed for points of the images, which correspond to the same point of the object. From a physical point of view, interpretation of this result is simple: Identical intensity distributions are observed where phase correlations between elementary waves are saved. Such interpretation is actually analogous to the considerations [66] conducted for the scheme where there is no optical system. It follows from (4.70) that if corresponding points $\frac{d_1}{p_1}\vec{X}_1 = \frac{d_2}{p_2}\vec{X}_2$ are in the image plane, then the correlation function alters due to a simple law:

$$\gamma_{\Delta I} = \frac{4(d_2/d_1)^2}{\left[1 + (d_2/d_1)^2\right]^2} \tag{4.72}$$

Decrease of the correlation under $d_2 \approx d_1$ can be explained by the fact that the resolution of the objective changes when the observation aperture changes, and other parts of the surface participate in the formation of a separate speckle. The resulting coherent sum of waves scattered by elementary diffusers will differ from the initial one and correlation will sink. When observation points, in which meeting condition in (4.71) lies in the planes P_1 and P_2 characterized by identical defocusing parameter $\varepsilon_1 = \varepsilon_2$, then decorrelation with α alteration comes much faster. If planes P_1 and P_2 are connected by the ratio $\varepsilon_1/\varepsilon_2 = \text{const}$, then for any defocusing $\varepsilon_{1,}$ there are two planes D_1 and D_2, for which intensities correlation does not alter, does not depend on the quantity, and its value is determined by the condition $d_2^2\varepsilon_2 = d_1^2\varepsilon_1$.

In conclusion, it is worth mentioning that (4.68) can be easily proved in experiments with speckle-photography and directly used for the development of new methods for determination of shifts and longitudinal velocities components.

4.6 Experimental Study of the Movement of Subjective Speckle-Fields Under Longitudinal Shift of the Diffuse Object

Methods of speckle-photography give the possibility to determine both longitudinal and transversal shifts of surfaces of biological and other diffuse objects [8, 44]. In one of the first works on speckle-interferometry [53], an efficient way of determination of axial shift was demonstrated, which is based on radial recession of the speckle-structure and obtaining of annular interferograms. Dependence of the contrast of such interferograms on the aperture was discussed in the works [44, 54], and in the work [84] correlation characteristics of the field were investigated, which was scattered by the moving along with the axis diffuser in the neighborhood of the plane image. According to these publications, the range of the determined shifts is limited by the depth of the focus of the objective.

In Sect. 4.5, speckle-fields correlation in two random points of image space is considered. In particular, it was shown that caused by the longitudinal shift, decorrelation of speckle-fields comes much more slowly if they are registered not in the fixed plane as it was before, but in planes conjugated with the initial and the shifted positions of the object. Dependencies of optimal distribution of registration planes under defocusing during the first exposition were obtained. The possibility of obtaining interferograms was experimentally investigated. The interferograms correspond to the longitudinal shifts of the object exceeding the depth of the field of the objective.

As follows from the work [118], radial shift of speckles is connected with the initial and the final position of the object d_1 and d_2 by the ratio

$$\Delta R = R\left(\frac{p_2}{d_2}\frac{d_1}{d_1} - 1\right) \tag{4.73}$$

where R is the distance from the center of the specklogram; p_1 and p_2 are the distance from the objective O_b to the registration plane during the first and the second exposures (Fig. 4.17).

In order to form speckle-interferograms, ΔR must exceed the average size of the speckles

$$\Delta R > \frac{1.22\lambda p}{q} \tag{4.74}$$

where $p = (p_1 + p_2)/2$; λ is the wavelength; q is the diameter of the diaphragm of the objective.

When $p_1 = p_2$ and $\frac{(d_2-d_1)}{d_2} \ll 1$, then (4.73) coincides with the result in the work [44]. In this case, the range of possible shifts $\Delta d = d_2 - d_1$ is determined by the ratio

$$\frac{1.22\lambda p d_2}{qR} < \Delta d < \frac{4\lambda d_1^2}{q^2}, \tag{4.75}$$

where $4\lambda d_1^2/q^2$ is the depth of the focus of the objective under shift at a value, which is more than the depth of the focus when decorrelation of speckle-structures occurs in the registration plane and a speckle-interferogram does not form.

However, the correlation between speckle-fields in the initial and the shifted states preserved much longer if a shift compensating of registration plane shift is made before the second exposure [119]. Such a shift gives the possibility to get speckle-interferograms, which correspond to the shifts of the depth of the focus of the objective, and, thus, to widen the range of the measured longitudinal shifts significantly using the method of speckle-interferometry and that was experimentally confirmed.

In the optical scheme (Fig. 4.17) with parameters $q = 17$ mm, $f = 120$ mm, $d_1 = 218$ mm, the subjective speckle-image was registered in the plane p_1 on film "Mikrat-N". The specklogram was exposed for the second time in the plane p_2 after the shift of the diffuser in the position D_2. Contrast interference rings were observed while illuminating of the specklograms with plane wave in such a way.

The interferogram is presented in Fig. 4.18a, which corresponds to the object shift $\Delta d = 4000$ μm that exceeds the depth of the focus almost in an order. Interference fringes did not form during illumination of the reference specklograms when the registration plane between the exposures was not shifted $\Delta d = 0$. The calculated ones according to the interferograms shifts of the object correspond to the actual ones with an accuracy of 1%. In the experiment [119], interference rings of tolerable quality were observed during shifts, which exceed the depth of the focus of the objective in 15–20 times. But significant shifts lead to decorrelation of speckle-structures, and contrast of interferograms decreases due to the regularity presented in the work [118].

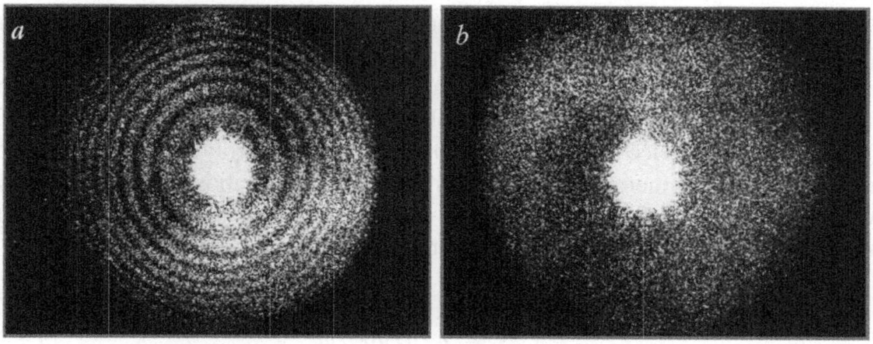

Fig. 4.18 Speckle-interferograms corresponding to longitudinal shift of the object exceeding the focal depth of the objective: **a**—$\Delta d = 4000$ μm; $\Delta p = 6000$ μm; **b**—$\Delta d = 4000$ μm; $\Delta p = 0$. Reprinted from [2] with permission

It is worth mentioning that the presence of transversal shift of the diffuser will lead to the shift of the center of the interference rings. Also an unexpanded laser beam can be used for determination of Δd by scanning the specklogram with its help [44].

4.7 Methods for Determining Diffuse Objects Deformations

Methods for determining diffuse objects deformations are based on the phenomenon of coherent addition of two random speckle-fields. The field scattered by the object in the initial state is recorded on the hologram. Then after some time, which is more than the time of the medium response, but less than the time of relaxation of the medium where the dynamic hologram was recorded, the second pulse reconstructs from the hologram speckle-field corresponding to the initial state of the object, and during this time the object, which has changed its state, forms the second speckle-field. Secondary illumination of the surface should be conducted with a wave, polarization of which is perpendicular to the polarization of the reconstructing wave. In this case, the hologram of the altered state is not recorded. The lens situated behind the specklogram forms a Fourier transformation in its focal plane. This Fourier transformation presents itself Yung's fringes. The direction and period of the fringes give the possibility to determine direction and value of shift during a small period of time τ, i.e., actually the velocity components of which are

$$\upsilon_x = \frac{\lambda f}{M\tau\Delta x}, \ \upsilon_y = \frac{\lambda f}{M\tau\Delta y}, \tag{4.76}$$

where f is the focal distance of the lens; Δx is the period of fringes along the x-axis; Δy is the period of fringes along the y-axis; M is the image system magnification.

It is suggested that a lens raster is used instead of one lens, which conducts Fourier transformation, to determine the shift of separate areas of the surface. And the relative aperture of a separate lens of the raster must be not less that the lens aperture, which creates the image of the surface on the hologram.

A registering device (e.g., a TV camera, a photodiode array) conjugated with a microcomputer or a memory block records Fourier transformation with the frequency ν and the obtained information is stored or processed in real time. To get shift of some area from time moment t_1 to moment t_2, it is enough to calculate the value

$$\vec{S} = \int_{t_1}^{t_2} \vec{\upsilon}(t)\mathrm{d}t \tag{4.77}$$

This process gives the possibility to follow the deformations of strongly changing surfaces, including structure-varying surfaces.

The Yung's fringes must register in the focal plane with a TV camera, which is conjugated with an automated system of interferogram processing, due to which the period and direction of fringes with the frequency of 2 Hz can be estimated. In its turn, this will give the possibility to measure the velocity of different areas of the plate and then to determine shift of these areas in different moments of time through microcomputer integration.

Also the described method can be implemented in the scheme presented in Fig. 4.19b. The laser *Ls* generates the first radiation pulse, which is divided on a mirror into an object and a reference beam. Passing through the optical shutter, the object beam incidents on the moving or deformed object *P*. The reflected from the object light passes through the objective L_1, which forms an image of the object covered with a speckle-structure on the dynamic hologram *DH*. The lasers Ls_1 and Ls_2 create radiation pulses after time τ. During this time, the electro optical shutter does not allow radiation with wavelength λ_2 passing onto the object. The object was illuminated with radiation with the wavelength λ_2, and the dynamic hologram is reconstructed with radiation with the wavelength λ_1. The lens and the lens raster conduct Fourier transform of the speckle-structure formed by the object and reconstructed from the hologram. In this case, the registering system should provide heterodyne detection of Yung's fringes. It is enough for this to use three photodiodes for each element of the raster as the input of the registering device.

The third scheme of the device for the probable implementation of this method is presented in Fig. 4.19c. During the first pulse, the radiation passes through the splitter Bs where it is divided into an object and reference beams. Then the object beam goes through electrooptical element E, which does not change its polarization during the first pulse, incidents on the object under study, passes through the image-forming system L1 and falls onto the cell DH where the dynamic hologram is recorded. During the second pulse, the electrooptical element turns the polarization plane of the object beam at 90° and thus the dynamic hologram is not recorded during the second pulse.

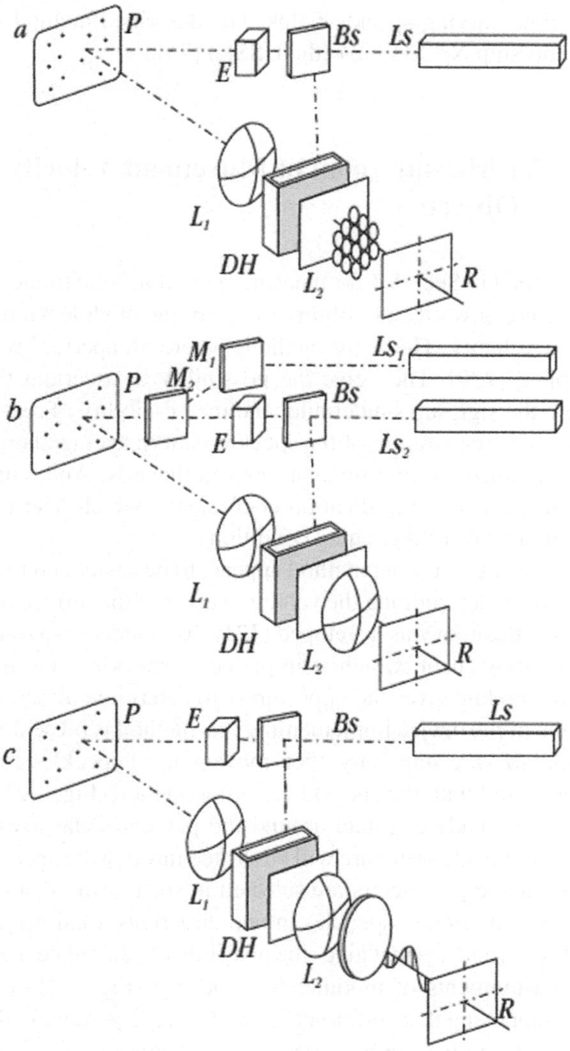

Fig. 4.19 Scheme of realization of determination method of deformation of diffuse objects. Reprinted from [2] with permission

The field of the hologram becomes an additional register, on which a reconstructed speckle-structure and formed by the object speckle-structure are registered.

An advantage of this method is the enhancement of the possibilities at the expense of measuring value and direction of fast-changing deformations and vibrations, changes of value and direction of velocity of movement of separate parts of the object under study or its movement on the whole, enlargement of the range of the measured deformations, deformations of the object with a varying surface microstructure, determination of degree of roughness of an object and mostly at the expense of composition of time-varying speckle-fields. The described method is protected by certificate of authorship No 1310624 (the USSR) [71].

4.8 Method for Measurement of Movement Velocity Vector of Diffuse Objects

The method described in Sect. 5.1 for measurement of muscle tissue velocity, which uses pinhole aperture, gives the possibility to determine module but not the direction of the movement velocity. There are methods where an aperture with preferential direction is applied [120]. They give the possibility to ascertain the direction of velocity, but here the sign stays indefinite, and the possibility of real-time measurements is lost. Besides registration of the speckle-pattern averaged on the larger area of the aperture minimizes signal fluctuations significantly. And amplifiers have to be used with much greater amplification coefficients, which lead to the increased nonlinear distortions under other equal conditions.

In this sense, speckle-counting method appears to be easier and more flexible. On its basis, a method for determining the velocity vector of the diffuse object as well as the device for its realization was developed [121]. Its essence consists in preliminary known additional movement, which is imparted to speckles, and measurement of velocity variation module gives the opportunity to determine direction of the movement. The scheme of the device implementing this method is presented in Fig. 4.20a. If mirrors M_1 and M_2 are stationary, then movement of speckles caused by object movement is registered with the speckle-counting method (Fig. 4.21).

If the mirrors are made to rotate around the perpendicular axes at an angular velocity ω, then the speckle-structure will be shifted through the aperture at a velocity of $\vec{\upsilon} + \delta\vec{\upsilon}$. The scheme parameters are selected in such a way that $|\delta\vec{\upsilon}| < |\vec{\upsilon}|$. After imparting additional motion to speckles in two directions, total velocities will have different modules $|\vec{\upsilon}_1|$ and $|\vec{\upsilon}_2|$ that gives the possibility to calculate the initial velocity vector. After measurements of modules $|\vec{\upsilon}_1|$ and $|\vec{\upsilon}_2|$ (Fig. 4.21) consequently or simultaneously, supposing that additions $\delta\vec{\upsilon}_1$ and $\delta\vec{\upsilon}_2$ are perpendicular and equal in size, the following formulae can be obtained for calculation

$$|\vec{\upsilon}| = \sqrt{\frac{|\vec{\upsilon}_1|^2 + |\vec{\upsilon}_2|^2}{2} - |\delta\vec{\upsilon}|^2}; \tag{4.78}$$

$$tg\alpha = \frac{|\vec{\upsilon}_2|^2 - |\vec{\upsilon}|^2 - |\delta\vec{\upsilon}|^2}{|\vec{\upsilon}_1|^2 - |\vec{\upsilon}|^2 - |\delta\vec{\upsilon}|^2}. \tag{4.79}$$

Accuracy of determining α-angle can be estimated according to the latter formula. Taking into account that $|\delta\vec{\upsilon}| \ll |\vec{\upsilon}|$ after some transformations, the dispersion of angle change can be found

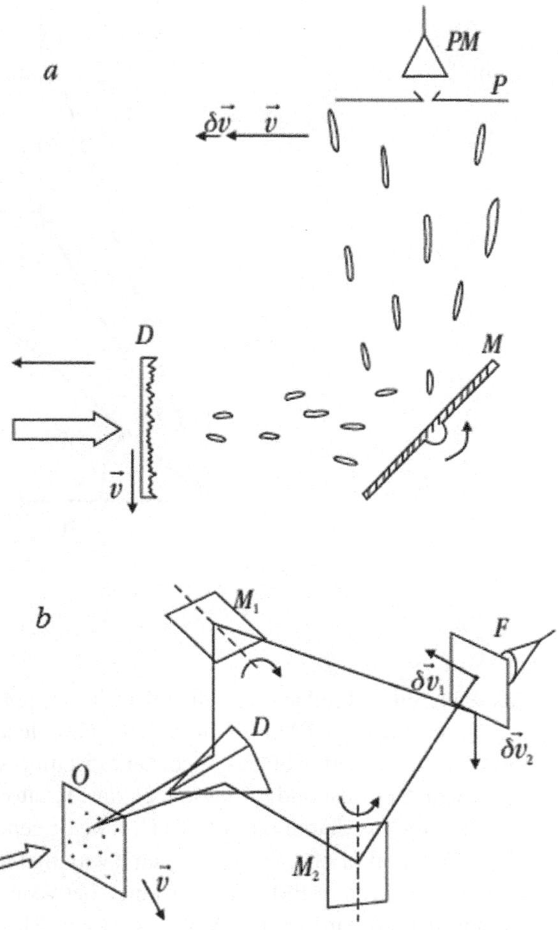

Fig. 4.20 Optical schemes for determining velocity sign (**a**) and measurement of velocity vector of diffusive objects (**b**). Reprinted from [2] with permission

$$\langle d\alpha^2 \rangle = \frac{2(|\vec{\upsilon}_2| - |\vec{\upsilon}|)(|\vec{\upsilon}_1| - |\vec{\upsilon}|)}{|\delta\vec{\upsilon}|^4} \langle d\upsilon^2 \rangle \tag{4.80}$$

where $\langle d\upsilon^2 \rangle$ is the dispersion of module determination $|\vec{\upsilon}|$, $|\vec{\upsilon}_1|$ and $|\vec{\upsilon}_2|$.

It was supposed that $|\delta\vec{\upsilon}|$ was determined with better accuracy than $|\vec{\upsilon}|$, $|\vec{\upsilon}_1|$ and $|\vec{\upsilon}_2|$.

A scheme was established for experimental check of a method for determining velocity sign of the diffuse object. In this scheme, a rotating with a frequency matt disk with a diameter of 15 cm was used as an object.

Unexpanded laser beam illuminated a small area at a distance $r = 18$ mm from the center, moving with a linear velocity of $\upsilon = \omega_0 r$.

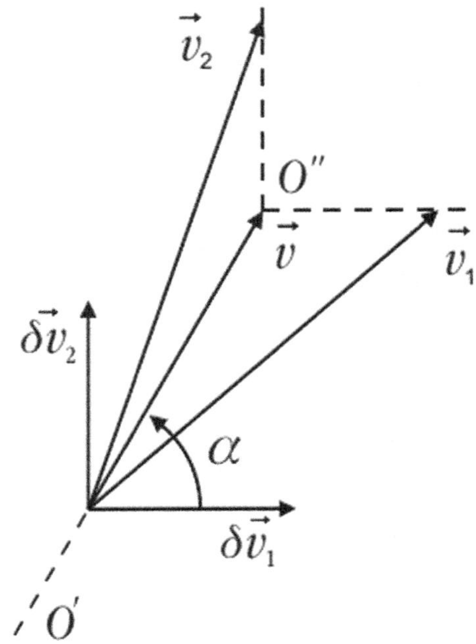

Fig. 4.21 Diagram of calculation of velocity vector. Reprinted from [2] with permission

The radiation was scattered and reflected by the mirror rotating at a frequency of two revolutions per minute and got on PMT 79 through the pinhole aperture. ChZ-33 frequency counter was used for counting pulses. Electronic circuit gave the possibility to form squared pulses, which correspond to single speckles. It automatically started when halo diffraction got onto the diaphragm PMT and time of generation of pulses was regulated within the interval of 60–500 ms in such a way that not to exceed the traveling time of a light spot through PMT. The distance between the diffuser and the mirror was 8 cm and from the mirror to PMT was 47 cm. Thus, the additional velocity was 9.84 cm/s. Direct current engine was used for disk rotation. The direction of the current was easily changed through switching the voltage sign. It should be mentioned that a one-dimensional method was realized in experiments that gives the possibility to determine the velocity sign. The results of the experiment for different rotation velocities of the object are presented in Table 4.1.

Table 4.1 Dependence of the modulated velocities υ_1, υ_2, actual velocity υ and frequency ω on voltage and power supply of the engine

U, B	5.0	6.0	7.0	8.0
ν_1, arbitrary units	70 ± 3.7	155.5 ± 8.3	242.5 ± 1206	317.7 ± 8.9
ν_2, arbitrary units	44.5 ± 4.6	125.2 ± 3.7	210.7 ± 18.5	292.0 ± 4.9
ν, arbitrary units	57.2	140.4	226.6	305.2
ω, Hz	3.03	5.55	7.69	10.0

All the results are reliable at significance level $p = 0.05$. It is seen from the comparison of υ_1 and υ_2 that velocity is codirectional with $\delta\upsilon$ in the first case, which was actually realized in the experiment. Average values $= (\upsilon_1 + \upsilon_2)/2$, which describe velocity module $|\vec{\upsilon}|$, lie well onto a straight line.

Thus, as a result of this study, a non-contact method of speckle-optics for determination of two-dimensional travel velocity vector of the diffuse object was suggested and developed, which is based on additional modulation of speckle-fields. An electronic device was created for the calculation of the number of speckles passing through PMT aperture per unit of time and for measurement of velocity module. The experimental check of this method on models provided a measurement of velocities in the range of 3–50 cm/s with an error of about 5%.

4.9 About Formation of Annular Speckle-Interferograms Emerging During Longitudinal Shift

It is known that double-exposure speckle-photography is mainly used for measurement of transverse shifts, and it has the greatest sensitivity to such shifts [44, 48]. Fourier transformation of the negative, on which there was registered image of the object covered with speckles, in initial and shifted states gives the opportunity to obtain equidistant interference fringes, which are known as Yung's fringes, according to which the required transversal shift is easily calculated. Along with this, speckle-photography allows determining the longitudinal shift of the diffuse object as well [58]. But speckle-interferograms themselves as well as their forming conditions considerably differ in both cases.

This work considers the process of formation of interference fringes corresponding to a longitudinal shift of the objective. Let two speckle-fields with intensity distribution $I_1(\vec{x})$ and $I_2(\vec{x})$ be linearly registered on photomaterial S_p, and its transmittance be described with expression

$$t = t_0 - t_1\left[I_1(\vec{x}) - I_2(\vec{x})\right] \tag{4.81}$$

where t_0 and t_1 are the constants; $\vec{x}$ is the two-dimensional vector in plane S_p.

We illuminate such a specklogram with a plane wave u_0 and behind it at distance d we place the optical system O_b with pulse response $K(\vec{x}, \vec{X}, d, p)$ where $\vec{X}$ is the coordinate in the observation plane P; p is the distance between the objective and P (Fig. 4.22). Part of the light t_0u_0 passed the specklogram without diffraction can be filtered by placing the minor blocking screen.

Intensity of a wave diffracted on speckle-structures will be equal

$$I_p\left(\vec{X}\right) = |u_0|^2 t_1 \int \left[I_1(\vec{x}) + I_2(\vec{x})\right]\left[I_1\left(\vec{x}'\right) + I_2\left(\vec{x}'\right)\right]$$
$$K\left(\vec{x}, \vec{X}, d, p\right) K^*\left(\vec{x}', \vec{X}, d, p\right) d^2\vec{x}\, d^2\vec{x}' \tag{4.82}$$

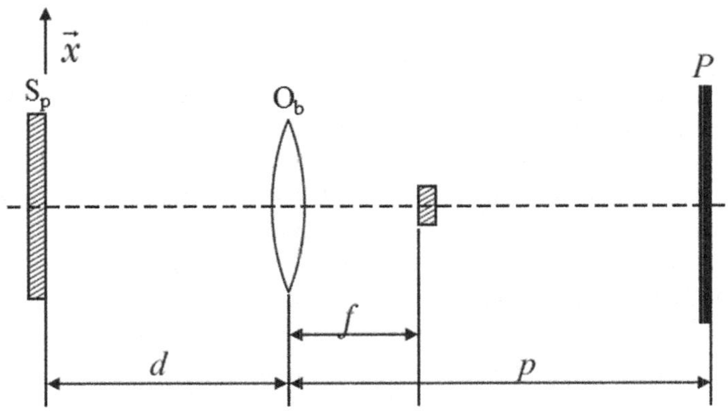

Fig. 4.22 Scheme for circular speckle-interferograms of longitudinal shear control. Reprinted from [2] with permission

Equation (4.82) decomposes into three components, two of which describe intensity of waves diffracted on speckles registered during the first and the second exposures, respectively, and the third one describes their interference. As $I_1(\vec{x})$ and $I_2(\vec{x})$ are the random functions, then the resulting distribution $I_p(\vec{x})$ in observation plane will be stochastic as well. For elimination of this noise and separation of interference fringes, we average (4.82) on the ensemble of specklograms. We analyze the latter compound in (4.82), which describes formation of interference fringes:

$$I\left(\vec{X}\right) = 2\text{Re}\iint I_1(\vec{x})\left\langle I_2\left(\vec{x}'\right)K\left(\vec{x}, \vec{X}\right)\right\rangle K^*\left(\vec{x}', \vec{X}\right)d^2\vec{x}d^2\vec{x}' \tag{4.83}$$

The shape of the speckle-interferogram is determined by the correlation between speckle-structures $I_1(\vec{x})$ and $I_2(\vec{x})$ and properties of the optical system. In fact, the expressions describe almost every speckle-interferogram. In the considered case, the object experiences longitudinal shift between exposures, and two speckle-fields $I_1(\vec{x})$ and $I_2(\vec{x})$ are alike and radially shifted. The common expression of correlation function $\left\langle I_1(\vec{x})I_2\left(\vec{x}'\right)\right\rangle$ is presented in the work [118]. But radial "recession" of speckles is most important for forming a speckle-interferogram of longitudinal shift and clarifying the main conditions of their observation. Neglecting decorrelation of speckles and their final size, correlation function of intensity can be written as

$$\left\langle I_1(\vec{x})I_2\left(\vec{x}'\right)\right\rangle = \langle I\rangle^2\delta\left(\vec{x} - \alpha\vec{x}'\right) \tag{4.84}$$

where α is the coefficient describing radial expansion of the speckle-image, the value of which is close to one in real conditions and is determined by longitudinal shift of the object; $\langle I\rangle$ is the average value of illumination of photomaterial during exposure.

For calculation of (4.83), we suppose that the optical system is a thin lens with apodizing Gaussian screen. It should be mentioned that the results of calculations using this model are in good agreement with the data of the experimental studies of speckle-fields [117]. Then, for example, according to the work [84]

$$K\left(\vec{x}, \vec{X}, d, p\right) = \exp\left\{ i\frac{k}{2d}\left(|\vec{x}|^2 + \frac{\left|\vec{X}\right|^2}{M}\right) - \frac{\left|\vec{x} - \frac{\vec{X}}{M}\right|^2}{\frac{4d^2}{k^2q^2} - i\frac{2d^2}{k}\left(\frac{1}{d} + \frac{1}{p} - \frac{1}{f}\right)} \right\} \tag{4.85}$$

where $M = p/d$, $k = 2\pi/\lambda$ is the wave number; q is the effective radius of the lens.

Substituting (4.85) and (4.84) into (4.83) and calculating normalized intensity distribution in registration plane, the contrast γ of the interference pattern can be obtained

$$\gamma = \left\{ \frac{\left(\alpha^2+1\right)^2}{4\alpha^2} + \left[\frac{\alpha^4 - 1}{2\alpha}\left(\frac{ka^2}{2d}(1+\beta^2) - \beta\right)\right]^2 \right\}^{\frac{1}{2}}$$
$$\times \exp\left\{ -\frac{(\alpha-1)^2}{(\alpha^2+1)} \frac{\left[1 + \frac{(\alpha+1)^2}{(\alpha^2+1)}\frac{ka^2}{d}\left(\frac{ka^2}{2d}(1+\beta^2) - \beta\right)\right]\left|\vec{X}\right|^2}{\left[1 + \frac{(\alpha^2-1)^2}{(\alpha^2+1)^2}\left(\frac{ka^2}{2d}(1+\beta^2) - \beta\right)^2\right]a^2M^2} \right\}$$
$$a = \frac{2d}{kq},\ \beta = \frac{kq^2}{2}\left(\frac{1}{p} + \frac{1}{d} - \frac{1}{f}\right) = \frac{\Delta p}{\delta}\frac{p_0}{p}, \tag{4.86}$$

where p_0 can be found from the condition

$$\frac{1}{p_0} + \frac{1}{d} - \frac{1}{f} = 0$$

where $\Delta p = p_0 - p$ characterizes the position of observation plane P; $\delta = \lambda p_0^2/\pi q^2$ is the depth of the field O_b is in the image space.

The physical sense of q is the distance in objects space resolved by objective O_b.

Intensity distribution of annular interference pattern is described by phase factor in (4.83):

$$\mathrm{I}\left(\vec{X}\right) = I_0\gamma \mathrm{Re}\left\{ \exp\left[\frac{i(\alpha-1)^2(\alpha+1)^2k\left|\vec{X}\right|^2}{\left(\alpha^2+1\right)^2 2dM^2} \times \frac{1 - \frac{(\alpha-1)^2}{(\alpha+1)^2}\beta\left(\beta - \frac{2d}{ka^2}\right)}{1 - \frac{(\alpha^2-1)^2}{(\alpha^2+1)^2}\left(\frac{ka^2}{2d}(1+\beta^2) - \beta\right)^2} \right] \right\} \tag{4.87}$$

where I_0 is the normalization factor.

Let us examine the contrast of the interference pattern more carefully. It is easy to show that the maximal value γ is determined by correlation

$$\frac{ka^2}{d}\beta - 1 = 0 \tag{4.88}$$

whence we find the condition for the plane of localization of interference rings:

$$p_L = \frac{2df}{2d - f} \tag{4.89}$$

Thus, the localization plane is optically conjugated with the plane, which is located twice farther from the objective, than the analyzed specklogram. The results of the experimental check of (4.89) are presented in Fig. 4.23.

The distance d was measured on the x-axis from the specklogram to the objective, and on the y-axis—distance p from the objective to the observation plane of rings with maximal contrast.

Having denoted the solution of (4.88) by β_0, $\Delta\beta = \beta - \beta_0$ and taking into account that α is close to unit, (4.86) can be rewritten in the form

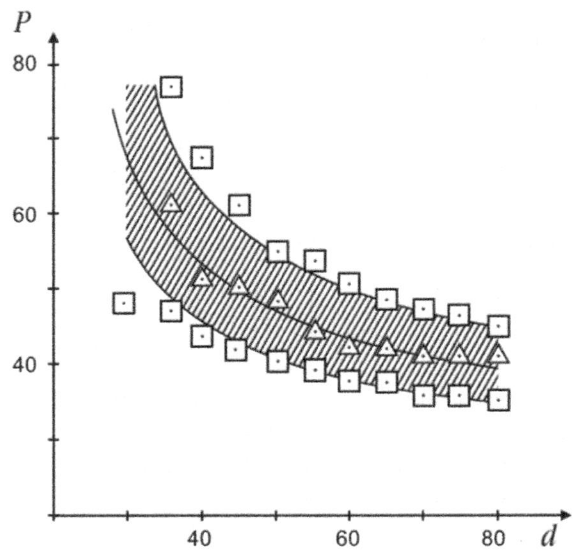

Fig. 4.23 Dependence of the position of the speckle-interferogram localization plane on distance d between the specklogram and the lens (solid curve—theoretical, triangles—experimental data). The experimentally observed localization region is bounded by squares, while the theoretically calculated one for the contrast $\gamma \geq 0, 1$ is shaded. Reprinted from [2] with permission

$$\gamma = \left\{1 + (\alpha - 1)^2\left(\frac{ka^2}{2d}\right)^2\left[\left(\frac{d}{ka^2}\right)^2 - \Delta\beta^2\right]^2\right\}^{-\frac{1}{2}}$$
$$\times \left\{-\frac{2(\alpha - 1)^2X^2}{q^2M^2} \times \frac{1 + \Delta\beta^2}{1 + (\alpha - 1)^2\left(\frac{ka^2}{2d}\right)^2\left[\left(\frac{d}{ka^2}\right)^2 - \Delta\beta^2\right]^2}\right\} \quad (4.90)$$

Analysis of (4.90) shows that contrast decreases on peripheral areas of the interference pattern. For observation with a good contrast of at least central rings, it is necessary that the multiplier before the exponent just slightly differs from the unit, whence it appears that

$$q^2 < \frac{2\lambda d}{\alpha - 1} \quad (4.91)$$

at that it was assumed that $\Delta\beta << d/\mathrm{ka}^2$ is the condition, which is usually fulfilled in real schemes. To obtain the best contrast $\Delta\beta = 0$ of all possible rings in the plane, the other condition should be fulfilled

$$(\alpha - 1)L \ll q \quad (4.92)$$

where L is the size of the specklogram; $(\alpha - 1)L$ is the maximal speckle divergence.

Equations (4.91) and (4.92) give the possibility to estimate the range of possible values of q. Rewriting (4.90) as a function from the objective aperture and determining maximum γ, an equation can be obtained for determination of optimal aperture q_0:

$$\left(\frac{\pi}{4\lambda d}\right)^2 q^6\left[1 + \left(\frac{(\alpha - 1)\pi}{4\lambda d}\right)^2 q^4\right]^{\frac{1}{2}} + 2L^2\left[1 + 3\left(\frac{(\alpha - 1)\pi}{4\lambda d}\right)^2 q^4\right] = 0 \quad (4.93)$$

For estimation of $q_{0,}$ we neglect the second compound in brackets in comparison with the unit taking into account (4.91). Then

$$q_0 \approx 1,48(L\lambda d)^{\frac{1}{3}} \quad (4.94)$$

Finally, we determine the area where the interference pattern will be localized. Taking into account (4.91) and (4.92), in the rough values p, under which rings can be observed with a contrast exceeding the given value γ_0 can be found from the condition

$$\left|\frac{1}{p} - \frac{1}{p_0}\right| \le \left(-\frac{2\ln\gamma_0}{k^2q^2(\alpha - 1)L^2} - \left(\frac{2}{kq^2}\right)^2\right)^{\frac{1}{2}} \quad (4.95)$$

The calculated using (4.95) localization area for $\gamma_0 \geq 0.1$ is shaded in Fig. 4.23. We use (4.87), which describes directly the interference fringes. Conditions for forming of an n-th ring

$$\frac{k(\alpha-1)^2X^2}{2dM^2}\frac{1-\frac{(\alpha-1)^2}{4}\beta\left(\beta-\frac{2d}{ka^2}\right)}{1-(\alpha-1)^2\left(\frac{k\alpha^2}{2d}\left(1+\beta^2\right)-\beta\right)^2} = \begin{cases} 2\pi n - \text{for bright,} \\ (2n+1)\pi - \text{for dark} \end{cases} \tag{4.96}$$

In localization plane, radius of the n-th bright ring is

$$X_n = M\sqrt{\frac{dn\lambda}{\alpha-1}} \tag{4.97}$$

The distance between the rings T_n from (4.97)

$$T_n = X_{n+1} - X_n = \frac{2M^2 d\lambda}{X(\alpha-1)} \tag{4.98}$$

where

$$\langle X\rangle = (X_{n+1} + X_n)/2 \tag{4.99}$$

Characteristics of the interference pattern in (4.97) and (4.98) agree with the results in the works [44, 58]. It is also worth mentioning that parameter $(\alpha - 1)$ included into (4.97) and (4.98) is determined by the shift of the object between exposures. Having measured the radii of the rings, the needed shift can be calculated.

Let us summarize the main results. There were theoretically examined and experimentally tested conditions of forming annular speckle-interferograms, which correspond to the longitudinal shift of the diffuse object using the lens system. The plane of best contrast and localization area of interferograms were found. Dependence of the contrast on the optical system of the aperture was obtained, and optimal diaphragm radii were found. Also it was investigated how parameters of the interference pattern change as the observation plane moves away from the plane of maximal contrast. It is also worth mentioning that the investigated lens systems for obtaining longitudinal speckle-interferograms have great advantage in comparison with lensless ones [8] and first of all in light-gathering power and possibility to control the sizes of the interference pattern. It seems that such an analysis can help to optimize more perfect schemes with annular apertures as well [44]. Though the work did not consider the questions of sensitivity of the method, the obtained results can help in the formation of speckle-photography of longitudinal shift as a non-contact method used in metrology and non-destructive testing, in medicine along with other methods of holographic and speckle-interferometry.

4.10 Results and Conclusions

1. Spectral and correlation characteristics of dynamic speckle-fields under longitudinal movement of diffuse objects were investigated in this chapter.

There were investigated statistical characteristics of intensity fluctuations of dynamic speckles during dispersion of the Gaussian laser beam on the rotating diffuser. It was investigated how the change of conditions of illumination and observation during longitudinal shift of the diffuser influences the spectral characteristics of the dynamic speckle-field, and it was established the dependence of amplitude of spectral harmonics and the area under intensity spectrum envelope on longitudinal oscillation of the object. Optimal conditions of illumination and observation were determined.

There was offered and realized determination method of longitudinal shift of rotating diffusely scattering object, and the frequency and velocity of which are unknown. A device for determination of linear longitudinal shifts of the surface was developed and experimentally tested.

2. Forming process of speckle-interferograms of longitudinal shift was theoretically and experimentally investigated. Optimal conditions of observation of the interference pattern were established, and its localization area, plane of maximal contrast from the aperture and optimal values of sizes of the filtering diaphragm were found. The obtained theoretical and experimental data agree well with each other.

Theoretical conclusions about three-dimensional character of speckle-fields trajectories were experimentally proved with the methods of speckle-interferometry of longitudinal shift. A ratio was ascertained, which determines sensitivity of speckle-interferometry of longitudinal shift, and it was shown that it can be controlled at the recording stage as well as at the stage of reconstruction of specklograms.

3. There were investigated statistical characteristics of dynamic speckle-field scattered by the diffuser in the longitudinal direction. The dependence was obtained, which interconnects relative intensity of the n-th harmonic and the amplitude of diffuser oscillations. A numerical simulation was conducted, and there was graphically obtained dependence of the amplitudes ratio of the first two harmonics on the oscillation amplitude of the object under different parameters, which determine the conditions of the diffuser illumination.

The possibilities of speckle-interferometry application for determining longitudinal components of shifts of diffuse objects were studied, dependence of interferograms on conditions of observation as well as optimal conditions of their obtaining was determined, and character of adequate spatial filtering was ascertained.

A speckle-optics method was suggested and experimentally realized, which allow determining not only the value, but also the direction of velocity of the diffusive object.

Research methods of diffusive surfaces of physical and medical–biological objects were developed. The methods give the possibility to determine deformations, velocity, shift and a number of other characteristics, which are not connected to them using a non-contact way due to the set amplitude correlation and intensity of the field diffraction on the object in different states.

Theoretical study was conducted aimed at development of speckle-optical methods for determination of such parameters as contraction velocity, change of form, deformations of diffusive objects, which are nerve trunks, muscles, epithelial tissues, etc.

In particular, double-pulse speckle-interferometry method was developed and experimentally implemented for determination of longitudinal shifts of the objects, which gives the possibility to uprange the shifts into two orders against the range of shifts of the known methods.

Behavioral peculiarities of speckles were investigated, which are formed with the objective used while studying such small preparations as nerve, separate muscle.

There were studied peculiarities of laser radiation transmission through several scattering media, which simulate complex biotissue.

References

1. J. Goodman, Laser speckle and related phenomena, **9** (1975)
2. L.V. Tanin, A.L. Tanin, *Biomedical and Resonance Optics: Theory and Practice* (Belaruskaya navuka, Minsk, 2011), 687p. (in Russian)
3. J. Goodman, *Statistical Optics* (Mir, Moscow, 1988), 527p. (in Russian)
4. F.P. Chiang, D.W. Li, Opt. Eng. **25**(5), 667 (1986)
5. J.M. Burch, J.H. Tokarski, Opt. Acta **15**(2), 101 (1968)
6. F.P. Chiang, P.W. Li, JOSA., A **3**(7), 1023 (1986)
7. E. Wolf, L. Mandel, UFN **87**(3), 491 (1965) (in Russian)
8. M. Franson, *Optics Speckles* (Mir, Moscow, 1980), 171p. (in Russian)
9. K. Hinsch, W. Schipper, D. Mach, Appl. Opt. **23**(4), 4460 (1984)
10. K. Hinsch, K. Brokopf, Opt. Lett. **7**(2), 51 (1982)
11. I. Yamaguchi, JOSA A **1**(1), 81 (1984)
12. V.V. Anisimov, S.M. Kozel, G.R. Lockshin, Opt. Spectrosc. **27**(3), 483 (1969). (in Russian)
13. V.V. Anisimov, S.M. Kozel, Technol. Electron. **15**(3), 539 (1970). (in Russian)
14. N. Takai, Jpn. J. Appl. Phys., **2025** (1947)
15. B. Ruffing, J. Fleiacher, JOSA A **2**(10), 1637 (1985)
16. T. Asakura, N. Takai, Appl. Phys. **25**(3), 179 (1981)
17. T. Iwai, N. Takai, T. Asakura, Opt. Coram. **44**(1), 13 (1982)
18. T. Iwai, N. Takai, T. Asakura, Appl. Opt. **22**(1), 170 (1983)
19. P.N. Pysey, J. Phys. **9**, 1399 (1976)
20. E. Jakeman, J.G. McWhirter, P.N. Pusey, JOSA **66**(11), 1175 (1976)
21. N. Nagamatsu, K. Nakegawa, T. Asakura, Opt. Quants Electron. **15**(2), 507 (1983)
22. G. April, H.H. Arsenault, JOSA **1**(7), 789 (1984)
23. T. Yoshimura, K. Nakegava, N. Wakabayashi, JOSA A **3**(7), 1018 (1986)
24. B. Crossignani, B. Daino, P. Porto, J. Appl. Phys. **42**, 399 (1971)
25. J.C. Leader, JOSA **67**, 1091 (1977)
26. H.A. Arsenault, G. April, JOSA **66**(11), 1160 (1976)
27. T. Consiorowski, J.C. Daint, JOSA **73**, 234 (1983)
28. T. Yoshimura, H. Doi, N. Wakabayashi, Opt. Comm. **48**(1), 17 (1983)
29. I.V. Markhvida, *Correlation Properties of Dynamic Speckle Fields And Their Use for Measuring the Velocities and Displacements of Diffuse Objects, Abstract of the Dissertation* (Univ, Minsk, Bel. State, 1989). (in Russian)

30. A.E. Ennos, Laser Speckle Related Phenom. **203** (1975)
31. K.A. O'Donell, JOSA **27**(11), 1459 (1972)
32. S.A. Akhmanov, Yu. E. Diyakov, A.S. Chirkin, *Introduction to Statistical Radiophysics and Optics* (Nauka, Moscow, 1981), 640p. (in Russian)
33. B. Ya. Zeldovich, N.F. Pilipetskii, V.V. Shkunov, *Treatment of Waterfront* (Nauka, Moscow, 1985), 247p. (in Russian)
34. N.B. Baranova, B. Ya, Zeldovich, V.V. Shkunov, Quant. Electron. **5**, 973 (1978). (in Russian)
35. J. Goodman, *Introduction to Fourier optics* (Mir, Moscow, 1970), 364p. (in Russian)
36. G.P. Weigelt, B. Stoffregen, Optik **48**(4), 399 (1977)
37. I.V. Markhvida, Proceeding of AS of the BSSR. Phys.-Math. ser. **4**, 86 (1986). (in Russian)
38. K.A. O'Donell, JOSA **72**(11), 1464 (1982)
39. D. Newman, JOSA **2**(1), 22 (1985)
40. N. Takai, T. Asakura, H. Ambar et al., JOSA A **3**(8), 1305 (1986)
41. H. Ambar, Y. Aoko, N. Takai et al., Optik **74**(1), 60 (1986)
42. H. Ambar, Y. Aoko, N. Takai et al., Optik **74**(1), 22 (1986)
43. T. Yoshimura, K. Nakagawa, Opt. Comm. **60**(3), 12 (1986)
44. I.S. Klimenko, *Holography of Focused Images and Speckle Interferometry* (Nauka, Moscow, 1985), 222p. (in Russian)
45. R. Jain, K. Singh, J. Optics (India) **9**(4), 74 (1980)
46. S. Lowenthal, H. Arsenault, JOSA **60**(11), 1478 (1970)
47. I. Yamaguchi, S.I. Komtasy, Opt. Acta **24**(7), 705 (1977)
48. R. Jones, K. Uaiks, *Holographic Speckle Interferometry* (Mir, Moscow, 1986), 327p. (in Russian)
49. T. Yoshimura, JOSA A **3**(7), 1032 (1986)
50. Y. Iwahashi, K. Zwata, K. Nageta, Appl. Opt. **24**(14), 2189 (1985)
51. L.L. Marty, *Investigation of The registration Process and Decrypt a Two-Beet-Exposure Photographs*, Abstract of the Dissertation, (Leningrad, 1985). (in Russian)
52. D.W. Li, P.P. Chiang, JOSA A **3**(7), 1023 (1986)
53. E. Archbold, A.E. Ennos, Opt. Acta **19**(4), 253 (1972)
54. I.S. Klimenko, V.P. Ryabukho, ZhTF **52**(5), 896 (1982). (in Russian)
55. I.S. Klimenko, T.V. Kuznetsova, A.N. Malov, ZhTF **56**(9), 1744 (1986). (in Russian)
56. J.B. Chen, F.P. Chiang, JOSA A **1**(6), 845 (1984)
57. E. Archbold, J.M. Burch, A.E. Ennos, Opt. Acta **17**(12), 883 (1970)
58. A.E. Ennos, Laser speckle and related phenomena, **203** (1975)
59. I.V. Volkov, I.S. Klimenko, ZhTF **50**(5), 1038–1043 (1980). (in Russian)
60. V.P. Ryabukho, *The Study of Methods for Obtaining and Interpreting Holo-Graphic Interogramms Based on Spatial Filtering*, Abstract of the Dissertation (Moscow, 1984). (in Russian)
61. V.K. Karppinen, Opt. Engineering **21**(1), 60 (1982)
62. P.M. Boone, Opt. Eng. **21**(1), 407 (1982)
63. P.K. Rastogi, P. Jacquot, Opt. Eng. **21**(3), 411 (1982)
64. I.V. Markhvida, L.V. Tanin, I.A. Utkin, *Proceeding of III All-Union Conference "The Use of Modern Physical Methods in Non-destructive Research and Control"*, Khabarovsk, the USSR, 1987, p. 211. (in Russian)
65. I.V. Markhvida, L.V. Tanin, I.A. Utkin, ZhTF **1** (1988). (in Russian)
66. Y. Dzialowski, M. May, Opt. Comm. **16**(3), 334 (1976)
67. J.A. Mendez, M.L. Roblin, Opt. Comm. **11**(3), 245 (1974)
68. J.Ch. Hagopian, M. May, J. Optics **9**(4), 217 (1978)
69. J.A. Leenderts, J. Phys. E. Soi. In-strum. **3**(3), 214 (1970)
70. Ch.S. Vikram, K. Vedam, Optik **64**(2), 171 (1983)
71. I.V. Markhvida, L.V. Tanin, Copyright certificate 1310624 (the USSR), (1984). (in Russian)
72. F.P. Chiang, Q.B. Li, Appl. Opt. **23**(24), 4469 (1984)
73. R. Murthy, R.K. Mohanty, R.S. Sirohi, Optik **67**(1), 85 (1984)

74. S.K. Dik, E.V. Zhuk, I.V. Markhvida, *Proceeding of the Conference "Holography in Industry and Scientific Study"*, (Grodno, the USSR, 1986), p. 16. (in Russian)
75. C.A. Shmidt-Harms, Appl. Opt. **23**(14), 2353 (1974)
76. D.E. Fitsjarrald, J. Phys. Isei. Instrum. **15**(9), 911 (1982)
77. K. Hinch, N. Schipper, Appl. Opt. **23**(24), 4460 (1984)
78. L.H. Tanner, Appl. Opt. **13**(9), 2026 (1974)
79. E. Inglestam, JOSA **66**(10), 1112 (1976)
80. F.P. Chiang, Opt. Eng. **21**(3), 377 (1982)
81. F.P. Chiang, J. Adashi, R. Anastasi et al., Opt. Eng. **21**(3), 379 (1982)
82. Ch.S. Vikrem, Optik **65**(3), 263 (1983)
83. K.A. Stetson, JOSA **66**(11), 1267 (1976)
84. J. Otsubo, Optic **57**(2), 183 (1980)
85. B.B. Gorbatenko, V.P. Riabuho, L.A. Maksimov, Tech. Phys. Lett. **30**(17), 68 (2004)
86. A. Assundi, Opt. Lett. **25**, 218 (2000)
87. N.A. Fomin, *Speckle Photography for Fluid Mechanics Measurements* (Springer-Verlag, Berlin, 1998), p. 264
88. E.I. Lavinskaya, E.F. Nogotov, N.A. Fomin, Phys. Eng. J. **72**(1), 96 (1999). (in Russian)
89. J. Goodman, *Statistical Optics*, ed. G.V. Skrotsskyi (Mir, Moscow, 1988), 527p. (in Russian)
90. P.A. Bakut, V.I. Mandrosov, Quantum Electron. **36**(3), 239 (2006). (in Russian)
91. H. Cummins, E. Pike, *Photon Correlation Spectroscopy and Velocimetry* (Springer, US, N.Y., 1977), p. 590
92. P.R. Jones, C. Wykes, *Holographic and Speckle Interferometry*, 2nd edn. (Cambridge University Press, Cambridge, 1989), p. 368
93. A.A. Kamshilin, E. Raita, K. Paivasaari, T. Jaaskelainen, Y. Kulchin, Appl. Phys. Lett. **73**(11), 1466 (1998)
94. D.N. Kozlov, B.A. Hemmerling, Appl. Phys. B: Laser and Opt. **71**(4), 585 (2000)
95. J.H. Churnside, H.T. Yura, Appl. Opt. **20**, 3539 (1981)
96. J.H. Churnside, J. Opt. Soc. Amer. **72**(11), 1464 (1982)
97. J.W. Goodman, J. Opt. Soc. Amer. **66**, 1194 (1976)
98. M. Born, E. Wolf, *Principles of Optics* (Nauka, Moscow, 1973), 719p. (in Russian)
99. M.H. Lee, J.F. Holmes, J.R. Kerr, J. Opt. Soc. Amer. **66**, 1164 (1976)
100. H. Kogelnik, T. Li, Appl. Opt. **5**, 1550 (1966)
101. S.K. Dik, I.V. Markhvida, L.V. Tanin, Proc. SPIE **1454**, 447 (1991)
102. S.K. Dik, L.V. Tanin, E.I. Rachkovsky, I. Markhvida, Copyright Certificate **176**, 576 (the USSR), 992. (in Russian)
103. L.Z. Dich, V.A. Zabelin, E.R. Malamed, G.N. Rassudova, Copyright Certificate A1093889 (the USSR), 23 May 1984. (in Russian)
104. O.A. Malyuntov, Copyright Certificate A1241062 (the USSR), 1986. (in Russian)
105. S.K. Dik, I.V. Markhvida, L.V. Tanin, *Proceeding of the International Congress on Optical Science and Engineering,* vol. 1508 (Hagne, Netherlands, 1991)
106. S.K. Dik, I.V. Markhvida, L.V. Tanin, A.S. Rubanov, Copyright Certificate 1810748 (the USSR) (1993). (in Russian)
107. N. Takai, T. Asakura, Appl. Opt. **17**(23), 3785 (1978)
108. H.J. Tiziani, Opt. Acta. **18**(12), 891 (1971)
109. L.M. Veselov, I.A. Popov, Opt. **68**(4), 953 (1990)
110. N. Takai, T. Asakura, J. Opt. Soc. Amer. **70**(7), 827 (1980)
111. J.Perina, Coherence of Light (Mir, Moscow, 1974), p. 367. (in Russian)
112. L.V. Tanin, S.K. Dik, I.V. Markhvida, A.S. Rubanov, *Proceeding of the International Symposium "Laser Optics '93"* (St. Petersburg, Russia, 1993), p. 554. (in Russian)
113. L.V. Tanin, S.K. Dik, V.I. Katsman, I.V. Markhvida, Proceeding SPIE 1914 (doc. 24) (1993)
114. B. Elliason, F.M. Mottier, JOSA **61**(6), 559 (1981)
115. N. Takai, T.A. Sutanto, JOSA **48**, 827 (1980)
116. N. Takai, Sutanto, T.A, Jap. J. Appl. Phys. **19**, L75 (1980)
117. J. Ohtsubo, Opt. Comm. **34**(2), 147 (1980)

118. I.V. Markhvida, L.V. Tanin, Optik **72**(4), 168 (1986)
119. I.V. Markhvida, L.V. Tanin, I.A.Utkin, Opti. Spectrosc. **63**(4), 941 (1987). (in Russian)
120. T. Iwai, N. Takai, T. Asakura, Opt. Acta. **28**(6), 857 (1981)
121. I.V. Markhvida, L.V. Tanin, L.I. Razhkovsky, L.L. Marti, Copyright Certificate 1474551 (the USSR) (22 Dec. 1988). (in Russian)
122. I.V. Markhvida, L.V. Tanin, I.L. Drobot, Proc. SPIE **1508**, 128 (1991)
123. L.V. Tanin, S.K. Dik, Proc. SPIE 2176(doc. 4) (1994)
124. L.V. Tanin, S.K. Dik, I.V. Markhvida, Proc. Russ. AS. Phys. ser. **59**(2), 200 (1995)
125. N. Takai, T.A. Sutanto, JOSA **70**(7), 827 (1980)

Further Reading

126. S.B. Dubin, J.H. Lunacek, G.B. Benedek, Proc. Nat. Acad. Sci. **57**(5), 1164 (1967)
127. J.D. Briers, Opt. Comm. **13**(3), 324 (1975)
128. R. Nossal, S.H. Chen, Opt. Comm. **5**(2), 117 (1972)
129. C.E. Riva, J.E. Grunwald, B.L. Petrig, Appl. Opt. **24**(5), 605 (1985)
130. M. Singh, G. Ramachandran, A. Periasamy, J. Opt. **15**(3), 63 (1986)
131. S.Yu. Kuzmin, S.S. Ulyanov, V.V. Tuchin, *Proceeding Proceedings of V International Conference "Laser Tehnologies'95"*, Shatura, the USSR, June 24–26, 1985. (in Russian)
132. R. Danliker, K. Weiss, Opt. Com. **1**(7), 297 (1970)
133. A. Luxmoore, J. Inst. Metals **101**, 208 (1973)
134. P.H. Deitz, J. Opt. Soc. Amer. **65**(3), 279 (1975)
135. K.A. Stetson, Opt. Eng. **14**(5), 482 (1975)
136. P. Kaminow, Appl. Opt. **15**(12), 3029 (1976)
137. E. Brinkmeyer, Appl. Opt. **16**(11), 2802 (1977)
138. T.G. Pavlopoulos, A.L. Lewis, H.N. McLandrich, Appl. Opt. **16**(6), 1466 (1977)
139. E. Brinkmeyer, Appl. Opt. **17**(1), 14 (1978)
140. M. Tateda, Appl. Opt. **17**(3), 475 (1978)
141. E.W. Powers, W.W. Flayer, Plast. Reconstr. Surf. **61**(9), 250 (1978)
142. N.G. Vlasov, S.G. Galkin, Y.P. Presnyakov, B.M. Stepanov, Holography and optical information processing: methods and apparatus, **143** (1980). (in Russian)
143. V.G. Kulkarni, Opt. Acta **28**(12), 1577 (1577)
144. P. Varman, C. Wykes, Opt. Laser Engin. **3**(2), 87 (1982)
145. O.J. Lokberg, *Proceeding of the International Conference "Optics in Biomedical Sciences"*, Graz, Austria, 7–11 September 1981, p. 144
146. I.S. Klimenko, S.N. Malov, Opt. Spectrosc. **52**(4), 745 (1982). (in Russian)
147. E.S. Wu, D.B. Geohegan, Amer. J. Phys. **51**(4), 315 (1983)
148. M. Hidetoschi, H. Masao et al., Jap. J. Appl. Phys. **23**(8), 1129 (1984)
149. B.Ya. Zeldovich, V.V. Shkunov, Quantum Electron. **11**(11), 2162 (1984). (in Russian)
150. V.I. Teleshevsky, Measur. Technol. **11**, 20 (1984)
151. M.P. Kothiyal, C. Delisle, Opt. Lett. **9**(8), 319 (1984)
152. K. Creath, Appl. Opt. **24**(18), 3053 (1985)
153. R.K. Mohanty, C. Joenathan, R.S. Sirohi, Appl. Opt. **24**(18), 3106 (1985)
154. P. Hariharan, Opt. Eng. **24**(4), 632 (1985)
155. N. Saga, K. Tanaka, Opt. Comm. **53**(1), 1 (1985)
156. G.P. Pyzin, S.B. Artemenko, ZhTF **56**(5), 868 (1986). (in Russian)
157. J. Shamir, Opt. Eng. **25**(5), 618 (1986). (in Russian)
158. M. Saillard, D. Maystre, J.P. Rossi, Opt. Acta **33**(9), 1193 (1986). (in Russian)
159. G.P. Costa, Manguni, P. Otonello, E. Piano, Rev. Sci. Instrum. **58**(1), 78 (1987)
160. I.S. Klimenko, T.V. Krivko, A.N. Malov, V.P. Ryabukho, ZhTF **58**(1), 182 (1988). (in Russian)
161. Ma. Shining, P. Xiang, Optik **80**(4), 184 (1988)

162. J.C. Danty, Prog. Opt. **14**, 3 (1976)
163. E. Ochoa, J.W. Goodman, JOSA **73**(3), 943 (1983)
164. N.G. Vlasov, G.V. Skrotsky, E.G. Soloviev, Probl. Holography **1**, 85 (1973). (in Russian)
165. N.G. Vlasov, U.P. Presnyakov, Recent developments in applied holography, **13** (1974). (in Russian)
166. T. Iwai, N. Takai, T. Asakura, Opt. Acta **30**(6), 759 (1983)
167. G.P. Lokshin, Methods and devices of optical holography, **125** (1983). (in Russian)
168. A.E. Ennos, Progress Opt. **16**, 233 (1978)
169. N. Takai, T. Iwai, T. Asakura, JOSA **70**(4), 450 (1980)
170. S. Komatsu, K. Osato, H. Ohau, Opt. Comm. **39**(6), 357 (1981)
171. M. Imai, M. Zida, T. Asakura, Optik **51**(4), 429 (1978)
172. M. Christopher, J. Haccot, F. Lawson, Opt. Acta **32**(5), 537 (1985)
173. T. Iwai, N. Takai, T. Asakura, Jap. J. Appl. Phys. **22**(10), 1523 (1983)
174. T. Iwai, N. Takai, T. Asakura, Jap. J. Appl. Phys. **22**(12), 1860 (1983)
175. R. Chislead, P. Sucin, I. Cucurezeann, Opt. Comm. **30**(6), 351 (1981)
176. M.F. Marechal, Structure of the Optical Image (Mir, Moscow, 1964), p. 295. (in Russian)
177. D. Ruth, W. Haind, Weidelich. Opt. Acta **31**(7), 759 (1984)
178. D. Ruth, W. Haina, Weidelich. Opt. Acta **30**(9), 1213 (1983)
179. R.L. Soloukhin, N.A. Fomin, Mixing Gasdynamic Lasers, Nauka i Technica (1984). (in Russian)
180. N.A. Fomin, W. Merzkirch, D. Vitkin, H. Wintrich, Exp. Fluids **20**, 476 (1996)
181. G.N. Blinkov, N.A. Fomin, M.N. Rolin, R.L. Soloukhin, D.E. Vitkin, N.L. Jadrevskaja, Exp. Fluids **8**, 72 (1989)
182. N.A. Fomin, R.L. Soloukhin, J. Revue, de Phys. Appl. **14**(2), 421 (1979)
183. A.S. Belsky, T.M. Korneychik, A.P. Khapalyuk, *The Spatial Structure of the Laser* (BSU Publisher, Minsk, 1982), 198p. (in Russian)
184. O.J. Lokberg, K. Hogmoen, O.M. Holja, Appl. Opt. **18**(6), 763 (1979)
185. E.V. Bulaenko, M.D. Prokoptsev, P.A. Friedman, Instr. Exp. Tech. **2**, 202 (1994). (in Russian)

Chapter 5
Laser Specklometer, Speckle-Optical Diagnostics and Laser Hemotherapy in Treatment of Diseases of Peripheral Nervous System

Development of modern medicine is closely connected with the achievements of science and technology. The tendency of medicine to obtain the needed amount of information about objects without damage or influence leads to the necessity of creation of highly sensitive non-contact methods. In the last decade, a number of interesting cross-disciplinary, medical and optical, scientifically applied schools appeared. Different optical devices and gadgets, for example, laser surgical scalpels, fiber gastroscopes, ophthalmologic devices for laser microsurgery, laser therapeutic units, etc., were developed and found wide clinical application. But the scope of optical possibilities is not limited by the developments named above. Introduction of speckle-optical methods, which use coherent characteristics of laser radiation, is rather prospective for medical instrument making.

During its short history (in fact from 1978), speckle-optics accumulated a large number of different methods of measurement in different technical areas of physics. But the factual material concerning medicine and biology is rather small.

As it was demonstrated in Chap. 2, the classical and holography interferometry, where the amplitude correlation participates, can be used only for the study of transparent phase microobjects: separate cells, fibers, microvessels [1]. More complex systems, for example, ensemble of cells, whole muscles, integumentary tissues, intensively scatter light. In this case, the measurements on the basis of intensity correlation, namely methods of speckle-interferometry and spectroscopy of optical shift, possess large possibilities of principal character. As it was shown in Chap. 4 (see Fig. 4.1) at radiation with coherent light of the rough surface diffusely scattered radiation forms in the observation plane typical "spot" structure of illumination called a speckle-image.

In this case, all the points of the illuminated by laser radiation diffusive surface send coherent interfering waves onto observer's retina. Image of each point of the surface on the retina is a diffraction pattern, which is determined by the optical system of the eye. Because of interference of these diffraction patterns, the surface of the object seems spotty to the observer, or covered with "speckles" as it is said

L. Tanin et al., *Biomedical and Resonance Optics*, Bioanalysis 11,
https://doi.org/10.1007/978-3-030-60773-9_5

today. Recently, the speckle-optical methods for measurement of real-time velocities and shifts have been developed intensively. This is determined by the combination of traditional organization of optical methods such as contactlessness, remoteness, high accuracy with the possibility of determination of movement parameters and coordinates of the object, which scatter the light significantly. If the object is stationary, then the observed speckle-structure is stationary too. The speckle-image detects pulsations of light spots (speckles) intensity, so called boiling, during oscillations of the object as a whole or during time alterations of its microrelief. As dynamics of these processes is connected with dynamics of the object, then certain judgments can be made about the character of movement or about alteration of the rough surface itself while studying frequency characteristics of intensity fluctuations in observation plane. If skin is considered as a rough surface, which suffers applied shock of muscle contractions, sanguimotions, etc., then abnormalities can be detected in the processes mentioned above according to features of intensity fluctuation spectra and can be connected with certain diseases.

The development of this direction is undoubtedly connected with studying nature and statistical characteristics of speckle-fields, which were investigated by a number of authors [2–8]. But this question was considered in detail only for the case when light is scattered on one surface. Along with this, in experimental practices, especially in the medical–biological one there are cases when light has to pass through one more scattering medium before the object or after it. The question of twice scattered radiation application in speckle-optics and in particular in speckle-interferometry requires more detailed studying. Consideration of statistical characteristics of dynamic speckles formed twice by scattered light is of theoretical interest as the field amplitude will have non-Gaussian statistics in this case.

During illumination of multilayer-structured skin, a dynamic speckle-structure is formed as a result of a coherent composition of many elementary waves scattered on moving particles, which are cells of dermis, epidermis, etc., which form fetal integument and blood cells, primarily erythrocytes.

For simple macroscopic homogenous systems, modeling of coherent dispersion has shown that average velocity of particle movement is proportional to fluctuation frequency [9]. But adequate models, which describe light dispersion from complex systems such as biological tissues in vivo [10, 11], have not been created yet.

In biological experiments while using speckle-optics, continuous mobility of objects under consideration must be mentioned, in particular, possible alteration of their microstructure that leads to speckle decorrelation. When speckle-interferometry is to be used, then a practical test of method independence is needed in every particular case. Useful information about the process dynamics can also be obtained during time-changing speckle-fields, which are of great interest today.

The distinctive feature of the study of many medical–biological preparations is their small sizes, and that requires the application of magnifying optics. Besides, being an integral part of many speckle methods, the optical system plays an important role in forming of speckle-fields and speckle-interferograms. Though its impact on contrast, size of speckles, etc., was studied in a number of works, questions such as the dependence of correlated intensity functions on movement of the object along the optical axis are slightly investigated.

The character of speckle decorrelation depends on the way of object movement, and it will differ from uniform transversal shift [4, 12], longitudinal [13, 14], three-dimensional [15] and periodical [5, 16, 17] ones.

Rather effective methods are developed for determination of: (a) transversal velocities due to the measurement of correlation time, shift of peak of correlation function [18, 19]; (b) amplitude and oscillation frequency due to ratio of harmonics of intensity fluctuation spectrum of the scattered radiation [20, 21]; (c) longitudinal coordinates due to interfrequency interval [22]; and (d) velocity of speckle movement [23].

If the diffusive object shifts across the optical axis, then the character of the formed and time-changing speckle-field depends greatly on conditions of illumination of the diffuser (radius of curvature of the wave front and sizes of laser beam) and on registration conditions [24–26]. Speckle lifetime is inversely proportional to the velocity of movement of the diffuser. The speckle lifetime determinates the character of time correlation function of intensity fluctuation and consequently power spectrum of intensity fluctuations of dynamic speckles connected with it according to the Wiener–Khinchin theorem. This fact underlies a number of methods of measurement of movement velocity [27, 28]. Much less attention was paid to the fact that proportionality coefficient depends on longitudinal parameters of the scheme [22]. If the velocity of object movement is known, then determination of proportionality coefficient gives the possibility to obtain information about longitudinal distances between the radiator and the diffuser, the diffuser and the observation plane that is undoubtedly interesting during the development of new methods and devices for measurements of longitudinal shifts.

Development of holographic and speckle-interferometry [29–34] for determination of shifts [35, 36], elastic responses of biotissues [37, 38], measurements of velocity of muscle contraction using speckle counting [39], etc., have shown the possibility and availability of development of new diagnostic coherent-optical methods and devices in medicine and biology as most objects of the study are diffusing.

The study of medical–biological objects is connected with great difficulties comparing to the objects of physical measurements. Mobility, complex structure, high fault probability, several simultaneous and often uncontrolled processes, bad repeatability of measurements and other factors cause the necessity of statistical methods of data processing and establishing correlation connections between the results of measurements and the functional state of the organism [15, 40, 41]. But due to contactlessness, non-invasiveness, high sensitivity, objectiveness, possibilities of speckle-optics do not raise doubts during diagnosing and obtaining new information about functional state of medical–biological objects [42–45].

First attempts of using laser anemometry [46, 47], optical mixing spectroscopy [48, 49], holographic methods for creation of archives and pathology classifiers [50], holographic and speckle-interferometry for studying neuro- and muscular preparations [51, 52], elastic responses of biotissues [53, 54], degree of roughness of human skin [55], polarization characteristics of radiation scattered by human skin in different functional states [56], as well as spectrometric methods of quantitative assessment of oxyhemoglobin concentration in cutaneous blood flow [57, 58], etc., have shown that with their help a new useful information about structure and functional state of living organisms can be obtained.

The coefficients of translational diffusion of biological macromolecules were first measured by the authors of the work [59]. Chloroplast mobility was determined through calculation of lifetime of dynamic speckles in different wavelengths. The mobility was 0.1–4 μm/s [60]. A number of works are dedicated to the determination of migration velocities of cell structures [61, 62]. T.I. Show, D. E. Newby adapted intensity correlation methods to the study of nerve tissues [63]. They observed considerable growth of intensity fluctuations of the scattered light while adding potassium ions. It was assumed that dispersion occurs on synaptic vesicles in ganglia mostly, which can move if nerve endings are depolarized. The same phenomenon was observed by R. W. Piddington [64] during current stimulation and the effect lessened by adding "nerve poison." Dynamic speckle-fields are also used for the study of microstimulation [65] and blood filling [66].

The analysis of the literature shows that on the one hand, studying intensity correlations becomes a standard laboratory method of measurement [48, 59, 67, 71] for determination of constant time of diffusion and average velocities in stationary systems of uniformly distributed simple-formed particles. On the other hand, great improvements of both theory and experimental equipment are required while studying correlation processes of specially shaped particles and other more complex systems.

In medical–biological study of comprehensive systems, great attention is given to spatial characteristics of the process that can be proved by success in tomography, thermography, NMR field forming.

The study, which was made on living tissues by other authors and us on isolated preparations [37, 39, 52, 68], shows the possibility of obtaining biotissues' deformation maps. The last work [69] gives rather encouraging results for optical diagnostics due to its non-damaging character and contactlessness. In this work, the correlation between chest deformation fields during functioning and heart diseases is established.

Complex biological systems (epithelial tissues and nerve trunks) scatter laser radiation rather heavily and cause forming of the developed speckle-structure, which can be used for studying parameters of these systems as well. Low requirements to vibroprotection in comparison with holographic interferometry, simple optical schemes and available equipment give hope that speckle-optical methods will become a new step in the development of research and diagnostic methods. Meanwhile, not only amplitude, half-width and shift of peak of correlation function of speckle intensity have informational value, as it was shown earlier, but also its derivatives. An important role is played by the second derivative of intensity correlation function in the recently developed speckle-counting method [70], which has a number of analogs [71, 72].

The researches of deformation fields of superficial tissues are of great practical and scientific interest in studying activity of neuromuscular system. Such a research gives the possibility to obtain detailed information about muscle contraction on great

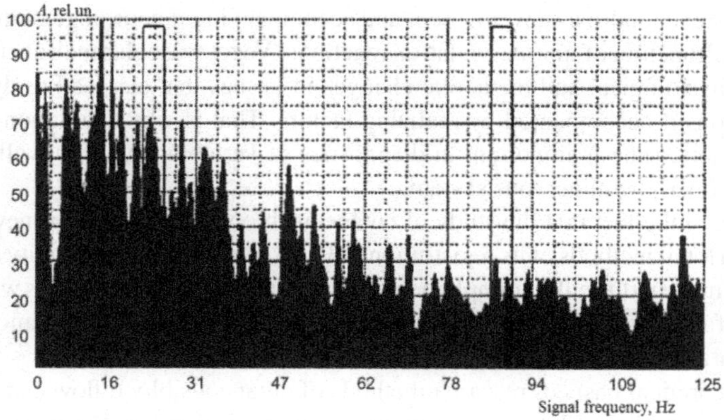

Fig. 5.1 Spectrum of intensity fluctuations of a dynamic biospeckle field formed by reflected radiation from the skin over the skeletal muscle of a patient. Reprinted from [87] with permission

areas unlike traditional electromyographic study, which gives integral information, or the speckle-counting method, which is characterized by measurement locality only in one point.

Biomedical applications of speckles are very variable, and the scope of their study is now determined mostly by enthusiasm and interest of studies. In a number of such fields as modeling in dentistry or orthopedics, some peculiarities of speckle-fields can be made due to specific characteristics of bone tissues, which scatter light (e.g., porosity or sizes of heterogeneity). In other respects, this study slightly differs from research of deformations of solid bodies. In other cases, speckles are formed during reflection of coherent light from living tissue. Complex three-dimensional macroscopic movement of the body surface and stochastic shift of separate elementary diffusers (blood cells and muscle fibers) cause temporary change of speckle-structure and lead to formation of dynamic biospeckles (Fig. 5.1).

Biospeckles are formed during reflection of coherent light, for example, He–Ne laser radiation from the surface of living tissue; in particular, it can be skin surface of a human, an animal, surface of a nerve, a muscle, in flora—surface of a leave, a stem, etc.

Recently, speckle-optical methods of determination of superficial blood flow have been developed intensively in Japan, Germany, Sweden and other countries [73–76]. A number of diseases, including diseases of peripheral nervous system such as traumatic nerve injury, vertebragenous radiculitis with vegetovascular manifestation and neuronopathies, are characterized with alteration of microcirculation. In turn, the decrease in the surface layer of skin (the depth is approximately 0.5 mm) of a number of erythrocytes and velocity of their movement causes the lowering of fluctuations of scattered laser radiation, which can be quantitatively registered by a photoreceiving equipment [73, 74]. At the same time, there are still many controversial and unsolved issues in vertebraneurological pathology, which require further experimental and clinical study. One of the important parts of pathogenetic mechanisms of spine osteochondrosis development and accompanying it neurological manifestations is malfunction of microcirculation reactions. Vegetovascular malfunctions lead to pathophysiological and metabolic changes in tissues that can lead to dystrophic malfunctions if accompanied by realizing factors. That is why estimation of tissue microdynamic state is important in diagnostics of vertebragenous as well as other diseases of peripheral nervous system.

In spite of the fact that the study of cutaneous blood flow is a rather new school, there are a lot of methods of its registration [73–77]. But small informativity and laboriousness make it difficult for implementation in practical medicine that is why development of new noninvasive informative methods of cutaneous microhemodynamic determination stays an urgent task.

The theoretical basis of research methods of cutaneous blood flow can be found in works by Asakura [71, 72], and its practical realization—in the works [73, 78, 79]. The study of the author and the workers of his research group of the Institute of Neurology, Neurosurgery and Physiotherapy of the Ministry of Health of the Republic of Belarus is of the greatest interest due to visualization of the examples and fundamentality of the approach [80–82].

It is also mentioned that three physiological parameters are used to describe microcirculation: physical movement, heat transport and oxygen content [66, 83]. Density of blood cells, which influences the heat transport, can be estimated according to average intensity of scattered and absorbed light, which is the subject of study of photoplethysmography [66].

Speckle-optics studies physical movement due to fluctuation spectra mostly, though the value of non-normalized spectrum at zero frequency is the average energy of scattered radiation [84]. In many studies, which are mostly connected with laser Doppler anemometry, it is assumed that cells move, and other tissues remain stable [85]. Thus, temporal change of speckles reflects the average velocity of erythrocyte movement.

But as the measurements have shown, if blood cells are isolated from the process of dispersion, for example, by putting adhesive tape on the part of the body under study, then qualitative dynamics of speckles will not change itself [82]. The point is that even motionless tissue constantly makes little involuntary movements because of contractions of separate muscle fibers near the area under study, activity of whole muscular groups (e.g., of the heart), pulsing transmission of blood in arteries and a number of other reasons. This movement is so significant that according to estimations of authors' works [69, 86], input into the total spectrum of light scattered on moving blood cells is 10%. Though this number is much greater than the part of light in backscattering, impact of shifts of the body is prevalent.

Despite such a complicated superposition of these processes and formed by them speckle-structures, a number of works were made connected with studying blood flow using dynamic speckle effect [73, 74].

Interpretation of measurement results is the most complicated question in speckle-optical measurements. Almost nobody of the authors calculated the average velocity of erythrocytes according to speckles. Speckles themselves have great variability, and their comparison is a rather difficult task. Various methods are offered for estimation of information about fluctuation [75].

A simpler parameter, which characterizes blood flow, is the value of time integral of the module of fluctuating signal. The value of the integral will be greater where the fluctuation frequency is higher if fluctuation amplitude is the same during the same time. Because of its simplicity, this integral characteristic suits well where a great value of information is processed in real time.

The authors of the works [69, 86] used it while constructing spatial maps of the blood flow. Skin surface with scattered coherent radiation was projected onto one-dimensional CCD matrix. Each of its pixels measured time fluctuations in its point. A computer added fluctuations, and a color-encoded pattern was displayed on the screen. A scanned image due to another coordinate was used for obtaining a two-dimensional pattern. Selection of informative parameters of intensity fluctuation spectra of the scattered radiation is also a rather difficult problem, because the task is multiparameter. It is due to the change of registration conditions as well as of the object of the study itself during changes. Skin suffers constant microvibrations caused by the contraction of separate fibers and a group of underlying skeletal muscles, tremor, etc. [41, 44]. Blood cells move not only with capillaries inside skin, but also they have their own velocity relative to them. Thus, the average velocity of erythrocytes appears to be much higher than velocity of shifts of cutaneous tissues [74].

The possibility of measurements in many points is a characteristic of speckle-interferometry. The works [37, 88–101] investigated maps of deformation distribution of isolated tissues as well as of living organisms using the method of two-dimensional speckle-photography. The question of speckle-photography method application in biological experiments was almost not considered, though it was successfully applied in nondestructive testing of solid bodies and testing of materials for construction of deformation fields. First of all, the following fact remained unclear, whether speckle-fields experience shifts at a value, which is greater than the speckle size preserving correlation, or deformation of living tissue, which is inevitably accompanied by microstructure changes, causes only speckle "boiling." That is why a number of experiments were conducted to determine the deformation field of living tissue using the method of double-pulse speckle-photography. According to the works [93–95], it is necessary to register two images covered with speckle-structure in the initial and excited states for determination of transversal shifts. The shift was conducted either artificially [95] or during real muscle contraction [37] in the plane of movement perpendicular to the optical axis. The obtained photography is enough to determine the shift field where two speckle systems were registered. Interference fringes were observed as a result of optical Fourier transformation of small areas. Their direction was perpendicular to the direction of shifts, and their period was inversely proportional to the shift [96, 97]. It is worth taking into consideration that broken speckles shift homogeneity of illuminated area of the specklogram leading to diminishing the contrast of interference fringes down to their disappearing. The diameter of the scanning beam had to be varied in experiments for obtaining maximal contrast interferograms for observation of considerably heterogeneous deformations occurring during contractions of living tissue.

Apart from useful physical information about mechanical properties of biotissues and amplitude–velocity regimes of motor organs, similar maps of deformations can give the information about the state of internal organs. The authors of the works [69, 102–105] developed speckle-photography methods for measurement of the map of chest surface deformation in a certain phase of heart contraction. As the length of elastic wave is more than the distance between the heart and the chest surface, then its deformations show the state of the heart muscle. According to them, there is a connection between this deformation field and the experienced heart disease. The obtained results show the possibility of effective application of speckle-photography in medical–biological study of living tissue contraction. Besides, it can be used in the study on biomechanics and supplement methods of holographic interferometry in other medical applications.

Biomechanical characteristics of human skeletal muscles are of great interest both for studies and for physicians. Constant contractions and relaxations of muscles occur in the body of a wakeful person, and they lead to respective shifts and vibrations of skin.

In the end, different abnormalities occurring in the human organism lead to alteration of degree of muscle contractility, which is connected with biomechanical parameters, and in turn, this changes amplitude and oscillation frequency of the muscle and the skin surface.

Meanwhile, the dynamics of these processes is connected with the dynamics of the object itself that gives the possibility to detect such abnormalities in muscle contractions according to the peculiarities of intensity fluctuation spectra and connect them with different diseases. In the work [106], Descherevsky's model [107, 108] is taken as a mathematic model of a skeletal muscle, which works in contraction regime. Amplitude and oscillation frequency of the muscle were calculated on its example at different values of biomechanical parameters. The expression for intensity of radiation scattered by the surface was obtained that gives the possibility to determine oscillation amplitude of the muscle. In particular, graphs were made, which describe the intensity of the scattered radiation depending on the load, which corresponds to the experimental data.

The progress of muscular physiology is determined by experimental equipment to a great extent. Speckle-optics gives different possibilities in this direction, and different methods can be adapted for measurement of deformations caused by motor activity of muscles, their contraction velocity and determination of tonic state of musculature.

It is known that muscles generate low-frequency sound oscillations, the intensity of which depends on the functional state of these muscles. Sounds of muscles are not connected with changes of the blood flow and have the same character during arbitrary contraction as well as during stimulation, and oscillation frequency does not depend on the surrounding temperature [109, 110]. The sound occurs as a result of transversal resonance oscillations of the muscle during its contraction [111]. Myoacoustic signals were used for analysis of muscular activity, in particular, for measurement of muscular fatigue [112, 113].

The study of muscular sounds of healthy people and patients who suffer from neuromuscular pathologies (myasthenia, polymyotoxia, polymyelitis, special muscle atrophies, Friedreich's ataxia, etc.) has shown reliable differences in average peak frequencies of sound oscillations in normal state and in pathology. The test people with neuromuscular pathologies generated signals, the amplitude of which was from 1/5 to ½ of the amplitude of signals detected in the group of healthy people [114].

Despite a number of interesting results obtained with the acoustic myography method, it is already clear that the study of muscle oscillations can be supplemented successfully with a rather simple method of speckle-optical myography due to a certain degree of noises in the low-frequency area (traffic noises, steps, etc.), filtration of which is a more complicated task, than filtration of high-frequency sounds (speech, etc.), and, correspondingly, a low special resolution. This method has a higher spatial resolution and protection in comparison with the method of acoustic myography. And advantages of optical methods such as contactlessness, remoteness and non-invasiveness give the possibility to overcome disadvantages of acoustic methods connected with the application of contact microphones, which reduce penetration of low-frequency background noises into the measurement section through the air [110].

One of the perspective directions is the development and implementation of coherent-optical research methods based on achievements of modern laser equipment. These research methods give the possibility to study processes of appearance and development of such a kind of diseases with high sensitivity and accuracy and high spatial resolution. The methods are based on the analysis of time–spatial correlation characteristics of laser radiation scattered by the living tissue and on determination of velocities, shifts, deformations, roughness and other characteristics of the objects under study with their help.

After the studies conducted by the author and his group, it was first found that changing of muscle tone, which is accompanied by changes of oscillations of muscle fibers and the surrounding tissues, deforms the spectrum of the speckle-field [80–82, 115–134]. These works were conducted in the Institute of Neurology, Neurosurgery and Physiotherapy of the Ministry of Health of the Republic of Belarus for 20 years on the initiative of L. V. Tanin, a Leading Scientific Researcher of the Institute of Physics of the NAS of Belarus, with the assistance and understanding the perspectives of this school by Professor I. P. Antonov, Academician, the Head of the Institute of Neurology, Neurosurgery and Physiotherapy of the Ministry of Health of the Republic of Belarus.

It was shown that measurements of spectral characteristics of dynamic laser speckle-field formed during reflection of the light from the tissue over the contracting muscle can serve as a serious alternative to sound myography considering high protection and significantly supplement the study of contractile function.

That is why the aim of the study was development of brand-new method of laser medicine in investigation of biomechanical characteristics of skeletal muscles, which was named laser speckle-optical myography, as well as development and creation of a new generation of devices, in particular, laser specklometer, which gives the possibility to implement this method in practice. The study was conducted by the author in collaboration with specialists of group of "coherent-optical research of medical–biological systems" created by him in the Institute of Neurology, Neurosurgery and Physiotherapy of the Ministry of Health of the Republic of Belarus. Using possibilities and advantages of the speckle-optical method, Academician I. P. Antonov set the task to investigate the functional state of skeletal musculature during diseases of peripheral nervous system and to conduct analysis of laser speckle-optical myograms in order to find informative and trustworthy parameters characterizing changes of muscle activity, which are difficult to obtain using the known methods.

The analysis of dependence of intensity fluctuation spectrum of dynamic speckle-field on geometric parameters of the scheme gives the possibility to substantiate methods and principles of devices for determining longitudinal coordinate according to the changes of the amplitudes of spectral harmonics. In this case, the author was interested in the possibility of development and implementation of laser specklometer (microhematomyograph) in the study of muscle contraction as well as of superficial blood flow.

Chapter 4 and the works [80–82, 123, 125, 129] inform about the results of earlier started theoretical and experimental study underlying development of laser specklometer. They describe the principle of the work and organization of the test sample of the device, software and the research results of vibration characteristics of skeletal muscles and microcirculation of human skin conducted with laser specklometer.

There is a serious problem connected with the development of speckle-optical methods and devices for studying biomechanical characteristics of skeletal muscles and microhemodynamics of human skin and their adaptation. This problem is studying the impact of a rather complicated movement of the surface of skin caused by contraction of the underlying skeletal muscle including longitudinal shifts, oscillations and stochastic shifts of separate elementary scatterers (blood cells and muscle fibers) onto statistic parameters of dynamic speckle-field formed by laser radiation scattered from skin surface. The change of fluctuation amplitude under load was detected using methods of speckle-optics [80–82]. Further study showed that spectra differ in healthy people and patients with inadequate motion activity, and these differences can be used in diagnostic practice [115–135].

The interpretation of speckle-optical results of study of biomechanical parameters of skeletal musculature and working out of diagnostic criteria of a neuromuscular pathology is possible if there is a dependence of these data on the processes, which influence the contractile muscle function (bioelectrical activity, biochemical shifts, structural changes, etc.). That is why the study of morphological and metabolic changes is important, which occur in the muscle tissue and in the structures connected with it (blood vessels, nerves and connective tissue) under vertebragenous pathology (muscular-tonic syndromes), experimental traumatic injury of somatic nerves (transaction with further neurorrhaphy and chronic nociceptive irritation) and under influence of physical factors (gas and hydrostatic hyperbary and laser radiation) and some medications.

5.1 Study of Velocity of Muscle Contraction Using the Speckle-Counting Method

The development of muscular physiology is essentially connected with the progress in experimental equipment: appearing of methods of irritation, holographic registration, calorimetry, etc. Microoptical visual observations gave the possibility to single out the main structural elements of muscle fibers. Observation of contraction of the muscle fiber and the whole muscle is complicated as it shifts beyond the limits of depth of field of the microscope.

The muscle tissue has a great number of diffusers, and at illumination with coherent radiation behind it a developed speckle-field is formed, the contrast of which is close to unit. Muscle contraction causes forward shift of all the diffusers as well as change of their relative location. In turn, this leads to the movement of separate spots–speckles and their decorrelation [136]. Also, the velocity of movement of the scattering object can be calculated after determining the time of speckle correlation. The velocity of muscle shortening is its main independent characteristic. Besides, it gives the possibility to calculate mechanical power of the muscle and to investigate the energy escape velocity [137]. We used the speckle-counting method for study of the dependence of velocity of muscle shortening during isotonic contraction [39, 138].

The device is offered for measurement of velocity of muscle tissue contraction, which is an important physiological parameter, which gives the possibility to ascertain power of muscle contraction and energy escape velocity. The device is based on counting the number of intensity fluctuations of dynamic speckle-field formed by diffusely moving scattering tissue. Also, the device can be used for investigation of biomechanical parameters of muscles, for determination of velocities of movement of diffusive objects with different types of scattering surfaces. The range of the measured velocities is from 1 to 100 cm/s. The spatial resolution is 10 mm^2. It is possible to measure time distribution of velocities in real time with temporal resolution of 10 ms.

This device has the following advantages in comparison with myograph traditionally applied for studying muscle contraction:

- The speckle-counting method is contactless and gives the possibility to study a muscle without isolating it.
- Horizontal position of the muscle gives the possibility to put it into physiological solution and stimulate it electrolytically without inserting electrodes, without cutting the preparation.
- It is possible to investigate different small areas of muscles.
- Zero lag,
- namely the contraction velocity of a certain area, but not changes of the total length of the muscle, is measured. The transition from velocity to dependence of the length on time, which is measured in direct myography, is connected with integration and reduction of test errors, while the reverse operation—differentiation greatly increases measurement errors.

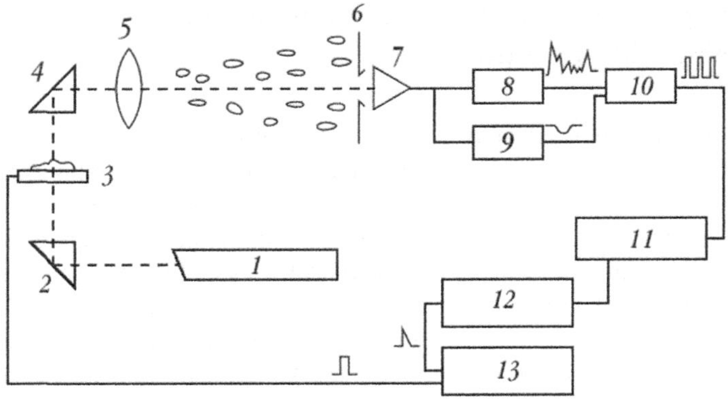

Fig. 5.2 Scheme of experimental device: 1—the laser, 2, 4—the rotating prisms; 3—the chamber for placement of muscle tissue preparation; 5—the objective forming an image in the plane of the diaphragm; 6, 7—the photomultiplier, 8—the active filter cutting off PMT noises; 9—the active filter, smoothing and intensifying the signal; 10—the trigger; 11—the frequency meter; 12—the delay line; 13—the electrostimulator. Reprinted from [87] with permission

A method [39, 138] based on counting the number of intensity fluctuations per unit time was developed for the study of velocity of muscle contractions. This method was implemented in the setup, the scheme of which is presented in Fig. 5.2. Laser radiation was directed by the prism 2 into the chamber 3 where the muscle under study was investigated. Some part of the radiation scattered by a small area of the muscle was turned by the prism 4 and directed into the objective 5, which formed an image covered with speckle-structure in the plane 6. If the preparation moves at the velocity v, its image together with the speckle-field shifted at a velocity

$$v_s = -mv,$$

where m is the magnification of the optical system. Measurement of speckle intensity due to movement in some point was registered with the photomultiplier 7 with a diaphragm, which is a little bit less than the average speckle size (in our experiments, 100 μ). Photocurrent PMT corresponded to spatial intensity resolution of speckle-structure passing through the aperture.

The average number of intersections n of the random value I with a set level I_s on a length unit was determined in the work [17] in the following way:

$$n(I_s) \sim \sqrt{\frac{I_s}{\langle I \rangle} \exp\left(-\frac{I_s}{\langle I \rangle}\right)} \tag{5.1}$$

where $\langle I \rangle$ is the average value of the speckle-field intensity.

If $I_s = \langle I \rangle / 2$ then the average distance between two cross points corresponds to the average speckle size σ and the number of shifts will be maximal:

$$n_{max} = n(\langle I \rangle 2) = R/\sigma, \quad (5.2)$$

where R is the parameter, which is determined by the geometry of the illuminating laser beam.

The electrical device compared a high-frequency random signal with a sliding signal, which corresponded to the average illumination of the speckle-pattern. When both signal crossed, the trigger transfer occurred. Thus, a sequence of right-angle pulses was formed on the trigger output. Their duration was determined by the width of a separate speckle. The electronic device produced $n_{max}|\vec{v}_s|$ pulses per unit time during the movement of the speckle-structure, which were read by frequency meter. Thus, the directly measured speckle-counting velocity is just proportional to the velocity of object movement.

In the conducted experiments, the counting time of the frequency meter was 10 ms and determined time resolution. Average velocity per every 10 s was practically measured. The frequency meter was launched in a set moment of time from the electrical stimulator through the delay line. The data were averaged on 10 measurements to increase the accuracy.

Biceps of a frog's thigh served as the research object. Its length was 20 mm, and the width was ~2 mm. It was placed into a physiological solution, and the size of the illuminated area was ~1 mm^2. Figure 5.3 illustrates the dependence of the relative velocity of contraction of the muscle area on time situated at a distance of 5 mm from the edge. A curve is dotted, which corresponds to external tensile force that is two and a half time more than in the case with a continuous curve. The first relatively short phase (~40 ms) corresponds to the active process of muscle contraction, and the second one—to passive relaxation. It is seen from the diagram that the process of contraction is faster than the process of relaxation.

The obtained results agree with the data obtained with direct myography traditionally applied for the study of muscle contraction. Along with this, the speckle-counting method has a number of advantages: It is non-contact and allows placing the preparation intentionally and conducting a study without separating it. This method is inertia-free (though it is worth mentioning that temporal resolution is limited by

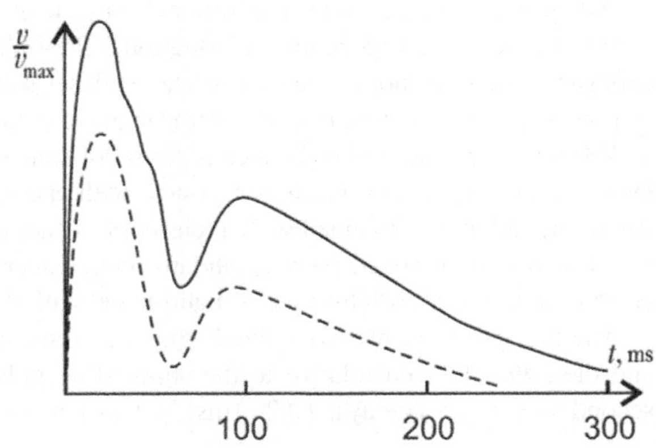

Fig. 5.3 Dependence of muscle contraction velocity on time. Reprinted from [87] with permission

minimal time of frequency meter). Using it, we can study separate small muscle areas by measuring velocity of their contraction.

While conducting this study, the dependence of velocity of muscle contraction on time was studied. The obtained results agree with the calculations of velocities due to implicit data. Duration of a single contraction is 200–300 ms, and about 40 ms of them goes to the process of active contraction. Maximal velocity of contraction is about 3 times higher than the velocity of relaxation, and it is developed during 20–25 ms. The increase of the external load leads to approximately proportional decrease of contraction velocity. The offered technique can be applied not only for studying muscle contraction, but for studying the movement of other diffusive objects.

5.2 Study of Deformations of Epithelial Tissues Using the Speckle-Photography Method

Majority of the data of electromyographic study give the possibility to estimate functioning of motor units or the whole muscle. Meanwhile, not only study of a local area, but also the detailed study of a huge surface of a tissue of a living organism participated in the process of muscle contraction, i.e., deformation fields, is of interest. Such an approach would give the possibility to obtain additional information about the activity of the muscle system of an organism along with the data of the already existing electromyographic methods.

Non-contact methods of speckle-photography exist in coherent optics for determining deformation fields and are successfully used in technical physics. Two images of the object in the initial and the excited states are registered on a photomaterial for determining transverse shifts. The images are covered with a spotted structure or speckles in coherent laser light. They obtained in such a way photography or specklogram is enough to determine the shift field. A posterior processing in the system of coherent information processing using spatial filtering gives the possibility to determine the areas of equal shifts or to determine shift at each point of the object [31].

A leg of a frog was used as a research object, in which contraction is characterized by very heterogeneous deformations. Preliminary test has shown that the exposure time must not exceed 1 ms. He–Ne laser usually used in such tests is not so powerful to provide registration of the speckle-pattern on films with a resolution of 100 lines/mm and higher in such a period of time. That is why a pulsed He–Ne laser was constructed as a radiation source with energy in free-running mode up to 0.8 J. The radiation expanded with a telescopic system was directed onto the object, which was photographed with a standard photocamera "Zenit" with magnification close to unit. Laser radiation was synchronized with the excitation of the object.

The first exposure was conducted when the tissue rested, then it was stimulated, and the pulse delayed relative to the stimulating pulse launched the laser for the second time (time of delay is 10:200 ms). After such a twofold exposure, the photofilm

("Micrat-N") was processed in a standard developer PMT. The energy of the illuminating laser pulse was usually selected in such a way that the specklograms were overexposed. They were bleached in a number of experiments. Nonlinear registration of specklograms enlarges the cone of the diffracted light during further reconstruction and leads to the appearance of a great number of interference fringes and increase of accuracy of deformations.

The posterior processing of speckle-photographs was made in the following way: A 24 × 35 mm^2 slide was scanned with a thin laser beam with a diameter of about 2 mm. A long-focal optical system $f = 300$ mm was placed behind it. The system conducts Fourier transformation of the light diffracted on the lighted area of the specklogram. Interference fringes were formed in the focal plane. Their direction and period reflect the shift of the surface under study, and

$$T = \frac{\lambda d}{\Delta m} \tag{5.3}$$

where Δ is the shift; m is the magnification of the optical system used during specklogram recording.

The focal distance f determines the scale of fringes. The characteristics of the interference pattern were measured on a special display that gives the possibility to filter that part of the light, which passed through the slide without diffraction, and has the possibility of angular attitude (with measuring the angle) and the measuring scale for determination of the fringe period.

Some typical speckle-interferograms are presented in Fig. 5.4a, b. They were made during filtering of non-diffracted light and reflecting deformation of separate areas of webbing. Also, a definite deformation field of the whole leg in 20 ms after the beginning of irritation is presented in the same figure.

Contractions of the elastic living tissue are accompanied by deformations and have considerable heterogeneity.

The diameter of the scanning beam had to be varied from 3 to 1 mm using a diaphragm for determining such deformations as, for example, in the right upper part of the figure in order to increase the contrast of bands and to change the resolution of the method. Deformations of the epithelial tissue during contractions were studied as a result of double-pulsed speckle-photography method ($\lambda = 694.3$ nm). Optimal conditions were determined for obtaining a speckle-photography of living tissue. The setup made on the basis of a pulse ruby laser gave the possibility to register the shift field on the area of about 6 cm^2 with spatial resolution of 2–4 mm^2.

A homogeneous deformation field of contracting epithelial tissue was investigated within the interval from 10 to 50 ms. After the beginning of contractions, vector diagrams of shifts were built, the range of which was from 40 to 205 μm.

Among the advantages of the method, the following ones should be mentioned: contactlessness, simple realization and equipment, and high accuracy (ranges of the measured fringe from dozens of micrometers to millimeters).

A disadvantage is a two-stage process of deformation determination—filming of specklograms and their posterior processing. Besides, the value of longitudinal shifts,

a

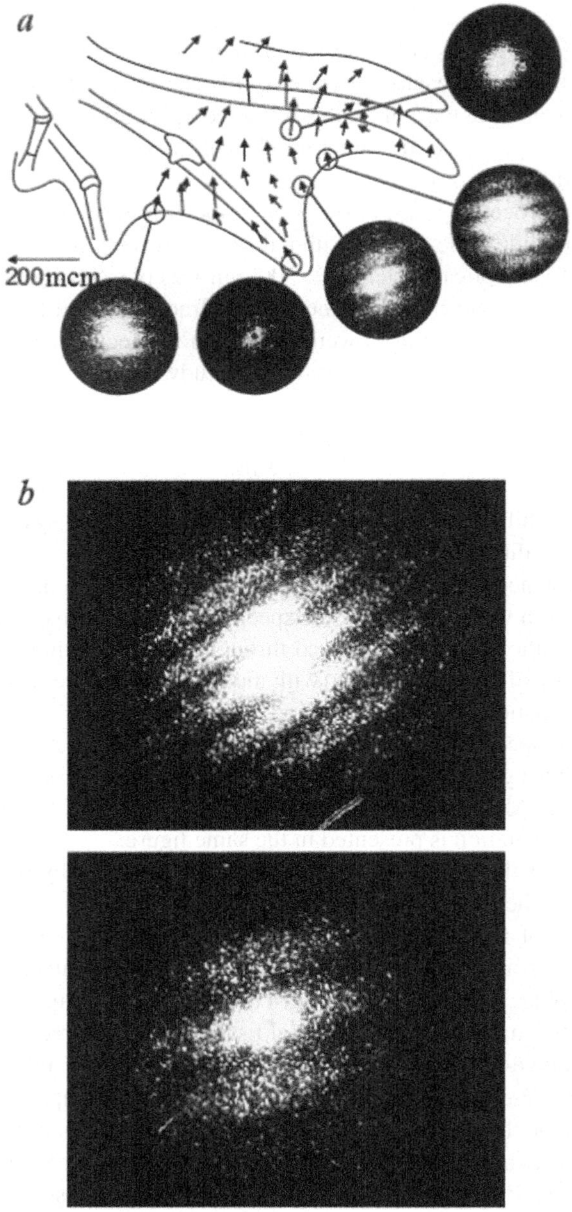

b

Fig. 5.4 Field of deformations of integument of a frog in 20 ms after the stimulation (a) and examples of speckle-interferograms with the help of the deformation field of frog's integument were determined (b). Reprinted from [87] with permission which usually accompany any tissue contraction, should not exceed the depth of the field of the objective, with the help of which a specklogram is recorded.

Medical–biological application of speckle-interferometry, in which it is necessary to register tissue deformations on the whole surface simultaneously, can be very diverse. Taking into account the results of the experiments, the described method can be recommended for determining shifts of the human larynx during speech production [139], studying contraction caused by electrical stimulation of nerves or smooth muscles of a lymphatic vessel [140]. In their studies, the authors of the works [139, 140] used mechanic–electrical sensors, which give the possibility to measure deformation in a separate point. A whole number of such sensors would be needed during estimation of human skin surface shift, and estimation of the surface shift while stretching such an object as a nerve is impossible. Meanwhile, this task of the study is interesting because of the information about mechanical changes in the nerve (swelling and changes of length) during action potential [141–143], about preserving the capability to transmit the action potential with a nerve stretched in 5 times against its initial length [144]. Only a number of specific possible applications of the speckle-photography method were enumerated above.

According to the authors of this monograph, the methods of speckle-optics can be used in restoration surgery and biomechanical study, and also add the methods of holographic interferometry in other medical applications.

5.3 Laser Specklometer (Microhematomyograph)

1. Purpose.

   1.1. The laser specklometer (further LS) (Figs. 5.5 and 5.6) is aimed at registration in relative units of time intensity fluctuations of a speckle-field formed as a result of dispersion of laser radiation by skin and calculation of their amplitude–frequency characteristics accompanied by a computer.

   1.2. The LS is used to study biomechanical characteristics of skeletal muscles and microhemodynamics of skin in order to estimate the functional state of separate systems (neuromuscular, blood circulation, etc.) and human organs.

   1.3. The LS is applied together with a computer (PC AT-286, 386, 486, monitors EGA, VGA, SVGA), which controls data input and processes and documents the obtained results.

   1.4. The LS power is from a one-phase alternating current, and voltage is 220 V $\pm$ 10% at a frequency of 50 $\pm$ 0.5 Hz.

2. Technical characteristics of the laser specklometer.

   2.1. The laser specklometer includes an optical-medical component, an electronic component and software.

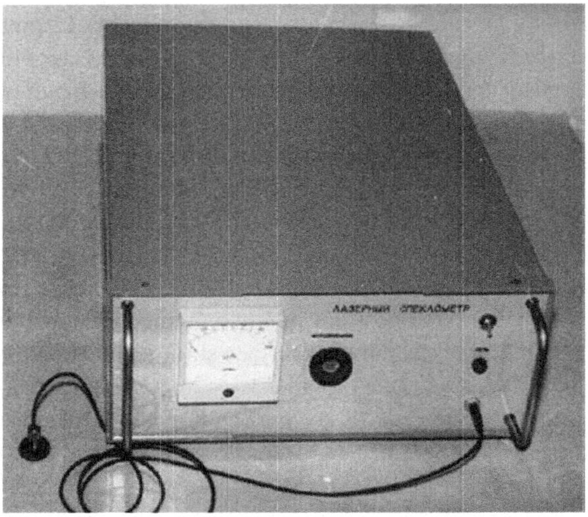

Fig. 5.5 Prototype of a laser specklometer. Reprinted from [87] with permission

2.1.1. The optical-mechanical component of the laser specklometer consists of He–Ne laser LGN-208 (with the following parameters: the wavelength is 632.8 nm, the radiation power is 2 mW, the beam divergence is 0.3 mrad, and the beam diameter is 0.3–0.8 mm); a receiving–transmitting light-fiber system; radiation input unit into the light fiber; a photoreceiver unit; a fixing unit; slots and the body of the power supply of the photoreceiver and the indicator of power of laser radiation; and a base plate and a dustproof housing.

2.1.2. The receiving–transmitting system includes a transmitting fiber, a receiving fiber, a head on the transmitting fiber for connection with the radiation input unit, a head on the receiving fiber for connection with a unit of the photoreceiver, an illuminating head and a safety cladding for the fibers.

2.1.3. The radiation input unit into the fiber provides input of radiation with minimal losses into fiber CB02-0.5-K2.GHO.735.IIFTC. Pointing of laser radiation to the face of the light guide is provided with adjusting screws with control on the maximum of the radiation on the fiber outlet.

2.1.4. Construction of the receiving–transmitting system provides replacement of the defective fibers.

2.1.5. The illuminating head forms a radiation zone with a diameter from 8 to 12 mm on the object (Fig. 5.7).

2.1.6. The distance between the faces of the fibers and the surface under study is from 8 to 12 mm.

2.1.7. The attachment point of the illuminating head provides its firm fixing on the object under study.

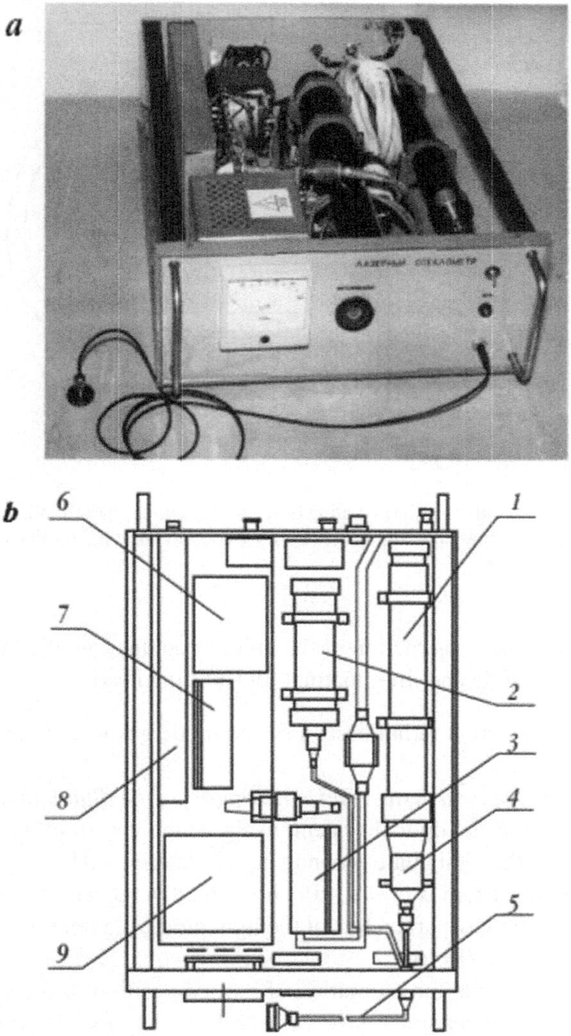

Fig. 5.6 General view of a prototype of laser specklometer with the cover removed (**a**) and mounting of the blocks of laser specklometer (**b**): 1—the laser LGN-208A; 2—PMT unit; 3—the laser power box; 4—the adjusting mechanism; 5—the light guide center; 6—the network transformer; 7—NP board; 8—the signal processing unit; 9—VP block. Reprinted from [87] with permission

2.1.8. The illuminating head has an easily removable protective glass as a variant, which should give the possibility to wipe it with an alcohol–ether mixture, but it is inadmissible to let the liquid inside the head.

2.1.9. A unit of the photoreceiver consists of a photoreceiver PMT-116: a voltage divider for PMT power, elements of fixing of the receiving fiber, an electromagnetic screen-casing.

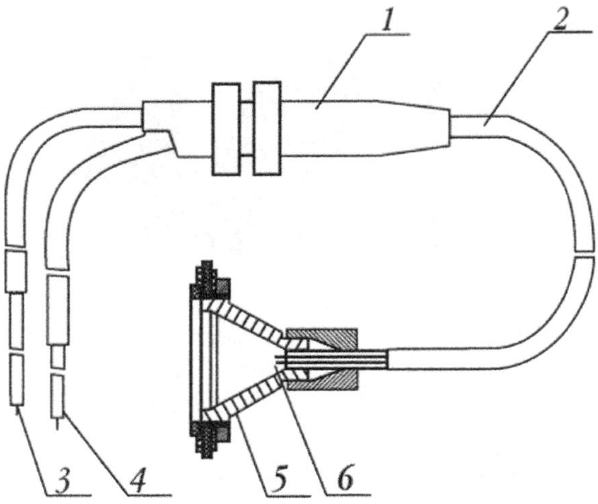

Fig. 5.7 Unit of a light guide: 1—the fastening bush; 2—the containment shell; 3—the illuminating guide light; 4—the receiving guide light; 5–the sensor body; 6—the light guide outputs. Reprinted from [87] with permission

2.1.10. The construction of the unit of conjunction of a laser with a light guide and illuminating–transmitting head.

The source of coherent radiation is a single-mode gas laser (wavelength is 0.63 μm).

The choice of the optical system and construction of the illuminating–transmitting system. The choice of the optical system, using which a most effective input of the optical energy into the light guide is conducted, is determined by the initial data: The diameter of the laser beam is $\varphi = 0.4{:}0.8$ mm, the divergence of the laser radiation is $\lambda = 0.6328$ μm, and by parameters of the light guide, diameter is $\varphi_w < 0.008$ mm, and the aperture is $\alpha = 0.146$.

The calculations made for estimation of mutual position of the light guides, the prism and the tissue under study gave the possibility to choose optimal sizes of the illuminating–transmitting sensor. The experimentally determined divergence of the laser radiation α_1 at the output of the single-mode light guide was ~6°. The size of the light spot on the surface under study is $D = 2S_1 tg\alpha \approx 1.4$ mm.

The face of the illuminated single-mode fiber is situated at a distance S_1 from the surface of the skin under study. This distance is

$$S_1 = a_1 + cn + b = 0.5 + 3 \cdot 1.515 + 2 = 7 \text{ mm}$$

where α_1 is the distance from the face of the illuminating light guide to the prism: c is the size of the prism face; n is the refraction index of the prism material; b is the distance between the output face of the prism and the surface of the skin under study.

To obtain the maximal contrast of the exit signal, the final size of a speckle in the plane of the input window of the receiving light guide should be ≈5 μm that is why the receiving light guide has to be put at a distance S_2 from the surface under study, which we determine from the expression

$$S_2 = \frac{Dd}{1.22\lambda} = \frac{1.4 \cdot 5 \cdot 10^{-3}}{1.22 \cdot 0.633 \cdot 10^{-3}} \approx 9 \text{ mm}$$

The light guides 1 (Fig. 5.8) are focused on the tips 2 with the help of the collets 3 and 4 and the nuts 5 and 6. The nut 7 serves for fixing the tips 2 in the block of the conjunction of the laser with the light guide (head) or on the PMT unit. Then, the light guides pass through the protective polyvinyl tubes 8, combine in the polyvinyl pipe 9 with the help of the ring 10 and enter the head 11. The prism 13 is fixed in the head on the mount 12. The head is fixed on the legs 14 on the surface under study

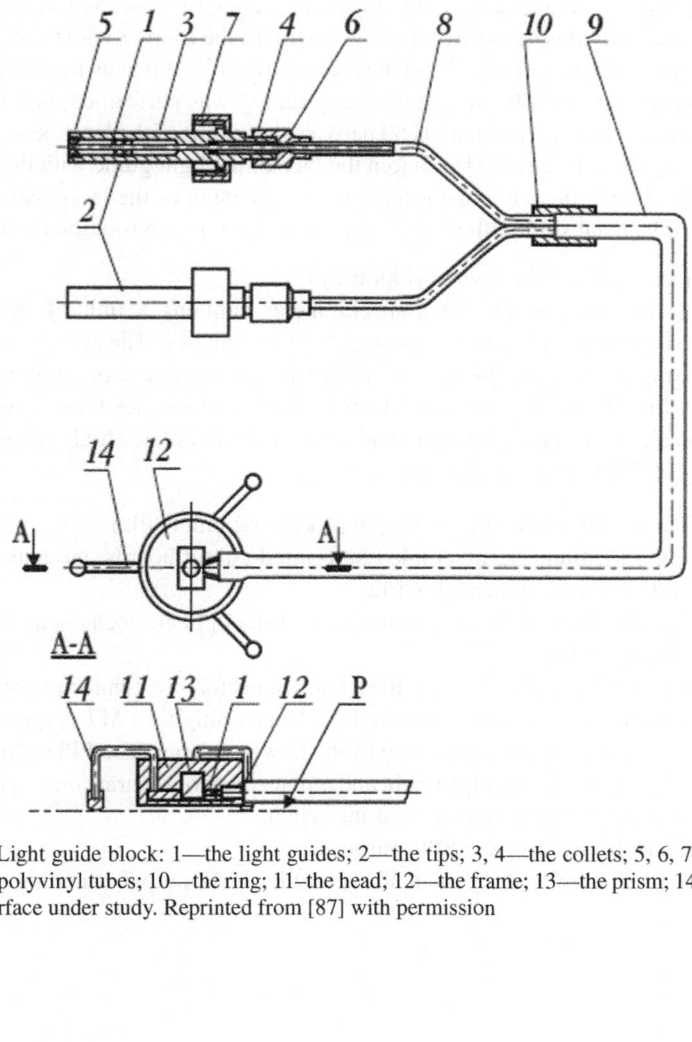

Fig. 5.8 Light guide block: 1—the light guides; 2—the tips; 3, 4—the collets; 5, 6, 7—the nuts, 8, 9—the polyvinyl tubes; 10—the ring; 11-the head; 12—the frame; 13—the prism; 14—the feet; P—the surface under study. Reprinted from [87] with permission

A. The position of the legs relative to the head can be modified due to the height and the angle.

In the process of operation, the laser radiation is brought into the light guide 1, which is established in the conjunction block (head), passes through it, then moves aside by the prism 13 and illuminates the surface under study A.

Partially reflected, scattered light passes through the prism 13 backward, comes into the second light guide 1 and goes through it to the tip 2, fixed on the PMT unit. Introduction of the beam to transmitting light guide is conducted in the block of conjunction of laser with the light guide. It consists of the body 1 (Fig. 5.9), in the borings of which the eccentrics 2 and the barrel 3 are placed. One of the tips 4 of the block of the light guides is fixed in the barrel. The eccentric axes 5 with the bearings 6 are fixed along the eccentric axis 2. The frame 8 with the objective 9 is pressed with the spring 7 to bearings. The frame 8 is placed in the gap between the body 1 and tablet 10. The nut 12 is put in the gap between the body 1 and cover 11. The nut is connected to the bushing.

During adjustment, the beam from prism goes along the objective axis and shifts perpendicular to its axis because of transverse shift of the frame 8 with the objective 9 during rotation of the eccentrics 2 and the eccentric axis 5. At that using the eccentrics 2 (their eccentricity is 500 μm), preliminary tuning was performed, and using the eccentric axes 5 (their eccentricity is 50 μm) accurate shift of the laser beam with the axis of the light guide was made. To join the face of the light guide with the focus of the objective 9 and, thus, to obtain maximal energy input of the laser radiation into the light guide, axial shift of the barrel 3 is conducted through rotation of the nut 12.

3. The electric part of the laser specklometer.
   The electric part of the laser specklometer contains a unit of preliminary processing and an indicator of power of the laser radiation. The unit of preliminary processing consists of a preliminary PMT signal intensifier: a programmed intensifier; a low-frequency filter 1; a detector and a low-frequency filter 2; connection with an electronic processing unit and a personal computer; a high-voltage power supply of PMT; and a power supply.

   3.1. The construction of preliminary processing unit (PPU).
        The PPU units are assembled on printed-circuit boards, mounted into the body and joined through wiring.
        The PPU is situated on a plate base of the optical-mechanical part of the specklometer.
        On the PPU body, there are fixed two connectors for connection of the electrical processing block and the PC; contact plug for PMT connection; and a contact plug for connection of the power voltage. The PPU construction provides access to adjustment and replacement of separate units, possibility to control their parameters during repair.
   3.2. Characteristics of the PPU purpose:
        (a) The preliminary intensifier provides PMT output matching.

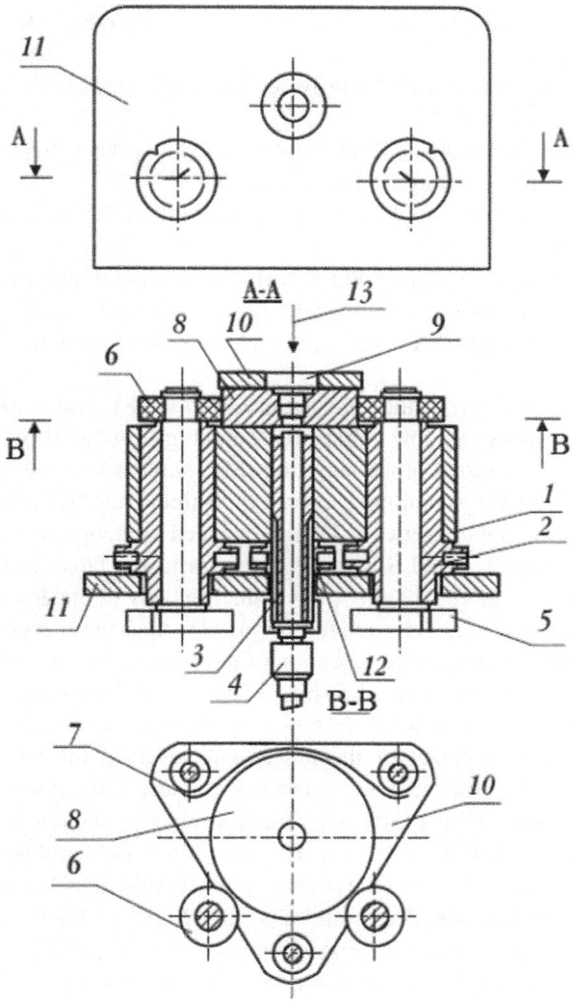

Fig. 5.9 Attachment: 1—the body; 2—the eccentrics; 3—the barrel; 4—the tip, 5—the eccentric axis; 6—the bearing; 7—the spring; 8—the frame; 9—the objective; 10—the table. Reprinted from [87] with permission

(b) The programmed intensifier has five fixed coefficients (1, 2, 4, 8 and 16), and the choice is made by the code, which is sent into the PPU through a serial line.
(c) The high-frequency filter 1 provides limitation of the spectrum signal, and the cutoff frequency is 2 kHz.
(d) The detector and low-frequency filter 2 provide obtaining a detected signal, the cutoff frequency is ~60 Hz, and switching of filters should

be conducted with a code, which is sent into the PPU through a serial line.
(e) The connection unit provides exchange through the serial line at a speed of 9600 bit/s.
(f) The high-voltage PMT power supply provides voltage of 1600 V, and adjustment range is lower than 10%.
(g) The power supply provides the operation of the above enumerated units.

3.3. Photodiodes PD 2A or PD 7 are used as a photoelement in the power indicator of laser radiation, and a needle indicator is used.
The needle indicator and the photoelements are put on the faceplate of the device.
The electric part of the LS is situated in the PPU and consists of units of signal processing, low voltage (LV) and high voltage (HV).
3.3.1. The signal processing unit (SPU) consists of the following units: a PMT signal intensifier, a high-frequency (HF) filter, an analog–digital converter (ADC), sequenced exchange with the PC.

3.4. Constructively, the LS is made in the form of a table device and has the dimensions of 168 × 340 × 154 mm. The body of the device has an upper removable cover, and when it is removed there is the access to the units and mounting of the device. Location of the LS units is shown in Fig. 5.9b.
On the front panel of the LS (Fig. 5.10a), there is a main-electrical breaker, an indicator "network," an outlet of the light guide, an indicator head and a photoreceiver of the power indicator; on the back panel of the LS (Fig. 5.10b), there are a network connector, a network safety device (common), a power indicator, a data receiving indicator, a ground clamp, a PPU network safety device, a connector to a PC serial port.
3.4.1. The capacity of the analog–digital converter is 12 bits. The range of the measured amplitudes is from 2048 to 2047 levels of the ADC.
3.4.2. The interface of connection with a PC is RS-232.
The exchange speed for data transmission

- into PC is 115,200 bit/s and
- from the PC is 9600 bit/s.

3.4.3. Programmable coefficients of PMT signal intensification have to be 1, 2, 4, 8, 16.

4. Organization and operation of the laser specklometer.
The operation of the laser specklometer is provided by hardware and software.

4.1. Duration of registration of time intensity fluctuations of the speckle-field: In low-frequency ones (1–62.5 Hz), it is not more than 8 s; in high-frequency ones (5–2000 Hz), it is not more than 1 s.

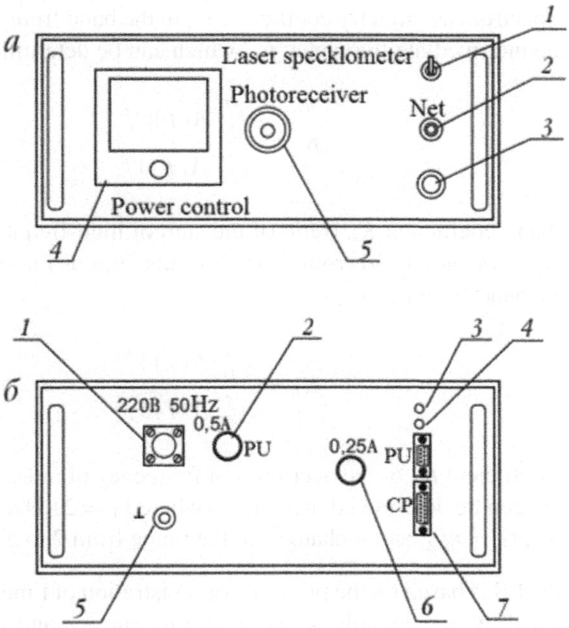

Fig. 5.10 General view of the front panel of a laser specklometer (**a**): 1—the main electric breaker; 2—the indicator "Net"; 3—the output of the light guide; 4—the indicator head; 5—the photoreceiver of power indicator. General view of the back panel of the laser specklometer (**b**): 1—the network connector; 2—the network safety device (general); 3—the power indicator; 4—the data receiving indicator; 5—the ground clamp; 6—the network PPU safety device; 7—the connector to a PC serial port. Reprinted from [87] with permission

4.2. The software gives the possibility to calculate the following spectra characteristics in the frequency range of 1–62.5, 1–125, 1–250, 1–500, 2–1000, 5–2000 Hz:

the average frequency $\langle f \rangle$ in the band from f_{L} to f_{U}:

$$\langle f \rangle = \frac{\int_{f_{\mathrm{L}}}^{f_{\mathrm{U}}} f A(f) \mathrm{d}f}{\int_{f_{\mathrm{L}}}^{f_{\mathrm{U}}} A(f) \mathrm{d}f} \tag{5.4}$$

where $A(f)$ is the amplitude at frequency f; f_{L}, f_{U} are the lower and upper frequencies of the frequency bands under study;
area S under the spectral curve in the band of frequencies from f_{L} to f_{U}:

$$S = \int_{f_{\mathrm{L}}}^{f_{\mathrm{U}}} A(f) \mathrm{d}f \tag{5.5}$$

the spectrum asymmetry coefficient K_a in the band from f_1 to f_2 relative to the intermediate frequency f_m, which can be determined as

$$K_a = \frac{\int_{f_m}^{f_2} A(f)\mathrm{d}f}{\int_{f_1}^{f_m} A(f)\mathrm{d}f} \tag{5.6}$$

the band coefficient K_b (ratio of the sum of high-frequency harmonics in the frequency band from f_3 to f_4 to the sum of low-frequency ones in the band from f_1 to f_2):

$$K_b = \frac{\int_{f_3}^{f_4} A(f)\mathrm{d}f}{\int_{f_1}^{f_2} A(f)\mathrm{d}f} \tag{5.7}$$

the coefficient K_b of the overtones of frequency of muscle contractions, which can be determined using (5.7) where $f_1 = 21$ Hz, $f_2 = 26$ Hz, $f_3 = nf_1, f_4 = nf_2$, and n changes in the range from 2 to 5.

The work of the LS is based on the principle of registration of time changes of the speckle-field formed by dynamically active scattering surface and further spectral analysis of the obtained data: $\langle f \rangle$, S, K_a, K_b.

Optical-electronic part of the LS consists of the following units: a single-mode gas laser LGN-208A; an adjustment mechanism; a light guide; PMT; and a power indicator. Irradiation of the surface under study and receiving of the scattered radiation are provided by two light guides with a diameter of the active part of the fiber of 4 μm. A single-mode gas laser (the wavelength is 0.63 μm) serves as a source of coherent radiation. Distance between the face of the irradiating light guide and the surface of the object under study is from 8 to 12 mm, and the size of the light spot on the skin at a distance of 10 mm to the face of the irradiating light guide is not more than 2 mm. Construction of the light guide unit is shown in Fig. 5.7.

In the preliminary processing unit (PPU), the registered PMT signal is intensified, filtered and converted into a digital form and transmitted through the serial line into the PC, which stores the necessary amount of the data in the memory. Then, the examination of the obtained time realization, the calculation of spectrum in the needed frequency range with giving of spectral characteristics and the data documentation can be organized. The structural scheme of the LS is in Fig. 5.11.

The main controlled parameter is the average frequency $\langle f \rangle$ of the main spectral harmonic, calculated due to the regulated intensity fluctuations of the speckle-field, which was formed during reflection of laser radiation from the diffuser, which is situated on the reference vibrator, oscillation frequency f_e of which is in the range of 15–50 Hz, and the oscillation amplitude does not exceed 5 μm. The average frequency $\langle f \rangle$ must coincide with f_e with the accuracy of 10%. Control should be conducted at three meanings f_e in the range of 15–50 Hz.

During determination of the value of the main controlled parameter, such a value of oscillation amplitude of the reference vibrator is set, at which the amplitude of

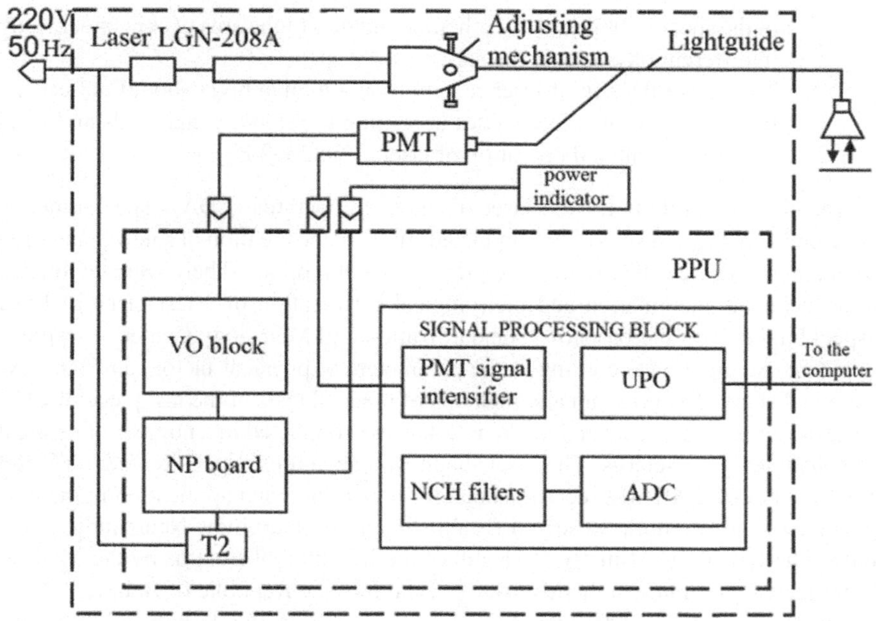

Fig. 5.11 Structural scheme of the laser specklometer. Reprinted from [87] with permission

the second spectral harmonics is not more than 10% from the amplitude of the main spectral harmonic in the amplitude–frequency spectrum of intensity fluctuations of the speckle-field.

The characteristics of speckles $\langle f \rangle$, S, K_a, K_b calculated on the basis of measurement results of intensity fluctuations of the speckle-field are used for analysis of data during conduction of the study of biomechanical characteristics of skeletal muscles and microhemodynamics of skin.

5. Medical requirements.

   5.1. The possibility of investigation of a different tone and contractive activity of skeletal muscles should be provided both for the state of rest and under the impact of different loads.
   5.2. The possibility of establishment of the correlation connection of spectra of intensity fluctuations of the speckle-field with biomechanical parameters of the skeletal muscles should be provided.
   5.3. The possibility of studying microhemodynamics of skin on the whole surface of the human body should be provided.
   5.4. The research area of microhemodynamics should have the diameter of not more than 5 mm.
   5.5. The processing of measurement results should give the possibility to establish a connection between the parameters of spectra of fluctuation intensity

of the speckle-field and microhemodynamics of the area of skin under study in different physiological states.

5.6. The power density of laser radiation on a human body should not exceed the maximum of safety values according to "sanitary standards and rules of arrangement and operation of lasers" No 2392-81.

Technical conditions for the laser diagnostic apparatus (LDA) "specklometer" were developed as a result of the conducted study on the creation of a new generation of diagnostic devices based on principles of coherent optics. There were conducted state technical examination and metrological certification of a test piece of LDA "specklometer" on the basis of "Standardization and Metrology Center" as well as hygienic examination according to the parameters of physical factors and sanitary-chemical study. The program and methods of medical tests of the test piece of LDA were developed. Hygienic and medical tests were conducted in a number of medical establishments of Belarus. The laser diagnostic apparatus "specklometer" TY RB 100103849.001-2001 was recommended for registration and application in medical practice of the Scientific Research Establishment of Neurology, Neurosurgery and Physiotherapy of the Ministry of Health of the Republic of Belarus by the Medical Equipment Committee of the Ministry of Health of the Republic of Belarus.

On the August 8, 2001, the Ministry of Health of the Republic of Belarus issued a registered card IMT No IM-7.2263 to the Scientific Research Establishment of Neurology, Neurosurgery and Physiotherapy of the Ministry of Health of the Republic of Belarus. Thus, the laser diagnostic apparatus "specklometer" is registered by the Ministry of Health of the Republic of Belarus (state registration number is MT-7.15047-0108) and today it is successfully applied in medical practice on the territory of the Republic of Belarus.

5.4 Theoretical Study of Biomechanical Characteristics of Skeletal Muscles and Determining the Ways of Optimization of the Optical Scheme of the Laser Specklometer

In this paragraph, the calculation of spatial distribution of the speckle-field intensity is depicted as well as the ways of optimization of the optical scheme.

One of the methods of studying muscle oscillation is the method of light diffraction on the oscillating under muscle impact skin surface. If it is assumed that the skin surface is a diffuser and the angle of light incidence onto the surface is rather small, then the amplitude distribution in the observation point after light diffraction is determined by the Debye integral [145]:

$$U(P) \equiv U(x, y, z) = -\frac{i}{\lambda} A \iint_{\Omega} \exp(-ik\vec{q}\vec{R}) \mathrm{d}\Omega \tag{5.8}$$

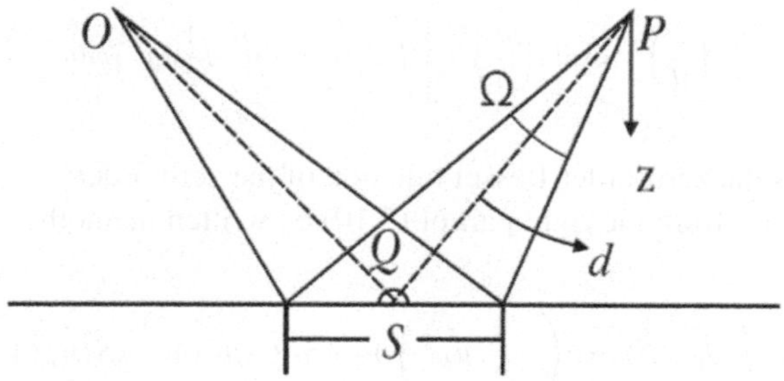

Fig. 5.12 To the question of optimization of the optical scheme of the laser specklometer. Reprinted from [87] with permission

where Ω is the element of the solid angle, at which the illuminated area *P* is seen (Fig. 5.12); *A* is the amplitude of the incident wave; λ is the wavelength of the incident light.

Let (*x*, *y*, *z*) be coordinates of (.) *P*, and (ξ, η, β) are coordinates of (.) *Q*. Suppose that $\xi = S\sin\theta$, $\eta = S\cos\theta$, $x = \rho\sin\psi$, $y = \rho\cos\psi$.

As (.) *Q* lies on the spherical wave front, then

$$\beta = -\sqrt{d^2 - S^2\rho^2} = -d\left[1 - \frac{1}{2}\frac{S^2\rho^2}{d^2} + \cdots\right]$$

that is why

$$\vec{q}\vec{R} = \frac{x\xi + y\eta + z\beta}{d} = \frac{S\rho\cos(\theta - \Psi)}{d} - z\left[1 - \frac{1}{2}\frac{S^2\rho^2}{d^2} + \cdots\right] \tag{5.9}$$

We will introduce dimensionless variables:

$$u = \frac{2\pi}{\lambda}\left(\frac{S}{d}\right)^2 z,\quad v = \frac{2\pi}{\lambda}\left(\frac{S}{d}\right)\rho = \frac{2\pi}{\lambda}\left(\frac{S}{d}\right)\sqrt{x^2 + y^2}$$

Let terms with ρ/d in the degree exceeding the second one be negligibly small. Then,

$$k\vec{q}\vec{R} = v\rho\cos(\theta - \Psi) - \left(\frac{d}{S}\right)^2 u + \frac{1}{2}u\rho^2,\quad d\Omega = \frac{S^2\rho d\rho d\theta}{d^2}$$

Taking into account the previously mentioned information, we obtain

$$U(P) = -\frac{i}{\lambda}\left(\frac{S}{d}\right)A\exp\left[i\left(\frac{d}{S}\right)^2 u\right]\int\limits_0^1\int\limits_0^{2\pi}\exp\left\{-i\left[v\rho\cos(\theta - \psi) + \frac{1}{2}u\rho^2\right]\right\}\rho d\rho d\theta$$

$$\equiv -\frac{2\pi i}{\lambda}\left(\frac{S}{d}\right)^2 A\exp\left[i\left(\frac{d}{S}\right)^2 u\right]\int_0^1 J_0(v\rho)\exp\left(-i\frac{1}{2}u\rho^2\right)\rho d\rho, \tag{5.10}$$

where $J_0(v\rho)$ is the zero-order Bessel function of the zero order.

Let the integral from the right part of (5.10) be written in another form:

$$2\int_0^1 J_0(v\rho)\exp\left(-\frac{1}{2}iu\rho^2\right)\rho d\rho \equiv C(u,v) - iS(u,v), \tag{5.11}$$

$$C(u,v) = 2\int_0^1 J_0(\rho,v)\cos\left(\frac{1}{2}u\rho^2\right)\rho d\rho \tag{5.12}$$

$$S(u,v) = 2\int_0^1 J_0(\rho,v)\cos\left(\frac{1}{2}u\rho^2\right)\rho d\rho$$

Having integrated equations (5.12) in parts for the case when $|u/v| < 1$ and having used the well-known ratio for Bessel function $\partial/\partial x\left[x^{n+1}J_{n+1}(x)\right] = x^{n+1}J_n(x)$, let $C(u, v)$ be written in the following form after integration in parts:

$$C(u,v) = \frac{2}{v}\int_0^1 [\rho J_1(v\rho)]\cos\left(\frac{1}{2}u\rho^2\right)d\rho$$
$$= \frac{2}{v}\left[J_1(v)\cos\left(\frac{1}{2}u\right) + u\int_0^1 \rho^2 J_1(v\rho)\sin\left(\frac{1}{2}u\rho^2\right)d\rho\right] \tag{5.13}$$

Having integrated in parts the integral in the right part of (5.13) again and continued this process, we obtain

$$C(u,v) = \frac{\cos\left[\left(\frac{1}{2}\right)u\right]}{\left(\frac{1}{2}\right)u}\left[\left(\frac{u}{v}\right)J_1(v) - \left(\frac{u}{v}\right)^3 J_3(v) + \cdots\right] + \frac{\sin[(1/2)u]}{(1/2)u}$$
$$\left[\left(\frac{u}{v}\right)^2 J_2(v) - \left(\frac{u}{v}\right)^4 J_4(v) + \cdots\right] = \frac{\cos\left[\left(\frac{1}{2}\right)u\right]}{\left(\frac{1}{2}\right)u}U_1(u,v)$$
$$+\frac{\sin\left[\left(\frac{1}{2}\right)u\right]}{\left(\frac{1}{2}\right)u}U_2(u,v) \tag{5.14}$$

Also, $S(u, v)$ are similarly found:

$$S(u,v) = \frac{\sin\left[\left(\frac{1}{2}\right)u\right]}{\left(\frac{1}{2}\right)u}\left[\left(\frac{u}{v}\right)J_1(v) - \left(\frac{u}{v}\right)^3 J_3(v) + \cdots\right] - \frac{\cos[(1/2)u]}{(1/2)u}$$
$$\left[\left(\frac{u}{v}\right)^2 J_2(v) - \left(\frac{u}{v}\right)^4 J_4(v) + \cdots\right] = \frac{\sin\left[\left(\frac{1}{2}\right)u\right]}{\left(\frac{1}{2}\right)u}U_1(u,v)$$
$$+ \frac{\cos\left[\left(\frac{1}{2}\right)u\right]}{\left(\frac{1}{2}\right)u}U_2(u,v) \tag{5.15}$$

where functions

$$U_n(u,v) = \sum_{S=0}^{\infty}(-1)^S\left(\frac{u}{v}\right)^{n+2S} J_{n+2S(v)}$$

are the Lommel functions for the case when $|u/v| < 1$. When $|u/v| > 1$, it is reasonable to implement expansions with positive degrees v/u. They are derived through integrating in parts relative to the trigonometric term. The following formula is obtained for this case after the first integration in parts for $C(u, v)$:

$$C(u,v) = \frac{2}{u}\int_0^1 J_0(vp)\frac{\mathrm{d}}{\mathrm{d}\rho}\left[\sin\left(\frac{1}{2}u\rho^2\right)\right]\mathrm{d}\rho$$
$$= \frac{2}{u}\left[J_0(v)\sin\left(\frac{1}{2}u\right) - v\int_0^1 J_1(v\rho)\sin\left(\frac{1}{2}u\rho^2\right)\rho\mathrm{d}\rho\right] \tag{5.16}$$

Like for the first case when the second term in the right part of (5.16) is integrated once more and the integrating process is continued, the following expression for $C(u, v)$ is obtained:

$$C\,(u,v) = \frac{\sin\left[\left(\frac{1}{2}\right)u\right]}{\left(\frac{1}{2}\right)u}\left[J_0(v) - \left(\frac{v}{u}\right)^2 J_2(v) + \cdots\right]$$
$$- \frac{\cos\left[\left(\frac{1}{2}\right)u\right]}{\left(\frac{1}{2}\right)u}\left[\left(\frac{v}{u}\right)J_1(v) - \left(\frac{v}{u}\right)^3 J_3(v)\right] + \frac{2}{u}\left[\frac{v^2}{2u} - \frac{1}{3}\left(\frac{v^2}{2u}\right)^3 + \cdots\right]$$
$$= \frac{2}{u}\sin\left(\frac{v^2}{2u}\right) + \frac{\sin\left[\left(\frac{1}{2}\right)u\right]}{\left(\frac{1}{2}\right)u}V_0(u,v) - \frac{\cos\left[\left(\frac{1}{2}\right)u\right]}{\left(\frac{1}{2}\right)u}V_1(u,v). \tag{5.17}$$

Also, $S(u, v)$ are similarly found:

$$S(u,v)=\frac{2}{u}\left[1-\frac{1}{2}\left(\frac{v^2}{2u}\right)^2+\cdots\right]-\frac{\cos\left[\left(\frac{1}{2}\right)u\right]}{\left(\frac{1}{2}\right)u}\left[J_o(v)-\left(\frac{v}{u}\right)^2J_2(v)+\cdots\right]$$
$$-\frac{\sin\left[\left(\frac{1}{2}\right)u\right]}{\left(\frac{1}{2}\right)u}\left[\left(\frac{v}{u}\right)J_1(v)-\left(\frac{v}{u}\right)^3J_3(v)+\cdots\right]=\frac{2}{u}\cos\left(\frac{v^2}{2u}\right)$$
$$-\frac{\cos\left[\left(\frac{1}{2}\right)u\right]}{\left(\frac{1}{2}\right)u}V_0(u,v)-\frac{\sin\left[\left(\frac{1}{2}\right)u\right]}{\left(\frac{1}{2}\right)u}V_1(u,v), \quad (5.18)$$

where the Lommel functions $V_n(u, v)$ are

$$V_n(u,v)=\sum_{S=0}^{\infty}(-1)^S\left(\frac{v}{u}\right)^{n+2S}J_{n+2S}(v).$$

The well-known ratio $\partial/\partial x\left[x^{-n}J_n(x)\right]=-x^{-n}J_{n+1}(x)$ for Bessel functions was used while deriving (5.17) and (5.18). Substitution of (5.14)–(5.15) and (5.17)–(5.18) into (5.10) gives the desired expression for the distribution of the amplitude in the observation point, and the intensity in observation point $I = |U(P)|^2$ is determined in the following way:

$$I(P)=\left(\frac{2}{u}\right)^2\left[U_1^2(u,v)+U_2^2(u,v)\right]I_0,\quad \left(\left|\frac{u}{v}\right|<1\right); \quad (5.19)$$

$$I(P)=\left(\frac{2}{u}\right)^2\left(1+V_0^2(u,v)+V_1^2(u,v)-2V_0(u,v)\cos\left[\frac{1}{2}\left(u+\frac{v^2}{u}\right)\right]\right.$$
$$\left.-2V_1(u,v)\sin\left[\frac{1}{2}\left(u+\frac{v^2}{u}\right)\right]\right)I_0(|u/v|>1) \quad (5.20)$$

If $v = 0$, then $V_0(u, 0) = 1$, and if $V_1(u, 0) = 0$, $U_1(u, 0) = U_2(u, 0) = 0$ then

$$I(P)\equiv I(u,0)=\frac{4}{u^2}\{2-2\cos[(1/2)u]\}I_0\equiv\left[\frac{\sin(u/4)}{u/4}\right]^2I_0, \quad (5.21)$$

where $I_0=\left(\frac{\pi S^2A}{\lambda d^2}\right)^2$; A is the amplitude of the radiation incident on the skin surface.

In order amplitude characteristics of the registered signal correctly reflected alterations of oscillation amplitude of the surface under study, it is necessary that the value of the amplitudes was within the limits of linear area of the distribution curve $I(P)$, which is reached at $z = 5$ µm (the wavelength of the incident radiation $\lambda =$ 633 nm, $r = 4$, $r = S/d$, where d is the distance from the illuminated surface to the receiver; S is the size of the illuminated spot of the skin surface).

The analysis of plots of muscle oscillation amplitudes shows, for example, in the contraction mode that the following condition $A < \Delta z$ is reached for loads up to 12 kg and time of registration up to 10 s. These dependencies of muscle oscillation amplitude on the applied load are illustrated in Fig. 5.13a, and dependencies of intensity distribution of the radiation scattered from the skin surface relative to z are shown in Fig. 5.13b. The obtained data gave the possibility to determine the ways of optimization of the optical scheme of the illuminating–receiving head.

The optical scheme of the illuminating–receiving head should be calculated in such a way that the angles of incidence and dispersion of radiation were small, and then muscle oscillation amplitude will be within the limits of the linear area of intensity distribution of the radiation scattered by the skin. The obtained results give the possibility to use the laser specklometer in the optimal way for studying vibration characteristics of muscles. Thus, both isometric and isotonic contraction and stretching of the skeletal muscle of vertebrates were theoretically studied. It was shown that the muscle oscillation amplitude reaches several microns [106]. The spatial intensity distribution of the scattered radiation was calculated for cases of small incident angles of the optical radiation onto the oscillating skin surface. It was shown that the muscle oscillation amplitude does not exceed Δz corresponding to the linear area of the intensity distribution.

5.5 Study on Optimization of Parameters of Measuring Path of the Laser Specklometer

As it was previously shown, biological tissues are more complex systems; in particular, these are whole muscles, integument tissues, ensemble of cells, etc. That is why before moving on to the application of the laser specklometer in medical–biological studies and clinical practice the researches were conducted on optimization of parameters of measuring path of the laser specklometer applied to the diagnostics of determination of vibration activity of technical issues. The speckle-optical method of studying vibration activity of the microinstrument of an ultrasonic welding system (USWS) was developed and introduced for these purposes.

This method can also be used for measurement of oscillation amplitudes of details and components of an engine, determination of resonance frequencies to control the destruction of a product due to changes in vibration amplitude, study of ultrasonic acoustic transducers for optimization of constructions of ultrasonic welding systems of microelectronic technique, etc. During the experiments, a laboratory prototype of the laser specklometer was assembled, with the help of which the study of vibration activity of the microinstrument of an ultrasonic welding system (USWS) was conducted.

It is known that ultrasonic microwelding is one of the most widespread technological operations during assembling products of electronic technique (integrated circuits, etc.). Loading force, time of welding and oscillations of the microinstrument are the main parameters of ultrasonic welding. But calibration and control of oscillation amplitude of the microinstrument during the process of welding are the most complicated processes.

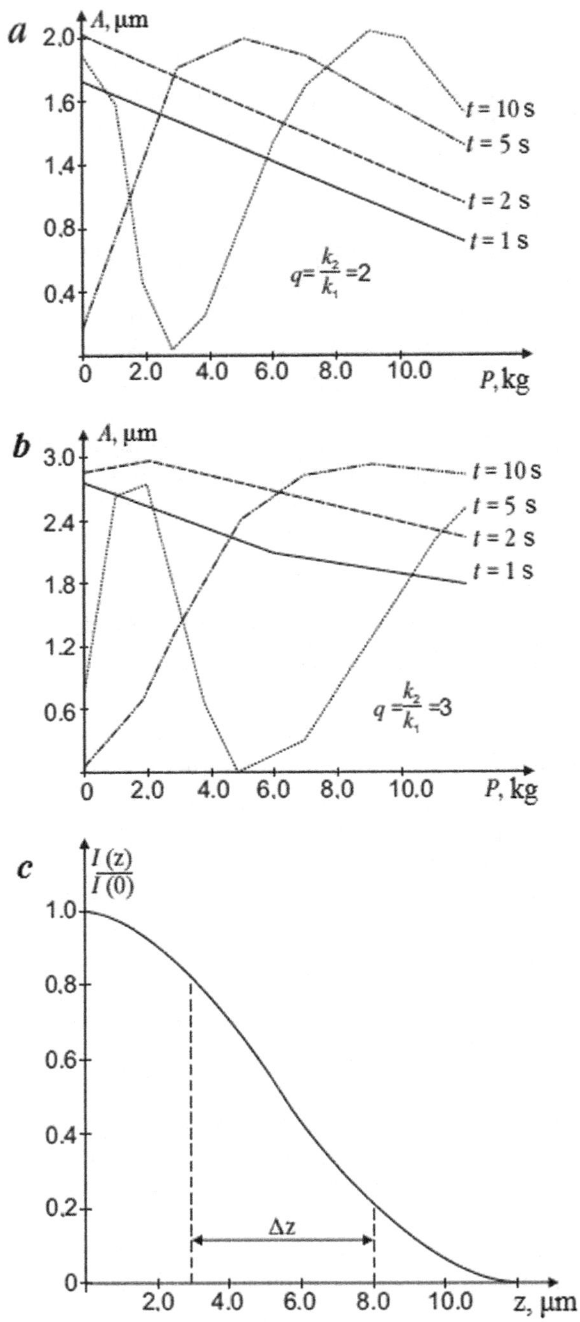

Fig. 5.13 Dependence of muscle contraction speed on applied load (**a**, **b**) and distribution of intensity of radiation diffused from the skin surface relative to z (**c**). Reprinted from [87] with permission

Piezoelectric, magnetic pickups and capacitive microphones, which are applicable of measuring of oscillation amplitude of ultrasonic transducer, are contact. They are sensitive to background impact in industry conditions and do not provide sufficient accuracy and spatial distribution during measurements.

Application of laser interferometers gives the possibility to avoid the above-mentioned disadvantages, but their usage for control of USWS oscillation amplitude requires additional treatment of the surface of the instrument to make it mirror. The structural–functional scheme of the laboratory prototype of the laser specklometer and the experimental scheme for research of the amplitude of the longitudinal component of USWS microinstrument vibration are shown in Fig. 5.14. The single-mode laser was used as a radiation source in the laboratory prototype of the laser specklometer. Its radiation was transmitted with interface optics of the matching device (MD) onto the face of the illuminating light guide (ILG), and then it went through the light guide into the illuminating–receiving sensor (IRS), using which the working face of the USWS microinstrument was illuminated.

The illuminating–receiving sensor IRS of the laser specklometer, light guides of which are oriented to the same area of the working face of the USWS microinstrument, was firmly fixed in such a way that the receiving light guide (RLG) was situated at a distance of 15 mm from the surface under study.

High optical-mechanical requirements are set to the illuminating–receiving sensor IRS, because the illuminating ILG and the receiving RLG should be directed to the same point of the surface under study. In a contrary case, the possibility of detecting the scattered radiation and IRS sensitivity decreases sharply.

In the laser specklometer, single-mode light fiber with a diameter of 4 mm was used as ILG and RLG of fiber optical light guides.

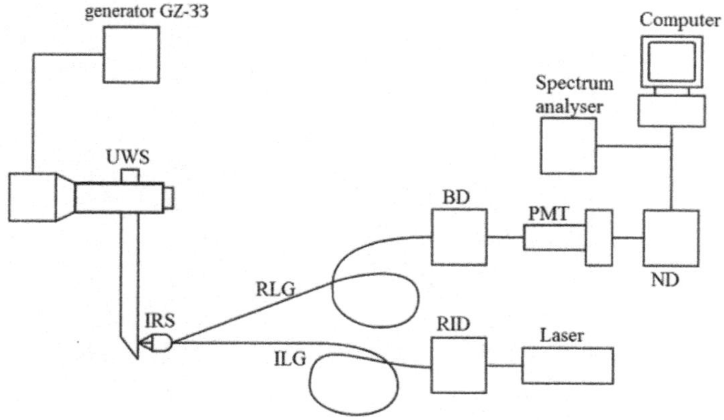

Fig. 5.14 Experimental scheme for studying the vibration of an ultrasonic welding system tool: USWS—the ultrasonic welding system; IRS—the illuminating–receiving sensor; RLG—the receiving light guide; ILG—the illuminating light guide; RID—the radiation input device; PMT—the photomultiplier tube; ND—the normalization device; BD—the bridging device. Reprinted from [87] with permission

It is known that the speckle lateral dimension was determined as $a = 1.22\lambda L/D$ where λ is the wavelength of the laser radiation and D is the illuminated plane. Using this expression, it is possible to calculate that $a \approx 4\ \mu\text{m}$. It is seen that the diameter of the light guide fiber and the speckle diameter almost coincide with each other. This makes it possible to detect a single signal, to provide maximal sensitivity of the measurement scheme and to observe the main oscillation harmonic only.

Sinusoidal electric signals are fed with set frequencies onto the USWS using the generator GZ-33. The frequencies were controlled by the frequency meter. The dynamic speckle-field scattered by the microinstrument surface was registered through the receiving light guide (RLG), and the radiation input device (ID) was registered and transformed into an electric signal using a photomultiplier tube (PMT). Then through a normalization device (ND), the signal went to the spectrum analyzer SK4-72 for processing. The amplitude A_1 of the first spectral harmonic of intensity fluctuation spectrum was registered. The spectrum was scattered by the surface of the working face of the USWS microinstrument, which makes harmonic oscillations. The value of the amplitude A_1 of the first spectral harmonic was converted to the value of vibration amplitude using the calibration curve obtained beforehand (Fig. 5.15). A normalization device of the signal received with the IRS was used for optimization of parameters of the measurement path of the laser specklometer. The structural–functional scheme of the normalization device of the receiving signal intensity is presented in Fig. 5.16.

Necessity of usage of the receiving signal normalization in the laser specklometer consists in maintaining a constant voltage at the input of the amplifier of spectrum

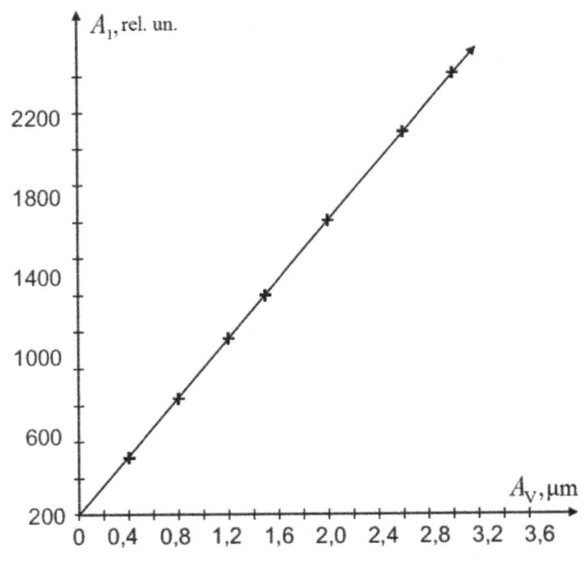

Fig. 5.15 Dependence of the amplitude of the first spectral harmonic A_1 on the amplitude of calibration vibrator oscillations A_V. Reprinted from [87] with permission

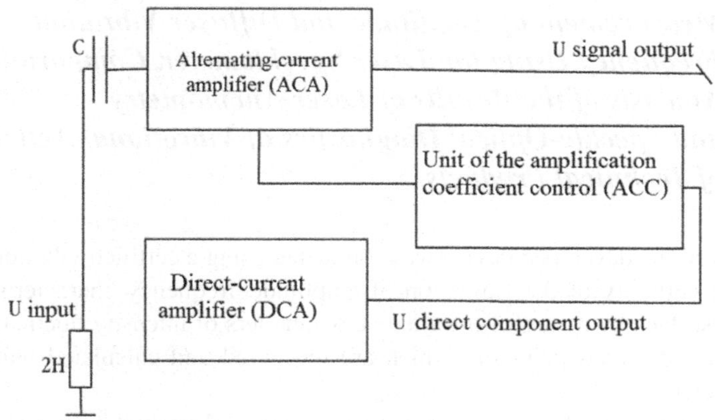

Fig. 5.16 Functional scheme of the device for input signal normalization in laser specklometer. Reprinted from [87] with permission

analyzer during scattered radiation intensity oscillations, which can occur in the processes of measurement due to the following reasons:

- change of the reflection power of the surface under study, changes of laser radiation power;
- change of geometry of the sensor-surface system, the inclination of the sensor, change of the distance between the face of the light guide and the surface under study during the process of measurement.

The receiving signal normalization device consists of three main functional units. An electric signal goes from the output of the photomultiplier tube (PMT) onto the input of the alternating component amplifier (ACA) with a regulated amplification coefficient through capacitor C (see Fig. 5.16), as well as onto the input of the direct current amplifier (DCA). The DC amplifier generates voltage for the unit of the amplification coefficient control (ACC) of the variable component amplifier. The normalized signal goes further from the output of the ACA onto the spectrum analyzer for processing.

The operating principle of the normalization device consists in the fact that the amplification coefficient is variable and changes depending on the intensity of the scattered radiation in the sensor reception area; i.e., there is a feedback. As a result, the output signal level remains almost unchangeable even under significant alteration intensity of the scattered radiation.

5.5.1 Measurement of Amplitude and Diffuser Vibration Frequency Using the Laser Specklometer. Comparative Analysis of the Results of Laser Anemometry and Speckle-Optical Diagnostics of Vibrational Activity of Technical Products

The study of the developed device was conducted using a certified vibrator for estimation of certainty of determination of amplitude–frequency characteristics of a vibrating surface by means of informative parameters of intensity fluctuation spectrum of the scattered radiation, which are registered and calculated with a laser specklometer.

The calibration of the specklometer was conducted through comparison of its results while measuring the amplitude of harmonic vibration of diffusing surface with the results of the model sensor in the form of a certified laser Doppler vibrometer. The electrodynamic vibrator RTF 11076 was used for the creation of harmonic vibration with a vibration exciter covered with a coating providing diffuse reflection of the laser radiation. The vibrometer was certified for the absence of transverse vibration components.

The experimental scheme of measurement of the amplitude and the frequency of the transverse component of diffuser harmonic vibration is shown in Fig. 5.17. The operating principle of the laser Doppler vibrometer is based on isolation of the Doppler shift of frequency of the laser radiation scattered by the vibrating object and on measuring the number of vibration periods.

Isolation of shift of frequency was conducted with a double-beam interferometer made according to a modified scheme of the Michelson interferometer.

The radiation scattered by object 5 was spatially superposed on the photoreceivers 7 and 8 with reference radiation reflected by the triple prisms 3. Detection of the Doppler signal was conducted with a balanced detector, which consists of the photoreceivers 7 and 8 and the differential amplifier 9 that gave the possibility to increase the signal/noise ratio at the expense of lowering the noise level due to the laser power fluctuations.

Alteration of vibroshift amplitude, which is determined with the method of interference fringes counting during vibration period according to the formula

$$A = \frac{1}{8}\lambda\frac{N}{n},$$

where λ is the wavelength of the laser radiation and N/n is the ratio of N periods of the Doppler signal to a certain number of vibration periods n.

This ratio was determined in the following way. The Doppler signal came from the output of the differential amplifier 9 to count the pulse former 10 build according to the Schmitt flip-flop circuit. The count pulses were fed from the outlet of the former 10 to one of the outlets of the frequency meter 13, which operates in the measurement mode of the ratio of two frequencies. Signal came from the generator

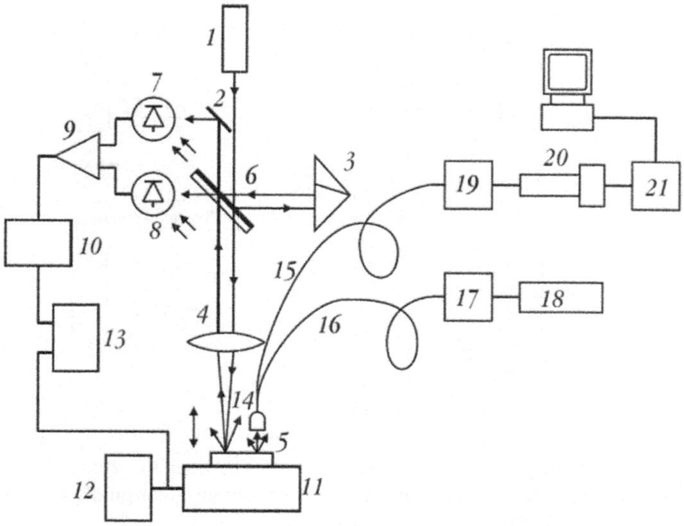

Fig. 5.17 Experimental scheme for measuring the amplitude–frequency characteristics of diffuser vibration: 1—the laser (M-207A); 2—the beam splitter; 3—the triple prism; 4—the focusing system; 5—the diffusely reflecting object; 6—the rotary mirror, 7, 8—the photoreceiver; 9—the differential amplifier; 10—the counting pulse shaper; 11—the electrodynamic shaker RTF 11076; 12—the signal generator (GZ-118); 13—the frequency meter (F 5137); 14—the light-receiving sensor, 15—the receiving light guide; 16—the illuminating light guide; 17—the conjugation device; 18—the laser (LGN-208A); 19—the device of radiation input; 20—the photomultiplier tube; 21—the normalization device; 22—PC. Reprinted from [87] with permission

12, which excites the vibrator 11 onto the second output of the frequency meter. The frequency meter 13 gives the possibility to determine the sought quantity N/n. The results of the frequency meter were multiplied by $\lambda/8$ for obtaining amplitude value A.

The estimated absolute error of measurement of amplitude vibration is 0.1 μm while measuring with the laser Doppler vibrometer. Amplitude–frequency characteristics of vibration were registered with a laser specklometer (Fig. 5.18) simultaneously with measurement using the laser Doppler vibrometer. The illuminating–receiving sensor 14 of the specklometer (Figs. 5.19 and 5.20) was firmly fixed in such a way that the receiving light guide 15 was at a distance of 5 mm from the illuminating light guide and at a distance of 15 mm from the diffusive covering of the vibration exciter surface. Laser radiation scattered by the surface under study was transmitted with the receiving fiber 15 through the radiation input device 19 to PMT 20, and signal from the multiplier came through the electronic path of the specklometer into PC 22 for processing and analysis. Fourier transformation of the obtained signals was conducted, and their amplitude–frequency characteristics (AFC) and power spectral density (PSD) were calculated.

Two kinds of illuminating–receiving heads were studied, which have some design peculiarities.

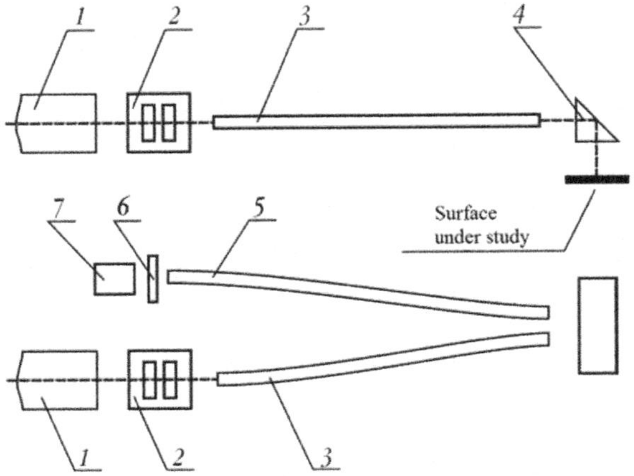

Fig. 5.18 Optical scheme of the laser specklometer: 1—the laser LG-208A; 2—the objective; 3—the transmitting light guide; 4—the prism; 5—the receiving light guide; 6—the light filter; 7—photomultiplier tube. Reprinted from [87] with permission

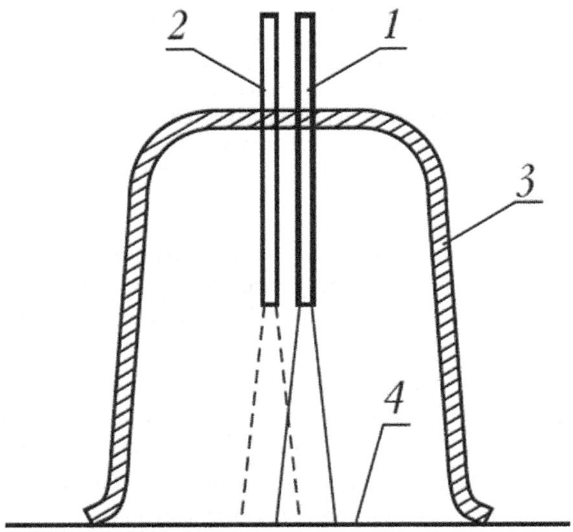

Fig. 5.19 Scheme of the first light-receiving head: 1—the illuminating light guide; 2—the receiving light guide; 3—the body; 4—the surface under study. Reprinted from [87] with permission

In the first illuminating–receiving head (Fig. 5.19), the illuminating–receiving light guides are approximately at the same distance from the facet of the head, and they are randomly oriented. In the other illuminating–receiving head, the receiving light guide is 5 mm farther in comparison with the illuminating one; moreover, the light guides are oriented to a point of the surface under study (Fig. 5.20).

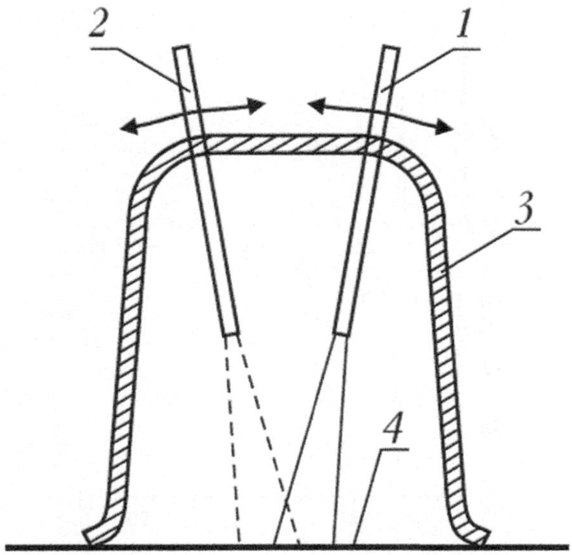

Fig. 5.20 Scheme of the second light-receiving head: 1—the illuminating light guide; 2—the receiving light guide; 3—the body; 4—the surface under study. Illuminating and receiving light guides have the ability to angular and longitudinal movements. Reprinted from [87] with permission

These heads were firmly fixed by turns at a minimal possible distance from the vibrating surface taking into account their design features. The measuring system was assembled on a massive basis for providing isolation from side sources of vibrations.

The first illuminating head was at a distance of ≈20 mm from the surface under study, and the receiving light guide of the second head was at a distance of ≈15 mm from the same surface.

As it is seen in Fig. 5.21, harmonics appear while using the first head in the spectrum apart from the main one. During the increase of the oscillation amplitude of the surface under study of more than 3 μm, the value of the amplitude of the main harmonic almost does not increase. Only redistribution between side spectral harmonics occurs.

Analogous PSD of vibrator oscillations with the frequency of 30 Hz and the same amplitudes for the second illuminating–receiving head are presented in Fig. 5.22. As it is seen in Fig. 5.22, in this case only one side harmonic appears under oscillation amplitude of 6 μm, but its value is much smaller than the value of the main spectral harmonic. And amplitude of the main harmonics increases monotonously with the growth of the oscillation amplitude of the surface under study; i.e., it adequately transmits information about changes of amplitude of vibrations under study.

The measurements were conducted for the following frequency oscillations of the surface under study of 25, 30 and 50 Hz. A series of amplitudes was set for each of these frequencies. And for these amplitudes, a vibrator was calibrated beforehand with the laser Doppler vibrometer.

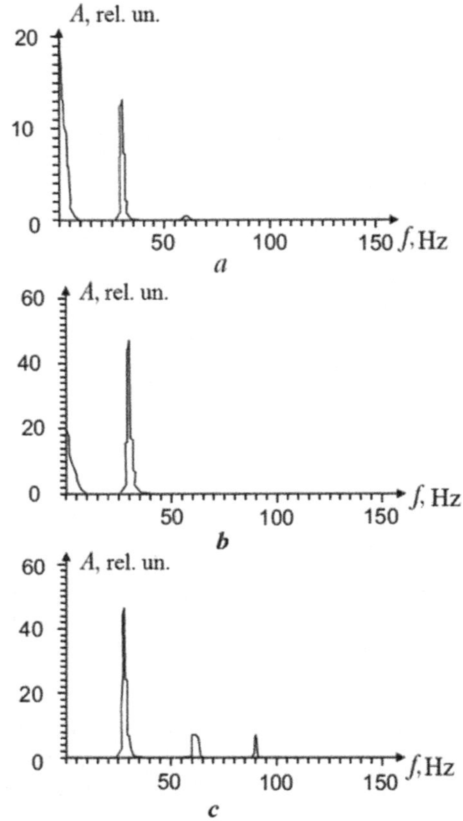

Fig. 5.21 PSD of vibrator oscillation with 30 Hz frequency and amplitudes: **a** 1.5 μm; **b** 3 μm, **c** 6 μm obtained with the first illuminating–receiving head. Reprinted from [87] with permission

Dependence of spectral density of the scattered radiation on vibrator oscillations on the frequency of 30 Hz with the amplitudes of 1.5, 3 and 6 μm is presented in Fig. 5.22. The dependence was obtained with the laser specklometer.

It is worth mentioning that if the oscillation amplitude of the surface under study increases then the amplitude of the main spectrum harmonic increases monotonously; i.e., it adequately transmits information about changes of amplitude of vibration.

Determination of vibration amplitude was conducted at different levels of the output signal of generator 12 (see Fig. 5.17), which corresponds to the determined ideal values A_V of vibration amplitude of the surface under study. It was determined through registration and calculation using the laser specklometer of amplitude of the first spectral harmonic A_1 in intensity fluctuation spectrum of dynamic speckle-field obtained by means of time averaging of a random signal over 16 realizations of this spectrum. The value of vibration amplitude measured with the certified laser Doppler vibrometer was taken for the ideal value A_V of the vibration amplitude.

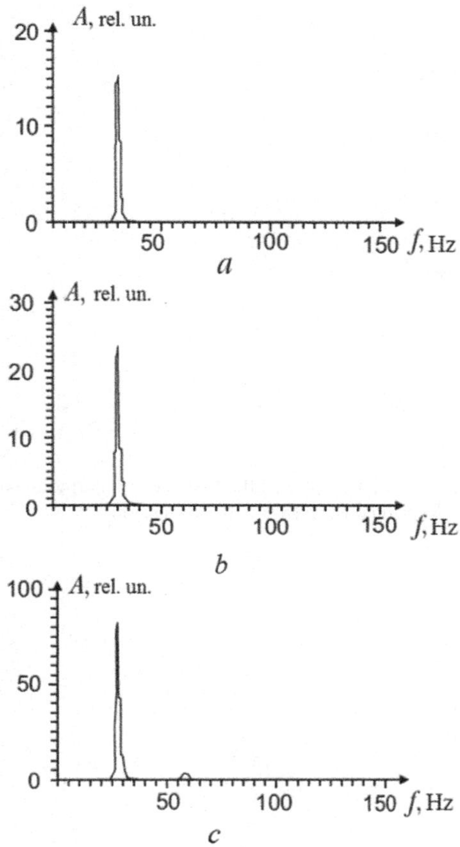

Fig. 5.22 PSD of vibrator oscillation with 30 Hz frequency and amplitudes: **a** 1.5 μm; **b** 3 μm, **c** 6 μm obtained with the second illuminating–receiving head. Reprinted from [87] with permission

Thus, 8 measurements were conducted. And the estimated relative error of amplitude measurement was not more than 5%. The conducted measurements gave the possibility to build a calibration plot shown in Fig. 5.23.

The experimentally obtained dependence of the amplitude value of the first spectral harmonic A_1 on the value of oscillation amplitude A_V of the calibration vibrator on the frequency of 30 Hz is presented in Fig. 5.24. The same dependencies were observed on frequencies of 25 and 50 Hz.

This illuminating–receiving sensor as a part of the laser specklometer can be used for analysis of oscillations with the amplitude up to 10 μm.

To broaden the range of linearity of the illuminating–receiving sensor, it is necessary to move the receiving light guide farther away from the surface under study in such a way that the maximal amplitude of the studied vibrations does not go beyond the limits of the linear area of intensity distribution in the longitudinal section of the

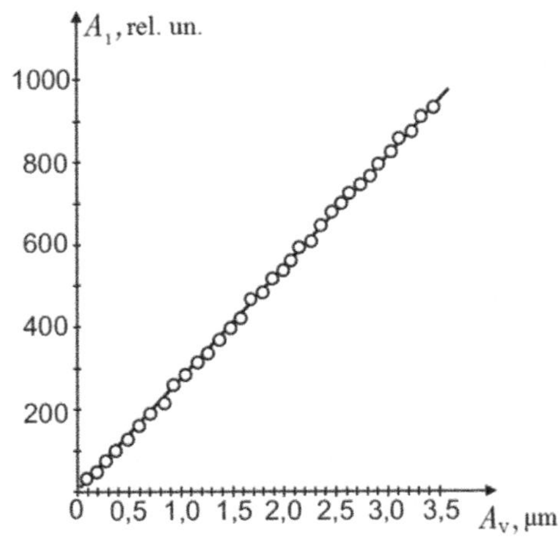

Fig. 5.23 Calibration curve of dependence of the first spectral harmonic amplitude A_1 on vibrator oscillation amplitude A_v. Reprinted from [87] with permission

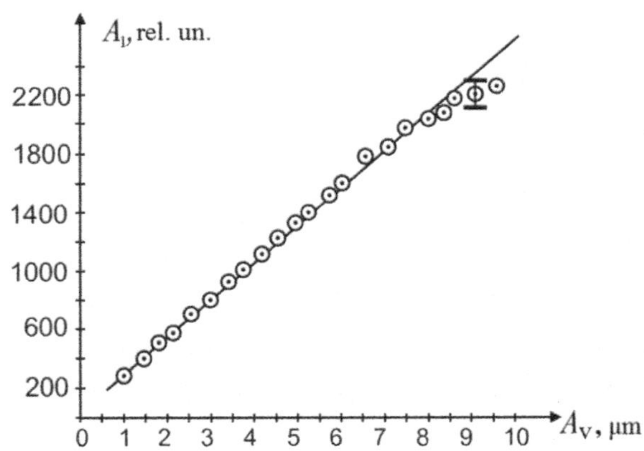

Fig. 5.24 Dependence of the oscillation of the first spectral harmonic A_1 from the oscillation amplitude of a calibrated vibrator A_8. Reprinted from [87] with permission

speckle. But meanwhile the intensity of the received signal decreases significantly that will lead to the necessity of using a more powerful radiation source.

Verification of the second controlled parameter f_0 is conducted due to the experimental scheme shown in Fig. 5.17. f_0 is the frequency of the main spectral harmonic calculated according to the speckle-field intensity fluctuations registered with a laser specklometer, and the speckle-field was formed during diffusion of the laser radiation from the diffusive reflector of the reference vibrator.

Sinusoidal electric signals were fed from the signal generator 12 onto the calibrated vibrator 11 serially on three fixed frequencies 4 in each of the set ranges of

15–50 and 100–2000 Hz with the amplitude that provide maximal amplitude of the diffusive surface of the vibrator 5 not more than 5 μm.

On each of the set frequencies of the calibrated vibrator, the registration and processing of the measuring signal were conducted with the laser specklometer and a computer and the f_0 value was determined. The calculation of the error of the measured frequency value f_0 from the frequency of the calibrated vibrator f_k showed that the relative error of measurement of vibration frequency in the range of 15–50 Hz did not exceed 4% and in the range of 100–2000 Hz was not more than 1.5%.

5.5.2 Study of Longitudinal Component Amplitude of Vibration of a Microinstrument of the Ultrasonic Welding System

The developed prototype of the laser specklometer gives the possibility to measure the longitudinal component of vibration of diffusive components that is of great importance for many tasks of vibrometry. Measurement of oscillation amplitude of details and units of aircraft engines, for example, of turbine buckets, with the laser interferometry method during their fatigue tests gives the possibility to study the vibration characteristics of the test specimens rather effectively: to determine the values of resonance frequencies, the beginning of product distortion according to changes of the longitudinal vibration amplitude.

Ultrasonic welding is one of the most widespread technological operations during assembling of products of electronic technology (integral schemes, hybrid microassemblies, etc.). Provision with high quality of connections during velocity welding on semiautomatic and automatic installations requires high accuracy of maintenance of parameters of the welding process. Load effort, welding time and oscillation amplitude of the microinstrument are the main parameters of the ultrasonic welding. But the greatest complication is the calibration and control of microinstrument oscillation amplitude during welding, and characteristics of the USWS (ultrasonic welding system) influence greatly its level.

Piezoceramic and magnetic pickups and capacitive microphones, which are applied for measurements of oscillation amplitude of the ultrasonic converters, are contactless and sensitive to background impacts in working conditions that does not provide sufficient accuracy and spatial distribution during measurements, and that is why they are not effective enough for calibration and control of the USWS.

Application of laser interferometers gives the possibility to avoid the above-mentioned disadvantages, but using them for control of USWS oscillation amplitude requires additional treatment of the surface of the instrument to make it a specular reflective one.

The presented laser Doppler vibrometer gives the possibility to make measurements of the USWS oscillation amplitude without preliminary treatment of the instrument. But the presence of angular or transverse vibration components can lead to diminution of amplitude of the interference signal and therefore to possibility that the signal amplitude can possess the values lower than the threshold level that will cause signal missing, will bear systematic error into the measurement result and will cause loss of information about the longitudinal vibration component. Laser specklometer application for measuring USWS oscillation amplitude made it possible to overcome this disadvantage of the laser vibrometer connected with measuring signal fading and to produce continuous measurements, because spectral features of intensity fluctuations of dynamic speckle-field are used as informative parameters. The working face of the microinstrument (Fig. 5.27) was illuminated with the coherent radiation from the laser through the interface unit (IU) and illuminating light guide 3. The dynamic speckle-field scattered by the surface of the microinstrument was registered through the headend light guide 5 and the radiation input device (ID), and it was converted into an electric signal with the photomultiplier tube (PMT). The signal came to the spectrum analyzer SK4-72 through the normalization device (ND) for processing. There were registered A_1 amplitudes of first spectral harmonic of intensity fluctuation spectrum of radiation scattered by the surface of the working face of the microinstrument that makes harmonic oscillations. The values of the amplitude A_1 of the first spectral harmonic were recalculated into values of vibration amplitude with a calibration curve obtained beforehand that is shown in Fig. 5.23.

The dependence of vibration amplitude of the working facet of the USWS microinstrument on vibration frequency in the area of working frequencies of 69–71 kHz was obtained with the laser specklometer.

Current of the ultrasonic transformer I_t was registered simultaneously with the measurement of vibration amplitude. The dependence of current on vibration frequency is of great importance as in the ultrasonic welding unit adjustment of the generator frequency to resonance frequency of transformer is conducted according to the maximal value of the current of the transformer. The results of measurements are presented in Fig. 5.25. The good coincidence of dependencies of transformer current and vibration amplitude on the frequency in the resonance region is observed that gives the possibility to make a conclusion about the appropriateness of the use of transformer current for adjustment of the USWS to resonance. Resonance frequency of vibration amplitude–frequency and transformer current is 70.2 kHz. Using the obtained amplitude–frequency characteristic of the USWS, the quality factor Q of the oscillation system can be determined:

$$Q = fp/2\Delta f \tag{5.22}$$

where $2\Delta f$ is the width of the resonance curve at the level of 0.707.

It is found from Fig. 5.25 that $2\Delta f = 0.17$ kHz, thus, $Q = 502$.

Dependence of the vibration amplitude of the working facet of the microinstrument on the transformer current is very important for designing the USWS. The obtained dependence is presented in Fig. 5.26. It is worth mentioning that it is linear. Vibration amplitudes were measured with the laser Doppler vibrometer [146–150].

Figure 5.26 shows dependencies of vibration amplitude on the frequency, which were obtained with the laser specklometer and the laser vibrometer, and give the

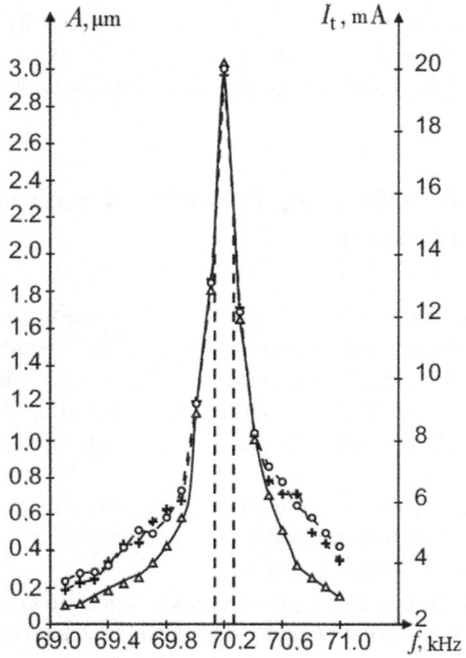

Fig. 5.25 Dependence of the amplitude A of working microinstrument threshold vibration and current of the transformer (I_t) vibration on frequency in resonance area: triangle—dependence I_t; multiplication symbol—dependence A obtained with the laser Doppler vibrometer; white circle—dependence A obtained with laser specklometer. Reprinted from [87] with permission

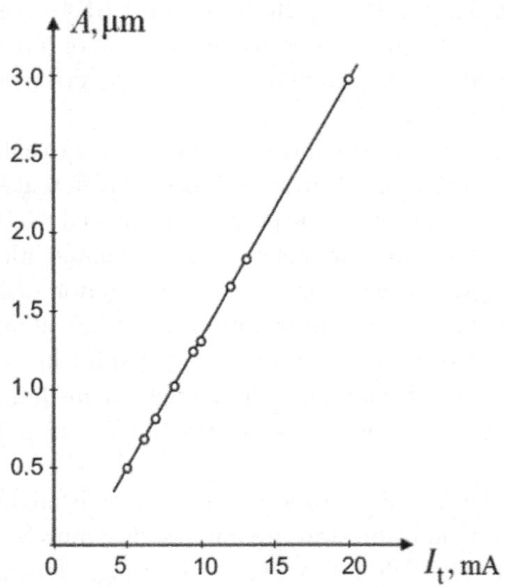

Fig. 5.26 Dependence of the amplitude A of working microinstrument threshold vibration on current of transformer (I_t). Reprinted from [87] with permission

possibility to judge about the good coincidence of the results. Besides unlike the laser vibrometer, the laser specklometer gave the possibility to register the vibration frequency of the working facet of the microinstrument along with the amplitude.

5.6 Technique for Obtaining Primary Information Using Laser Specklometer

Intensity fluctuation spectra of the scattered laser radiation from the human body surface are registered using the scheme (see Fig. 5.27b) for study of functional state of muscles using the laser specklometer (see Fig. 5.27a). The illuminating–receiving head (IRH) is laid on the studied body surface with a soft strap. Meanwhile, minimal pressure on the muscle is reached so that a normal physiological behavior is not broken in it.

The illuminated area was covered with the elastic white diffuser to achieve homogeneous conditions of diffusion in different experiments without the influence of blood flow surface while studying muscles.

The scattered radiation, which forms a speckle-field, is received by the receiving light guide, the aperture of which matches the speckle size and is transmitted into the photodetector. The amplifying signal is proportional to the random time measurement of intensity and corresponds to the functional state of the muscle under study (surface blood flow). The signal is analyzed with low-frequency spectrum analyzer (Fig. 5.28a) having the range between 0.02 and 2000 Hz (with further registration on a recorder) or with an electronic processing unit, or, finally, with a PC with an interface block (Fig. 5.28b). The spectrum analyzer SK4-72 gives the possibility to determine the full spectrum of intensity fluctuations of a dynamic speckle-field formed by the scattered radiation from the reflective covering on the skin above the muscle (Fig. 5.28c).

The analysis of the full spectrum gives the opportunity to calculate, for example, parameters such as deviation of spectral amplitudes at different frequencies and area under the spectral curve. The electronic processing unit and the PC with the interface block allow obtaining the same parameters, but in the automatic mode.

The muscle tone research program includes registration of intensity fluctuation spectra of the scattered laser radiation from the surface above the paravertebral muscles of coupling and thorax from both sides while sitting at rest and during tonic reactions: inspiration, head rotation to the right, to the left, down, bend over of the body back and forth for 30° and to the maximal degree, of gastrocnemius muscles of both legs by turn.

As an example in the standing position, there were registered intensity fluctuation spectra of the speckle-field from paravertebral muscles, muscles of the front surface of the lower one-third of a thigh (Fig. 5.29a), back surface of lower one-third of a shin (Fig. 5.29b), area of a hallux at rest and during isometric contraction of extremity muscles (Fig. 5.29c).

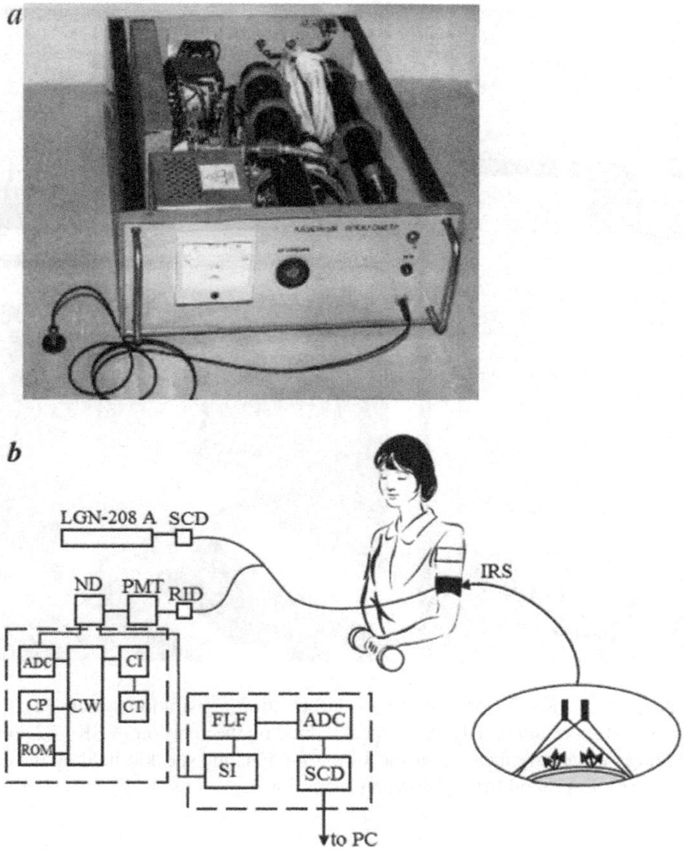

Fig. 5.27 Exterior view of laser specklometer (**a**) and structural and functional diagram of laser specklometer (the study of muscle tone) (**b**): LGN-208 A—He–Ne laser; IRS—the illuminating-receiving sensor, RID—the radiation input device, PMT—the photomultiplier tube, ND—the normalization device; ADC—the analog–digital converter; CP—the central processor, ROM—the read-only memory; CW—the common width, CI—the console interface, CT—the console terminal; FLF—the filter of low frequencies; SI—the signal intensifier; SCD—the sequential change device. Reprinted from [87] with permission

Tonic state of the muscles under study at rest on the sore and the healthy side as well as presence and degree of manifestation of the nearest and separate synergies were compared.

There were developed methods of research and analysis of the studied intensity fluctuation spectra of dynamic speckle-fields during operation of the laser specklometer both with a spectrum analyzer and with PC of AT286 type. Informative parameters of spectra were marked out, which reflect some biophysical processes, in particular alteration of microdynamics of skin and vibrotonic characteristics of muscles. These parameters became a basis for development of separate software variants for the laser specklometer.

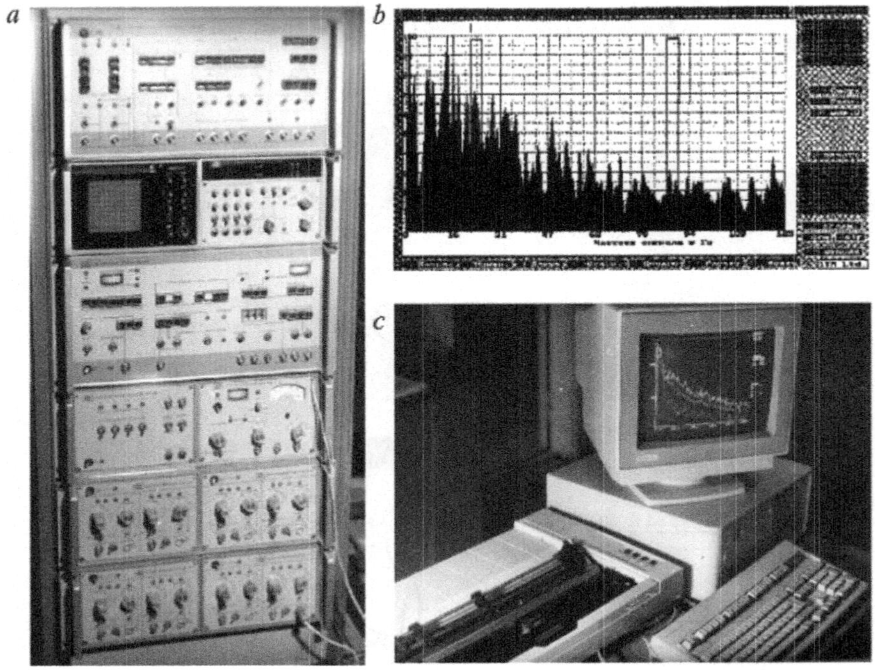

Fig. 5.28 Analyzer of SK4-72 spectrum type—information registration system (**a**), recording of intensity fluctuation spectrum of a dynamic speckle-field on the analyzer of SK4-72 spectrum type (**b**) and recording of intensity fluctuation spectrum of a dynamic speckle-field on the computer of PC AT 286 type (**c**). Reprinted from [87] with permission

5.7 Study of Intensity Fluctuation Spectra of Dynamic Speckle-Fields of Skeletal Muscles of Healthy People Obtained with the Laser Specklometer and the Speckle Analyzer

The calculations conducted in paragraph 5.4 showed that longitudinal oscillation amplitudes of a muscle do not exceed 5 μm. This gives the possibility to investigate longitudinal vibrations of human skeletal muscles during contraction.

In this paragraph, there are presented the investigation results of biomedical characteristics of human skeletal muscles. The experiments were conducted on a developed prototype of the laser specklometer. A series of investigations of contractive activity of human skeletal muscles was conducted in order to find most sensitive parameters of intensity fluctuation spectrum of a dynamic speckle-field, which bear information about the functional state of the muscles.

Motion (oscillation) of skin above the skeletal muscle during its constipation or movement of blood cells in the capillary network causes shift and boiling of speckles formed by the scattered laser radiation, and it represents just a pattern of intensity

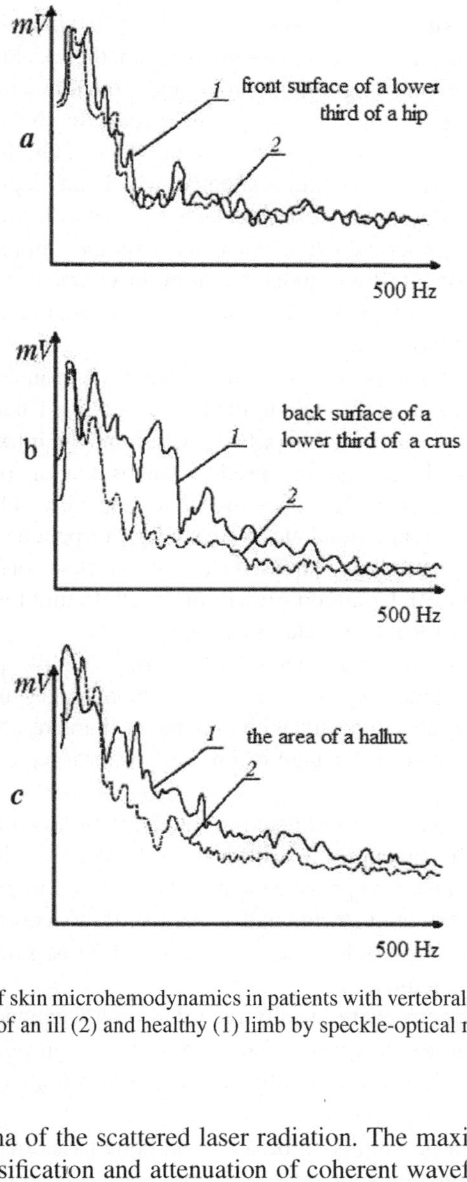

Fig. 5.29 Condition of skin microhemodynamics in patients with vertebral radiculitis S_I registered in symmetrical points of an ill (2) and healthy (1) limb by speckle-optical method. Reprinted from [87] with permission

maxima and minima of the scattered laser radiation. The maxima and minima are the results of intensification and attenuation of coherent wavefront with a random phase distribution.

A reflective coating is brought onto the surface of skin during the study of biomechanical parameters of skeletal muscles. The coating gives the possibility to estimate selectively contribution of oscillation movement of muscles to the total of intensity fluctuation spectrum of the speckle-field form while scattering the laser radiation from skin and microcirculation of blood in the skin. The scattered dynamic speckle-field bears information about the intensity of oscillations of the skeletal muscles, and thus about the degree of contractility, muscle tone, etc., as well as about the change of the velocity and intensity of the blood flow in each studied point of the surface with a distribution of 2–5 mm. Through a rotary microprism, the speckle-field gets onto the outlet face of the receiving single-mode light guide, the diameter of which matches the size of a solitary spectrum and is ~ 5 μm. The receiving light guide is connected with a sensitive photosensor by means of an input device. On the output window of the photosensor, an interference filter is situated, which intercepts spurious radiation (incandescent lamps, day light, etc.). An electric signal is formed on the output of the photoelectric multiplier, which is connected with the amplifier. The analysis of the signal gives the possibility to judge about biomechanical parameters of human skeletal muscles (degree of contractility and tone) and about velocity alterations and intensity of the skin blood flow.

Determination of signal speckles was conducted through analog processing with a spectrum analyzer as well as through digital PC processing. Changes of the speckle-field characteristics were analyzed in order to determine the informative parameters, which reflect the functional state of muscles. Values such as power of fluctuation spectrum, average frequency, frequency dispersion, half-breadth of a spectrum and amplitude ratio on different frequencies were studied. Dependence of maximal value of the spectral curve caused by a stressed state of muscles from the applied load is shown in the work [132]. Common growth of spectral amplitudes is accompanied by natural extension of the area under the spectral curve.

Most sensitive spectrum parameters, which have simple physical interpretation, are frequency characteristics, in particular, the average fluctuation frequency. Frequency parameters are connected with movement characteristics in a most simple way as frequency of intensity change is directly proportional to the velocity of the diffuser.

The average frequency was estimated in the range between 0 and 150 Hz as on higher frequencies the signal did not differ from noise during linear measurement. Moreover, the program was to provide the possibility of recording a random process into a file and organization of storing and activation of the recorded files, possibility of PC input of 1024 or 4096 dots and conduction of a fast Fourier transformation due to a full or partial realization.

Low-frequency spectra were usually recorded on a frequency band to 65.5 Hz at time of realization accumulation of about 8 s. The main purpose of the above-mentioned changes initially was to obtain reactive quantitative information about muscle tone.

Besides, there is a possibility to broaden the boundaries of method application and to use it for estimation of muscle relaxation time after removal of static load. This mode gives the possibility to record up to 12 spectrograms with different muscle loads and as a result to obtain graphic numerical dependencies of integral parameters of spectrograms on the exposure factor.

Preliminary investigation and most vivid distinctions in low-frequency spectrograms on different levels of muscle activity appear when frequency $f = 26$–26.5 Hz is chosen. As for the upper frequency f, its value is equal to the cutoff frequency of a low-frequency filter (50 Hz). A signal of intensity fluctuations of the speckle-field is recorded in a LF mode, which passes through the filter. The first two recorded spectrograms of 12 were obtained with a tense muscle of the right biceps, the other ones with a relaxed one.

The obtained spectrograms give the possibility to determine the time of muscle relaxation, as the last one was excited with a natural stress, but not with a large static load. Meanwhile, two important peculiarities can be mentioned. First, greatest contrasts display two characteristics—the average spectrum frequency and its asymmetry coefficient. Second (and what is most interesting), the process of muscle relaxation after its tension has a pulsating (wavy) character, and all the characteristics point at it.

Besides, the works [80–82, 115–117, 122, 123] showed the possibility of determination of relaxation time of biceps muscle when the measurement was conducted as an exception in the frequency range to 125 Hz. But the signal of intensity fluctuations from the multiplier ran through a low-frequency filter with a cutoff frequency of 50 Hz, so that the harmonics over the above-mentioned frequency suffered additional relief. Such a bit inconvenient experiment was explained by a desire to reduce the time of realization accumulation from 8 to 4 s. The calculation of specklogram frequency was also conducted in the band up to 125 Hz. Oscillation spectra of a biceps were investigated by healthy people at different measured isometric loads from 0 kg to the maximum for each person with discreteness of 2.5 kg. The subjects held the load of the needed mass in the hand in such a way that the angle between forearm and the shoulder was $\approx 115°$. The measurements were repeated many times to provide the credibility of the results (Fig. 5.30).

Let us examine the peculiarities of muscular spectrograms under different loads from 0 to 15 kg with a step of 2.5 kg. The study was conducted in the frequency band up to 62.5 Hz. The methods of measurements consisted in sequential load of a half-bent right arm and recording of corresponding specklograms of the biceps. It was shown in paragraph 5.3 that $\langle f \rangle$, K_a and K_b show different characters of load dependency. As far as we can see, it can be caused by two reasons: either by biophysical peculiarities of work of the muscle itself or by the nonlinearity of the device mentioned above. For the latter case, a rather clear explanation can be

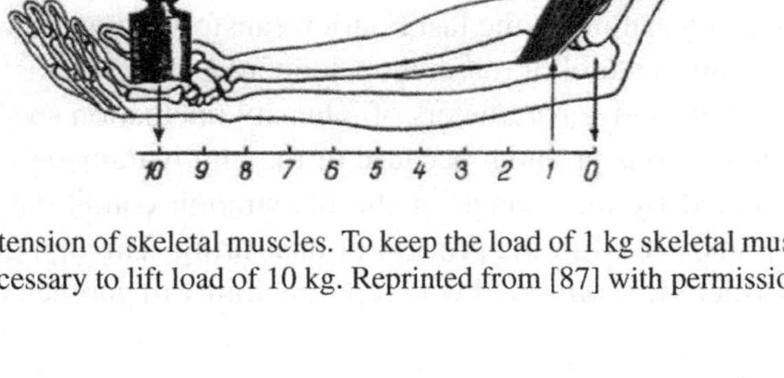

Fig. 5.30 Static tension of skeletal muscles. To keep the load of 1 kg skeletal muscle must contrast with the force necessary to lift load of 10 kg. Reprinted from [87] with permission

suggested if the muscle is presented as a simple model of a string system. In this case, if the skin oscillation amplitude does not exceed the limits of the longitudinal radius of the speckle with load increment, then gradual tension of the string muscle should lead to smooth increase of frequency of muscle "humming" and as a result to smooth dependence of the average frequency. But histograms of all the three integral characteristics examined above point to surges of muscle tone under some loads. In accordance with the accepted model, it can be assumed that load increment must cause increase of the number of involved muscle fibers that leads to enhancement of the oscillation amplitude of the resonant skin above the longitudinal average speckle radius. But if the amplitude of skin contractions is equal, for example, to a doubled average longitudinal speckle radius, then pulsations of the light field (and thus of the electrical signal from the photoreceiver) will occur with a double frequency of skin oscillations; with the amplitude, which is equal to three longitudinal radii; with a tripled one, etc. These frequencies are named "nodal" for definiteness.

It should be mentioned that in our case consideration of harmonics above the third one does not have any sense as the recorded signal passes through a LF filter with the cutoff frequency of 50 Hz first. Pulsation amplitudes of the electrical signal, which is concentrated on the mentioned nodal frequencies, are maximal and proportional to the contrast of the speckle-structure, and they amount to maximal contributions into the average frequency during loads corresponding to the occurrence of these nodal frequencies. During all the intermediate loads, the electrical signal has intermodal frequency values, but along with this it has a strongly pronounced phase modulation. The latter causes spectrum smearing of its energy with the width of signal frequency order, which will appear on the graph of the average frequency in the form of a dip between two surges. If the considered presentation is true, then the presence of maxima on dependences of the average frequency on the load is evidence of firmly fixed amplitudes of skin oscillations, which are easily calculated basing on the longitudinal size of a speckle. But this aspect requires additional study. It is worth mentioning that analysis of the informative content of spectrograms and search for informative parameters is very complicated [151, 152]. It is known that traditional methods of estimation of spectral characteristics of processes are based on the application of Fourier transformation either directly to their realizations (Cooley–Tukey's method) or to statistic estimates of autocorrelation functions of time dependencies (Blackman–Tukey's method). Meanwhile, resolution and accuracy of the obtained estimations are usually limited with the finite duration of the used realizations of the process or with a finite interval of values of autocorrelation functions, respectively. Special smoothing functions [153] are usually used for decreasing the negative impact of process truncation. Namely, the rational choice of the latter ones is of the greatest interest as algorithms of the fast Fourier transformation are well known and have become a traditional tool of researchers long ago.

The choice of informative parameters of intensity fluctuation spectra of the scattered radiation is a serious problem because of the multiparameter character of the current task specified by the change of the registration conditions as well as by the object of the study itself in the process of measuring. The first group of factors should include difficulties connected with reproducibility of geometry of experiment during repeated mounting of the illuminating–receiving sensor into the point under study (another area of skin and therefore distance from its surface to the face of the light guides) as well as changes of geometry of the experiment at different degrees of contraction of the skeletal muscle, etc. The second large group of conditions includes distinctions of reflectivity of different skin areas, impact of the physiological state of a patient, etc.

As the given measurements of medical–biological objects in vivo are unstable and have a great confidence interval, special attention should be paid to the maximal possible repetition of identity of conditions of the experiment as well as the set of the needed statistics. Also, this reason explains the usage of integral informative parameters mostly. The structure of the curves obtained in the same conditions is not preserved; meanwhile, their more conservative integral characteristics are significantly more resistant.

Taking into account the stated uncertainties of measurements as the informative ones, it is reasonable to choose not absolute, but relative indices of spectrograms, which are normalized to the zero harmonic, maximal spectrogram amplitude or to the area of the spectral curve, etc., and to base methods of diagnostics on the comparative principle.

One of rather evident-registered experimental facts is the increase of amplitude of spectral harmonics; moreover, this change is observed in the area of low as well as higher frequencies.

The dependence of maximal value on the applied load is shown in particular in the works [80–82]. The common growth of spectral amplitudes is accompanied by the growth of the area under the spectral curve, respectively.

The calculation of the area was conducted for the primary spectra as well as for the smoothed spectra. Simultaneously, the same study was conducted with an electromyography (EMG). The calculation of the spectrum was conducted in the frequency range of 0–62.5 Hz due to the realization of 1024 dots (sampling time is 8 s).

The increase of the width of spectral harmonic (sectionalization) was applied to enhance stability of the obtained results of measurements during spectrum calculation, and also the averaging of spectra of several consecutive signal selections was conducted.

The use of the enumerated methods requires the length either of the selection or of their number that finally leads to the increase of time of measurement. But as the results showed time of measurement is limited by the capability of a testee to hold the maximal loads. If load P was 10 kg, this time did not exceed 1.5–2 min for such muscles as a biceps. Meanwhile, the number of realizations was chosen to be 10 in the range of 62.5 Hz. The previous chapters have shown that calculations and model experiments for longitudinally oscillating diffuser gave the possibility to use the area under the spectral curve (spectrum power) as an informative parameter of the spectrum, which characterizes oscillating activity of human skeletal muscles:

$$S = \int_{f_\mathrm{L}}^{f_\mathrm{U}} A(f)\mathrm{d}f, \tag{5.23}$$

where $A(f)$ is the spectral amplitudes, and f_L and f_U are the lower and the upper frequencies of the analyzed spectral range.

The changes of amplitudes of spectral harmonics caused by longitudinal vibrations of the muscle during contractions are accompanied by the changes of the area under the spectral curve, respectively. The areas under the primary and the smoothed curves were almost equal.

The frequency parameters of oscillations of the studied skeletal muscle are determined by movement characteristics, as the frequency of change of intensity fluctuation is directly proportional to the change of the velocity of the diffuser movement. The average spectrum frequency in the band from f_L to f_U was calculated as

$$\langle f \rangle = \frac{\int_{f_\mathrm{L}}^{f_\mathrm{U}} f A(f)\mathrm{d}f}{\int_{f_\mathrm{L}}^{f_\mathrm{U}} A(f)\mathrm{d}f}, \tag{5.24}$$

where $A(f)$ is the spectral amplitude on frequency f.

The band coefficient K_b (ratio of the sum of high-frequency harmonics in the frequency band from f_3 to f_4 to the sum of low-frequency ones in the band from f_1 to f_2) was calculated as

$$K_\mathrm{b} = \frac{\int_{f_3}^{f_4} A(f)\mathrm{d}f}{\int_{f_1}^{f_2} A(f)\mathrm{d}f}. \tag{5.25}$$

As a result of the study, there were determined spectrum parameters, which are the most sensitive to changes of the physiological state of muscle; they are power and the average frequency of the spectrum. It was ascertained that power and the average frequency of the spectrum grow up to 80% from the maximal effort of a patient with the increase of the isometric load applied for the biceps of the upper arm. Further load increment leads to decrease of the muscle activity that is proved by lessening of power and the average frequency of the spectrum. The same results were described in the literature for sound myography [111, 154–156].

The changes of intensity fluctuation spectra of the speckle-field, i.e., the area under the spectral curves during different isometric loads, are shown in Fig. 5.31a, b.

The changes of power spectral density (PSD) and spectrum power (SP) from the value of the applied isometric loads are presented in Figs. 5.32 and 5.33. To

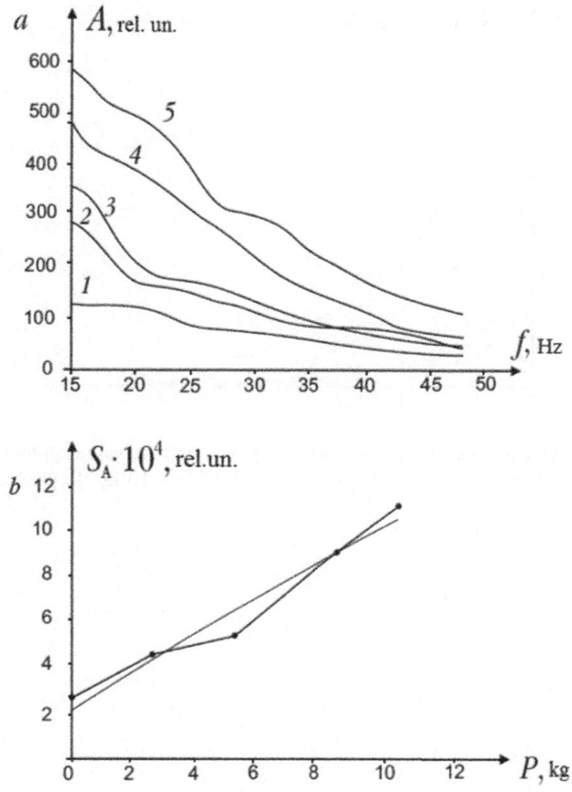

Fig. 5.31 Dependence of intensity fluctuation spectra of the speckle-field on load value, applied to biceps muscle of arm (**a**: 1—without load; 2—2.5 kg; 3—5 kg; 4—7.5 kg; 5—10 kg) and dependence of area under spectral curve on the load value, applied to biceps muscle of arm (**b**). Reprinted from [87] with permission

improve reliability of results of the experiment 10, intensity fluctuation spectra of the speckle-field were registered for each measured value of the isometric load.

Simultaneously with the conduction of this research using the electromyograph, bioelectric activity of the biceps of the upper arm under study was registered. The dependence of the bioelectric activity of the muscle on the applied load is presented in Fig. 5.34. It is seen in the figure that bioelectric activity of the muscle (electromyogram) grows in proportion to increment of the isometric load up to maximal muscle force.

Thus, as the study showed, the laser specklometer gives the possibility to obtain additional information about oscillatory activity, which cannot be obtained with the help of the electromyograph.

The analysis of spectra made it possible to determine a number of biomechanical characteristics of human skeletal muscles. The growth of amplitudes of spectral harmonics is observed in the given frequency range with increment of the applied

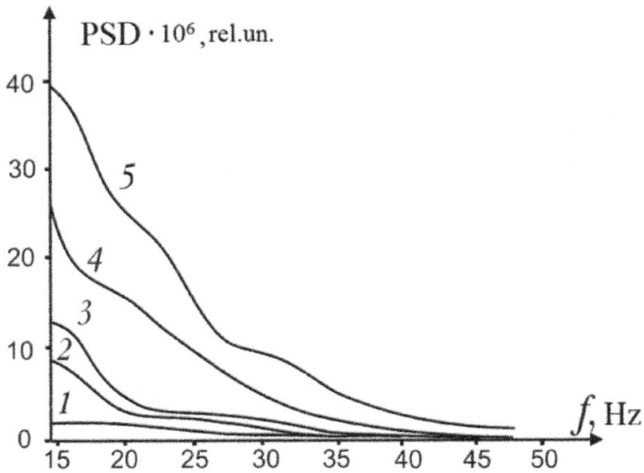

Fig. 5.32 Dependence of spectral density and PSD power on load value applied to biceps muscle of arm: 1—without load; 2—2.5 kg; 3—5 kg; 4—7.5 kg; 5—10 kg. Reprinted from [87] with permission

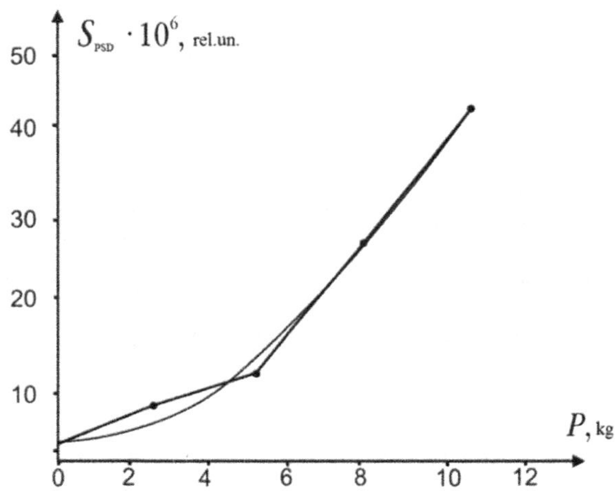

Fig. 5.33 Dependence of spectrum power S_{PSD} on load value, applied to biceps muscle of arm. Reprinted from [87] with permission

load, which determines tonic state of the biceps of the upper arm, i.e., degree of its contractility. The increase of harmonics is accompanied by extension of the area under the spectral curve (power of the spectrum), respectively.

As a result of the conducted study, it was ascertained that the power of the spectrum increases monotonically up to 80% from the maximal muscular power for each of the subjects. Further growth of isometric load leads to decrease of oscillatory activity of the muscle that seems to be connected with fatigue that is proved by lessening of spectral amplitudes and spectrum power.

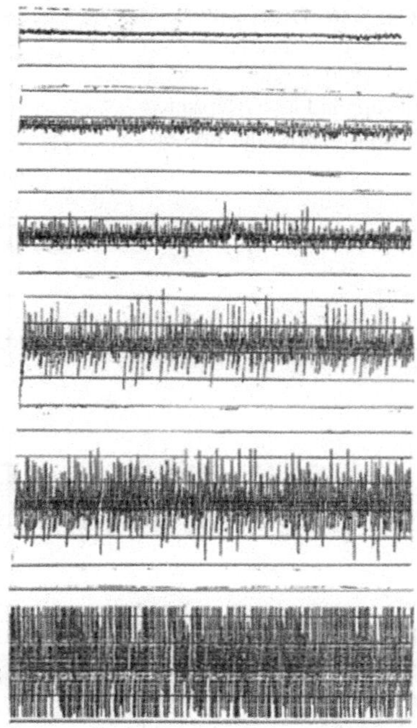

Fig. 5.34 Dependence of on the applied load. Reprinted from [87] with permission

Typical plot of dependence of spectrum power of speckle-field intensity fluctuations on the applied load is presented in Fig. 5.35.

For comparison, the dependence of the power of the registered electromyograms on the value of the isometric load is presented in Fig. 5.36 that corresponds to the results obtained for sound myography [113, 114].

Estimation of the average frequency $\langle f \rangle$ of the spectrum was conducted in the selected frequency range of 8–62.5 Hz.

Dependence of the average frequency of fluctuations on the load is a sensitive characteristic of speckle-optical measurements, which is presented in Fig. 5.37. As it

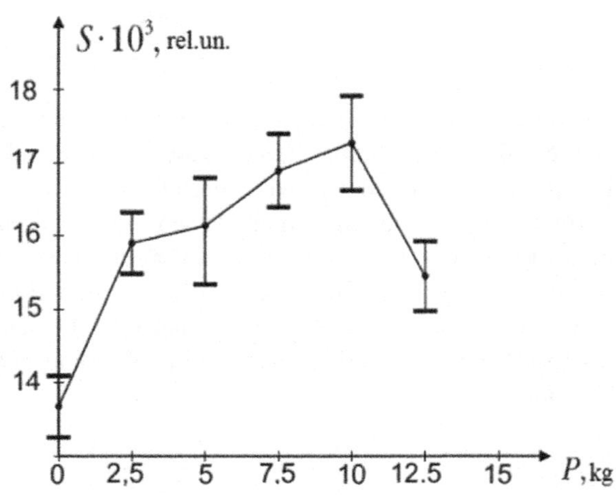

Fig. 5.35 Dependence of spectrum power of speckle-field intensity fluctuations on the load rate applied to biceps muscle of arm. Reprinted from [87] with permission

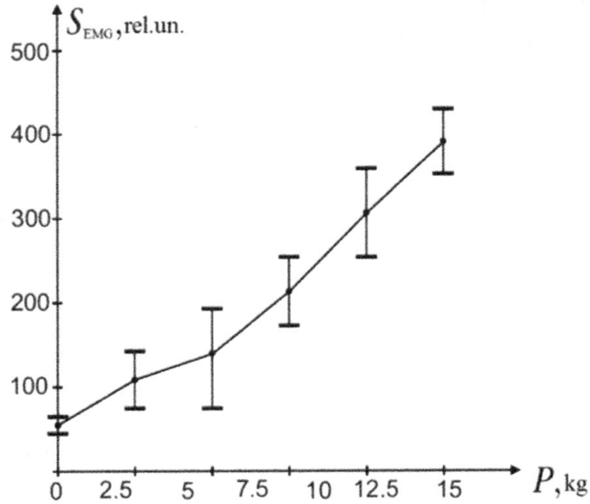

Fig. 5.36 Dependence of power of registered electromyograms S_{EMG} on the applied load. Reprinted from [87] with permission

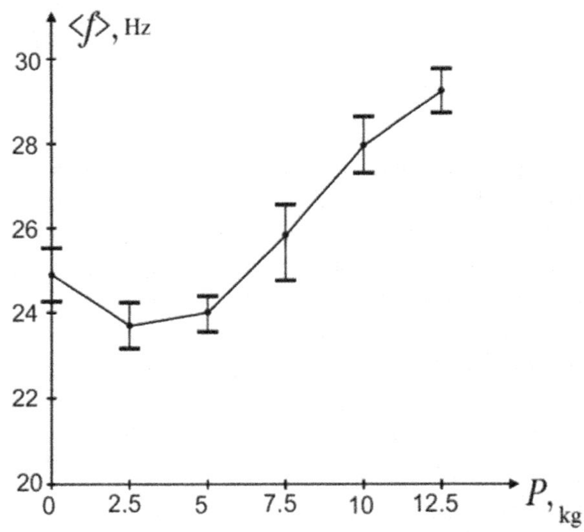

Fig. 5.37 Dependence of average frequency $\langle f \rangle$ of intensity fluctuation spectra of the speckle-field on load value applied to biceps muscle of arm. Reprinted from [87] with permission

has already been mentioned, it is obvious that while using the informative parameters connected with amplitude characteristics of the signal, the value of the signal should be carefully controlled to provide reproducibility of results, for example, to introduce normalization. To do this, a method was offered and a device for normalization of the intensity of the signal received from the sensor was developed, which can be used for the following intensity changes of the scattered radiation as a result of changes of the reflection power of skin, geometry of the sensor-surface system, power of laser radiation, etc., during investigation of biomechanical characteristics of human skeletal muscles.

Besides, the ratio of the estimated value of the oscillations under study and intensity distribution in longitudinal section of the speckle should be taken into consideration.

If the estimated maximal amplitude of the fluctuations under study does not go beyond the linear area of intensity distribution in longitudinal section of the spectrum, then the given informative parameters will describe the amplitude characteristics of oscillations taking into account corresponding frequencies of the spectrum. For example, spectrum power used during the study of contractive activity of human skeletal muscles showed with a sufficient degree of reliability and credibility the changes of amplitude characteristics of the vibrating biceps by the testee at changes of the value of the applied isometric load. Relaxation time characterizing the fatigue characteristics of muscles can also serve as an informative parameter.

This parameter was determined through registration and analysis of spectra after relieving stress from the muscle caused by the applied physical load.

Along with this, it is possible to determine informative parameters $\langle f \rangle$, S, Q not for the whole spectral curve, but for its separate specified parts, for example, local maxima, i.e., to change limits of integration. Search for most optimal conditions of selection and determination of informative parameters was conducted in the process of the study. It will be shown further during description of the experiments how these informative parameters work during the study of dynamics of the surface blood flow and contractive activity of muscles in different conditions. In particular, it is reasonable to use integral characteristics of spectra: square S, average frequency F, etc. It is obvious that it is also reasonable to complement the mentioned integral characteristics with moments of higher orders $\int f^n A(f)\mathrm{d}f$ as well where $n = 2.3$.

As an additional parameter, the dispersion can be used as well (or standard deviation of time dependence of the signal, which characterizes the average power of intensity fluctuation in the observation plane). Besides, it is also sensible to add to the above-mentioned characteristics some indicators of nonlinearity of transformation with a device of pulsation spectrum of the skin into intensity fluctuation spectrum. They can be in the form of coefficients of overtones of the main frequency of muscle contractions. It is known that the muscle "buzzing" effect is observed at frequencies of about 23 Hz, so the overtone coefficients can be taken into consideration as well. When the applied isometric load is intensified up to the maximal effort developed by the biceps of the upper arm, the increase of the average fluctuation frequency within the limits of 23.6–23.9 Hz is observed. This dependence can be explained by representing the muscle as a simple model of a string system [111, 154–156].

Thus, the conducted studies gave the possibility to develop a noninvasive method of determining biomechanical characteristics of human skeletal muscles—a method of speckle-optical myography. This allows obtaining additional information about oscillatory activity of muscles and degree of their contractility through the registration of amplitude–frequency parameters of intensity fluctuation spectra of radiation scattered by skin. This information cannot be obtained with traditional method of electromyography.

That is why the developed method can be rather informative for diagnostics of neuromuscular diseases, as it is known that disturbing factors of any genesis, both endo- and exogenous ones, realize their pathological action in different ways and influence first of all the contractile apparatus of somatic muscle fibers [157].

The developed methods and the determined informative parameters became a basis for the development of some variants of software for the laser specklometer.

5.8 Development of Diagnostic Speckle-Optical Criteria for Estimation of Skin Microhemodynamics

Contemporary methods of coherent optics give the possibility to visualize the surface blood flow without contact. Estimation of the skin blood flow is important while ascertaining the diagnostics as well as during individual therapy of a number of diseases, for example, polyneuropathy, angiopathy, arterial thrombosis and other disorders.

In clinical practice, the role of non-contact methods cannot be overestimated as they do not require sterilization. Such works are conducted in Germany, Japan, the USA and a number of other industrialized countries and directed first of all to ascertaining qualitative characteristics of the blood flow [44, 75, 84, 158–160], which are necessary for the development of diagnostic criteria.

An overwhelming number of researchers conduct experiments on determination and measuring of microcirculation in one single point. The only exception is the work [75] where two-dimensional pictures of the blood flow are made with electromechanical scanners in combination with registration with a one-dimensional DAS rule.

Development of the technology of photoreceivers and telemetry intensified with the power of contemporary PCs gives the possibility to begin development of sensitive noninvasive systems of determination of the surface blood flow. The systems can visualize great areas and make tests in all the necessary points. Observation of distribution of microcirculation and following the dynamics of this distribution under a number of physiotherapeutic means on the one hand will definitely favor the development of new methods of diagnostics, and on the other hand it is aimed at optimization and acceleration of the process of treatment. In a number of cases, observation of the skin blood flow is also complicated as it is sensitive to changes in temperature and depends on the state of skin and pressure on the skin. That is why the determination of the actual blood flow without any influence of the measurement itself can be possible while using the non-contact noninvasive speckle-optical method.

The use of the speckle-optical method for studying microhemodynamics in skin excludes sterilization, allows probing any place on the skin with obtaining objective results and enables to make measurements in patients with tactile hypersensitivity and skin vulnerability who suffer from peripheral circulatory disorders, different dermatoses and skin allergic onsets, and eliminates the risk of contracting both transcutaneously transmitted diseases and the human immunodeficiency virus (HIV).

The use of optics of speckles for studying processes in living tissues gave the possibility to obtain information about mobility of the scattering particles and their velocities. Laser radiation penetrates into skin, which is full of capillaries into the depth of 500 μm and scatters back. It forms a dynamic speckle-pattern, and its velocity of change depends on the intensity of microcirculations [44, 75, 84]. Though at such depth, there are many small blood vessels, in particular the mobile blood cells (first of all erythrocytes), which take part in forming of the dynamic speckle-structure. Small depth of penetration gives the possibility to estimate selectively the surface blood flow in comparison, for example, with ultrasonic study, which estimate the data averaged from the depth of 1 cm [84].

He–Ne laser radiation of several milliwatts is transmitted through a single- or multimode light guide to provide the flexibility of the experiment. The skin area under study is illuminated either with focused or with divergent radiation, and in the latter case the fiber is situated close to the surface so that illuminance is too small (Fig. 5.38). The measurements were conducted from the backside of a palm, in the nail bed, finger bones, surface of the neck, legs, etc. [84]. The scattered radiation is received directly by a photomultiplier tube or through the receiving light guide united with a transmitting one [44, 86] into a pair and in further experiments with the DAS rule with a scanning device [75].

The electrical signal, which corresponds to random time change of intensity, is intensified and investigated with spectrum analyzer CK4-72. The spectra differ slightly for different parts of the body.

The mechanism of this spectrum formation is rather complicated. As a result of a small difference in the index of refraction of cells and their rather big sizes compared

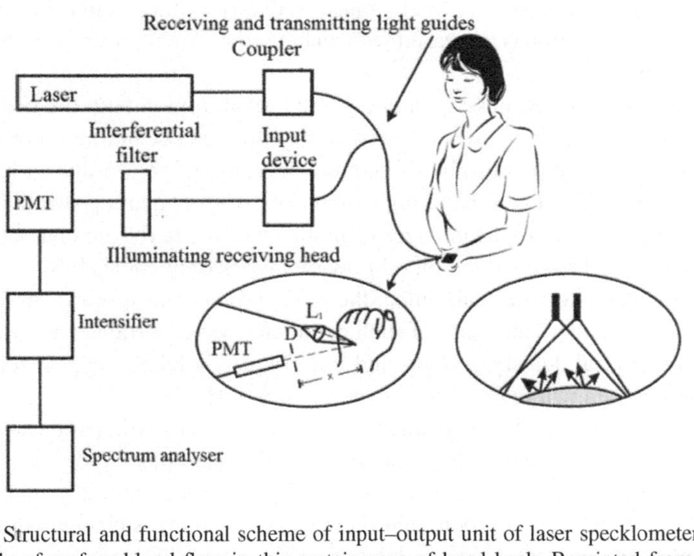

Fig. 5.38 Structural and functional scheme of input–output unit of laser specklometer radiation in the study of surface blood flow in this certain case of hand back. Reprinted from [87] with permission

to the wavelength in the tissues, forward scattering prevails and only a small amount of the light (0.1%) is diffused backward with red blood cells [44, 45, 84, 161].

In this case, the light scattered from deeper areas of skin does not reach the inverse skin surface as a result of high absorption. This is the main reason why the measurements were conducted with He–Ne laser as it gives the possibility to estimate namely the capillary blood flow [84]. Movement of erythrocytes illuminated with a coherent light source emits fluctuations of the light scattered from the capillaries. Frequency of fluctuations of the scattered irradiation is shifted relative to frequency of irradiations of the laser at the value proportional to the velocity of the blood flow and depends on the orientations of capillaries relative to the direction of the inclined and scattered light. The light scattered from the motionless tissues is not shifted in frequency [162]. Typical boiling frequencies of the speckle-structure formed by the laser irradiation scattered of the surface tissues are in the range from 0 to 10 kHz and change at a different blood filling [44].

As far as we are concerned, there are two peculiarities, which give this possibility to divide motions of speckles caused by the blood flow and shift of the body.

Firstly, the study of microcirculations is usually concentrated not on determination of absolute indices of the blood flow, but on their temporal changes as a result of any tests. Meanwhile, the muscular tone and other factors, which influence involuntary body movements, are supposed to be stationary and do not influence the measured value. A typical example of such dynamics is the procedure of reactive hyperemia. Its essence is in the fact that parameters of a normal blood flow are estimated from the spectrum, then the artery is compressed, the blood flow stops, measurements are conducted (in this case, the spectrum is practically the same as the spectrum formed by the shift of the body), and finally the cuff is removed. The blood flow increases dramatically and then sinks gradually to the normal level. In other studies, microcirculation is increased with an injection of tuberculin or with warming ointments [163].

Secondly, blood cells make common translational motion with the body surface as well as they have their own velocity and on the whole, unlike other diffusers, their velocity distribution must be shifted into the range of great values. Fluctuations formed by this diffuser will have high frequency. Experiments prove that spectra, which correspond only to the changes of mobility of the body, are characterized by lower frequencies [164] and the blood flow by higher frequencies [78].

Resulting characteristics can reflect the inclination of the spectral curve in some way [44]. In particular, changing of such parameter was ascertained as a ratio of the level of the spectral density at high- and low-frequency $\text{HLR} = f_{\text{L}}^2 f_{\text{U}}^2$ for a test of reactive hyperemia where $f_{\text{L}} = 40$ Hz and $f_{\text{H}} = 640$ Hz.

The study was conducted on change of blood flow intensity in patients who suffer from arterial occlusion before and after an operation [163].

The use of multi-channel registration and digital processing of signal gives the possibility to obtain the microcirculation map, which reflects changes in the blood flow intensity in the area of skin under study [75, 86].

Methods of measurements. The measurement was conducted with an experimental setup shown in Fig. 5.38. In a series of conducted experiments, laser irradiation illuminated a small area of skin of 2×2 mm^2 through a light guide path.

The scattered irradiation was gathered with the second light guide and transmitted to the photomultiplier. The photoelectrical signal from PMT, which is proportional to the irradiation intensity, was digitalized with a 12-discharge ADC and recorded in the form of integral numbers into a file, which includes 4096 numbers. Signal measurement accuracy was 1/2000 of the average signal level. Methods of measurements and working principle of the device were described in more detail in the works [80, 81]. Essential distinction is that the frequency range for blood flow determination is broadened considerably into the high-frequency area as the blood flow is characterized by a signal of higher frequency [78].

The measurements were conducted 10 times in each point. Thus, the total number of measurements in each point was 40,960. Then, the sensor was shifted with a step of 10 mm in accordance with the net marked on the inside of the hand. In such a way, the area of 300 mm^2 was scanned.

Two series of experiments were conducted, in one of which the blood flow was normal and in the other one it increased as a result on therapeutic impact (warming ointment "Finalgon" was rubbed into).

Random characteristics of the speckle signal. The speckle signal is formed as the result of coherent composition of many wave elements scattered on moving particles. In our case, such elements are cells, which form skin, and blood cells, and erythrocytes are in the first place. Modeling of coherent dispersion for simple macroscopic homogeneous systems shows that the average velocity of particle movement is proportional to fluctuation frequency. Today, there are no adequate models, which describe light dispersion from the tissue of complex models such as biological tissues. Muscle tissue along with the adjoining skin suffers constant non-regular microvibrations conditional on the tremor, contraction of separate fibers and groups of muscles, but along with this they also have their own velocity relative to vessels. The average velocity of erythrocyte movement appears to be higher than the velocity of shift of covering tissues. Part of the light scattered by the blood cells is smaller than the light scattered by skin as the density of erythrocytes is less than the density of covering tissues.

However, both theoretical estimation [78, 84] and experimental data [75, 84, 132–134] obtained by foreign authors and by us give the possibility to assert with confidence that movement of blood cells affects greater in the area of high-frequency fluctuations.

Thus, processing of a random signal, which reflects intensity of a scattered speckle-field, should be directed to the analysis of change of high-frequency components in the spectrum. From this point of view, the analysis of the whole spectrum is optimal. Informative parameters of the spectrum can be diverse [41, 44, 85, 132–134, 163] though as it mentioned above none of them reflects directly either the number of blood particles or their velocity because of complexity of the physical model. This connection is of correlation character and is often ascertained during tests. The following parameters can be mentioned: the average frequency, half-breadth of the spectrum, frequency dispersion and moments of higher orders, average fluctuation power, ratio of spectral harmonics from different frequency ranges, correlation time, median, density maximum and distribution modality, correlation time, number of surges, density of maxima and minima, characteristic distribution function, coefficients of asymmetry, excess, measure of skewness, etc.

The choice of parameters, which are optimal and most sensitive to changes of surface blood flow, is a rather laborious computational task. Analogous studies conducted for determination of the tone of the muscle tissue showed that fluctuation power suffers the greatest alteration. But the high-frequency spectrum tail, which is changeable during amplification/abatement of the blood flow, embraces a small area. Dynamic range of the change of the spectrum area is about 10% of the whole spectrum area. As characteristic frequencies of boiling the speckle-structure of the scattered irradiation change under different blood filling along with μ parameter, which represents the ratio of the spectral density on high and low frequencies [44, 164], in our case it was calculated as ratio of areas of small zones $\Delta f = 10$ Hz near the set frequencies $f_1 = 50$ Hz and $f_2 = 610$ Hz then

$$\mu = \frac{\int_{f_2-\Delta f}^{f_2+\Delta f} A(f)\mathrm{d}f}{\int_{f_1-\Delta f}^{f_1+\Delta f} A(f)\mathrm{d}f} \tag{5.26}$$

The following parameters were suggested for estimation of dynamics of blood flow changes:

1. The average spectrum frequency

$$\langle f \rangle = \frac{\int_{f_L}^{f_U} f A(f)\mathrm{d}f}{\int_{f_L}^{f_U} A(f)\mathrm{d}f} \tag{5.27}$$

   where f is the frequency; *A*(*f*) is the spectral amplitudes; f_L and f_H are the smallest frequency (40 Hz) and the biggest frequency (2000 Hz) of the analyzed spectral range, respectively.
2. Coefficient of asymmetry relative to the average frequency

$$A_s = \frac{\int_{f_L}^{\langle f \rangle} A(f)\mathrm{d}f}{\int_{\langle f \rangle}^{f_U} A(f)\mathrm{d}f} \tag{5.28}$$

3. Correlation of the average spectrum frequency to asymmetry

$$Q = \frac{\langle f \rangle}{A_s} \tag{5.29}$$

The study of microhemodynamics of skin in different functional states was conducted using the developed sample of the laser specklometer.

5.9 Investigation of Microhemodynamics of Human Skin Using the Speckle-Optical Method and Obtaining Microhemodynamic Maps

The study of microhemodynamics of skin in different functional states was conducted with the laser specklometer. The scheme of testing is presented in Fig. 5.38.

The studied skin area on the back of the hand was illuminated with the laser irradiation. The blood flow intensity was changed with an inflatable cuff, which was preliminary put on the forearm. The spectrum was calculated in the range of 40–2000 Hz with the realization on 4096 dots. The intensity fluctuation spectra of the speckle-field reflecting the state of the microhemodynamics of skin in different functional states are shown in Figs. 5.39 and 5.40. Spectrum A corresponds to the normal blood flow. Spectrum B was registered in 120 s after stopping the blood flow using the inflatable cuff.

Lowering of the blood flow intensity leads to suppression of high-frequency components in the registered spectra. Spectrum B corresponds to the state of the

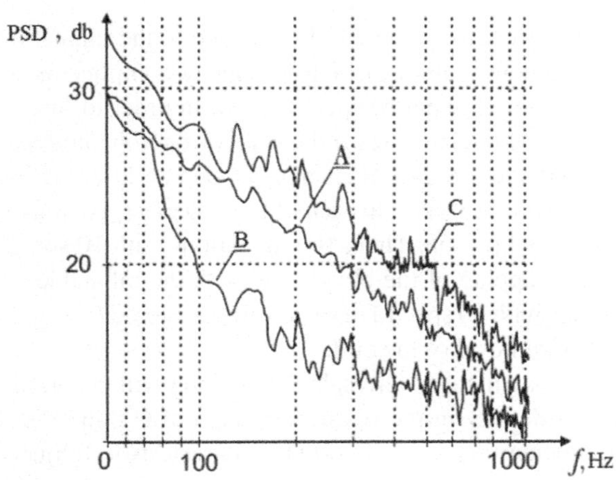

Fig. 5.39 Intensity fluctuation spectra of the speckle-field at different functional states of skin: A—normal, B—cross-clamping with inflatable cuff; C—after removal of clamping cuff. Reprinted from [87] with permission

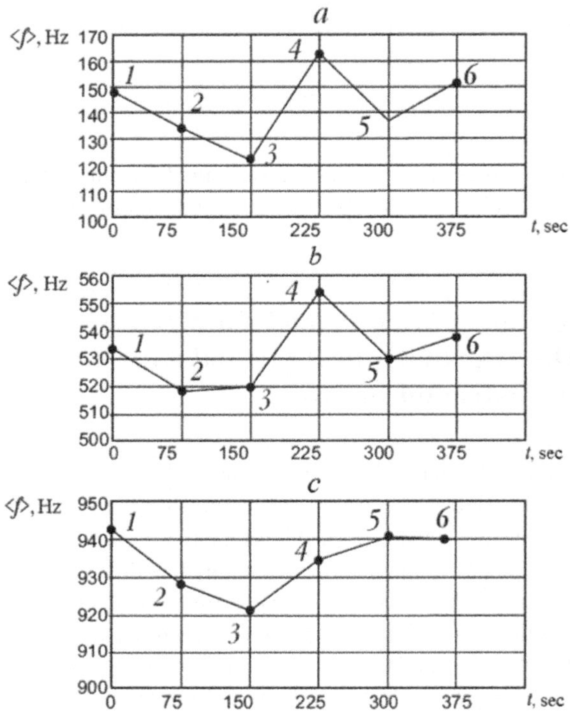

Fig. 5.40 Frequency characteristics of skin blood flow (⟨f⟩ intensity fluctuations of the speckle-field) at hypoxic probe: 1—normal; 2, 3—cross-clamping with inflatable cuff; 4—compression removal; 5, 6—blood flow recovery; frequency range is 0–2000 Hz (**a**) 125–2000 Hz (**b**); 500–2000 Hz (**c**). Reprinted from [87] with permission

microhemodynamics from the moment of removing of the inflatable cuff. A significant increase of the contribution of high-frequency components was observed in the fluctuation spectrum of dynamic speckle-field. In order to find new informative parameters, which reliably characterize the state of microhemodynamics of human skin, there were conducted studies of changes of skin at different fluctuation times.

Ten realizations with 10 spectra in each of them were registered during 180 s that corresponds to a normal person. Then, 10 realizations with 10 spectra in each were fixed in 120 s after stopping of the blood flow with the inflatable cuff ($t = 180$ s). Then, 20 realizations with 10 spectra in each were registered ($t = 360$ s) immediately from the moment of removal of the cuff.

New informative parameters were calculated along with a parameter such as the ratio of the spectral density in the frequency range of 600–620 Hz to the spectral density in the frequency range of 40–60 Hz (characteristic frequencies of boiling of the speckle-structure of the scattered irradiation at normal blood filling): average spectrum frequency $\langle f \rangle$, spectrum asymmetry A_s relative to the average frequency $\langle f \rangle$ and ratio of the average frequency $\langle f \rangle$ to asymmetry A_s, i.e., Q.

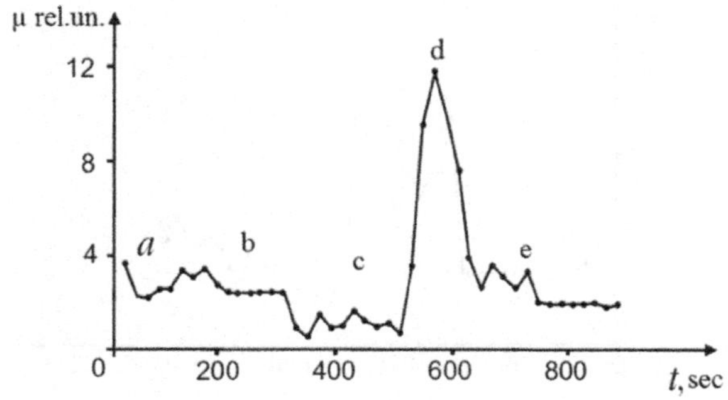

Fig. 5.41 Dependence of μ parameter on surface blood flow at different functional states of skin: a—normal; b—inflatable cuff pumping; c—cross-clamping; d—cuff removal; e—blood flow recovery. Reprinted from [87] with permission

The changes of the suggested informative parameters depending on the state of microhemodynamics of skin in different functional states can be seen in Figs. 5.41, 5.42, 5.43, 5.44 and 5.45. The following fact should be mentioned that while lowering intensity of the skin blood flow lessening of parameters $\langle f \rangle$, μ and Q is observed and the value of parameter A_s grows.

Maximal relative error of measurements does not exceed 10% for each of the suggested parameters. While registering μ, $\langle f \rangle$ and Q, immediately after removal of the inflatable cuff there were observed a sharp increase of the mentioned parameters and a decrease of A_s due to blood emission into capillaries of skin. Then, gradual lowering of quantities $\langle f \rangle$, μ and Q and growth of A_s was observed to quantities, which correspond to a normal blood flow. Thus, one can make a conclusion that application of the given parameters gives the possibility to estimate a certain degree of efficiency of the state of microhemodynamics of human skin.

The study was conducted on making microhemodynamic maps of the skin blood flow. The studied area of the skin of 40 × 50 mm^2 on the back of the patient's hand

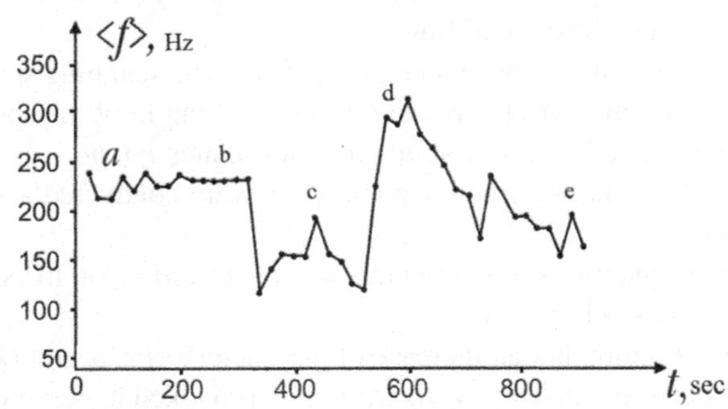

Fig. 5.42 Dependence of average frequency $\langle f \rangle$ spectrum on blood flow change at different functional states of skin: a—normal; b—inflatable cuff pumping; c—cross-clamping; d—cuff removal; e—blood flow recovery. Reprinted from [87] with permission

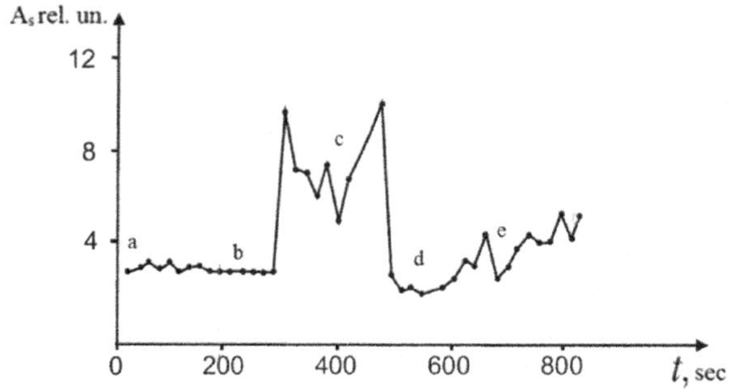

Fig. 5.43 Dependence A_S spectrum skewness on surface blood flow change at functional states of skin: a—normal; b—inflatable cuff pumping; c—cross-clamping; d—cuff removal; e—blood flow recovery. Reprinted from [87] with permission

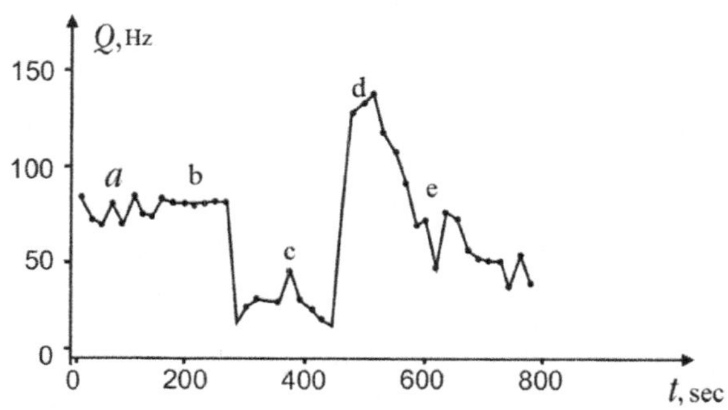

Fig. 5.44 Dependence of Q parameter on surface blood flow change at functional states of skin: a—normal; b—inflatable cuff pumping; c—cross-clamping; d—cuff removal; e—blood flow recovery. Reprinted from [87] with permission

(Fig. 5.45) was scanned with the receiving–illuminating head of the laser specklometer. And the suggested informative parameters $\langle f \rangle$, μ, Q and A_s were registered, which reflect the state of the blood flow.

The scanned area of skin is presented in Fig. 5.45. The scanning step was 10 mm. Then, artificial hyperemia was provoked through rubbing in of the special ointment "Finalgon" on the studied areas of skin with coordinates *bb*, *bd*, *cb*, *cd*, *db*, *dc*, *dd*. And measurements of the informative parameters were conducted according to the mentioned method.

Distributions of quantities of parameters μ, $\langle f \rangle$, Q and A_s on the studied area of the skin are presented in Fig. 5.46.

It is worth mentioning that an increase of parameters $\langle f \rangle$, μ and Q and decrease of parameter A_S occur on the areas with artificially provoked hyperemia. Quantity of changes of the informative parameters depends considerably on skin hyperemia, i.e., on the state of microhemodynamics of skin on the given area. The parameter Q has the greatest sensitivity of skin that can be connected with multidirectional character

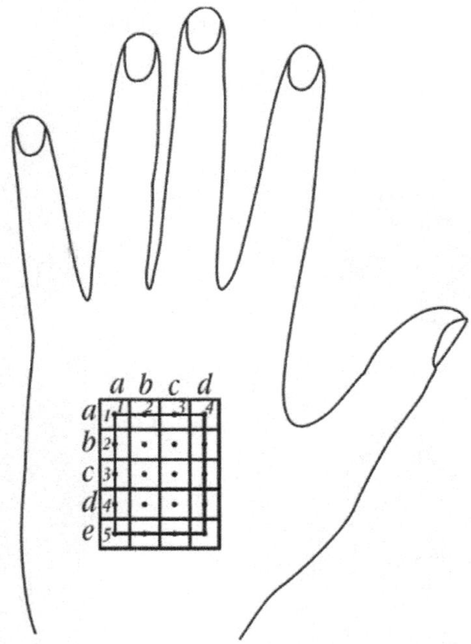

Fig. 5.45 Experimental skin area ($40 \times 50\,\text{mm}^2$) on hand back. Reprinted from [87] with permission

of changes of quantity $\langle f \rangle$ and A_s during changes of microhemodynamics of skin. Maximal relative error of parameter changes did not exceed 10%.

The values of the calculated informative parameters, which correspond to the studied dots on the area of skin, gave the possibility to estimate rather effectively the state of the surface blood flow in the normal state and during artificially provoked hyperemia of the skin area of $40 \times 50\,\text{mm}^2$.

5.10 Study of Biomechanical Parameters of Skeletal Muscles in Patients with Diseases of Peripheral Nervous System

Muscle-tonic syndromes, which are accompanied by vertebragenous diseases of peripheral nervous system, cause lingering algesias and disorders of motion activity and lead to limitations of the ability to work. Regarding muscle-tonic syndromes by neurologic onsets of lumbar osteochondrosis (NOLO), medical policy is directed either to stimulation of muscles, which realize local muscular fixation and perform the duties of a muscular collar, or to elimination of muscle-tonic disorders at the stage of progressing. The following items are not enough examined: topographical diagnostics of groups of muscles, which provide this overtone, quantitative assessment of

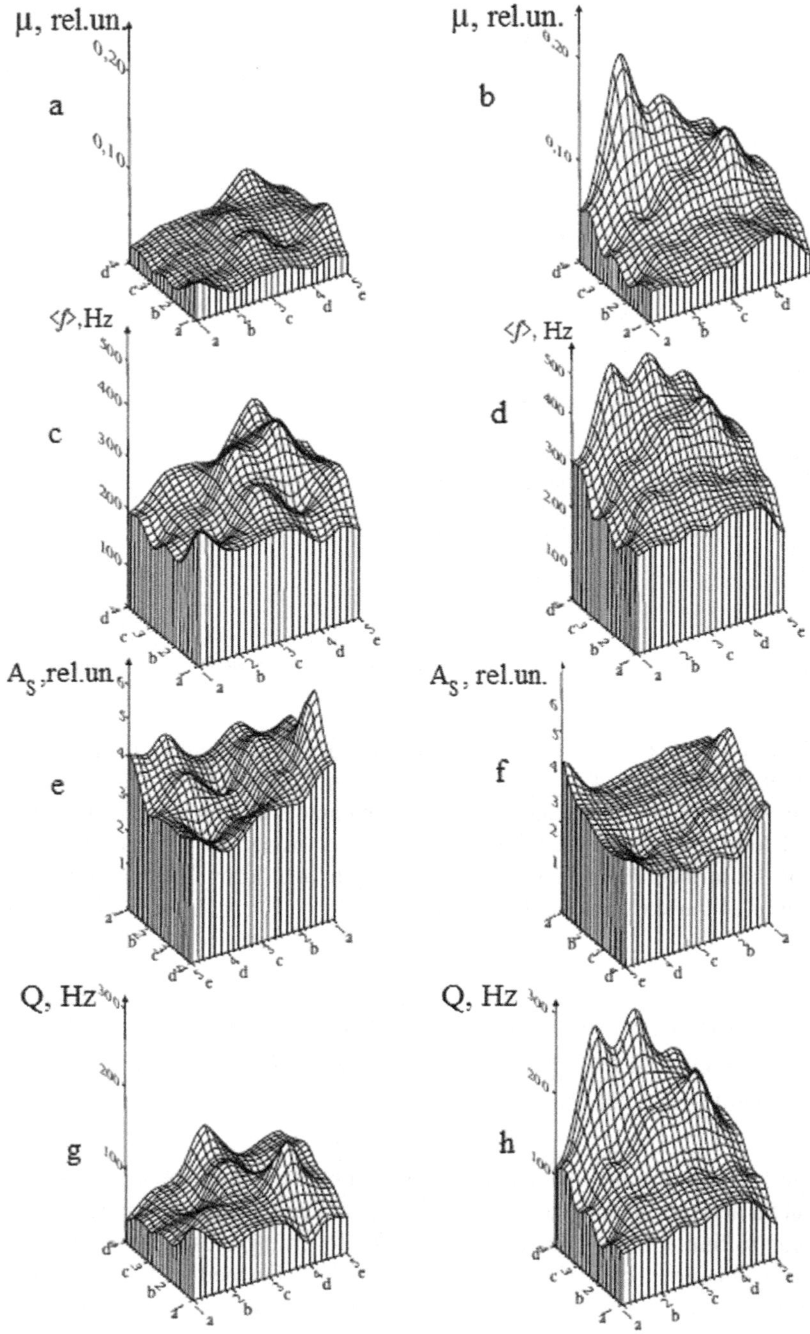
μ, rel.un.
a
b
⟨f⟩, Hz
c
d
A$_S$, rel.un.
e
f
Q, Hz
g
h

◂**Fig. 5.46** Dependence of μ parameter on state of experimental skin area surface blood flow: a—normal; b—hyperemia. The dependence of average frequency $\langle f \rangle$ spectrum on the state of experimental skin area blood flow; c—standard; d—hyperemia. The dependence of A_s spectrum skewness on surface blood flow of experimental skin area: e—standard; f—hyperemia. The dependence of Q parameter on the state of surface blood flow of experimental skin area; g—standard; h—hyperemia. Reprinted from [87] with permission

degree of its manifestation as well as objectification of its dynamics in the process of treatment. Traditional methods of investigation of muscle pathophysiology are either imperfect like myotonometry or connected with nociceptive impact on a patient like stimulating isometric mechanomyography. And electromyography (EMG), which is most widely applicable in investigation of patients with neurological diseases, gives a limited image about contractive muscle activity and characterizes only indirectly the main function of muscles—the contractile one.

The study of these questions with speckle-optical laser myography method and comparison of the obtained results with clinical, electrotonometric and electromyographic data give the possibility to conduct a deeper analysis and a more adequate interpretation of changes of tonic and contractive functions of muscles by diseases of peripheral nervous system.

At angiopathies of a different etiology and vegetovascular form of neurological manifestations of spine osteochondrosis, the study of microhemodynamics of skin with noninvasive speckle-optical methods gives the possibility to assess the degree of manifestation of trophic disorders and to objectify efficiency of treatment according to blood flow dynamics.

Seven hundred and fifty speckle-optical myograms were registered while investigating the muscle overtone in patients with neurological manifestations of spine osteochondrosis accompanied by a moderate or a full-blown pain or muscle-tonic syndromes. The registration of intensity fluctuation spectra of scattered laser irradiation was conducted from the surface over the paravertebral muscles of lumbar and thoracic spines from both sides, muscles of the back surface of a thigh and gastrocnemius muscle of the right and the left legs. The study was conducted with a sitting and standing person, at rest and at voluntary contractions of extremities and at tonic reactions. The tonic condition of the observable muscles at rest and at voluntary contractions on the diseased and sound sides was compared; besides, the presence and the degree of manifestation of the nearest and remote synergies were detected. For the mechanical measuring of the tone, mechanical myotonometer and electrical myotonometer were used. Electromyographic activity was registered with the electromyography "Diza." In this process, there were taken into account the value of the amplitude of the oscillations, asymmetry of this indicator, qualitative assessment: presence of the activity of fasciculation type at rest and during tonic reactions, appearance of the pathologic electromyograms (II a and II b type), short grouping of the increased amplitude "volleys" (electromyogram of III type), appearing of the increased synchronization. The state of microhemodynamics of cutaneous integument was detected in symmetrical points of diseased and healthy limbs; these results were compared with clinical trial data.

The mathematic processing of 200 specklograms was carried out in order to develop diagnostic speckle-optical criteria of the muscle tone condition changes and the disorder of dermal microdynamics. For the solution of these problems in previous studies, the amplitude of the fluctuation spectra of the intensity of sparse laser radiation and the coefficient defining the correlation of amplitudes of fluctuation spectra and amplitudes of the intensity of the speckle-field on frequencies 30 and 10 Hz were estimated. But the informativity of these indices, however, did not provide a clear view of the investigated phenomenon. Further search for informative criteria led us to the attempt of assessment of the functional condition of muscles on the area under the spectral curve and also on the coefficient $A_i(f)$ (normalized amplitude), to the equal ratio of the initial amplitude to the area under spectral curve taking into account coefficients α_1, α_2, etc., and calculated by the formula

$$A_i(f) = \langle A_i(f) \rangle + \sum_k \alpha_{ik} . \varphi_k(f), \tag{5.30}$$

where $\langle A_i(f) \rangle$ is the average normalized amplitude and αk is the coefficient of the resolution of vector $\varphi_k(f)$.

The data of the registration of speckle-optical myograms of gastrocnemius muscle of patients with NOLO showed that the amplitude of the fluctuation spectra of the speckle-field at rest on the side of pain syndrome was lower than on the healthy limb in the cases when the decrease of the tone of gastrocnemius muscle were clinically registered. The amplitude increase of the spectra of the fluctuations on both limbs was observed under the maximum clonus of gastrocnemius muscle, but less vivid on the affected side (Fig. 5.47). These data most often had unidirectional character with the results of electromyogram studies. However, in patients with long clinical course and evidence of its clinical development by isometric contractions of gastrocnemius muscle not the increase was registered, but the reduction of the amplitude of the fluctuation spectra in comparison with the condition at rest. The comparison of this data with the results of electromyogram of gastrocnemius muscles of diseased and healthy limbs was carried out. With the help of electromyogram powerful fasciculations were registered from both sides. The appearance and disappearance of these fasciculations during muscle tension was different at the injured side and at the health limb. The registered fasciculations are the reflections of irritations of motoneurons of the anterior horns. It is known that the lesion of motoneurons leads to abnormal changes of muscle fibers and causes the disorder of their contractive activity that can result in the development of observable reactions as a result of a maximal effort in patients of this group. In situations like this for the estimation of functional possibilities of muscles, the results of speckle-optical myography appear to be more informative than traditional electromyogram.

At the study of paravertebral muscles of patients with clinical manifestation of moderate and evident muscle tone syndromes, where the direction of electromyogram

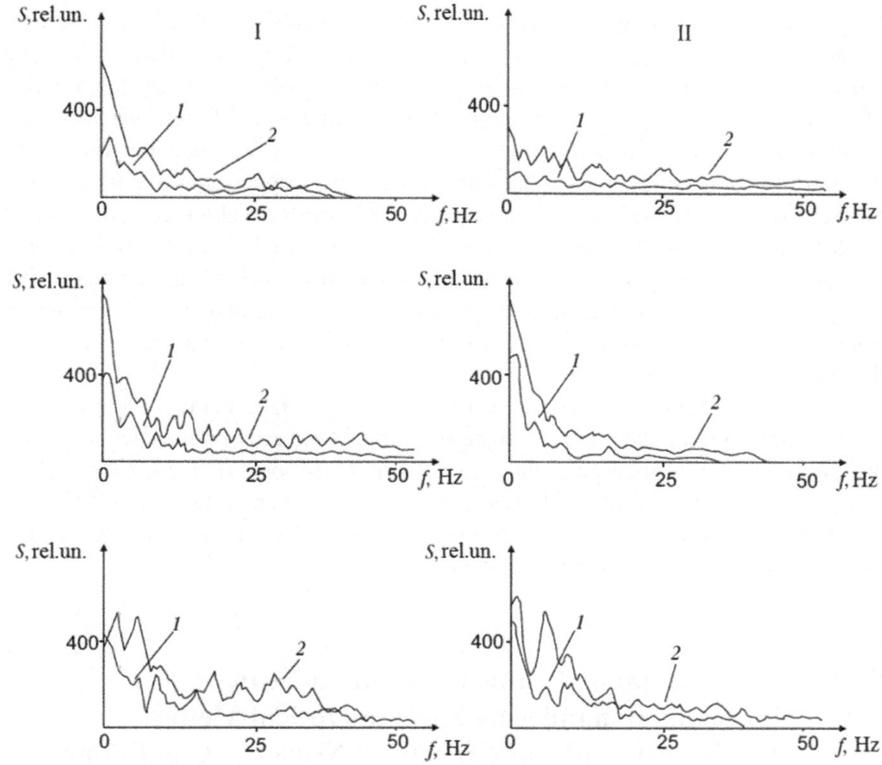

Fig. 5.47 Speckle-optical myogram of gastrocnemius muscles in 3 patients with NOLO on a healthy I and injured II sides, in relaxed condition 1 and at maximal contraction 2. Reprinted from [87] with permission

changes and speckle-optical indicators coincide, the asymmetry of muscle tension on the diseased and health side at rest was detected. The exclusion was the single cases, when under bilateral pathologic hypertonus of muscles under study these regularities were not registered and the results of applicable methods of research were multidirectional. Perhaps, this can be explained by the duration (17 and 47 years) and evidence of the pathologic process: the presence of pathologic pulsations in the form of volleys on the electromyogram, single and multiple fasciculations at rest and during the relaxation period, low-frequency rate, presence of increased synchronization. The data of tonometry in these cases testified different degrees of tone evidence of paravertebral muscles on its length.

In our studies, the degree of participation of different groups of muscles provided the specific postural pose was important. In this way, in the sitting position the tone of paravertebral muscles at the injured side changed insignificantly and in the majority of cases oscillated within the limits of ±11% in comparison with healthy side. In staying position because of the increase of the load on the muscles supporting this position, the indicator of tonic tension of muscles was 45–70% higher at the diseased side.

In cases when bilateral hypertonus of muscles under study was observed in patients in staying position, in consequence of protective muscle defense, the tonus of the muscles of the back at the healthy side was two times more tensed than the muscles of the back on the opposite side, although the picture was different in the sitting position, and muscle tone at the diseased side increased by 80% in comparison with the condition of muscles on the other side. In such a way, hypertonus of paravertebral muscles of the healthy side had mioadaptive mechanism of development, conditioned to postural overloads of certain muscles. Quite often in such cases, multidirectional changes of electromyogram indicators were observed, as well as various qualitative changes of electromyograms up to the appearance of single and multiple fasciculations at rest and at the period of relaxation, the presence of increased synchronization, the slowering of the frequency.

The conducted studies allowed revealing segmental differences in muscle defense. Thus, while studying the prevalent hypertension of paravertebral muscles we found differences in their condition at the affected vertebral-motor segment L_5, where in comparison with the healthy side it was 65.5% higher than at the level of thoracic vertebrae T12, where the figure was lower, and only 18% higher than in voltage muscles on the affected side than on the healthy one.

5.11 Experimental and Clinical Studies of Skin Microhemodynamics by Speckle-Optical Method After Neurorraphy of Peripheral Nerves in Conditions of Intravenous Laser Blood Irradiation (ILBI) in Patients with Compressive–Ischemic Neuropathies and Neurological Manifestations of Lumbar Osteochondrosis

Traumatic injuries of peripheral nerves (PNs) are characterized by severe flow with expressed disturbances of motor, sensitive and vegetative functions. In this regard, experimental and clinical studies aimed at pathogenetic, pharmacological correction of arising disorders and search for new therapies are of particular importance.

With this purpose, there were conducted the study of changes in the degenerated tissue of limbs at full transection of the sciatic nerve with simultaneously suturing it with seams and the investigation of the effects of pharmacological agents and intravenous laser blood irradiation, as well as their combinations on trophic processes in the tissue conditions of degenerated limbs.

Experimental studies were carried out on non-pedigreed adult rabbits with the weight of 4 kg. Transection of the sciatic nerve was carried out under intravenous rabbit hexenal anesthesia. One group of animals in the postoperative period was injected a dose of edadenum of 1 ml per animal intramuscularly daily during 10–15 days. Two courses were conducted with intervals of 10 days. The sessions of intravenous laser blood irradiation were conducted with the second series of animals.

The influence of laser radiation on the regenerative processes of somatic nerves after injury was investigated, and the study of the dynamics of recovery microhemocirculatory bed sciatic nerve and its regeneration of nerve fibers after reconstructive operations on the injured nerve trunk under the influence of low-intensity laser irradiation was conducted.

In addition, the studies were carried out in chronic experiments on 18 white rats with the weight of 250–300 g in a model of sciatic nerve transection on the border of the upper and middle third, followed by neurorraphy. The laser beam with a wavelength of 632.8 nm and a power density of 22.5 mW/cm^2 (He–Ne laser LG-75) focused on the area of post-operational scar (1 min of irradiation exposure for 8 days). Application as a therapeutic factor in low-intensity laser irradiation contributed to the reduction of the inflammatory tissue response in areas of trauma and reduction of swelling of nerve damage. Adhesions in the area of traumatic injury of the sciatic nerve were expressed significantly less, and scar remained rather loose. This created favorable conditions for germination in the connective tissue of the emerging regeneration neuroma of the newly formed blood vessels, both central and peripheral segments of the sciatic nerve. Most of the thin regenerating nerve fibers, which grow through the damaged area in the distal segment of nerve, accompanied by newly formed vessels, repeating their course, and by the end of the month evenly permeated the loose scar.

Thus, the use as a curative factor of low-intensity laser irradiation in the experiments with animals having an injury of sciatic nerve did not only reduce the inflammatory response in the areas of trauma, but also led to the formation of loose connective tissue scar more, thus creating favorable conditions for recovery of microhemocirculatory bed in the area of damage and improved nerve regeneration of nerve fibers.

In clinical studies using speckle-optical method, it has been found that the most sensitive to the action of laser radiation and informative indicator of changes in surface microcirculation in patients with traumatic peripheral nerve is the average frequency $\langle f \rangle$ of fluctuations in the intensity of the speckle-field scattered by the skin, in the range of 0–2000 Hz. It is shown that preoperative mean frequency $\langle f \rangle$ reflecting the state of the cutaneous blood flow in the area of innervation of the damaged nerves is significantly lower than the corresponding parameter in the symmetric points of healthy limbs. According to the speckle-optical studies in patients with peripheral nerve neurorraphy, intravenous laser blood irradiation causes the activation of cutaneous microcirculation, more pronounced after the exchange rate effect, which leads to a decrease in the asymmetry mean frequency $\langle f \rangle$ to the healthy and injured limbs. It is obvious that the revealed biostimulatory effect of laser radiation is associated with the mobilization of potential reserves of microhemocirculatory channel as intact [165–169] and traumatic lesions of peripheral nerves [170]. It was established experimentally that the activation process of post-traumatic reparative regeneration of growing axons and axo-muscular synapses correlates with the reactive rearrangement associated with blood vessels [171] and accompanied by improved blood flow in the surrounding tissues due to arteriolar enlargement and the inclusion of reserve capillaries [172, 221]. This effect of microcirculation system on laser impact is associated with photoactivated suppression of tone of smooth myocytes in large arterioles [172], the stabilization of the histohematogenous barrier and functional state of the wall of micro- and macrovessels, including its endothelium, reduction of regional vascular resistance, the development of collaterals, that improves the tropism of denervated tissues [173] and creates favorable conditions for regeneration of nerve fibers in the early stages of restoration and regenerative process [176].

Speckle-optical studies of skin microhemodynamics found that the use of intravenous laser blood irradiation creates in patients with compressive–ischemic neuropathy of the peripheral nerves more optimal, compared with the control group, conditions of blood supply to damaged tissues, causing an increase in the power spectrum and the increase in the frequency $\langle f \rangle$ of the speckle intensity fluctuation field scattered by the skin. By the medicamentous therapy, only a tendency to increase the average frequency $\langle f \rangle$ and reduction in power spectrum is noted.

The activating effect of intravenous laser blood irradiation on the power spectrum of intensity fluctuations of the speckle-field of the limb muscles with compressive–ischemic neuropathy of the peripheral nerves and the contralateral limb is revealed, whereas medicament therapy by activating the muscle tone of healthy limbs does not have positive effect in the early stages of study on the functional state of muscle fibers innervated by the nerve. It is established that both methods of treatment do not cause positive changes of the amplitude parameters of contractile function of the myons, but optimize their frequency characteristics.

Further, the study of the diagnostic features of methods of coherent optics made it possible to identify another speckle-optical indicator for an objective assessment of the cutaneous microcirculation, and in particular the ratio of the area of small regions around the two selected frequencies of the spectrum, such as the ratio of the area under the spectral curve in the frequency range of 600–640 Hz to the area under the spectral curve in the frequency range of 40–80 Hz. It is shown experimentally that the study of blood flow to this parameter was very sensitive and can be used as a criterion for the degree of violation of microhemocirculatory processes.

Based on experimental studies, there was installed the effectiveness of low-level radiation of He–Ne and semiconductor laser in ischemic injury and neurorraphy peripheral nerves, which helps to activate the immune processes, improves microcirculation of tissues by increasing the degree of oxygenation of the blood and improves oxygen function. A comparative analysis of various methods of low-intensity laser irradiation showed higher efficiency of intravenous laser blood irradiation as compared with the experimental data of laser puncture in immunological studies.

By speckle-optical study of the biomechanical parameters of the functional activity of the muscles and skin blood flow conditions, there was revealed a positive therapeutic effect of intravenous laser blood irradiation in patients with neurorraphy and compressive–ischemic neuropathy of the peripheral nerves, resulting in normalization of the cutaneous microcirculation in the damaged tissues, improving the state of tonic muscle fibers. However, this positive effect influences only some of the speckle-optical indices of contractile properties.

The conducted studies deepen the impact of low-intensity laser irradiation on some indicators of immune status, the processes of microcirculation, oxygen transfer–blood function that improves the functional activity of the muscles and demonstrates the feasibility of intravenous laser blood irradiation in complex treatment of patients with traumatic and ischemic injury of peripheral nerves.

The dynamics of skin blood flow of the lower extremities in patients with neurologic manifestations of lumbar osteochondrosis in our experiments corresponded to the changes in high-frequency spectra of intensity fluctuations of the speckle-field. In this case the regional blood flow changes of lower limbs, which were detected by the method of choralography and which reflect anti-spasmodic processes, were accompanied by a drop of skin temperature and decrease in the frequency spectra of fluctuations. On the contrary, at invariable regional blood flow (on the intact limb) and at the absence of autonomic changes in the form of disturbances in thermoregulation the frequency shift of the spectra of fluctuations in the higher frequency region was observed. Consequently, the growth of the intensity of blood supply to the integumentary tissue and the increase of the blood flow in the microvasculature of the skin are reflected in increasing contribution of high frequencies to the range of fluctuations of the speckle-field. This is also evidenced by the decline in the asymmetry of the spectrum relative to the average frequency.

In the result of the studies, diagnostic speckle-optical criteria of violations of the surface blood flow were developed, and microhemodynamic cards were experimentally obtained. It was established that for the assessment of dermal microhemodynamics the following parameters can be used: the average frequency of spectrum; the area covered by the spectrum; the coefficient of asymmetry of the spectrum; the ratio of the power spectral density levels at high and low frequencies; and the ratio of the average frequency spectrum to the asymmetry. It is shown that different parameters can characterize various links of the microcirculatory process. It is assumed that the average frequency of the spectrum to a greater extent reflects the blood flow, and the value of the area under the spectral curve reflects the capacity of the capillary bed, the degree of capillaries opening.

Thus, the identified parameters of the spectra of fluctuations in the intensity of dynamic speckle-field are characterized by both physiological and pathological changes in the tissue and were used to assess the functional state of skin blood flow.

The studies of the microhemodynamics of human skin with artificially induced ischemia and hypertension showed that as the diagnostic criteria, which characterize the state of skin blood flow, the following indicators are the most informative: the average frequency spectrum in the frequency range of 40–2000 Hz, the ratio of the spectral density in the frequency range of 600–620 Hz to the spectral density in the range of 40–60 Hz, the asymmetry of the spectrum relative to the average frequency, the ratio of the average frequency spectrum to the coefficient of asymmetry, at that the last parameter is the most sensitive indicator of microhemodynamics of the skin.

The values calculated by informative parameters corresponding to the point in the area of the skin made it possible to construct microhemodynamic maps and effectively evaluate the state of the surface of blood flow in normal and artificially induced hyperemia in the skin area of 40×50 mm ^{2}.

Thus, summarizing all of the studies, we can say that the methods of coherent optics can be successfully used in neurological practice. The developed methods allow us to register in patients with lesions of the peripheral nervous system of various etiologies the microhemodynamic processes of denervation and changes in muscle activity with the identification of local hypertonus, and objectively assess the dynamics of the patient's condition and treatment effectiveness.

As follows from the data presented, the use of coherent light for the treatment of diseases of the peripheral nervous system is especially effective when it is exposed on the internal environment. The absence of adverse effects from the use of intravenous laser blood irradiation in the treatment of a number of lesions of the peripheral nervous system, which are investigated in our study, testifies the feasibility of its wider adoption in neurological practice. However, it is necessary to bear in mind that at extending the range of application of ILBI in neurology it seems to be necessary to study the remote effects of therapeutic intervention, as well as to analyze its effects on neuromotor apparatus, the state of endothelial cells and other pathogenetic links of diseases of the nervous system [177, 178].

5.12 Speckle-Optical Diagnostics of Muscle Activity and Microhemodynamics of Human Skin in Patients with Diseases of Peripheral Nervous System

Below, there are some of the results of medical studies carried out using the laser specklometer. The purpose of the investigation is to study the functional disorder of the neuromuscular system and skin microhemodynamics in ischemic and traumatic lesions of peripheral nerves, as well as muscular-tonic syndromes of neurological manifestations of lumbar degenerative disk disease through the development of new methods of optical speckle evaluation and possible use of intravenous laser blood irradiation in the treatment of damage to peripheral nerves and cerebral ischemia.

The clinical trials were conducted in 116 patients (95 patients with neurologic manifestations of lumbar degenerative disk disease, 21—with compressive–ischemic neuropathies). The control group consisted of 30 healthy individuals.

Experimental studies were performed on 144 pubertal rabbits of both sexes. Speckle-optical indicators of skin microhemodynamics (MHD) and functional activity of muscles were studied under intravenous hexenal narcosis after modeling ischemia of a hind limb (ISH) (39 animals), traumatic injuries of sciatic nerve with further neurography (40 rabbits) and local ischemia of brain (LIB) (28 animals).

Rabbits (37) with the simulation of a similar pathology served as controls. The rabbits had not been treated with intravenous laser blood irradiation. The biochemical parameters of the calf muscles were studied at rest condition and at their passive contraction.

The speckle-optical estimation of functional activity of gastrocnemius muscles of a rabbit showed that the spectrum power (SP) reduced by 50% in 7 days after modeled ischemia of a hind limb and in 22–24 days—by 61.2%, compared with the level of tonic muscle tension at rest in non-operated animals (Fig. 5.48), and can be considered as a more informative characteristic. The deterioration of the contractive activity of muscles at the violation of their blood supply was found.

The study of microcirculatory changes in the skin blood flow found that the power of the spectrum was reduced by 41% in 7 days after the modeling of ischemia of the hind limb on the operated limbs, and in 22–24 days after surgery by 67%, in comparison with the level of skin microhemodynamics of the limb before operation (Fig. 5.49). At the same time, the asymmetry power spectrum of intensity fluctuations of the speckle-field between the operated and intact limbs during all periods of

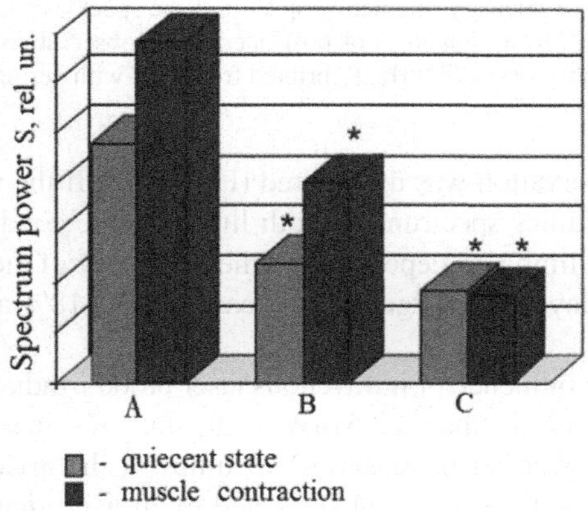

Fig. 5.48 State of myotonus and contractile activity of muscles (according to SP data): A—in intact rabbits; B—in 7 days after modeling hind limb ischemia; C—in 22–24 days after surgery; asterisk—differences are true with respect to the values of SP in intact rabbits. Reprinted from [87] with permission

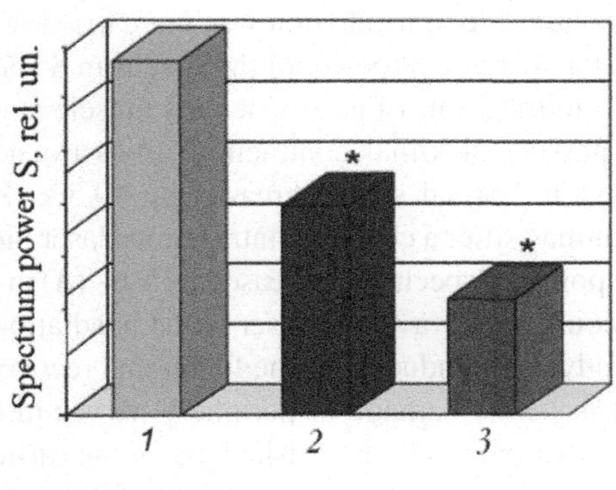

Fig. 5.49 Characteristics of skin blood flow (according to SP data in frequency range of 0–2000 Hz): 1—in intact rabbits; 2—in 7 days after modeling hind limb ischemia; 3—in 22–24 days after surgery; asterisk—differences are true with respect to the values of SP in intact rabbits. Reprinted from [87] with permission

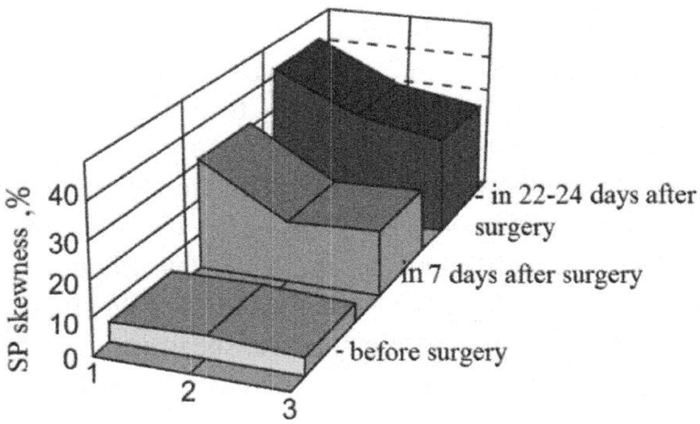

Fig. 5.50 SP skewness of intensity fluctuation spectra of the speckle-field A_S, scattered by skin, of intact and damaged limbs in dynamics of post-operational observation in different frequency ranges: 1–0–2000 Hz and 3–500–2000 Hz. Reprinted from [87] with permission

post-operational observation was discovered (Fig. 5.50). If the symmetry of power skin microhemodynamics spectrum on both limbs before conducting modeling of ischemia of the back limb was dependent on the frequency of the transmission band of 4–6%, then in 7 days after surgery, it increased to 31–18% and in 22–24 day to 42–27%.

The studies of the influence of intravenous laser blood irradiation of He–Ne laser with power at the end of the fiber of 2.5 mW on the state of skin microhemodynamics of intact animals showed an increase in 2–3 times of the area under the spectral curve in comparison with rabbits, not subjected to laser irradiation. It is found out that course application of intravenous laser blood irradiation in the intact animal has little stimulating effect on the background tonic and contraction muscle activity.

After the influence of intravenous laser blood irradiation, the average vibration frequency $\langle f \rangle$ of muscle fibers at rest in animals, which had not been operated, did not differ from the average vibration frequency $\langle f \rangle$ in normal rabbits, in regard to which the intravenous laser blood irradiation was not carried out, and the tendency was only observed to the increase of power of the spectrum S and average frequency $\langle f \rangle$ of speckle-optical mimograms of gastrocnemius muscle at rest (Fig. 5.51a, b).

In conditions of passive maximal contraction of gastrocnemius muscles, the average frequency $\langle f \rangle$ in normal state increased by 40.5% (*A*) compared to the background, and in animals after a course of intravenous laser blood irradiation—by 47.8% (*B*), at that the power of spectrum increased by 59% (*A*) in the group of normal animals, and after a course of intravenous laser blood irradiation to intact rabbits—by 47.8% (*B*). The study was conducted of the functional recovery of neurovascular tissue interactions at ischemic damage of periphery nerves in conditions of intravenous laser blood irradiation. It was established that at the earlier stage of the study of regeneration–reinnervation process after modeling of ischemia of a hind limb the intravenous laser blood irradiation has a moderate positive impact on the indicators of skin microcirculation and contractile activity of muscles, without affecting the

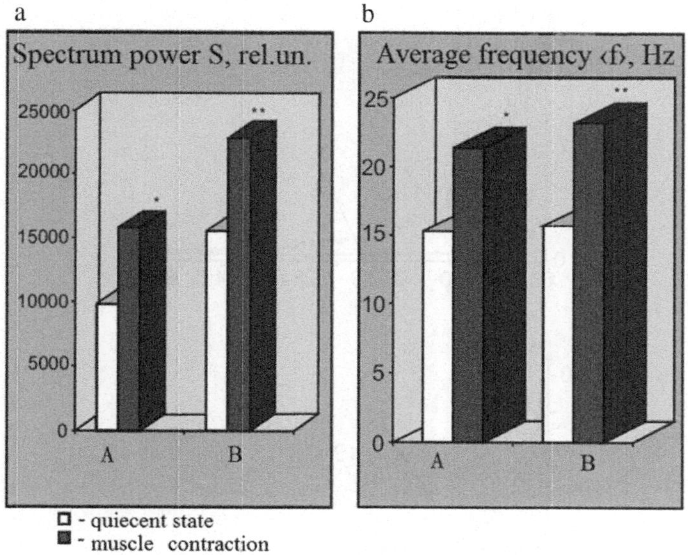

Fig. 5.51 Speckle-optical characteristic: spectrum power (**a**) and average frequency (**b**) of gastrocnemius muscle myotonus in intact rabbits (A) and in conditions of using ILBI (B): * $p < 0.05$ in comparison with quiescent state; ** $p < 0.05$ in comparison with intact rabbits. Reprinted from [87] with permission

myotonus at rest and not leading to full restoration of the functional state of the neuromotor apparatus [159–162, 168–175].

The spectrograms of spectrum power of skin blood flow in rabbits after modeling of ischemia of a hind limb without treatment (A) and after a course of intravenous laser blood irradiation (B) are shown in Fig. 5.52. A significant difference in the level of skin blood flow of non-operated and operated limb in control group and the reduction of this skewness after the intravenous laser blood irradiation is seen. It was found out that in the rabbits with traumatic damages of sciatic nerve with the following neurography in 3.5 months after operation the contractive activity of gastrocnemius muscles was reduced (B) (Fig. 5.53). During the same period of post-operational observation, considerable reduction of the intensity of skin blood circulation was indicated: SP was 30% lower at the limb after surgery, then on the intact one.

This is confirmed by the data about developing at the traumatic damage of the nerve hemodynamic disorders, which deteriorate reinnervation process. The tendency to the salvage of contractive activity of gastrocnemius muscle was observed after the conduction of the course of intravenous laser blood irradiation. However, the degree of the muscle contraction was considerably lower, than in intact rabbits (B). Also, a 51% increase of the skin blood flow at the damaged limb compared to the control animals [180] was observed. According to the information derived from angiographic and reovasographic studies, the surface and intravascular laser irradiation influence

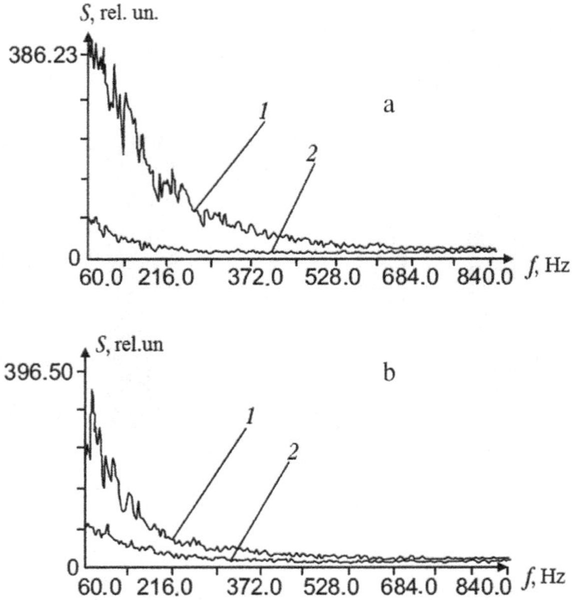

Fig. 5.52 Skin microhemodynamic in operated rabbits: **a** without treatment; **b** course ILBI influence in 22–24 days after modeling of ischemia of the back limb; 1—intact limb, 2—operated limb. Reprinted from [87] with permission

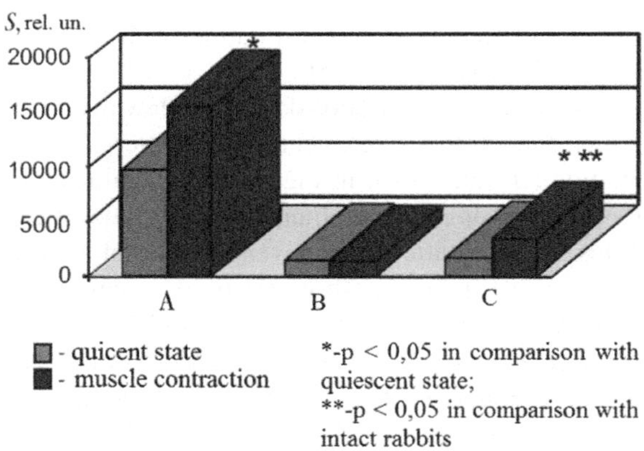

Fig. 5.53 Speckle-optical characteristic of gastrocnemius muscle myotonus in intact rabbits (A) and in 3.5 months after nerve suture (B) in conditions of using ILBI. Reprinted from [87] with permission

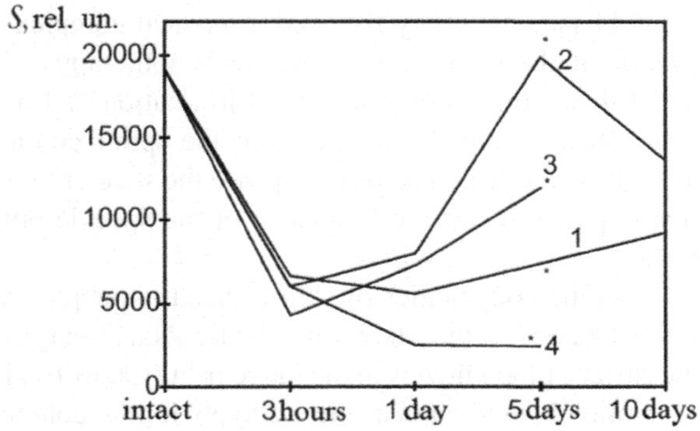

Fig. 5.54 Characteristic of skin blood flow when modeling local brain ischemia in intact rabbits (1) after ILBI course (632 nm; 2.5 mW) (2); infrared laser (860 nm; 2.0 mW) (3); and infrared laser (8.5 mW) (4). Reprinted from [87] with permission

the intensity of the blood filling of tissues under study as a result of spasm removal, vasodilator reaction of the vessels, mobilization of the collaterals, which did not function previously.

In the experiments based on the study of cerebral ischemia, it was found out that local ischemia of the brain causes phase changes of speckle-optical parameters of skin microhemodynamics: the deterioration of the ischemia during the first day and the improvement in a longer time period (on the 5th–10th day) without restoration of the level of blood flow to the normal indicators that testifies the development of compensatory reactions (Fig. 5.54). The use of intravenous laser blood irradiation by the irradiation of infrared laser with a power of 2 mW during the early post-operational period at rabbits with ischemia of brain has positive effect on microhemodynamics processes in tissues compared with control (Fig. 5.54).

The course effect of intravenous laser blood irradiation with a power of 8.5 mW does not have a favorable impact on the state of microhemodynamics processes, causing deterioration of blood supply to the tissues, associated perhaps with the violation of the rheological properties of blood with the activation of hemostasis (Fig. 5.54). The changes of the speckle-optical indices of the skin microhemodynamics at rabbits with local cerebral ischemia during treatment of intravenous laser blood irradiation ($\lambda = 632.8$ nm, power of 2.5 and 4.5 mW) reflect the activation of processes of microhemocirculation in the skin under the influence of laser effect on the 5th day, when their level reaches normal level. Application of 10 sessions of intravenous laser blood irradiation weakens the severity of this effect by worsening the condition of microhemodynamics, as is shown in Fig. 5.54 [180].

The conducted experimental speckle-optical studies made it possible to justify pathogenetically the use of intravenous laser blood irradiation in the treatment of compressive–ischemic neuropathies. Depending on the conducted treatment, patients were divided into two groups: Group I (control group) consisted of 10 patients. They received the following medical complex: nifedipine, multivitamins, massage, mechanotherapy. The treatment was performed for 10 days. Group II (experimental group) consisted of 11 patients. They received a medical complex of intravenous laser blood irradiation, multivitamins, massage, mechanotherapy.

It was found out that intravenous laser blood irradiation had a more apparent positive therapeutic effect than medication, causing the optimization of microcirculation in the damaged tissues, which helps to improve the state of tonic muscle fibers and has a positive impact only on the frequency of the speckle-optical indices of contractile properties.

The studies of microhemodynamics of skin of healthy people were carried out at different functional states. At the same time, the frequency range of the intensity fluctuations of the surface blood flow is in the range from 100 to 10 kHz. As a model of disorders of skin microhemodynamics, here a hypoxic test, cold test (application of ice to the skin of the back of the hand to reduce the skin temperature up to 19 °C) and infriction of "Finalgon" ointment, which causes congestion, were used. For the purpose of conducting a hypoxic test in the tonometer cuff on the top of the shoulder, the pressure was raised up to 200 mm Hg; during this process in the range of frequencies of 125–2000 Hz the reduction of all amplitude–frequency characteristics of a spectrum was set. During the restoration of blood flow, the speckle-optical indicators increased, reaching the norm values. The changes of spectrum power under conducting the test "norm—the cold—hyperemia," the values of which decreased under hypothermic test and increased under hyperemia, are presented in Fig. 5.55.

The tonic tension and contractile function of skeletal muscles were studied at 30 healthy people using methods of speckle-optics. The increase of the average frequency of spectrum of various muscle groups at their maximal voluntary contraction in comparison with the condition at rest was observed (Fig. 5.56). The power of spectrum connected with the spectrum amplitude also appeared to be a rather informative parameter of functional muscle condition. The method of speckle-optical myography made it possible to discover and assess objectively the peculiarities of functional condition of muscles during the conduction of test using isometric

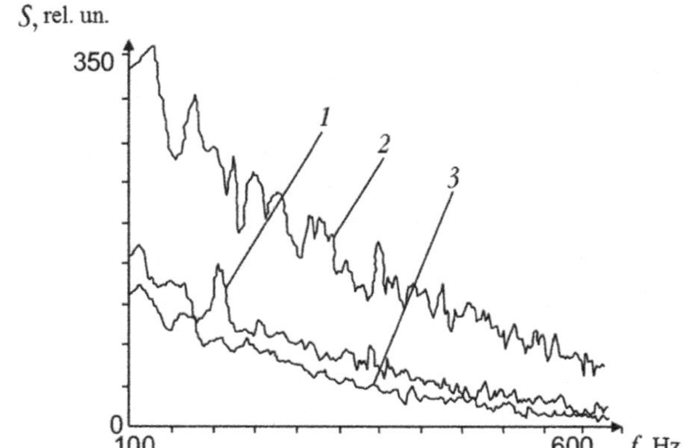

Fig. 5.55 Skin microhemodynamic of the back surface of a hand: 1—normal; 2—at artificial hyperemia; 3—at cooling. Reprinted from [87] with permission

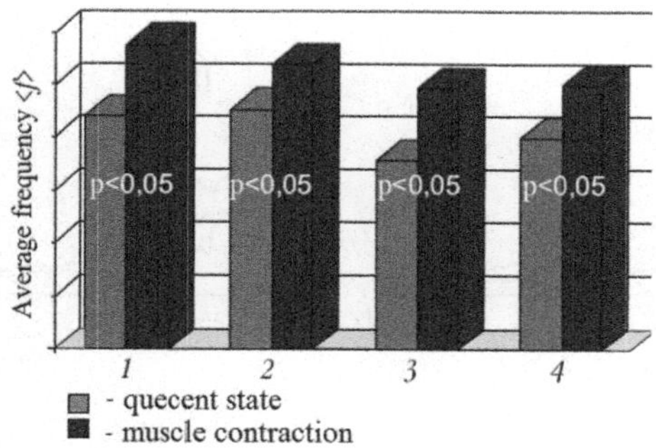

Fig. 5.56 Characteristic of muscle contraction activity: 1—paravertebral muscles; 2—gluteus maximus muscle; 3—rectus muscle of thigh; 4—gastrocnemius muscle. Reprinted from [87] with permission

exercise. During the parallel study of speckle-optical parameters and bioelectrical activity of biceps muscle of arm, the increase of average frequency of spectrum $\langle f \rangle$ and amplitude indicator of global electromyogram in proportion to the increase of isometric exercise was observed (Fig. 5.57). However, the increase in the spectrum power lasted only till 80% from the maximal effort developed by the muscle, and then the spectrum power decreased that is obviously connected with the reduction

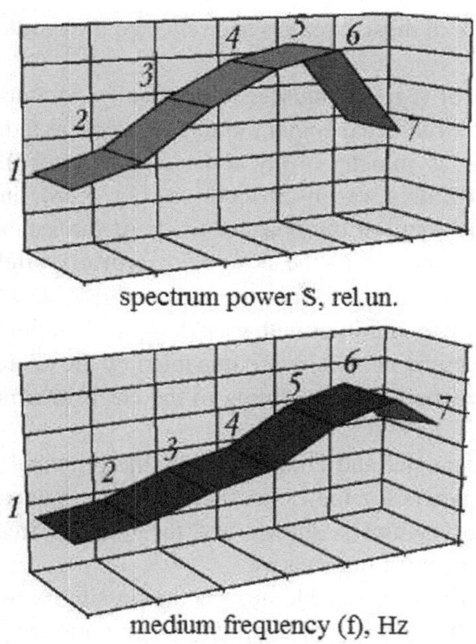

Fig. 5.57 Dependence of power and medium frequency of biceps muscle of arm spectrum under isometric contraction on the size of the load: 1—without load; 2—2 to 2.5 kg; 3—5 kg; 4—7.5 kg; 5—10 kg; 6—12.5 kg; 7—15 kg. Reprinted from [87] with permission

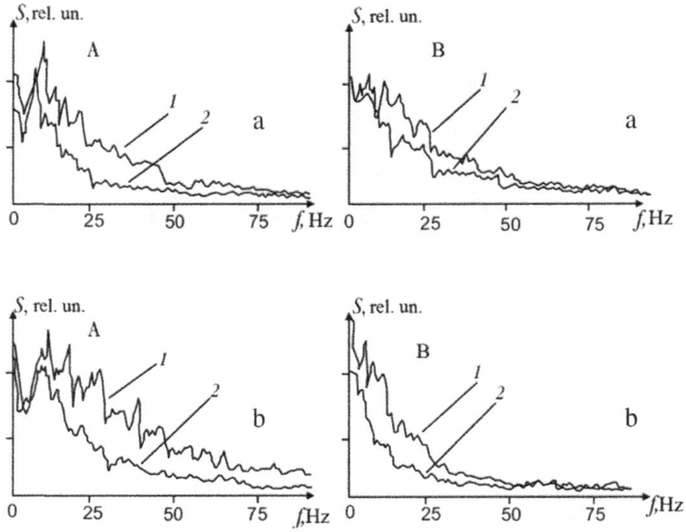

Fig. 5.58 Speckle-optical myograms of paravertebral muscles on the level of spinal motion segments L_V—A and L_1—B from the right side—1 and from the left side—2 in sitting position (**a**) and in standing position (**b**) in patient with right-side radiculitis L_5–S_1. Reprinted from [87] with permission

of vibration amplitude of muscle tissues and reflected the development of processes of muscle fatigue.

During the study of regional muscle imbalance in patients with muscle-tonic syndrome in the area of affected spinal motion segment (at radiculitis), as opposed to healthy people, the asymmetry of myotonus of paravertebral muscles according to speckle-optical indicators was discovered (see Fig. 5.58), and its size fluctuated from 12 to 46% depending on the postural pose: At the standing position, it was larger and was not connected to the nosologic form of vertebral pathology, and it reflected only the dynamics of tonic muscle condition. During the study of the microdynamics of lower extremities with speckle-optical method, the proximo-distal gradient of speckle-optical indicators accompanied by the decrease of skin temperature at distal areas of both legs according to the electrothermometry and thermal study was detected.

Thereby, in experimental and clinical studies the principal possibility of using speckle-optical parameters for the estimation of biomechanical muscle properties and skin microhemodynamics is shown, and the medium frequency $\langle f \rangle$ and the spectrum power are the most informative.

The advantage of the laser specklometer is the possibility to conduct simultaneously the multidisciplinary study of a patient. Once registering fluctuations of the intensity of the speckle-field scattered by the skin, it is possible to derive the information about the state of skin microdynamics, muscle tonus and other biomechanical characteristics of tissues in this area, tremor, in particular. Further differentiated analyses of the derived information include the calculation of spectra registered in different frequency ranges: For the evaluation of the degree of the tremor intensity 1–16 Hz and for the diagnostics of the disorders of muscle tonus 10–62 changes of skin microdynamics are detected in the ranges of 1–1000 and 50–1000 Hz. This approach to the processing of speckle-optical indicators was further developed in the experimental works and is successfully implemented in the clinical practice. The wide application of speckle-optical method of the study of skin microdynamics and myotonus made it possible to derive the objective characteristics of microhemocirculatory processes in patients with compressive–ischemic, toxic and traumatic injuries of peripheral nerves [165, 169], also in patients with acute and chronic ischemia of the brain [193–203, 206, 209].

With the help of noninvasive speckle-optical methods, the tonic muscle condition and their contractile activity were studied, and the objective estimation of tremor and skin microhemodynamics (MHD) was provided in patients with the diseases of central and peripheral nervous system. Speckle-optical myography made it possible to provide objective estimation of changes in functional state of skeletal muscles in patients with axonal and demyelinating polyneuropathy (PNP), facial dyskinesias and also during Parkinson's disease.

During the process of the study of functional disorders of neuromotor apparatus, speckle-optical diagnostics of muscle tonus considerably completes the results of electroneuromyographic (ENMG) testing of peripheral nerves and electromyography (EMG) and extends the possibilities of objective assessment of processes under study. As it was shown in the works [181–215], it was used for the study of patients with traumatic and compressive–ischemic damages of nerves. Further speckle-optical estimation of disorders of contractile function of muscles at demyelinating polyneuropathy of different genesis was carried out. In patients with alcohol and demyelinating polyneuropathy due to the skin microhemocirculation disorders, particularly obvious in distal areas of lower extremities, there were detected insignificant changes in muscle tonus at rest and worsening of their contractile properties with the disorder of contractile function, mainly, in the distal areas of upper and lower extremities with certain predominance of pathologic changes in patients with chronic inflammatory demyelinating polyneuropathy [183, 184]. The received data make it possible to estimate the degree of postsynaptic changes developing directly in the muscle fibers and conditioned by the pathology of neutralizing component of peripheric neuromotor apparatus at polyneuropathy of demyelinating and axon genesis. Altogether with the results derived from ENMG testing, the speckle-optical estimation of tonic and contractile muscle activity quite clearly defines electrophysiological and biomechanical patterns of functional disorders of peripheral neuromotor apparatus.

The question of the possibility of applying speckle-optical indicators as differential diagnostic criteria with the estimation of biomechanical characteristics of facial muscles in patients with facial dyskinesias is discussed [185–187]. The unidirectionality was detected of changes of speckle-optical parameters of myotonus of orbicular muscle of the eye in patients with focal form of muscle dystonia—blepharospasm and facial hemispasm with establishment of the most essential differences in the frequency range of 1–62 Hz. Zygomatic muscles are the optimal area of registration [187, 188].

Patients with facial hemispasm had less obvious tonic disorders. Consequently, these indicators can be applied as differential diagnostic criteria for the estimation of functional muscle condition, which is important for prescribing the differentiating treatment.

Noninvasive speckle-optical methods of registration have been developed, and informative parameters of tremor of facial muscles have been defined in patients with facial dyskinesia, which are used as diagnostic criteria and also for the estimation of the effectiveness of the treatment [189–191]. It was shown that after treatment the tremor and the intensity of forced closing of one's eyes in patients decreased that was objectively characterized by the change of speckle-optical myogram and tremorogram with the restoration of the parameters under study up to the indicators in healthy people.

The possibility of using speckle-optical indicators for the objective estimation of the muscle rigidity and tremor in patients with Parkinson's disease is shown. It made it possible to derive considerably new objective information about biomechanical disorders during this pathology, which can be used at neurological practice with the aim to provide diagnostics and control of the efficiency of the conducted therapy [186, 192].

During the experimental studies, the unidirectionality of changes of speckle-optical indicators of skin and cerebral microhemodynamics during the modeling of local ischemia of brain was found out [193–203]. These studies had been taken as principal ones for the development of the method of the diagnostics of cerebral microhemodynamic changes in patients with acute and chronic disorders of brain blood flow using registration and estimation of speckle-optical indicators of skin blood flow [204–206]. It was established that in patients with the more favorable flow of dyscirculatory encephalopathy and its atherosclerotic form there are registered an increase of power spectrum, a reduction of spectrum-medium frequency and asymmetry coefficient, which characterize the capacity of microvasculatory channel that reflect the development of compensatory processes. In patients with less favorable flow of the disease and mixed form of encephalopathy (atherosclerotic and hypertensive), the worsening of skin microhemodynamics has been found out [207].

The dynamics of changes in speckle-optical parameters of skin microhemodynamics has a diagnostic value for differentiating various variants of acute disorder of brain blood circulation on the basis of detection of lateralization of medium frequency and asymmetry coefficient of the spectrum and reduction of the indicators in patients in comparison with healthy faces. While comparing speckle-optical characteristics of microhemodynamics of skin in temporal region and over mammillary tubercle at the side of lesion of brain during atherothrombotic variant of brain with parameters of microhemodynamics in the analogous zones in healthy people, mainly the reduction of the values in patients at the side of brain in infarction was observed [208, 209].

Unlike patients with atherothrombotic infarction, patients with lacunar brain infarction did not show the asymmetry of speckle-optical indicators of blood flow, which is connected with systematic affection of microhemocirculatory channel that is due to the formation of microangiopathy and atherosclerosis of small cerebral arteries [191, 210–214]. This phenomenon is supported by the data of magnetic resonance tomography, in the result of which some patients, along with localization of the process at the one side, have the multi-focal focuses on different brain and cerebellum structures that become obvious with worsening the condition of microhemocirculatory channel in general, which is registered with the help of speckle-optical method.

Thereby, the presented original results of the studies demonstrated the diagnostic possibilities of coherent methods of the study of skin microhemocirculation, speckle-optical myography and tremorography that makes it possible to refer them to the highly informative methods of estimation of functional disorders of microhemodynamics, tonus and biomechanical indicators of muscle activity, which are developed during the diseases of central and peripheral nervous system.

In the works [221, 223] at modeling the traumatic damage of the sciatic nerve in the experiment based on the clinical, speckle-optical and pathologic methods of estimation of degenerative-reinnervational process, trophic disorders, myotonus, contractile and bioelectrical muscle activity, and also skin microhemodynamics, the character of the disorders of neuromotor apparatus was detected. In 3 months after neurography of the sciatic nerve of rabbits there was determined a considerable lowering of electroneuromyographic (amplitude and area M-answer, speed of the conduction velocity on the motor fibers) and speckle-optical (medium frequency of fluctuations of intensity of speckle-field and power of the spectrum) indices during clinical picture of obvious paresis of the damaged limber with considerable trophic disorders. The most informative speckle-optical indicators of contractile muscle activity and skin microhemodynamics, which make it possible to objectify the flow of regenerative-reinnervational process, were defined.

In the works [216, 221, 223], the stimulating effect of ILBI with He–Ne laser ($\lambda = 632.8$ nm, power of 2–2.5 mW) was shown on the indexes of skin microhemocirculation, tonic and contractile muscle activity at intact rabbits. The positive effect of laser hemotherapy on the flow of regenerative-reinnervational process was registered after neurography of sciatic nerve at rabbits that becomes evident by the increase of background tonic and contractile muscle activity, improvement of the indicators of neural conductivity, increase of blood flow in skin microhemocirculatory channel and, as a result, in enhancement of blood supply of the tissues of damaged limb.

In the works [216, 217] there was experimentally substantiated the reasonability of using ILBI in complex post-operative treatment of patients with traumatic injuries of peripheral nerves in early post-operational period, starting from the second day after neurography, in order to create optimal conditions for treatment of regenerative-reinnervational processes occurring in a damaged neuromuscular system.

As it was shown in the works [218, 219, 222], clinical and speckle-optical studies at trauma of all nerves examined in patients, which was conducted in the presurgical period, testify considerable disorder of pain sensitivity, trophic and motional function, neural conductivity and contractile properties of muscle fibers at damaging of nerves, which innervate them. Biomechanical muscle disorders are characterized by the absence of difference in the medium frequency of fluctuation intensity of speckle-field scattered by skin over the examined muscles at their maximal arbitrary contractions and in the condition of functional rest. In patients with traumatic damages of peripheral nerves using the speckle-optical method of diagnostics, the character of disorders of skin microdynamics was found with the shift of the medium frequency of fluctuation intensity of speckle-field in the area of low frequency and reduction of spectrum power more distal than the place of damage.

On the grounds of the analysis of clinical, ENMG and speckle-optical data of the study results, it was found out that using the ILBI method by irradiation of He–Ne laser ($\lambda = 632.8$ nm, power of 4–8 mW) after neurography of damaged nerves in patients, it accelerates the appearance of initial clinical, ENMG and speckle-optical signs of regeneration of peripheral nerves and provides quicker and sterling functional recovery of damaged nerves, improving the results of surgical treatment of this pathology [215, 218–220, 224, 225].

Thereby, the studies, which we have conducted in the Institute of Neurology, Neurosurgery and Physiotherapy of the Ministry of Health of the Republic of Belarus (today: the State Establishment "The Republican Theoretical and Practical Center of Neurology and Neurosurgery") together with other workers of this institution in the period from 1996 till 2008, of the influence of ILBI on patients with the diseases of peripheral and central nervous system with the help of the developed fundamentally new noninvasive, non-contacting, highly sensitive speckle-optical method and laser specklometer—the device of a new generation in medicine [80–82, 115–134, 165, 166, 179–264]—convincingly showed the perspective of their usage in treatment and diagnostics of these diseases.

Besides in works [56–58, 265–269], the studies of changes of concentration of oxyhemoglobin of surface blood flow at various functional skin conditions were conducted, and a new spectrometric method for the determination of the level of ischemia of the organs and tissues of a human being was proposed. The dependence was established of local extremums of the spectral curve of the coefficients of reflections of skin on the level of concentration of oxyhemoglobin of the surface cover and the level of ischemia.

5.13 Analysis of Spectral Characteristics of Radiation Scattered by Human Skin and Development of Non-contact Noninvasive Optical Method of Blood Flow Study

At the transmission of the radiation of optical range through skin, it considerably scatters and is absorbed by structural skin layers and chromophores contained in them [270, 271]. Skin is an extremely dynamic organ, and its optical properties must be studied in vivo. For the normally incident beam in a spectral range of 250–3000-nm mirror component, scattered light by human skin for the light as well as dark skin [272, 273] amounts to approximately 4–7% of the power of incident radiation. Consequently, in the inner skin layers, approximately 93–96% of the incident on its radiation is absorbed and scattered. The diffuse component of scattered radiation is a function of the processes of absorbing and scattering from the inner skin layers, which are reached by the incident radiation.

The results of the measurements of the radiation reflected from the skin on certain selected wavelengths can be used for the quantitative estimation of the changes of pigments of skin. As the penetration depth of optical radiation into the tissue depends on the wavelength of the incident radiation, the allocation of pigments in various skin layers influences the spectral composition of the radiation, which reaches this layer.

Absorbing regions corresponding to each of the main dermal chromophores can be determined as minima of spectral reflection in vivo [274–276]. The possibilities of this method include the determination of vasodilation conditions, oxygen saturation of skin blood flow, levels of bilirubin and melanin pigmentation.

The measurements of bilirubin levels, skin chromophores, hemoglobin and oxyhemoglobin in skin blood flow have great importance for the diagnostics of the peripheral vascular diseases and, vasomotor disorders [270–273]. The studies of the absorption areas of oxyhemoglobin and hemoglobin at the registration of spectra of radiation scattered by skin allow estimating the concentration of given chromophores in skin blood flow.

5.13.1 Measurement of Spectral Reflection Coefficients of Human Skin

The experimental studies of spectral reflection coefficient of skin preparation (epidermis and dermis) in vitro and skin in vivo showed considerable differences in curves of the reflection coefficient. In particular, for the skin in vivo the presence of typical absorption peaks at wavelengths of 542.577 nm is registered [277, 278].

We have carried out the studies of spectral reflection coefficient of skin in vivo on the inner side of a thumb of 8 people (4 men and 4 women) at the age from 20 to 40. For more precise measurements of the reflection coefficient, the spectrophotometer "Specord M-40" with the attachment for the measurement of reflection with photometric sphere was used, as the reflection of skin is diffusive. Beforehand, the reproducibility of the results of measurements was studied. Repeatedly spectral ratio of reflection coefficients K and mean-square deviation σ for various spectral intervals (Table 5.1) were recorded.

In the works of foreign researchers [279], it was stated that the tendency of measurement of curves of reflection coefficients of skin of fingers in vivo of men and women is the same; however, the curves of reflection coefficient of women have bigger amplitude than the corresponding curves of men.

With the aim of conducting the comparative analysis of spectral curves (Fig. 5.59) at wavelengths of 400, 450, 500, 560, 600, 650, 750 nm, the calculations of average

Table 5.1 Results of determining of reproducibility of measurements

Parameter	Wavelength (λ), nm						
	400	450	542	560	577	650	800
$\overline{K}$	9.6	14.6	16	17.4	15.2	121.4	150
σ	1.02	1.36	0.89	1.02	0.75	2.94	5.55

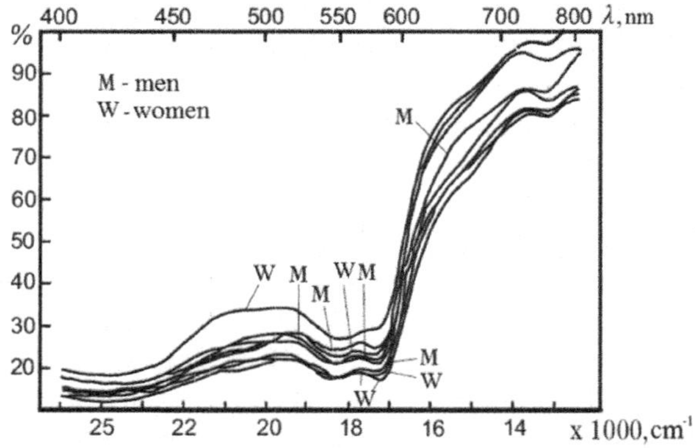

Fig. 5.59 Spectral changes of human skin reflection coefficient. Reprinted from [87] with permission

values of reflection coefficients $\overline{K}$ and mean-square deviation σ separately for men and women were conducted (Table 5.2).

The obtained results indicate the absence of differences in reflection coefficients of human skin depending on the sex. The difference in amplitudes of spectral curves of reflection coefficient for men and women presented in the work [279] is obviously connected with the absence of reliable statistical information. The changes of amplitude of spectral curves of reflection coefficient, derived by us, do not depend on the

Table 5.2 Results of the measurements of spectral reflection coefficients of human skin

λ, nm	Male		Female	
	$\overline{K}$, %	σ	$\overline{K}$, %	σ
400	30.25	3.63	32.50	4.92
450	34.75	2.95	38.50	5.12
500	49.50	5.22	51.75	10.01
560	45.50	4.61	45.25	7.92
600	81.50	9.55	79.0	13.6
660	141.75	11.80	146.25	17.64
750	171.5	13.09	177.0	12.98

sex of the respondents and are explained by the different degrees of pigmentation of skin of every individual.

5.13.2 Absorption Spectra of Blood Preparation at Different Oxyhemoglobin Concentrations

The pigments contained in blood, such as hemoglobin, oxyhemoglobin, β-carotene and bilirubin, in considerable degree distinguish the absorption of the radiation of the visible area of spectrum in dermis in vivo. Hemoglobin and oxyhemoglobin have a discrete area of absorption at the range of 400–1000 nm [272, 279]. Spectral reflection in the area from 400 to 510 nm can be used for the noninvasive assessment of the level of bilirubin in blood [280], and shortwave part of the spectrum of visible radiation corresponds to the main area of absorption of β-carotene [272].

With the aim of a contactless noninvasive method of defining the degree of blood oxygenation, absorption spectra of blood preparation at different concentrations of oxyhemoglobin were studied. The studies were conducted with the help of special attachment “Specord M-40” for the measurement of transmission and absorption.

The absorption spectrum of blood preparation at normal air pressure of 760 mm m.c. was registered. The degree of saturation of blood preparation O_2 corresponded to the partial pressure O_2, which equals to 20.9% from normal air pressure. The curve 1 for the given concentration of oxyhemoglobin in blood preparation is presented in Fig. 5.60. The characteristic peaks in absorption spectrum at wavelengths of 542, 560, 577 nm are observed. With the aim of changing the concentration of oxyhemoglobin, the blood preparation was bubbled by clear oxide during 30 min. The degree of saturation of the blood preparation O_2 reduced to 5% from normal air saturation of

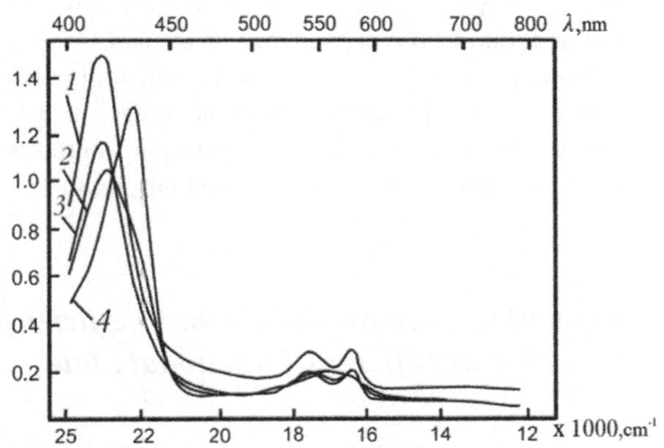

Fig. 5.60 Absorption spectra of blood preparation at different concentrations of oxyhemoglobin: 1—partial pressure O_2 is maximal 20.9% of normal atmospheric pressure; 2—pressure O_2 is 5% of normal atmospheric pressure; 3—pressure O_2 is 0; 4—in 15 min after O_2 saturation. Reprinted from [87] with permission

the blood pressure saturation after this procedure. In the line 2 corresponding to the given spectrum of absorption of blood preparation, the lowering of the amplitude of characteristic peaks at wavelengths of 542, 560, 577 nm is observed. Na_2SO_4 was added to the cuvette studied by the preparation, which made it possible to bring the concentration of oxyhemoglobin practically to zero. The line 3 shows complete disappearance of characteristic peaks for wavelengths of 542 and 577 nm, and also the changes of the sign of curvature of spectral curve for a wavelength of 560 nm. Then, the cuvette with blood preparation was depressurized and stayed under normal air pressure during 15 min. At this time, oxygen absorption occurred to a certain level. In the line 4 corresponding to the absorption spectrum of blood preparation and at the given degree of oxygenation, the characteristic peaks for wavelengths of 542 and 577 nm appear, and the sign of curvature for the given spectral curve at a wavelength of 560 nm changes.

The obtained results allow concluding that change of the amplitudes of characteristic peaks of spectral curve at wavelengths of 542, 560 and 577 nm was connected with the degree of oxygenation of the blood preparation and gives the possibility to produce contactless noninvasive method of determination of the concentration of oxyhemoglobin in blood. The measurement of spectral reflection coefficients from human skin in vivo for three waves of 542, 560, 577 nm constitutes the essence of this method. Let us introduce two coefficients Q_1 and Q_2, which correspond to the differences between amplitudes of spectral curve at wavelengths of 542, 577 nm and amplitude of spectral curve for a wavelength of 560 nm:

$$Q_1 = A_{542} - A_{560}, \quad Q_2 = A_{577} - A_{560}.$$

The results of the calculations of the coefficients Q_1 and Q_2 in the relative units taking into account the sign of curvature are the following: for the line 1—$Q_1 = 0.07$, $Q_2 = 0.08$, for the line 2—$Q_1 = 0.02$, $Q_2 = 0.02$, for the line 3—$Q_1 = -0.02$, $Q_2 = -0.03$, for the line 4—$Q_1 = 0.05$, $Q_2 = 0.05$. The derived information allows building a calibrating diagram of the dependence of Q_1 and Q_2 from the degree of saturation of O_2 of blood preparation (Fig. 5.61). The diagram allows determining the degree of saturation of blood preparation with the oxygen. For instance, for the condition determined by the curve 4 of blood preparation saturation spectrum, the degree of saturation of O_2 equals to 13% from normal air pressure.

5.13.3 Measurement of Spectral Reflection Coefficients of Skin In Vivo at Different Functional States

On the results of the studies of the same group of respondents (see Fig. 5.59), coefficients Q_1 and Q_2 have been calculated, which characterize, as stated above, the degree of oxygenation of skin blood flow. The results of the calculations are provided in Table 5.3.

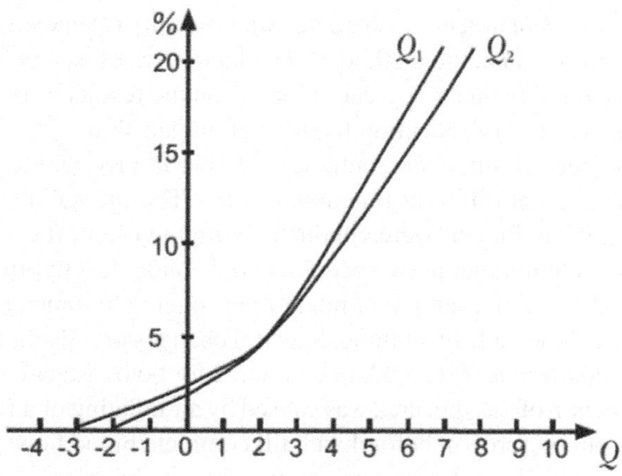

Fig. 5.61 Calibrating diagram of coefficients $Q_1 \times 10^{-2}$ and $Q_2 \times 10^{-2}$ dependence on blood oxygenation degree. Reprinted from [87] with permission

Table 5.3 Coefficients Q_1 and Q_2 for groups of male and female

Respondents	Q_1	Average measurements and errors	Q_2	Average measurements and errors
Female (group)				
1	5	$Q_1 = 3.5$	5	$Q_1 = 3$
2	3	$D = 0.75$	2	$D = 1.5$
3	3	$\sigma = 0.87$	3	$\sigma = 1.23$
4	3	$\delta = 21\%$	3	$\delta = 50\%$
Male (group)				
1	4	$Q_1 = 3.75$	4	$Q_1 = 3.75$
2	4	$D = 0.19$	3	$D = 0.69$
3	3	$\sigma = 0.43$	3	$\sigma = 0.83$
4	4	$\delta = 5\%$	5	$\delta = 18\%$

The average values of groups of respondents (Q), dispersion (D), root-mean-square deviation (σ) and relative error (δ) are presented in the table.

Regardless of the sex of respondents and reflecting features of the skin, the coefficients Q_1 and Q_2 remain almost stable. Their size is distinguished by the functional state of the human organism and corresponds to a certain level of concentration of oxyhemoglobin in blood.

Some authors [281] point to the possibility of diagnostics of ischemia by measuring the intensity of reflected from skin radiation with a wavelength $\lambda = 632.8$ nm. However, the experimental data derived by us at registration of spectral reflection coefficient for the respondent groups (Fig. 5.59 and Table 5.2) showed considerable differences in the amplitudes of spectral coefficients of reflection regardless of sex, determined by individual reflecting characteristics of skin of every single respondent. Thus, according to our data, the method of the diagnostics of ischemia suggested by the authors is not reliable due to a great influence on the results of measurements of individual characteristics of reflecting features of human skin.

The study of spectral reflection coefficient of skin in vivo on the inner side of a thumb was conducted at different functional states. The measurements of spectral curves of reflection coefficient were conducted with the use of the attachment with integrating sphere photometer at two people. People under test tightly applied to the aperture for installation of a sample of integrating sphere photometer the area under study of skin at the inner side of a thumb. Spectral changes of reflection coefficient of skin at the same conditions (Fig. 5.62, lines 1 and 2 for both testees) were registered.

Artificial ischemia of the skin area was caused by air inflating of air-inflated collar, which had been put on forearm beforehand, till complete blood flow stops of a limb. Later, the changes of spectral reflection coefficients of skin were registered, which are presented in Fig. 5.62 (the line 3 for the first testee and the line 4 for the second one). Then, a special cuvette with blood preparation was installed into the aperture for the installation of a sample of integrating sphere photometer (Fig. 5.62, line 5). In lines 1 and 2 corresponding to the spectral curves of the reflection coefficient of skin at wavelengths of 542, 560, 577 nm, characteristic peaks were observed, which coincide with the peaks of spectral curve of the blood preparation reflection (Fig. 5.62, line 5) and the spectrum of blood absorption (Fig. 5.60, line 1).

Artificially caused ischemia leads to the decrease of the amount of oxyhemoglobin practically to zero. Complete disappearance of characteristic peaks of spectral reflection curve for wavelengths of 542 and 577 nm is registered, and the sign of curvature of spectral reflection curve at a wavelength of 560 nm. Averaged values of coefficients Q_1 and Q_2 at normal condition equal to 3 and 2, at ischemia—3 and 7, respectively. Thereby, introduced coefficients Q_1 and Q_2, being independent

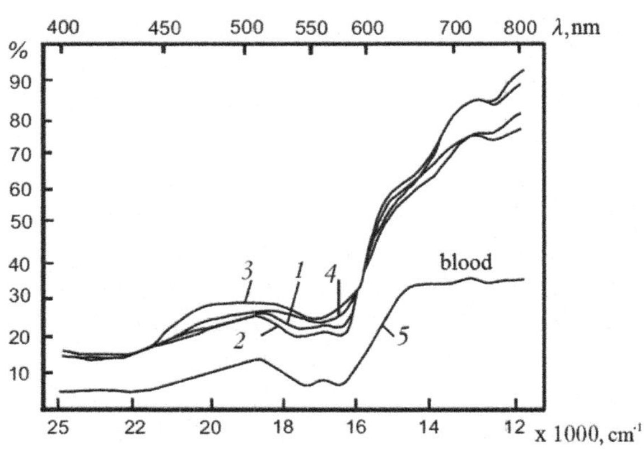

Fig. 5.62 Spectral changes of reflection coefficient of skin and human blood. 1, 2—normal in the first and second testees; 3, 4—ischemia in the first and second testees, respectively. Reprinted from [87] with permission

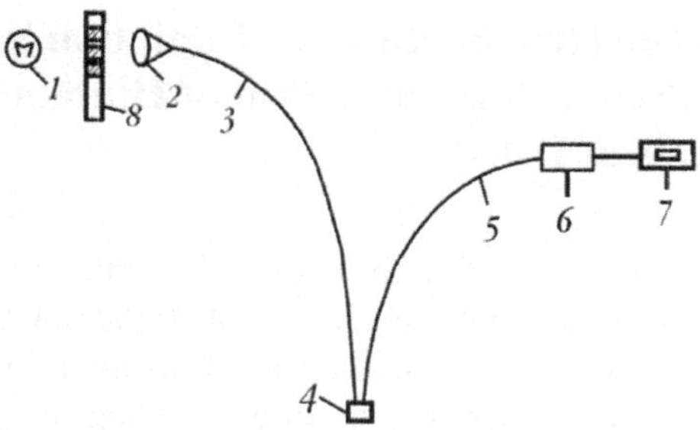

Fig. 5.63 Scheme of the device for determination of skin oxygenation degree: 1—the light source (filament lamp); 2—the optical system for radiation input into the illuminating fiber; 3—the illuminating light guide; 4—the illuminating and receiving head; 5—the receiving light guide; 6—the photoreceiver; 7—the analyzer determining Q_1 and Q_2; 8—turret head with two or three inferential filters. Reprinted from [87] with permission

of individual reflecting features of skin, change considerably at the disorder of microhemodynamics of skin, ischemia in particular.

The obtained results let us develop contactless noninvasive method of defining the concentration of oxyhemoglobin in skin blood flow in vivo and the degree of ischemia by analogy with the method of defining the degree of oxygenation of blood preparation.

Having directed the radiation with wavelengths of 542 and 560 nm to the area of skin of a person under study, having registered corresponding spectral coefficients of reflection and having calculated coefficient Q_1, using calibrating diagram it is possible to determine the degree of ischemia for the given area. Additionally, coefficient Q_2 is being determined in order to increase credibility.

This method can have rather simple realization with the help of the device (Fig. 5.63). Mounting the illuminating–receiving head 4 at different areas of skin and conducting local calculations, it is possible to develop a map of the allocation of the degree of oxygenation of skin of human body, and in the combination with speckle methods of the study of microhemodynamics of skin it is possible to determine the quantity and speed of oxyhemoglobin transfer.

Thereby, the developed method can be widely applied in practical medicine at the diagnostics of different functional disorders, connected with the changes of microhemodynamics of skin (ischemia, for instance), and also at conducting different scientific studies.

5.14 Studies of Spectral Features of Radiation Scattered by Blood Preparations and Skin of Human Being and Animals at ILBI

In our studies [56–58, 265–269], it has been experimentally established that at traumatic and ischemic damages of peripheral nerves ILBI therapy causes normalization of prooxidant reactions, enzymatic and hormonal disorders, increase of the level of immune active proteins and the activity of antioxidant of system, improvement of functional state of neuromotor apparatus and microhemocirculation that leads to the enforcement of compensatory adaptive reactions of organism in the conditions of simulating pathology [165, 166]. Consequently, ILBI influencing the circulating blood has biostimulating effect, which normalizes the condition of the main functional systems and organs that allows applying this kind of non-medicament therapy at different pathologies, including the diseases of nervous system: multiple sclerosis, cerebrovascular pathology, disease of peripheral nervous system [169].

The information about the level of skin chromophores, hemoglobin and oxyhemoglobin in skin blood flow has importance for diagnostics of hypoxic changes in tissues at damages of periphery nerves [83, 265, 266, 278, 282, 283, 284].

Spectral analysis is one of the effective methods of getting this information. The studies of the absorption spectra of blood preparation of animals with the help of three-wavelength spectrophotometric method at ILBI allow determining the influence of low-intensity laser irradiation (LILI) on such a significant index of oxygen transfer–blood function, as a degree of oxygenation, and using this criterion as a test for the objectification of the disorders of oxygen-transfer blood function, at different variants of damages of peripheral nerves (traumatic damages of nerve trunks and partial ischemia of back limb of animals with the development of ischemic neuropathy of sciatic nerve), also they allow studying the possibility of correction of this indicator via course of ILBI.

5.14.1 Study of Influence of ILBI on Spectral Properties of Radiation Scattered by Blood Preparation of Animals with the Help of Three-Wavelength Spectrophotometric Method

The experiments based on the studies of the changes of spectral properties of radiation scattered by blood preparation of rabbits in a range of 200–900 nm before and after conducting the course ILBI were carried out on 37 rabbits of both sexes at norm, at traumatic damage of peripheral nerves and partial ischemia of peripheral nerve of animals. The studies were conducted with the help of special attachment to the spectrophotometer “Specord M 40” for the measurement of transmission and absorption.

For conducting experiments, blood preparation of rabbits was used, which was preliminary prepared by adding heparin to it for preventing coagulability and subsequent centrifugation with the aim of elimination of plasma, white blood cell, etc. Hermetically closed cuvette with homogeneous aqueous solution of red blood cells was placed into the compartment for the test of probes of spectrophotometer, and cuvette with distilled water—on the place of comparison probe.

The spectra of blood absorption were registered, on which characteristic local extremums on three wavelengths of $\lambda_1 = 542$ nm, $\lambda_2 = 560$ nm and $\lambda_3 = 577$ nm were observed, and their amplitudes depended on the degree of blood oxygenation.

Simultaneously with conducting of these experiments, the protein concentration in probes under study was determined by biuret method.

The conducted experimental studies of absorption spectra of blood preparation during changes of concentration of oxyhemoglobin allowed concluding that changes of amplitudes of typical local extremums of spectral curve at wavelengths of 542, 560, 577 nm are connected with the level of blood oxygenation. Registration of spectral curves of reflection coefficient at different degrees of artificially caused ischemia of skin allowed stating the fact that the increase of the degree of ischemia leads to gradual distinction of local extremums of spectral curve at wavelengths of 542 and 577 nm and change of a sign of curvature at a wavelength of 560 nm.

Correspondingly, there takes place an increase of suggested coefficients, which are responsible for the differences between the amplitudes of local extremums at wavelengths of 542 and 547 nm and the amplitude for a wavelength of 560 nm and depending on the concentration of oxyhemoglobin in blood.

For the estimation of influence of ILBI on the change of the concentration of oxyhemoglobin, laser radiation of blood via placing optical fiber into the auricle of a rabbit was conducted. The exposure time during this process amounted to 10 min at the radiation power at the exit of fiber that equals to 4 mW. He–Ne laser of LG-79 ($\lambda = 632.8$ nm) type was used for the radiation. The course of ILBI was conducted for 10 days (10 procedures, 10 min each).

Spectral properties of radiation scattered by the preparations of blood of experimental animals were studied after 1 month after conducting the course of ILBI therapy. Animals with damages of peripheral nerve served as control, and the course of ILBI therapy was not conducted on them.

Based on the described technique, blood preparation after ILBI was prepared and absorption spectrum with the help of spectrophotometer “Specord M 40” was registered. At the spectral curves, there were fixed the changes of amplitudes of characteristic local extremums on three wavelengths of 542, 560 and 577 nm, which were obviously caused by the influence of laser radiation.

Spectral curves of absorption spectra of native preparations of blood of experimental animals before and after the course of ILBI with the help of He–Ne laser are presented in Fig. 5.64.

For the quantitative estimation of concentration of oxyhemoglobin in blood preparation of experimental animals before and after conducting ILBI and for the normalization of the results of the measurements, the difference of the calculated ratios of the amplitudes of characteristic local extremums, registered spectral curves of

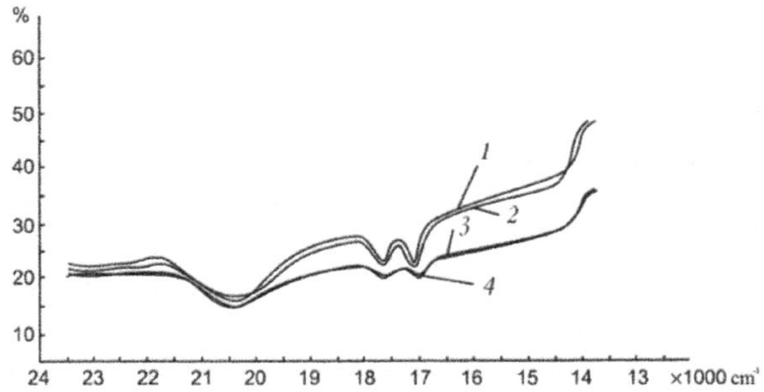

Fig. 5.64 Absorption spectra of native preparation of blood of experimental animals (rabbits) before (curves 3–4) and after (curves 1 and 2) ILBI course with He–Ne laser. Reprinted from [87] with permission

absorption to the concentration of protein in the blood probes under study, was used, which also corresponds to the coefficients Q_1 and Q_2 that equal to

$$Q_1 = \text{Ka}_{\lambda_1} - \text{Ka}_{\lambda_2}; \quad Q_2 = \text{Ka}_{\lambda_3} - \text{Ka}_{\lambda_2}. \tag{5.31}$$

The results of the conducted measurements and calculations of the coefficients Q_1 and Q_2, suggested for the estimation of the degree of oxygenation of blood preparations, are presented in Tables 5.4, 5.5 and 5.6.

Thereby, on the basis of the experimental studies, which were conducted using methods of spectrophotometry, it is important to highlight that ILBI causes changes in spectral characteristics in absorption spectra of native blood preparation and, consequently, the concentration of oxyhemoglobin in native red blood cells.

It was found out that conduction of ILBI course causes 36.7% change of the calculated coefficients Q_1, and Q_2—35.9%; consequently, it also causes an increase in the degree of blood oxygenation of experimental animals.

Table 5.4 Amplitude of the extremums of spectra of blood absorption on three wavelengths before and after conducting ILBI

Probe No.	Experimental group of animals (standard)							
	Before conducting ILBI				After conducting ILBI			
	Amplitude of the extremums of blood absorption at three wavelengths			Amount of the protein in the probe, mg	Amplitude of the extremums of blood absorption at three wavelengths			Amount of the protein in the probe, mg
	λ_1	λ_2	λ_3		λ_1	λ_2	λ_3	
1	16.77	18.92	16.68	0.982	19.47	22.85	19.41	0.973
2	15.41	17.68	15.35	0.964	17.49	20.96	19.42	0.996
3	18.27	20.81	18.32	0.920	20.39	23.63	19.43	1.015
4	19.56	22.31	19.44	1.057	21.38	24.79	19.44	0.987

Table 5.5 Ratio of extremums of the amplitude of spectrum of blood absorption to the amount of protein in a probe at three wavelengths before and after conducting ILBI

Probe No.	Experimental group of animals (standard)							
	Before conducting ILBI				After conducting ILBI			
	Ratio of amplitude of extremums of a spectrum to the quantity of protein in a probe at three wavelengths Ka (1/mg)			Average value Ka_{λ_i} av	Ratio of amplitude of extremums of a spectrum to the quantity of protein in a probe at three wavelengths Ka (1/mg)			Average value Ka_{λ_i} av
	λ_1	λ_2	λ_3		λ_1	λ_2	λ_3	
1	17.08	19.27	16.99	Ka_{λ_1} av 17.8 ± 0.9	20.01	23.48	19.95	Ka_{λ_1} av 19.8 ± 0.9
2	15.99	18.55	15.92	Ka_{λ_2} av 20.4 ± 1.1	17.59	21.24	17.52	Ka_{λ_2} av 23.3 ± 0.9
3	19.85	22.62	19.91	Ka_{λ_3} av 17.8 ± 0.9	20.38	23.29	20/01	Ka_{λ_3} av 17.8 ± 0.9
4	18.52	21.11	18.39		21.66	25.12	21.57	

Table 5.6 Difference of the ratio of blood absorption spectra amplitude extremums to the quantity of protein in the probe Q_1 and Q_2 before and after conducting ILBI

Probe No.	Experimental group of animals (standard)					
	Before conducting ILBI			After conducting ILBI		
	Q_1	Q_2	Q_1 av, Q_2 av	Q_1	Q_2	Q_1 av, Q_2 av
1	2.19	2.28	Q_1 av	3.47	3.53	Q_1 av
2	2.56	2.63	2.53 ± 0.14	3.68	3.72	3.46 ± 0.11
3	2.77	2.71	Q_2 av	3.21	3.28	Q_2 av
4	2.60	2.72	2.59 ± 0.12	3.46	3.54	3.52 ± 0.10

Analyzing the derived experimental information from the point of view of possible mechanism of laser photoactivation of hemoglobin in blood, it is possible to assume the following: Chromophore hemoglobin group, hem, obviously sustains intramolecular redistribution of electronic levels under the influence of monochromatic radiation that results in significant enforcement of affinity of porphyrin ring of hem to oxygen. The studies of blood preparations of a rabbit showed that at ILBI absorption spectra aspect the changes typical for the oxygenated form of this protein arise.

As far as the influence of laser radiation led to the increase of the level of oxygenation, it can be assumed that the mechanism of photosensitization of reaction of formation of oxyhemoglobin at ILBI is due to intramolecular conformational changes in the structure of protein part of the hemoglobin molecule along with direct influence of laser light on the molecule of chromophore (hem).

We should not exclude the fact that the formation of the process of photosensitization of the formation of oxyhemoglobin, and consequently the reaction of oxygen transfer is due to the activity of biologically active compounds, the emission of which into the bloodstream is stimulated by the activity of laser light. A change of the bioenergetics of red blood cell can be the consequence of the activity of such endogenous bioregulators, the final result of which is higher level of oxygenation of endoglobular hemoglobin in animals with ILBI that has been found out by the experiment.

Based on the results of the conducted work, it is possible to conclude that the further development of the ideas about the mechanism of photochemical activation of oxyhemoglobin compounds in condition in vivo can be reached in the studies with the use of model systems with direct radiation of laser light as a structural form of hemoglobin (solid red blood cell) and also the homogenous aqueous solution of this protein.

5.14.2 Studies of Influence of ILBI on Absorption Spectra of Blood Preparation of Animals with Traumatic Damages of Peripheral Nerves During the Usage of Radiation of He–Ne Laser

This paragraph is devoted to the study of spectral properties of radiation in a range of 200–900 nm scattered by native blood preparations of experimental animals during traumatic damage of peripheral nerves (PN) before and after the course of ILBI therapy (He–Ne laser $\lambda = 632.8$ nm).

The studies were conducted at blood preparations of nine experimental animals. With the help of spectrophotometer "Specord M-40," absorption spectra of blood preparation were registered, on which characteristic local extremums were registered at wavelengths of 542, 560 and 577 nm, the amplitude of which according to the previously conducted studies was determined by the degree of blood oxygenation. Protein concentration in the blood probes under study was determined by the biuretic method.

The studies of absorption spectra of blood preparation were conducted for the estimation of ILBI influence on the change of the concentration of oxyhemoglobin at traumatic damages of sciatic nerve; also, the protein concentration in the probes under study in nine experimental animals before and after conducting ILBI course (10 sessions during 10 days) was determined. The exposure time amounted to 10 min at the power of radiation at the output of light guide equal to 2–2.5 mW.

ILBI course therapy at rabbits with the trauma of PN was started in 10 days after operation. There was one course (10 procedures).

Spectral properties of radiation scattered by blood preparation of experimental animals were studied in one month after conducting ILBI therapy.

Along with this, absorption spectra (1st and 2nd probes) were registered twice and protein concentration in blood preparation in seven experimental animals, which constitute control group, was determined. Animals with the damage of PN served as control, and they did not undergo the course of ILBI therapy.

With the aim of developing a criterion for the quantitative estimation of concentration of oxyhemoglobin in blood and normalizing the results of measurements, coefficients Q_1 and Q_2 were calculated that correspond to the ratio of amplitudes of characteristic local extremums, of registered spectral curves at wavelengths of 542, 560, 577 nm and concentration of protein in the probes under study

$$Q_1 = \mathrm{Ka}_{\lambda_1} - \mathrm{Ka}_{\lambda_2}; \quad Q_2 = \mathrm{Ka}_{\lambda_3} - \mathrm{Ka}_{\lambda_2},$$

where $\mathrm{Ka}_{\lambda_1} = A_{542}/B$; $\mathrm{Ka}_{\lambda_2} = A_{560}/B$; $\mathrm{Ka}_{\lambda_3} = A_{577}/B$, where $A_{542}, A_{560}, A_{577}$ are the amplitudes of spectral curve at wavelengths of 542, 560 and 577 nm; B is the concentration of protein in probes under study.

The results of the measurements and calculations of coefficients Q_1 and Q_2, suggested for the estimation of the degree of oxygenation of blood preparations, are presented in Tables 5.7, 5.8 and 5.9.

On the basis of the experimental studies conducted with the help of three-wavelength spectrophotometric method, it was found out that ILBI at traumatic damages of PN in experimental animals leads to the change of 30 and 37%, respectively, to the calculated coefficients, which correspond to the difference between amplitude ratio of local extremums of absorption spectra of native blood preparation at wavelengths $\lambda_1 = 542$ nm, $\lambda_2 = 560$ nm and $\lambda_3 = 577$ nm to the protein concentration in the probes under study, in comparison with the analogous coefficients of animals of the control group.

5.15 Studies of Spectral Properties of Radiation Scattered by Blood Preparation of Animals at Partial Ischemia of Sciatic Nerve Before and After ILBI with the Help of Semiconductor Laser

The study was conducted on the spectral properties of radiation in the range of 200–900 nm scattered by native blood preparations of experimental animals (rabbits) at partial ischemia of sciatic nerve before and after the course of ILBI therapy (semiconductor laser with $\lambda = 850$ nm).

The studies were conducted on blood preparation of 17 experimental animals.

With the help of spectrophotometer "Specord M-40," absorption spectra of blood preparation were registered, on which characteristic local extremums at wavelengths of 542, 569 and 577 nm were recorded, the amplitude of which was determined by the degree of blood oxygenation. The concentration of protein in the blood probes under study was defined by biuretic method.

For the estimation of the influence of ILBI on the change of concentration of oxyhemoglobin at partial ischemia of sciatic nerve, spectra of absorption of blood preparation were studied and protein concentration was defined in the probes under study of ten experimental animals (rabbits) before and after conducting the course

Table 5.7 Amplitude of the extremums of blood absorption spectra at three wavelengths before and after conducting ILBI

Probe No.	Experimental group of animals (trauma of peripheral nerves)								Control group (the 1st and 2nd probes)			
	Before conducting ILBI				After conducting ILBI							
	Amplitude of the extremums of blood absorption spectra at 3 wavelengths			The amount of protein in the probe, mg	Amplitude of the extremums of blood absorption spectra at 3 wavelengths			The amount of protein in the probe, mg	Amplitude of the extremums of blood absorption spectra at 3 wavelengths			The amount of protein in the probe, mg
	λ_1	λ_2	λ_3		λ_1	λ_2	λ_3		λ_1	λ_2	λ_3	
1	16.70	18.96	16.59	1.349	19.68	22.18	19.61	0.735	10.49 11.12	12.72 13.67	10.54 11.03	1.21 1.487
2	11.24	13.34	11.25	1.510	16.99	19.26	17.09	0.909	10.17 10.30	12.67 12.97	10.28 10.54	0.992 0.985
3	10.02	12.01	9.95	1.451	18.31	20.73	18.27	0.973	11.32 9.83	13.43 12.31	11.34 9.79	1.29 0.937
4	11.94	14.00	11.98	0.925	18.95	21.22	18.98	0.858	12.98 14.38	12.67 16.49	15.36 14.32	1.37 1.316
5	10.60	12.74	10.41	0.973	14.67	16.97	14.71	0.962	9.56 9.47	11.17 11.83	9.52 9.45	0.914 0.918
6	9.75	11.89	9.77	1.085	16.61	19.03	16.65	1.035	10.29 11.31	12.34 13.74	10.38 11.26	1.042 1.263
7	10.41	12.32	10.38	1.303	14.76	17.54	15.21	1.061	11.07 10.12	13.11 12.43	11.14 10.07	1.11 1.104
8	9.66	11.97	9.66	0.909	15.26	17.49	15.21	1.061				
9	10.87	13.08	10.82	1.223	12.49	14.77	12.57	1.145				

Table 5.8 Ratio of the amplitude of the extremums of blood absorption spectra to the amount of protein at three wavelengths before and after conducting ILBI

Probe No.	Experimental group of animals (trauma of peripheral nerves)								Control group (the 1st and 2nd probes)			
	Before conducting ILBI				After conducting ILBI							
	The ratio of the amplitude of the extremums of blood absorption spectra to the amount of protein at three wavelengths Ka, 1/mg			Average value of Ka	The ratio of the amplitude of the extremums of blood absorption spectra to the amount of protein at three wavelengths Ka, 1/mg			Average value of Ka	The ratio of the amplitude of the extremums of blood absorption spectra to the amount of protein at three wavelengths Ka, 1/mg			Average value of Ka
	Ka_{λ_1}	Ka_{λ_2}	Ka_{λ_3}		Ka_{λ_1}	Ka_{λ_2}	Ka_{λ_3}		Ka_{λ_1}	Ka_{λ_2}	Ka_{λ_3}	
1	12.38	14.10	12.29		26.81	30.20	26.73		7.48	9.19	7.42	$Ka_{\lambda_1}av_1$ 9.64 ± 0.29
2	7.42	8.83	7.45	$Ka_{\lambda_1}av$ 9.667 ± 0.76	18.72	21.26	18.89	$Ka_{\lambda_1}av$ 17.666 ± 1.7	10.46	13.17	10.70	$Ka_{\lambda_1}av_2$ 9.69 ± 0.50
3	6.90	8.28	6.86		18.84	21.33	18.82		11.32 9.83	13.43 12.31	11.34 9.79	$Ka_{\lambda_2}av_2$ 11.5 ± 0.4
4	12.91	15.14	12.95		22.13	24.72	22.14		10.93	12.53	10.88	$Ka_{\lambda_2}av_2$ 11.8 ± 0.6
5	10.91	13.09	10.70	$Ka_{\lambda_2}av$ 11.5 ± 0.7	15.31	17.57	15.28	$Ka_{\lambda_2}av$ 20.1 ± 1.7	10.32	12.89	10.29	$Ka_{\lambda_3}av_2$ 9.69 ± 0.29
6	8.99	10.96	9.00		16.04	18.39	16.09		8.95	10.88	8.92	$Ka_{\lambda_3}av_2$ 9.68 ± 0.51
7	7.99	9.46	7.89		15.55	18.49	15.43		9.17	11.26	9.12	
8	10.63	13.17	10.63	$Ka_{\lambda_3}av$ 9.63 ± 0.75	14.40	16.48	14.34	$Ka_{\lambda_3}av$ 17.6 ± 1.6				
9	8.89	10.70	8.85		10.90	12.90	10.98					

Table 5.9 Difference in the ratio of the blood absorption spectra amplitude extremums to the quantity of protein in the probe Q_1 and Q_2 before and after conducting ILBI

Probe No.	Experimental group of animals (trauma of peripheral nerves)						Control group (the 1st and 2nd probes)		
	Before conducting ILBI			After conducting ILBI					
	Q_1	Q_2	Q_1 av, Q_2 av	Q_1	Q_2	Q_1 av, Q_2 av	Q_1	Q_2	Q_1 av, Q_2 av
1	1.72	2.28		3.39	3.47		1.71	1.77	$Q_1 \text{cp}_1 \pm S_x = 1.90 \pm 0.18$
2	1.41	2.63	$Q_1 \text{cp} \pm S_x = 1.86 \pm 0.14$	2.54	2.37	$Q_1 \text{cp} \pm S_x = 2.52 \pm 0.15$	2.71	2.47	$Q_1 \text{cp} \pm S_x = 2.18 \pm 0.18$
3	1.38	2.71		2.49	2.51		2.65	2.69	
4	2.23	2.72		2.59	2.58		1.61	1.65	$Q_2 \text{cp}_2 \pm S_x = 1.85 \pm 0.11$
5	2.18	2.39	$Q_2 \text{cp} \pm S_x = 1.89 \pm 0.15$	2.26	2.29	$Q_2 \text{cp} \pm S_x = 2.52 \pm 0.17$	2.57	2.60	$Q_2 \text{cp} \pm S_x = 2.18 \pm 0.17$
6	1.97	1.96		2.35	2.30		1.93	1.96	
7	1.47	1.48		2.94	3.06		2.09	2.14	
8	2.54	2.54		2.08	2.14				
9	1.81	1.85		2.00	1.92				

of ILBI (10 sessions within 10 days). Radiation was conducted with the help of the device for ILBI (semiconductor laser with $\lambda = 850$ nm). Power of radiation amounted to 2 mW.

Along with these experiments based on the given technique, absorption spectra were registered twice and concentration of protein was defined in blood preparation in seven experimental animals, which constituted control group.

For the quantitative estimation of the concentration of oxyhemoglobin in blood preparation of experimental animals before and after the conduction of ILBI and normalization of the results of measurements, there were used the calculated ratios of amplitudes of characteristic local extremums of the registered spectral curves of absorption to the concentration of protein in the blood probes under study, and also coefficients Q_1 and Q_2, corresponding to:

$$Q_1 = \mathrm{Ka}_{\lambda_1} - \mathrm{Ka}_{\lambda_2}; \quad Q_2 = \mathrm{Ka}_{\lambda_3} - \mathrm{Ka}_{\lambda_2},$$

where $\mathrm{Ka}_{\lambda_1} = A_{542}/B$; $\mathrm{Ka}_{\lambda_2} = A_{560}/B$; and $\mathrm{Ka}_{\lambda_3} = A_{577}/B$ are the ratios of amplitudes of local extremums of absorption spectra at the wavelengths of 542 and 560 and 577 nm to the protein concentration in the blood probe under study, respectively.

The results of the conducted measurements and calculations of the coefficients Q_1 and Q_2, suggested for the estimation of the degree of oxygenation of blood preparation, are presented in Tables 5.10, 5.11 and 5.12.

With the help of three-wavelength spectrophotometric method, it was found out that the conduction of ILBI therapy by the radiation of semiconductor laser at partial ischemia of sciatic nerve in experimental animals leads to 55% change of the calculated coefficient Q_1 that corresponds to the difference between the ratios of amplitudes of local extremums of absorption spectra of blood preparation at wavelengths of 542 and 560 nm to the concentration in the probes under study, and to the 57% change of coefficient Q_2, corresponding to the difference between ratios of amplitudes of local extremums of spectral curves at wavelengths of 542 and 577 nm to the concentration of protein in the probes under study, in comparison with the analogous coefficients in animals of control group.

Thereby, as a result of conducted studies it was found out that the size of spectral intensity changes considerably as a result of distinctive reflective features of skin of different people. However, the relative value of local extremums for the wavelengths $\lambda_1 = 542$ nm, $\lambda_2 = 560$ nm and $\lambda_3 = 577$ nm remains practically stable for all the respondents.

The conducted studies of spectral characteristics of human blood preparation in vivo for the range of 200–700 nm at different levels of oxygenation showed that there are the same local extremums $\lambda_1 = 542$ nm, $\lambda_2 = 560$ nm and $\lambda_3 = 577$ nm at spectral curves of blood preparation, relative values of which considerably change during the change of the level of oxygenation. The curve of dependence between oxygenation level of human blood preparation and relative values of local extremums at corresponding spectral curves was made, by which it is possible to determine the blood preparation oxygenation level.

Table 5.10 Amplitude of the extremums of the blood absorption spectrum at three wavelengths before and after conducting ILBI

Probe No.	Experimental group of animals (partial ischemia of a sciatic nerve)								Control group (the 1st and 2nd probes)			
	Before conducting ILBI				After conducting ILBI							
	Amplitude of the extremums of the blood absorption spectra at three wavelengths			The amount of protein in the probe, mg	Amplitude of the extremums of the blood absorption spectra at three wavelengths			The amount of protein in the probe, mg	Amplitude of the extremums of the blood absorption spectra at three wavelengths			The amount of protein in the probe, mg
	λ_1	λ_2	λ_3		λ_1	λ_2	λ_3		λ_1	λ_2	λ_3	
1	10.16	12.03	10.09	1.213	18.86	21.61	18.81	0.832	10.38 11.43	12.21 13.64	10.46 11.47	1.27 1.192
2	9.82	11.52	9.80	0.984	19.47	22.13	19.41	0.907	9.87 10.04	11.70 12.15	9.84 10.06	1.063 1.087
3	11.74	13.61	11.78	1.065	17.54	20.57	17.63	0.764	11.41 10.43	13.26 12.32	11.54 10.41	1.156 0.958
4	10.94	13.44	10.88	1.302	17.87	20.81	17.82	1.034	13.18 12.16	14.88 13.87	13.23 12.21	1.416 1.429
5	10.64	12.52	10.74	1.122	20.26	23.38	20.31	0.937	10.21 11.30	12.17 11.61	10.16 11.33	1.032 0.996
6	11.91	13.87	11.94	0.971	15.77	18.73	15.73	0.798	11.58 13.21	13.41 15.14	11.64 13.18	1.128 1.216
7	11.26	13.17	11.23	1.217	16.52	19.39	16.49	0.987	10.08 9.63	11.73 11.3	10.21 9.61	0.986 0.924
8	14.71	16.91	14.67	1.423	19.17	22.23	19.21	0.892				
9	9.72	11.46	9.76	0.992	16.43	19.74	16.41	1.096				
10	10.24	12.06	10.21	1.176	15.87	18.81	15.81	1.115				

Table 5.11 Ratio of the amplitude of the extremums of blood absorption spectra to the amount of protein at three wavelengths before and after conducting ILBI

Probe No.	Experimental group of animals (partial ischemia of a sciatic nerve)								Control group (the 1st and 2nd probes)			
	Before conducting ILBI				After conducting ILBI				The ratio of the amplitude of the extremums of blood absorption spectra to the amount of protein at three wavelengths Ka, 1/mg			Average value of Ka
	The ratio of the amplitude of the extremums of blood absorption spectra to the amount of protein at three wavelengths Ka, 1/mg			Average value of Ka	The ratio of the amplitude of the extremums of blood absorption spectra to the amount of protein at three wavelengths Ka, 1/mg			Average value of Ka				
	Ka_{λ_1}	Ka_{λ_2}	Ka_{λ_3}		Ka_{λ_1}	Ka_{λ_2}	Ka_{λ_3}		Ka_{λ_1}	Ka_{λ_2}	Ka_{λ_3}	
1	8.38	9.92	8.32		26.67	25.97	22.61		9.59	11.44	9.62	$Ka_{\lambda_1} av_1$ 9.57 ± 0.30
2	9.98	11.71	9.96	$Ka_{\lambda_1} av$ 9.676 ± 0.41	21.47	24.40	21.40	$Ka_{\lambda_1} av$ 19.3 ± 1.1	9.24	11.18	9.25	$Ka_{\lambda_1} av_2$ 10.1 ± 0.4
3	11.02	12.78	11.06		22.96	26.92	23.08		10.89	12.86	10.87	
4	8.40	10.32	8.36		17.28	20.12	17.23		8.51	9.70	8.54	$Ka_{\lambda_1} av_1$ 11.2 ± 0.4
5	9.48	11.16	9.57	$Ka_{\lambda_2} av$ 11.5 ± 0.4	21.62	24.95	21.68	$Ka_{\lambda_2} av$ 22.2 ± 1.2	11.35	13.66	11.38	$Ka_{\lambda_2} av_2$ 11.9 ± 0.5
6	12.27	14.28	12.30		16.76	23.47	19.71		10.86	12.45	10.84	$Ka_{\lambda_3} av_1$ 9.53 ± 0.30
7	7.99	9.46	7.98	$Ka_{\lambda_3} av$ 17.6 ± 1.6	16.55	19.65	16.71	$Ka_{\lambda_3} av$ 19.3 ± 1.1	10.42	12.22	10.40	$Ka_{\lambda_3} av_2$ 10.1 ± 0.4
8	10.63	13.17	10.63	$Ka_{\lambda_3} av$ 9.63 ± 0.75	14.40	16.48	14.34					
9	8.89	10.70	8.85		10.90	12.90	10.98					

Table 5.12 Difference of the ratio of the blood absorption spectra amplitude extremums to the quantity of protein in the probe Q_1 and Q_2 before and after conducting ILBI

Probe No.	Experimental group of animals (partial ischemia of a sciatic nerve)						Control group (the 1st and 2nd probes)		
	Before conducting ILBI			After conducting ILBI					
	Q_1	Q_2	Q_1 av, Q_2 av	Q_1	Q_2	Q_1 av, Q_2 av	Q_1	Q_2	Q_1 av, Q_2 av
1	1.54	1.60		3.30	3.36		1.85	1.82	
2	1.73	1.75		2.93	3.00		1.94	1.93	$Q_1av_1 \pm S_x = 1.81 \pm 0.40$
3	1.76	1.72	$Q_1av \pm S_x = 1.71 \pm 0.05$	3.96	3.84	$Q_1av \pm S_x = 3.21 \pm 10.13$	1.97	1.99	$Q_1av_2 \pm S_x = 1.81 \pm 0.14$
4	1.92	1.96		2.84	-2.89		1.19	1.16	
5	1.68	1.59		3.33	3.27		2.31	2.28	$Q_2av_1 \pm S_x = 1.87 \pm 0.21$
6	2.01	1.98	$Q_2av \pm S_x = 1.70 \pm 0.05$	371	3.76	$Q_2av \pm S_x = 3.22 \pm 0.13$	1.59	1.61	$Q_2av_2 \pm S_x = 1.87 \pm 0.14$
7	1.57	1.59		2,91	2.94		1.80		
8	1.55	1.52		3,43	3.38				
9	1.75	1.72		3.02	3.04				
10	1.54	1.57		2.63	2.69				

The studies of spectral properties of radiation scattered by human skin during artificially caused ischemia in the part of the limb under study showed that with an increase of the degree of ischemia, that is with the decrease of the degree of blood oxygenation, the value of two relatively local extremums decreases to zero and after becomes negative. The curve of the dependence of the change of relative value of second local extremum (λ_2, λ_3) on the degree of compressing the limb was made.

During the removal of air-inflated collar, abrupt increase of relative quantity of extremum happens. According to the conducted studies, the given spectral method could be used for the estimation of the degree of ischemia and the level of oxygenation of human skin.

5.16 Results and Conclusions

The studies dedicated to the development of speckle-optical methods for determination of parameters such as velocity of contraction and deformations of diffuse objects, which are nerve stems, muscles, epithelial tissues, were conducted.

1. The technique of the study of the contraction process of muscle tissue, based on speckle counting, was developed. The dependence of the velocity of contraction of biceps muscle of thigh of a frog on time and mechanical load was experimentally studied. The conducted studies showed two phases of contraction: The first, relatively short (10–80 ms), corresponds to the active process of muscles contraction, and the second (150–200 ms)—to the passive process of relaxation. Absolute values of the maximal velocity of contraction depend on the size of outer stretching force and are different for various preparations (reach several dozens of meters within a second). It was shown that the maximal velocity of muscle contraction develops in 20 ms at duration of all phase of muscle contraction of 80 ms. During the relaxation, maximal velocity values are approximately two times smaller. Derived information coincides with the results of direct myography traditionally applied for the studies of muscle contraction. At the same time, the used method possesses a number of advantages: It is contactless, allows conducting studies by placing the preparation arbitrary and not even extracting it, is zero inertia and allows studying the movement of separate small areas of preparation.
2. The method of double-pulsed speckle-interferometry in real time for the determination of longitudinal shift of an object (muscle), which allows increasing the limit of measurable shifts by two orders in regard to the shift range determined by the known techniques, was suggested and experimentally realized. The principal possibility of using speckle-interferometry for the study of surface deformations of human tissues was shown. On the setup developed with the use of monopulse ruby laser $\lambda = 694$ nm, with the help of double-pulsed speckle-photography method with the accuracy of ~10 μm the field of deformation of surface tissues at the process of muscle contraction from the surface (2 cm^2) at spatial resolution of 2–4 mm^2 in 20 s after the beginning of contraction was studied. The vector

diagram of shifts in 10–15 ms, the range of which amounted to from 40 to 250 μm, was made.

Besides, the possibility of study using the method of speckle-optics of elasto-mechanical properties of an isolated nerve stem, which preserves its viability at stretching, was shown. The contactlessness of such a method in combination with the possibility of determining the whole deformation field can make it a unique instrument in the study of elastic properties of a nerve and the homogeneity of their deformations depending on the force of stretching.

3. The method of speckle-optics was suggested and experimentally studied, which allows defining not only the size, but also the vector of velocity of a diffuse object due to additional modulation of shift of speckle-fields.

Both isometric and isotonic contraction and stretching of skeletal muscle of vertebrata were theoretically studied. It was shown that amplitude of muscle oscillations can reach several microns. For the case of small incidence angles of optical radiation at oscillating skin surface, spatial distribution of the intensity of scattered radiation was calculated. It was shown that the amplitude of muscle oscillation does not exceed Δ_z situated on the line section of the intensity distribution.

4. The complex of scientific research and instrument-making works providing the implementation of principally new contactless coherence-optical methods of diagnostics, treatment and prevention of atherosclerosis and other diseases, which are connected with the disorders of vascular and metabolic processes, was conducted. During these studies, physical principles of a new class of medical–biological devices based on holography and speckle-optics for remote noninvasive diagnostics of diseases and preventive examination with high precision, sensitivity and big spatial resolution were developed, and that allowed increasing the objectivity, information value and credibility of derived results. A laser-holographic complex, namely holographic cardiograph (Chap. 3), laser speckle-optical microhematomyograph for contactless defining of muscle tone and surface blood flow with defining the components of this integration process, was created on the basis of these developments. The field of application of the data from coherent-optical devices can spread not only on the measurement of muscle tone and skin blood flow, but also on the studies of other tissues and organs. It is supposed that these developments may be used in the sphere of cardiosurgery in clinical practice, along with electrocardiography, angiocardiography and other methods.
5. An experimental model of laser specklometer consisting of a measuring block, a signal processing module, a module of low and high voltage, was developed, produced and tested. With its help, medical–biological experiments were conducted, which allowed developing informative parameters for the analysis of information, derived during the studies of biomechanical characteristics of skeletal muscle and microhemodynamics of skin.

The laser specklometer is equipped with software and is used together with computer. Software of experimental model of laser specklometer is assigned for the automatization of spectro-metrical measurements: information collection, Fourier analysis, statistical processing and graphic means of visualization of the derived results. The possibility of using the majority of classical methods of spectral estimation based on the algorithm of quick Fourier transformation is realized in it.

The laser specklometer is protected by the certificate of authorship of the USSR No. 1620037 (11.07.08) and does not have analogs on the territory of the former USSR and countries of Eastern Europe. The device possesses novelty in regard to countries of the former USSR, USA and Germany.

6. The preliminary studies of characteristics of reconstructing of illuminating–receiving head and the whole laser specklometer with the help of calibrated vibrator were conducted. It was shown that for the frequency of oscillations of 50 Hz during the increase of the amplitude of oscillation from 4 to 10.7 μm and during the usage of reconfigurable illuminating–receiving head only one main harmonic (others were more than 100 times smaller in their amplitude) was observed in the spectrum. During this, the dependences of amplitude changes of spectral harmonic and the area under spectral curve on the oscillation amplitude of the surface of the vibrator under study were practically linear. This fact during deciding the issue of normalizing the signal allows using laser specklometer for the quantitative estimation of not only the frequencies, but also the amplitudes of vibrations of the surface under study, and also its little movements. Additional studies of laser specklometer with the help of calibrating vibrator showed that the measurements conducted with the use of reconfigurable illuminating–receiving head are practically linear in the range from 4 to 11 μm, and consequently it gives the possibility to conduct qualitative estimation of the amplitudes and frequencies of oscillations under study.
7. With the aim of development of speckle-optical technique, the studies of amplitude–frequency characteristics of ultrasonic acoustic transformer of YZPYA2M3.507.178 type were conducted.

Resonance USWS frequency was determined, and the dependence of oscillation amplitude of the working end of USWS microinstrument on frequency of vibrations and current of transformer near resonance was obtained.

These characteristics are necessary for the optimization of calculations and construction of USWS. The derived information was compared with the results of the studies with the help of the laser Doppler vibrometer.

In these studies, the following was used: the sample of laser specklometer—device designed for registration in relative units of temporary fluctuations of intensity of speckle-field, formed as a result of scattering by vibrating diffuse objects of radiation and calculation of their amplitude–frequency characteristics during collaboration with computer.

The laser specklometer allowed measuring the amplitude in the range of 1–15 μm and the frequency of vibration in the range of 1–2000 Hz with the resolution power of 1 μm.

Determining the vibration amplitude was conducted by registration with the help of laser specklometer of the amplitude of the first spectral harmonic of fluctuation spectrum of the radiation intensity, scattered by diffuse surface, which makes harmonic vibration, with further recalculation of amplitude values according to the derived beforehand calibration curve.

The calibration of experimental model of laser specklometer was carried out with the help of certified model of the laser Doppler vibrometer. During this, relative error of the measurement of the amplitude of longitudinal component of harmonic vibration amounted to not more than 5% and frequency of vibrations did not exceed 4%.

The developed methods and means of the defining amplitude–frequency characteristics of vibrations of diffuse objects based on the analysis of time fluctuations of the intensity of speckle-field formed as a result of the dispersion of laser radiation by vibrating objects.

The derived results of the work testify the effectiveness of the usage of the developed method and device for the vibrations of a wide range of objects in real-time scale.

Principal importance of the derived results consists in the fact that on their basis in industry the possibility of the usage of remote, contactless sensors of location and positioning appeared, study with the help of experimental model of laser specklometer of vibrating activity of the elements of technical items by relatively simple and sufficiently cheap method.

The results of the work can be used for the studies of ultrasonic fields of acoustic converters and oscillating systems, which are used in technological processes at the enterprises of electronic industry, definitions of vibrating activity of the elements of technical items, motor and tractor technology, developments of the location and positioning sensors in the technology and robotics.

8. Contactless method for the estimation of muscle tone depending on functional state of the organism was developed. The results of the studies of muscle-tonic condition during the measurements of temporary spectra of fluctuations of intensity of laser radiation scattered on them were derived. The method of effective removal of the influence of surface blood flow on the speckle-optical indicators during the study of biomechanical characteristics of muscles was developed.

It is known that a muscle at its contraction creates low-frequency sound vibrations, the intensity of which depends on its functional condition. However, acoustic measurements are hindered due to considerable noises in the low-frequency area. The usage of laser radiation gives a wide range of advantages: first of all, contactless, remote, insensitive toward noises in the area of sound oscillations, high spatial resolution.

It was found out that it is preferable to use integral parameters of spectra due to their larger degree of stability. Besides, it is also reasonable to choose relative, but not absolute parameters of spectra, and to create techniques of the study of biophysical processes based on the comparative principle. The derived results can be applied during conducting of myographic studies in neurosurgery, restoration surgery, muscle tone determination.

As a result of the conducted studies the most sensitive to the changes of muscles, physical condition spectrum parameters, namely its power and medium frequency, were detected. It was found out that with an increase of isotonic load applied to the biceps muscle of arm, power and medium frequency of a spectrum increase up to 80% from maximal effort developed by a patient. Further increase of the load leads to the decrease of muscle vibrational activity, which is confirmed by the decrease of power and medium frequency of a spectrum. Analogous results were described in the literature for sound myography. Bioelectrical muscle activity grows proportionally to the increase of isotonic load up to maximal effort developed by a muscle. Thereby, as the studies showed, the laser specklometer allows receiving additional information about vibrational function of a muscle, which cannot be obtained with the help of electromyography.

A series of studies of vibrational characteristics of human skeletal muscles at different physical exercises was conducted. For every given constant load, 10 spectra were registered in the same conditions. Statistical processing of the results was carried out. The dependences of informational parameters of spectra, namely the area under the curve and average frequency of the physical load intensity, were made. It was shown that in average the obtained curves increase monotonically with the growth of loads up to 80% from effort maximally developed by a patient. With the help of speckle-optical methods, it was also found out that at different degrees of manifestation and distribution of muscle fixation in patients with reflectory and radicular syndromes of lumbar osteochondrosis there is observed the difference of speckle-optical indicators of muscle activity of paravertebral muscles both at the level of damaged segment at the sore and intact side and in cases of spread myohypertonicity at overlying areas of chest.

9. The informative parameters of fluctuation spectra of intensity of dynamical speckle-field, which reflect biophysical processes occurred in the skin blood flow, which can also characterize pathophysiological changes in the tissues under studies, were detected. The possibility of study of surface blood flow in the limited zone allowed determining its segmental differences along the spine in patients with vertebral pathology. The increase of the frequency of spectra fluctuations of the intensity of speckle-field in the zone of damaged vertebral–locomotive segment is accompanied by the growth of the temperature in this area and indicates the intensification of skin microhemodynamics. This phenomenon can be the reflection of dyscirculatory processes, which occur in located deeper peridisk tissues, disks, when sanogenetic reactions prevail also in patients with dystrophic changed disk in order to reduce the disorders of metabolism, microcirculatory processes are intensified.

The dynamics of skin blood flow of lower limbs in patients with neurologic manifestation of lumbar osteochondrosis corresponded to the change of high-frequency fluctuation intensity spectra of speckle-field. Here shifts of regional blood flow on the lower limbs, which were detected by reovasography and which reflect angiospastic processes, were accompanied by the lowering of skin temperature and change of the frequency of fluctuations spectra. On the contrary, during unchangeable regional blood flow (at the intact limb) and absence of vegetative changes such as disorders of the thermoregulation, a shift of frequency spectra of fluctuations into high-frequency area was observed. Consequently, the increase of the intensity of blood supply of covering tissues and growth of the volume of sanguiferous microcirculatory blood channel find its reflection in the increase of the impact of high frequencies into the spectrum of fluctuations of the intensity of speckle-field. The same fact is indicated by the lowering asymmetry coefficient of spectrum relative to the average frequency. The proof of this data was found in the results of the experiment with artificially caused hyperemia of limited areas of skin of a hand, where in the area of applying irritative agent a burst of the most high-frequency fluctuation spectra, which correspond to the points of the most intensive microhemodynamic changes, was observed.

It is important to point out that during the study of skin microhemodynamics proximo-distal gradient of speckle-optical indicators in patients with polyneuropathies, which correlates with the changes of skin temperature and the data of thermal imaging observation, was detected.

It was found out that in patients with neurological diseases and in animals with damages of peripheral nervous system the course application of the method of intravenous laser blood irradiation mobilizes the system of neurohumoral regulation, stimulating hypophysis and adrenals, thyroid and thyroid glands; has immunomodulatory effect on the organism; affects the structure of membranes of erythrocytes and white blood cell part of the blood positively; and increases the level of hemoglobin and the number of erythrocytes. The clinical effect of ILBI correlates with the lowering of the activity of system of lipid peroxidation, and antioxidant system activates; the evidence of vertebral syndrome decreases, and rebuilding of sensory and motional function of nervous system accelerates. It is shown that ILBI stimulates the regeneration of damaged peripheral nerve after neurography (in the experiment).

It was detected that conducting the course of intravenous laser blood irradiation by emission of He–Ne laser in the early post-operational period in patients with traumatic damages of peripheral nerves activates skin microhemodynamics that is manifested by the increase of average frequency of fluctuations of the intensity of speckle-field, formed by the scattered radiation of skin. It was shown that the therapeutic ILBI effect is more evident on the damaged limb; however, low-intensity laser irradiation caused the increase of skin microhemodynamics also at the healthy limb.

Differences of fluctuation intensity spectra of laser radiation on diseased and healthy limbs were detected. Specklograms were presented, which had been detected in patient suffered from vertebrogenit radiculitis with the damage of a root. In the zone of innervation of mainly this root at the back surface of lower third of a crus, the most obvious asymmetry of indicators under study is observed, while the difference in the conditions of surface blood flow on the diseased and healthy side in other points is either less obvious (the area of a thumb) or is absent at all (the front surface of lower third of a hip).

Thereby, the revealed physical parameters of spectra of fluctuations of intensity of dynamic speckle-field are characterized as physiological and pathologic changes occurred in the tissues under study and can be used for the estimation of functional state of skin blood flow.

10. Complex theoretical and experimental studies of patho- and sanogenetic processes occurred during certain pathologic conditions of peripheral nervous system by the methods of coherent optics were conducted.

The technique of clinical application of intravenous laser blood irradiation of patients during certain pathologic conditions of peripheral nervous system was developed. It was shown that during the treatment of patients with peripheral nervous system ILBI stimulates the immune system. It was found out that ILBI does not influence the activity of complement and content of immune complexes. It was found out that during the treatment of patients with vertebrogenti damages of peripheral nervous system ILBI has positive dynamics in the clinics and does not cause any side and negative effects.

The possibilities of application of methods of speckle-optics for the detection of the intensity of skin blood flow during different diseases of peripheral nervous system were studied. Segmental disorders of surface microhemodynamics along the spine and on the skin of limbs in the areas of innervation of roots were detected. According to the results of speckle-optical study and functional condition of limb vessels according to the data from reovasography at vertebral neurological pathology in patients with mono- and poly-neuropathies, the correlated connection of disorders of surface microhemodynamics was detected. The influence of hyperbaric oxygenation on the skin blood flow in patients with neurologic developments of lumbar osteochondrosis was shown. The results of speckle-optical studies can be used for the estimation of the condition of surface microhemodynamics and the effectiveness of conducted therapy.

Using ILBI during the treatment of patients with damaged peripheral nervous system leads to the decrease of leukocytosis and normalization of the stability of red blood cells to the hemolysis. At the same time, the stimulation of immune system and oxidative reaction are detected.

During the treatment of patients with vertebrogenit damages of peripheral nervous system with the method of intravenous laser irradiation, positive dynamics in the clinics was derived. The given method does not have any side and negative effects. Laser radiation stimulates the proliferation of ill people a little bit more than timalin that testifies that laser radiation affects similarly to timalin—stimulator of T-system of immunity—and can be applied for the correction of immunodeficiency states.

Application of low-intensity laser irradiation does not have any damaging influence on blood formation and cells of peripheral blood that is especially urgent during the stimulation of common protective forces of organism and also local resistibility.

It was found out that dealing with rabbits application of intravascular laser irradiation during experimental transection of sciatic nerve with its further sewing has positive effect on the flow of pathology process. Most probably that it is connected with the fact that ILBI through special mechanisms hinders the development of inflammatory process and activation of inhibitors of growing of fibrous tissue at the same time that creates the most favorable conditions for the regeneration of fibrous tissues.

High sensitivity of speckle-optical methods of regeneration of hemodynamic disorders in microcirculatory channel of skin was registered. Dealing with animals suffering from occlusion of femoral artery progressive decrease of power of spectrum of fluctuation intensity of speckle-field was observed in comparison with the norm by 41–67% depending in terms of the operation that is connected with considerable vasomotor disorders.

A considerable decrease of power spectrum of skin blood flow at the operated limb during traumatic damage of sciatic nerve with further neurography by 67.3% in the transmission band of 0–2000 Hz in comparison with intact group was detected. According to the results of the conducted studies on animals, the application of intravenous laser blood irradiation ILBI at traumatic damages of peripheral nervous system decreases trophic disturbances, postponing the time of their appearing in general. Morphological studies showed that ILBI contributes to the development of connective tissue, causes earlier normalization of pulse activity and quickens regeneration processes of a damaged nerve.

Besides, on the basis of in-depth study of the mechanisms of the influence of intravenous laser blood irradiation on the principal manifestations of atherosclerotic process depending on the degree of its manifestation there will be developed special techniques, which allow determining the regimes of the radiation of whole blood and vascular walls (intensity and duration of laser radiation, frequency and time of influence), and provide the best therapeutic effect. It was shown that the developed speckle-optical methods reflecting biophysical processes in capillary skin channel are applicable for the estimation of microhemodynamics of human skin during vertebral pathology. The studies of microhemodynamics of human skin in different functional states both in separate points and in skin areas of $40 \times 55\ \text{mm}^2$ were conducted with the help of laser specklometer. Microhemodynamic maps of skin areas under study were built on the basis of the measurements.

11. The studies of the change of the concentration of oxyhemoglobin of surface blood flow in different functional states of skin were conducted, and a new spectrometric method of determining the degree of ischemia of human organs and tissues according to the change of the concentration of oxyhemoglobin of skin was suggested. The dependence of the local extremums of spectral curve of the reflection coefficients of skin on the level of concentration of oxyhemoglobin of surface blood flow and the degree of ischemia was found.

Spectral properties of radiation scattered by human skin in the range of 200–700 nm were studied.

On the basis of experimental studies conducted with the help of three-wave long spectrophotometer method, it was found out that during the traumatic damage of peripheral nervous system in experimental animals ILBI leads to the changes of the calculated coefficients to 30 and 37% that correspond to the difference between ratios of amplitudes of local extremums of absorption spectrums of native blood preparations at wavelengths of 542, 560 and 577 nm to the concentration of protein in the probes under study in comparison with the same coefficients in animals of control group. With the help of three-wavelength spectrophotometric method, it was found out that the conduction of the course of ILBI therapy by the radiation of semiconductor laser during partial ischemia of sciatic nerve in experimental animals leads to the 55% change of the calculated coefficient Q_1, that corresponds to the difference between ratios of amplitudes of local extremums of absorption of blood preparation at wavelengths of 542 and 560 nm to the concentration in the probes under study to the change up to 57% of the corresponding coefficient Q_2 at wavelengths of 542 and 577 nm in comparison with the same coefficients in animals of control group.

On the basis of conducted studies, it was found out that the given spectral method can be used for the estimation of the oxygenation degree and defining the degree of ischemia of human skin.

References[1]

1. G.M. Frank, *With the Eyes of a Scientist* (Academy of Sciences of the BSSR, Moscow, 1963), 579 p. (in Russian)
2. J.W. Goodman, JOSA **66**(11), 1145 (1976)
3. I.A. Popov, L.M. Veselov, Opt. Commun. **105**(3–4), 167 (1994)
4. T. Alexander, J. Harvey, A. Weeks, Appl. Opt. **33**(35), 8240 (1994)
5. J.H. Churnside, H.T. Yura, Appl. Opt. **20**, 3539 (1981)
6. M.N. Lee, J.F. Holmes, J.R. Kerr, JOSA **66**, 1164 (1976)
7. I. Yamaguchi, S.I. Komatsy, Optica Acta **24**(7), 705 (1977)
8. E. Ochoa, J.W. Goodman, JOSA **73**(3), 943 (1983)
9. Y. Aizu, T. Asakura, Opt. Laser Techol. **23**, 205 (1991)
10. M. Gemert, G. Schets, M. Bishop, Laser Life Sci. **2**(1), 1 (1988)
11. G. Yoon, A. Weich, M. Motamedi, IEEE J. Quantum Electron. **23**(10), 1721 (1987)
12. T. Asakura, N. Takai, Appl. Phys. **25**(3), 179 (1981)
13. N. Takai, Sutanto, T. Asakura, JOSA A **70**(7), 827 (1980)
14. J. Ohtsubo, Optik **57**(2), 183 (1980)
15. I. Markhvida, Correlation properties of dynamic speckle fields and their use for measuring the velocities and displacements of diffuse objects, Abstract of the dissertation, Minsk, 1989. (in Russian)
16. N. Takai, T. Asakura, Appl. Opt. **17**(23), 3785 (1978)
17. B. Ruth, D. Haina, W. Waidelich, Optica Acta **30**(6), 841 (1983)
18. S. Komatsu, K. Osato, H. Ohzu, Opt. Commun. **39**(6), 357 (1981)
19. Y. Aizu, T. Asakura, Opt. Commun. **68**(5), 329 (1988)
20. L. Veselov, I.A. Popov, Opti. Range **68**(6), 1155 (1990). (in Russian)
21. L. Veselov, I. Popov, Opt. Spectr. **74**(6), 1155 (1993). (in Russian)
22. M. Giglio, S. Musazzi, U. Perini, Appl. Opt. **20**(5), 721 (1981)

[1]We provide the main [see references 82–84, 119–122, 126, 127, 131, 134–138, 168, 170, 189–198, 208–210, 213, 219, 220, 224, 234, 235, 240, 241, 245, 248, 250, 252, 253, 259, 260, 262, 264, 268, 270] and below additional list of scientific works carried out in the "National Research Center for Neurology and Neurosurgery," of the Ministry of Health of the Republic of Belarus using the diagnostic device "Laser specklometer" scientific basis and structural developments of which are presented in this monograph (Chap. 5).

23. A. Hayashi, Y. Kidagava, Opt. Commun. **49**(2), 91 (1984)
24. N. Takai, T. Iwai, T. Asakura, Appl. Opt. **22**(1), 170 (1983)
25. J. Goodman, *Laser Speckle and Related Phenomena* (1975), p. 9
26. I. Yoshimura, JOSA **3A**(7), 1032 (1986)
27. S. Komatsu, K. Osato, H. Ohzu, Pot. Commun. **39**(6), 357 (1981)
28. J.H. Churnside, JOSA **72**(11), 1464 (1982)
29. N.A. Fomin, *Speckle Interferometry of Gas Flows* (Science and Technology, Minsk, 1989), 166 p. (in Russian)
30. A.E. Ennos, Prog. Opt. **16**, 233 (1978)
31. I.S. Klimenko, *Holography of Focused Images and Speckle Interferometry* (Nauka, Moscow, 1985), 222 p. (in Russian)
32. Ch.I. Vest, *Holographic Interferometry* (Mir, Moscow, 1982), 504 p. (in Russian)
33. Yu.I. Ostrovsky, M.M. Butusov, G.V. Ostrovskaya, *Holographic Interferometry* (Nauka, Moscow, 1977), 336 p. (in Russian)
34. Yu.I. Ostrovsky, L.V. Tanin, ZhTF **1**, 1756 (1975). (in Russian)
35. V.S. Siniyakov, Bull. Exp. Biol. Med. **11**, 495 (1986). (in Russian)
36. H.E. Hoyer, J. Dorheide, *Biomechanics: Basic and Applied Research*, vol. 3 (1987), p. 135
37. I.V. Markhvida, L.V. Tanin, V.V. Dubovik, Archive AGE (5), 73 (1988). (in Russian)
38. O.P. Bolshakov, T.A. Iliinskaya, N.E. Protsenko, *Proceedings of the International Conference "Advances in the Biomechanics of Medicine"*, Riga, Latvian SSR, 1986. (in Russian)
39. I.V. Markhvida, L.V. Tanin, Proc. BSSR Biol. Ser. **2**, 115 (1986). (in Russian)
40. A.V. Sokolov, Coherent optical methods of studying of biological microorganisms, Abstract of the dissertation, Leningrad, 1981. (in Russian)
41. A.E. Fercher, J.D. Briers, Opt. Commun. **37**, 326 (1981)
42. A. Oulamara, G. Tribillon, J. Duvernoy, J. Mod. Opt. **36**(2), 165 (1989)
43. A.V. Priezzhev, V.V. Tuchin, L.P. Shubochkin, Proc. AS BSSR Phys. Ser. **53**(8), 1490 (1989). (in Russian)
44. H. Fujii, T. Asakura, K. Nohira et al., Opt. Lett. **10**(3), 104 (1985)
45. H. Tasawa, T. Hiraguchi, T. Asakura, Med. Big End Comput. **89**(27), 580 (1989)
46. E.M. Boer, D.R. Bezencer, Dermatol. Beruf Umwelf **37**(2), 58 (1989)
47. E. Okada, H. Minamitani, Y. Fukuoka, *Proceedings of the Annual International Conference of IEE Engineering in Medicine and Biology Society*, New Orleans, USA, 4–7 Nov 1988
48. H.Z. Cummins, E.R. Pike (eds.), *Spectroscopy of Optical Mixing and Correlation of Photons* (Springer, Berlin, 1978). (in Russian)
49. E.V. Bulaenko, M.D. Prokoptsov, P.A. Friedman, Dev. Tech. Exp. **2**, 202 (1994). (in Russian)
50. I.P. Antonov, S.D. Bezzubik, S.K. Dik, V.V. Dubovik, I.V. Markhvida, L.V. Tanin, Physical factors and tools in medicine, 32 (1986). (in Russian)
51. S.K. Dik, E.V. Zhuk, I.V. Markhvida, *Proceedings of the Conference "Holography in Industry and Research"*, Grodno, 1986, p. 133. (in Russian)
52. I.P. Antonov, A.V. Goroshkov, V.N. Kalyunov, I.V. Markhvida, A.S. Rubanov, L.V. Tanin, ZhPS **39**(1), 103 (1983). (in Russian)
53. D.A. Zimnyakov, V.V. Tuchin, G.E. Brill, *Abstracts of the Conference "Photonics West'95"*, San Jose, 1–28 Feb 1995
54. T. Higgins, Laser Focus Word **28**(8), 67 (1992)
55. F. Dalmases, R. Cibrian, M. Buendia et al., Phys. Med. Biol. **K33**(8), 913 (1988)
56. L.V. Tanin, S.K. Dik, S.A. Aleksandrov, Proc. Acad. Sci. Phys. Ser. **59**(6), 84 (1995). (in Russian)
57. S.A. Aleksandrov, S.K. Dik, M.M. Loiko, L.V. Tanin et al., Patent 2061514RU C1 (Russia), 1996. (in Russian)
58. L.V. Tanin, S.A. Aleksandrov, S.K. Dik, Opt. Prot. Man. Env. Nat. Tech. Dis., 239 (1993)
59. S.B. Dublin, J.H. Lunacek, Proc. Nat. Acad. Sci. **57**, 1164 (1967)
60. J.D. Briers, Opt. Commun. **13**(3), 324 (1975)
61. R. Nossal, S.H. Chen, Opt. Commun. **5**, 117 (1972)
62. S. Fufime, M. Maruyama, S. Asakura, J. Mol. Biol. **68**, 347 (1972)

63. T.I. Shaw, B.J. Newby, Biochim. Biophys. Acta **255**(1), 411 (1972)
64. R.V. Piddington, *Spectroscopy of Optical Mixing and Correlation Photons* (Mir, Moscow, 1987). (in Russian)
65. C.E. Riva, J.E. Gruhwald, B.L. Petriy, Appl. Opt. **24**(5), 605 (1985)
66. I.D. Swain, L.I. Grant, Phys. Med. Bid. **34**(2), 151 (1989)
67. N. George, Opt. Eng. **25**(6), 754 (1986)
68. V.S. Voronov, A.D. Drogin, Mech. Compos. Mater (6), 520 (1985). (in Russian)
69. M. Singh, G. Ramachandran, A. Periasamy, J. Opt. **15**(3), 63 (1986)
70. B. Ruth, D. Haina, W. Weidelich, Optica Acta **30**(9), 1213 (1983)
71. N. Takai, T. Iwai, T. Asakura, JOSA **70**(4), 450 (1980)
72. T. Iwai, N. Takai, T. Asakura, Opt. Acta **28**(6), 857 (1981)
73. B. Ruth, D. Abendroth, L. Sunder-Plassmann, W. Waidelich, *Proceedings of 7th Congress International Society for Laser Surgery and Medicine in Connection with Laser'87*, Berlin, 1987
74. B. Ruth, J. Schmand, D. Abendroth, Lasers Surg. Med. **13**(2), 179 (1993)
75. H. Fujii, K. Nohira, Y. Jamamoto, Y. Ikawa, T. Ohura, Appl. Opt. **26**(24), 5321 (1987)
76. A.F. Fercher, M. Peukert, E. Roth, Opt. Eng. **25**(6), 731 (1986)
77. I.D. Swain, L.J. Yrant, Biophys. J. **30**, 119 (1980)
78. T. Asakura, L. Yrant, Anritsu News **38**(38), 152 (1988)
79. Y. Airu, T. Asakura, Opt. Laser Technol. **10**(7), 827 (1991)
80. L.V. Tanin, S.K. Dik, I.P. Antonov, I.V. Markhvida, L.I. Rachkowsky, L.A. Vasilevskaya, **14**, 48 (1992). (in Russian)
81. L.V. Tanin, A.S. Rubanov, I.Y. Markhvida, S.K. Dik, L.I. Rachkovsky, *Proceedings SPIE Holography'89*, vol. 1183 (1989), p. 425
82. S.K. Dik, I.V. Markhvida, D. Rachkowsky, I. Antonov, L.V. Tanin, Copyright certificate 1620037, 11 July 1988. (in Russian)
83. R.R. Anderson, I.A. Parrish, J. Invest. Dermatol. **77**(1), 13 (1981)
84. B. Ruth, J. Mod. Opt. **34**(2), 257 (1987)
85. E. Okada, H. Minamitani, Y. Fukuoka et al., *Proceedings of the Annual International Conference of the IEEE Engineering in Medicine and Biology Society*, New Orleans, LA, USA, 4–7 Nov 1988, p. 1321
86. Z. Yi, X. Ni, Y. Xu, The development at clinical practice of microcirculatory blood flow monitor. Paper presented at the annual international conference of the IEEE Engineering in Medicine and Biology Society, 1987
87. L.V. Tanin, A.L. Tanin, *Biomedical and Resonance Optics: Theory and Practice* (Belaruskaya navuka, Minsk, 2011), 687 p. (in Russian)
88. I.S. Klimenko, *Holography of Focused Images and Speckle Interferometry* (Nauka, Moscow, 1985), 224 p. (in Russian)
89. H.E. Hoyer, J. Dorheide, in *Biomechanics: Basic and Applied Research* ed. by G. Bergmann, R. Kölbel, A. Rohlmann. Developments in Biomechanics (Springer, Dordrecht, 1987), p. 135
90. S.K. Dik, E.V. Zhuk, I.V. Markhvida et al., *Proceedings of the Conference "Holography in Industry and Research"*, Grodno, 1986, p. 133. (in Russian)
91. E. Archbold, J.M. Burch, A.E. Ennos, Opt. Acta **17**(12), 883 (1970)
92. A.E. Ennos, *Laser Speckle and Related Phenomena* (1975), p. 203
93. I.V. Volkov, I.S. Klimenko, ZhTF **50**(5), 1038 (1980). (in Russian)
94. E. Archbold, A.E. Ennos, Opt. Acta **19**(4), 253 (1972)
95. I.S. Klimenko, V.P. Ryabukho, ZhTF **52**(5), 896 (1982). (in Russian)
96. I.S. Klimenko, T.V. Kuznetsova, A.N. Malov, ZhTF **56**(9), 1744 (1986). (in Russian)
97. G.R. Lockshin, *Methods and Devices of Optical Holography* (FTI, Leningrad, 1983), pp. 125–146. (in Russian)
98. I.V. Markhvida, L.V. Tanin, I.A. Utkin, *The Use of Modern Physical Methods in Non-destructive Research and Control* (Khabarovsk, 1987), p. 211. (in Russian)
99. I.V. Markhvida, L.V. Tanin, I.A. Utkin, ZhTF **1**, 121 (1988). (in Russian)
100. I.V. Markhvida, L.V. Tanin, I.A. Utkin, Opt. Spectrosc. (1987). (in Russian)

101. I.V. Makrhvida, L.V. Tanin, Copyright certificate 1310624 (USSR), 1984. (in Russian)
102. D.W. Li, P.P. Chiang, JOSA A **3**(7), 1023 (1986)
103. I. Yamaguchi, JOSA A **1**(1), 81 (1984)
104. N. Takai, T. Asakura, H. Ambar, JOSA A **3**(8), 1305 (1986)
105. F.P. Chiang, Q.B. Li, Appl. Opt. **23**(24), 4469 (1984)
106. L.V. Tanin, V.A. Dmitriev, *Proceedings of the International Conference on Laser Applications in Life Sciences*, Minsk, 1994
107. V.I. Descherevsky, *Mathematical Models of Muscle Contraction* (Nauka, Moscow, 1977), 160 p. (in Russian)
108. V.I. Descherevsky, Biophysics **13**(5), 928 (1986). (in Russian)
109. G. Oster, World Sci. (5), 62 (1984). (in Russian)
110. G. Oster, J.S. Jaffe, Biophys. J. **30**, 119 (1980)
111. J.V. Frangioni, T.S. Kwan-Gett, L.E. Dobrunz, T.A. McMahon, Biophys. J. **51**, 775 (1987)
112. B. Diemont, *Proceeding 7th Nordic Meeting on Medical and Biological Engineering*, Trondheim, 1987
113. M.R. Shorten, D.J. Kerwin, *Biomechanics: Basic and Applied Research*, ed. by G. Bergmann, R. Kölbel, A. Rohlmann. Developments in Biomechanics (Springer, Dordrecht, 1987), p. 141
114. B.A. Rhatigan, K.S. Mylrea, IEEE Trans. Biomed. BME **33**(10), 967 (1986)
115. S.K. Dik, I.V. Markhvida, L.I. Rachkovsky, A.S. Rubanov, L.V. Tanin, *Proceedings of the Conference on CLEO'89*, Baltimore, 1989, p. 54
116. L.V. Tanin, S.K. Dik, I.V. Markhvida, L.I. Rachkovsky, A.S. Rubanov, *Proceedings of the International Conference «Optics in Life Sciences» ICO 15 SAT*, Garmish-Partenkirchen, 12–15 Aug 1990, p. 23
117. L.V. Tanin, Speckle-optics and holography in investigations of nervous-muscular tissue. Paper presented at the international conference «Optics in life sciences» ICO 15 SAT, Garmish-Partenkirchen, 12–15 Aug 1990
118. L.V. Tanin, A.A. Kumeisha, I.V. Markhvida, M.M. Loiko, S.A. Aleksandrov, S.K. Dik, *Proceedings SPIE*, vol. 2083 (1993), p. 280
119. S.K. Dik, L.V. Tanin, L.I. Rachkovsky, I.V. Markhvida, A.S. Rubanov, L.A. Vasilevskaya, *Proceedings of VI All-Union Conference on Holography*, Vitebsk, 1990, pp. 157–158. (in Russian)
120. S.K. Dik, L.V. Tanin, I.V. Markhvida, *Proceedings SPIE*, vol. 1454 (1991), p. 447
121. S.K. Dik, I.V. Markhvida, L.V. Tanin, *Abstracts of the International Congress of Optical Science and Engineering*, Hague, 1991, v.1508
122. L.V. Tanin, I.V. Markhvida, A.S. Rubanov, S.K. Dik, L.I. Rachkovsky, *Optics in Medicine, Biology and Environmental Research* (Elsevier, Amsterdam, New York, 1993), p. 149
123. L.V. Tanin, *Optics in Medicine, Biology and Environmental Research* (Elsevier, Amsterdam, New York, 1993)
124. L.V. Tanin, S.K. Dik, I.V. Markhvida, A.S. Rubanov, *Laser Optics'93: Proc. Reports*, vol. 2 (1993), p. 554. (in Russian)
125. S.K. Dik, I.V. Markhvida, L.V. Tanin, A.S. Rubanov, Copyright certificate 1810748. (in Russian)
126. S.K. Dik, L.V. Tanin, L.I. Rachkowsky, I.V. Markhvida, Copyright certificate 1765768 (USSR). (in Russian)
127. L.V. Tanin, A.A. Kumeisha, M.M. Loiko, S.A. Aleksandrov, I.V. Markhvida, S.K. Dik, L.A. Vasilevskaya, *Abstract Book of BiOS Europe'93, International Symposium on Biomedical Optics, Budapest*, Hungary, 1–5 Sept 1993, p. 49
128. L.V. Tanin, S.K. Dik, I.V. Markhvida, Proc. Rus. Acad. Sci. Phys. Ser. (1994). (in Russian)
129. L.V. Tanin, S.K. Dik, *Proceedings SPIE*, vol. 2176 (1994), p. 332
130. L.V. Tanin, S.K. Dik, S.A. Aleksandrov, M.M. Loiko, L.A. Vasilevskaya, *Abstracts of BiOS Europe'94. International Symposium on Biomedical Optics*, Lille, 1994
131. L.V. Tanin, S.A. Aleksandrov, M.M. Loiko, S.K. Dik, L.A. Vasilevskaya, A.A. Kumeisha, *Optical Method in Bio-Medical and Environmental Sciences* (Elseiver, Tokyo, Amsterdam, 1994), p. 76

132. L.V. Tanin, S.A. Aleksandrov, A.A. Kumeysha, M.M. Loyko, I.V. Markhvida, S.K. Dik, L.A. Vasilevskaya, Peripheral Nerv. Syst. **17**, 50 (1994). (in Russian)
133. L.V. Tanin, S.K. Dik, S.A. Aleksandrov, M.M. Loiko, A.A. Kumeisha, I.V. Markhvida, L.A. Vasilevskaya, *Laser Optics ICONO'95*, vol. 11 (1995), p. 304
134. L.V. Tanin, S.K. Dik, S.A. Aleksandrov, M.M. Loiko, L.A. Vasilevskaya, I.V. Markhvida, *Abstracts of International Conference "Light and Biological Systems"*, Wroclaw, Poland, 3–6 July 1995
135. G.A. Ivanichev, Clinic, diagnostics, mechanisms of development and treatment of miofascic-ulatory hypertensive syndromes (local muscular hypertonicity), Abstract of the dissertation, Kazan State Institute of Advanced Training of the Doctors, 1986. (in Russian)
136. M.A. Franson, *Optics of Speckles* (Mir, Moscow, 1980), 171 p. (in Russian)
137. V.S. Gurfinkel, U.S. Levik, *Skeletal Muscle. The Structure and Function* (Nauka, Moscow, 1985), 144 p. (in Russian)
138. I.V. Markhvida, L.V. Tanin, *Proceedings of the UE Respiratory Conference. Young Scientists*, Minsk, 1986, p. 133. (in Russian)
139. D.P. Danilova, J. Neurol. Psychiatry (1979). (in Russian)
140. T. Ohhashi, N.G. Mcnale, I.C. Roddie, Pflugers Arch. **388**(3), 221 (1980)
141. I. Tasaki, K. Iwasa, J. Physiol. **77**(9), 1055 (1981)
142. I. Tasaki, K. Iwasa, J. Upsala, J. Med. Sci. **85**(3), 211 (1980)
143. I. Tasaki, K. Iwasa, Biol. Bull. **159**(2), 494 (1980)
144. A. Fulton, *Cytoskeleton. Architecture and Choreography of a Cell* (Moscow, 1987). (in Russian)
145. M. Born, E. Wolf, *Principles of Optics* (Nauka, Moscow, 1973), p. 719. (in Russian)
146. S.K. Dik, L.V. Tanin, V.M. Aranchuk, M.M. Korol, A.S. Rubanov, *Proceedings of Optics Within Life Sciences OWLS IV*, Munster, Germany, 1996, p. 6
147. V.M. Aranchuk, Opt. Spectr. **78**(2), 284 (1995). (in Russian)
148. Y.F. Zastrogin, *Precision Measurements of Motion Parameters Using a Laser* (Mashinos-troenie, Moscow, 1986), 272 p. (in Russian)
149. V. Klyuyev (ed.), *Devices and Systems for the Measurement of Vibration, Noise and Impact: Handbook*, vol. 2 (Mashinostroenie, Moscow, 1978), 439 p. (in Russian)
150. V.M. Aranchuk, Methods and means of laser interferometry for control of multicomponent vibration of solids, Abstract of the dissertation, Minsk, 1989. (in Russian)
151. V.S. Gutnikov, *Filtration of Measurement Signals* (Energoatomizdat, Leningrad, 1990), 192 p. (in Russian)
152. M.F. Ponomarenko, V.M. Karpa, V.D. Tsidelko, Informational spectral estimation (measure-ment, control, automation) 3, 55 (1987). (in Russian)
153. J. Max, *Methods and Techniques of Signal Processing in the Physical Dimensions* (Mir, Moscow, 1983), 312 p. (in Russian)
154. C. Orizio, R. Perini, A. Velctelnas, Enr. J. Apple Physiol. **58**, 528 (1989)
155. D.T. Barry, Biophys. J. **51**, 769 (1987)
156. D.T. Barry, N.M. Cole, Biophys. J. **53**, 899 (1988)
157. I.P. Antonov, G.K. Nedzved, Peripherals Nerve Syst. **5**, 129 (1982). (in Russian)
158. Y. Aizu, T. Asakura, Opt. Laser Technol. **23**, 205 (1991)
159. B. Ruth, D. Haina, W. Waidlich, Opt. Acta **31**, 759 (1984)
160. J. Aizu, K. Ogino, *Laser Anemometry in Fluid Mechanics*, vol. 3 (1987)
161. B. Ruth, Int. J. Mocricir. Clim. Exp. **12**(3), 227 (1993)
162. R.I. Utyamishev, I.M. Arefief, A.G. Mazovetsky, Copyright certificate AS 975004 (the USSR). (in Russian)
163. L.A. Vasilevskaya, Functional disorders of neuromotor apparatus, their speckle-optical diag-nostics and correction with the means of low intensity laser radiation in certain diseases of the peripheral system, Abstract of the dissertation, Minsk State Medical Institute, 1999. (in Russian)
164. J. Aizu, T. Asakura, Opt. Laser Technol. **23**(4), 398 (1991)
165. L.A. Vasilevskaya, Peripheral Nerve Syst. **21**, 6 (1998). (in Russian)

166. N.I. Nechipurenko, A.R. Gavrilova, R.M. Tanina, G.G. Petrovsky, L.V. Tanin, *Proceedings of the Third Congress of the Belarusian Society Photobiology and Biophysics*, Minsk, 1998. (in Russian)
167. V.V. Skupchenko, T.G. Makhovskaya, G.N. Handurina, *Proceedings of the Conference 'Low-Intensity Laser Radiation in Medical Practice"*, Khabarovsk, 1990, p. 148. (in Russian)
168. S.E. Rovdo, A.R. Gavrilova, N.U. Chumakova, Peripheral Nerv. Syst. (21), 39 (1998). (in Russian)
169. E.M. Yurah, Questions of morphology in theor. and clinical medicine 91, 43 (1982). (in Russian)
170. A.A. Novozemtseva, V.A. Levitsky, S.E. Rokhind, Morphogenesis and regeneration, 25 (1977). (in Russian)
171. E.P. Melman, V.A. Levitsky, B.M. Mytskan, *Proceedings of IX All-Union Congress of Anatomists, Histologists, Embryologists*, Minsk, 1981. (in Russian)
172. G.A. Azizov, Laser photocorrelation of microcirculatory disorders in patients with chronic arterial ischemia of lower limbs, Abstract of the dissertation, Moscow, 1991. (in Russian)
173. B.J. Saligeperova, Formation of roundabout routes of blood flow in the pelvic limbs under laser irradiation, Abstract of the dissertation, Kharkov, 1991. (in Russian)
174. V.L. Zelyak, V.A. Levitsky, I.I. Gerzanich, Fiziol. J. **33**(5), 87 (1987). (in Russian)
175. V.I. Eliseenko, Usage of lasers for diagnostics and treatment of diseases, 3 (1996). (in Russian)
176. M.Z. Kreyman, I.R. Udalyi, *Low-Energy Laser Therapy* (Tomsk, 1992). (in Russian)
177. R.Z. Losev, O.A. Tsarev, V.P. Romankin, Low-intensity laser radiation in medical practice, 100 (1990). (in Russian)
178. V.N. Shabalin, T.V. Skokova, A.J. Olshansky, Immunology **6**, 30 (1990). (in Russian)
179. N.I. Nechipurenko, A.R. Gavrilova, R.M. Tanina, P.A. Vlasyuk, L.V. Tanin, I.P. Antonov, Eur. J. Neurol. **4**(1), 117 (1997)
180. N.I. Nechipurenko, L.A. Vasilevskaya, A.R. Gavrilova, *Proceedings of the 3rd International Conference* **6**(2) (1998), p. 55. (in Russian)
181. N.I. Nechipurenko, I.P. Antonov, G.V. Sherstyuk, L.V.Tanin, Peripheral Nerv. Syst. Sci. Works (21), 13 (1998). (in Russian)
182. A.L. Tanin, Recovery of function of peripheral nerves after neurography under the influence of laser hemotherapy, Abstract of the dissertation, Institute of Neurology, Neurosurgery and Physiotherapy, 2003. (in Russian)
183. L.A. Vasilevskaya, N.I. Nechipurenko, Actual problems of neurology and neurosurgery (8), 20 (2006). (in Russian)
184. N.I. Nechipurenko, E.A. Korotkevich, L.A. Vasilevskaya, L.I. Matusevich, N.U. Shcherbina, *Proceedings of the Conference "Current Approaches and the Introduction of New Techniques in the Diagnostics"*, 25 Mar 2005, p. 147. (in Russian)
185. S.A. Likhachev, E.V. Veevnik, L.A. Vasilevskaya, Zh. of Grodno State Medical University 3, 111 (2006). (in Russian)
186. S.A. Likhachev, E.V. Veevnik, L.A. Vasilevskaya, *Guidance for Doctors Based on Materials of National Congress* (Moscow, 2008), p. 307. (in Russian)
187. S.A. Likhachev, E.V. Veevnik, L.A. Vasilevskaya, Patent 11237 (Belarus), 22 July 2008. (in Russian)
188. S.A. Likhachev, L.D. Vasilevskaya, E.V. Veevnik, Actual problems of neurology and neurosurgery **8**, 78 (2006). (in Russian)
189. S.A. Likhachev, E.V. Veevnik, L.A. Vasilevskaya, Actual problems of neurology and neurosurgery **10**, 127 (2008). (in Russian)
190. S.A. Likhachev, L.A. Vasilevskaya, E.V. Veevnik, *Proceedings of the Conference "Current Approaches and the Introduction of New Techniques in the Diagnostics"*, 25 Mar 2005, p. 122. (in Russian)
191. L.A. Vasilevskaya, S.A. Likhachev, G.V. Veevnik, Patent 9310 (Belarus), 07 Feb 2007. (in Russian)
192. S.A. Likhachev, S.K. Dik, L.A. Vasilevskaya, V.V. Vashchilin, Actual problems of neurology and neurosurgery **6**, 66 (2004). (in Russian)

193. N.I. Nechipurenko, L.A. Vasilevskaya, I.D. Pashkovskaya, Actual problems of neurology and neurosurgery 7, 103 (2005). (in Russian)
194. N.I. Nechipurenko, T.V. Griboyedova, L.A. Vasilevskaya, N.U. Shcherbina, *Proceedings of IV International Conference "Medical and Social Ecology of Personality: Status and Prospects"*, Minsk, Belarusian State University, 2006, p. 191. (in Russian)
195. N.I. Nechipurenko, L.A. Vasilevskaya, T.V. Griboyedova, N.Y. Shcherbina, Y.I. Musienko, Bull. Exp. Biol. Med. Annex D, 224 (2006). (in Russian)
196. N.I. Nechipurenko, L.A. Vasilevskaya, Y.I. Musienko, I.D. Pashkovskaya, N.U. Shcherbina, Molecular, membrane and cellular bases of functioning of biosystems **2**, 224 (2006). (in Russian)
197. N.I. Nechipurenko, L.A. Vasilevskaya, Y.I. Musienko, I.D. Pashkovskaya, *Proceedings of Scientific and Practical Conference "Recent Advances in Laser Medicine and Their Application in Medical Practice"*, Moscow, 2006, p. 176. (in Russian)
198. N.I. Nechipurenko, L.A. Vasilevskaya, Y.I. Musienko, Actual problems of neurology and neurosurgery **9**, 99 (2007). (in Russian)
199. N.I. Nechipurenko, L.A. Vasilevskaya, I.D. Pashkovskaya, Proc. Natl. Acad. Sci. Med. Ser. (3), 76 (2007). (in Russian)
200. N. Nechipurenko, L. Vasilevskaya, Y. Musienko, G. Maslova, SPIE-OSA 6632, R-OK-1 (2007)
201. L.A. Vasilevskaya, M.I. Nechipurenko, L.V. Tanin, *Secondary Reactions of Organism in the Laser Hemotherapy of Cerebral Ischemia* (Minsk, 2002). (in Russian)
202. L.A. Vasilevskaya, Actual problems of neurology and neurosurgery 6, 30 (2004). (in Russian)
203. L.A. Vasilevskaya, *Proceedings of the International Conference "Problems of Integration of Functions in Physiology and Medicine"*, Minsk, 15–16 June 2004, p. 61. (in Russian)
204. S.A. Likhachev, N.I. Nechipurenko, L.A. Vasilevskaya, *Proceedings of the IV International Scient.-Tech. Conference "Means of Medical Electronics and New Medical Technologies"*, Minsk, 2006, p. 181. (in Russian)
205. L.N. Anatskaya, N.I. Nechipurenko, E.I. Matusevich, L.A. Vasilevskaya, N.U. Shcherbina, G.K. Nedzved, *Proceedings of the Conference "Endothelial Dysfunction: Experimental and Clinical Studies"*, Vitebsk, 2006, p. 115. (in Russian)
206. N.I. Nechipurenko, U.I. Musienko, L.A. Vasilevskaya, L.I. Alekhnovich, Advances of medical science of Belarus, in *Practical Yearbook*, vol. 11 (2006), p. 140. (in Russian)
207. S.A. Likhachev, N.I. Nechipurenko, L.A. Vasilevskaya, Proc. Natl. Acad. Sci. Belarus Med. Ser. **4**, 69 (2007). (in Russian)
208. N.I. Nechipurenko, L.A. Vasilevskaya, E.K. Sidorovich, G.N. Shilov, V.V. Evstigneev, Recept **41**(3) (2005). (in Russian)
209. N.I. Nechipurenko, E.K. Sidorovich, I.P. Antonov, L.A. Vasilevskaya, Purines and nitrogen monoxide. Regulatory function in the body, 86 (2003). (in Russian)
210. N.I. Nechipurenko, L.N. Anatskaya, L.I. Matusevich, L.A. Vasilevskaya, N.U. Shcherbina, Med. Panorama **6**, 52 (2006). (in Russian)
211. L.N. Anatskaya, N.I. Nechipurenko, L.I. Matusevich, N.Yu. Shcharbina, L.A. Vasilevskaya, Int. J. Stroke **1**, 50 (2006)
212. L.N. Anatskaya, N.I. Nechipurenko, L.I. Matusevich, L.A. Vasilevskaya, J. Neurol. Psychiatry Special Issue, 229 (2007). (in Russian)
213. L.A. Vasilevskaya, Actual problems of neurology and neurosurgery 10, 37 (2008). (in Russian)
214. L.A. Vasilevskaya, N.I. Nechipurenko, L.N. Anatskaya, *Proceedings of V Scientific and Practical Conference "Endothelial Dysfunction: Experimental and Clinical Studies"*, Vitebsk, 2008. (in Russian)
215. A.L. Tanin, Actual problems of neurology and neurosurgery 1, 134 (1999). (in Russian)
216. A.L. Tanin, Actual problems of neurology and neurosurgery 2, 171 (2000). (in Russian)
217. A.L. Tanin, Actual problems of neurology and neurosurgery 3, 152 (2001). (in Russian)
218. A.L. Tanin, L.A. Vasilevskaya, A.F. Smeyanovich, Actual problems of neurology and neurosurgery 1, 138 (1999). (in Russian)
219. A.L. Tanin, Med. News (7), 55 (2000). (in Russian)

220. A.L. Tanin, V.I. Hodylev, V.I. Tanin, Actual problems of neurology and neurosurgery 4, 138 (2002). (in Russian)
221. N.I. Nechipurenko, L.A. Vasilevskaya, A.L.Tanin, The therapeutic effect of intravenous laser blood irradiation after experimental traumatic peripheral nerve injury, in *Pathological Physiology and Experimental Therapy* (Meditsina, Moscow, 1999), 13 p. (in Russian)
222. A.L. Tanin, *Proceedings of the 3rd Congress of Neurosurgeons of Russia*, St. Petersburg, 2002, p. 538. (in Russian)
223. N. I. Nechipurenko, *Proceedings of IV Bel. Soc. of Union of Photobiology and Biophysics*, Minsk, 2000, p. 230. (in Russian)
224. A. Smeyanovich, A. Tanin, N. Nechipurenko, Eur. J. Neurol. **7**(3), 1079 (2000)
225. A.F. Smeyanovich, M.I. Nechipurenko, A.L. Tanin, Patent 19980400 (Belarus), 1998. (in Russian)
226. A.L. Tanin, U.G. Shanko, *Diagnostics and Surgical Treatment of Traumatic Lesions of Peripheral Nerves of Limbs* (BelMAPO, Minsk, 2008), 60 p. (in Russian)
227. Y.G. Shanko, A.L. Tanin, A.N. Naledko, *Brain Injury* (Belprint, Minsk, 2008), 96 p. (in Russian)
228. L.V. Tanin, N.I. Nechipurenko, L.A. Vasilevskaya, G.K. Nedzved, S.E. Rovdo, A.L. Tanin, *Laser Hematherapy in the Treatment of Diseases of the Peripheral Nervous System* (LLC "Magic Book", Minsk, 2004), 148 p. (in Russian)
229. L.V. Tanin, I.V. Markhvida, A.A. Kumeysha, S.K. Dik, S.A. Aleksandrov, M.M. Loyko, I.P. Antonov, L.A. Vasilevskaya, Adv. Med. Sci. Belarus **2**, 17 (1997). (in Russian)
230. N.L. Nechipurenko, L.V. Tanin, I.P. Antonov, L.A. Vasilevskaya, P.A. Vlasyuk, *Abstracts of the European Biomedical Optics Week BIOS Europe'96*, Vienna, 1996
231. R.M. Tanina, T.V. Gavrilina, L.V. Tanin, S.V. Rovdo, G.V. Sherstyuk, *Abstracts of the International Conference on New Technologies in the Humanities and Fourth International Conference on Optics Within Life Sciences (OWLS IV)*, Munster, Germany, 1996
232. L.V. Tanin, G.G. Petrovsky, R.M. Tanina, *Proceedings of European Biomedical Optics Week, BIOS Europe'96*, Vienna, 1996, paper 2922-75
233. N.I. Nechipurenko, A.R. Gavrilova, R.M. Tanina, P.A. Vlasyuk, L.V. Tanin, I.P. Antonov, Eur. J. Neurol. **4**, 117 (1997)
234. L.V. Tanin, L.A. Vasilevskaya, A.V. Astapenko, *Proceedings SPIE*, vol. 3193 (1998), p. 157
235. L.A. Vasilevskaya, L.V. Tanin, S.E. Rovdo, *Proceedings of the International Symposium SPIE's BiOS'98—International Biomedical Optics Symposium*, San Jose, CA, 24–30 Jan 19, paper N3251-33
236. N.I. Nechipurenko, A.R. Gavrilov, R.M. Tanya, G.G. Petrovsky, L.V. Tanin, *Proceedings of the Third Congress "Belarusian Society Photobiology and Biophysics"*, Minsk, 1998. (in Russian)
237. V.A. Lapina, R.M. Tanina, *Proceedings of the International Symposium on Biomedical Optics Europe'93*, Budapest, Hungary, 1993, p. 80
238. R.M. Tanina, P.A. Vlasyuk, S.D. Bezzubik, *Optical Methods in Biomedical and Environmental Sciences: Selected Contributions to the Third International Conference on Optics Within Life Sciences, OWLS III*, Waseda University, Tokyo, Japan, 10–14 Apr 1994
239. L.V. Tanin, M.M. Loyko, S.K. Dik, S.A. Aleksandrov, A.A. Kumeysha, Patent application 93028293 (Belarus), 01 June 1993. (in Russian)
240. L.V. Tanin, I.V. Markhvida, I.L. Drobot, *Proceedings SPIE*, vol. 1508 (1991), p. 128
241. L.V. Tanin, S.K. Dik, *Proceedings SPIE*, vol. 2248 (1994), p. 427
242. S.K. Dik, L.V. Tanin, L.A. Vasilevskaya, P.N. Bagrov, T.D. Lebedeva, *Proceedings SPIE*, vol. 4515 (2000), p. 185
243. N.I. Nechipurenko, I.P. Antonov, G.V. Sherstyuk, L.V. Tanin, Peripheral Nerv. Syst. **21**, 13 (1998). (in Russian)
244. N.I. Nechipurenko, L.V. Tanin, I.P. Antonov, L.A. Vasilevskaya, Lasers Med. Sci. **15**, 101 (2000)
245. L.V. Tanin, *Proceedings of the International Conference "Lasers in Biomedicine"*, vol. 1 (2003), p. 73. (in Russian)

246. L.V. Tanin, *Proceedings of IX International Conference on Laser Applications in Life Sciences*, Vilnius, Lithuania, 7–11 July 2002, p. 73
247. S.E. Rovdo, A.R. Gavrilova, I.P. Antonov, L.V. Tanin, Peripheral Nerv. Syst. **19**, 79 (1996). (in Russian)
248. L.V. Tanin, I.P. Antonov, L.A. Vasilevskaya, *Proceedings SPIE*, vol. 2623 (1995), p. 321
249. I.P. Antonov, L.V. Tanin, N.I. Nechipurenko, A.R. Gavrilova, L.A. Vasilevskaya, Adv. Med. Sci. Belarus **3**, 5 (1998). (in Russian)
250. L.A. Vasilevskaya, M.I. Nechipurenko, L.V. Tanin, I.P. Antonov, *Abstracts of I Caucasian Symposium on Life Sciences*, Tbilisi, 1999. (in Russian)
251. N.I. Nechipurenko, L.A. Vasilevskaya, R.M. Tanina, P.A. Vlasyuk, in *Abstracts of the Conference "The Effectiveness of Low-Intensity Laser Radiation in Experimental Cerebral Ischemia"*, Moscow, 2001. (in Russian)
252. L.A. Vasilevskaya, L.V. Tanin, P.A. Vlasyuk, Actual problems of neurology and neurosurgery 3, 26 (2001). (in Russian)
253. L.A. Vasilevskaya, L.V. Tanin, A.N. Fillipovich, N.F. Fillipovich, S.M. Zubritsky, Actual problems of neurology and neurosurgery 4, 32 (2002). (in Russian)
254. N.I. Nechipurenko, L.A. Vasilevskaya, G.T. Maslova, Y.I. Musienko, A.L. Tanin, *Proceedings of the Conference "Experimental Study of Laser Hemotherapy in Cerebral Ischemia"*, Grodno, 2002. (in Russian)
255. L.A. Vasilevskaya, N.I. Nechipurenko, L.V. Tanin, E.A. Korotkevich, Patent of Belarus No. 9238, 25 Jan 2007. (in Russian)
256. L.A. Vasilevskaya, N.I. Nechipurenko, L.V. Tanin, *Proceedings of the Conference "Medelektronika 2002"*, Minsk, 2002, p. 262. (in Russian)
257. L.V. Tanin, L.A. Vasilevskaya, *Speckle-Optical Diagnostics of Skin Changes and Cerebral Ischemia in Experimental Microhemodynamics the Brain and the Laser Irradiation of Blood* (St. Petersburg, 2003). (in Russian)
258. N.I. Nechipurenko, L.V. Tanin, L.A. Vasilevskaya, *Proceedings of the International Conference "Eazerno-Optical Technologies in Biology and Medicine"*, Minsk, 2004. (in Russian)
259. L.A. Vasilevskaya, P.A. Vlasyuk, L.V. Tanin, Actual problems of neurology and neurosurgery 4 (2002). (in Russian)
260. L.A. Vasilevskaya, L.V. Tanin, J.I. Musienko, *Proceedings of the International Conference on Optics Within Life Sciences*, Lucerne, 2002, p. 72
261. L.A. Vasilevskaya, L.V. Tanin, *Proceedins of the International Conference "Laser Physics and Laser Applications"* Minsk, 2003, p. III-26u. (in Russian)
262. V.A. Lapina, R.M. Tanina, L.I. Matusevich, *Abstracts of Optics of Lasers'93*, vol. 2, St. Petersburg, 1993, p. 60. (in Russian)
263. L.V. Tanin, S.A. Aleksandrov, M.M. Loyko, S.K. Dik, L.A. Vasilevskaya, A.A. Kumeysha, *Proceedings of III International Conference "Optical Methods in Biomedicine"*, Tokyo, 10–14 Apr 1994. (in Russian)
264. L.V. Tanin, S.K. Dik, *International Conference of Electronic Images. Science and Technology*, San Jose, (1994). (in Russian)
265. L.V. Tanin, S.A. Aleksandrov, V.A. Lapina, R.M. Tanina, *Proceedings SPIE*, vol. 1887 (1993), p. 158
266. S.A. Aleksandrov, S.K. Dik, L.V. Tanin, R.M. Tanina, V.A. Lapina, Peripheral Nerv. Syst. (16), 13 (1993). (in Russian)
267. L.V. Tanin, S.K. Dik, S.A. Aleksandrov, Investigation of polarization properties of radiation scattered by human skin. Paper presented at the conference "Laser tehnologies'93", Shatura, 1993, Report number 21. (in Russian)
268. S.A. Aleksandrov, L.V. Tanin, S.K. Dik, G.G. Peter, R.M. Tanina, M.M. Loyko, Application for invention 93028292/028706, 1 June 1993. (in Russian)
269. L.V. Tanin, S.A. Aleksandrov, S.K. Dik, R.M. Tanina, V.A. Lapina, *Selected Scientific Works of International Society. Optics in the Life Sciences* (1993), p. 239. (in Russian)
270. R.R. Anderson, J.A. Parrish, J. Invest. Dermatol. **77**, 13 (1981)

271. J.D. Regan, J.A. Parrish, in *The Science of Photomedicine*, ed. by I.D. Regan (Springer, New York, 1982), pp. 148–192
272. J.A. Parrish, R.R. Anderson, F. Urbach, D. Pilts, *Biological Effects of Ultraviolet Radiation with Emphasis on Human Responses to Longwave Ultraviolet* (Springer, New York, 1978), p. 262
273. L. Balovitch, M. Avery, Biol. Neonate **15**, 348 (1981)
274. M.A. Edwards, S.Q. Duntley, Am. J. Anat. **1**(65), 1 (1969)
275. H. Baden, C. Pearliman, J. Invest. Dermatol. **48**, 71 (1974)
276. J.W. Feather, M. Halfzadch-Saffar, G. Leslie J. Dawson, Phys. Med. Biol. **34**(7), 807 (1989)
277. J.A. Jaquer, H. Kuppenheim, J. Appl. Physiol. **8**, 212 (1956)
278. J.W. Feather, D.J. Ellis, G. Leslie, Phys. Med. Biol. **33**(6), 711 (1988)
279. J. Tang, Ch. Fu, Laser J. **7**(4), 213 (1986)
280. R.A. Bruce, Noninvasive estimation of bilirubin and hemoglobin oxygen saturation in the skin by reflection spectrophotometry, Ph.D. thesis, Duke University, Durham, 1978
281. R.T. Madzhidov, S. Aturbekov, Science—Practical Health, 48 (1985). (in Russian)
282. L.V. Tanin, S.K. Dik, S. Aleksandrov, *Abstracts of Optics of LASERS'93*, St. Petersburg, vol. 2, 1993, p. 627. (in Russian)
283. J. Zundorf, D. Tauschek, K. Frank, *Proceedings of 19th Annual Meeting. Int. Soc. Oxyg. Tran. Tiss*, 1991, p. 222
284. J.W. Feather, M. Hajizadeh-Saffar, G. Leslie, J.B. Dawson, Phys. Bed. Biol. **34**(7), 807 (1989)

Further Reading

285. D.T. Barry, S.R. Geiringer, R.D. Ball, Muscle Nerve **8**, 189 (1985)
286. F. Dalmases, R. Cibrian, M. Buendia et al., Phys. Med. Biol. **33**(8), 913 (1988)
287. H.J. Tiziani, Opt. Acta **18**(12), 891 (1971)
288. V.A. Khanov, Optoelectronics **4**, 103 (1987). (in Russian)
289. B. Ruffing, J. Fleiscker, JOSA A **2**(10), 1637 (1985)
290. G. April, H.H. Arsenault, JOSA A **1**(7), 738 (1984)
291. T. Yoshimura, K. Nakagava, N. Wakabayashi, JOSA A **3**(7), 1018 (1986)
292. T. Iwai, N. Takai, T. Asakura, Opt. Acta **30**(6), 759 (1983)
293. H. Kogelnik, T. Li, Appl. Opt. **5**, 1550 (1966)
294. B.Ya. Zeldovich, N. Pilipetsky, V. Shkunov, *Phase Conjugation* (Nauka, Moscow, 1985), p. 247. (in Russian)
295. H. Fujii, T. Okamoto, T. Asakura, JOSA A **4**(8), 1366 (1987)
296. N. Takai, T. Asakura, Appl. Opt. **27**(3), 557 (1988)
297. J. Goodman, *Statistical Optics* (Mir, Moscow, 1988), 527 p. (in Russian)
298. J.C. Danty, Prog. Opt. **14**, 3 (1976)
299. F.P. Chiang, D.W. Li, Opt. Eng. **25**(5), 667 (1986)
300. V.V. Anisimov, S.M. Kozel, G.R. Lokshin, Opt. Spectrosc. **3**, 483 (1969)
301. V.V. Anisimov, S.M. Kozel, G.R. Lokshin, Radio engineering and electronics **15**(3), 539 (1970). (in Russian)
302. J.C. Leader, JOSA **67**, 1091 (1977)
303. N. Takai, Jpn. J. Appl. Phys. **13**, 2025 (1974)
304. T. Iwai, N. Takai, T. Asakura, Opt. Commun. **44**(1), 170 (1982)
305. P.N. Pysey, J. Phys. **9**, 1399 (1976)
306. J. Ohtsubo, T. Asakura, Opt. Commun. **14**, 30 (1975)
307. J. Ohtsubo, T. Asakura, Nouv. Rev. Opt. **6**, 189 (1975)
308. H.M. Pedersen, Opt. Acta **22**, 15 (1975)
309. H.M. Pedersen, Opt. Commun. **16**, 63 (1976)
310. E.G. Rawson, A.B. Neffarate, R.E. Norton, J.W. Goodmann, J. Opt. Soc. Am. **66**, 1290 (1976)

311. F. Roddier, Opt. Commun. **10**, 103 (1974)
312. C.P. Grover, Appl. Opt. **12**, 149 (1973)
313. J.B. Chen, F.P. Chiang, JOSA A **1**(6), 845 (1984)
314. R. Murthy, R.K. Monahty, R.S. Sironi, Optik **67**(1), 85 (1984)
315. T. Iwai, N. Takai, T. Asakura, Jpn. Appl. Phys. **22**(10), 1523 (1983)
316. T. Iwai, N. Takai, T. Asakura, Jpn. Appl. Phys. **22**(12), 1860 (1983)
317. Y. Ichioka, K. Yamamoto, T. Suzuki, Jpn. J. Appl. Phys. **20**, 239 (1975)
318. M. Christopher, J. Haccot, F. Lawson, Opt. Acta **32**(5), 537 (1985)
319. ChS Vikram, K. Vedam, Optik **64**(2), 171 (1983)
320. R.D. Bahuguna, K.N. Gupta, JOSA **70**(7), 874 (1980)
321. S. Mallick, M.L. Roblin, Opt. Commun. **6**, 45 (1972)
322. T.S. McKechnie, Opt. Commun. **13**, 29 (1975)
323. E. Marom, I. Kasher, J. Opt. (Nouv. Rev. Opt.) **8**, 1 (1977)
324. H.M. Pedereen, JOSA A **1**(8), 850 (1984)
325. E. Jakeman, E.R. Pike, G. Parry, B. Saleh, Opt. Commun. **19**, 359 (1976)
326. V.K. Karppinen, Opt. Eng. **21**(1), 60 (1982)
327. P.M. Boone, Opt. Eng. **21**(1), 407 (1982)
328. P.K. Rastogi, P. Jacquot, Opt. Eng. **21**(3), 411 (1982)
329. K. Nagata, N. Yoshida, J. Nishiwaki, Jpn. J. Appl. Phys. **9**, 505 (1970)
330. Y. Dzialowsky, M. May, Opt. Commun. **16**(3), 334 (1976)
331. A. Kozma, C.R. Christensen, J. Opt. Soc. Am. **66**, 1257 (1976)
332. A.W. Lohmann, G. Weilgeilt, Opt. Commun. **14**, 252 (1975)
333. J.A. Mendez, M.L. Roblin, Opt. Commun. **11**(3), 245 (1974)
334. JCh. Hagopian, M. May, J. Opt. **9**(4), 217 (1978)
335. O.J. Lokberg, K. Hogmen, O.M. Holja, Appl. Opt. **18**(6), 763 (1979)
336. J.A. Leenderts, J. Phys. E: Sci. Instrum. **3**(3), 214 (1970)
337. B.J. Gausman, A.J. Zakharchenko, M.I. Kataev, A.S. Larshin, Health Med. **1**, 42 (1990)
338. V.V. Skupchenko, G.G. Makhovskaya, G.N. Handurina, L.P. Mironov, Actual problems of the clinic and treatment of neurogenic pain sinbromov, 112 (1989). (in Russian)
339. U.N. Sirenko, S.Y. Savitsky, S.S. Krasnitsky, A.V. Shabilnov, L.I. Popova, J. Sov. Med. **3** (1990). (in Russian)
340. O.G. Kogan, S.N. Arsenov, A.I. Nebotkip, The use of lasers in medicine and the clinic **4**(2), 172 (1989). (in Russian)
341. A.A. Degtyareva, J.M. Chmielevski, New laser medicine and surgery **4**(1), 24 (1990). (in Russian)
342. G.G. Schwalb, A.L. Zakharchenko, A.A. Sigaevl, M.I. Kataev, J. Sov. Med. **3** (1990). (in Russian)
343. N.M. Korochkin, E.V. Babenko, J. Sov. Med. **3** (1990). (in Russian)
344. B. Beutler, A. Cerami, *The Role of Cells and Cytokines in Immunity and Inflammation* (Oxford University Press, New York, 1990), p. 226
345. E. Panasyuk, A.M. Frost, B. Gatalyak, The use of lasers in medicine, 100 (1985). (in Russian)
346. I.I. Yakovtchenko, L.I. Lebedeva, A.S. Sergievsky, A.M. Karaevkov, The action of low-intensity laser radiation, 50 (1989). (in Russian)
347. V.V. Udut, V.E. Prokopiev, A.B. Karpov, S.A. Naumov, B.N. Zyryakov, Bull. Tomsk Sci. Cent. Acad. Med. Sci. **2**, 65 (1990). (in Russian)
348. V.M. Inyushin, I.B. Beklemishev, A collection of papers on the resorts of Kazakhstan, 235 (1975). (in Russian)
349. P.D. Schwalb, M.H. Kataev, A.J. Zakharchenko, Modern methods of laser therapy 96, 3 (1989). (in Russian)
350. I.B. Laprun, Experiment in modern biology and medicine, 33 (1975). (in Russian)
351. M.S. Zhumankulov, M.S. Pluzhnikov, L.V. Shabunevich, B.S. Ivanov, L.I. Basiladze, V.E. Kholmogorov, The action of low-level laser treatment on the blood, 15 (1989). (in Russian)
352. N.F. Gamaleya, V.Y. Stadnik, Vrach. Delo **9**, 67 (1988). (in Russian)

353. S.I. Gorodenko, A.A. Novitski, I.P. Korolik, On the question of hormonal shifts in the body during whole-body irradiation of helium-neon laser in the chronic experiment, Abstract of the dissertation, Kiev, (1992). (in Russian)
354. S.I. Gorodenko, M.V. Uglova, V.V. Rossinskaya, Use the methods and means of laser technology in biology and medicine, 122 (1981). (in Russian)
355. A.K. Cheban, R.E. Golovach, A.A. Chumak, D.A. Badyka, J.N. Minchenko, The action of low-intensity laser radiation on blood, 167 (1989). (in Russian)
356. V. Rodionov, The action of low-intensity laser radiation on blood, 147 (1989). (in Russian)
357. J.A. Grinevich, I.F. Laounets, I.P. Benkovskaya, The action of low-intensity laser radiation on blood, 150 (1989). (in Russian)
358. V.L. Stadnik, A.G. Fedorchuk, L.M. Skivka, G.A. Karasevskaya, The action of low-intensity laser radiation on blood, 160 (1989). (in Russian)
359. J.A. Zozulya, P.V. Spasichenok, G.M. Oleinik, O.F. Ponomarev, G.M. Yakhnenko, Z.M. Rudykh, The action of low-intensity laser radiation on blood, 92 (1989). (in Russian)
360. V.I. Prohorenkov, A.V. Plotnikov, J. Dermatol. **1**, 34 (1991). (in Russian)
361. S.A. Shishkin, V.I. Boborov, J. Ear Dis. **29** (1991). (in Russian)
362. Y. Kushpiruk, A. Neros, The action of low-intensity laser radiation on blood, 118 (1989). (in Russian)
363. N.I. Belis, Intravascular laser blood irradiation in treatment of female infertility of inflammatory origin, Abstract of the dissertation, Kiev, 1992. (in Russian)
364. S.V. Dmitrepko, E.E. Korobkov, V.V. Osipov, K.P. Vorobyev, Modern methods of laser therapy (1989). (in Russian)
365. V.A. Voronina, et al., The use of lasers in medicine, 16 (1985). (in Russian)
366. V.E. Kuzmina, Y. Varizhnikov, The biological action of laser radiation, 51 (1984). (in Russian)
367. T. Gavrilin, I.P. Antonov, G.K. Nedzved, J. Neuropathy Psychiatry **4**, 492 (1985). (in Russian)
368. I.P. Antonov, G.K. Nedzved, T.V. Gavrilina, *Proceedings of IV All-Union Congress of Gerontology and Geriatrics*, Kiev, 1982, p. 140. (in Russian)
369. G.K. Nedzved, T.V. Gavrilina, Peripheral Nerv. Syst. **8**, 95 (1985). (in Russian)
370. A.G. Mrochek, T.A. Alyushina, Fiziol. Exp. Ther. **5**, 56 (1990). (in Russian)
371. P.M. Lapetva, S.A. Baksheva, T.S. Frazineva, Immunopathology and allergy, 181 (1991). (in Russian)
372. O. Molotkov, M. Gudkov, M. Mishenkov, O. Rubakov, *Low-Energy Lasers in Medicine*, chap. 1 (Obninsk, 1991), p. 66. (in Russian)
373. T.A. Bagrova, L.P. Antonov, *Proceedings of Constituent Congress International Society for Pathophysiology*, Moscow, 1991, p. 70
374. G.K. Abasheva, O.T. Prasmytsky, D.A. Budko, Curr. Probl. Biol. Med. **2**, 297 (1996). (in Russian)
375. J.I. Azhipa, *Trophic Function of the Nervous System* (Nauka, Moscow, 1990), p. 672. (in Russian)
376. I.P. Antonov, *Diseases of the Nervous System. Selected Papers and References* (Science and Technology, Minsk, 1992), 415 p. (in Russian)
377. I.P. Antonov, H.H. Gitkina, Zh. Neuropathol. Psychiatry **87**(4), 481(1987). (in Russian)
378. G.A. Azizov, Lazernaya photocorrection microcirculatory disorders in patients with chronic arterial ischemia of lower limbs, Abstract of the dissertation, Moscow, 1991. (in Russian)
379. L.J. Alperin, G.I. Bulgakov, N.F. Gamaleya, V. Gromov, in *Low-Energy Lasers in Experiment and Clinics*, ed. by M.G. Maslova, V.M. Chertok (Far Eastern University Publisher, Vladivostok, 1991), 220 p. (in Russian)
380. A. Akzam, Intravascular laser blood irradiation in complex treatment of peritonitis, Abstract of the dissertation, Institute of Laser Medicine, Moscow, 1991. (in Russian)
381. I.P. Antonov, E.N. Ponomarev, I.A. Gonchar, Peripheral Nerv. Syst. **19**, 53 (1996). (in Russian)
382. I.P. Antonov, E.P. Titovets, N.I. Nechipurenko, Peripheral Nerv. Syst. **2**, 15 (1979). (in Russian)
383. V.B. Gavrilov, V.M. Mazhul, G.G. Gauko, S.V. Konev, Copyright certificate 1179225 (the USSR), 15 Sept 1985 (Bull. No. 34). (in Russian)

384. A.R. Gavrilova, N.F. Khmara, Copyright certificate 1282003 (the USSR), 07 Jan 1987 (Bull. No. 1). (in Russian)
385. I.K. Akhmedov, *Nerve Regeneration in Acute Blood Loss and Purulent Wounds and Its Correction* (Meditsina, Tashkent, 1989), 79 p. (in Russian)
386. K.M. Babina, Clinical efficacy of intravenous laser blood irradiation in chronic bronchitis and its effect on lipid peroxidation, Abstract of the dissertation, Ryazan State Medical University, Ryazan, 1995. (in Russian)
387. L.O. Badalian, I.A. Skvortsov, *Clinical Electroneuro-Myography* (Meditsina, Moscow, 1986), 386 p. (in Russian)
388. I.M. Baibekov, A.K. Kasimov, V.I. Kozlov, *Morphological Bases of Low-Intensity Laser Therapy* (Ion Sina Publisher, Tashkent, 1991), 223 p. (in Russian)
389. A.E. Baranovski, Regional blood flow in patients with traumatic nerve damage limbs in the pre-and postoperative period, Abstract of the dissertation, Bel. Inst. of Impro. of Doct., Minsk, (1993). (in Russian)
390. V.P. Belyaev, A.S. Fedorov, B.N. Malyshev et al., in *Lasers in Clinical Medicine: A Guide for Physicians*, ed. by S.D. Pletnev (Meditsina, Moscow, 1996), 428 p. (in Russian)
391. G.D. Berdyshev, The influence of laser radiation on blood, 178 (1989). (in Russian)
392. T.A. Berezovskaya, General patterns of morphogenesis and regeneration, 33 (1972), p. 33. (in Russian)
393. T.A. Berezovskaya, Microvascularization value of the lumbosacral plexus and its major nerve trunks in the preservation of the morphofunctional integrity of the innerved by them muscles, Abstract of the dissertation, Medical Institution, Lvov, 1973 p. (in Russian)
394. V.P. Bersnev, Diagnostics and surgical treatment of nerve injuries of extremities, Abstract of the dissertation, Leningrad Scientific-Research Neurosurgeon Institute named after A.L. Polenova, Leningrad, 1986. (in Russian)
395. G.E. Brill, O.V. Proshina, V.N. Zhigalina, L.S. Filimonovskaya, Low-energy lasers in the experiment and clinic, 26 (1992). (in Russian)
396. D. Buga, Irradiation of blood low-energy radiation from a helium-neon laser in the complex treatment of patients with obliterating diseases of lower limb arteries, Abstract of the dissertation, Kiev Institute of Cardiovascular surgery, Kiev, 1995. (in Russian)
397. V.P. Veselovsky, F.A. Khabirov, *Muscular Syndromes of Osteochondrosis (Clinic, Diagnostics, Treatment): The Method, the Recommendations* (Kazan, 1989), 19 p. (in Russian)
398. A. Volotovskaya, N. Nechipurenko, N. Slobozhanina, *Proceedings of the Conference "Molecular, and Cellular Membrane Basis of Functioning of Biosystems"*, Minsk, 2002, p. 109. (in Russian)
399. A.R. Gavrilova, Clinical and diagnostic assessment of lipid peroxidation in neurological manifestation of lumbar degenerative disc disease, Abstract of the dissertation, Institute of Neurology, Neurosurgery and Physiotherapy, Minsk, 1990. (in Russian)
400. A.R. Gavrilova, L.I. Matusevich, P.A. Vlasyuk, Peripheral Nerv. Syst. **14**, 48 (1992). (in Russian)
401. O.P. Galeev, G.A. Dmitriev, T.P. Songolo, Topical issues of reconstructive and restorative surgery: proceedings of the final work 1, 29 (1989)
402. N.F. Gamaleya, *Proceedings of the Conference "The Action of Low-Energy Laser Radiation on Blood"*, Kiev, 1989, p. 180. (in Russian)
403. B.M. Gecht, *Theoretical and Clinical Electromyography* (Nauka, Leningrad, 1990), 229 p. (in Russian)
404. S.P. Gladkih, Y.V. Alekseev, N.P. Istomin, The use of lasers to diagnose and treatment of the diseases, 7 (1996). (in Russian)
405. F.S. Govenko, Surgery of nerve injury in children, Abstract of the dissertation, Leningrad Scientific-Research Neurosurgeon Institute named after A.L. Polenova, Leningrad, 1991. (in Russian)
406. M.D. Golub, *Regeneration of Nerves and Reinnervation of Organs* (Nauka i tehnika, Minsk, 1981), 112 p. (in Russian)

407. M.G. Gonchar, Morphological and functional changes in limb injury it MAGIS neutral arteries and their surgical, Abstract of the dissertation, Kiev Institute of Advanced Training of the Doctors, Kiev, 1986. (in Russian)
408. M.G. Gonchar, B.M. Mytskan, V.I. Fuchko, J.I. Bogaychuk, *Proceedings of the All-Union Conference "Topical Issues of Violations of the Regulation of Hemodynamics and Microcirculation in Clinical and Experimental"*, Minsk, 1984, p. 127. (in Russian)
409. O. Gorgon, Laser therapy in complex treatment of injuries of peripheral nerve trunks, Abstract of the dissertation, Kazakhstan Institute of Clinical and Experimental Surgery named after A.N. Syzganov, Alma-Ata, 1992. (in Russian)
410. T.V. Griboedova, Features of enzymatic disorders in peripheral nerves and in the zone of denervation in ischemic and traumatic lesions of the sciatic nerve, Abstract of the dissertation, Academy of Sciences of Belarus, Institute of Radiobiology, Minsk, (1996). (in Russian)
411. K.A. Grigorovich, *Surgical Treatment of Nerve Injury* (Meditsina, Leningrad Division, Leningrad, 1981), 302 p. (in Russian)
412. E. Grunina, Lipid peroxidation in the evaluation of the effectiveness of intravascular laser therapy of rheumatoid arthritis, Abstract of the dissertation, Yaroslavl State Medical Institute, Yaroslavl, 1994. (in Russian)
413. E. Danilova, I. Potozkaya, N. Kolesova, Violation of regulatory mechanisms and their correction 1, 119 (1989). (in Russian)
414. S. Dmitrienko, E.E. Korobkov, V.V. Osipov, E.P. Vorobyeva, Modern methods of laser therapy, 38 (1989). (in Russian)
415. V.V. Evstigneev, Health **3**, 44 (1996). (in Russian)
416. A.P. Eliseev, Microsurgery in the treatment of isolated and associated injuries of peripheral nerves, Abstract of the dissertation, Moscow Medical Academy named after I.M. Sechenov, Moscow, 1993. (in Russian)
417. V.I. Eliseenko, The use of lasers to diagnose and treat disease, 3 (1996). (in Russian)
418. M.S. Zhumankulov, M.S. Pluzhnikov, E.V. Shabunevich, B.S. Ivanov et al., *Proceedings of the All-Union Conference "Effect of Low-Energy Laser Radiation on Blood"*, Kiev, 1989, p. 15
419. E. Zaitsev, K.A. Erigorovich, Surgery of the nerves 2, 280 (280–286). (in Russian)
420. J.A. Zozulya, S.A. Romodanov, V. Rozumenko, *Laser Neurosurgery* (Zdorovie, Kiev, 1992), 167 p. (in Russian)
421. F. Zusmanovich, B. Bunov, T. Berkutskaya, A. Naumov, Surgery **6**, 34 (1985). (in Russian)
422. E.I. Zlotnik, E.A. Korotkevich, *Microsurgery of Peripheral Nerves* (Minsk, 1981), 19 p. (in Russian)
423. V.I. Zyablov, *Problematic Issues of Regeneration of the Nervous System* (Simferopol, 1986), 38 p. (in Russian)
424. I.N.P. Istomin, I.V. Petukhov, O.D. Mishnakh et al., Circulation **12**(2), 30 (1979). (in Russian)
425. I.I. Kanus, P.M. Nazarenko, *Proceedings of the Conference "Anesthetic Management and Intensive Therapy of Critical States"*, Minsk, 1995, p. 52. (in Russian)
426. G.I. Klebanov, *Proceedings of the Conference "Molecular and Cellular Basis for the Functioning of Biological Systems"*, Minsk, 2000, p. 255. (in Russian)
427. V.I. Korepanov, *Laser Therapy in Neurology, Arthrology and Treatment of Pain Syndrome* (Moscow, 1995), 43 p. (in Russian)
428. E.A. Korotkevich, Surgery of traumatic lesions of the limb nerves, Abstract of the dissertation, Bel. State. Inst of Advanced Training of the Doctors, Minsk, 1985. (in Russian)
429. V.A. Kostrov, The clinical and hemorheologic effectiveness of intravascular laser blood irradiation in treatment of hypertension, Abstract of the dissertation, Nizhegorskaya Medical Academy, Nizhny Novgorod, 1994. (in Russian)
430. M.Z. Kreyman, I.F. Udalyi, *Low-Energy Laser Therapy: Pract. Manual* (Publishing House of the University, Tomsk, 1992), 110 p. (in Russian)
431. G.N. Kryzhanovsky, *General Pathophysiology of the Nervous System: A Guide* (Meditsina, Moscow, 1997), 352 p. (in Russian)

432. V.V. Krylov, I.N. Kamalov, M.M. Shudlo, *Proceedings of Rep. Scientific-Practical Conference "Modern Methods of Diagnostics and Treatment of Patients in the Clinic"*, Kazan, 1996, p. 191. (in Russian)
433. V.V. Kupriyanov, J.E. Karaganov, V.I. Kozlov, *Microcirculatory Bed* (Meditsina, Moscow, 1975), 214 p. (in Russian)
434. S.D. Pletneva (ed.), *Lasers in Clinical Medicine: A Guide for Physicians* (Meditsina, Moscow, 1996), 427 p. (in Russian)
435. V.B. Laskov, Differentiated treatment of injuries of the nerves of extremities (Experimental and clinical studies), Abstract of the dissertation, Moscow. Med. Inst. named after N. I. Pirogov, Moscow, 1981. (in Russian)
436. B.C. Lobzin, V.B. Laskov, N.M. Zhulev, *Injury of Nerves* (Voronezh Univ Publisher, Voronezh, 1989), 190 p. (in Russian)
437. I.G. Lyandres, A.P. Shkadarevich, Modern technology and equipment in the surgery, the prospects of development. Paper presented at the international conference "Laser physics and laser applications", Minsk, 14–16 May 2003. (in Russian)
438. A.V. Marochkov, *Intravascular Laser Blood Irradiation. Interaction Mechanisms and Clinical Application* (Minsk, 1996), 85 p. (in Russian)
439. N.V. Matveev, Clinic, diagnostics and treatment of postoperative wound infection in neurosurgery, Abstract of the dissertation, Military. Med. Acad., St. Petersburg, 1996. (in Russian)
440. N.A. Melnikova, Effect of ultraviolet and laser radiation on the structure and function of membranes of blood cells, Abstract of the dissertation, Ogarev Mordovia State University, Saransk, 1994. (in Russian)
441. V.A. Mostovnikov, G.R. Mostovnikova, V.Y. Plavsky et al., *Proceedings of Symposium "Low-Intensity Lasers in Medicine"*, Obninsk, 1991, Part 1, p. 67. (in Russian)
442. V.A. Mostovnikov, G.R. Mostovnikova et al., The influence of laser radiation on blood, 193 (1989). (in Russian)
443. V.A. Mostovnikov, G.R. Mostovnikova, V.Y. Plavsky, L.G. Plavskaya, *Proceedings of International Conference "Laser Physics and Laser Application"*, Minsk, 2003, p. (in Russian)
444. L.V. Musikhin, M.L. Avrutsky, D.C. Katkovskaya, T.U. Guseinov, Anesthesiology and intensive care 5, 74 (1991). (in Russian)
445. G.K. Nedzved, Some of the etiology, pathogenesis and treatment of neurological manifestations of lumbar degenerative disc disease: clinical and genet., immunogenet. studies, Abstract of the dissertation, Academy of Medical Sciences: Institute of Neurology, Minsk, 1991. (in Russian)
446. N. I. Nechipurenko, Pathophysiological mechanisms and issues of pathogenetic therapy of traumatic lesions of peripheral nerves (experimental and clinical study), Dissertation, St. Petersburg Institute of Advanced Training of the Doctors, Minsk, 1992. (in Russian)
447. N.I. Nechipurenko, Acute issues of neuropathology and neurosurgery 2, 164 (1995). (in Russian)
448. S.S. Nikitin, Pathophysiological mechanisms of movement disorders in lesions of different levels of short-muscular system (clinical-neurophys. and morpholog. studies), Abstract of the dissertation, Medical Sciences Institute of Neurology, Moscow, 1994. (in Russian)
449. V.V. Nikiforov, V.V. Babichenko, Diagnostics and treatment of lesions of the peripheral nervous system, 72 (1989). (in Russian)
450. M.A. Nikulin, Effects of physical fields in the tissue, and their use in treating diseases of the lower extremities: theory, experiment, the clinic, Abstract of the dissertation, St. Petersburg Institute of Emergency Care., St. Petersburg, 1995. (in Russian)
451. J.P. Paltsev, G.I. Zheltov, A.A. Komarova, Proc. Acad. Med. Sci. **1**, 32 (1992). (in Russian)
452. T.L. Panchenko, E.I. Kharkov, A.L. Egorov, G.I. Nechipurenko, Coronary heart disease, 73 (1989). (in Russian)
453. G.K. Nedzved, I.P. Antonov, S.E. Rovdo, Patent RB 1604, 22 Oct 1996. (in Russian)

454. L.E. Perminova, Clinical and physiological characteristics of patients dyscirculatory encephalopathy in intravenous laser therapy, Abstract of the dissertation, Nizhegorodskaya Medical Academy, N. Novgorod, 1994. (in Russian)
455. G.G. Petrovsky, Biochemical mechanisms of development and metabolic correction of the local neurodistrofies, Abstract of the dissertation, Academy of Sciences of the BSSR, Institute of Radiobiology, Minsk, (1991). (in Russian)
456. S.S. Pogosyan, The role of the syndrome gemovaskudyary imbalance in the dynamics of ischemic vascular lesions in the lower extremities and their surgical correction, Abstract of the dissertation, Yerevan State. Med. Univ. named after Mkhitar Eeratsim, Yerevan, 1996. (in Russian)
457. C.L. Popel, Morphofunctional state of microcirculatory race and the nerve fibers of the facial nerve in normal and in experimental neuropathy in the conditions of laser irradiation, Abstract of the dissertation, Ukr. Med. Univ named after A.A. Bogomoletz, Kiev, 1993. (in Russian)
458. Ya.Yu. Popelyansky, *Diseases of the Peripheral Nervous System* (Meditsina, Moscow, 1989), 462 p. (in Russian)
459. A.R. Rahishev, V.N. Coi, Methodological, theoretical and methodological aspects of modern neuromorphology, 128 (1987). (in Russian)
460. S. Rovdo, Clinical evaluation of the effectiveness and status of lipid peroxidation in patients with neurological manifestations of osteochondrosis in the treatment of intravenous laser blood irradiation and fotoblokades, Abstract of the dissertation, Bel. State. Inst. of Adv. Train. of the Doctors, Minsk, 1998. (in Russian)
461. S.E. Rovdo, A.R. Gavrilova, I.P. Antonov et al., Peripheral Nerv. Syst. **9**, 79 (1996). (in Russian)
462. A.N. Rubinov, A.A. Afanasiev, *Proceedings of the International Conference on "Laser Physics and Laser Applications"*, Minsk, 2003, p. 8. (in Russian)
463. A.N. Rubinov, A.A. Afanasiev, *Proceedings of the International Conference "Lasers in Biomedicine"*, Grodno, 2002, p. 22. (in Russian)
464. Z.Sh. Sadikova, Post-traumatic peripheral nerve regeneration in acute blood loss and the imposition of tow, Abstract of the dissertation, First Tashkent. State. Med. Inst., Tashkent, 1995. (in Russian)
465. S. Seleznev, G.I. Nazarenko, V.C. Zaitsev, *Proceedings of the Conference "Topical Issues of Violations of the Regulation of Hemodynamics and Microcirculation in Clinics and Experiments"*, Moscow, 1984, p. 50. (in Russian)
466. V.V. Skupchenko, Low-intensity laser radiation in medical practice, 3 (1990). (in Russian)
467. V.L. Stadnik, L.G. Fedorchuk, N.M. Skivka, N.A. Karasevskaya, *Proceedings of the Conference "Effect of Low-Energy Laser Radiation on Blood"*, Kiev, 1989, p. 160. (in Russian)
468. A.A. Tereshchenko, Morphological features of nerves of leg muscles (microscopic and experimental-morphological studies), Abstract of the dissertation, Kharkov. Med. Inst., Kharkov, 1991. (in Russian)
469. V.C. Ulashchik, Health **5**, 3 (1998). (in Russian)
470. V.I. Fuchko, M.L. Gonchar, *Proceedings of Symposium "Acute Ischemia of Organs and the Response to Postischemic Disorders"*, Moscow, 1973, p. 35. (in Russian)
471. I.V. Harlap, A.G. Ovcharenko, *Proceedings of the International Conference "New Developments of Laser Medicine"*, Moscow, St. Petersburg, 1993, pp. 562–563. (in Russian)
472. N.F. Khmara, A.R. Gavrilov, P.A. Vlasyuk, Peripheral Nerv. Syst. **10**, 50 (1987). (in Russian)
473. V.P. Tsymbalyuk, N.N. Suly, M.F. Kvasha, *Proceedings of the Conference "New Technologies in Neurology and Neurosurgery"*, Samara, 1992, p. 191. (in Russian)
474. I.I. Shamelaptvili, The clinic and surgical treatment of closed injuries of the brachial plexus conduction disturbance, Abstract of the dissertation, Scientific and Research Neurosurgery Institute named after A.L. Polenov, St. Petersburg, 1996. (in Russian)
475. P.G. Schwalbe, M.I. Kataev, A.J. Zakharchenko, Col. Sci. Works **96**, 3 (1989). (in Russian)
476. I.R. Schmidt, Neurologic manifestations of spinal osteochondrosis: (origins of multifactorial model of rehabilitation and prevention). Abstract of the dissertation, Moscow Ped. Acad. named after I.M. Sechenov, Moscow, 1991. (in Russian)

477. A. Iovichich, S. Petkovic, V. Ivanisevic, J. Neuropathol. Psychiatry **4**, 67 (1996). (in Russian)
478. N.R. Yulish, Dynamics of damage and recovery of the sciatic nerve with prolonged compression of tissues, Abstract of the dissertation, Kiev Medical Institute, 1978. (in Russian)
479. E.M. Yurah, Morphological and functional studies of neurovasal components of the sciatic nerve in normal and regenerating under de influence of laser radiation, Abstract of the dissertation, Kiev. Med. Inst. named after A.A. Bogomoletz, Kiev, 1990. (in Russian)
480. N.V. Yakovenko, F.S. Govenko, V.B. Semenyutin et al., Diagnostics and treatment of lesions of the peripheral nervous system, 87 (1989). (in Russian)
481. M.M. Asimov, R.M. Asimov, A.N. Rubinov, *Proceedings of International Symposium SPIH on Biomedical Optics Europe'97*, San Remo, 199, Paper 3198-20
482. V. Hanser, in *Holography in Medicine and Biology*. Springer-Series in Optical Sciences 18 (1979), p. 27
483. H. Kasparak, H. Podbielska, G. von Bally, G. Fechner, Int. J. Lehal. Med. **106**(3), 132, 228 (1993)
484. H. Millesi, Int. Surg. **65**(6), 503 (1980)
485. K. Piwernetz, in *Holography in Medicine and Biology*. Springer-Series in Optical Sciences 18(7) (1979), pp. 15, 272
486. L.A. Vasilevskaya, N.I. Nechipurenko, E.N. Anatskaya, Actual problems of neurology and neurosurgery (11), 27 (2009). (in Russian)
487. S.A. Likhachev, L.A. Vasilevskaya, E.V. Veevnik, N.I. Nechipurenko, L.N. Anatskaya, Functional diagnostics **1**, 61 (2009). (in Russian)
488. L.A. Vasilevskaya, Actual problems of neurology and neurosurgery 12, 51 (2009). (in Russian)
489. S. Likhachev, L. Vasilevskaya, V. Vashchilin, *Proceedings of III Ukrainian Scientific-Practical Conference*, Kiev, 2009, pp. 67–69. (in Russian)
490. S.A. Likhachev, L.A. Vasilevskaya, E.V. Veevnik, J. Neurol. **14**(5), 35 (2009). (in Russian)
491. N.I. Nechipurenko, V.I. Hodylev, E.A. Korotkevich, L.A. Vasilevskaya, G.V. Zabrodets, Neurol. Neurosurg. (2), 18 (2010). (in Russian)
492. L.A. Vasilevskaya, N.I. Nechipurenko, L.N. Anatskaya, Patent 12852, 04 Nov 2009. (in Russian)
493. S.A. Likhachev, M.I. Nechipurenko, A.I. Veres, T.V. Griboyedova, L.A. Vasilevskaya, I.D. Pashkovskaya, L.A. Tishina, Collection of scientific articles "New and diagnostic technology", 7 (2009). (in Russian)
494. V.D. Rybakova, L.A. Vasilevskaya, Actual problems of neurology and neurosurgery 13 (2010). (in Russian)
495. N.I. Nechipurenko, E.A. Korotkevich, Hodylev et al., Actual problems of medical rehabilitation, 195 (2009). (in Russian)
496. N.I. Nechipurenko, S.A. Likhachev, T.V. Griboyedova et al., J. Grodno State Med. Univ. **2**, 79 (2009). (in Russian)
497. S.A. Likhachev, A.I. Veres, N.I. Nechipurenko, *Proceedings of V International Scientific-Practical Conference "Hypertension in the Aspect of Solving the Problem of Population Security"*, Vitebsk, 2009, p. 134. (in Russian)
498. N.I. Nechipurenko, I.D. Pashkovskaya, L.A. Vasilevskaya, G.T. Maslova, T.V. Griboyedova, *Proceedings of the International Conference "The Patterns of Development of Pathological States and Their Correction"*, Minsk, 2009, p. 167. (in Russian)
499. I.D. Pashkovskaya, N.I. Nechipurenko, L.A. Vasilevskaya, Health **2**, 74 (2010). (in Russian)
500. S.A. Likhachev, L.A. Vasilevskayal, N.I. Nechipurenko, *Proceedings of VI Scientific Conference "Endothelial Dysfunction: Experimental and Clinical Studies"*, Vitebsk, 2010, p. 16. (in Russian)
501. I.D. Pashkovskaya, N.I. Nechipurenko, L.A. Vasilevskaya, Public Health **2**, 74 (2010). (in Russian)
502. S.A. Likhachev, L.A. Vasilevskaya, V.V. Vashchilin, Patent BY13502 (Belarus), 30 Aug 2010. (in Russian)

503. N.I. Nechipurenko, V.I. Hodylev, G.V. Zabrodets, L.A. Vasilevskaya, G.V. Zobnina, M.A. Schurevich, *Abstracts of the International Scientific-Practical Conference "Actual Problems of the Medical Examination and Rehabilitation of Sick and Disabled"*, Minsk, 2010, p. 94. (in Russian)
504. N.I. Nechipurenko, Y.I. Stepanov, L.A. Vasilevskaya, I.D. Pashkovskaya, Laser hemotherapy in ischemic cerebrovascular diseases (experimental and clinical aspects) (Biznesofset, Minsk, 2010), 192 p. (in Russian)
505. L.N. Anatskaya, G.K. Nedzved, N.I. Nechipurenko, L.A. Vasilevskaya, Yu.I. Stepanova, Diagnostics and comprehensive treatment of lacunar infarcts of the brain using low-intensity laser radiation. Instructions for use (GU "RSPC of Neurology and Neurosurgery", Minsk, 2007) (Reg. No. 086-1107). (in Russian)
506. S.A. Likhachev, L.A. Vasilevskaya, E.V. Veevnik, Speckle-optical diagnostics of the functional activity of facial muscles during face dyskinesia. Instructions for use (GU "RSPC of Neurology and Neurosurgery", Minsk, 2007) (Reg. No. 090-1107). (in Russian)
507. V.I. Hodulev, N.I. Nechipurenko, G.V. Zabrodets, G.V. Zobnina, L.A. Vasilevskaya, E.A. Korotkevich, E.N. Apanel, Diagnostics, treatment and rehabilitation measures at compression-ischemic neuropathies, using methods of laser hemotherapy. Instructions for use (GU "RSPC of Neurology and Neurosurgery", Minsk, 2009) (Reg. No. 109-1109). (in Russian)
508. V.D. Rybakova, L.A. Vasilevskaya, Actual problems of neurology and neurosurgery 13, 133 (2010). (in Russian)
509. L.A. Vasilevskaya, L.N. Anatskaya, V.K. Zabarovskaya, Actual problems of neurology and neurosurgery 13, 28 (2010). (in Russian)
510. N.I. Nechipurenko, L.A. Vasilevskaya, L.A. Tishina, Actual problems of neurology and neurosurgery 13, 105 (2010). (in Russian)

Afterword

My main mentor was mother who formed my attitude toward medicine by her personal example: her attitude to work and her capacity to work, ability to communicate with people, to always render help, to be useful to others no matter what kind of problems they have. Her advice, lessons and instructions helped me to define my purpose in life. Being a teacher of the Russian language and literature, wittingly or unwittingly she instilled in me first and then in his grandson (my son) Andrew the feeling of love and respect to medicine, which determined the future of our professions.

Medical studies have always attracted me. But I knew, or rather felt, that medicine is one of the most complex sciences, and it requires special training, and, for some reasons, particularly deep knowledge of physics and mathematics. From time to time, I had hesitations. First I entered the Physics Department of the Leningrad State University, after the first year, I suddenly decided to transfer to the First Medical Institute, and when the question with my transferring had almost been decided, I told myself that I would move to medicine only after the graduation from the Physics Department. When I successfully graduated from the Physics Department and received a diploma in physics, I suddenly realized that I had to come to medicine only after defending my doctoral thesis. And only after that I would be able to comprehend the mysteries of medicine.

At the same time, the voice from above suggested me: "Think about holography," a scientific direction, about which at that time—it was in 1966—little was known, especially at the undergraduate level.

Having defended the thesis in the field of holography, I immediately began to develop the sphere—holography in medicine and biology. I was so much fascinated with this problem that on the basis of the results of my own works and studies of foreign authors, I prepared a well-illustrated (with hologram format of $30 \times 40\ cm^2$) report on the theme "Holography in Biology and Medicine," which I presented to public several times in the USSR and abroad. Like many young scientists, I believed that at the meeting point of physics and medicine, I would soon be able to discover something inaccessible to others. That was the main goal of my work. It was very

L. Tanin et al., *Biomedical and Resonance Optics*, Bioanalysis 11,
https://doi.org/10.1007/978-3-030-60773-9

difficult and important for me at that time to properly search and select the areas in medicine where I could maximally realize the unique features of optical holography, holographic interferometry and microscopy, speckle-optics and laser physics. I was lucky: in 1975–1977 years, the fate brought me with Belarusian scientists, endlessly in love with their profession. We can say today about these people that they are outstanding scientists who created in medicine, each in his own direction, the whole epoch. The topic of the report "Holography in Medicine and Biology" so much interested and intrigued them that I was invited to present it first at the Institute of Photobiology of the Academy of Sciences of the BSSR, where I had a long conversation with Aleksander Arkadievich Shlyk, Sergei Vasilyevich Konev, Yevgeniy Aleksandrovich Chernitskiy and Vladimir Mikhailovich Mazhul. After I presented my report at the Institute of Physiology of the Academy of Sciences of the BSSR, where I met Ivan Andreevich Bulygin and Galina Vasilievna Abramchik, which also left a pleasant mark in my life. Sometime later, my report was made at the Institute of Oncology of the Ministry of Health of the BSSR, and so much interested Director of this Institute Nikolai Nikolaevich Alexandrov that he offered me a job at his institute. But fate had prepared me another meeting at the Institute of Neurology, Neurosurgery and Physiotherapy of the Ministry of Health of the BSSR, where I also presented my report "Holography in Medicine and Biology" together with Ignat Petrovich Antonov, Iosif Abramovich Sklut and Georgy Konstantinovich Nedzved. The prospect of the study, which Ignatiy Petrovich opened to me, finally defined my choice for many years. As a result, the main object of my research became neuromuscular system and later on cardiovascular system, the study of which was conducted jointly with the Institute of Cardiology of the Ministry of Health of the BSSR directed by George Ivanovitch Sidorenko. In order to prepare young professionals for future scientific work, I gave lectures—special course "Holographic Interferometry and Microscopy" on the Physics Faculty (Department of Spectroscopy and Quantum Electronics) from 1976 to 1983. During those years, there were many skeptics among the majority of physicists who assured me that the issue related to the study of a biological object was unpromising. However, as a physicist, I did not see then and do not see now the difference, whether I explore the vibration modes of blades of compressors and turbines engines, elasticity, fatigue, residual stress, aging of materials, flow and thermal gradients, plasma, laser resonators, whether I study vibrational modes of blades of compressors and engine turbines, elasticity, fatigues, residual stress, aging of materials, flows and thermal gradients, plasma, laser resonators, create laser systems or study cells, organs, tissues, processes of the contraction of a heart muscle, deformations inside and outside of the skull, stress concentrators in the jaw, the elasticity of the vessels, the speed of red blood cells, neural networks, pulsation of the brain, etc. Moreover, today I can say for sure that a medical and biological object is more complex for study than a physical one, its state and properties are multivariate and multi-functional.

I gratefully and warmly recall the period of my study at High School in Leningrad from 1965 to 1975, especially the years of study in the graduate school. After finishing work on my thesis devoted to the studies using resonant processes in gaseous media, which are based on the phenomenon of anomalous dispersion, Yuri Isaevich Ostrovsky said: "Remember, Leonid, that you were my graduate student, I was a graduate student of Nikolai Petrovich Penkin, and Nicholai Petrovich Penkin was a graduate student of Dmitriy Sergeevich Rojdestvenskiy, so you are a grand student of Dmitry Sergeyevich Rojdestvenskiy, the founder of optical school in the USSR" [14] (see Chap. 1, p. 69).

The parting words of the mentor turned into active scientific searches at the meeting point of laser physics, holography and medicine. In 1978, I created the scientific group "Coherent-optical studies of medical and biological systems" at the Institute of Neurology, Neurosurgery and Physiotherapy of the Ministry of Health of the BSSR with the active support of Director of this Institute, Academician, Doctor of medical sciences, Professor I. P. Antonov. This group was assembled of highly qualified experts in the field of coherent-optics and holography, laser physics, medicine and biology: physicist–optician, biophysicists, biologists, physiologists, physicians (neurologists, pathophysiologists, cardiologists), electronics, programmers, mechanics, engineers, designers and technicians who made up a single scientific team.

In order to develop new and modern scientific approaches, methods, instruments and devices for the treatment and diagnostics of diseases of the peripheral neuromuscular and cardiovascular systems in the period from 1978 to 1999, this group of scientists and engineers with my participation as one of the supervisors carried out a range of scientific and research works on the topics:

1. Topic—code "Neuron" (1980) (principle investigator: Candidate of Physical and Mathematical Sciences L. V. Tanin),
2. Development and improvement of methods of holographic interferometry and microscopy for solving the biomedical problems (1983) (principle investigator: Candidate of Physical and Mathematical Sciences L. V. Tanin),
3. Development of holographic methods for the study of the structural and functional changes of peripheral nerves at norm and pathology (1983–1987) (principle investigator: Candidate of Physical and Mathematical Sciences L. V. Tanin),
4. The study of the impact of a powerful pulsed magnetic field on nerve–muscle preparation by the coherent-optical electrophysical methods (1985) (principle investigators: Academician, Doctor of Medical Sciences I. P. Antonov, Candidate of Physical and Mathematical Sciences L. V. Tanin);
5. The development of coherent-optical methods and devices for the study of neurobiological objects and making recommendations on the study of the peripheral nervous system and the treatment of its diseases (1988) (principle investigators: Academician, Doctor of Medical Sciences I. P. Antonov, Candidate of Physical and Mathematical Sciences L. V. Tanin);
6. The development of coherent-optical, holographic, electronic systems and measurement techniques for the conducting the studies of structural and functional states of nervous and muscles tissues (1988), code «Pressure» (principle investigators: Academician, Doctor of Medical Sciences I. P. Antonov, Candidate of Physical and Mathematical Sciences L. V. Tanin);

7. The study of functional state of nervous and muscle tissue at norm and at pathology of peripheral nervous system by coherent-optical methods (1991), (principle investigators: Academician, Doctor of Medical Sciences I. P. Antonov, Candidate of Physical and Mathematical Sciences L. V. Tanin);
8. Development and creating of laser-holographic complex for the determining of the condition of the human blood circulation system (1992–1995), (principle investigators: Academician, Doctor of Medical Sciences I. P. Antonov, Candidate of Physical and Mathematical Sciences L. V. Tanin);
9. Development and creating of the prototype of laser specklometer (microhematomyograph) (1993), (principle investigators: Candidate of Physical and Mathematical Sciences Science L. V. Tanin);
10. Coherent-optical study of biomechanical conditions of muscles and microhemodynamics of skintissues and the development on its basis contactless methods of the diagnostics of peripheral nervous system diseases, Republican scientific research program (1993), (principle investigators: Academician, Doctor of Medical Sciences I. P. Antonov, Candidate of Physical and Mathematical Sciences L. V. Tanin);
11. To study certain questions of pathogenesis, diagnostics and treatment of the diseases of PNS by the methods of coherent-optics and to develop the appliances and devices on their basis (1994), (principle investigators: Academician, Doctor of Medical Sciences I. P. Antonov, Candidate of Physical and Mathematical Sciences L. V. Tanin);
12. To study the therapeutic effectiveness of intravenous laser blood irradiation (ILBI) at the traumatic damage and ischemia of peripheral nerves by noninvasive coherent-optical methods (1998), (principle investigators: Academician, Doctor of Medical Sciences I. P. Antonov, Candidate of Physical and Mathematical Sciences L. V. Tanin);

It is necessary to point out that at the times of carrying out the mentioned topics, I was not only the coordinator, but also the responsible executer who participated in the conducting of the majority of the experiments, as well as in the development of the theoretic basis of the studies. All this helped me to sum this complex study together with the co-authors and create a scientific work—monograph "Biomedical and Resonance Optics: Theory and Practice."

One of the ideas that I suggested and realized was the creation of an experimental model of a diagnostic apparatus "Laser specklometer," in the creation of which took part a lot of scientists and specialists from different areas, institutes and companies of the former USSR, in particular, such establishments as the Institute of General Physics of the Russian Academy of Science (Moscow), GOI named after S. I. Vavilov (St. Petersburg), the Institute of Physics of the NAS of Belarus, the Specialized Design and Technology Bureau of the Institute of Physics of the NAS of Belarus, the Central Design Bureau of the NAS of Belarus, HMTI of the NAS of Belarus, LLC "Information technologies and measurements," the Institute of Applied Physics of the NAS of Belarus, the Belarusian Research Institute of Neurology, Neurosurgery and Physiotherapy of the Ministry of Health of the Republic of Belarus and other organizations.

Today, "Laser specklometer" is successfully used in the State Establishment "Scientific and Practical Center of Neurology and Neurosurgery" of the Ministry of Health of the Republic of Belarus for the development of a number of new scientific schools. A number of fundamental and applied studies in medicine (mainly the studies of peripheral and central nervous system) have been carried out with the help of "Laser specklometer." I would like to express my gratitude to Ludmila Aleksandrovna Vasilevskaya for the specially presented for the monograph results of the clinical studies of patients derived with the help of the given microhematomyograph, for the list of the scientific works carried out with her participation and the titles of the conferences. I would like to point out that as it can been seen from the list of scientific publications (the main and additional ones, see pp. 501-502), the important summary result of the multi-aspect studies carried out in the State Establishment "Scientific and Practical Center of Neurology and Neurosurgery" of the Ministry of Health of the Republic of Belarus for us, authors of this monograph, is the fact that our device "Laser specklometer" has successfully passed through long years (about 10 years) of clinical tests and today benefits to the practical medicine. The main significance of the device is its ability to measure noninvasively practically simultaneously the muscles tone and surface blood flow at any area of human body by the means of switching the device to different frequency ranges. As a great scientist in the sphere of development and using of speckle-optical methods and devices in the studies of the surface blood flow, Professor T. Asakura from the Research Institute of Applied Electricity Hokkaido University (Japan) said, there are no analogs of this device today. All this allows us, authors, to recommend this unique device of a new generation "Laser specklometer" to the Ministry of Health of the Republic of Belarus for a series manufacture and wide implementation both for operational remote diagnostics during the treatment of patients with different kinds of diseases, as well as for contactless noninvasive control of fatigue states of muscles and changes in the conditions of skin microhemodynamics in the process of sportsmen training. Besides, it is possible to carry out further detailed study of the mechanism and restoration means and regulating the physical state of a sportsmen with the help of the given device (microhematomyograph) without any medical interference.

It is also important to point out that the possibilities of the given device have not yet been fully discovered. From the point of view of the search and defining of the new informative parameters of dynamic speckle-fields formed during the diffuse reflection of life tissue from the surface, the laser specklometer contains a great potential of highly sensitive diagnostic criteria of the estimation of the degree of the evidence of different kinds of diseases. In particular, the area of application of the developed by us speckle-optical device is spread not only on the measurement of the skin blood flow and muscle activity, but also on the studies of functional states of other organs and tissues.

And as result, you have our work before you, the main result of which is the development of the perspective school in science–biomedical optics. The prospects of the given school are confirmed by the fact that a few publications of the 80s years of twentieth century turned out to be today as grandiose international scientific congresses and conferences. A number of reports have been presented in different countries of the world, including international, all-union and republican conferences. Today physicists and medics defend a number of theses on this theme.

The main results of this monograph have been tested during:

VI All-union Conference on non-linear optics (the BSSR, Minsk, 1972),
All-union Conference "Optical holography and its application" (RSFSR, Leningrad, 1974),
IV All-Union Conference on Low Temperature Plasma Physics (USSR, Kiev, 1975),
I All-Union Conference "Lasers based on complex organic compounds" (the BSSR, Minsk, 1975),
II All-Union Conference on Holography (USSR, Kiev, 1975),
VIII All-Union Conference on Coherent and Nonlinear Optics (GSSR, Tbilisi, 1976),
II All-Union Conference "Lasers based on complex organic compounds and their application" (TSSR, Dushanbe, 27—September 30, 1977),
IV Republican Conference of Young Scientists on Physics (the BSSR, Minsk, 1977),
I All-Russian Conference "Laser Optics" (RSFSR, Leningrad, 4—January 8, 1977),
All-Union School on Holography (BSSR, Minsk, 1978),
The International Symposium on Applied Holography «Interkamera» (Czechoslovakia, Prague, 1978),
IV All-Union Conference on Holography (Yerevan, 1982),
The International Conference "Holography in non-destructive studies" (SFRU, Dubrovnik, 1983),
XV All-Union School on Holography (the BSSR, Minsk, 1984),
VIII Republican Conference of Young Scientists on Physics (the BSSR Minsk, 1984),
II All-Union Conference of Young Scientists "Theoretical and Applied Optics" (RSFSR, Leningrad, 1986),
The Republican Conference "Physical factors and technical means in medicine" (BSSR, Minsk, 1986),
All-Union Conference "Use of modern physical means in non-destructive research and control" (RSFSR, Khabarovsk, 1987),
All-Union Seminar "Holography in biology and medicine" (the BSSR, Minsk, 1982),
All-Union workshop "Holography in industry and scientific studies" (the BSSR, Minsk, 1984, 1986),
In scientific reports presented in the Center of Scientific Instrument Making (GDR, Berlin, 1986),
Graduate School of Medicine (GDR, Suhl, 1986) and the University of Bologna (Italy, Bologna, 1987),

All-Union Scientific-practical conference "Actual issues of Radio Electronics in Medicine" (RSFSR, Moscow—Kuibyshev, 1988),
Conference on lasers and electrooptics (USA, Baltimore),
The International Conference "Holography'89" (Bulgaria, Varna, 1989),
I International Conference "Optics in Life Sciences" (Germany, Garmisch-Partenkirchen, 1990),
VI All-Union Conference on Holography (Belarus, Vitebsk, 1990),
The International Symposium "Electronic Imaging: Science and technology" (USA, San Jose, 1991),
II International Conference "Optics in Life Sciences" (Germany, Munster, 1992),
The International Symposium "Biomedical optics'93" (USA, Los Angeles, 1993),
The International Symposium "Laser optics'93" (Russia, St. Petersburg, 1993),
The European international symposium "Biomedical optics'93" (Hungary, Budapest, 1993),
The International Symposium "Electronic Imaging: science and technology" (USA, San Jose, 1994),
III International Conference "Optics in Life Sciences" (Japan, Tokyo, 1994),
The European international symposium "Biomedical optics'94" (France, Lill, 1994),
The European Symposium "Optics in Industry" (Germany, Frankfurt-am-Main, 1994),
The International Conference "Laser optics'95" (Russia, St. Petersburg, 1995),
The International Conference "Light and biological systems" (Wroclaw, 1995),
The European international symposium "Biomedical optics'95" (Spain, Barcelona, 1995),
IV International Conference "Optics in Life Sciences" (Munster, 1996),
The European international symposium "Biomedical optics" (Austria, Vienna, 1996),
III International Scientific and Technical Conference "Physics and radio electronics in medicine and biotechnology" (Russia, Vladimir, 1998),
The Republican scientific seminar "Laser Technologies in Medicine" (Belarus, Minsk, 1999),
IV congress of the Belarusian public association of photo biologists and biophysicists(Belarus, Minsk, 2000),
The Congresses of the Belarusian Society of Physiologists (Belarus, Minsk, 2001, 2006),
IX International Conference "Application of lasers in life sciences" (Lithuania, Vilnius, 2002),
The International scientific-practical conference "Dysfunction of endothelium: experimental and clinical studies" (Belarus, Vitebsk, 2002, 2006, 2008),
The International Conference "Lasers in biomedicine" (Belarus, Grodno, 2002),
The International scientific-technical conference "Medelektronics" (Belarus, Minsk, 2002–2004, 2006, 2008),
The Congress of neurologists and neurosurgeons of the Republic of Belarus (Belarus, Minsk, 2003),

International Conference "Laser physics and laser applications" (Belarus, Minsk, 20 the 03),
The International Conference "Laser and optical technologies in biology and medicine" (Belarus, Minsk, 2004),
The International Conference "Problems of integration of functions in physiology and Medicine" (Belarus, Minsk, 2004),
The Scientific-practical conference "Modern approaches and the introduction of new techniques in the diagnostics" (Belarus, Vitebsk, 2005),
The International Conference on Biomedical Optics (Germany, Munich, 2005),
The International scientific-practical conference "Modern achievements of laser medicine and its application in practical health care" (Russia, Moscow, 2006),
IV International Conference "Medical and social ecology of personality: condition and prospects" (Belarus, Minsk, 2006),
The International Conference "Molecular, membrane and cellular basis for the functioning of biological systems" (Belarus, Minsk, 2006), Scientific Conference "Interaction of laser radiation with the substance" (Belarus, Gomel, 2006),
II Russian International Congress "Cerebrovascular diseases and a stroke" (Russia, St. Petersburg, 2007),
The Conference on completed in 2007 assignments under the "Modern technologies in medicine",
International scientific-practical conference "Actual problems of neurology",
"The successes of neuroprotection—from theory to everyday practice" (Belarus, Minsk, 2008),
I National International Congress on Parkinson's disease (Russia, Moscow, 2008),
The Republican scientific-practical conference "Actual problems of Neurology and Neurosurgery" (Belarus, Minsk, 2008),
VII International Conference "Laser physics and optical technologies" (Belarus, Minsk, 2008),
The Symposium on "Regional hemodynamic and microcirculation" in the framework of the IV All-Russian Conference on clinical hemostaziology and hemorheology in cardiovascular surgery (Russia, Moscow, 2009),
The Scientific-practical conference "Fundamental and applied aspects of physiology" (Belarus, Grodno, 2009),
III Ukrainian scientific-practical conference "Extrapyramidal diseases and the age" (Ukraine, Kiev, 2009),
The International Scientific Conference "Regularities of development of pathological states and their correction" (National academy of sciences, Department of Medical Sciences, Institute of Physiology, Belarus, Minsk, 2009),
The International scientific-practical conference "Actual problems of medical examination and rehabilitation of sick and disabled people" (Belarus, Minsk, 2010),
The Republican scientific-practical conference dedicated to the Day of Science (Belarus, Minsk, 2010),
VI International Scientific and Practical Conference "Endothelial dysfunction: experimental and clinical studies" (Belarus, Vitebsk, 2010),

7th International Conference “HoloExpo – 2010”,
“Holography. Science and Practice” (Russia, Moscow, 2010).

I am sure that those, whom we will pass our knowledge and experience, will discover new pages in the prospective sphere of biomedical optics, which has a great future.

Academician of the International engineering academy,

Doctor of physical and mathematical sciences  L. V. Tanin

Index

L. Tanin et al., *Biomedical and Resonance Optics*, Bioanalysis 11,
https://doi.org/10.1007/978-3-030-60773-9

The manufacturer's authorised representative in the EU is Springer
Nature Customer Service Centre GmbH, Europaplatz 3, 69115 Heidelberg,
Germany. If you have any concerns regarding our products, please
contact ProductSafety@springernature.com

Printed and bound by CPI Group (UK) Ltd, Croydon, CR0 4YY
28/09/2026
02224062-0010